Artemisinin-Based and Other Antimalarials

Artemisinin-Based and Other Antimalarials

Detailed Account of Studies by Chinese Scientists Who Discovered and Developed Them

Editors Chinese Edition

Li Guoqiao, Li Ying, Li Zelin, Zeng Meiyi

Editors English Edition

Muoi Arnold, Keith Arnold, Zeng Meiyi

Chinese to English Translation

Yuexin Rachel Lin

Postdoctoral Researcher,
Mahidol-Oxford Tropical Medicine Research Unit

Muoi Arnold

Academic Press is an imprint of Elsevier
125 London Wall, London EC2Y 5AS, United Kingdom
525 B Street, Suite 1800, San Diego, CA 92101-4495, United States
50 Hampshire Street, 5th Floor, Cambridge, MA 02139, United States
The Boulevard, Langford Lane, Kidlington, Oxford OX5 1GB, United Kingdom

Library of Congress Cataloging-in-Publication Data
A catalog record for this book is available from the Library of Congress

British Library Cataloguing-in-Publication Data
A catalogue record for this book is available from the British Library

ISBN: 978-0-12-813133-6

For information on all Academic Press publications visit our website at https://www.elsevier.com/books-and-journals

Publisher: Mica Haley
Acquisition Editor: Glyn Jones
Editorial Project Manager: Katie Chan
Production Project Manager: Mohana Natarajan
Designer: Vicky Pearson

Typeset by TNQ Books and Journals

Contents

The Development of Qinghaosu (Artemisinin)—A View From the Outside

The story of the discovery of qinghaosu is remarkable by any standards. The "outside world" first learned of this extraordinary Chinese malaria treatment in the late 1970s with the publication in English of a landmark paper in the Chinese Medical Journal. Six small pages reported the discovery, structure, physicochemical characterization, pharmacology, and the laboratory and clinical evidence of antimalarial activity. A plant-derived traditional Chinese medicine seemed about to challenge the products of bark of the cinchona tree (quinine), which has been preeminent in medicine for the past 350 years. The speed of antimalarial action claimed for these new compounds was unprecedented, and they reportedly lacked significant toxicity. Like many malaria researchers when I read this paper I was astounded and eager to learn more and to obtain these compounds for clinical testing. Southeast Asia where I worked was experiencing a rapid decline in the efficacy of the existing antimalarial drugs and a surge in malaria morbidity and mortality. Furthermore, our Chinese colleagues were willing to provide us with the drugs. But the World Health Organization was concerned—the quality of the Chinese products was uncertain—we were strongly advised to wait until WHO could provide us "high-quality material" for clinical testing. But this never came. Instead WHO TDR embarked on a separate course to develop their own artemisinin derivative—arteether—a compound we learned later had already been synthesized and rejected by the Chinese.

By the late 1980s malaria was causing havoc in rural areas of Southeast Asia, the historical epicenter of antimalarial drug resistance. The available antimalarial drugs (chloroquine, sulfadoxine–pyrimethamine, amodiaquine) had fallen to resistance, and the countries of the region decided they could wait no longer. They decided that they had no option but to import or make the artemisinin derivatives themselves. Trials started and they very soon confirmed all the Chinese claims—artemisinin and its derivatives were indeed the most rapidly acting antimalarial drugs ever. Patients with uncomplicated malaria got better a day earlier, and the responses in severe malaria were equally impressive. Furthermore, the drugs were simple to administer and there was no evident toxicity. Later they were combined with more slowly eliminated antimalarial drugs to form "artemisinin combination therapies" or ACTs, but despite excellent results from clinical trials with ACTs there was reluctance from international agencies to support these new drugs, and change from the inexpensive and increasingly ineffective treatments they supported. The global malaria death toll continued to rise.

It is sometimes claimed that WHO recommended a policy switch from the failing antimalarial drugs to ACTs in 2001, but this recommendation was very weakly supported, and there was little immediate policy change in Africa, which bears the brunt of the malaria death toll. Eventually the rising tide of antimalarial drug resistance associated suffering and death, and the clear advantages of ACTs over existing drugs which were demonstrated in WHO TDR coordinated multinational trials, forced a change. From 2006, some 35 years after the original discovery of qinghaosu, the new WHO malaria treatment guidelines unequivocally recommended that the world should change to ACTs for the treatment of falciparum malaria. National policies then changed rapidly, and donor support to purchase and deploy the drugs increased. Millions of lives, particularly those of African children, have been saved as a direct consequence.

In the treatment of severe malaria the first large trials outside China were with intramuscular artemether as 30 years ago WHO was interested primarily in a parenteral treatment that could be given easily at a village or health center level. In hindsight this was a mistake. Artemether, which is absorbed slowly and erratically after intramuscular injection, did prove better than quinine—but it was not superior enough to effect a policy change. Eventually very large randomized trials with the water-soluble parenteral artesunate were conducted first in Asia and then in Africa. These showed a large and unequivocal survival benefit with parenteral artesunate, and so artesunate has now replaced quinine as the treatment of choice for severe malaria.

Meanwhile many of the earlier toxicity concerns have been progressively allayed. Neurotoxicity had been shown in beagle dogs after high doses of intramuscular arteether or artemether. This was found to have a pharmacokinetic explanation. There is no evidence for neurotoxicity with current doses of the artemisinin derivatives. From the beginning qinghaosu was shown to affect fetal outcomes in experimental animals adversely. However, careful clinical and experimental studies have shown no evidence of iatrogenic toxicity affecting fetal development or birth outcomes in pregnant women. Today artesunate is the drug of choice for pregnant women with severe malaria at any stage of pregnancy. ACTs are actively recommended for uncomplicated malaria in the second and third trimesters of pregnancy, and the restriction on their use in the first trimester may be lifted cautiously in the near future. Increasingly ACTs are also being used to treat vivax and other human malarias. But there is a darkening cloud on the horizon and that is resistance. It proved very difficult to change susceptibility to qinghaosu and its derivatives in the laboratory, but decades of unregulated use in Southeast Asia did eventually select for resistance, and artemisinin-resistant *Plasmodium falciparum* parasites are now spreading across the Greater Mekong subregion.

Qinghaosu has been one of the most important antiinfective drugs of the past century, it has saved millions of lives, and it has allowed us again to plan for malaria elimination. It is a discovery worth celebrating. In this comprehensive volume we can now learn for the first time from different perspectives of the details of the early discovery and the extensive and far-reaching research in China that followed.

Professor Sir Nicholas J White, FRS
Chairman, Oxford University – Wellcome Trust
South East Asian Tropical Medicine Research Units

Foreword

Malaria is the most serious parasitic disease worldwide. In the 1960s and 1970s, annual morbidity reached 30–50 million, with around 5 million—mostly children—dying from the disease. The history of mankind's struggle with malaria is a long one and, in the prevention and treatment of the disease, antimalarial drugs play an extremely important role. The appropriate use of such medications can result in good therapeutic outcomes, as well as eliminate the sources of infection, and prevent its transmission. Beginning with the successful isolation of quinine from cinchona bark in the 1880s, chloroquine was developed in the 1940s, ushering in an era in which chloroquine was the drug of choice for malaria treatment and prophylaxis. In the 1960s, *Plasmodium falciparum* began to develop resistance toward chloroquine. Morbidity rates soared and mortality was high, especially in Southeast Asia, South America, and tropical and subtropical areas of Africa. This became a severe threat to life and health.

During the Vietnam War, both Americans and North Vietnamese suffered greatly from malaria. Noncombat casualties sometimes outnumbered combat losses. To support North Vietnam, China's leaders considered this request urgent and called for a task force to research for new antimalarial drugs to aid North Vietnam. The relevant authorities across China formed a Project 523 leading group and set up an administrative body, the Project 523 Head Office, to organize the cooperative efforts of ministerial, military, provincial, metropolitan, and autonomous district research units. They devoted themselves to the research and development of a new antimalarial. At first, compound therapies based on old medicines were devised to meet the pressing need for malaria treatment and prophylaxis. Simultaneously, the researchers directed their energies toward synthesizing new antimalarial compounds and screening remedies in the Chinese *materia medica*.

In the 1970s, Chinese scientists unearthed and developed artemisinin from the treasure trove of Chinese medical classics. Three artemisinin derivatives with even higher efficacy—artesunate, artemether, and dihydroartemisinin—were subsequently produced. Research into new compounds yielded lumefantrine and naphthoquine phosphate while developing piperaquine further. From the 1990s, these drugs formed the basis of a series of combination therapies designed to raise the cure rate and delay the onset of drug resistance: artemether–lumefantrine, artemisinin–naphthoquine phosphate, and artemisinins–piperaquine, etc. Domestic and international collaboration brought artemisinin-type antimalarials to the world. Today, China's innovative artemisinin-based combination therapies have become one of the main drugs used worldwide for the prevention and treatment of drug-resistant falciparum malaria. They have been taken up rapidly and broadly to treat tens of millions of cases worldwide, saving the lives of millions of critically ill patients. The discovery of artemisinin and the research outcomes that followed were a significant breakthrough in global antimalarial research and a milestone in the history of the struggle against malaria. They represent an important contribution in this area by Chinese scientists and a model of success for modern medical research in China. In the first 5 years of Project 523, I took part in the screening of antimalarial drugs and in clinical trials in malaria-endemic areas. I experienced the great tribulations of antimalarial research and felt its significance deeply. For what Project 523 has accomplished, I am immeasurably happy.

New drug research is a form of basic science and a medical engineering project that crosses disciplines and specializations. In more than a decade of research under Project 523, scientific, medical, pedagogical, manufacturing, and other units undertook large-scale collaboration. They not only achieved important outcomes in multiple fields but also established a system and management style based on "unity in will, goals, direction, and action." With little regard for their personal gain, several hundred scientists and technicians forged ahead cooperatively, maintaining timely and open channels of communication. This spirit, combined with the capable and effective coordinating role of the Project 523 Leading Group and Head Office, was a high point in the world of science and technology. After the conclusion of Project 523, research into artemisinins gradually came to adhere to international standards under the stewardship and coordination of the China Qinghaosu Steering Committee. Research standards improved time and again, and the research and development of new antimalarials attained greater heights.

The new antimalarials developed by Chinese scientists have won multiple high-profile awards and garnered international recognition. Both at home and abroad, researchers have published a large volume of scholarly and scientific works on the artemisinins. The discovery of artemisinin and invention of artemisinin-based antimalarials have received wide-ranging attention internationally. Until now, however, a systematic, comprehensive account from the discoverers and those personally involved in the research is unfortunately still lacking. The publication of *Artemisinin-Based Antimalarials* fills this gap.

The authors of *Artemisinin-Based Antimalarials* are all elderly scientists from different units and specialties who took part in the research and development first-hand. The book is divided into 10 chapters, which include a history of the research endeavor and a scholarly scientific record and discussion. In the history section, the events surrounding China's discovery and research into artemisinins, the scientific standards at different points in time, scientific reasoning and methodology, as well as its administrative, managerial, and collaborative qualities are described extensively and objectively. The scientific achievements reported present the key findings of half a century of antimalarial research in China and a full and accurate summary of scientific data. This is a valuable academic reference work on medical and pharmaceutical research that is of great scientific significance. I hope that its publication will inspire and aid in the research and development of new drugs.

Sun Manji
Member of the Chinese Academy of Sciences

Foreword

This book is a definitive scientific account of *qinghaosu* (artemisinin) antimalarial drugs, written and edited by the original scientists involved, and is a major contribution to the academic literature and should serve as a reference source for those interested in antimalarial drug development. It supplements and adds more thorough research background information to the earlier publication, *A Detailed Chronological Record of Project 523 and the Discovery and Development of Qinghaosu (Artemisinin)*, edited by Zhang Jianfang and translated into English (Chinese version published 2006, English version published 2013). This earlier book was written and edited by administrators of the project more as a general history and to correct some false claims being made as to who was involved in *qinghaosu*'s discovery. Both books give due credit to the Chinese scientists involved in a remarkable achievement, especially since this was achieved under the very adverse conditions of the Cultural Revolution and with much outdated and obsolete equipment.

Artemisinin, its derivatives, and its combinations with other antimalarial drugs are the first-line and most widely used drugs in the world for malaria. This demonstrates how important and amazing this discovery and development has been to the decrease in morbidity and mortality for millions of patients suffering from this disease.

The breadth and depth of the scientific studies described in the following pages clearly demonstrate the competence, hard work, and unreserved commitment to their professional calling of the contributors. These contributors range from basic laboratory specialists in chemistry, physics, botany, pharmacology, and pharmaceutical formulation, to crystallographers, electron microscopists, microbiologists, toxicologists, biologists for animal studies, and clinicians treating patients. But certainly not to be neglected for credit are the contributions to the organization and management of the project by the large numbers of equally competent administrators, such as Zhang Jianfang, Zhou Keding, and colleagues in Project 523. Scientists had to collaborate among themselves and among many geographically widely separated research centers, as well as with the administrators in different locations, who were overseeing the whole undertaking. That this was accomplished in such a short time among such a diverse group of people and under such adverse conditions is to be highly commended.

It has been my privilege and honor to contribute in a minor way to *qinghaosu*'s development and introduction to the western world by publishing, with Chinese authors, the first clinical research paper in the western medical literature in the *Lancet* in 1982. And also to be involved in the translation on the artemisinin book written by the administrators, mentioned above and to contribute to this book by the original researchers in the translation and editing of this English edition.

My association with one of the editors and contributors, Li Guoqiao, extends over nearly 40 years, from the occasion when he first introduced me to *qinghaosu* in 1979 in his laboratory at Guangzhou University of Chinese Medicine. In subsequent years, we have worked together in China and Vietnam to further document the efficacy and safety of *qinghaosu*, its derivatives and combinations. This collaboration continues today as we work to study and promote his concept of Fast Elimination of Malaria by Source Eradication.

Keith Arnold MD FACP

Preface

In the 1960s, as the war between the United States and North Vietnam raged ever more fiercely, both sides often needed to reinforce their troop strength. Drug-resistant falciparum malaria was highly endemic in the country. Morbidity and mortality were seriously reducing the number of combat-ready troops. China's leaders responded to North Vietnam's call for aid and initiated Project 523, a research program to find a new drug capable of preventing and treating drug-resistant falciparum malaria. They assembled a project leading group and management organization, named Project 523 Head Office. It generated a large-scale collaborative effort involving both military and civilian research units. Based on their experience in the Chinese *Materia Medica*, scientists discovered the active antimalarial component of the *Artemisia annua* plant, artemisinin. From this discovery they synthesized and created a series of new antimalarials especially suited to the treatment of drug-resistant falciparum malaria and referred to as artemisinins.

Research into these drugs took place in the 1970s, a troubled time in China. As in a relay event, multiple ministries, disciplines, units, and personnel worked in concert, overcoming one obstacle after another and making progress through extensive collaboration. Of most importance was the finding that the chemical structure of artemisinin was completely different from that of all known antimalarials. The discovery of this compound and the formulation of more artemisinins and the discovery of other drugs were a major milestone in the global history of antimalarial research following the introduction of quinine several hundred years earlier. It also represented an important scientific breakthrough that combined Chinese and Western medicine and advanced the legacy of Chinese medicine itself.

To deal with the relative insolubility of artemisinin, Chinese scientists sought out suitable preparations and, at the same time, changed the structure of the compound. They chemically reduced artemisinin and used its product, dihydroartemisinin, to synthesize artemether and artesunate. These were first developed into four new formulations: artemisinin suppositories, injectable artesunate and artemether, and dihydroartemisinin tablets. Since the 1990s, to delay the onset of drug resistance, improve cure rates, and shorten the course of treatment, artemisinin-type drugs were combined with other new drugs synthesized or invented in China, such as lumefantrine, naphthoquine, and piperaquine. These combinations, termed artemisinin combination therapies (ACTs) by the WHO, have become the first-line worldwide treatment against falciparum malaria. ACTs have been lauded as the hope for the global containment of malaria in the 21st century, and artemisinin-type drugs have already saved the lives of hundreds of thousands, even millions, of critically ill patients.

As a group and as individuals, the scientists involved in this research have won many prizes both in China and abroad; from the State Technological Invention Award, Second Class in 1979—awarded for the "New Antimalarial, Artemisinin"—other laurels include the State Technological Invention Award, Third Class for artesunate (1989), and artemether (1996); the "Outstanding Technological Achievements Group, Artemisinin" (1996), from the Qiu Shi Science & Technologies Foundation; and the State Technological Invention Award, First Class for lumefantrine (1990), and Second Class for naphthoquine phosphate (1996). In 2004, the King of Thailand awarded the country's top prize for medicine to researchers of artemisinin-type drugs. The piperaquine (2005) and artemether (2007) ACTs also won the State Technological Progress Award, Second Class. In 2009 came the European Inventor Award for non-European countries, followed by the Lasker Award in 2011 and the Nobel Prize in 2015.

When Project 523 was first launched, it was a secret program to deal with urgent matters of military preparedness. Most research reports and data were only rarely publicly disclosed, especially those dealing with the history of this research and clinical studies; although they have long attracted the interest of medical historians in China and abroad. Therefore, all this detailed research had never been collected together and made public. Unfortunately as time went on and the people involved aged, this period of arduous struggle was gradually being forgotten or even fabricated and distorted. A detailed and accurate academic work on research on the artemisinins and the other drugs has not been published until now.

In March 2012, a few scientists and leaders who had dedicated their early years to this great work gathered together in the belief that, in their twilight years, it was their responsibility to organize and publish their experiences and their cumulative research data. This would be a legacy to future generations and perhaps provide food for thought for future drug research and development. It would also be the most fitting memorial to Project 523. Forty years ago, when they were in the prime of their lives, these scientists had dedicated themselves wholeheartedly into finding a new drug to prevent and cure drug-resistant falciparum malaria under difficult circumstances. Now in their old age, for neither glory nor gain, they wish to uphold the spirit of cooperation, very prominent at that time, by using their remaining energy to set down a truthful record of those precious years of combined experiences.

Artemisinin-Based Antimalarials is published after three plus years, through joint effort and after several revisions. The editors and writers came from different units and worked in different disciplines. They were either the inventors who had carried out this innovative research first-hand or were key participants and leaders and organizers. The book reflects the relentless creativity of their research and methods, as well as some unique organizational, management, and collaborative techniques. It strives to comprehensively and objectively describe the history of artemisinin-based antimalarial research in China and its technological achievements. The book's main focus is to introduce China's research findings, including specialized and detailed data from chemistry, pharmacology, toxicology, and clinical studies. Because of concerns over patents and international collaboration, a portion of this material has not been previously published and is presented here for the first time. Furthermore, the latest international technical and clinical data are also included where appropriate.

Artemisinin antimalarial research has not ceased in China. Beginning in 2003, ACTs were used to study Fast Elimination of Malaria by Source Eradication in Cambodia and Comoros, which yielded promising results. It could provide a new path for the control and eradication of malaria worldwide. Some of the most recent data on this research are assembled in the book.

This book is the first, comprehensive summary of the important achievements made in research into artemisinin-based antimalarials, which has taken place in China for almost half a century. It is a valuable, scientific publication. It can serve as a reference work for new drug research in China, medical education, and for historians of medicine, as well as finding a place in academic libraries and archival collections.

Naturally, as times change and society develops, and as science and technology continue to progress, the research facilities and conditions of the past and our understanding of the unknown are nothing like those of today. Discussion and criticism from different academic viewpoints, and suggestions for improvement on any of the book's shortcomings, are welcomed.

The Editors
2015

List of Contributors

Editorial Board Members and Contributors

Chen Chang, National Institute of Parasitic Diseases, Chinese Center for Disease Control and Prevention, Shanghai, China

Guo Xingbo, Institute of Tropical Medicine, Guangzhou University of Chinese Medicine, Guangzhou, Guangdong Province, China

Jiao Xiuqing, Institute of Microbiology and Epidemiology, Academy of Military Medical Sciences, Beijing, China

Li Guoqiao, Institute of Tropical Medicine, Guangzhou University of Chinese Medicine, Guangzhou, Guangdong Province, China

Li Ying, Shanghai Institute of Materia Medica, Chinese Academy of Sciences, Shanghai, China

Li Zelin[†], Institute of Chinese Materia Medica, China Academy of Chinese Medical Sciences, Beijing, China

Liang Li, Institute of Biophysics, Chinese Academy of Sciences, Beijing, China

Liu Tianwei, CITIC Group Corporation, Beijing, China

Liu Xu, Guilin Pharma, Guilin City, Guangxi Zhuang Autonomous Region, China

Luo Zeyuan, Yunnan Institute of Materia Medica, Kunming, Yunnan Province, China; Sichuan Institute of Materia Medica, Chengdu, Sichuan Province, China

Ning Dianxi, Institute of Microbiology and Epidemiology, Academy of Military Medical Sciences, Beijing, China

Shi Linrong, National Project 523 Head Office, Academy of Military Medical Sciences, Beijing, China

Wang Jingyan, Institute of Microbiology and Epidemiology, Academy of Military Medical Sciences, Beijing, China

Wu Yulin, Shanghai Institute of Organic Chemistry, Chinese Academy of Sciences, Shanghai, China

Ye Hechun, Institute of Botany, Chinese Academy of Sciences, Beijing, China

Ye Zuguang, Institute of Chinese Materia Medica, China Academy of Chinese Medical Sciences, Beijing, China

Yuan Shoujun, Institute of Radiology and Radiation Medicine, Academy of Military Medical Sciences, Beijing, China

[†]Deceased.

Zeng Meiyi, Institute of Chinese Materia Medica, China Academy of Chinese Medical Sciences, Beijing, China

Zhou Zhongming, Institute of Chinese Materia Medica, China Academy of Chinese Medical Sciences, Beijing, China

Additional Contributors to Chinese Edition

Chen Lin, Antimalarial Laboratory, Institute of Parasitic Diseases, Second Military Medical University, Shanghai, China

Ding Deben, Institute of Microbiology and Epidemiology, Academy of Military Medical Sciences, Beijing, China

Hu Sihua, Artemisia annua L. Research Center, Institute of Tropical Medicine, Guangzhou University of Chinese Medicine, Guangzhou, China

Li Fulin†, Institute of Microbiology and Epidemiology, Academy of Military Medical Sciences, Beijing, China

Li Guofu, Institute of Microbiology and Epidemiology, Academy of Military Medical Sciences, Beijing, China

Li Linna, Institute of Radiology and Radiation Medicine, Academy of Military Medical Sciences, Beijing, China

Li Tao, Institute of Chinese Materia Medica, China Academy of Chinese Medical Sciences, Beijing, China

Ma Lina, Institute of Chinese Materia Medica, China Academy of Chinese Medical Sciences, Beijing, China

Wang Hong, University of Chinese Academy of Sciences, Beijing, China

Wang Yiwei, Institute of Chinese Materia Medica, China Academy of Chinese Medical Sciences, Beijing, China

Yang Weipeng, Institute of Chinese Materia Medica, China Academy of Chinese Medical Sciences, Beijing, China

Zhang Huihui, Institute of Chinese Materia Medica, China Academy of Chinese Medical Sciences, Beijing, China

Zhang Xiuping, State Institute of Pharmaceutical Industry, Shanghai, China

Zhang Zhixiang, Institute of Microbiology and Epidemiology, Academy of Military Medical Sciences, Beijing, China

Zhong Jingxing, Institute of Microbiology and Epidemiology, Academy of Military Medical Sciences, Beijing, China

†Deceased.

Individuals Provided Assistance with Archival and Source Data

Gu Haoming, Shanghai Institute of Materia Medica, Chinese Academy of Sciences, Shanghai, China

Huang Heng, Yunnan Institute of Materia Medica, Kunming, Yunnan Province, China; Sichuan Institute of Materia Medica, Chengdu, Sichuan Province, China

Li Runhong, Institute for Medical Humanities, Peking University, Beijing, China

Lou Songnian, Shandong Academy of Chinese Medicine, Jinan City, Shandong Province, China

Tian Ying, Shandong Academy of Chinese Medicine, Jinan City, Shandong Province, China

Wu Zhaohua, Shanghai Institute of Organic Chemistry, Chinese Academy of Sciences, Shanghai, China

Zhan Eryi, Yunnan Institute of Materia Medica, Kunming, Yunnan Province, China

Zhang Shugai†, Shanghai Institute of Materia Medica, Chinese Academy of Sciences, Shanghai, China

Contributors to English Edition

Richard K. Haynes, Center of Excellence for Pharmaceutical Sciences, North-West University, South Africa; Department of Chemistry, Hong Kong University of Science and Technology, Hong Kong

Tran Tinh Hien, Oxford University Clinical Research Unit, Ho Chi Minh City, Vietnam; Nuffield Department of Medicine Oxford University, Oxford, England

†Deceased.

Introduction

Eight years ago, my colleagues and I were seeking out drafts and opinions for our edited volume, *A Detailed Chronological Report of Project 523 and the Discovery and Development of Qinghaosu (Artemisinin)*, from the specialists involved in artemisinin research. Shen Jiaxiang, a member of the Chinese Academy of Sciences familiar with the history of artemisinin, suggested that our book should be accompanied by a scholarly volume on the research into artemisinin-based antimalarials. Now that *Artemisinin-Based Antimalarials* has been published, jointly written and edited by the scientists involved in the research and development of these drugs, an important milestone has been reached in the history of Project 523. For this, I am sincerely happy.

Project 523 was an urgent assignment handed down by China's leaders to aid North Vietnam's war effort. It was a complex and important task, in which time was of the essence. Ministries, commissions, and the military high command adopted the same spirit that had characterized Chairman Mao's note on the "recipe" for the atom bomb: "active coordination to achieve this task". With the approval of the State Council and the Central Military Commission, the National Leading Group on Research into Malaria Prevention and Treatment (also known as the Project 523 Leading Group) was formed. A head office was also established to organize and coordinate the work. They brought together 50–60 research units across the country, assembling 500–600 scientists and technicians into one formation to carry out research into combating drug-resistant falciparum malaria. Based on China's traditional materia medica, artemisinin and its derivatives—artemether, artesunate, and dihydroartemisinin—were invented. Using chemical synthesis, a series of new drugs was created and developed, such as pyronaridine, lumefantrine, and naphthoquine. After Project 523 came to an end, the Ministry of Health and the State Administration of Medicine set up the China Qinghaosu Steering Committee to administer and coordinate the further advancement of research into artemisinin-based drugs.

The cooperative efforts undertaken under Project 523 greatly accelerated the Project's own research work and the development of artemisinin-based antimalarials as a whole. Back then when China's technical personnel and facilities were relatively lacking, only the mustering of skills and equipment scattered across various departments and units ensured that their research tasks could be achieved in a short span of time. Active collaboration rallied the energies of the entire group, unified plans and programs, allowed for the division of labor, and made up for the deficiencies in any one specialty or facility. If any unit encountered difficulties, many others came to their aid, creating strength in unity and a wholehearted pursuit of their research goals. With the discovery of artemisinin and development of artemisinin-based antimalarials, for example, many different specialties and units were involved, from screening of traditional and regional herbal drugs, discovery of qinghao and its effective extract portion to isolation of qinghaosu (artemisinin), production of a therapeutically effective crystal, and the fashioning of a new drug from the crude traditional medicinal extract. The Institute of Chinese Materia Medica, China Academy of Chinese Medical Sciences led the way by obtaining a preparation from *Artemisia annua* L., which yielded good clinical results, and isolating the effective antimalarial component. This was an important first step, but the clinical trials on the crystals isolated by the Institute did not provide sufficient proof of their antimalarial effects. Research into *huanghuahao* (the local name for *A. annua*) by the Yunnan Institute of Materia Medica and Shandong Institute of Chinese Medical Sciences began somewhat later, but they progressed rapidly and an effective crystal was isolated. Shandong lacked the necessary patients to establish the crystal's efficacy against falciparum malaria. With the Project 523 Head Office playing a coordinating role, Yunnan's *huanghaosu* (artemisinin) was transferred to the Guangzhou University of Chinese Medicine's Project 523 research unit, which was then working on the treatment of cerebral malaria in the Yunnan-Myanmar border region. In 2 months, a basic assessment of the extract could be made. It was highly effective against falciparum malaria and had low toxicity. The discovery of artemisinin as a new antimalarial drug, from crude extract of *A. annua* (*huanghuahao*) to the crystal isolated and subsequent clinical evaluation, took only a little more than 2 years. Only 3 years passed before a small amount of artemisinin was manufactured and provided for clinical trials outside China. And only 6 years were needed before batches of the drug were produced on an industrial scale for military use.

The invention of new antimalarial drugs of artemisinin, its many derivatives, and artemisinin-based combination therapies are all the fruits of the Project 523 units' concerted efforts. Artemisinin itself was appraised at an evaluation conference organized by the National Project 523 Leading Group. The drug was brought before the awards committee of the State Science and Technology Commission by the Ministry of Health and, in September 1979, the Commission issued certifications of invention to six research units: the Institute of Chinese Materia Medica of China Academy of Chinese Medical Sciences, the Shandong Institute of Chinese Medical Sciences, the Yunnan Institute of Materia Medica, the Institute of Biophysics of the Chinese Academy of Sciences, the Shanghai Institute of Organic Chemistry of the Chinese Academy of Sciences, and the Guangzhou University of Chinese Medicine. At the same time, the *People's Daily* issued a formal announcement to both domestic and foreign audiences that Chinese scientists had invented a new antimalarial, artemisinin. After artemisinin suppositories, artesunate for injection, and artemether injection had been successfully developed, the Ministry of Health held a press conference on

September 11, 1987 to introduce these new preparations. It announced that the suppositories, artemether-oil injections, and artesunate-water injections were especially effective artemisinin derivatives for the treatment of uncomplicated falciparum malaria, and cerebral malaria, and formed the fruits of successful research and development.

In December 2006, in an international conference on research into artemisinins held in Guangzhou, a representative from the WHO said to colleagues from the Project 523 Head Office, "Without Project 523, there would have been no artemisinin. Without artemisinin, there would have been no means of saving the lives of millions of people living in malaria-endemic areas worldwide. Thank you!"

Artemisinin-Based Antimalarials is a record of the various scientists and technicians who took part in Project 523 and worked tirelessly for more than a decade to aid their brothers and friends in a war of counterinvasion, to enhance the military strength of China's border defense forces, and to save the lives of people across the globe living with malaria. It is proof of their humanity, their disregard for personal reward, and their bravery in the face of danger. As a member of the Project 523 Head Office, I witnessed the entire development process behind artemisinin and other antimalarials. With various comrades and friends, we toiled hand in glove for 10 years, creating deep friendships. Thinking back on those uncommon times, the invention of artemisinin and other antimalarial drugs has been lauded throughout the world, especially by the people of Asia, Africa, and Latin America, and has brought glory to China. I would like to share this special honor with the authors from Project 523 and their many readers.

Zhang Jianfang
Former principal administrator,
National Project 523 Head Office

Chapter 1

Discovery of *Qinghaosu* (Artemisinin)—History of Research and Development of Artemisinin-Based Antimalarials

Chapter Outline

Artemisinin-Based and Other Antimalarials. https://doi.org/10.1016/B978-0-12-813133-6.00001-9

Malaria is a mosquito-borne disease caused by the parasite *Plasmodium* and is widely prevalent in tropical and subtropical areas. The most severe form is falciparum malaria, which can readily lead to severe illness in those with no immunity, such as children, and those living in malaria-free area. Morbidity and mortality are high: In Africa alone, more than a million children per year lost their lives to the disease in the past, but mortality has significantly decreased recently to around 500,000.

Since the 1960s, *Plasmodium falciparum* has developed resistance to chloroquine and other commonly used antimalarials. This has spread throughout Southeast Asia, South America, and Africa, posing an imminent threat to the control and treatment of malaria worldwide.

Malaria is also an important problem for military medicine. Already in the 1960s, during the Vietnam War, malaria decimated troop strength on both sides. Consequently, the United States of America established a specific Malaria Research Unit under the Walter Reed Army Institute of Research to intensify the study for new drugs. The search for new antimalarials was also conducted in other research institutes around the world including the United Kingdom, France, Australia, and in established pharmaceutical companies in Europe.

To a request from the leader of the North Vietnamese government, Chinese Communist Party Chairman Mao Zedong and Prime Minister Zhou Enlai decided that solving the malaria problem was critical to supporting North Vietnam's combat capabilities. Thus, China's State Science and Technology Commission (SSTC) and the General Logistics Department of the People's Liberation Army (PLA) convened a national collaborative meeting on malaria drug research in Beijing on May 23, 1967. In attendance were officials from the relevant ministries and military organizations and from the provinces (both ordinary and autonomous), cities, and military districts. The top-priority military project that resulted was termed "Project 523," after the date of this meeting. The National Project 523 Leading Group was created, with the SSTC at its head and the General Logistics Department of the PLA as deputy. Other members of the Group came from the Ministry of Health; Ministry of Chemical Industries; Commission for Science, Technology and Industry for National Defense; and the Chinese Academy of Science. It was headquartered at the Academy of Military Medical Sciences (AMMS), with corresponding offices at the provincial, city, and military district level. These offices directed the formation, management, and coordination of plans relevant to Project 523.[1]

After the development and use of antimalarial combinations for the North Vietnamese troops—malaria prevention and treatment tablets no. 1 (pyrimethamine–dapsone), no. 2 (pyrimethamine–sulfadoxine), and no. 3 (piperaquine–sulfadoxine)—Chinese scientists discovered *qinghaosu* (artemisinin) from studies on traditional Chinese medicine. Because artemisinin had a novel chemical structure, could effectively and quickly treat falciparum and vivax malaria, and had no cross-resistance to chloroquine, it solved the difficult problem of drug-resistant falciparum malaria. To improve the efficacy and solubility of artemisinin, derivatives such as artesunate, artemether, and dihydroartemisinin were synthesized. Moreover, to delay the emergence of drug resistance, improve cure rates, and shorten treatment time, artemisinin and its derivatives were combined with other antimalarials discovered in China—such as lumefantrine, naphthoquine, and an imitated piperaquine. The combination with pyronaridine, another Chinese creation, was by a Korean pharmaceutical company. The results were a series of artemisinin-based combinations which the World Health Organization (WHO) has termed artemisinin-based combination therapies (ACTs). They have had a significant impact on the global Roll Back Malaria initiative.

This chapter introduces the discovery of artemisinin, the creation of artemisinin derivatives, and the innovation of artemisinin-based antimalarials (ACTs) by Chinese scientists. Under a unified national plan, these researchers worked in tandem and shared resources, with each unit playing important unique roles. In so doing, the National Project 523 Head Office and its successors—the Chinese Qinghaosu (artemisinin) Steering Committee and SSTC—were critical in planning and coordinating the research process.

1.1 DISCOVERY OF ARTEMISININ AND EARLY ANTIMALARIAL RESEARCH

Zeng Meiyi[1], Li Guoqiao[2], Li Ying[3], Li Zelin[1†]
[1]Institute of Chinese Materia Medica, China Academy of Chinese Medical Sciences, Beijing, China; [2]Institute of Tropical Medicine, Guangzhou University of Chinese Medicine, Guangzhou, Guangdong Province, China; [3]Shanghai Institute of Materia Medica, Chinese Academy of Sciences, Shanghai, China

From its beginnings in 1967, the search for a new antimalarial was two-pronged. First, a large number of existing chemical compounds were selected for screening and new chemical compounds were synthesized, resulting in the creation of pyronaridine phosphate, lumefantrine, and naphthoquine phosphate. Second, Chinese traditional medicine was reviewed looking for natural compounds. The 3-year research plan drawn up that year proposed seven traditional medicinal plants as priority subjects for research. *Qinghao* (sweet wormwood) was among them.[2]

At that time, some research units had carried out preliminary screening on *qinghao* but failed to detect antimalarial properties. This may have been due

†Deceased.

to differences in the species and habitat of the *qinghao* studied or the extraction methods used. It was subsequently confirmed that only medicinal *qinghao* derived from the *huanghuahao* plant (*Artemisia annua* Linnaeus, of the Asteraceae family) had antimalarial properties. Its active component, *qinghaosu* (artemisinin), varies significantly depending on whether the plant was obtained from northern or southern China and when it was harvested. The content of *qinghaosu* is also highest in the leaves of *huanghuahao* at the growing stage before flower budding. These key factors were only recognized during later research.

A. THE DISCOVERY OF *QINGHAOSU* (ARTEMISININ)

1. An Investigation into Traditional Antimalarials: Survey and Screening of Herbs Used in Endemic Areas for Treating Malaria

Chinese herbal medicine may be broadly divided into two categories: remedies described in the historic *materia medica* or medicine classics, and regional folk herbalism. A specialized collaborative team dealing with Chinese medicine was established by the Project 523 Leading Group. While reviewing traditional Chinese *materia medica* and folk herbal knowledge, the team also visited areas such as Hainan, the Yunnan border, Guangdong, Guangxi, Sichuan, Jiangsu, and Zhejiang, where malaria is endemic. There, it made extensive inquiries among the local people, gathering their various antimalarial remedies. Samples of medicinal herbs were collected and preliminary preparations made in the laboratory for efficacy and safety tests on animals. These studies were followed by tests on patients at pilot sites to decide which remedies should be explored further.

From 1967 to 1969, for example, an investigative team on folk herbs staffed by members from the Chinese Academy of Medical Sciences' Institute of Materia Medica (IMM of CAMS), the AMMS, and researchers from Yunnan, Guangdong, and Jiangsu learned of a traditional antimalarial in Guangdong: *yingzhao* (*Artabotrys uncinatus* [Lam.] Merr.). After tests and a preliminary clinical trial, it was found to be effective. Under the auspices of Yu Dequan from the IMM of CAMS, the Sun Yat-sen University and its Zhongshan School of Medicine, and the South China Institute of Botany of the Chinese Academy of Sciences (CAS), the active component in *A. uncinatus*—*yingzhaosu* A ($C_{15}H_{26}O_4$)—was isolated. Liang Xiaotian, Yu, and others confirmed that this was a new sesquiterpene peroxidic compound.[3] Because of the plant's limited resource supply and the poor stability of *yingzhaosu* A it was not developed further. Nevertheless, the discovery of *yingzhaosu* A greatly inspired the subsequent identification of *qinghaosu's* chemical structure, which was also of a sesquiterpene peroxide type.

2. Screening of Remedies in Chinese Folk Recipes

In January 1969, the Institute of Chinese Materia Medica (ICMM) of the China Academy of Chinese Medical Sciences (CACMS) in Beijing joined Project 523.

The CACMS put Tu Youyou in charge of its 523 Research team, with Yu Yagang as the only team member. Both had degrees in pharmaceutical sciences from Beijing Medical University (now the Peking University School of Pharmaceutical Science) and were trained in modern scientific research methodology. They had also completed a training course at the CACMS, integrating western and Chinese medicine, and understood the historical development of Chinese *materia medica.*

That April, the department of general affairs of CACMS compiled a mimeographed *Collection of Simple and Secrete Antimalarial Remedies* with 640 recipes selected from thousands of letters sent by the public over several years.[2] At the same time, it sent extracts from a few herbal remedies out of the collection to the AMMS for rodent malaria tests. They found that pepper and chilli-alum extracts had an over 80% malaria inhibition rate on rodent malaria. From July to October, Tu Youyou, Yu Yagang, and Lang Linfu carried out on-site clinical trials in Hainan using these two extracts. Out of 44 cases, only one case in each extract group showed parasite clearance. Eventually, these poor results terminated the trial on these extracts.[4] Then, in early 1970, the National Project 523 Head Office dispatched Gu Guoming from the AMMS, where he was researching herbal antimalarials, to work with Yu to produce herbal extracts and test them on rodent malaria.[5] Because of her other responsibilities, Tu did not continue with the team but remained at its head.[6]

3. First Indications of Qinghao's Antimalarial Properties

Yu and Gu continued combing through the Chinese medical literature, preparing and sending herb extracts to AMMS for rodent malaria tests. As his main resources, Yu used the *Special Compilation on Malaria*—a collection of ancient Chinese antimalarial prescriptions edited by the Shanghai Literature Institute of Traditional Chinese Medicine—and the malaria section of the Qing-dynasty *Complete Medical Works of the Library Collection, Ancient and Modern* compiled in 1723. He concluded after in-depth study that "black plum, aconite root, shell of fresh-water turtle *Carapax Trionycis*, *qinghao*, and others" should be singled out, believing that these remedies "had been used in isolation and appeared frequently in combined prescriptions and are worthy of multiple animal tests."[2]

With "Project 523 Team" as the author, Yu edited a manuscript titled *Malaria Remedies in Chinese Medical Literature*.[2] Yu's manuscript noted that *qinghao* remedy was first recorded in *A Handbook of Prescriptions for Emergencies* written by Ge Hong (AD 284–364), prepared thus: "One bunch of *qinghao* in two *sheng* of water, mash it and administer the juice." The extracts prepared by him and Gu were tested against rodent malaria by Jiao Xiuqing of the AMMS. Gu reported that "Herbs that appeared more frequently among the traditional antimalarial remedies were selected as objects of study. Among them were *qinghao*, fresh-water turtle shell etc. Extracts were obtained via boiling in water or ethanol extraction, and sent to the screening group for rodent malaria tests. It was found that *qinghao* (dried aerial part of *Artemisia annua* L. purchased) had definite antimalarial properties, with an inhibition rate of around 60–90%."[7]

It was Yu and Gu who were the first to experimentally establish *qinghao*'s antimalarial properties, providing a valuable reference point for their successors. Yu reported *qinghao* extract's high–rodent malaria inhibition rate to Tu Youyou.[2] In September 1970, Tu added *qinghao* to the screening list.[2] After that, Yu was assigned other work and left Project 523. Gu also returned to his original unit. From September 1970 to May 1971, ICMM's antimalarial herbal research was suspended.[4,8]

4. Discovery of the Effective Extract of Qinghao Against Malaria

In late May 1971, a Malaria Prevention and Control Research Symposium was held in Guangzhou. The Ministry of Health and National Project 523 Head Office again emphasized the importance of Project 523, prompting the ICMM to redouble its research efforts.[9]

Continuing in her role as team leader, Tu and her team member Zhong Yurong resumed preparation of Chinese medicine extracts.[10] After studying over 100 samples, however, they had to consider selecting new remedies and revisiting those which had already been screened and showed relatively higher antimalarial properties.[11] They had previously tested a 95% ethanol *qinghao* extract, but the rodent malaria inhibition rate was only 40%. The team then turned to ether extraction with better results: Rodent malaria inhibition rates reached 99%–100%, but toxicity was high. After removal of its acidic part, the neutral portion of ether extract obtained had high efficacy and low toxicity.[12] With this, research into *qinghao* as an antimalarial gained new emphasis.

By March 1972, at a Nanjing meeting of the Project 523 Chinese medicinal herb research group, Tu represented the ICMM in reporting that the neutral portion of *qinghao* ether extract could achieve a 100% inhibition rate against rodent malaria.[12] This drew the interest of the National Project 523 Head Office and other research units at the meeting.

After toxicology studies in animals and healthy volunteers showed no obvious side effects, Tu and Dai Shaode went to Hainan from August to October 1972 to carry out clinical trials with the neutral portion of *qinghao* ether extract (coded No. 91). Three treatment dosages were tested and all were effective against malaria, with the high-dose group—total dosage 36 g—yielding the best results. It was effective in treating all 11 cases of vivax malaria, including one case of mixed-species infection. Of nine falciparum malaria cases, seven were effectively treated.[13] With the Project 523 Head Office playing a coordinating role, the PLA's Hospital 302 also carried out clinical trials on nine vivax patients; all were cured. The results obtained in all 29 cases were relatively satisfactory.[14]

Through animal experiments and clinical trials, ICMM confirmed that the neutral portion of *qinghao* ether extract had antimalarial properties, especially in treating falciparum malaria. This was a touchstone for further research into *qinghao*. Armed with this news, the Shandong Institute for Parasitic Disease Control and Prevention tested their own neutral portion of ether extract produced from a local plant, *huanghuahao*, in October 1972. It was also effective against rodent malaria.[15] In 1973, the Institute used this *huanghuahao* extract to treat vivax malaria patients on-site, again with good results.[15]

5. Discovery of the Active Antimalarial Component, Crystal II, From Qinghao

In August 1972, as Tu was running clinical trials with the neutral portion of *qinghao* ether extract in Hainan, her colleague Ni Muyun attempted further purification of this neutral portion using aluminum oxide column chromatography but failed to get any significant substance. Zhong Yurong used silica gel column chromatography and gradient elution with ethyl acetate petroleum instead. On November 8, 1972, she isolated three types of crystals numbered I, II, and III.[4,8] Ye Zuguan, of the pharmacology department, carried out rodent malaria tests and found that crystal II was the only one with antimalarial properties. Parasite clearance could be achieved when 50–100 mg of crystal II was fed to mice. Subsequently, ICMM renamed crystal II as *qinghaosu II*.[4,8]

6. Initial Signs of Qinghaosu II's Efficacy Against Vivax Malaria

The quantity of *qinghaosu II* produced by ICMM of CACMS via the chromatography method was limited. Hence, right after acute toxicity tests in mice, experiments were run on healthy volunteers. Three of the researchers took *qinghaosu II*, with one receiving a total dose of 3.5 g over 3 days and the other two receiving 5 g. ECGs, EEGs, and routine liver function, renal function, blood, and urine tests showed no significant changes after medication.[16]

Then, from September to October 1973, Li Chuanjie took a team to Changjiang, Hainan, to carry out clinical trials on *qinghaosu II*. The trial report stated: "Three non-local vivax cases were given a total dose of 3–3.5 g in capsules. Average parasite clearance time was 18.5 hours, fever reduction time 30 hours. Follow-up for three weeks, two cases were cured and one was partially effective (parasites recurred at Day 13). Of five non-local falciparum malaria cases, one case was again partially effective (parasite over 70,000/mm^3, total tablet dosage 4.5 g, fever reduction after 37 hours, parasite clearance after 65 hours, but parasites reappeared on Day 6). Medication was stopped in two cases due to the appearance of premature heart contractions (one was a first-time infection, parasites over 30,000/mm^3, fever subsided at 32 hours after 3 g dosage, but parasites reappeared and temperature rose one day after medication stopped). Two cases were ineffective."[17]

This clinical trial provided the first signs that *qinghaosu II* had some effect on vivax malaria but could not confirm its efficacy against falciparum malaria. From 1974, however, ICMM could no longer isolate *qinghaosu II* for various reasons and could not run further clinical trials.

7. Huanghaosu's Efficacy Against Falciparum Malaria in Clinical Trials

Inspired by the results of ICMM of CACMS' clinical trials with the neutral portion *qinghao* ether extract, Luo Zeyuan from the Yunnan Institute of Materia Medica (YIMM) tried to find plants from the *Artemisia* genus with antimalarial properties. From March to April 1973, she collected *kuhao* ("bitter" wormwood) and tried extraction with four different solvents: petroleum ether, ether, ethyl acetate, and methanol. Several types of crystal were obtained. The one extracted using ether was dubbed kuhao crystal III, and its efficacy against rodent malaria was demonstrated by Huang Heng of the institute. Later, botanist Wu Zhenyi from the Kunming Institute of Botany, CAS confirmed that *kuhao* was a variant of *huanghuahao* with a large flower head, commonly known as "large-headed *huanghuahao*" (*A. annua* L. forma *macrocephala* Pamp.). Kuhao crystal III was therefore renamed *huanghaosu* after the plant's original name.

In June 1973, due to a local shortage of raw materials, researchers bought medicinal *qinghao*—obtained from *huanghuahao* plants grown in the Youyang area—from Sichuan Medicinal Materials Corporation. The content of *huanghaosu* in these plants was 10 times higher than in Yunnan's large-headed *huanghuahao*. The artemisinin they yielded furnished ample material for pharmacological, chemical research, and clinical trials. Therefore, in October, efficacy and toxicity tests with *huanghaosu* were completed, showing good results against rodent malaria and no obvious toxicity.[18] The YIMM's Zhan Eryi, Luo, and others went on to establish petroleum solvent extraction as the basis for large-scale production of *huanghaosu*.[18]

In autumn of that year, Zhou Keding of the National Project 523 Head Office and Zhang Yanzhen from ICMM of the CACMS went to the YIMM to observe the progress of research into large-headed *huanghuahao* and the Institute's extraction techniques. Luo introduced the extraction of *huanghaosu* in detail, sending some samples to ICMM. Then, in 1975, Liu Jingming and Pan Jiongguang from ICMM of CACMS traveled to the YIMM to learn its extraction process and bought the high-quality *huanghuahao*. This enabled ICMM to solve its artemisinin supply problem,[19] a manifestation of the mutual support among research units.

The Yunnan clinical team carried out trials in the province's Yun County from September to October 1974 to determine *huanghaosu's* efficacy against falciparum malaria. Because patients were scarce, however, the trial was inconclusive. During a field visit, director of the Project 523 Head Office Zhang Jianfang made a snap decision to assign *huanghaosu* clinical research to the Project 523 team at the Guangzhou University of Chinese Medicine (Guangzhou UCM). The team, led by Li Guoqiao, was then carrying out research into cerebral malaria treatments in Gengma County Hospital, Yunnan.[20]

With the first three cases, Li found that ring-form falciparum parasites ceased development and rapidly decreased after oral administration of *huanghaosu*. He felt that the rapid action of *huanghaosu* on the parasites far exceeded that of quinine and chloroquine. To quickly evaluate its efficacy against falciparum malaria, Li immediately drew up a treatment protocol for nasogastric administration of *huanghaosu* to cerebral malaria patients. He sent a team to establish an additional test site at the Nanla Community Hospital in Cangyuen County on the China–Burma border, seeking cerebral malaria patients there and from the Gengma County People's Hospital. Between October and December, a total of 18 cases were entered in the trial. 1 had cerebral falciparum malaria, 2 had falciparum malaria with jaundice, and 11 had uncomplicated falciparum malaria. The remaining four had vivax malaria; all were rapidly cured.[21] This was the first clinical confirmation of the efficacy of *huanghaosu* against falciparum malaria and of its rapid action and low toxicity. It was an important discovery of *huanghaosu*'s clinical value.[22] Li's team suggested that injectable artemisinin be developed as soon as possible to treat severe malaria.

Meanwhile, at the Shandong Institute of Chinese Medicine (SICM), now as Shandong Academy of Chinese Medicine (SACM), Wei Zhengxing was also spurred on by ICMM of CACMS' clinical work on the *qinghao* ether extract. In November 1973, Wei extracted the active component from local *huanghuahao* plants, calling it *huanghuahaosu*. From May to October 1974, his team used *huanghuahaosu* to treat 19 cases of vivax malaria in Shandong's Juye County: "Parasites rapidly cleared, symptoms quickly controlled, no apparent side effects."[23]

As mentioned in the previous sections, there were significant differences between the 1973 clinical trials of *qinghaosu II* and the 1974 clinical trials of *huanghuahaosu* or *huanghaosu*. ICMM of CACMS faced many drawbacks in extracting and isolating *qinghaosu II* for clinical trials, making it difficult to avoid impurities. Moreover, as knowledge of *qinghaosu I, II, and III* was still in its infancy, it was mistakenly thought that *qinghaosu II* would still be effective at half the dosage if combined with *qinghaosu III*.[24] The *qinghaosu II* used in Beijing's clinical trials may therefore have contained a large amount of *qinghaosu III*.

The discrepancies in the *qinghaosu II* and *huanghaosu* trials involved their therapeutic efficacy against falciparum malaria. Against vivax malaria, however, there were also clear differences in dosage and effectiveness. As we have seen, ICMM's *qinghaosu II* trial used a 3-day regimen and a total dose of 3–3.5 g. Of the three patients cured, one showed recrudescence at day 13. Shandong's *huanghuahaosu* trial, by contrast, had a 3-day regimen and total dosage of 0.6–1.2 g. All 19 cases were rapidly cured. Yunnan's *huanghaosu* trial gave its four vivax cases a total dosage of 0.2–0.3 g over only 1 day. Again, all were rapidly cured. This clinical data indicated that the content of pure substance in ICMM's *qinghaosu II* was too low.

In sum, ICMM of CACMS was the first to isolate the active principle, *qinghaosu II*, in 1972. Its 1973 clinical trial produced initial signs of *qinghaosu II*'s efficacy against vivax malaria but not against falciparum malaria. Then, the 1974 *huanghaosu* trials in Yunnan and *huanghuahaosu* trials in Shandong—with 14 cases of falciparum malaria and 23 cases of vivax malaria—took the first step in confirming the therapeutic value of *huanghaosu* and *huanghuahaosu*. It was especially effective against falciparum malaria, could treat cases of cerebral malaria effectively, and had low toxicity. The National Project 523 Head Office concluded that the results of the Shandong and Yunnan trials had "greatly prompted further research into *qinghao*."[25] It provided the policy impetus for expanding collaborative research across the country.[26]

After 1978, the names *qinghaosu II*, *huanghaosu*, and *huanghuahaosu* were unified under the term *qinghaosu* in Chinese. Its English name, *artemisinin*, first appeared in the 1980s and entered the Chinese Pharmacopoeia in its 1995 edition.

B. NATIONWIDE COLLABORATION ON ARTEMISININ RESEARCH IN TREATING MALARIA

In April 1975, to accelerate research into *qinghaosu*/artemisinin, the National Project 523 Leading Group held a Project 523 Traditional Chinese Medicine Group Meeting in Chengdu. More than 70 officials were present, including leaders and representatives from the relevant ministries and military organizations and technical personnel from dozens of Project 523 research groups from 10 provinces and autonomous regions, cities, and military districts. At the meeting, reports were delivered on the Yunnan *huanghaosu* trial against uncomplicated and severe falciparum malaria and on the Shandong *huanghuahaosu* trial against vivax malaria. The meeting drew up a road map for artemisinin research. First, a basic evaluation of its speed of action, toxicity, and cross-resistance to other drugs had to be carried out. Second, its chemical structure had to be determined so that it could be synthesized or modified. Third, combination therapies and simple dosage forms had to be researched. These should be easy to implement on a mass scale with greater efficacy in both military and civilian contexts, using readily available raw materials. A nationwide "united front" was organized around these key points.[27]

From 1975 to 1977, participating units across the country worked in tandem to answer a range of research questions. They looked at the distribution of *huanghuahao* (*A. annua* L.), established a technical process for production of artemisinin, studied its pharmacologic and toxicological characteristics, identified and modified its chemical structure, established a quantitative assay and quality control standard, and expanded clinical trials. With the last few tasks, technical institutes were mobilized nationwide in a joint research offensive.

1. Expansion of Clinical Trials

The main endeavor throughout 1975 was to complete clinical trials of suitable scale to demonstrate the effectiveness and properties of artemisinin. That year, Li Guoqiao's team continued with their nasogastric *huanghaosu* trial on cerebral malaria patients in Hainan's Ledong County Hospital. Working with the Dongfang County Hospital, Li's team treated 36 cases of cerebral malaria, achieving a cure rate of 91.7% with nasogastric *huanghaosu*. This further confirmed artemisinin's effectiveness in treating cerebral and other forms of severe malaria.[28] In July, under the auspices of the National Project 523 Head Office, Li held a workshop on clinical trials at Ledong County Hospital and established uniform clinical trial protocols for the whole country.

After the Chengdu meeting, various formulations of artemisinin were tested in Hainan, Yunnan, Henan, Shandong, Jiangsu, Hubei, Sichuan, and Guangxi, as well as in Laos and Cambodia. A total of 2099 patients were treated, 588 with falciparum and 1511 with vivax malaria. Fever reduction and parasite clearance times were clearly faster than those of chloroquine. In oral and injectable forms, the short-term cure rate of artemisinin reached 100%. The recrudescence rate of a 3-day oral regimen was 50%, as opposed to 10% for an oil suspension injection. In 141 cases of cerebral malaria given artemisinin via nasogastric intubation or oil- or water-based injection, the cure rate was 91.5%. No adverse effects were noted in pregnant patients. In Hainan, Yunnan, and other areas, artemisinin proved effective in 143 cases of falciparum malaria that had failed to respond to chloroquine. This indicated that it had no cross-resistance to chloroquine and could be used to treat chloroquine-resistant falciparum malaria. Meanwhile, simple preparations of *qinghao* extract were used to treat 4456 cases of malaria with good results.[29]

2. Determining the Chemical Structure of Artemisinin

After *qinghaosu II* (artemisinin) was isolated by ICMM of CACMS, its melting point, specific optical rotation, infrared spectrum, and elemental analysis were measured. However, the result of the elemental analysis was not in accordance with the molecular formula $C_{15}H_{22}O_5$. Then, in 1973, as suggested by Professor Lin Qishou of the Beijing Medical College's Faculty of Pharmaceutical Sciences, ICMM started the determination of functional groups in artemisinin, a necessary step for later creation of its derivatives. ICMM attempted to detect ketone or aldehyde carbonyl groups by using potassium borohydride to react with artemisinin. The results of the chemical reaction were compared against the infrared spectrum of artemisinin, showing that a carbonyl peak was lost and a hydroxyl peak gained. Back then, it was thought that artemisinin might contain a ketone group.[30]

That same year, to speed up its investigation into the chemical structure of artemisinin, the ICMM collaborated with the Shanghai Institute of Organic Chemistry (SIOC of CAS) to work in this area. From 1974 to 1976, ICMM successively dispatched Ni Muyun, Fan Jufen, and Liu Jingming to the SIOC, where they worked with Wu Zhaohua, Wu Yulin, and Zhou Weishan. In early 1975, because the supply of artemisinin was inadequate, the National Project 523 Head Office arranged for both the SICM and YIMM to step in and provide relatively purer artemisinin crystals (called *huanghuahaosu* in Shandong and *huanghaosu* in Yunnan, respectively) for this research.[31]

In early 1974, after Ni of ICMM had arrived at the SIOC, Wu Zhaohua repeatedly recrystallized artemisinin using the standard procedures of structural analysis. When a constant melting point was reached, an elemental analysis was performed and the results were in concordance with $C_{15}H_{22}O_5$. With hydrogen- and carbon–proton nuclear magnetic resonance spectrometry and high-resolution mass spectrometry, Wu identified the molecular formula and weight of artemisinin and the types of its 15 carbon atoms.

Using chemical and color reactions and quantitative analysis, Wu worked with Ni, Liu, and Fan and established the presence of a lactone group in artemisinin. Then, with infrared spectroscopy, they found that what was previously thought to be a ketone peak was a lactone peak. Thereafter, a series of chemical reactions confirmed that the artemisinin molecule did not have a double bond, epoxy, and free carbonyl group of aldehyde or ketone.[32]

In April 1975, inspired by the fact that the chemical structure of *yingzhaosu* contained a peroxide group (see Section 1.1, A, 1), Wu Zhaohua and Wu Yulin used qualitative and quantitative analysis to show that artemisinin also had such a group. Subsequently, they reduced artemisinin with zinc borohydride or sodium borohydride and obtained the same reduced product but with much higher yield than using potassium borohydride previously. Qualitative analysis of this product showed that the peroxide group was retained. Nuclear magnetic resonance spectrometry indicated that the lactone group had been reduced to form a pair of lactol epimers. They named the product *reduced artemisinin*. This is a very rare lactone reduction.[33]

Together with the existing data, it was possible to deduce some segments of artemisinin's molecular structure. After consulting the chemical structure of arteannuin B, isolated in 1972 from *A. annua* L. by chemists in Yugoslavia, Wu's group proposed a probable chemical structure in the latter half of 1975. This included a peroxyl-lactone ring.[33]

In March 1975, the Project 523 Head Office had also invited the Institute of Biophysics of the CAS to determine the chemical structure of artemisinin. The single crystal of artemisinin and its molecular formula of $C_{15}H_{22}O_5$ were provided by the Qinghaosu Research Group of ICMM of CACMS. Li Pengfei and

Liang Li took charge of the assignment. In November 1975, they successfully established the preliminary chemical structure of artemisinin by X-ray crystal analysis. The structural characteristic of the artemisinin molecule is the presence of a trioxane ring (a six-membered ring comprising an oxygen bridge, a peroxide bridge, and three carbon atoms). This finding rejected the structure with peroxyl-lactone ring—the peroxy group and the lactone group in the molecule located in the same ring—proposed by Wu's group at SIOC mentioned above. Li Pengfei passed away at the end of 1976. In early 1978, Liang Li et al. determined the absolute configuration of artemisinin using anomalous X-ray scattering of CuKa radiation based on a fine-structure model of artemisinin.[34] The stereostructure established that artemisinin is a novel type of sesquiterpene lactone with a peroxyl group. It was a new antimalarial with a chemical structure totally different from all other known antimalarials. By determining the chemical structure of artemisinin, the development of a new generation of antimalarials was given a scientific basis.

In February 1976, the National 523 Head Office announced the goal of developing artemisinin derivatives. By the end of 1977, Li Ying from the Shanghai Institute of Materia Medica (SIMM of CAS) and Liu Xu from the Guilin Pharmaceutical Factory (GPF) had rapidly developed artemether and artesunate injection, which had a higher efficacy than artemisinin.

3. Establishing Assay Methods and Quality Control

Ascertaining the chemical structure of artemisinin was a necessary first step toward establishing quality standards and methods for its modification. In February 1977, the Project 523 Head Office's Zhou Keding chaired a technical workshop on establishing an assay method for detecting artemisinin, at the SICM.[35] Participants came from the ICMM of CACMS, the SICM, the YIMM, and 15 research units from institutes and pharmaceutical factories in Shanghai, Guangdong, Guangxi, Jiangsu, Henan, Sichuan, and Hubei. They carried out on-the-spot demonstrations to share their individual assay techniques, comparing the pros and cons of each method.

In September, another such workshop was held, this time at ICMM of CACMS. Discussions centered on improving the ultraviolet spectrophotometric method developed by the Nanjing Pharmaceutical University and Guangzhou UCM. After comparing the measurement errors resulting from different instruments, it was unanimously decided that the specificity, precision, and accuracy of this method rendered it suitable for determination of the pharmaceutical substance artemisinin and artemisinin preparations. It was also easy for grassroots laboratories to use. Finally, a draft on the use of the ultraviolet spectrophotometric method was written by Shen Xuankun (Guangzhou UCM), Yen Kedong (National Institute for the Control of Pharmaceutical and Biological Products, NICPBP), Luo Zeyuan (YIMM), Tian Ying (SICM), and Zeng Meiyi (ICMM of CACMS). Zeng's draft for the quality standard and the specification of quality control for artemisinin was adopted with input from the Yunnan and Shandong Institutes. This formed a unified quality standard for *qinghaosu* (artemisinin) nationwide.

C. PROJECT VALIDATION AND THE NATIONAL INVENTION AWARD

By the end of 1977, the various research agendas relating to artemisinin had basically been completed. On November 23–29, 1978, the National Project 523 Leading Group held a conference in Yangzhou, Jiangsu, to evaluate these results. More than 100 participants attended from the Ministry of Health, SSTC, General Logistics Department of the PLA, Project 523 regional offices, and principal and collaborative research units. The data submitted for evaluation were divided into 12 topical reports, including *qinghao* plant resources, the chemistry of artemisinin, pharmacology, quality standards, formulations, production technology, clinical trials, and research into derivatives.[36] The specialist group at the conference endorsed these results and approved the Certificate of Validation.

In September 1979, at the recommendation of the Ministry of Health, the SSTC awarded Second Class Certificate of Invention for a New Antimalarial: Artemisinin, to six research units. These were the ICMM of CACMS, the SICM, the YIMM, Institute of Biophysics of CAS, the SIOC of CAS, and the Guangzhou UCM.

The path to the discovery of artemisinin was long but successful. At a time when China was technologically underdeveloped, Chinese scientists of different specialties worked under a unified national research plan and a competent project administration, each made their own contribution and together led to the discovery of new antimalarials, artemisinin and its derivatives, a historical breakthrough in the last century in the treatment of drug-resistant falciparum malaria.

ACKNOWLEDGMENTS

This section is based on a large volume of original archival documents and memoirs provided by the organizers and numerous participants of Project 523. Editorial committee members Shi Linrong, Liu Tianwei, Ning Dianxi, Jiao Xiuqing, and others made corrections and suggestions after completion of the manuscript. Li Runhong also offered ideas and opinions. We would like to express our gratitude to all of them.

1.2 THE DEVELOPMENT OF ARTEMISININ AND ITS DERIVATIVES AS ANTIMALARIALS

The discovery of artemisinin initiated a new era in the history of antimalarials. In 1985, China established "Provisions for New Drug Approval." Therefore, a fresh application was made for artemisinin and its derivatives and the relevant technical data submitted.

The first new drugs approved by the Ministry of Health were artemisinin (qinghaosu) and artemisinin (qinghaosu) suppository, which had been developed by the ICMM of CACMS and the Guangzhou UCM. Approval was then given to artesunate, the brainchild of eight research units: the GPF, Guangxi Medical University (GMU), Guangxi Institute of Parasitic Disease Control and Prevention, Guangzhou UCM, Shanghai Institute of Pharmaceutical Industry (SIPI), Institute of Microbiology and Epidemiology (IME) of the AMMS, the IMM of CAMS, and the ICMM of CACMS. It was followed by artesunate injection, developed by the Guilin No. 2 Pharmaceutical Factory (GPF No. 2), SIPI, and Guangzhou UCM. Next were artemether and artemether oil injection, by the SIMM of CAS and the Kunming Pharmaceutical Factory (KPF) (now Kunming Pharmaceutical Co. Ltd. (KPC)). Having been evaluated, these new drugs were all certified and the necessary production permits issued. They were the first three Class I drugs to be approved in China. Dihydroartemisinin tablets were developed later.

For these new China-made antimalarials to rapidly enter the international market, the quality of artemisinin and its derivatives began to approach international drug registration requirements. This brought the antimalarials up to international standards.

A. ARTEMISININ

Li Zelin[1†], Li Guoqiao[2]
[1]Institute of Chinese Materia Medica, China Academy of Chinese Medical Sciences, Beijing, China; [2]Institute of Tropical Medicine, Guangzhou University of Chinese Medicine, Guangzhou, Guangdong Province, China

In 1981, the Scientific Working Group on the Chemotherapy of Malaria (CHEMAL-SWG) and the group in the Special Programme for Research and Training in Tropical Diseases (TDR) from WHO held their fourth TDR/CHEMAL-SWG meeting in China. The conference advocated "prioritizing the development of intravenous artesunate injection to meet the urgent needs of patients with severe malaria." However, the drug had to be produced in factories adhering to Good Manufacturing Practices (GMP) and approved by the WHO. Fresh pharmacology, toxicology, and clinical studies had to be conducted in research institutes and hospitals with Good Laboratory Practices (GLP) and Good Clinical Practices (GCP). WHO experts inspected Chinese pharmaceutical factories producing artemether and artesunate, but not one of them met GMP standards. Therefore, the WHO did not approve China-made artesunate for direct use in overseas clinical trials. Instead, it asked China to supply the artesunate active ingredient, which would be prepared in the United States for further research. China staunchly maintained its independence and would not allow this raw material out of the country. Further cooperation came to a temporary halt.

[†]Deceased.

Chinese medical and pharmaceutical researchers recognized that China's formulations of artemisinins were far below international benchmarks for the manufacture of new drugs. Their own research and production standards would have to be raised if they wished to venture abroad. However, with China's production technologies and capabilities in their current state, it would be difficult to meet international criteria in a short time. Intravenous injection standards especially would not reach GMP levels quickly. Using his experience with nasogastric and rectal administration in severe malaria patients, Li Guoqiao suggested a twofold approach. First, pharmaceutical factories making injections should be renovated in accordance with GMP standards. Fresh laboratory and clinical trials should be conducted on existing artesunate and artemether injection to GLP and GCP criteria. Second, artemisinin suppositories should be studied. This would allow artemisinin to reach malaria-endemic areas as soon as possible for treating cerebral and other severe forms of the disease.

In 1982, Liu Jingming of the chemistry laboratory and Li Zelin of the pharmacology laboratory from the ICMM of CACMS reasoned that artemisinin suppositories could be readily used on critical malaria cases and children in impoverished areas. Therefore, they discussed the development of such suppositories with Shen Lianci from the pharmaceutical preparation laboratory. When they learned that the Nanjing Medical College's newly synthesized suppository matrix—polyoxyethylene monostearate or S-40—had been approved for production as a substitute for imported cocoa butter, they used it as the matrix for their own artemisinin suppositories.

Liu, Li, and Shen were the principal investigators in the research into artemisinin suppositories. The ICMM's pharmaceutical preparation laboratory was responsible for the manufacturing technology and supplied the drug for clinical use. Its pharmacological laboratory was responsible for pharmacodynamic, pharmacokinetic, and toxicology studies, including three special toxicity tests. Finally, the analytical laboratory had to establish an assay method for artemisinin content in suppositories and body fluids. It cooperated with the pharmacology laboratory in the pharmacokinetic tests. Clinical trials were under the Guangzhou UCM. From 1982 to 1984, artemisinin suppositories were given to 358 cases of falciparum malaria (including 32 severe cases) in Hainan and 105 cases of vivax malaria in Zaoyang City, Hubei,[37] Shenzhen, and Hainan. The total dose for a 3-day treatment course was 2800 mg. The results were very good (see Section 6.1, B).

The Project Validation on Research into Antimalarial Artemisinin Suppositories was held in October 1984 in Guangzhou. The meeting learned that the Ministry of Health would soon issue the Provisions for New Drug Approval, effective from July 1985. Artemisinin suppositories, artesunate injection, and artemether injection—whose development had been prioritized by the Chinese Qinghaosu Steering Committee—would be the first group of Class I drugs to be evaluated. Hence, the meeting was asked to submit an application for artemisinin suppositories in accordance with the new requirements. In June 1985, the Chemical Subcommittee of the Ministry of Health's Drug Evaluation Committee held its first pilot drug evaluation meeting. It considered the artemisinin suppositories application, testing the evaluation process at the same time. Based on the evaluation report, the research group was asked to submit further data.

Official evaluation of artemisinin suppositories took place in March 1986. The research group had 3 months to submit more data in two areas. First, rectal administration should be used in animal pharmacodynamic tests. Second, the pharmacokinetic studies should include results from Phase I trials on healthy volunteers. After the completion of the relevant studies, the Ministry of Health agreed to a 2-year trial production of the artemisinin suppositories as a new Class I drug.[38]

The rectal administration animal tests were successfully completed by Li Zelin, Yang Lixin, and Dai Baoqiang, with the help of Jiao Xiuqing and Ning Dianxi of the AMMS' IME. Zhou Zhongming and others ran the Phase I pharmacokinetic study on healthy volunteers. To measure the concentration of artemisinin in the blood, Zhou used the high-performance liquid chromatography method developed at the analytical laboratory by Zhao Shishan and Zeng Meiyi.

When submitting its application for the suppositories, the research group learned that it had to supply the relevant data on the pharmaceutical substance artemisinin at the same time. Therefore, the group supplemented its dossier with information on artemisinin's physical properties, chemical structure, standards, and so on. The data had been made publicly available by the ICMM of CACMS and other collaborating research units, while Project 523 was still active. Some of these had been published under the authorship of the partner research units. On October 3, 1986, the ICMM was granted a new drug certificate for artemisinin suppositories. It was also the sole recipient of the new drug certificate for artemisinin. This contradicted the National Certificate of Invention for Artemisinin, which had been awarded in September 1979 to the ICMM and five partner research units (see above). This unfortunate event meant that only ICMM of CACMS and Guangzhou Baiyun Mountain Pharmaceutical Factory had the rights to produce artemisinin in China. They would have to apply for a transfer of certificate to others in case of necessity. It was unforeseen by the research group.

Because of the lack of falciparum malaria patients in China, the Guangzhou Baiyun Mountain Pharmaceutical Factory—the manufacturer of artemisinin suppositories—neither promoted its product domestically nor sought to enter international markets. Finally, in the late 1980s, Dr. Keith Arnold brought the suppositories to Vietnam. There, he and Dr. Tran Tinh Hien successfully treated children[39] and adults[40] with uncomplicated and severe malaria patients with the new product.[41] Subsequently, Vietnam produced its own suppositories and used them widely in the country.

B. ARTEMETHER

Li Ying
Shanghai Institute of Materia Medica, Chinese Academy of Sciences, Shanghai, China

1. The Invention of Artemether

In February 1976, the National Project 523 Head Office assigned the SIMM of CAS the task of modifying the chemical structure of artemisinin to find a new injectable antimalarial with an even higher efficacy. The Institute divided the work among its Project 523 research groups. This included the synthetic chemistry laboratory, which was in charge of making minor modifications to the structure of artemisinin; the phytochemistry laboratory, which looked into larger modifications and metabolism of artemisinin; and the pharmacology laboratory, responsible for testing the modified compounds on animals.

Based on existing studies on the chemical reactions of artemisinin, the synthetic chemistry laboratory initiated research into the relationship between chemical structure and antimalarial properties. It found that artemisinin's peroxide group was essential for its antimalarial activities. Studies on rodent malaria also showed that dihydroartemisinin, a reduction product of artemisinin, was twice as effective as artemisinin itself. Because of its hemiacetal structure, however, it was unstable. Its solubility was also not ideal. Hence, Li Ying and other chemists in the synthetic chemical laboratory synthesized its ethers, carboxylic esters, and carbonates.[42] Gu Haoming and colleagues in the pharmacology laboratory tested these derivatives using intramuscular injection on chloroquine-resistant rodent malaria. When their SD_{90} (the required dosage for 90% parasite suppression) was measured, it was found that the efficacy of dozens of derivatives was higher than that of artemisinin.[43] Among them, SM224 (later named artemether) was selected as the candidate for further development because of its greater oil solubility, stability, and efficacy against rodent malaria—six times that of artemisinin.

2. Early Development of Artemether Injections

Chen Zhongliang and other chemists of the phytochemistry laboratory studied the production technology of artemether. They developed a one-step process in which artemisinin was reduced with potassium borohydride and directly etherified without separation of dihydroartemisinin. After its pharmacology, toxicology, and preclinical tests were completed, the National Project 523 Head Office approved the first artemether clinical trial in Hainan. From July to September 1978, the Project 523 clinical research group of the Guangzhou UCM obtained good results with artemether injection.

The KPC in Yunnan was made responsible for the pilot production of artemether for extended clinical trials. In the early summer of 1980, Zhu Dayuan and others from the SIMM of CAS went to the KPC to help expand production. Having completed the pilot production of artemether and artemether oil injection, KPC became the manufacturer of artemether and its injection.

From 1978 to 1980, expanded clinical trials of the artemether oil injection took place in Hainan, Yunnan, Guangxi, Hubei, and Henan. A total of 1088 malaria cases were treated, with 829 falciparum malaria (including 56 cases of cerebral malaria and 99 chloroquine-resistant cases) and 259 vivax malaria patients. The total dosage over a 3- or 4-day course of treatment was 600–640 mg. Short-term cure rates were 100%; patients experienced fever reduction and parasite clearance times of 24–48 h, faster than with chloroquine. A follow-up on 354 patients showed that the 1-month recrudescent rate was 7.8%. In January 1981, an Artemether Antimalarial Project Validation was held in Shanghai, which validated the research results. Thus, the Yunnan Health Department approved artemether for trial production and sale.

3. Injections and Other Preparations

With the Provisions for New Drug Approval introduced in 1985, the Chinese Qinghaosu Steering Committee arranged for participating research units to carry out carcinogenic, teratogenic, and mutagenic testing on artemether. Guangzhou UCM's Sanya Institute of Tropical Diseases ran Phase I and II clinical trials according to GCP standards. These had 315 falciparum malaria patients, including seven cases of cerebral malaria. All cases were cured, with a 28-day recrudescence rate of 7.4%. This set the course of treatment at 5 days and the total dosage at 480 mg (see details in Section 6.3). In September 1987, the SIMM of CAS and KPC were jointly granted new-drug certification for the pharmaceutical substance artemether and artemether injection.

Thereafter, the KPC developed two more formulations: capsules and tablets. In October 1992, artemether capsules received new-drug certification. In 1996, schistosomiasis prevention was added to the approved indications of artemether capsules. They were now a Class V drug and given fresh certification. Finally, artemether tablets certificate was granted in April 1997.

In 1996, artemether was awarded the SSTC's State Award for Inventions, third Class. Later, through international cooperation and exchange, it was included in the ninth edition of the WHO Essential Medicines list in 1997. Artemether capsules, tablets, and injection were recorded in volume V of the third edition of the *International Pharmacopoeia* in 2003 and in its fourth edition in 2013.

4. International Good Manufacturing Practices Certification for Two Lines at the Kunming Pharmaceutical Factory

The first production line to synthesize artemether at the KPF was granted GMP certification by the Australian Therapeutic Goods Administration (TGA) in

2002 and by the WHO in 2005. A newly built synthesis production line obtained this certification from the American Food and Drug Administration (FDA) in 2008 and from the TGA in 2013.

5. Artemether Injections Enter the International Market

The Kunming Factory's artemether products entered the international market in the 1990s. It became a supplier to UNICEF and other international organizations, with artemether injection as an emergency drug. These products have been listed in the national antimalarial or health insurance guidelines of Sudan, Nigeria, Ghana, and dozens of other countries. They have also been registered or marketed in Myanmar, Tanzania, Pakistan, Yemen, and 33 other countries.

ACKNOWLEDGMENTS

The historical content of this segment was based on records provided by Zhang Shugai, the former Technology Director of the Shanghai Institute of Materia Medica, and by Kunming Pharmaceutical Co. Ltd.

C. ARTESUNATE

Liu Xu
Guilin Pharma, Guilin City, Guangxi Zhuang Autonomous Region, China

1. The Invention of Artesunate

The National Project 523 Head Office convened the Combined Western and Traditional Medicine Specialty Group Meeting on Antimalarial Drugs in Nanning, Guangxi, in April 1977. A representative from the SIMM introduced the synthesis of artemisinin derivatives and their efficacy against rodent malaria. Because Guangxi was one of three provinces where *huanghuahao* (*A. annua*) plants were abundant and had higher artemisinin content, the province's Project 523 District Office swiftly set up a Qinghaosu Derivatives Research Collaboration Group.

In June, Liu Xu of the GPF attended a meeting on antimalarial compounds in Shanghai. A representative from SIMM of CAS reported on the synthesis and experimental results of SM224/artemether (see Section B) and other artemisinin derivatives and on the research plans to modify the structure of artemisinin. On his return, Liu immediately started synthesizing artemisinin derivatives. To reduce artemisinin, he successfully used the factory's existing stock of potassium borohydride in place of sodium borohydride. Beginning in August, Liu and others synthesized dihydroartemisinin hemisuccinate (coded 804). It was three times more effective than artemisinin against rodent malaria if given intragastrically and seven times more effective if given intravenously.[44] Its sodium salt (coded 804-Na) could be used to formulate intravenous injection (artesunate) for rapid and convenient treatment of severe malaria.[45]

2. Early Research Into Artesunate Preparations

In July 1978, the GPF No. 2 produced the 804-Na powder injection, while the GPF prepared the 804 tablets. Pharmacology, toxicology, and other tests were conducted on 804-Na by the GMU, Guangxi Institute for Parasitic Disease Control and Prevention (GIPDCP), Guangzhou UCM, and Radiochemical Research Laboratory of Beijing Normal University. No obvious toxicity was detected. That September, the GIPDCP used 804-Na intramuscular injection to treat 32 malaria patients in Ningming, Guangxi. The total dose was 300 mg over 3 days. All cases showed rapid fever reduction and parasite clearance.

Over the next 2 years, the GIPDCP and Guangzhou UCM treated 284 malaria patients in Hainan with 804-Na intramuscular and intravenous injection, 804 tablets, and an 804-sulfadoxine oral combination. These proved effective, fast-acting, and had low toxicity.

In November 1980, the Guangxi Zhuang Autonomous Region Science and Technology Committee, Medical Administration, and Department of Health held the 804 Project Validation meeting in Guilin. The drug was approved and labeled "artemisinin ester" (renamed "artesunate" in 1987). In March the next year, the Department of Health approved the production and marketing of artesunate and artesunate tablets by the GPF.

3. Further Development of Artesunate and Artesunate Injections

In the WHO/TDR conference of October 1981 in Beijing (see Section A), the Chinese delegation reported on the good results obtained with artesunate intravenous injection and artemether intramuscular oil injection in treating cerebral malaria. WHO experts said that the development of artesunate intravenous injection should be prioritized to meet the urgent needs of critically ill patients.

In March 1982, the Science and Education Division of the State Administration of Medicine (SAM) discussed the standardization of artemisinin-based injection at a meeting with the SIPI, GPF, GPF No. 2, and KPF. In May, the GPF received a grant from the Guangxi Economic and Trade Commission for the construction of an artesunate production line. The SAM arranged for the Shanghai Pharmaceutical Design Institute to plan this line according to GMP standards. Construction was completed and production began in 1987.

In August 1982, Jin Yunhua, standing deputy director of the Technical Committee at the State Drug Administration, worked with Uppsala University's pharmacokinetic laboratory on artesunate pharmacokinetics research.[46]

Because the stability of freeze-dried sodium artesunate powder for injection was poor, the SIPI attempted to overcome the substance's unstable hygroscopic property with a double-ampule preparation. An ampule containing a sodium bicarbonate solution would be injected into an ampule of artesunate microcrystals to dissolve them before use. In June 1983, Shi Guangxia from the GPF No. 2 successfully produced artesunate microcrystals using a "wet method." This preparation method received the national invention patent in 1988.

From 1984 to 1986, the Qinghaosu Steering Committee assigned the pharmaceutical, pharmacology–toxicology, and clinical research units to repeat works for artesunate in accordance with the Ministry of Health's Provisions for New Drug Approval of July 1985 and the requirements of the WHO. Between 1985 and 1986, the Guangzhou UCM ran clinical trials with intravenous artesunate injection in Hainan and Xishuangbanna, Yunnan. These were conducted according to international standards. A total of 500 cases were treated, including 443 falciparum malaria (of whom 31 had cerebral malaria), 56 vivax, and 1 malaria cases. The results were good (see Section 6.2, A).

Later, from 1988 to 1989, 180 cases were given intramuscular artesunate injection over 3, 5, and 7 days to compare their efficacy. Recrudescence rates at day 28 were 52% for the 3-day, 9.8% for the 5-day, and 2.5% for the 7-day course. One hundred falciparum malaria cases were given oral artesunate. Recrudescence rates at day 28 were 51.2% for a 3-day and 4.4% for a 5-day course. In the late 80s and early 90s, intravenous artesunate was studied in Vietnam[47] following research into artemisinin suppositories.[41]

Artesunate and artesunate for injection obtained new-drug certification and production licenses in April 1987; artesunate tablets followed in 1988. In March 1989, the GPF and Thailand's Atlantic Laboratories Corp. carried out artesunate tablets clinical trials on the Thai–Burmese border. The Thai Ministry of Health registered artesunate tablets in July 1991. They were subsequently registered and marketed in 43 countries, including Burma (now Myanmar), Pakistan, Vietnam, India, Peru, and some African countries. Also in 1989, artesunate was awarded the State Technological Invention Award, third Class by the SSTC. In 1991, its invention patent won the Chinese Patent Award of Excellence. And in 1995, it won the SAM's Outstanding Medicine Award, first Class.

In 1993, the Ministry of Health placed artesunate and artesunate tablets among the country's list of essential drugs. Artesunate tablets were later recorded in the second section of the *Pharmacopoeia of the People's Republic of China*, 1995 edition.

Furthermore, open and randomized clinical trials were conducted in areas with multidrug-resistant falciparum malaria in 1994. One hundred patients with uncomplicated falciparum malaria were given 600 mg artesunate tablets over 5 days or 800 mg over 7 days. The 5-day course produced a recrudescence rate of 19.1%, compared to 4.2% for the 7-day course. No adverse clinical reactions were seen for both groups. The results showed that the 7-day, 800 mg regimen had a high cure rate.[48] In 1996, a WHO/TDR meeting in Manila proposed this regimen as the standard treatment course (as determined by in vivo sensitivity).

The Ministry of Health approved the marketing of 100 mg artesunate tablets for preventing schistosomiasis, which was a new indication for the drug, in 1996. And in 2002, the scheme to optimize and promote artesunate regimens for schistosomiasis prevention was awarded the National Science and Technology Progress Award, second Class. That same year, artesunate tablets and injection were listed in the 11th edition of the WHO's List of Essential Medicines. They were recorded in Volume V of the 2003 *International Pharmacopoeia* (third edition) and again in 2013 (fourth edition). In December 2005, they obtained WHO prequalification. The WHO's 2006 *Guidelines for the Treatment of Malaria* rated artesunate injection as the drug of choice for severe malaria.

In 2011, the suppliers of artesunate for injection received prequalification from the WHO. At present, Guilin Pharmaceutical Co. is the main production center for artesunate for injection.

D. DIHYDROARTEMISININ

Zeng Meiyi[1], *Li Guoqiao*[2]

[1]*Institute of Chinese Materia Medica, China Academy of Chinese Medical Sciences, Beijing, China;* [2]*Institute of Tropical Medicine, Guangzhou University of Chinese Medicine, Guangzhou, Guangdong Province, China*

Early on dihydroartemisinin was known as "reduced artemisinin." Because of its low solubility and unstable chemical property, it was used as a reaction intermediate in the synthesis of artemisinin derivatives such as artemether, developed by the SIMM, and artesunate, by the GPF.

1. The Chemical Structure of Dihydroartemisinin

From October to December 1973, with the guidance of Professor Lin Qishou, Ni Muyun from the ICMM of CACMS reduced artemisinin with potassium borohydride and measured its melting point and infrared spectrum. In 1974, Wu Zhaohua of the SIOC of CAS used zinc borohydride to reduce artemisinin. Elemental analysis showed that the molecular formula of the product was $C_{15}H_{24}O_5$. The next year, Wu Yulin of the SIOC carried out the same reaction with sodium borohydride and obtained an identical product. It was found that the lactone group in artemisinin had been reduced to form a pair of lactol epimers and was called "reduced artemisinin." Only after the absolute configuration of artemisinin had been confirmed by Liang Li of the Institute of Biophysics of CAS in 1978[49] could the structure of its reduction product finally be deduced.[32]

2. Development of Dihydroartemisinin Tablets

The efficacy of "reduced artemisinin" was double that of artemisinin itself. In the human body, it was also an effective artemether and artesunate metabolite. This was further proof of the importance of this substance.

In 1985, the ICMM of CACMS applied for funds from the Ministry of Health to study the treatment of chloroquine-resistant falciparum malaria with "reduced artemisinin." The principal investigators were Tu Youyou of the chemistry laboratory, Li Zelin of the pharmacology laboratory, and Zeng Meiyi of the analytical chemistry laboratory. A 2-year research plan was drawn up, including the preparation of dosage forms; establishing method for quantitative determination of "reduced artemisinin" content and its quality standards; pharmacodynamic, toxicology, and pharmacokinetic studies (under Zhou Zhongming); and Phase I–II clinical trials with the cooperation of relevant units. The Chinese Qinghaosu Steering Committee provided start-up funding for this research.

That same year, Zeng and colleagues completed their research and developed an assay method for "reduced artemisinin."[50] This became the methodology of choice for dihydroartemisinin and dihydroartemisinin tablets and was detailed in the 2000, 2005, and 2010 editions of the *Pharmacopoeia of the People's Republic of China* (see Section 3.7). For various reasons, however, the pharmacology and other research failed to proceed as planned.

Thereafter, the ICMM made Tu the group leader and commissioned the Beijing University of Chinese Medicine to carry out tests on rodent malaria. The IME at the AMMS was asked to run similar tests with simian malaria, whereas the parasitic diseases research group at the Beijing Union Medical College was assigned in vitro testing with *P. falciparum*. Fu Hangyu's team at the ICMM was in charge of subacute and reproductive toxicity tests in rats and dogs. NICPBP performed the three mutagenic tests: micronucleus, Chinese hamster lung cell chromosomal aberration, and Ames test.[51] The radioimmunoassay method developed by the IMM of CAMS was used in rabbit and dog pharmacokinetic studies[52] and in a Phase I clinical pharmacokinetic study.[53] Tu's group adopted the findings on absorption, distribution, excretion, and metabolism in mice from studies by Zhou Zhongming and the Biology Department of Beijing Normal University.[54] The established method of measuring dihydroartemisinin content by Zeng and Zhao Shishan was also used.[51]

The Sanya Institute of Tropical Medicine of the Guangzhou UCM was in charge of clinical trials. Phase I trials and blood sampling for pharmacokinetic studies were carried out by Xiyuan Hospital of the CACMS. Clinical trials were conducted in six hospitals in Hainan, in areas where chloroquine- and piperaquine-resistant falciparum malaria was endemic. A total of 349 cases of falciparum malaria were treated, all successfully, with rapid action and low side effects. Of these, 239 cases received a total dosage of 480 mg over 7 days. After a 28-day follow-up, 205 cases showed a recrudescence rate of 2.0%. This regimen was proposed as a standard treatment.

"Reduced artemisinin" was renamed as dihydroartemisinin and its oral formulation named dihydroartemisinin tablets. In July 1992, dihydroartemisinin and dihydroartemisinin tablets obtained new-drug certification and received the Chinese Medicine Science Technology Progress Award from the State Administration of Traditional Medicine. The ICMM transferred the tablets to the Beijing No. 6 Pharmaceutical Factory for production. They were promoted worldwide by the Beijing Cotecxin Technical Co. under the "Cotecxin" brand name.

ACKNOWLEDGMENTS

This section was based on data and memoirs supplied by the original researchers involved. It was reviewed and supplemented by members of the editorial board, including Wu Yulin and Liu Tianwei. We would like to express our thanks to all of them.

1.3 INNOVATION OF ARTEMISININ-BASED COMBINATIONS

Beginning in the 1980s, China started combining artemisinins with its other antimalarial inventions—lumefantrine and naphthoquine—and also with the imitated piperaquine. Five effective artemisinin-based combinations were successively developed in the 1990s. They were promoted outside China through various channels. Some have become mainstream drugs for the treatment of falciparum malaria globally. The WHO has recommended that these ACTs should be the first-line treatment in areas where falciparum malaria is endemic.

To enhance efficacy, delay the emergence of resistance, and prolong the therapeutic life span of artemisinins, the malaria research team at the IME of AMMS explored the synergism between artemisinins and lumefantrine or naphthoquine to create artemisinin-based combinations with characteristics superior to each of the partner drugs. In keeping with China's Provisions for New Drug Approval, it developed the artemether–lumefantrine tablets (Coartem) and the artemisinin–naphthoquine tablets (ARCO). The malaria team at the Guangzhou UCM, with its lengthy clinical research experience and preclinical studies, combined artemisinins with piperaquine phosphate and later with piperaquine base. Tablets of dihydroartemisinin–piperaquine phosphate (DHA–PQP)–primaquine–trimethoprim combination (CV8), DHA–PQP (Artekin, Duo-Cotexin), and artemisinin–piperaquine (Artequick) were developed in succession.

At the time the ACTs developed overseas were mostly using artemisinins together with old antimalarials, such as artesunate + mefloquine, artesunate + amodiaquine, and artesunate + sulfadoxine–pyrimethamine. South Korea's Shin Poong Pharmaceutical Co. also created Pyramax using artesunate and another Chinese invention, pyronaridine.

A. ARTEMETHER–LUMEFANTRINE COMBINATION (COARTEM)

Ning Dianxi
Institute of Microbiology and Epidemiology, Academy of Military Medical Sciences, Beijing, China

Artemether–lumefantrine combination was created by the antimalarial research team at the IME of AMMS and jointly developed with KPC in accordance with the Chinese regulations for new drug approval and obtained the new drug certificate under the name of "Compound Artemether" in 1992. CITIC Technology Ltd. united with AMMS and KPC then entered a partnership with Swiss Ciba–Geigy Limited (now Novartis Pharma AG), to develop the combination in accordance with the international standards of GLP, GCP, and GMP with Novartis finally marketing the drug internationally. The WHO gave this compound the English name "Artemether and Lumefantrine" (AL) and placed it in the *List of Essential*

Medicines, the first and second editions of the *Guidelines for the Treatment of Malaria*, and the fourth edition of the *International Pharmacopoeia* (2013) under the name of "Artemether and Lumefantrine Tablets."

Between 1981 and 1982, the antimalarial research team of IME of AMMS proposed the idea of creating a combination and placed it on the agenda. Through basic research, combination tests, new-drug development, and international cooperation, artemether–lumefantrine combination was promoted worldwide. Its development can be divided into three phases.

1. Independent Innovation and Development (1981–92)

At the beginning of the 1980s, monotherapies with artemisinins (artemisinin, artesunate, artemether) showed high efficacy, quick action, and low toxicity in treating falciparum malaria. However, 3-to 5-day regimens of these drugs could not totally eliminate the parasites, easily resulting in recrudescence. In the second half of 1982, Zhou Yiqing and Teng Xihe asked the Chinese Qinghaosu Steering Committee to look into the use of combination therapies to delay the emergence of artemisinin resistance. They obtained approval and research grants.

Despite being fast-acting, highly effective, and of low toxicity, artemisinin was quickly metabolized and had a short half-life. Hence, Zhou suggested that if it were combined with another drug that was metabolized more slowly and had a longer half-life, the resulting therapy would thoroughly kill the parasites in the blood. This would achieve the goals of greater efficacy, shorter treatment times, and delayed resistance. It could even generate new concepts in combination therapy. Benflumetol (later named lumefantrine)—which had been invented by Deng Rongxian, Teng, and others from AMMS—had low toxicity and long-lasting effects. This met the requirements of the proposed partner drug. Moreover, lumefantrine had not been marketed at the time, making it an ideal choice for a combination therapy. Therefore, research was done on an artemisinin and lumefantrine combination. A large amount of experimental data showed that the combination of long- and short half-lives, as well as complementary efficacies, was indeed feasible. Both drugs did boost the others' efficacy and hopefully would slow down the emergence of resistance.

Ning Dianxi joined in this research in 1984, reviewing the experimental results of the artemisinin and lumefantrine combination. On analyzing the best ratio of artemisinin to lumefantrine (20:0.625), the research group felt that the dosage of artemisinin was too high and the gap in the relative dosages too large. It called into question the stability and practicality of this combination. Finally, the group decided to start afresh by exploring a combination of AL (A + B, with B for benflumetol, the original name of lumefantrine). Ning was the principal investigator. Together with Zhou and Wang Yunling, they began studying a new artemether antimalarial combination.

In early 1987, pharmacodynamic studies on rodent and simian malaria, initial research into formulations and assay methods (in collaboration with Zeng Meiyi of the ICMM of CACMS), acute toxicity animal experiments, and pharmacological tests were completed. Because of a lack of funding, however, the Chinese Qinghaosu Steering Committee facilitated a joint venture between the AMMS and KPC. According to this Agreement on the Cooperative Development of Artemether Combinations, the KPC pledged 300,000 yuan (about US$50,000) in installments. It was responsible for pilot test and for supplying the drug for clinical trial use. For its part, the AMMS had to complete new-drug research and apply for certification within the agreed timeframe. If the combination failed evaluation, the AMMS had to return the 300,000 yuan. In the end, all the necessary studies were performed on time and the drug successfully underwent clinical trials and the evaluation process. This collaborative research project was a relatively early example of a joint venture between research institution and pharmaceutical enterprise in China. Both parties weathered the risks of new-drug development.

Experiments showed that the optimal ratio of artemether to lumefantrine in treating rodent malaria was 2:0.75. It then had to be confirmed in simian malaria tests. After reviewing each drug's efficacy against rodent, simian, and human malaria, Ning inferred that the best drug ratio for simian malaria was around 1:5.5. Therefore, he tested two different ratios—1:4 and 1:6—on simian malaria. Results showed that the ratio 1:6 yielded a higher cure rate, but both ratios had similar parasite clearance times. In 1987, clinical trials were conducted in Hainan with patients taking the two component drugs together at ratios of 1:5 and 1:6 in parallel. The 1:6 artemether to lumefantrine combination was found to be the most suitable.

Further clinical trials were carried out between 1987 and 1991 in areas of Hainan and Yunnan where chloroquine-resistant falciparum malaria is endemic. In 1988, Ding Deben and Shi Yunlin also conducted a clinical trial with the artemether–lumefantrine combination in southern Somalia. In the 24 falciparum malaria patients treated, average fever reduction time was 25.7 h. Average parasite clearance time was 29.9 h. A follow-up after 28 days showed a cure rate of 100%. This early, influential trial of the artemether–lumefantrine combination in Africa was a first glimpse of its efficacy. By 1992, all studies had been completed and the drug passed evaluation. Artemether–lumefantrine combination was given new-drug certification under the name of "compound artemether" and a production permit to be manufactured by KPC.

2. Sino-Swiss Joint Research and Development (1990–94)

In August 1988, the Ministry of Science and Technology's Department of Social Development (DSD) called for tenders to export artemisinins or entrust domestic companies to cooperate with potential partners overseas to exploit the global market. One of the 14 successful firms was the China International Trust Investment

Company (CITIC), via its subsidiary CITIC Technology Company. At the DSD's recommendation, Swiss company Ciba–Geigy (now Novartis Pharma AG) began talks with CITIC Technology and Zhou Keding. In March 1990, it signed a confidentiality agreement with CITIC Technology Company, AMMS, and KPC.

In April, Ciba–Geigy sent an assessment delegation under overseas licensing manager Jean Heimgartner to China for an inaugural meeting. One of the members was Dr. Anton Poltera. The Chinese delegation included experts and researchers from the DSD, CITIC Technology Company, KPC, and AMMS. Cong Zhong and Zhou acted as spokespeople. Ciba–Geigy emphasized that it was only interested in drugs that could obtain international patents. Drugs with high recrudescence rates or prone to resistance would not be selected. All the drugs proposed by the Chinese—artemisinin, artemether, and artesunate—were off the table. When Dr. Poltera asked Zhou if "there is anything else on offer that we do not know about," Zhou and Cong, after consultation with AMMS experts, revealed that the Chinese were conducting Phase III trials on an ACT.

The Ciba–Geigy delegation returned to Beijing in October 1990 with a new member, patent affairs expert Dr. Hanspeter Schluep. The Chinese side included Jiao Xiuqing, Zhou Yiqing, Deng Rongxian, Teng Xihe, and Zeng Meiyi. The Swiss were informed of the research into artemether–lumefantrine combination. After investigating the possibility of an international patent for this combination, Ciba–Geigy's definite reply was: "The artemether–lumefantrine combination is novel, can be patented internationally and may represent a breakthrough in malaria treatment. Ciba–Geigy would like to select this combination as a basis for cooperation with the Chinese."

In April 1991, Ciba–Geigy signed a Phase One Agreement with the Chinese. Annex A of the agreement was signed specifically to underscore Chinese interests. Then, in August, it was approved by the DSD. The Agreement, lasting from 1991 to 1994, laid out three issues to be addressed by both parties: the patent application, validation of clinical results, and pharmacokinetic studies of lumefantrine. Failure in any of these areas could terminate the collaboration.

International Patent Application

There were several problems regarding the patent application. First, in 1990, inventions granted priority patent rights in China had to be made available to the public after 1 year. This left only half a year for both sides to sign a formal cooperation agreement. Time was short. Second, the public content of the text of the Chinese patent was too detailed. Not only was the ratio of the component drugs clearly stated but also no scope of protection had been set out. Third, an article on the assaying of the combination had been submitted to a journal. It was finalized and about to be printed. If the article was to be published, it would be impossible to obtain an international patent for the combination. Therefore, it was necessary to ask the journal to withdraw the article manuscript, to which it agreed.

Ciba–Geigy's experts assisted the Chinese with redrafting the patent application. According to China's new patent laws, given the initial patent application in 1990, the final deadline for an international application was in just 45 days time. A series of procedures had to be completed, including the completion and inspection of the application documents. The WHO was asked to issue a statement clarifying the validity of the patent during the application period. Moreover, where a combination therapy was concerned, it was necessary to show that the efficacy of the two active components was higher when used together than when applied singly. Therefore, Professor W.H. Wernsdorfer—Ciba–Geigy's technical consultant—conducted experiments showing the synergistic effects of AL. Through close cooperation with the Swiss and working night and day, the necessary materials were submitted to 64 different countries and territories 2 days before the deadline. Patent protection for the artemether–lumefantrine combination was secured in 49 countries. Most of the work outside China was accomplished by Dr. Schluep and Prof. Wernsdorfer.

Good Clinical Practices–Standard Clinical Trials

Ciba–Geigy invited Prof. Wernsdorfer to be a consultant on this project. For 3 years, both sides conducted a comprehensive evaluation of the artemether–lumefantrine combination based on international GLP and GCP standards. The reasons for the lengthy timeframe were, firstly, that both parties had their own internal issues and, secondly, that lumefantrine was insoluble in water and artemether was a fat-soluble compound. Foreign experts more familiar with water-soluble compounds had their doubts. Therefore, this combination was low on the product selection priority list of Ciba–Geigy's Treatment Review Committee and the Global Research and Development Committee. In addition, the high cost of toxicology studies drove some to reject this unplanned project. Ciba–Geigy decided to run a clinical trial on 100 falciparum malaria patients in China according to GCP standards. If the results proved similar to those already obtained by the Chinese, the project would continue. If not, it would be terminated.

Dr. Poltera and Prof. Wernsdorfer designed the experimental protocol and patient record forms. In the summer of 1993, Jiao Xiuqing conducted a 28-day inpatient trial in Sanya, Hainan, with Ciba–Geigy's inspector supervising throughout. The study showed that the combination had a 96% cure rate and was very well tolerated. A 1994 clinical trial observed changes in *P. falciparum* gametocytes before and after treatment. Its findings, that the number of gametocytes fell after treatment, were confirmed in subsequent studies. This trial also proved that the artemether–lumefantrine combination was more effective than each drug in isolation and that both drugs worked in synergy. The Swiss finally gave this project the green light.

Lumefantrine Blood Concentration Determination

For the pharmacokinetics study, Prof. Wernsdorfer arranged for healthy people to be given oral lumefantrine. Two blood samples were drawn at intervals and sent for independent testing by the AMMS and the WHO's Clinical

Pharmacological Test Center in Penang, Malaysia. Teng Xihe was in charge of the pharmacokinetics study at the AMMS, and Zeng Meiyi was invited to design a method to determine lumefantrine blood concentration. In 1992, both laboratories compared results. Although they found similar trends in blood concentration over time, the concentration values from the Penang laboratory were consistently lower, even zero. Those from the AMMS were 14 times higher on average. The Swiss suspected that the AMMS' methodology was incorrect. The Chinese suggested that Ciba–Geigy Headquarters organize a meeting at which both laboratories could report on their experimental procedures. After discussions at the meeting, the Penang laboratory admitted that they had errors in methodology. To be fair, Ciba–Geigy formed a secret team to prepare blood samples with lumefantrine at various concentrations. These were sent to both laboratories for testing and AMMS' results were completely validated.

This collaborative process increased the Swiss experts' confidence in their AMMS counterparts. The Chinese experts were also deeply impressed by the pragmatic, scientific attitude of the Swiss. Measuring the blood concentration of artemether and dihydroartemisinin required equipment in strict controlled condition. Ciba–Geigy was therefore in charge of establishing methods and performing tests for these two drugs. With the success of the patent application and research process, the collaboration reached a new height.

In September 1994, Ciba–Geigy signed a formal License and Development Agreement with the Chinese. It authorized the use of a Chinese patent to develop a drug of international standards. In December, a joint press conference announced "Chinese–Swiss collaboration to develop a new-generation artemisinin-based antimalarial." The artemether–lumefantrine combination thus entered the stage of collaborative international development.

3. Novartis and the Global Arena (1995–2011)

In 1996, Ciba–Geigy merged with Sandoz Laboratories and became Novartis Pharma AG. It continued to work with the Chinese and focused on opening up international markets, undertaking out further research into the effectiveness, safety, compliance, and acceptability of the artemether–lumefantrine ACT. According to international drug registration requirements, Novartis had to run supplementary studies, such as international multicenter clinical trials spanning Asia, Africa, and Europe. These monitored trials took in 3599 children and adults of different ethnic origins. They included observations on young children (soluble, dispersible tablets suitable for infants were formulated) and pregnant women and comparisons with the individual component drugs and other antimalarials. The trials amply showed that the combination drug was effective and safe.

The drug was marketed under two brand names, Coartem and Riamet, at different prices. Coartem was sold to the WHO, Global Fund, and international charity organizations at cost and to developing countries at low prices.

By contrast, Riamet was sold to the private sector in developed countries at a higher price. Novartis also collaborated with the WHO to improve on the treatment method, reduce prices, and build factories to increase production to rapidly meet demand from countries or organizations.

In 2009, the drug was approved by the American FDA and registered in 86 countries and regions. Almost 700 million treatment doses had been sold by 2014. In 2007, the development, internationalization, and industrialization of the artemether–lumefantrine combination won the National Science and Technology Progress Award, second Class. In 2009, Zhou Yiqing, Ning Dianxi, and their team were given the European Inventor Award, and in 2010 Coartem won the Prix Galien USA for best drug.

ACKNOWLEDGMENTS

This section was based on the "Firsthand Accounts" section of the book *The Road to the Artemether Combination*, written by the antimalarial team at the AMMS.

B. ARTEMISININ–NAPHTHOQUINE PHOSPHATE COMBINATION (ARCO)

Wang Jingyan, Jiao Xiuqing
Institute of Microbiology and Epidemiology, Academy of Military Medical Sciences, Beijing, China

After Coartem, artemisinin–naphthoquine phosphate combination tablets (ARCO) were another ACT invented by the antimalarial research group at the AMMS. It combined naphthoquine phosphate and artemisinin at a ratio of 1:2.5. Like Coartem, ARCO contained complementary components and could delay the emergence of drug resistance. Its most outstanding feature, however, was convenience. A single oral dose could cure drug-resistant falciparum malaria, making it suitable for patients in impoverished areas and children. It was especially appropriate for the preventive treatment of troops and travelers entering malaria-endemic areas. ARCO is currently the most easy-to-use ACT worldwide.

1. Background

Naphthoquine phosphate was a new drug invented by the antimalarial team at the AMMS. It was designed and synthesized by Li Fulin in the 1980s. Li and Jiao Xiuqing's team ran pharmacology, toxicity, and clinical studies and obtained new-drug certification in 1993.

Although naphthoquine and chloroquine were both 4-aminoquinolines, naphthoquine's efficacy was greater and it had no cross-resistance to chloroquine. Against chloroquine-resistant rodent malaria, the effective dose of chloroquine eliminating 90% of the parasites (ED_{90}) was 291.6 mg/kg. However, the ED_{90} of naphthoquine phosphate was only 1.2 mg/kg. The resistance index (I_{90}) of chloroquine was 89.7 as opposed to 1.7 for naphthoquine phosphate. Naphthoquine phosphate was prone to emergence of resistance: After mice were fed naphthoquine phosphate, blood samples were drawn weekly to cultivate for drug resistance. At the 30th and 40th passages, the I_{90} was 83.6 and 164.5, respectively. Clinical tests found that although naphthoquine phosphate could eliminate parasites and had a high cure rate, it acted relatively slowly. Due to its long half-life ($t_{1/2}$)—over 300 h after a single oral dose in healthy volunteers—it also had the advantage of short treatment duration.

To build on its strengths and overcome its weaknesses and to protect the new drug and prolong its use, research was conducted to delay the emergence of resistance to naphthoquine phosphate even before it had entered the market. Artemisinin antimalarials were fast-acting and highly effective but had a short half-life, and 3-day treatment courses showed high recrudescence rates. But they had mutual complementary relationship with naphthoquine phosphate.

The research group tested combinations of naphthoquine phosphate with artemisinin, artemether, and artesunate. The results showed that artemisinin–naphthoquine phosphate and artemether–naphthoquine phosphate had clear synergistic effects, with synergistic indices higher than that of artesunate–naphthoquine phosphate. Artemisinin was selected because of its availability, stability, and cost. In 1991, research on artemisinin-naphthoquine phosphate combination began in earnest. These experiments found that a ratio of 1:2.5 of naphthoquine phosphate to artemisinin was synergistic. Not only could the dosage of each component drug be reduced but also a single administration could be used. This was a major innovative break from the traditional treatment course of at least 3 days.

2. Research and Development

In 1991, research into the artemisinin–naphthoquine phosphate ACT was made mandatory by the General Logistics Department of the PLA. First, orthogonally designed tests on rodent malaria confirmed the synergistic effect of artemisinin and naphthoquine phosphate. The combination's synergistic index against normal strains of rodent malaria was 4.2. By contrast, it was 8.2 against chloroquine-resistant strains. Cultivation of drug-resistant parasites showed that, at the 20th passage, I_{90} of naphthoquine phosphate was 32.8 (moderate resistance), as opposed to 1.5 for the combination (sensitive). The combination significantly delayed the emergence of single-drug resistance.

In experiments with simian malaria, two groups of three monkeys were given a single oral dose of 10 mg/kg naphthoquine phosphate or 30 mg/kg artemisinin. None was cured. However, all three monkeys in a third group given 10 mg/kg naphthoquine phosphate and 25 mg/kg artemisinin were cured, proving the combination's synergy. Acute and long-term toxicity tests comparing the combination to its component drugs were also carried out on mice, rats, and dogs, according to GLP standards. The results indicated that the combination did not increase toxicity and no changes to its target organs were observed. In 1998, the China Food and Drug Administration approved clinical trials for the artemisinin-naphthoquine phosphate combination.

From 1998 to 2000, AMMS—together with the Hainan Tropical Diseases Research Institute, Nandao Farm Hospital, and Tianan Hospital in Dongfang—ran clinical trials on the ACT, with the component drugs as controls. Thirty patients received artemisinin tablets four times over 3 days, the total dosage being 2500 mg. The 28-day cure rate was 66.7%. Another 100 patients were given naphthoquine phosphate tablets twice in 2 days, total dosage containing 1000 mg naphthoquine base. Their 28-day cure rate was 100%. A final group of 100 patients took a single dose of 1400 mg artemisinin–naphthoquine phosphate tablets (1000 mg artemisinin and 400 mg naphthoquine base). These had a 28-day cure rate of 97% (see Table 8.59). The combination had the rapid action of artemisinin and the short treatment course and high cure rate of naphthoquine phosphate.

Expanded clinical trials in Hainan treated 320 falciparum malaria patients. After receiving a single dose of the combination tablets, they showed a 28-day cure rate of 97%. Preclinical and clinical studies both supported the scientific basis for the combination treatment.

Artemisinin-naphthoquine phosphate combination tablets obtained a national patent in 2001 and new-drug certification in 2003. They were transferred to the KPC in 2004 and approved for production the next year. International, multi-center clinical trials organized and completed by Kunming Pharmaceutical in 13 countries all showed that a single oral dose of the combination was effective and safe. Average cure rates for falciparum malaria were over 96% in endemic areas of Asia and Africa. By May 2013, the artemisinin-naphthoquine phosphate combination tablets had been registered in or approved by 22 countries.

This project was funded by Initiatives 85 and 95 of the PLA's General Logistics Department and by the Ministry of Science and Technology's key strategic program for new-drug research and industrialization. In 2009, the Development and International Marketing of a New Artemisinin–Naphthoquine Phosphate Antimalarial Drug (ARCO) won the Military Science and Technology Progress Award, second Class.

ARCO was a new-generation artemisinin combination developed by the AMMS antimalarial research group following its invention of Coartem. It is currently the only single-dose oral ACT in the world for drug-resistant falciparum malaria. Unfortunately, despite ARCO's excellent clinical trial results (see Chapter 8), it has not yet been accepted by certain authorities. Their document states: "An ACT requires a three-day regimen to cover three developmental cycles of the asexual parasite." Nevertheless, we argue that the 28-day or 42-day cure rate is the sole criteria for judging the efficacy of artemisinin combinations. There is reason to believe that ARCO, with its single-dose efficacy, should play a major role in global malaria prevention and treatment.

C. ARTEMISININS–PIPERAQUINE COMBINATIONS (CV8, ARTEKIN, DUO-COTEXIN, ARTEQUICK)

Shi Linrong[1], *Li Guoqiao*[2]

[1]*National Project 523 Head Office, Academy of Military Medical Sciences, Beijing, China;* [2]*Institute of Tropical Medicine, Guangzhou University of Chinese Medicine, Guangzhou, Guangdong Province, China*

In the early 1980s, the antimalarial team of the Guangzhou University of Chinese Medicine carried out clinical studies on a combination of artemisinins and piperaquine with relatively good results. In 1992, when the team was invited to Vietnam to treat cerebral malaria with artemisinin, a decision was made to thoroughly research new combinations.

1. The Birth of the First Artemisinins–Piperaquine Combination: CV8

In the late 1980s, Vietnam was reforming its economy. With internal migration at a high level, malaria outbreaks were rife and many patients with cerebral and other severe forms of the disease died. Dr. Keith Arnold—formerly a researcher of antimalarials with the US Walter Reed Army Institute of Research—was working for the Roche Asian Research Foundation at this time. Together with Tran Tinh Hien, he had been conducting clinical trials in Ho Chi Minh City on artemisinin, its derivatives, and in combination with mefloquine. Arnold had cooperated with Li Guoqiao's team[55] in Hainan in the late 1970s and early 1980s to observe the efficacy of mefloquine and artemisinin[56] and was intimately familiar with the team's early research on artemisinin and cerebral malaria. To Professor Trinh Kim Anh, director of the Cho Ray Hospital in Ho Chi Minh City, Arnold suggested that Li Guoqiao be invited to Vietnam to participate in clinical work. Around the same time, he arranged for the Wellcome Trust under Dr. Nicholas White to establish a research unit in Cho Quan Hospital, Ho Chi Minh City, to research the comparative efficacy of artemisinin derivatives and quinine on cerebral malaria.

A few doctors from Li's team were therefore invited to Ho Chi Minh City in 1991, to offer guidance on the use of artemisinin in treating severe malaria. Previously, quinine was used and malaria deaths across Vietnam stood at over 5000 annually. With such a severe situation, Li proposed to introduce artesunate—produced by the GPF—widely and immediately in all malaria-endemic areas. The team had brought artesunate injection with them from China, which was used in demonstrations on cerebral malaria patients in Cho Ray Hospital. Together with the Chinese doctors' strictly enforced rule of "never leave a comatose patient unattended," which had been set by Li's team, the mortality rate for cerebral malaria was soon brought below 10%. This was far superior to the results obtained with quinine.

Prof. Trinh Kim Anh appealed to hospital staff to learn the Chinese doctors' techniques. Then, with chief physician Dr. Huynh An Binh, a graduate from the Guangzhou Sun Yat-sen Medical College, Trinh took Li to Hanoi to meet the Minister of Health, Professor Pham Song. Pham said, "Malaria outbreaks are prevalent throughout Vietnam. The most critical problem now is how to reduce deaths." Li was moved by the gratitude of the families of cerebral malaria patients and, with the Minister's request for urgent help, his determination to study new artemisinin-based combinations was confirmed.

At the time, both artesunate and artemether had been commercially available for many years but had not yet been widely used in malaria-endemic areas. One reason was the long course of treatment needed. In comparative studies of treatment duration in 1988, Li found that a 7-day regimen was necessary to achieve a cure rate of over 95% for artemisinin and its derivatives. However, patient compliance with this regimen was relatively poor, which increased the burden of dosage and cost. Also, because artemisinin was fast-acting, patients experienced fever reduction and relief of symptoms the day after medication. They were able to work by the third day and the remaining medication was frequently left for the next time the patient became ill. The result was recrudescence and potential drug resistance.

Given this situation, a practical solution was to use existing effective antimalarials together to quickly develop a combination with short treatment duration and high efficacy, able to block malaria transmission at a low cost. It could be widely distributed in community hospitals in malarial areas to lower the occurrence of the disease, reduce the incidence rate, and greatly decrease the cases of severe malaria. Morbidity and mortality would drop. It was a better way of achieving this goal than concentrating only on treating severe malaria patients.

In 1992, therefore, the team tested a single-dose regimen with artesunate and piperaquine phosphate given together. Previous clinical trials conducted from 1984 to 1988 in Hainan showed that this regimen had a relatively high cure rate. The Hainan study followed 47 patients over 28 days and found no recrudescence. In Vietnam, however, the recrudescence rate was 30%. An analysis indicated that multidrug-resistant falciparum malaria was endemic in Vietnam. The parasitic strain might also have been different. Thus, the team repeated clinical trials with various combination regimens.

From 1991 to 1995, while helping with the treatment of cerebral malaria in Vietnam, the team also began studying the effects of artemisinin-based antimalarials on *P. falciparum* gametocytes and the dose–response relationship of primaquine on the infectivity of gametocytes in the *Anopheles* mosquito vector. They found that although artemisinin-based antimalarials could kill early-stage gametocytes quickly, their inhibitory effect on infectious mature gametocytes was slow, with 100% losing their infectivity 14 days after treatment. Patients could still be sources of infection for 14 days after being cured. In the primaquine study, a single oral dose of 7.5 mg primaquine in adults could produce 100% infectivity loss in mature gametocytes within 24 h; even a dose as low as 3.75 mg could reduce infectivity. When used with artemisinin, both drugs had synergistic effects. Li realized that artemisinin's ability to kill gametocytes could be combined with a low dose of primaquine to rapidly block transmission and reduce malaria occurrence.

Comparative clinical trials on 10 different regimens with various combinations of partner drugs were carried out consecutively at the research unit in Xuan Loc Hospital between 1993 and 1994. Because the combinations all contained artemisinin, their rapid action and low toxicity made them superior to the quinine and sulfonamide–pyrimethamine (Fansidar) currently in use. They were being researched for Vietnam and the clinical trials had the support of Trinh and the director of Xuan Loc. With low recrudescence, lack of side effects and rapid loss of gametocyte infectivity as the criteria, Combination No. 8 (DHA–PQP–primaquine–trimethoprim), was finally chosen.

Preclinical studies on Combination No. 8 were conducted in the laboratory of the Guangzhou UCM according to China's new-drug research regulations. Later, Vietnam's National Institute for Malaria and Parasitic Disease Research repeated animal toxicity tests and confirmed that the combination was safe. Clinical trials were performed in multiple hospitals in central and southern Vietnam. The combination was renamed "CV8 malaria tablets," as the eighth combination studied in the China–Vietnam collaboration research project.

In June 1997, CV8 malaria tablets were approved for registration by the Vietnamese Ministry of Health, to be manufactured by the Ministry's No. 26 Pharmaceutical Factory. On December 9, 1999, the Ministry designated CV8 as the first-line drug to be given free of charge to patients in areas where falciparum malaria is endemic. Since then, the Ministry has distributed over several hundred thousand doses of CV8 per year in this manner. Because CV8 contains a low dose of primaquine, clinical studies demonstrated that it could cause *P. falciparum* gametocytes to lose infectivity and block transmission without inducing hemolysis. It played an important role in controlling malaria in Vietnam. In March 2007, to honor the contributions of Li's team toward malaria control, the Ministry of Health awarded Li the People's Health Medal. In June 2011, the Vietnamese government also presented Li with the Friendship Medal.

2. Artekin

CV8 was registered, manufactured, and widely used in Vietnam, drawing the attention of the WHO's representative in Vietnam, Cambodia, and Laos, Dr. Allan Schapira. In 1998, he paid a special visit to the Guangzhou UCM for a detailed look into the development and clinical results of CV8.

At the beginning of 2000, the WHO invited Li Guoqiao and Wang Xinhua to attend a WHO/TDR meeting in Chiang Mai, Thailand. Wang reported on the research into CV8. At the WHO/TDR's request, Li signed a confidentiality agreement with its representative after the meeting and submitted all the technical data on CV8 research for evaluation. Thereafter, Schapira was in frequent contact with Li, providing suggestions on improving CV8.

In May 2001, Schapira and Dr. Jeremy Farrar of the Wellcome Trust-Oxford University Tropical Diseases Clinical Research Unit in Cho Quan Hospital, Ho Chi Minh City went to Guangzhou to discuss specific ideas for improving the combination. Li presented Artekin, a new ACT with a revised ratio of dihydroartemisinin and piperaquine phosphate and with no primaquine or trimethoprim. Schapira and Farrar brought 200 treatment doses back to the Wellcome-Oxford research center in Vietnam for clinical trials.

Artekin is composed of dihydroartemisinin and piperaquine phosphate, to be taken four times over 2 days. Clinical trials were completed in 2002, giving a cure rate of over 97% at 28 days. Considering its efficacy, low level of side effects, cost, and ease of use, Artekin was a relatively well-rounded antimalarial. In January 2003, Artekin was registered and granted new-drug certification in China and was produced by Holley Pharmaceutical Co. Ltd. In March 2010, it was listed in the second edition of the WHO's *Guidelines for the Treatment of Malaria*.

It should be mentioned that Artekin was a legacy of Project 523. Made up of dihydroartemisinin and piperaquine phosphate, which had been used in China for many years, its cost was relatively low. Schapira was very interested in Artekin and hoped that it could be accepted by public health organizations, replace chloroquine, and resolve the problem of growing drug resistance and worsening morbidity and mortality in Africa. Later, he was appointed WHO/TDR representative for the Western Pacific region, becoming a principal member of the WHO headquarters responsible for malaria control. In November 2001, after organizing a WHO/TDR conference in Shanghai on the development of antimalarials, he also planned a June 2002 WHO/TDR meeting in Guangzhou to discuss bringing DHA–PQP tablets to international standards more quickly. This was yet another meeting in China dedicated to artemisinin-based antimalarials, following on from the two Beijing conferences organized by WHO/SWG-CHEMAL in 1981 and 1989 (see Section 1.4). It solicited 12 international authorities to help have Artekin registered worldwide and entered into the WHO's List of Essential Medicines, making it a first-line antimalarial globally.

To bring Artekin research to international standards, Schapira proposed that Li collaborate with Oxford's Professor Nicholas White and the WHO/TDR. Together, they applied for and obtained Medicines for Malaria Venture grants. On this basis, Holley Pharmaceutical Co. Ltd. embarked on a joint venture with Italy's Sigma-Tau to reevaluate the research in line with international standards. With the Wellcome Trust–Oxford University Unit, Holley-Cotec Pharmaceuticals Co. carried out international multicenter clinical trials with more than 5000 cases and a cure rate of over 97%. In December 2009, the WHO's *World Malaria Report* included DHA–PQP in its recommended medications list.

In 2010, the WHO published ACT procurement regulations and the second edition of its *Guidelines for the Treatment of Malaria*. DHA–PQP was among its recommended combinations. It was later produced and marketed by Holley-Cotec under the brand name Duo-Cotecxin and was one of the antimalarials donated by China's leaders to African countries.

3. Artequick

European Union (EU) aid to Cambodia for malaria control reached its sixth year in 2003. During this time, a large number of insecticide-treated bed nets were sent to Cambodia. An artesunate–mefloquine combination (Malarine), made in Cambodia, was the first-line drug for malaria treatment and distributed to community health centers nationwide for free. According to statistics from the Cambodian National Malaria Control Program, malaria morbidity fell from 170,387 cases in 1997 to 110,762 in 2002, a reduction of 35% in 5 years. In 2003, however, it rose to 132,571 cases. Hence, morbidity only declined 22.2% from 1997, an average annual decrease of 4.1%. This was also the final year of the EU aid project. The director of Cambodia's Malaria Control Program asked Li Guoqiao for a solution to reduce morbidity. It encouraged Li once again to improve artemisinin combinations.

The piperaquine phosphate content in Artekin was the same as that of CV8. In early CV8 clinical trials, the drug was taken in three doses, with a significantly lower rate of adverse reactions—such as nausea and vomiting—than chloroquine. Nevertheless, this rate still stood at 5%–10%. Therefore, the regimen was changed to four doses over 3 days (second dose at 6–8 h, third at 24 h, and fourth at 48 h). This reduced each individual dose by 25%. The rate of adverse reactions dropped to 3%–6%, while the 28-day cure rate was maintained at over 95%. The Artekin regimen was therefore also changed to four doses over 2 days (twice per day, 6–8 h apart). It still met the goals of high cure rates, few side effects, and shorter treatment duration. Patient compliance was also better. However, this regimen was inconvenient to apply *en masse*.

Cambodia in 2003 was determined to reduce malaria morbidity as quickly as possible, so Li tapped into his years of practical experience controlling malaria in China. The best method of rapidly bringing down morbidity or controlling outbreaks was to swiftly eliminate sources of infection via mass drug administration (MDA). Therefore, any newly developed artemisinin combination should have MDA as its aim. Such a drug had to have a high cure rate and few side effects. It should be cheap and easy to use and be able to block transmission by quickly inhibiting gametocyte infectivity. In this respect, the 1993–96 study in Vietnam had shown that a single low dose of primaquine (7.5 mg), when combined with artemisinin, could cause 100% infectivity loss in *P. falciparum* gametocytes 24 h after medication (see Section C, 1). The new ACT would require the addition of just a low dose of primaquine at the first administration to achieve the same effect.

To reduce the frequency of medication and gastrointestinal side effects, piperaquine was used instead of piperaquine phosphate in the new combination. This was based on the promising results obtained during a pre-1973 Project 523 study, in which such side effects significantly decreased after the phosphate radical was removed. Another change was the replacement of dihydroartemisinin with artemisinin, at a third of the original cost. As for the required dose of artemisinin in the regimen, a repeat, comparative, dose-finding clinical study was needed. In 2003, two groups of falciparum malaria patients were given daily doses of either 100mg artemisinin or 100mg dihydroartemisinin orally. They were medicated when the parasites were at the ring-form stage. Blood smears were tested every 6h to calculate the number of parasites in the blood. The results showed that both drugs reduced parasitemia by 95% 24h after medication, with 99% in 36h. A serum pharmacology study in healthy volunteers showed that 2mg/kg oral artemisinin was maintained at a concentration effective for parasite inhibition for 12h. This formed the basis for a suitable dosage of artemisinin in the combination.

Finally, the daily dose of artemisinin to piperaquine was set at a ratio of 125–750mg, which greatly reduced the dosage of artemisinin. The new ACT, Artequick, was administered twice in 24h, making its regimen rather shorter than other combinations. Later, after comparative clinical trials with Artekin and Coartem tablets, it was found that there were no significant differences in parasite clearance among the three drugs. An international multicenter clinical trial of Artequick, with nearly a 1000 uncomplicated falciparum malaria cases, showed a cure rate of over 95% at 28days. It demonstrated that Artequick had high efficacy, fast action, low medication frequency, and few side effects. It was stable and cheap and therefore suitable for widespread use in poor areas.

By 2005, all new drug studies had been completed for Artequick, and it was certified and registered in April 2006. In 2009, it was listed as an essential drug for falciparum malaria by the Chinese Ministry of Health. In 2010, the Ministry of Commerce placed it among the drugs sent to Africa for malaria aid. Artequick now has patent protection in 38 countries, is registered and marketed in 16 malaria-endemic countries, and is one of the key antimalarials gifted by China to Africa.

4. Artequick in Fast Elimination of Malaria by Source Eradication

The checkered, decades-long history of malaria control in Hainan and worldwide showed that, given the prevalence of the mosquito vector in the wild, methods based on mosquito elimination were too slow. Therefore, a conceptual change was needed in malaria control to allow for a radical departure from the traditional approach. This was a strategy focusing on active, rapid, and complete removal of the source of infection, namely the malaria parasite and especially the gametocyte. Such methods could swiftly control and even eradicate malaria.

The principle was simple. The parasites can survive in the human body for 1–2 years or longer, whereas the mosquito's life span is normally around 1 month. Parasites in the *Anopheles* vector disappear with the death of the mosquito, and if a new generation of mosquitoes cannot feed on blood containing parasitic gametocytes, they will not transmit malaria. Measures for malaria prevention and control should thus emphasize the destruction of those gametocytes in the human body. In principle and in practice, this approach was fundamentally different from the policy previously advocated in malarial areas around the world, i.e., killing mosquitoes through spraying insecticide and using insecticide-treated bed nets.

In early 2004, together with Cambodia's Ministry of Health and National Malaria Control Program, Artequick clinical trials were conducted in Kampong Speu, where falciparum malaria is severely endemic. To see if malaria could be quickly controlled, MDA of Artequick plus 9 mg primaquine was conducted twice in 40 days, taking in 28,000 people in three areas. The first site was Aoral District. Its population of 7000 saw a reduction in population parasite carrier rate in the first year, from 52.3% down to 13.2%. In particular, falciparum parasite carrier rate fell from 35.9% to 5%, a decline of 86.1%. Following some improvements, another study was carried out in April 2005 in Sprin, Kampot, with a population of 3000. After 6 months, falciparum parasite carrier rate decreased from 20.8% to 0%. With further refinements to the technical protocol, this policy was termed Fast Elimination of Malaria by Source Eradication (FEMSE).

To commend his contributions to malaria control, in June 2006, the Cambodian government awarded Li Guoqiao the Royal Order of Monisaraphon Medal, Knight Grand Cross Class. This was also recognition for the efforts of the China–Cambodia joint antimalarial team.

In September 2006, the antimalarial team at the Guangzhou UCM recommended FEMSE to the African island country of Comoros. The results of FEMSE in Cambodia and Comoros—where it was implemented nationwide—showed that if a mosquito-based policy of malaria control was replaced by a source elimination one, concerted action could rapidly turn high-morbidity areas into low-morbidity ones (see Section 10.6). It could reduce malaria deaths, shorten the period necessary to control or eradicate malaria from decades to a few years, and greatly reduce its cost. This would swiftly relieve the social and economic burdens caused by malaria.

1.4 CHINA'S RESEARCH PROGRAM UNDER THE CHINESE STEERING COMMITTEE FOR RESEARCH OF QINGHAOSU AND ITS DERIVATIVES

Liu Tianwei
CITIC Group Corporation, Beijing, China

A. WHO INTEREST IN CHINA'S ARTEMISININ ANTIMALARIALS

In October 1978, China's Minister of Health Jiang Yizhen and WHO director-general Dr. H. Mahler signed the "Memorandum on Technical Cooperation on Health between the Ministry of Health of the PRC and the WHO" in Beijing. This was a milestone in bilateral cooperation.[57]

The following March, a WHO/TDR delegation visited the Shandong Institute of Parasitic Disease Control and Prevention, where TDR director Dr. A.O. Lucas reviewed the results of artemisinin research.[58] Before the visit, the Ministry of Health had proposed a plan for collaboration with the WHO on parasitic diseases, including research into the mechanisms behind artemisinin's antimalarial properties.[59] In practice, this included the effects of artemisinin on phospholipid metabolism in the cell membranes of rodent plasmodium; metabolic and pharmacokinetic research on artemisinins; and the relationship between the configuration of artemisinin and the cell membrane structure of plasmodium. The research units would be the ICMM of CACMS and SIMM of CAS.[60]

Professor Wallace Peters of WHO/TDR's SWG-CHEMAL was invited to visit by the Health Ministry in March 1980.[61] Peters proposed, first, to organize a conference in Beijing under the auspices of SWG-CHEMAL and, second, to help send a Chinese scholar (Li Zelin) to his laboratory at the London School of Hygiene and Tropical Medicine for joint research into artemisinin's antimalarial activity and mechanism of action in animal models.[62]

Prompted by WHO Assistant Director Professor Chen Wenjie, Mahler sent a letter to Health Minister Qian Xinzhong in December 1980. It stated, "WHO/SWG-CHEMAL believes that the next scientific conference should discuss artemisinin and its derivatives. Due to the urgent necessity of developing new antimalarials, we suggest that the conference be held at the beginning of April 1981 in Beijing or another location, with a preference for China."[63] Indeed, in April 1981, the State Council approved an international academic conference on artemisinin in Beijing, based on a submission from the Ministry of Health, SAM, and SSTC.

Although Project 523 had been terminated in March 1981, the Ministry of Health considered that further research into artemisinin antimalarials was necessary. It also had to prepare for the WHO/SWG-CHEMAL meeting and thereafter to expand cooperation with the WHO. Therefore, a conference preparatory group was formed based on the original National Project 523 Head Office. From late April 1981, following a WHO proposal, the relevant Project 523 researchers were mobilized to write 14 reports on the chemistry, pharmacology, toxicology, and clinical research into artemisinin and its derivatives. A comprehensive summary with seven themes was produced out of these, discussed, and refined. At the same time, technical protocol for joint research with the WHO was drafted.[64] In July, Deputy Minister for Health Huang Shuze oversaw the establishment of a Preparatory Leading Group.

WHO/TDR held its fourth SWG-CHEMAL conference in Beijing on October 6–10, 1981, around the theme of Research into the Artemisinin Antimalarial and its Derivatives. Chinese representatives delivered seven reports, which were discussed in specialist groups. Suggestions for further research were made in these groups. WHO experts rated the significance of the discovery of artemisinin highly. The conference concluded, "Any ideal new drug should have a new chemical structure and mechanism of action… In addition, an easily tolerated, safe and fast-acting formulation to treat severe falciparum malaria, especially cerebral malaria, is needed. Artemisinin conforms to these key requirements because it has a new chemical structure. Its mechanism of action seems to differ from those of existing schizontocidal drugs, and it is fast-acting."[65] At this meeting, Chinese researchers came to understand GLP, GCP, GMP, and other international standards, which was a great boost to their subsequent work.[66]

WHO/SWG-CHEMAL convened this meeting to evaluate the existing data on these drugs, identify gaps in the research, and draw up relevant plans to allow these drugs to be used in malaria control.[67] It was followed by a discussion on how Chinese research institutes and SWG-CHEMAL could collaborate on the research of artemisinin and its derivatives. TDR experts pointed out that the main problem was insufficient pharmacokinetic and toxicology data. At the same time, they inquired about prioritizing joint research as a basis for future international registration of artemisinin and its derivatives. They suggested that China set up a steering committee to implement this plan and ensure effective coordination of the groups involved.[68]

The talks concluded by emphasizing that the relevant Chinese ministries and WHO/SWG-CHEMAL would cooperate only on the scientific aspects and development of artemisinin and its derivatives as antimalarials, particularly on artesunate, and not on commercial or manufacturing activities.[68] After the meeting, the China Academy of Chinese Medicine's report to the Ministry of Health proposed the initial establishment of a leading group for artemisinin development and research.[69]

B. THE CHINESE STEERING COMMITTEE FOR QINGHAOSU RESEARCH AND ITS TASKS

In January 1982, a joint meeting of the Ministry of Health and the SAM was held in Beijing on research problems in artemisinin and its derivatives and the establishment of a committee to oversee the program. On March 20, 1982, the Chinese Steering Committee for the Research of Qinghaosu and its Derivatives, or the Chinese Steering Committee for Qinghaosu Research in short, was formed. Chen Haifeng, director of the Science and Technology Department (STD) at the Ministry of Health, headed the Committee. Wang Pei, deputy dean of the CACMS, and She Deyi, manager of Science and Education at the SAM, were appointed deputy heads. Leaders and experts of the principal research units also held Committee posts. To strengthen communication with the WHO, chief engineers Jin Yunhua and Shen Jiaxiang of the SAM were asked to serve as consultants. (Shen was later invited to work as a consultant for WHO/SWG-CHEMAL.) In the Qinghaosu Steering Committee's secretariat, the AMMS' Zhou Keding—previously responsible for the National Project 523 Head Office—was both a committee member and full-time secretary. Other member secretaries were Wang Xiufeng from the STD, Ministry of Health; Zhang Kui and Li Zelin from the ICMM; Zhu Hai from the SICM; and Yang Shuyu from SAM. The secretariat was located at the CACMS. Zhou was in charge of the daily work of the Committee. Four specialty groups—chemistry, pharmacology, clinical matters, and drug formulation—were set up. That same month, the secretariat issued its first work report.[70]

The report laid out the tasks of the Committee. They were: "To organize the drafting and coordination of research plans, producing a blueprint for foreign technical cooperation; to organize scholarly exchange and discussion of findings; to make proposals on the allocation of funds and training of personnel; and to bring together China's experience in researching and developing innovative drugs." The Committee should collaborate with the WHO/TDR, while maintaining its independence and self-reliance. It particularly emphasized the following two issues.

First was centralized authority. Research into artemisinin and its derivatives was a critical state project. Routine work relating to joint research would be dealt with by the CACMS, but important issues should be submitted to the Ministry of Health and SAM for approval. All research results were the achievements of multiple departments and units working in concert throughout the country. To safeguard national interests, all findings, pharmaceutical ingredients, or preparations, which must be handed over to the WHO or abroad, or any collaborative talks, should henceforth be handled by the Bureau of Foreign Affairs at the Health Ministry. The Bureau would negotiate with the relevant departments or units on an ad hoc basis or pass requests on to superiors for approval.

Second, the ethos of nationwide cooperation and mutual aid must be maintained. The inaugural meeting had agreed that to bring technical collaboration with the WHO to a successful conclusion, domestic collaboration had to be effective first. Artemisinin and its derivatives were Chinese innovations, but if they were to reach international registration requirements, enter the global market, and be used widely, there was still a long way to go. This could not be accomplished by a single department or unit. It required national cooperation and the mutual support of each department and unit, furthering the spirit of unity, looking to the bigger picture, and tackling problems in unison.

This meeting also formulated 2-year research priorities, namely: "To prioritize preclinical pharmacology and toxicology experimental data on sodium artesunate injection, artemether oil injection and oral artemisinin formulations (later changed to artemisinin suppositories) in accordance with the requirements of international drug registration standards. This will form the basis for the commercialization and international registration of these three drugs."

C. JOINT PROJECTS BETWEEN THE CHINESE STEERING COMMITTEE FOR QINGHAOSU RESEARCH AND WHO/TDR

Taking its cue from the October 1981 WHO/SWG-CHEMAL conference in Beijing, which expressed the desire to promote artemisinins antimalarials globally, the January 1982 collaborative meeting drafted a joint research blueprint for 1982–83. These laid out suggestions for collaboration with the WHO.

In February 1982, SWG-CHEMAL secretary Dr. P.I. Trigg, drug policy advisor Dr. M.H. Heiffer (head of pharmacology at the Walter Reed Army Institute of Research), and toxicology expert Dr. C.C. Lee (Li Zhenjun) visited Beijing, Shanghai, Guangzhou, and Guilin. They agreed to select seven projects from the Chinese research blueprint and submit them to WHO/SWG-CHEMAL. A preliminary consensus was reached on the 2-year research projects, technical requirements, funding issues, and the training plans submitted to the WHO/TDR for consideration—namely, for five scientists to study abroad and hold pharmacokinetic and drug metabolism workshops—as well as on clinical trials for sodium artesunate injection in Thailand. Table 1.1 lists the topics studied in

TABLE 1.1 Topics of Collaboration Between the Chinese Steering Committee and TDR

No.	Topic	Work Units Involved
1	Special toxicity studies of sodium artesunate and artemether, including teratogenic, mutagenic, and effects on male reproduction	Institute of Chinese Materia Medica, China Academy of Chinese Medical Sciences
2	Sensitive assay methods for artemisinin and sodium artesunate in biological samples; pharmacokinetic studies	Institute of Materia Medica, Chinese Academy of Medical Sciences
3	Acute and subacute toxicity studies of sodium artesunate and artemether in dogs	Institute of Microbiology and Epidemiology, Academy of Military Medical Sciences
4	Acute and subacute toxicity studies of sodium artesunate and artemether in rodents (rats and mice)	Shanghai Institute of Materia Medica, Chinese Academy of Sciences
5	Sensitivity of the drug-resistant ANKA strain of rodent malaria parasite *Plasmodium berghei* to artemisinin, sodium artesunate, and artemether	Shandong Academy of Chinese Medicine
6	Clinical trials of sodium artesunate and artemether	Guangzhou University of Chinese Medicine
7	Standard formulations of sodium artesunate and artemether	Shanghai Institute of Pharmaceutical Industry

the Steering Committee and WHO/SWG-CHEMAL joint research project and the main research units responsible (the funds were provided by China).

Other themes not included in the agreement but in need of further exploration—such as oral artemisinin formulations, its antimalarial mechanisms, aspects of its systemic pharmacology, delaying drug resistance, and *A. annua* L. resource distribution[71]—were placed on the domestic research agenda.[72]

1. Protracted Negotiations on Sodium Artesunate

In 1982, WHO/SWG-CHEMAL formulated a development plan for sodium artesunate. Its summary stated, "For sodium artesunate to obtain international registration, a batch of this drug should be manufactured as a chemical substance reference to set and maintain standards in line with the US FDA's current GMP criteria… This should be carried out in China, with FDA officials inspecting and confirming the plant material and approving all equipment involved. It will determine if it is possible to carry out the various procedures according to current GMP standards." The areas inspected included the synthesis of the pharmaceutical substance of the new drug, appropriate equipment and procedures for freeze drying and encapsulation, packaging and labelling, analysis of pharmaceutical substance and finished products, as well as stability, sterilization, and pyrogen tests. The production process and technology of Guilin No. 2 Pharmaceutical Factory, manufacturer of the injection, had to be "inspected on-site by qualified experts before clinical trials can be conducted in China and other countries."[73]

In March 1982, Trigg wrote to China to confirm that the development of intravenous sodium artesunate injection in the treatment of cerebral malaria was a priority, with GMP manufacture as a prerequisite.[74] He wrote again in June, informing China of the TDR and FDA personnel who would be arriving in September for the inspection, as well as of their travel arrangement.[75] China requested that KPF, which produced artemether injection, also be included in the tour. To accommodate the FDA inspection, the Qinghaosu Steering Committee organized its own reviews and improvements of the SIPI, GPF, GPF No. 2, SIMM, and other units. These took in the preparation of pharmaceutical ingredient for sodium artesunate injection, manufacturing process flow, and quality control. In July, a study conference on the preparation of artemisinin derivatives was held in Beijing, focusing on evaluating the production technology of sodium artesunate. It concluded that the use of sodium bicarbonate solution and improved freeze drying were the next main research targets.[76]

In September to October of that year, Dr. D.D. Tetzlaff of the US FDA's international inspections department carried out his appraisal. He was accompanied by Trigg, Committee secretaries Zhou Keding and Li Zelin, and Wang Dalin from the SIPI.[77] They inspected the production of sodium artesunate in the GPF and GPF No. 2 and of artemether injection in the KPF. However, they had a negative view of all three factories' attempts to reach GMP status. At China's request, they also reviewed the Shanghai Xinyi Pharmaceutical Factory, which had the highest production standards in China at the time. Nevertheless, it also failed to obtain GMP approval from the FDA inspector.

At the end of September, Zhou Minjun, deputy director of the Foreign Affairs Bureau at the Health Ministry, presided over the concluding meeting. The TDR felt that the factories producing intravenous sodium artesunate injection were "not in accordance with GMP criteria. Therefore, the sodium artesunate formulations provided by China cannot be used in experimental research or supplied to overseas clinical trials. The TDR suggests that preclinical and clinical trials do not commence as yet."[78] Trigg recommended, first, that China build a new production line to GMP standards and, second, that China work with other countries to process a batch of the drug that met GMP criteria. The TDR could help in selecting a suitable foreign organization for joint production.[78]

In November, Zhou Minjun wrote to Trigg on behalf of the Chinese Steering Committee, setting out the plan to construct a GMP-compliant production line. At the same time, he raised the possibility of cooperating with the Walter Reed Army Institute of Research to process a batch of the drug and a more general collaborative effort relevant to this task. Specific proposals relating to this were included.[79]

Trigg replied to Zhou in January 1983, saying that WHO/SWG-CHEMAL had decided to encourage the Walter Reed Institute to work with China.[80] In March, another of Trigg's letters said that terms of cooperation had been drafted.[81] Then, a telephone call from Trigg disclosed that since the Walter Reed Institute was under the US Department of Defense, Colonel R. Young would travel to China to negotiate this matter.[82] The Qinghaosu Steering Committee decided to postpone negotiations,[83] wishing to see the agreement first.[84] Over the next 2 years,[85] a flurry of letters passed between both sides,[86] which basically accepted the draft terms.[87]

In September 1985, WHO/TDR Director Dr. A.O. Lucas (see above) wrote to Xu Wenbo, then director of the Steering Committee. He said, "Because a Chinese scientist in Dr. Brossi's laboratory at the National Institutes of Health (NIH) has found that artesunate is unstable under certain conditions, the TDR considers that the development of artesunate should not be a priority."[88] Thus the two and a half years of negotiations came to an end. Yet, without notifying China, the TDR invited tenders worldwide for six research projects on artemisinin and its derivatives, including techniques for extracting artemisinin from *A. annua* L. and the synthesis of new derivatives.[89] In April 1986, the TDR drew up a comprehensive development plan for arteether, which had been synthesized by China first, but was abandoned because of its inferior efficacy in comparison with artemether.

In February 1987, WHO/SWG-CHEMAL secretary Prof. W.H. Wernsdorfer and WHO Western Pacific Region representative Dr. A. Shirai went to China to discuss cooperative research into artemisinin derivatives with the Committee. Nominally, the discussions were on artesunate and artemether, but their main purpose was to obtain supplies of raw artemisinin from China to support arteether development.[90]

2. WHO Workshops in Assay Methods for Determining Artemisinin and Its Derivatives in Body Fluids

In October 1983, the CACMS hosted a training course on assay methods for artemisinin and its derivatives in body fluids. It had been jointly organized by the Ministry of Health and WHO. The WHO had invited Professors M.G. and E.C. Horning of Houston's Baylor College of Medicine; Professor M. Rowland of the Pharmacology Department at the University of Manchester; Professor W. Bertsch from the Chemistry Department of the University of Alabama; and Professor V.P. Butler from Columbia University's School of Medicine. Fifteen Chinese technicians and researchers attended.[91] The course content included drug metabolic pathways and the extraction, isolation, and identification of metabolites; the use of gas chromatography–mass spectrometry to determine the quantity of drug and metabolites in body fluids; capillary gas chromatography and its application; establishing a radioimmunoassay method; and pharmacokinetic studies.

The course provided an important impetus for the later development of artemisinins and assay methods in body fluid in China. In 1985, Song Zhenyu of the IMM of CAMS used radioimmunoassay to determine the concentration of artesunate and artemisinin in body fluids.[92] Yang Shude did the same for artesunate and dihydroartemisinin using high-performance liquid chromatography and electrochemical polarographic detection methods.[93] In 1986, Zhao Shishan and Zeng Meiyi from the ICMM of CACMS measured the concentration of artemisinin in plasma via high-performance liquid chromatography with pre-column reaction and UV detection. An article on the subject soon followed.[94]

3. The Second Beijing WHO/SWG-CHEMAL Conference

In March 1987, the Committee assembled to hear a report from Shen Jiaxiang, who had participated in the WHO/SWG-CHEMAL meeting. With China-made artesunate and artemether in mind, the meeting's chairman Dr. C.J. Canfield maintained the view that "any drug that is not up to standard cannot be promoted on an international scale." Shen made strenuous counterarguments. TDR Director Dr. T. Godal arranged to meet with him, and Shen brought up six suggestions to which Godal responded enthusiastically. These centered on WHO entrusting China with large-scale production of artemisinin and its derivatives. Godal also said that TDR/SWG-CHEMAL would hold a meeting in China in 1989.[95]

On April 25–29, 1989, WHO/SWG-CHEMAL held its second meeting in Beijing. Its main focus was to evaluate the progress of research into China's antimalarials.[96] Because the Qinghaosu Steering Committee had been disbanded, a conference preparatory leading group was set up, headed by Huang Yongchang, director of the Science and Education Department at the Ministry of Health. A preparatory group was established, with Wang Xiufeng, Zhou Keding, and Zhang Kui in charge of practical matters. The Chinese reports were delivered by Li Zelin, Teng Xihe, Song Zhenyu, An Jingxian, Zeng Yanlin, Li

Guoqiao, Deng Rongxian, Zhou Yiqing, and Chen Chang. They spoke on the progress on artemisinins and synthetic drugs such as pyronaridine, piperaquine, lumefantrine, and artemether–lumefantrine combination, showing that China's antimalarial research agenda had moved from single drugs to ACTs.[97]

4. Other Support From WHO/TDR

The WHO provided its own published material on GLP and GMP, calling on experts to review the English language application documents for artesunate injection and artemether injection. These experts gave advice based on their knowledge of international registration requirements. It arranged for the FDA inspectors to conduct GMP appraisals in China and sponsored seven Chinese researchers to go abroad for specialist training courses and technical observation. Finally, it provided eight beagles, some instruments, and reagents. Together with a French firm, it supported artemether toxicity experiments and Phase III clinical trials in Africa.

D. RAISING THE STANDARDS OF NEW-DRUG RESEARCH AND PRODUCTION SPECIFICATIONS

During its lifetime, the Chinese Qinghaosu Steering Committee drafted three artemisinin research and development plans for 1982, 1983, and 1985. A total of 2 million yuan (around 340,000 USD; 1.33 million USD in 1982) was disbursed by the Finance Ministry for the development of artemisinin-based drugs. Among these, artemisinin, artesunate, artemether, dihydroartemisinin, and the artemether–lumefantrine combination were all projects supported financially by the Committee. Other projects that received funding included a Hainan clinical research base, a facility for the AMMS to house laboratory animals, a nationwide resource survey of *A. annua*, and installations for the industrialization of artemisinin production. Thirteen units were responsible for these research tasks in 1982 (see Table 1.2).

The Qinghaosu Steering Committee emphasized that international cooperation had to be combined with self-reliance. Therefore, it raised the following practical demands[98]:

> To strive for support or funding from the WHO that will benefit China but not to wait for, rely on, depend on, or be restricted by it;
> To accelerate the work within China and construct a GMP production line. The path to new-drug development would have to be taken alone, sooner or later. Foreign aid would always be temporary and cannot replace one's own work;
> To actively seek out other channels for processing drug formulations or for assistance with international registration overseas, protecting China's rights in the manufacture and sale of the drugs internationally; and
> To actively pursue bilateral technological exchange and open the way toward marketing these drugs in countries where malaria is endemic.

TABLE 1.2 Development of Artemisinin Antimalarials and Artemisinin Derivatives: 1982 Projects

Work Unit	Project
Guilin Pharmaceutical Factory	Production technology for artesunate and artemether formulations
Guilin No. 2 Pharmaceutical Factory	Production technology for sodium artesunate formulation
Kunming Pharmaceutical Factory	Production technology for artemether formulation
Shanghai Institute of Pharmaceutical Industry	Production technology for sodium artesunate formulation
Institute of Chinese Materia Medica, China Academy of Chinese Medical Sciences	Artemisinin derivatives: Carcinogenic, teratogenic, and mutagenic studies; pharmacology; formulary research
Institute of Microbiology and Epidemiology, Academy of Military Medical Sciences	Artemisinin derivatives: Acute and chronic toxicity tests on large animals
Guangzhou University of Chinese Medicine	Artemisinin derivatives: Clinical and clinical pharmacology studies
Shanghai Institute of Materia Medica, Chinese Academy of Sciences	Artemisinin derivatives: Acute and chronic toxicity tests on small animals
Institute of Materia Medica, Chinese Academy of Medical Sciences	Assay methods for ultramicroanalysis of artemisinin
Shandong Academy of Chinese Medicine	Resistance to and oral preparations of artemisinin and its derivatives
Guangxi University of Chinese Medicine	Systems pharmacology of artemisinin derivatives
Guangxi Medical University	Systems pharmacology of artemisinin derivatives
China Academy of Chinese Medical Sciences; and Academy of Military Medical Sciences	Raising and breeding of beagles
Xiamen Institute of Medical Sciences	Survey of *Artemisia annua* botanical resources

On the one hand, the Committee did its utmost to win WHO support and take the chance to collaboratively develop sodium artesunate injection. It actively introduced the regulations, standards, and technical methods of international new-drug research to China. On the other hand, it carried out the work on artemisinins antimalarials with China as its focus, bringing openness to a project that had previously been insular. It accelerated the process of aligning China's drug research with the world and cultivated a team of scientists for new drug registration internationally. The team gained an understanding of international procedures and familiarized itself with the GLP, GCP, and GMP requirements of new-drug development. Based on WHO research protocol, it reassessed and supplemented the preclinical pharmacology and toxicology data of artesunate injection and artemether injection. It carried out fresh clinical trials, introduced purebred beagles and white mice as experimental animals, and improved toxicology test conditions. Purebred animal models—i.e., rodents and monkeys—were refined, and carcinogenic, teratogenic, and mutagenic special toxicology studies were run. Finally, an assay method was established for artemisinin, artesunate, and dihydroartemisinin in body fluids by Chinese scientists.

To learn more about international drug registration requirements, Prof. Shen Jiaxiang obtained a grant from the WHO in 1988. With Zhou Keding, he organized a specialist group to prepare and translate the registration documents relating to artesunate injection and artemether injection. The submission material for these two drugs was revised and supplemented according to international standards by Zhao Xiuwen of the Health Ministry's New-Drug Evaluation Office. Teng Xihe, Deng Rongxian, Zeng Meiyi, and others were responsible for the English translation, while Song Shuyuan and Shen reviewed the draft in both languages. The English text was entrusted to a WHO expert for appraisal, who in turn commissioned the Pharmaceutical Systems Inc. (PSI) to assess the dossiers according to the Organization for Economic Cooperation and Development and US FDA criteria. In October 1990, the specialist registration group sent the WHO/SWG-CHEMAL's assessment of the English dossier to the research and production units and editorial staff working on both drugs.[99] This was a great help to the work of Chinese researchers and their understanding of international registration and became a reference for companies exploring international markets.[99]

From 1985—when China began to implement its Provisions for New Drug Approval—to 1995, 14 new Class A drugs were approved in the country. This included five artemisinin-based antimalarials. It showed the significance of these drugs in the history of China's pharmaceutical research and development and in its path toward international involvement.

E. SURVEYS INTO *ARTEMISIA ANNUA* LINNAEUS RESOURCES AND A BASE FOR ARTEMISININ PRODUCTION

Soon after its establishment, the Qinghaosu Steering Committee quickly placed a survey of *A. annua* L. plants and centralized industrial production of artemisinin on its agenda.[100] Funds were allocated to the Xiamen Institute of Medical Sciences for a survey.[101] In August 1983, a conference on the development and use of *A. annua* L. was held in Xiamen, Fujian. It considered that "to produce artemisinin on a large scale, there must be planned cultivation and development of high-quality *A. annua*."

In 1984, Tan Shixian of the Chongqing Nanchuan Medicinal Plant Research Institute at the Sichuan Academy of Chinese Medical Sciences and Feng Tianjiong of the Institute for Drug Control in Fuling, Sichuan, were put in charge of the survey in the province's Youyang County. In July the next year, they completed their investigative *Report into the A. annua resources of Youyang, Sichuan, its ecobiology and artemisinin content.*[102] It stated that the volume of *A. annua* L. purchased from Youyang was around 500,000 kg/year. There was over 550,000 mu (equivalent to 90,610 acres) of land with dry soil suitable for the plant, and average annual yield could be kept at 2,000,000 kg, with a maximum of 7,000,000 kg. The average artemisinin content was 0.59%; few of the coexisting substances contained in the leaves of Youyang *A. annua* interfered with the extraction of artemisinin and hence purity could be easily obtained. Considering Youyang's remote and mountainous location, transporting the leaves to be processed elsewhere was less cost-effective than investing in a new pharmaceutical factory to extract or process the intermediate artemisinin product.[103]

In April 1986, an evaluation conference in Chengdu on Youyang's *A. annua* L. resources confirmed that the region's plants met the requirements for industrial production of artemisinin. The Youyang County government hoped that the Committee would site an extraction plant there. In June, Zhou Keding wrote in response to Feng that the Committee would support the residents of that mountainous area in using local resources, developing their industry and enabling them to rise out of poverty. He suggested that they start by training technical personnel.[104]

In August, the Committee commissioned Professor Wei Zhenxing from the Shandong Academy of Chinese Medicine to go to Youyang to investigate the varieties and artemisinin content of *A. annua* L. and to compare extraction methods. In September, Zhou Keding made his first visit to Youyang, negotiated the construction of an artemisinin extraction production line, and obtained financial support from Sichuan's Science and Technology Commission.

In 1987, Wei took charge of the technical aspects, while Wang Taoxian and another four veterans were dispatched from the Great Wall Pharmaceutical Factory of the PLA's General Logistics Department to assist in the construction of the artemisinin production plant in Youyang. With the Committee's "guarantee that artemisinin will enter international markets," the newly established Three Gorges Corporation invested 1.25 million yuan (about 200,000 USD; 336,000 USD in 1982) into resettlement assistance for poverty-stricken residents—including some in Sichuan—who were affected by the dam project. The KPF chose the Youyang facility as its raw material production site. The artemisinin produced there would be packaged and sold by KPF. KPF also invested 300,000 yuan (50,000 USD; 64,500 USD in 1982) into construction as advance payment for the raw materials. Building of the artemisinin production line at Sichuan's Youyang Wulingshan Pharmaceutical Factory began in 1988. Completed in 1990, it was the world's first plant to produce artemisinin by the ton. By end-September 1994, the Wulingshan Pharmaceutical Factory had produced nearly 8000 kg of artemisinin.

F. PREPARING FOR THE GLOBAL MARKET

The main precondition for entering global markets was that the research and production of new drugs had to satisfy international GLP, GCP, and GMP requirements. Given China's technological level and production facilities in the 1980s, this was a tall order. Therefore, the primary task of the Committee at that time was to prepare for artemisinin drugs' first foray into international markets.

In May 1988, SSTC Deputy Director Guo Shuyan returned from a visit to Africa and organized a forum on the production, marketing, and internationalization of artemisinin-based antimalarials. In July, the SSTC, SAM, Ministry of Foreign Economic Relations and Trade, Ministry of Agriculture, and Ministry of Health jointly issued a notice on accelerating the promotion and export of research into artemisinin-based drugs. Competitive methods would be used to entrust dynamic and qualified companies with international cooperative projects on artemisinin. By then, the research and production of artesunate, artemether, and other antimalarials had gradually improved and GMP production lines established. After 4 years of effort, some drugs had already been registered in over 20 countries in Southeast Asia, Africa, and Latin America and exported as well. International technological and industrial collaboration were proceeding apace.

G. TERMINATION OF THE CHINESE STEERING COMMITTEE FOR QINGHAOSU RESEARCH

In 1987, the CACMS submitted a request to dissolve the Qinghaosu Steering Committee. Health Minister Chen Minzhang, considering the Committee's unfinished tasks and joint ventures with foreign organizations, indicated on 18 May that this should be temporarily postponed.[105]

In June 1988, the Ministry of Health and State Administration of Medicine formally issued a notice on the dissolution of the Committee. It stated that the Committee had completed the research and development of three new antimalarial formulations—artesunate, artemether, and artemisinin suppositories—in June 1987 as planned, and its projects had come to a successful conclusion. Subsequent research into artemisinin-based drugs would be managed by the relevant ministries.

Over its 7-year life span, the Qinghaosu Steering Committee fostered the development and internationalization of artemisinin-based antimalarials, perfected the research and development of a few drugs, and explored international cooperation. In its early days, because of the authority and influence of its leaders, the Committee became a temporary management organization with a certain degree of authority. From 1986, with a change in leadership, full-time members Zhou Keding and Zhang Kui moved over to Prof. Shen Jiaxiang's Chinese Medicine Research and Development Center. They helped in the preparations for the 1989 TDR conference, participated in the exploration of international markets under the auspices of the SSTC, and actively assisted in coordinating the various work units. After the Committee's dissolution, the SSTC continued to support collaboration between the AMMS and Swiss pharmaceutical company Novartis (former Ciba–Geigy) to develop an artemether ACT, which appeared on the market in 1998 as Coartem.

1.5 MANAGEMENT OF ARTEMISININ ANTIMALARIAL PROJECTS

Shi Linrong
National Project 523 Head Office, Academy of Military Medical Sciences, Beijing, China

Looking back on the development of artemisinin-based antimalarials, which took place during a difficult and turbulent age with few scientific and technical personnel, poor equipment, and limited time, one can see that the research team took many a road less traveled. They went from *qinghao* in the Chinese *materia medica* to extracts from *A. annua* L., to artemisinin, to artemisinin derivatives, and finally to the invention of ACTs. The creation of a series of excellent antimalarial drugs validates China's achievement with appropriate pride.

Project 523 and the research into artemisinin-based antimalarials took place under unified leadership and careful planning, with clear research direction, goals, and division of labor. They were managed using a systems engineering model. The National Project 523 Head Office and its successors, the Chinese Qinghaosu Steering Committee and the SSTC, headed the entire process of developing China's new antimalarials. In strategic thinking, the assignment of tasks, securing of human and material resources, and coordination and management, all played a crucial role.

Project 523 began during the Cultural Revolution (1966–76). Yet it was mostly free of disturbances and interruptions, for it was an urgent military assignment in which Mao Zedong and Zhou Enlai had taken a direct interest, one which they had specifically ordered to aid the north Vietnamese government in their war with the United States of America. Mao personally reviewed Project 523's situation reports. Zhou, President Li Xiannian and other state leaders gave important instructions to Project 523 on multiple occasions. With top-level support, Project 523 was highly regarded by central state institutions down to provinces and autonomous regions, cities, and military districts. Despite the unusual circumstances of the Cultural Revolution, when scientific research seemed to be at a standstill, Project 523 was guaranteed ample staffing, material, and financial support from various departments and units from top to bottom. Those scientific and technical personnel lucky enough to be a part of the Project were honored and impressed with their responsibilities. It inspired the highest degree of enthusiasm and dynamism and strongly propelled Project 523 and artemisinin research toward success.

A. CLEAR DIRECTION AND GOALS

With the National Project 523 Head Office and the succeeding Chinese Qinghaosu Steering Committee at the helm, the development of artemisinin and other new antimalarials followed a clear research direction and goals throughout.

The research plan drawn up at the first Project 523 conference in 1967 clearly aimed at prevention and treatment of drug-resistant falciparum malaria in Southeast Asia, with special attention to troop movement. Prior to this, field inspections had been carried out and documents inspected and compiled. Expert opinions were sought. In the drafting and implementation of each year's work thereafter, research into the prevention and treatment of drug-resistant falciparum malaria was always at the top of the agenda.

A two-pronged approach to antimalarial research—combing through China's antimalarial remedies on the one hand and synthesizing new chemical compounds on the other—had already been decided with that first plan. The result was that an active antimalarial substance, artemisinin (qinghaosu), was discovered in the traditional medicinal herb *A. annua* L., and new artemisinin-based antimalarials were invented. At the same time, other new antimalarials such as lumefantrine, naphthoquine, and pyronaridine were synthesized, and the existing drug piperaquine was replicated and developed. Both paths reaped many rewards and laid an important material foundation for the subsequent creation of ACTs.

In the nearly 14 years of Project 523's existence, three general research plans were laid out. Goals were established in a timely manner based on the progress of research. A total of 21 years had passed under the management of the National Project 523 Leading Group and its Head Office, as well as the Chinese Qinghaosu Steering Committee and its Secretariat. Specific research agendas and plans for division of labor were handed down to the work units each year, with the existing direction and goals in mind. Their progress was summed up or exchanged through expert meetings or specialized discussions. Some necessary adjustments were made but, from start to finish, they did not deviate from the original goal.

The Qinghaosu Steering Committee and its Secretariat were established after the termination of Project 523. With the achievements of artemisinin and its derivatives in hand, the Committee took aim at the opportunity to develop them internationally. Seizing the moment, it drafted a series of research plans in accordance with international requirements for new drug registration. This was a step forward in realizing the fundamental goal of commercializing and internationalizing China's artemisinin antimalarials. Over seven plus years, the Committee organized many pharmaceutical research and production units, completing the development of artemisinin suppositories, artesunate, and artemether formulations. At the same time, it conducted a domestic survey of *A. annua* plants and built an artemisinin production site. It introduced international criteria to raise China's own standards of drug research and production and played an important role in the initial preparations for entering global markets.

B. FORWARD-LOOKING INNOVATION

After Project 523 had been mandated, based on the situation then, it was necessary to rapidly produce an effective drug for military use that could reduce malaria morbidity and mortality. Back then, the solution to this emergency was to "buy a horse while riding a donkey," i.e., to replicate or find new uses for old drugs, turn these into combinations, and promptly address the troops' urgent malaria problem. Nevertheless, it was acknowledged that using combinations of old drugs would lead to drug resistance in time. If a long-term answer to drug-resistant malaria was to be found, one had to seek for a drug with a new type of structure, forging a new path.

On the chemical synthesis front, new compounds were developed while existing ones were screened. The criteria for new chemicals were strict. Many of the tens of thousands of newly synthesized or selected compounds produced relatively positive results in animal malaria tests. However, if their chemical structures were similar to those of existing antimalarials—4-aminoquinolines (e.g., chloroquine, to which there already was resistance), sulfonamides, chloroguanides, for example—comparative drug resistance studies were required before they were studied further.

On the Chinese medicine front, research focused on ease of use, exploiting local resources and the extraction of active components. The isolation of the effective monomer was essential for finding an antimalarial with a new structure. Artemisinin was discovered in the Chinese medicinal herb *A. annua* L. and, when clinical trials confirmed its antimalarial properties, forces were immediately organized to determine and modify its chemical structure. This would overcome the obstacle of artemisinin's large required dosage and short treatment duration, which led to a high rate of recrudescence. Hence artesunate, artemether, and dihydroartemisinin were invented in succession, with high efficacy. To protect these new drugs, delay the onset of resistance and increase efficacy, the AMMS and Guangzhou College of Traditional Chinese Medicine (now the Guangzhou UCM) initiated research into new drug combinations. Of these, artemether–lumefantrine, DHA–PQP, artemisinin–piperaquine, and artemisinin–naphthoquine phosphate have become increasingly important in global malaria control.

Artemisinin is a natural compound. From the perspective of production costs, a wholly synthetic version still cannot replace natural extraction even after several decades of research. Zhou Keding, Zhang Kui, and Zhu Hai of the Chinese Qinghaosu Steering Committee Secretariat foresaw the future prospects of artemisinin. They felt that the construction of an artemisinin production site was an urgent project and immediately organized a survey of wild *A. annua* L. plants. They arranged for Wei Zhenxing of the SICM to look into improving the production technology of artemisinin and seek out a site for its manufacture. Finally, with support from provincial leaders in Sichuan, an artemisinin production plant was built in Youyang with annual output in tons, promptly resolving the raw material supply issue.

C. NEW CENTRALIZED MANAGEMENT

The momentous achievements of Project 523 and the research into artemisinin-based drugs were made possible by a strategy devised by core authoritative leaders. Executive organizations drew up plans and agendas based on this strategy, dividing it up, allocating the work, and putting it in practice. The high productivity achieved by each work unit was created and brought about by this systems engineering form of management.

From 1967 to 1981, as Project 523 and artemisinin research were being implemented, the National Leading Group and its Head Office were responsible for the formulation of plans, supervision, and inspection. They played a central role in command and coordination. Below them were the Project's regional leading groups and offices, in charge of implementing these plans, coordination, and management. These offices resolved the issues surrounding the realization of the Project. Leading groups and work units on both national and regional levels assigned liaison officials to maintain contact with the Head Office. Their tasks were to make timely reports and convey leaders' instructions, ensuring that implementation of the Project was always under effective, centralized, and unified management.

In 1981, the Project had been completed and its structures dissolved. To meet the new circumstances and strengthen cooperation with the WHO and other bodies, the Health Ministry and the State Administration of Medicine jointly approved the establishment of the Chinese Qinghaosu Steering Committee for Research on Qinghaosu and its Derivatives and its Secretariat officially on March 20, 1982. Over the six plus years to June 1988, the Committee followed Project 523's management model and continued to strengthen its oversight on the research into artemisinin and its derivatives. It guided and coordinated the development of three new antimalarial drugs: artemisinin suppositories, sodium artesunate injection, and artemether injection. The Committee also submitted an application to the Ministry of Finance for 2 million yuan (around 330,000 USD; 1.33 million USD in 1982) in research funding. This supported 15 units studying artemisinin, artemisinin derivatives and ACTs, and the construction of the artemisinin production plant in Youyang, Sichuan. Some of the money went toward a clinical research base in Dongfang, Hainan, or was used to introduce GLP,

GMP, and GCP standards in new-drug research and production according to international requirements. This promoted the construction of GMP-compliant factories and further refined China's system of new-drug evaluation.

In the 21 years from the first Project 523 meeting in 1967 to the dissolution of the Chinese Qinghaosu Steering Committee in 1988, the Project's Head Office and Committee Secretariat inspected the execution of plans every year according to overall objectives. They gained a deep and practical understanding of promising new leads in each specialty, exploring each one thoroughly and tenaciously. Tough issues arising between the different units and specialties were promptly mediated and resolved by the Head Office and Secretariat. After the Committee was terminated, the Social Development and Technology Department at the SSTC took over the management and coordination of artemisinin promotion overseas. It played a leading role in bringing artemisinin-based antimalarials to the world.

Project 523 and the artemisinin research agenda had strict management systems. Multiple departments and units worked in tandem to ensure that their research would progress swiftly toward the defined goal.

D. MUTUAL AID AND STRENGTH IN NUMBERS

To find an innovative new drug, especially one—such as artemisinin—whose scientific import was of international standards, the cooperation of numerous academic specialists, departments, and units was needed. It involved chemical, pharmaceutical, pharmacological, and clinical work, as well as the technical aspects of drug formulation and production. It was an integrated research project. Approximately 500 people and 60 research units became involved.

In the mid-20th century, China's scientific research conditions and drug development capabilities were poor. Given the state of the research groups and technology then, it would have been difficult to complete the Project in a relatively short time with only one department or unit. Be it the discovery of artemisinin or the synthesis of its derivatives, all the achievements took place with the coordination of a unified management structure. Widely dispersed technical resources and facilities were pulled together following a single plan and based on tight organizational principles. Laboratory experiments, clinical trials, and production technology research were linked together to form a coherent whole. The talent and material strengths of the military, various departments, regions, units, and specialists were all brought into play. They shared everything and worked closely together, frequently conducting technical exchanges and supplementing any lack of equipment. If one party encountered difficulties, others were ready to help. All energies were directed forward, such as in a relay race. Swift, effective, and high-quality progress on all research fronts was assured.

The ICMM had taken the first steps in artemisinin research and made a good start but then ran into difficulties and setbacks. The Shandong and Yunnan Institutes of Materia Medica were inspired by the ICMM's study of *qinghao* extract and investigated their local *huanghuahao* (*A. annua* L.) plants. Although they had begun later, progress was smooth. However, there were no falciparum malaria cases in Shandong's clinical research site. The Yunnan clinical trial group had difficulties finding patients, as the peak of the malaria season would soon pass. To exploit the advantages of each clinical trial group, the National Project 523 Head Office immediately decided to assign clinical testing of *huanghaosu* to the team from the Guangzhou UCM, which was studying the treatment of cerebral malaria in Gengma, Yunnan. Within 2 months, the Guangzhou team showed that artemisinin—then called *huanghaosu*—was effective, fast-acting, and had low toxicity in the treatment of falciparum and severe malaria. On this evidence, the Project 523 Leading Group resolved to launch a nationwide scientific research offensive. The Head Office instantly mobilized the Project 523 teams in the various regions to carry out a comprehensive, collaborative research agenda, including plant resources, the elucidation of artemisinin's chemical structure, pharmacology and toxicity, expanded clinical trials, extraction technology, quality standards, and improvement on pharmaceutical formulations. It rapidly brought forward the success of artemisinin research.

After the discovery of artemisinin, the development of artesunate was another instance of successful collaborative research integrating multiple units, specialists, facilities, and technical know-how. The WHO proposed to prioritize the development of artesunate out of all the artemisinin derivatives. But because China's laboratories, clinical trials, and pharmaceutical factories did not meet international GLP, GCP, and GMP regulations, there were difficulties in promoting artesunate internationally. The Chinese Qinghaosu Steering Committee promptly applied for state funding and brought together experts from eight research units at the IMM of CAMS and AMMS to carry out research according to the WHO's international criteria for new drugs. High-quality work on artesunate was thus swiftly concluded.

E. RESPECTING KNOWLEDGE AND TALENT

While Project 523 and artemisinin research were under way, the Leading Group and the Steering Committee held scientific knowledge and talent in high regard. They supported, defended, and intensified the enthusiasm of their personnel, putting great emphasis on giving scope to experts and technicians. Such people were front and center of the research projects and both Leading Group and Committee fostered their dynamism and creativity. Especially during the turbulent Cultural Revolution, when some harassment occurred, the large number of technical personnel fortunate enough to participate in Project 523 and artemisinin research was given encouragement. In that oasis, they were able to work with only limited disturbance. With hearts and minds at rest, they were able to put themselves fully into urgent research. It was a rare unique situation.

Again, during this time, the selection of emergency medicines to be studied and formulated, or the drafting of research plans and important projects, was all suggested, evaluated, and confirmed by experts. Their innovative ideas were supported and scholarly creativity respected. Senior technical personnel made up the specialty groups, such as chemical synthesis, Chinese medicine, on-site clinical trials, and so on. Regular seminars were held to resolve planning or technical issues, based on the overarching goals and needs of the Project and the practicalities of each specialty. They tackled thorny questions in the research and exercised democratic scholarship, bringing many minds together in search of solutions for the common good. Specialists from different units visited, exchanged information with, and inspired each other. In this atmosphere of lively scholarship, they exploited their own knowledge and powered the progress of research. Each specialty group and their seminars provided first-hand information to the managing organizations. Therefore, they not only mobilized scientists and technicians but also served as a bridge to the leadership. Their role in the management and strategy of the Project was indispensable.

In those years, when daily necessities were in critically short supply, the Project's leaders did more than energize and encourage the participants intellectually. They went some way toward helping participants meet their personal needs. With food and other goods rationed, supplementary food supplies and living expenses were given to researchers working on-site or in rural areas. Therefore, although units and personnel were scattered all over the country, the shared project and good internal environment allowed them to form a lively team capable of an energetic campaign. This was a rare event in China, at a time of nothing but "uncontrolled revolution." The sincere friendships forged during those times are unforgettable. Some of the scientific and administrative staff working in specific regions still organize meetings on 23 May to relive that memorable time and recall the common struggles and friendships of the past.

Under the direction and coordination of the Leading Group and Steering Committee, a world-class scientific achievement was attained. Artemisinin and its derivatives, as well as innovative ACTs, have won renown across the globe and are proudly labeled: Made and discovered in China.

Drawing together those unforgettable experiences, one may underscore Health Ministry Qian Xinzhong's assessment of a Project 523 summary report: "During the Cultural Revolution, Project 523 produced a number of scientific achievements, cultivated a number of scientific and technical talents, and also protected a number of research units and intellectuals." Through the implementation of Project 523 and the work of the Chinese Qinghaosu Steering Committee, a series of artemisinin-based antimalarials were developed and a group of experts in the various aspects of drug research were trained. Beyond that, an effective management technique for cooperative research was successfully applied. These self-evident and valuable achievements steeled the resolve and the efforts of the people in charge. Zhou Keding of the AMMS, who had been part of the Project 523 Head Office and the Chinese Qinghaosu Steering

Committee, was involved in the entire research management process from the beginning to the end. He was one of the administrators praised by all and represents the many unsung heroes of China's artemisinin project.

Today, with China's reform process under way, the country's economic strength, scientific technology, and facilities are well ahead when compared to the past. But these lessons and understandings of the past may serve as a reference for the mobilization and management of current and future scientific projects.

REFERENCES

1. Zhang JF, editor. *A detailed chronological report of project 523 and the discovery and development of Qinghaosu (artemisinin), English ed*. Houston: Strategic Book Publishing and Rights Co.; 2013. p. 7–10.
2. Li RH. Project 523 and the rediscovery of *Qinghaosu's* antimalarial efficacy. *Chin J Hist Sci Technol* 2011;**32**(4):488–500. [黎润红."523任务"与青蒿抗疟作用的再发现. 中国科技史杂志, 2011, 32(4): 488–500.]
3. Liang XT, Yu DQ, Wu WL, et al. The chemical structure of *Yingzhaosu* A. *Acta Chim Sin* 1979;**37**(3):215–30. [梁晓天,于德泉, 吴伟良, 等. 鹰爪甲素的化学结构. 化学学报, 1979, 37(3): 215–230.]
4. Institute of Chinese Materia Medica, China Academy of Chinese Medical Sciences. *Key events in the discovery of Qinghaosu at the China Academy of Chinese medical sciences (1969–1973)*. Compiled. Held at the Center for the History of Medicine, Peking University; 2012. [中国中医科学院中药研究所. 中国中医科学院发现青蒿素的主要历程(1969-1973年), 2012.]
5. Zhang JF, editor. *A detailed chronological report of project 523 and the discovery and development of Qinghaosu (artemisinin), English ed*. Houston: Strategic Book Publishing and Rights Co.; 2013. p. 17.
6. Li YM. The discovery of artemisinin: the whole story. *Tianjin Ribao* 14 October, 2011:15. [李雅民. 青蒿素发现始末. 天津网-数字报刊2011-10-14.]
7. (a) Note from Gu Guoming on his participation in qinghaosu research, June 5, 2004. Held at the Center for the History of Medicine, Peking University. [顾国明. 关于参加部分青蒿研究工作的回顾. 2004.6.5.]

   (b) Note from Jiao Xiuqing on his participation in qinghaosu research, December 8, 2012. Held at the Center for the History of Medicine, Peking University. [焦岫卿.情况说明. 2012.12.8.]
8. Research Group on the History of the Academy, China Academy of Chinese Medical Sciences. *Events in the discovery of Qinghaosu by the Institute of Chinese Materia Medica (1969–1973)*. Compiled. Held at the Center for the History of Medicine, Peking University; 2012. [中国中医科学院院史研究组. 2012, 中药研究所发现青蒿素的历程(1969-1973年) 2012.]
9. Zhang GZ. *Accelerating the work on project 523 in accordance with chairman Mao's important instructions on theoretical questions*. Drafted. Held at the Center for the History of Medicine, Peking University; May 16 and July 16, 1975. [章国镇. 学习毛主席关于理论问题的重要指示. 把五二三工作促上去. 1975年5月16日, 7月16日稿.]
10. Tu YY, editor. *Qinghao and Qinghaosu drugs*. Beijing: Chemical Industry Press; 2009. p. 1–2. [屠呦呦编著. 青蒿及青蒿素类药物. 北京：化学工业出版社， 2009：1–2.]

11. Li RH, Rao Y, Zhang DQ. An inquiry into project 523 and the history of the discovery of *qinghaosu*. *J Dialect Nat* 2013;**35**(1):107–21. [黎润红，饶毅，张大庆. "523任务"与青蒿素发现的历史探究.自然辩证法通讯, 2013, 35(1): 107–121.]
12. Malaria Prevention Team, China Academy of Chinese Medical Sciences. *Breaking new ground in Chinese herbal medicine research through the guidance of Maoist thought*. Held at the Center for the History of Medicine, Peking University; March 4, 1972. [中医研究院疟疾防治小组.用毛泽东思想指导发掘抗疟中草药的工作. 1972, 3. 4.]
13. Project 523 Clinical Research Group, Institute of Chinese Materia Medica, China Academy of Chinese Medical Sciences. *Summary of clinical trial No. 91*. Dated. Held at the Center for the History of Medicine, Peking University; 1972. [中医研究院中药研究所"523"临床验证小组. 91# 临床验证小结. 1972.]
14. Zhang JF, editor. *A detailed chronological report of project 523 and the discovery and development of Qinghaosu (artemisinin), English ed.* Houston: Strategic Book Publishing and Rights Co.; 2013. p. 22–3.
15. Zhang JF, editor. *A detailed chronological report of project 523 and the discovery and development of Qinghaosu (artemisinin), English ed.* Houston: Strategic Book Publishing and Rights Co.; 2013. p. 26.
16. Pharmacology Laboratory at the Institute of Chinese Materia Medica, China Academy of Chinese Medical Sciences. Pharmacology research into *Qinghao*. *N Med J* 1979;**1**:23–33. [中医研究院中药研究所药理研究室. 青蒿的药理研究. 新医药杂志, 1979, (1): 23–33.]
17. Institute of Chinese Materia Medica, China Academy of Chinese Medicine. *Research into the antimalarial Qinghao (1971–1978)*. 1978. p. 26–34. [中医研究院中药研究所. 青蒿抗疟研究(1971~1978): 26-34.]
18. Zhang JF, editor. *A detailed chronological report of project 523 and the discovery and development of Qinghaosu (artemisinin), English ed.* Houston: Strategic Book Publishing and Rights Co.; 2013. p. 27–9.
19. Xu TS. *Brief report on the Beijing Institute of Chinese Materia Medica's Scientific Research* Issue 3. Held at the Center for the History of Medicine, Peking University; 1975. [徐天生.中药研究所科研工作简报1975年第三期.]
20. Zhang JF, editor. *A detailed chronological report of project 523 and the discovery and development of Qinghaosu (artemisinin), English ed.* Houston: Strategic Book Publishing and Rights Co.; 2013. p. 36.
21. Project 523 Team, Guangzhou University of Chinese Medicine. *Report on 18 malaria cases treated with huanghaosu*. Held at the Center for the History of Medicine, Peking University; 1975. [广州中医学院523小组.黄蒿素治疗疟疾18例总结（原稿.） 1975.]
22. Zhang JF, editor. *A detailed chronological report of project 523 and the discovery and development of Qinghaosu (artemisinin), English ed.* Houston: Strategic Book Publishing and Rights Co.; 2013. p. 37–9.
23. Shandong Institute of Chinese Medicine. Preliminary observations on vivax malaria patients treated with a simple preparation of *Huanghuahaosu* and *Huanghuahao* acetone extract. In: *Shandong Institute of Chinese Medicine, "materials on research into traditional Chinese medicine: special collection on Huanghuahaosu*. Held at the Center for the History of Medicine, Peking University; 1980. p. 57. [山东省中医药研究所.黄花蒿素及黄花蒿丙酮提取物简易剂型治疗间日疟现症病人初步观察（见: 山东省中医药研究所中医药研究资料"黄花蒿素专辑"1980: 57).]
24. Institute of Chinese Materia Medica. *Antimalarial research into the traditional medicine, Qinghao*. Issue 11. Held at the Center for the History of Medicine, Peking University; 1975. p. 6–7. [中医研究院中药研究所,中药青蒿抗疟研究. 1975, 11: 6–7.]

25. Project 523 Head Office. *The state of antimalarial research into Qinghao*. Held at the Center for the History of Medicine, Peking University; October 1977. [全国523办公室. 关于青蒿抗疟研究的情况. 1977.10.]
26. Zhang JF, editor. *A detailed chronological report of project 523 and the discovery and development of Qinghaosu (artemisinin), English ed.* Houston: Strategic Book Publishing and Rights Co.; 2013. p. 29–30.
27. Project 523 Traditional Chinese Medicine Group. 1975 research plan for the project 523 group on traditional Chinese medicine. In: *Appendix of the "report of the project 523 meeting on traditional Chinese medicine*. Held at the Center for the History of Medicine, Peking University; 1975. [523中医中药专业组. 523中医中药专业座谈会简报（附件: 1975年523中医中药研究计划表).]
28. Li GQ, Guo XB, Jin R, et al. Clinical trial on Qinghaosu and its derivatives in treating cerebral malaria. *J Tradit Chin Med (Engl Ed)* 1982;**2**(2):125–30. [李国桥，郭兴伯，靳瑞，等. 青蒿素及其衍生物治疗脑型疟疾的临床研究. 中医杂志(英文版), 1982, 2(2): 125–130.]
29. National Malaria Prevention and Treatment Research Directorate. *Certification for Qinghaosu*. Held at the Center for the History of Medicine, Peking University; 1978. [全国疟疾防治研究领导小组. 青蒿素鉴定书. 1978.]
30. Zhang GZ. *Summary of the Beijing Institute of Chinese Materia Medica's research work for 1973*. Held at the Center for the History of Medicine, Peking University; 1974. [章国镇.中药研究所一九七三年科研工作总结. 1974.]
31. Zhang JF, editor. *A detailed chronological report of project 523 and the discovery and development of Qinghaosu (artemisinin), English ed.* Houston: Strategic Book Publishing and Rights Co.; 2013. p. 48.
32. Liu JM, Ni MY, Fan JF, et al. The structure and reactions of Qinghaosu. *Acta Chim Sin* 1979;**37**(2):129–43. [刘静明，倪慕云，樊菊芬，等. 青蒿素的结构和反应. 化学学报, 1979, 37(2): 129–143.]
33. Li Y, Wu YL. Advances in research into the medicinal chemistry and pharmacology of artemisinins. In: Donglu B, Kaixian C, editors. *Advances in medicinal chemistry*. 2005. p. 435. [李英，吴毓林. 青蒿素类化合物的药物化学和药理研究进展. 白东鲁，陈凯先. 药物化学进展. 北京：化学工业出版社.]
34. Qinghaosu Research Group, Institute of Biophysic, Academia Sinica. Crystal structure and absolute configuration of *Qinghaosu*. *Sci Sin* 1980;**23**(3):380–96.
35. Office of the National Malaria Prevention, Treatment Directorate. Report on the exchange training course on quantitative assay techniques for *Qinghaosu*. In: *Report on the research into malaria prevention and treatment, issue 1*. Held at the Center for the History of Medicine, Peking University; 1977. [全国疟疾防治研究领导小组办公室.青蒿含量测定技术交流学习班简报（见《疟疾防治研究工作简报1977年第一期》).]
36. Zhang JF, editor. *A detailed chronological report of project 523 and the discovery and development of Qinghaosu (artemisinin), English ed.* Houston: Strategic Book Publishing and Rights Co.; 2013. p. 65–7.
37. Li YQ, Xie GL, Zhang M. Observations on the efficacy of rectally administered artemisinin in the treatment of vivax malaria. *Chin J Parasitol Parasit Dis* 1984;**2**(4):279. [李应庆，谢贵林，张明. 青蒿素直肠给药治疗间日疟的效果观察. 寄生虫学和寄生虫病杂志, 1984, 2(4): 279.]
38. Drug Evaluation Committee Office, Ministry of Health. The evaluation of new drugs: anshuping tablets and artemisinin suppositories. *Chin J Clin Pharmacol* 1986;**1**:31. [卫生部药品审评委员会办公室. 新药安舒平片和青蒿素及其栓剂审评情况. 中国临床药理学杂志, 1986, (1): 31.]

39. Hien TT, Tam DTH, Cuc NTK, Arnold K. Comparative effectiveness of artemisinin suppositories and oral quinine in children with acute falciparum malaria. *Trans R Soc Trop Med Hyg* 1991;**85**:210–1.
40. Arnold K, Hien TT, Chinh NT, et al. A randomised comparative study of artemisinin suppositories and oral quinine in acute falciparum malaria. *Trans R Soc Trop Med Hyg* 1990;**84**:499–502.
41. Hien TT, Arnold K, Vinh H, et al. Comparison of artemisinin suppositories with intravenous artesunate and intravenous quinine in the treatment of cerebral malaria. *Trans R Soc Trop Med Hyg* 1992;**86**(6):582–3.
42. Li Y, Yu PL, Chen YX, et al. The synthesis of ethers, carboxylic acid esters and carbonic acid esters of reduced artemisinin. *Acta Pharm Sin* 1981;**16**(6):429–39. [李英, 虞佩琳, 陈一心, 等. 青蒿素类似物的研究 I . 还原青蒿素的醚类、 羧酸酯类和碳酸酯类的合成. 药学学报, 1981, 16(6): 429–439.]
43. Gu HM, Lü BF, Qu ZQ. The antimalarial activity of artemisinin derivatives on chloroquine-resistant strains of *Plasmodium berghei*. *Acta Pharmacol Sin* 1980;**1**(1):48–50. [顾浩明，吕宝芬，瞿志祥. 青蒿素衍生物对伯氏疟原虫抗氯喹株的抗疟活性. 中国药理学报, 1980, 1(1): 48–50.]
44. Liu Xu. Research into artemisinin derivatives. *Chin Pharm Bull* 1980;**15**(4):183. [刘旭. 青蒿素衍生物研究. 药学通报, 1980, 15(4): 183.]
45. Yang QC, Gan J, Li PS, et al. The antimalarial activity and toxicity of an artemisinin derivative, artesunate. *J Guangxi Med Coll* 1981;**4**:1–6. [杨启超，甘俊，李培寿，等. 青蒿素衍生物——青蒿酯的抗疟活性与毒性. 广西医学院学报, 1981, (4): 1–6.]
46. Edlund PO, Westerlund D, Jin YH, et al. Determination of artesunate and dihydroartemisinin in plasma by liquid chromatography with post-column derivatization and UV-detection. *Acta Pharm Suec* 1984;**21**:223–34.
47. Vinh H, Huong NN, Ha TTB, et al. Severe and complicated malaria treated with artemisinin, artesunate or artemether in Vietnam. *Trans R Soc Trop Med Hyg* 1997;**91**(4):465–7.
48. Fu LC, Li GQ, Guo XB, et al. A Re-examination of the dosage of artesunate tablets in the treatment of falciparum malaria. *J Guangzhou Univ Tradit Chin Med* 1998;**15**((2):81–3. [苻林春, 李广谦, 郭兴伯, 等. 青蒿琥酯片治疗恶性疟的剂量再探索. 广州中医药大学学报, 1998, 15(2): 81–83.]
49. Qinghaosu Research Group. *Crystal structure and absolute configuration of Qinghaosu*. 1980. p. 380–96.
50. Zeng MY, Zhao SS. UV spectrophotometry in the measurement of dihydroartemisinin content. *Chin J Pharm Anal* 1986;**6**(3):135–8. [曾美怡，赵世善. 紫外分光光度法测定双氢青蒿素含量. 药物分析杂, 1986, 6(3): 135–138.]
51. Tu YY, editor. *Qinghao and Qinghaosu drugs*. Beijing: Chemical Industry Press; 2009. p. 187–224. [屠呦呦编著. 青蒿及青蒿素类药物. 北京：化学工业出版社， 2009, pp. 187–224.]
52. Zhao KC, Chen QM, Song ZY. A *in vivo* study of the pharmacokinetics of artemisinin and two of its active derivatives in dogs. *Acta Pharm Sin* 1986;**21**(10):736–9. [赵凯存，陈其明，宋振玉.青蒿素及其两个活性衍生物在狗体内药代动力学的研究. 药学学报, 1986, 21(10): 736–739.]
53. Zhao KC, Song ZY. A comparison of the pharmacokinetics of dihydroartemisinin and artemisinin in humans. *Acta Pharm Sin* 1993;**28**(5):342–6. [赵凯存，宋振玉. 双氢青蒿素在人的药代动力学及与青蒿素的比较. 药学学报, 1993, 28(5): 342–346.]
54. Pharmacology Laboratory at the Institute of Chinese Materia Medica. *Pharmacology research into Qinghao*. 1979. p. 23–33.

55. Jiang JB, Li GQ, Guo XB, et al. Antimalarial activity of mefloquine and Qinghaosu. *Lancet* 1982;**2**(8293):285–8.
56. Li GQ, Arnold K, Guo XB, et al. A randomised comparative study of mefloquine, qinghaosu and pyrimethamine-sulfadoxine in patients with falciparum malaria. *Lancet* 1984;**2**(8416):1360–1.
57. *Memorandum on technical cooperation on health between the Ministry of health of the PRC and the WHO*. October 5, 1978. http://www.law-lib.com/law/law_view.asp?id=75742.
58. Shandong Institute of Parasitic Disease Control and Prevention, "List of Delegates received since 1979". http://www.sdipd.com/newsshow.asp?id=580&newsclass=13. [山东省寄生虫病防治研究. 1979年以来我所接待来访人员一览.]
59. World Health Organization. *Making a difference: 30 years of research and capacity building in tropical diseases*. 2007. p. 15. http://www.who.int/tdr/publications/documents/anniversary_book_phase1.pdf?ua=1.
60. *Draft proposal on talks dealing with technical cooperation with the WHO on parasitic diseases*. Held at the Center for the History of Medicine, Peking University; February 28, 1979. [我部拟与世界卫生组织在寄生虫病方面进行技术合作会谈方案（草稿）》, 1979年2月28日.]
61. Dalrymple DG. *Artemisia annua, artemisinin, ACTs & malaria control in Africa: tradition, science and public policy*. 2012. p. 17–9. http://www.mmv.org/sites/default/files/uploads/docs/publications/Malaria_Book_3.pdf.
62. World Health Organization. *Making a difference: 30 years of research and capacity building in tropical diseases*. 2007. p. 14. http://www.who.int/tdr/publications/documents/anniversary_book_phase1.pdf?ua=1.
63. Preparatory Group for the Conference on the Development of Qinghaosu and its Derivatives. In: *Collection of documents on the development of Qinghaosu and its derivatives for the 4th TDR-CHEMAL-SWG Meeting*. Held at the Center for the History of Medicine, Peking University; August 1981. [青蒿素及其衍生物的发展会议筹备组编印, 联合国计划开发署/世界银行/世界卫生组织 热带病研究和培训特别规划疟疾化疗科学工作组第四次会议《青蒿素及其衍生物的发展会议文件汇集》, 1981年8月.]
64. Conference Preparatory Group. In: *Summary on the preparation work for the conference on the development of Qinghaosu and its derivatives*. Held at the Center for the History of Medicine, Peking University; July 10, 1981. [筹备小组编, 《青蒿素及其衍生物的发展会议筹备工作简报 (第一期)》, 1981年7月10日.]
65. Zhang JF, editor. *A detailed chronological report of project 523 and the discovery and development of Qinghaosu (artemisinin), English ed*. Houston: Strategic Book Publishing and Rights Co.; 2013. p. 93–5.
66. Zhou TC, Song ZY, Zhou KD. The WHO conference in Beijing on artemisinin and its derivatives. *Acta Pharm Sin* 1982;**17**(2):158–9. [周廷冲，宋振玉，周克鼎. 世界卫生组织在北京召开青蒿素及其衍生物学术讨论会. 药学学报, 1982, 17(2): 158–159.]
67. *UNDP/World Bank/WHO special Programme for research and training in tropical diseases, fourth SWG-CHEMAL Meeting. "The development of Qinghaosu and its derivatives as antimalarial drugs"*. Held at the Center for the History of Medicine, Peking University; October 6–10, 1981.
68. Trigg PI. Key points from the meeting on the cooperation between Chinese research institutions and CHEMAL on the study of artemisinin and its derivatives. In: *Collected materials on project 523 and Qinghaosu (1981–1988)*. Held at the Center for the History of Medicine, Peking University; October 12, 1981. [Trigg博士. 《中国研究机构与疟疾化疗科学工作组之间在抗疟药青蒿素及其衍生物研究 的合作》会谈记录要点, 1981年10月12日（五二三与青蒿素资料(1981~1988).]

69. China Academy of Chinese Medical Sciences. A situation report on the international conference on *Qinghaosu* and suggestions for strengthening this research. In: *Collected materials on project 523 and Qinghaosu (1981–1988)*. Held at the Center for the History of Medicine, Peking University; October 20, 1981. [中医研究院, 关于国际青蒿素会议的情况报告和对加强这项研究工作的建议, 1981年10月20日（五二三与青蒿素资料(1981~1988).]
70. Secretariat of the Qinghaosu Steering Committee. Report No. 1. In: *Collected materials on project 523 and Qinghaosu (1981–1988)*. Held at the Center for the History of Medicine, Peking University; January 10, 1982. [青蒿素及其衍生物研究开发指导委员会秘书处, 简报, 第一期, 1982年1月10日.（五二三与青蒿素资料(1981~1988年).]
71. Qinghaosu Steering Committee. Report on the state of discussions on *Qinghaosu* development and cooperation with the WHO. In: *Collected materials on project 523 and Qinghaosu (1981–1982)*. Held at the Center for the History of Medicine, Peking University; February 26, 1982. [青蒿素及其衍生物研究开发指导委员会, 关于青蒿素发展研究与WHO合作问题讨论情况的报告, 1982年2月26日.（五二三与青蒿素资料汇集1981~1982年).]
72. Secretariat of the Qinghaosu Steering Committee. Report No. 2. In: *Collected materials on project 523 and Qinghaosu (1981–1988)*. Held at the Center for the History of Medicine, Peking University; March 24, 1982. [青蒿素及其衍生物研究开发指导委员会秘书处, 简报, 第二期, 1982年3月24日 （五二三与青蒿素资料(1981~1988年).]
73. Outline of a plan for sodium artesunate development. Translated. In: *Collected materials on project 523 and Qinghaosu (1981–1982)*. Held at the Center for the History of Medicine, Peking University; 1982. [蒿酯钠开发计划概要（译稿）. 1982 (五二三与青蒿素资料汇集(1981~1982年).]
74. "Letter from P.I. Trigg to Xue Songzhuo, director of the foreign affairs bureau with the Ministry of health". Translated, March 26, 1982. In: "Collected materials on project 523 and *Qinghaosu* (1981–1982)". Held at the Center for the History of Medicine, Peking University. [P.I. Trigg 致卫生部外事局薛公焯局长的信(译稿). 1982年3月26日（五二三与青蒿素资料汇集(1981~1982年).]
75. "Letter from P.I. Trigg to Chen Haifeng". Translated, June 29, 1982. In: "Collected materials on project 523 and *Qinghaosu* (1981–1982)". Held at the Center for the History of Medicine, Peking University. [P.I. Trigg 致陈海峰信, 1982年6月29日（五二三与青蒿素资料汇集(1981~1982年).]
76. Secretariat of the Qinghaosu Steering Committee. Bulletin No. 3. In: *Collected materials on project 523 and Qinghaosu (1981–1988)*. Held at the Center for the History of Medicine, Peking University; July 25, 1982. [青蒿素及其衍生物研究开发指导委员会秘书处, 简报第三期, 1982年7月25日.]
77. Ministry of Health. *Plan for the reception of the WHO/SWG-CHEMAL Group*. Held at the Center for the History of Medicine, Peking University; August 28, 1982. [卫生部. 接待世界卫生组织疟疾化疗科学工作组计划, (82)卫外字第916号, 1982年8月28日.]
78. "P.I. Trigg's Note for the Record". Handwritten summary of the meeting by Zhou Keding. 30 September 1982. In: "Collected materials on project 523 and *Qinghaosu* (1981-1982)". Held at the Center for the History of Medicine, Peking University. [P.I. Trigg. Note for the Record. 1982年9月30日（会谈纪要，周克鼎手迹）.五二三与青蒿素资料汇集(1981~1982年).]
79. Reply from deputy director of the foreign affairs bureau Zhou Minjun to P.I. Trigg". November 16, 1982. In: "Collected materials on project 523 and *Qinghaosu* (1981–1982)". Held at the Center for the History of Medicine, Peking University. [周敏君副局长给Trigg博士复函. 1982年11月16日（五二三与青蒿素资料汇集(1981~ 1982年).]
80. "Letter from P.I. Trigg to Zhou Minjun". January 4, 1983. In: "Collected materials on project 523 and *Qinghaosu* (1983–1986)". Held at the Center for the History of Medicine, Peking University. [P.I. Trigg. Letter to Zhou Minjun, 4 January 1983.（五二三与青蒿素资料汇集(1983~1986年).]

81. "Letter from P.I. Trigg to Prof Zhou Minjun, T16/83/M2/2". March 11, 1983. In: "Collected materials on project 523 and *Qinghaosu* (1983–1986)". Held at the Center for the History of Medicine, Peking University.
82. "Briefing Note from P.I. Trigg, subject: visit of Colonel Robert young to China for discussion of production of artesunate". Held at the Center for the History of Medicine, Peking University.
83. "Materials from Zhou Minjun". May 24 , 1983. In: "Collected materials on project 523 and *Qinghaosu* (1983–1986)". Held at the Center for the History of Medicine, Peking University. [周敏君有关资料. 1983年5月24日（五二三与青蒿素资料汇集1983~1986年).]
84. An Agreement for the Cooperative Study of Artesunate Among The Ministry of Public Health, The World Health Organization and The U.S. Department of Defense. In: "Collected materials on project 523 and *Qinghaosu* (1983–1986)". Held at the Center for the History of Medicine, Peking University.
85. "Note for the Record on the Meeting of March 14–16, 1984 amprdquosemicolon. In: "Collected Materials on Project 523 and *Qinghaosu* (1983–1986)". Held at the Center for the History of Medicine, Peking University. [1984年3月14~16日会谈纪要（Note for the Record）（五二三与青蒿素资料汇集1983~ 1986年).]
86. "WHO letter from H. Mahler to the minister of public health of the PRC, T16/83/M2/2". March 12, 1985. In: "Collected materials on project 523 and *Qinghaosu* (1983–1986)". Held at the Center for the History of Medicine, Peking University.
87. Handwritten note by Zhou Keding, "Views on the WHO's letter on Revisions to the draft agreement on artesunate cooperation". May 20, 1985. In: "Collected materials on project 523 and *Qinghaosu* (1983–1986)". Held at the Center for the History of Medicine, Peking University. [周克鼎手迹. 对WHO来信关于修改'青蒿酯合作协议'（草稿）的复核意见. 1985年5月20日（五二三与青蒿素资料汇集(1983~1986年).]
88. "Memorandum from A.O. Lucas to Dr Lu Rushan, subject: development of qinghaosu and its derivatives, T16/83/M2/2". August 20, 1985. In: "Collected materials on project 523 and *Qinghaosu* (1983–1986)". Held at the Center for the History of Medicine, Peking University.
89. World Health Organization. *WHO newsletter, special issue: TDR workplan*. Issue 22. August 1986. p. 5.
90. "Note for the record. Meeting with representatives of Chinese steering committee for development of qinghaosu and its derivatives, Beijing, February 26, 1987". In: "Collected materials on project 523 and *Qinghaosu* (1987–1993)". Held at the Center for the History of Medicine, Peking University.
91. Handwritten draft by Zhou Keding, "Secretariat of the qinghaosu steering committee, report No. 6". December 5, 1983. Held at the Center for the History of Medicine, Peking University. [中国青蒿素及其衍生物研究开发指导委员会秘书处，简报（第六期)，1983年12月5日(周克鼎起草手迹).]
92. Song ZY, Zhao KC, Liang XT, et al. Radioimmunoassay of artesunate and artemisinin. *Acta Pharm Sin* 1985;**20**(8):610–4. [宋振玉，赵凯存，梁晓天，等. 青蒿酯和青蒿素的放射免疫测定法.药学学报, 1985, 20(8): 610–614.]
93. Yang SD, Ma JM, Sun JH, et al. Assay of artesunate and dihydroartemisinin in human plasma using reduced electrochemical polarographic detection and high-performance liquid chromatography. *Acta Pharm Sin* 1985;**20**(6):457–62. [杨树德, 马建民, 孙娟华,等. 还原型电化学极谱检测高效液相色谱法测定人血浆中青蒿酯和双氢青蒿素.药学学报, 1985, 20(6): 457–462.]
94. Zhao SS, Zeng MY. Application of precolumn reaction to high-performance liquid chromatography of qinghaosu in animal plasma. *Anal Chem* 1986;**58**:289–92.

95. Handwritten notes by Zhou Keding, "Secretariat Aide-Memoire for the steering committee on discussions on cooperation with CHEMAL". March 20, 1987. In: "Collected materials on project 523 and *Qinghaosu* (1987–1993)". Held at the Center for the History of Medicine, Peking University. [秘书处备忘录，指导委员会讨论与CHEMAL合作事, 1986年3月20日 (周克鼎手迹) (五二三与青蒿素资料汇集1987~1993年).]
96. "Letter from P.I. Trigg to C.C. Shen, M20/83/2", July 12, 1988; "Letter from WHO director-general Hiroshi Nakajima to the Ministry of public health of P.R. China", October 5, 1988. In: "Collected materials on project 523 and *Qinghaosu* (1987–1993)". Held at the Center for the History of Medicine, Peking University. [五二三与青蒿素资料汇集1987~1993年).]
97. Shen JX, editor. *Antimalarial drug development in China*. Beijing: National Institutes of Pharmaceutical Research and Development; 1991. *passim*. [沈家祥. "中国抗疟药物研究的进展". 北京，中国医药科技出版社1989.]
98. Handwritten notes by Zhou Keding, "Qinghaosu steering committee: report on guiding principles for negotiations on the cooperative development of artesunate". February 21, 1984. Held at the Center for the History of Medicine, Peking University. [青蒿素及其衍生物研究开发指导委员会. 关于合作开发青蒿酯谈判方案的请示报告. (周克鼎手迹二页, 1984年2月21日).]
99. Editorial and Translation Specialist Group for China's Dual-Drug Registration. "TDR experts' comments on China's registration materials". October 1990. "Contractual Services agreement exchange of letters from Tore Godal and Dr E.B. Doberstyn to Dr C.C. Shen, HQ/88/058952". April 20, 1988. In: *Collected Materials on Project 523 and Qinghaosu (1987–1993)* Held at the Center for the History of Medicine, Peking University. [中国"两药"注册文件编译专家组.（TDR专家）对中国注册文件的评论1990年10月.五二三与青蒿素资料汇集(1987~1993年).]
100. Secretariat of the Qinghaosu Steering Committee. "Report No. 2". [青蒿素及其衍生物研究开发指导委员会秘书处，简报，第二期，1982年3月24日 （五二三与青蒿素资料1981~1988年).]
101. Secretariat of the Qinghaosu Steering Committee. Funding allocation for 1982. In: *Collected materials on project 523 and Qinghaosu (1981–1988)*. Held at the Center for the History of Medicine, Peking University; July 27, 1982. [青蒿素指导委员会秘书处. 1982年经费分配方案, 1982年7月27日.（五二三与青蒿素资料1981~1988年).]
102. Nanchuan Medicinal Plant Research Institute, Chengdu Institute of Traditional Chinese Medicine; and Fuling Institute for Drug Control, Sichuan. *Report into the A. annua resources of Youyang, Sichuan, its ecobiology and artemisinin content*. Held at the Center for the History of Medicine, Peking University; July 1985. [成都中医药研究院南川药物种植研究所，四川省涪陵地区药品检验所，四川省酉阳县青蒿资源及其化学生态调查研究报告， 1985年7月.]
103. Handwritten notes by Wang Meisheng, "A few arrangements undertaken by the project 523 head office [sic, should be Qinghaosu steering committee] after the report on Qinghaosu in Youyang". June 3, 1984. Held at the Center for the History of Medicine, Peking University. [523办公室对酉阳青蒿素工作汇报后的几点安排, 1984年6月3日（王美胜手迹).]
104. "Handwritten letter from Zhou Keding and Zhang Kui to Feng Tianjiong". June 20, 1986. Held at the Center for the History of Medicine, Peking University. [周克鼎, 张逵. 1986年6月20日致冯天炯的信 （周克鼎手迹).]
105. "Collected materials on project 523 and Qinghaosu (1981–1988)". Held at the Center for the History of Medicine, Peking University.

Chapter 2

Agronomics and Biology of *Artemisia annua* L.

Chapter Outline

2.1 PROPAGATION AND CULTIVATION OF *ARTEMISIA ANNUA*

Hu Sihua

Artemisia annua *L. Research Center, Institute of Tropical Medicine, Guangzhou University of Chinese Medicine, Guangzhou, China*

A. BOTANICAL AND MEDICINAL PROPERTIES

1. Name and Morphology

Artemisia annua is an annual herbaceous plant of the Asteraceae family, aromatic, and either hairless or with scattered, dense patches of fine hair.

Artemisinin-Based and Other Antimalarials. https://doi.org/10.1016/B978-0-12-813133-6.00002-0

The stem is erect, ridged, and of green, yellow–green, or violet–green; in the late growth stage, it becomes brown or violet–brown. In the wild, it reaches heights of 30–150cm. Cultivated specimens can reach 300cm. The leaves are green or yellow–green, with dotted glands distributed on the surface in a honeycomb pattern. They have short stems, are oval in shape, and are tripinnately divided into small, elliptical, lanceolate leaflets. These leaflets are short and pointed, with entire margins that may have one or two teeth. The cauline leaves on the midsection are bipinnately divided, with the leaves on the upper section having no stems and relatively smaller and fewer pinnae. Leaves of the uppermost section are bracteolate, simple, and with a small number of lobes on each side.

The flowers are yellow, with globelike flower heads. They appear in large numbers, dispersed or drooping, on short pedicels that form long, tiered, cone-shaped panicles. The involucre is smooth, hairless, and green, with linear, oval outer bracts. Inner bracts are oval or rounded and glossy, with a broad membranous margin. The receptacle is raised, hairless, with 10–20 filiform female flowers on the outside. These have spotlike glands and a thin stigma with a blunt top, which projects out of the smooth, tubelike corolla. The florets in the disc are bisexual, numbering 10–30. The anther is thin, with long pointed appendages. Appendages on the base are extremely short, with slight points. The style is shorter than the stamen, with a thin, slightly bisected stigma, which has cilia on the top. The fruit are long, elliptical, yellow–brown achenes, with glossy surfaces and clear longitudinal furrows. The endosperm is white and oily.[1,2]

Based on the botanical and medicinal characteristics listed in the pre-Ming *Materia Medica*, the *qinghao* mentioned in these works refers to *huanghuahao* or *A. annua*.[3] The Ming physician Li Shizhen listed *qinghao* and *huanghuahao* as two different herbs. In 1753 and 1852, Linnaeus and Henry Hance identified *A. annua* L. and *Artemisia apiacea* Hance, respectively, both of which are associated with *qinghao*. Chinese and other scholars subsequently confirmed *qinghao* as *A. apiacea* and *huanghuahao* as *A. annua*. There have therefore been differences in the standard names used in botanical classification and in Chinese herbal medicine.[4] In the *Pharmacopoeia of the People's Republic of China*, the plant containing artemisinin is termed *qinghao*, described as "the dry, aboveground portions of *huanghuahao* (*A. annua* L.) of the Asteraceae family."[5]

2. Medicinal Properties

In traditional Chinese medicine, the medicinal portion of *A. annua* is, as stated above, the dry parts found above ground. It is harvested in autumn when the plant is in full bloom. Old stems are removed and the plant is air dried. The stems used in medicine are cylindrical, the upper sections of which have many branches. They are 30–80cm in length and 0.2–0.6cm in diameter, appearing yellow–green or tan on the surface, with longitudinal ridges. The texture is slightly hard and brittle, and the pith can be seen when broken. The brittle leaves grow in an alternating pattern and are curled, dark green, or brown. When flattened out, complete leaves are tripinnately divided, with oblong or elliptical lobes and lobules covered in short hairs. The aroma is characteristic and the taste is slightly bitter.[5]

The dry leaves of *A. annua* form the raw material and are used for the extraction of artemisinin. These are brittle; when dried under different conditions, they can be green, brown, or tan. They have the characteristic aroma of *A. annua*, with a slightly bitter and minty taste. The artemisinin content is 0.7% and above.[1]

A. annua is used very sparingly in traditional Chinese medicine. The Chinese *Pharmacopoeia* does not stipulate the artemisinin content of medicinal *A. annua*.[5] Wild *A. annua* was sufficient for the needs of traditional medicine. Therefore, this section deals only with the extraction of artemisinin from the dry leaves of *A. annua* and the production of high artemisinin yields. The artemisinin content referred to here relates to the content found in these leaves, except where otherwise specified.

B. RESOURCE DISTRIBUTION AND ORIGIN

1. Geographic Distribution

A. annua originates in China.[1] In its wild form, it is found throughout the country at altitudes of 1500m above sea level and below in the eastern regions, at 2000–3000m above sea level in the north- and southwestern regions, and at 3650m above sea level in Tibet. It is adaptable, growing by the roadside and in wastelands, hillsides, and forest margins in the eastern and southern regions. In other areas, it can also be found in grasslands, forest meadows, dry river valleys, semiarid climates, and rocky slopes. It can be seen in salinized soil and in certain areas can become the dominant or main accompanying plant species. Wild *A. annua* is also widespread in the temperate, cold-temperate, and subtropical regions of Asia and Europe, occurring most frequently in the central, eastern, and southern parts of Europe and the northern, central, and eastern parts of Asia. Its southernmost distribution extends to the Mediterranean, North Africa, and southern and southwestern Asia. From northern Asia, its range has spread to North America, throughout Canada, and the United States.[6]

After the 1975 Chengdu meeting of the Project 523 Head Office, experts in botany and traditional Chinese medicine from Guangdong, Yunnan, Jiangsu, Sichuan, Guangxi, Fujian, and Beijing swiftly embarked on a nationwide investigation of wild sources of *A. annua*. They confirmed that the plant was widely distributed in both the northern and southern provinces of China. In general, the artemisinin content of wild *A. annua* was higher in the south than in the north. Plants with high artemisinin content were found south of the Nanling and Wuyi mountains. Those in Guangxi, Guangdong, and the north of Hainan had the highest content. Intermediate artemisinin concentrations occurred to the north of Nanling and south of Qinling. Content was low in north of Qinling, and plants in the northeastern regions had only trace amounts of artemisinin. The investigation discovered that *A. annua* grew in large amounts in Youyang, Pengshui, and Xiushan in Chongqing; Xiangxi and the Wuling Mountains in Hunan; and the Guizhou, Yunnan, Guangxi, and Hunan border. Not only could production be high but also these plants had good artemisinin content. These regions are the main sources of wild *A. annua*.[7]

The artemisinin content of wild *A. annua* was measured. In Feng County, Baoji, and Pucheng in Shaanxi, it was 0.13%–0.23%[8]; in Xilin, Guangxi, it was 0.66%–0.98%[9]; in 13 regions of Guangxi, the range was 0.50%–1.01%.[10] Concentrations in the Wuling Mountains (southeastern Sichuan, western Hubei, western Hunan, and northeastern Guizhou) were 0.48%–0.89%. Chongqing, Wuhan, Huarong in Hunan, and Beijing had concentrations of 0.40%, 0.57%, 0.42%, and 0.11%, respectively.[11] Inner Mongolia had a concentration of 0.12%–0.17%.[12] It was 0.50% in Tianjin,[13] 0.70%–1.11% in Changde, Hunan,[14] and 0.01%–0.09% in Heilongjiang.[15] Outside of China, concentrations were higher in Vietnam, reaching 0.86%.[16] Plants in some regions of India had concentrations of 0.42%.[17] Mostly, however, artemisinin content was very low (0.1% or less).[11] These data were based on single or multiple specimens of *A. annua*. Within China, many large-scale artemisinin production companies in the main growing regions purchase dry leaves from wild *A. annua*, which have artemisinin content below 0.7%, mostly 0.4%–0.6%.

In 1987, the first artemisinin production enterprise—Wuling Mountain Pharmaceutical Factory in Sichuan—announced the commencement of its operations. The Jishou Pharmaceutical Factory in Hunan started up in 1992. Until 2001, these two companies accounted for over 90% of total artemisinin production. Raw *A. annua* was sourced from Youyang and the surrounding areas in the Wuling Mountains, as well as western Hubei. Most were wild plants, of which 1500 tons were purchased annually. The artemisinin extraction ratio was around 0.4%.

Under the influence of international markets, the artemisinin industry has rapidly expanded since 2002. From a handful of production enterprises, more than 100 have sprung up, converting or creating new artemisinin extraction lines. They rely on local administration or the "company-peasant household" model for the large-scale expansion of *A. annua* cultivation. In 2006, the area devoted to *A. annua* cultivation in the main growing regions of Chongqing, Hunan, Guangdong, Guangxi, Guizhou, and Sichuan reached 500,000 mu (equivalent to 335,000 ha or 80,000 acres), leading to severe overproduction. The artemisinin industry rapidly entered a downturn, with serious losses for both enterprises and farmers. From 2007 to 2008, the growing area declined sharply; in 2009, the industry began to rationalize.

Because of the rapid expansion of the artemisinin industry, wild *A. annua* had been overharvested for many years, resulting in a dramatic fall in the plant's population. Competition between enterprises was intense and, at the same time, demand for artemisinin-based drugs in international markets gradually increased. Purchases of wild *A. annua* and the plant's low artemisinin content could not meet the needs of artemisinin extraction. At present, raw *A. annua* is mainly cultivated by farmers, with the main growing areas in Youyang, Chongqing, and its surrounding areas, the Three Gorges Reservoir area in northeastern Chongqing, western Hunan, northeast Guangxi, eastern and northern Guangdong, southwestern Hubei, and eastern Guizhou. The total area cultivated is 100,000–200,000 mu (equivalent to 67,000–134,000 ha or 16,000–32,000 acres). Artemisinin content in large-scale production is above 0.7%, mostly 0.8%–1.0%.

2. Optimal Growth Conditions

A. annua is highly adaptable, favoring sunny environments. It germinates at a temperature of above 7°C.[1] Experience has shown that the optimal temperature for growth is 20–25°C or an average year-round temperature of 13.5–17.5°C. It grows well in mountains, forest margins, and wastelands, at various altitudes: 600–800 m in China, 50–500 m in Vietnam, and 1000–1500 m in Tanzania and Kenya.[1] The plant flourishes in warm weather, above 10°C, and the accumulated annual temperature should be 3500–5000°C depending on individual situation. Annual sunshine duration should be around 1000 h.[1] It is now known that the best growth environment is in humid, subtropical monsoon climates. During the growth period, the average temperature in these zones is 17.6–28.4°C. Annual precipitation is 1150–1350 mm, whereas the precipitation needed in the growing season is 600–1000 mm.[1] *A. annua* can grow in most types of soil as long as the soil pH is between 4.5 and 8.5, the top soil is deep, and drainage is good.[18] The plant cannot withstand waterlogged soil, which easily leads to delayed growth, root rot, and death of the seedlings.

When selecting ground for planting, field experiments should be carried out with different environmental conditions. Artemisinin content and yield could be greatly affected by different microenvironments. Each *A. annua* cultivar or line also has its optimal growth environment. The plant is cultivated only to extract artemisinin, with maximum artemisinin yield per unit area as the sole aim. This is very different from the cultivation principles of traditional Chinese medicine.

Wild and cultivated *A. annua* have different optimal growing conditions. Areas where artemisinin content is higher in wild specimens, and where the plant is more widely distributed, may not be the most suitable for cultivated varieties.

C. PROPAGATION

1. Sexual Reproduction

A. annua is a short-day plant. In the branching season, it is very sensitive to short periods of light stimulus. If induction occurs at this time, budding commences about 2 weeks later. The plant's photoperiod is around 13.5 h.[18,19] From budding to flowering, the plant's vegetative growth slows significantly, young leaves gradually thicken and shrink, and old leaves slowly turn yellow and fall off. Sunshine duration varies in different regions, and hence it is important to cultivate the plant at the right time or the yield of dry leaves and artemisinin content will decrease. The number of hours of sunlight influences all growth phases, with longer durations causing the plant to flower prematurely. Shorter durations delay flowering.[1] Different wild *A. annua* strains and cultivars have varying photoperiods and degrees of sensitivity to light duration. This is manifested in contrasting budding and flowering phases. In Guangdong, flowering begins on 12 September at the earliest and on 15 October at the latest. Under the same cultivation conditions, the flowering period is more consistent in purer sexually propagated cultivars and identical in single-strain, asexually propagated specimens. This indicates that light sensitivity and photoperiod are subject to polygenic inheritance.

As a typical cross-pollinating plant, *A. annua* is self-incompatible. The structure of each inflorescence head is highly conducive to self-pollination, but the self-pollination rate is relatively low and produces seed only with difficulty.[20] In September 2000, during the flowering period, Zhu Weiping selected 50 specimens of *A. annua* and tagged and bagged one branch from each. The bag was removed after the flowering period. The fruit was collected and husked in December. Only five seeds were found out of all 50 branches.[21] According to research by Zhang Long, the self-pollination rate is less than 0.05%.[22] Finally, in 2005, Tang Qi strictly quarantined and self-pollinated a specimen of *A. annua*, after which only seven seeds were found. In 2007, bagged self-pollinated seeds from 2006 were germinated indoors. The germination rate was extremely low or absent. Those which did germinate grew poorly and could not be transplanted. Observation of the stigma showed that both bisexual and female flowers had dry stigmas. This confirmed that *A. annua* had sporophytic self-incompatibility.[23]

In 2006, Hu Sihua et al. from the Artemisia Research Center at the Guangzhou University of Chinese Medicine cut half the branches of three specimens of *A. annua* in Fengshun, Guangdong. The specimens were strictly quarantined and self-pollinated and the resulting seeds harvested and sown. Each specimen yielded 210, 357, and 1034 seedlings, respectively. Given that cross-pollinated specimens produce 40 g of seeds on average, the self-pollination rates were 0.05%, 0.08%, and 0.24%, respectively, for each plant. Then in 2007, an *A. annua* cutting was used to asexually produce nine other specimens. These were quarantined together and the seeds harvested for observation. Their appearance, self-pollination rate, and germination results were similar to those yielded by self-pollination of a single specimen. Seeds were small and light and did not separate from the tubelike buds after maturing. They were nonlustrous and of the same color as the fragments of bud and disk flowers, being indistinguishable by the naked eye. Hence, it was difficult to pick out the pure seeds. When the seeds were sown in early March, they germinated in 7–10 days, 4–7 days later than seeds produced by cross-pollination. The germination rate was very low, the offspring were dwarfed, internode spaces were shortened, and leaves were shrunken and denser, with less disease resistance.

The sexual reproduction coefficient of *A. annua* is extremely high, reaching 10^4–10^5. A small area for seed production can provide sufficient seeds for the planting of a large area. This has a positive impact on the propagation of the plant.

2. Asexual Propagation

Asexual reproduction depends on the totipotency of plant cells and their capacity to dedifferentiate and return to a meristematic state, manifested in the strong regenerative ability of the vegetative organs. The genotypes of parent and offspring are identical, and the characteristics of the parent are maintained. This has great significance for research into the cultivation and propagation of *A. annua*.

Asexual propagation can be carried out via cutting, grafting, and tissue cultures. Xu Chengqiong et al. made cuttings of the base, middle, and top sections of *A. annua* branches in May, June, July, and August 1995, before the flowering period. The best time for cuttings was in July and August. Top sections were optimal, with a strike rate of 86.0%. These top sections were cultivated in fertile soil (silt), ash soil, and sandy soil. It was found that the strike rate was related not only to the level of nutrients in the medium but also to its looseness. Soil contains a certain nutrient load and is sufficiently loose, yielding a strike rate of up to 96.0%. Silt has more nutrients but is not loose enough, with a lower strike rate of 84.0%. Sand is loose but, containing almost no nutrients, had the lowest strike rate of 78.0%.[24]

In March 2007, Hu Sihua et al. conducted an experiment on *A. annua* cuttings in Fengshun, Guangdong, as aforementioned. Loose soil was used for the base of the medium, followed by 4 cm of fine loess. The strike rate was 64%–100%; the average strike rate for stem tips (90%) was slightly lower than that of the main part of the stem (96%). The former was more fragile and prone to infection before a callus could form over the cut, leading to decay. However, they grew more quickly and produced more roots. After striking, the speed at which terminal buds grew and underwent tillering was faster, with shoots appearing after 26 days (height 15 cm). This was around 10 days earlier than with the main stem sections. After replanting, they grew more slowly than seedlings produced from seed, with fewer branches. At the peak of vegetative growth, specimen height was only 1.2–1.5 m, as opposed to 1.7–2 m for plants grown from seed. From 2008 to 2015, a large number of *A. annua* clones were used in field experiments. The strike rate was 80% and above.

The artemisinin content of an *A. annua* specimen produced via cutting was measured. The artemisinin content correlation coefficient between the parent and offspring was 0.818, which was significant on the 0.025 level. Cuttings had basically the same growth conditions and field management. Hence, differences in artemisinin content across strains resulted from the parent genotype. Genetic studies could be used to cultivate plant strains with high artemisinin content.

Many studies on tissue culture techniques for *A. annua* have taken place in China and abroad. In 1983, He Xichun et al. first successfully induced *A. annua* callus formation and regeneration of the plant.[25] The explant can be the stem tips, stems, leaves, flower buds, flowers, or seeds. They are usually disinfected using alcohol, sodium hypochlorite solution, and mercuric chloride solution. In general, the culture conditions include a light intensity of 1500–2000 lux and duration of 14–16h a day, with a temperature of 25°C.

Growth regulators usually have significant effects on the development of the plant. He Xichun et al. used *A. annua* leaves as explants in an improved N6 medium, cultivated at 25°C for 5–10 days. Calluses were produced, which were placed in a B5 medium for further cultivation. Later, an enhanced Murashige–Skoog medium (MS)+BA medium was used and yielded small plant specimens.[25] Yang Yaowen et al. showed that flower explants had a proliferation rate two to three times higher than that of stem sections. Differentiated seedlings rapidly produced roots in 1/2MS+IAA 0.5mg/L+1-naphthylacetic acid (NAA) 0.5mg/L mediums. After transplantation, the strike rate was high.[26]

With pedicels, flowers, and leaves as explants and MS as a base culture medium, Wang Mengqiong induced callus formation and plant regeneration under different hormone combinations. For the pedicels, MS+6-benzylaminopurine (6-BA) 1mg/L+2,4-dichlorophenoxyacetic acid (2,4-D) 0.5mg/L induced calluses, MS+IAA 0.1mg/L+kinetin (KT) 1mg/L produced differentiation into shoots, and IAA 0.1mg/L+6-BA 1mg/L+NNA 1mg/L caused differentiation into roots. Regenerated *A. annua* could thus be obtained at relatively low cost, following simple methodology with a short cultivation period.[27] Wu Xiaoli et al. used young *A. annua* leaves as explants and found that MS+6-BA 0.1mg/L+2,4-D 0.3mg/L was ideal for inducing calluses, MS+6-BA 0.1mg/L+IAA 0.3mg/L for differentiation, and MS+IAA 0.1mg/L+NAA 0.3mg/L for root formation.[28] A study by Li Xiaojuan showed that MS+6-BA 0.05mg/L+NAA 0.01mg/L was the best culture medium for stem and leaf induction. MS+6-BA 1mg/L+NAA 0.2mg/L+KT 0.5mg/L was best for proliferation.[29]

Since 2005, Li Nuo et al. of the Artemisia Research Center at the Guangzhou University of Chinese Medicine used young stem tips, flower buds, and pedicels as the explants in yearly tissue culture tests. For disinfection, the explants were first washed in soapy water and then soaked in 75% alcohol for 20s. Saturated sodium hypochlorite solution was then used to disinfect the explants for 10min, followed by 0.05% mercuric chloride solution for 8min. Culturing conditions included a light intensity of 1500 lux, duration of 14h a day, and a temperature of 25°C. Media used were MS+6-BA 0.4mg/L+IAA 0.5mg/L to induce sprouting, MS+6-BA 1.5mg/L+IAA 0.3mg/L for differentiation, and MS+NAA 0.1mg/L to produce root growth. The induction time needed for different explants and inoculation methods and the location and number of shoots, all varied. Stem tips inserted into the medium could produce a single shoot at the original node in around 20 days, which developed very rapidly. Flower buds could be inserted into the medium or placed on the surface, with many shoots produced in 10 days. Insertion yielded only single shoots near to the culture medium, but surface placement created multiple shoots at all nodes. In strong light, some bud explants became leaf buds, forming disk-shaped, green leaf shoots, which then produced sprouts. Pedicels required a longer induction time, with sparse shoot clusters appearing after around 30 days. Their location was similar to those in the flower buds. Pedicels on the verge of blooming did not become leaf buds under strong light and turned brown easily.

Asexual propagation—such as cuttings, tissue culture, grafting, etc.,—can be used to produce clones of *A. annua* for field experiments, as well as to preserve and produce seeds, but it is not possible to use the resulting seedlings directly in large-scale cultivation. The financial, technical, and manufacturing demands of such seedlings are higher, overriding the benefits of these clones.

3. Selective Propagation

Selective propagation of *A. annua* refers to the cross-breeding of existing cultivars, strains, or types. Individual, multiple, or mass selection is used to produce new, pure cultivars and strains with similar genotypes.

A. annua is cross-pollinating, with a very low rate of self-pollination. Offspring of self-pollination also easily fall prey to natural selection. In a natural state, the plant is highly heterozygous. Specimens in the same cluster have varying appearances and artemisinin content, with high genetic diversity. This provides plenty of raw materials for selective propagation.

Beginning in 1992, Ding Derong initiated a major research project at the Sichuan Science and Technology Commission on the selective propagation of *A. annua* and artificial cultivation methods. One variant, termed "Sichuan Yellow 1," was produced out of the wild types growing in Youyang, Chongqing. Its artemisinin content was 1.0% and yield was 80 kg/mu (equivalent to 0.12 kg/m^2). The Science and Technology Commission of Chongqing approved it in 1998.[30]

Chen Herong, Ding Derong, Luo Rongchang, and Yang Pinghui et al. conducted selective propagation of *A. annua* in Youyang. After continuous selection and preservation of seeds by many research units and scientists, strains suitable for cultivation in Youyang were produced. Termed the "Youyang cultivar," these had a tiered shape, were relatively tall, and had purplish stems or internode spaces and sparsely lobed leaves. They were resistant to disease, yielded around 160 kg/mu (equivalent to 0.24 kg/m^2) of dried leaves, and had an artemisinin content of 0.8%–1.0%.

In 2004, Ma Xiaojun et al. compared the germplasms of 72 *A. annua* specimens from various regions. They found that the germplasm from Jingxi,

Guangxi, had generally higher artemisinin content. Individual specimens were analyzed before flowering. Those with lower artemisinin content were rejected. Specimens with contents above 1% were numbered, quarantined, and their seeds preserved. After 3 years, improved bulk selection was used to produce the "Guihao No. 3" cultivar, with an average artemisinin content of 1.08%. The specimen with the highest content came in at 1.62%. Medicinal yield was 229.5 kg/mu (equivalent to 0.34 kg/m^2). In December 2006, it was registered by the Crop Cultivar Approval Committee of the Guangxi Zhuang Autonomous Region. The cultivar is robust, has tripinnate compound leaves, and has a tiered shape. The stem and primary branches are purple–green. Plants have a height of 180–305 cm, canopy breadth of 80–135 cm, and internode spaces of 2.8–4.5 cm. Primary branches number 60–105, with a branch angle of 40–65 degrees. Leaves are 6.0–9.8 cm long and 4.8–9.5 cm wide and are deep green in color. The bud stage begins in mid-August, with budding in early September and blooming in late September. This strain can be grown in northern and western Guangxi or in similar climates.[31]

Xiang Jixian et al. collected and evaluated wild *A. annua* in the Wuling Mountains in 2001. They found seven main phenotypes. From 2002 to 2004, in Enshi Tujia and Miao Autonomous Prefecture in Hubei, they carried out directed selection via derived line method and group mixing method. Specimens of the same type were hybridized to form a pure line. Then, comparative tests between different cultivars were conducted from 2005 to 2006. The best result was the "Hubei *A. annua* No. 1" cultivar, approved by Crop Cultivar Approval Committee of Hubei Province in March 2007. It is tiered, relatively more compact, with a white–green stem, light-yellow and sparsely lobed leaves, and a height of 180 cm. Each individual plant has 105 branches. It grows for 244 days, yielding 280 kg of dried leaves per mu (equivalent to 0.42 kg/m^2) with an artemisinin content of 1.07%. In Enshi, it can be planted at altitudes below 1200 m.[32,33]

From 2000 to 2001, Li Longyun et al. collected cultivated and wild *A. annua* from Youyang and the Wuling Mountains area. These were selectively propagated from 2001 to 2006 in Youyang and Fengdu, using the matrilineal and bulk selection system. Field tests were run in 2007–2008, from which a new cultivar, "Chongqing Green No. 1," was selected. The Crop Cultivar Approval Committee of Chongqing approved it in April 2009. It is compact, with purple or purple–red stems and strong resistance to disease. Yield is 150 kg and above of dry leaves per mu (equivalent to 0.22 kg/m^2); highly productive specimens can reach 200 kg/mu (equivalent to 0.30 kg/m^2) and above. The artemisinin content is 0.82%–1.81%. This cultivar can be grown in Chongqing at altitudes of 800 m and below.[34]

In 2003–2004, Chen Herong et al. of the Artemisia Research Center at the Guangzhou University of Chinese Medicine assembled *A. annua* seeds and plants from Qionghai and Tongshen in Hainan, Baise in Guangxi, Youyang in Chongqing, Shaoshan in Hunan, and Fengshun, Lianping, Wengyuan, and Shaoguan in Guangdong. They were planted in 2004 in Fengshun. The resulting plants had different appearances and characteristics. Stems were purple, green, or yellow; most strains were tiered, with relatively long internode spaces and sparse leaves. Their artemisinin contents were 0.73% (Qionghai), 0.63% (Tongshen), 0.61% (Baise), 0.51% (Youyang), 0.61% (Shaoshan), 0.47% (Fengshun), 0.50% (Lianping), 0.58% (Wengyuan), and 0.32% (Shaoguan). Then, from 2005 to 2011, Hu Sihua et al. systematically and selectively propagated these specimens, picking those which had similar phenotypic traits and high yield. Individual offspring were then tested for artemisinin content, and seeds with high content were harvested according to the number needed that year. The specimens with the highest artemisinin content were strictly isolated, and their seeds were used to produce the next generation, undergoing the same selection process in the second year. This process produced a new strain, "Nanfeng No. 1." The highest artemisinin content found in a single strain increased from 0.99% to 2.47%. Average content per year in the parent specimens increased from 0.82% (2005) to 0.83% (2006), 1.22% (2007), 1.42% (2008), 1.53% (2009), 1.79% (2010), and 1.97% (2011).

"Nanfeng No. 1" is a mass-selection strain, shaped like a column, with green or green–yellow stems. The leaves are green, broadly lobed, and dense. Internode spaces are relatively short. It is midmaturing, moderately resistant to disease, and produces around 150 kg of dry leaves per mu (equivalent to 0.22 kg/m^2). Random sampling of dry leaves showed an artemisinin content of 1.20%. From 2006 to 2012, more than 60,000 mu (equivalent to 40,200 ha or 9600 acre) were devoted to the cooperative cultivation of this strain in Guangdong (Fengshun, Qingyuan, Lechang), Hunan (Qidong, Linwu), Chongqing (Liangping, Youyang), and Sichuan (Gao County).

4. Combination Breeding

Combination breeding involves the artificial hybridization of different parents with desirable characteristics. Using the principles of selective propagation, the offspring are weeded out over many generations, compared, and evaluated. The result is a genotypically pure or almost-pure strain of *A. annua*. This process is also termed sexual hybridization. Key aspects of the methodology include cultivation of specimens, preservation and isolation of parent plants, regulation of flowering time and preservation of pollen, and management of the hybrids. Systematic selection was also carried out on the offspring.

In Switzerland, Delabays et al. hybridized clones of Chinese *A. annua* in 1993, using in vitro propagation. These were pollinated with specimens from Italy, the former Yugoslavia, and Spain, producing hybrid strains. Dry leaves yield was 133.3 kg/mu (about 0.20 kg/m^2), with an average artemisinin content of 0.64%–0.95%.[35] In 1994, Delabays et al. created a Chinese–Yugoslavian hybrid (Chine × Yougoslavie 49) with an artemisinin yield of 1.776 kg/mu (equivalent to 0.0027 kg/m^2).[36] From two groups of *A. annua* in South Vietnam, Debrunner selected five karyotypes with high artemisinin content in 1996. They were hybridized with Chinese strains with high artemisinin content, producing offspring with yields of 2.53 kg artemisinin/mu (about 0.0038 kg/m^2).[37] De Magalhaes et al. also selected and hybridized Chinese and Vietnamese strains in 1999, which created a strain suitable for cultivation in Brazil. Its artemisinin yield could reach 1.43 kg/mu (about 0.0021 kg/m^2).[38]

In 2006, Hu Sihua et al. isolated and self-pollinated a single specimen with desirable characteristics. The offspring were selectively propagated in 2007. Tissues from specimens with high yield and artemisinin content were cultured. In 2008, they were asexually propagated and planted for testing. Then, in 2009, those specimens with the most stable karyotypes relative to the original specimen were asexually propagated to produce seed. Finally, in 2010, these seeds were planted and tested for artemisinin content and yield. Some groups had very stable karyotypes, but their phenotypes needed to be improved. Therefore, desirable specimens among their offspring were backcrossed to the parents, successfully creating the "Guangzhong No. 1" cultivar. The strain is column-shaped, with green stems. Some branch nodes have purple spots. Leaves are green, broadly lobed, and dense. Internode spaces are short. The strain has moderate resistance to disease and is late maturing, with a yield of 160 kg/mu (about 0.24 kg/m^2) of dry leaves. Average artemisinin content from a sample of the leaves was around 1.30%. From 2010 to 2013, cultivation of this strain was expanded to 40,000 mu (equivalent to 26,800 ha or 6400 acre) in Guangxi (Rongan and surrounding areas), Guizhou (Wuchuan district), Hunan (Yongchuan of Qidong district), and Guangdong (Shaoguan of Fengshun district).

5. Hybrid Propagation

Hybrid propagation utilizes the hybrid advantages common in biology, with selective breeding conducted on the hybrid offspring. Compared to conventional strains produced by systematic and mass selection, first-generation hybrids have significantly higher yields, disease resistance, and uniformity. Because dominant or recessive characteristics often emerge in second-generation hybrids, this gives the first generation a broad industrial application prospect. Hybrids now make up 80% of world crops. Apart from developing countries, which still use conventional varieties, hybrid strains dominate the market in seeds.

Cross-pollinating plants are highly suitable for hybrid propagation. For such plants, including *A. annua*, continuous self-pollination results in inbreeding depression, manifested in slower growth, dwarfism, lowered disease resistance, reduced yield, etc. As cross-pollinating plants propagate, unfavorable recessive genes are more likely to be preserved in the hybrids. Once these are self-pollinated, the offspring may be homozygous for these genes and hence show reduced fitness. Studies have shown that *A. annua* has marked inbreeding depression. Hybridization of different self-pollinated strains can restore high genetic heterozygosity and the corresponding desirable characteristics. As we have seen, the self-pollination rate is extremely low in *A. annua*, and it can be discounted in the production of hybrid seeds. There is no need to remove male plant parts or specimens. This makes it easier to pursue hybrid propagation.

Effective cultivation and preservation of inbred lines of *A. annua* are critical for hybrid breeding. Such lines have been selected over generations of self-pollination to have pure or almost-pure genotypes, producing a uniform phenotype and stable inheritance. Target phenotypes are preserved to the greatest possible extent via selection. Normally, hybrids reach this stage of purity only after many generations. No reports have yet emerged, either in China or abroad, on the cultivation of inbred lines of *A. annua*.

In 2005–06, Hu Sihua et al. studied the cultivation of self-pollinated *A. annua*. Cycles of selective propagation and tissue culturing began in 2006 to preserve different inbred types. A large number of such lines had been obtained by 2010. After two to five generations, they all had different characteristics. Tests showed relatively large differences in the self-pollinating compatibility of the various types. By the third generation, some types could no longer produce seeds capable of germination. Others could be self-pollinated to the fifth generation. Tissue cultures of inbred specimens became more difficult to preserve as the number of self-pollinated generations increased. In 2011, tissues from representative specimens of each type were cultured for hybridization experiments. A large number of combinations were cross-pollinated in strict quarantine. The F1 generation of some combinations successfully produced hybrid seeds, which were planted in 2012. Many hybrids had high compatibility, and in some sampled plants, the dried leaves had artemisinin content of 2.0% and above. In 2013, a high-yield, high-content hybrid was planted in larger fields and given the initial name of "Guangzhong No. 2." It was expanded into large areas in 2014, with dry leaves containing up to 1.50% artemisinin content and yields of 170 kg/mu (about 0.25 kg/m^2). Subsequently types with higher compatibility were then used to produce hybrid seeds, new inbred lines were cultivated, and large-scale compatibility tests were run. This may create a virtuous cycle for the production of desirable *A. annua* seeds.

6. Mutation and Ploidy

Mutation breeding refers to the use of physical and chemical factors to induce genetic mutation and then selectively cultivating new strains from the mutated offspring. Physical mutagenesis utilizes physical factors such as various kinds of radiation, ultrasound, lasers, space environments, etc. Chemical mutagenesis relies on pharmaceutical treatment. Ploidy-based propagation is based on research into ploidy changes in the chromosomes. The number of chromosomes in the plant is artificially modified. From there, new strains can be selectively bred.

In 1989, Chen Herong treated *A. annua* seeds with colchicine, selectively breeding a strain with high artemisinin content and significantly greater vegetative mass. Specimens were tall and strong, adaptable, stable, and had strong resistance. Dubbed "Jingxia No. 1," it had an artemisinin content of 1.0%.[39]

Wallaart et al. used colchicine treatment to obtain tetraploid *A. annua*. Its artemisinin content was 38% higher than that of diploid specimens and the leaves were much larger, but it was small and yields were lower by 25%. Nevertheless, cultivated tetraploid plants grew rapidly, and their high artemisinin content made them good raw materials.[40]

Xun Xiaohong et al. soaked callus tissue containing buds in 0.05% colchicine solution for 48 h. The induction rate was 60%, and $2n=4x=36$ polyploid test-tube seedlings were obtained (i.e., tetraploid plant with 36 chromosomes). The vegetative sections and petals of these specimens were markedly larger and heavier.[41] Mu Shengyu soaked clustered buds in colchicine and also added colchicine to the culture medium. The tetraploidy induction rate was 69.2%, with 10 such specimens produced. After transplantation, their leaves were significantly larger than those of diploid plants. The average length and width of diploid leaves were 5 cm and 2.5 cm, respectively, whereas those of tetraploid leaves were 7 cm and 3 cm, respectively. Leaves were also thicker, with a darker color; stems and roots were markedly larger. The artemisinin content was higher, reaching up to 0.96% (as opposed to the diploid control, with 0.64%).[42] Finally, Li Xiaojuan conducted research on the best experimental protocol for using colchicine solution to induce chromosome mutation. Seeds should be soaked in a 0.1%–0.5% solution for 5–7 days; a 0.25% or 0.5% solution should be used for stems over 4–5 days. Better results were obtained if a 60 mg/L colchicine solution was added to the solid culture medium. The chromosome number was $2n=4x=36$.[29]

Li Nuo et al. began research into polyploidy breeding in 2005. From 2005 to 2008, seeds were soaked in 0.05%–0.2% colchicine for 24 h and incubated at different temperatures. They were then sown and tested. In 2009, the seeds were cultured in a sterile medium after colchicine treatment, producing polyploid seedlings. From there, three tetraploid lines were created—DB3, DB4, and BX69—which were dwarfed, with thicker leaves and larger pores and pollen. The chromosome number of the roots was established as $2n=4x=36$. At present, studies are being conducted on the hybridization of tetraploid and diploid plants and creating triploid *A. annua*.

In 1996, Li Guofeng et al. from the Chinese Academy of Sciences' Institute of Botany placed *A. annua* seeds in a recoverable satellite, "Jianbin No. 1." On return, the seeds were propagated asexually, cultured, and screened to produce a postspace, high-yield strain, "SP-18." The strain was grown in greenhouses, in a suburb of Beijing and in Youyang of Chongxing. All had long internode spaces, strong stems, long and robust branches, and wide leaves. Specimens were tiered, with yellow–green leaves. This was in contrast to the "001" control strain, which was irregularly shaped and had deep green leaves. In the vegetative growth and early budding stages, SP-18 had twice the height and number of branches as the 001 strain. When planted in Youyang, SP-18 could produce up to 1.89 kg/mu (about 0.0028 kg/m^2) of artemisinin, higher than the 1.35 kg/mu (about 0.0020 kg/m^2) for the 001 strain.

7. Seed Production

Currently, the propagation of *A. annua* as a whole is in its early stages when compared with other important crops. Conventional seeds are usually produced directly by breeders. Alternatively, directed line or bulk selection based on type may be used, where seeds are produced based on adherence to type and consistency. Once the seed is obtained, they must be further propagated because of their limited numbers, with further selection of desirable specimens. The seeds produced are supplied to farms.

Hybrids of self-pollinated lines are relatively easier to produce because the plant's rate of self-pollination is so low as to be negligible. Parent plants may be grown in alternating rows, and mature seeds may be harvested together. If first cross or reciprocal cross hybrids have obvious differences in phenotype, they can be collected separately. To improve seed production and quality, attention must be paid to ensure synchronization of flowering.

One mu (equivalent to 667 m^2 or 0.67 ha or 0.16 acre) of land used for seed production (containing 1000 specimens) can usually produce 25 kg of seed and above. Assuming that 1 g of seed is required to sow 1 mu of land, this fulfills the requirements for 25,000 mu (equivalent to 16,750 ha or 4000 acre) of land. However, the propagation coefficients of different seed production methods, conditions, and plant varieties vary widely. Hence, the planting area should be based on empirical conditions.

Seeds are normally assessed based on water content, mass per 1000 grains, purity, germination potential, and germination rate. In late December 1999, Zhu Weiping collected the fully developed wild *A. annua* seeds from three areas in Hunan. The seeds were germinated in an illuminated incubator at 25°C and exposed to light for 12 h a day. The germination rate was around 90%.[21] Dong Qingsong et al. took the first step toward setting out germination techniques: top of paper germination bed, temperature 15–25°C (or 30°C), first count at day 7, and last count at day 14.[43] Li Hongli et al. established a preliminary classification standard for seed quality. Water content had to be less than or equal to 11.0%; 1000-grain weight was ≥0.06 g for Grade 1, ≥0.04 g for Grade 2, and ≥0.03 g for Grade 3. Seed purity was ≥60% for Grade 1, ≥50% for Grade 2, and ≥40% for Grade 3. Germination rate was ≥90% for Grade 1, ≥70% for Grade 2, and ≥50% for Grade 3.[44]

Based on the *Inspection Procedures for Crop Seeds*, Hu Sihua et al. checked the *A. annua* seeds produced in fields from 2006 to 2011. After this extensive analysis and based on the criteria for seed preservation, the following grading standard for seeds was created (Table 2.1).

A. annua seeds have no dormant period and, when mature, can sprout on the branch after several days of rain. Their shelf life depends on the temperature and humidity of the preservation conditions and the moisture and purity of the seeds. A study by Xu Chengqiong et al. showed that the germination rate was highest if seeds were stored at low temperatures, followed by storage at room

TABLE 2.1 Grading Standards for *Artemisia annua* Seeds

	Seed Grading Reference		
Item	**Grade 1**	**Grade 2**	**Grade 3**
Moisture	≤9%	≤9%	≤9%
Purity	≥90%	≥80%	≥70%
1000-grain weight (g)	≥0.03	≥0.025	≥0.02
Germination potential	≥50%	≥40%	≥30%
Germination rate	≥70%	≥60%	≥50%

temperature in dry conditions, and finally by storage at room temperature in wet conditions. Regardless of the above, however, the germination rate was higher after storage in bottles than in cloth bags.[24] Li Hongli et al. found that seeds kept at low temperatures (4–5°C) for 420 days had a germination rate only 2.2% lower than before storage. After freezing (−10 to 4°C) for 420 days, the seeds' germination rate fell by 15.6%. The germination rate fell the fastest for seeds stored at room temperature, losing viability after 180 days.[45] Seed respiration and metabolism were very low at low temperatures. Seeds should be stored in sealed containers and frozen.

D. CULTIVATION

1. Planting

Many studies on growing methods for *A. annua* have appeared in China and abroad. Chen Herong et al. conducted an early experiment on the effects of different sowing times on artemisinin content. It showed that in Xiamen, Fujian Province, plants had significantly higher artemisinin content if the seeds were sown on 14 February and 14 March, rather than on 20 April and 14 May. After putting the plants in shade, it was found that longer light duration increased artemisinin content. Individual plants should not block each other; hence, a suitable sowing density should be used. A north–south row direction was ideal.[46] Field experiments by Cao Youlong et al. proved that longer growth periods (up to 118 days) resulted in greater production of dry leaves and a gradual increase in artemisinin content. They suggested bringing forward the sowing and delaying the harvest.[47] Lu Hongshun found that early sowing produced much higher artemisinin content. March sowing resulted in an artemisinin content of 0.68%, which fell to 0.25% for May sowing.

A. annua grows well in warm, sunny climates with no shade, good drainage, and fertile loose clay loam. In such conditions artemisinin content is high and pests are few.[48] Wang Sangen et al. studied the effects of sunshine on artemisinin content. These effects were marked in both wild and cultivated plants. Plants grown in dark, damp places had low artemisinin content, as opposed to a content of 0.95% in plants without shade.[49]

Liang Huiling et al. investigated *A. annua* pests and diseases in Guangxi from 2005 to 2006 and proposed preventive measures. There, *A. annua* is susceptible to three diseases: stem rot, damping-off, and powdery mildew. It is affected by 11 types of pest: leaf beetles, leafworm moths (*Spodoptera litura*), red ants, green peach aphids (*Myzus persicae*), black cutworms, chrysanthemum midges, froghoppers (spittlebugs), yellow peach moths, inchworms, gray weevils, and cabbageworms. Of these, stem rot, damping-off, green peach aphids, black cutworms, and chrysanthemum midges were widely distributed throughout the area. Powdery mildew occurred sporadically in Xiangzhou. Red ants fed on seedlings in the autonomous regions of Longlin. The aphids were a serious pest in Jingxi.[50]

In 2006, the WHO summarized the existing literature in its *Good Agricultural and Collection Practices*.[1] Based on the *Good Agricultural Practices for Chinese Traditional Medicine*, published by the China Food and Drug Administration, Jiang Yunsheng et al. studied the distribution, environment, germination and cultivation techniques, harvesting, processing, storage, quality control, packaging, and transportation of *A. annua* in Guangxi. They determined the best indicators and methods of each stage, formulating standard production criteria and operating procedures suitable for *A. annua* cultivation in Guangxi.[51]

From 2005 to 2014, Huang Ronggang et al. from the Artemisia Research Center at the Guangzhou University of Traditional Chinese Medicine commenced research into the production of *A. annua* in Guangdong, Hunan, Chongqing, Guangxi, and Guizhou, cooperating to expand the growing area to almost 100,000 mu (equivalent to 67,000 ha or 16,000 acre). Empirical data from farmers were used to establish the following planting techniques. However, variations in microenvironment mean that planting must be adjusted to local conditions.

Sowing Phase

Seed Selection

The seed is obtained from production units, which meet technical criteria, and then inspected. These seeds should be capable of yielding 150 kg of dry leaves per mu (about 0.22 kg/m^2) and above, with an artemisinin content of at least 1.0%.

Seedbed Preparation

Soil should be deep, loose, and fertile, conveniently irrigated and drained, with relatively few weeds. It should be sunny, with sandy loam preferred. Weeds, plant residue, stones, and other foreign matter are removed. Then, 1200 kg/mu (about 1.80 kg/m^2) of manure and 25 kg/mu (equivalent to 0.037 kg/m^2) of compound fertilizer (15-15-15) are spread evenly. The land is plowed at a depth of 20 cm and above and divided into furrows 1.4 m wide and not more than

6 m long. The top soil in each furrow should be fine and flattened. For each mu (667 m^2 or 0.67 ha or 0.16 acre), a mixture of 1.5 kg of 50% carbendazim wettable powder and 30 kg of soil is spread evenly on the furrows. These are thoroughly watered before sowing.

Sowing

Sowing takes place in January or February. In seedbeds not earmarked for transplanting, 1 g of seeds per 15 m^2 can be used, as opposed to 1 g per 5 m^2 in fields for transplanting. Seeds are mixed well with plant ash or thin soil at a ratio of 1:500 and uniformly spread on the surface of each furrow. A structure is then erected and covered with agricultural film, the corners of which are held down by soil.

Management of Seedlings

Seedbed

Attention must be paid to weather. In high temperatures and dry conditions, the moisture level of the seedbed must be observed and supplemented promptly to ensure a high germination rate and good uniformity. When one or two true leaves appear, two ends of the film are uncovered every morning on sunny days to aid growth. This is replaced in the evening. After 3–5 days, the film can be completely removed, to be replaced during heavy rain. The seedbed should be watered when temperatures are high and the weather is dry. Weed control begins after the film is uncovered; they are removed immediately when still small and few in number to reduce damage to the plants and make elimination easier. When three to four true leaves are present, a 0.3% compound fertilizer–water mixture is used every 5 days.

Thinning

Thinning is carried out when four true leaves appear, with seedlings left in according to need. Diseased and weak seedlings are removed first before adjusting for density, to maintain consistent growth.

Transplanting

Transplant bed requirements are essentially the same as the seedbed preparation in the sowing phase. When seedlings are around 5 cm tall, they are transplanted at intervals of 10 cm and watered for a few consecutive days to ensure their survival. Once striking is achieved, they are given a 0.5% compound fertilizer–water mixture once in every 5 days. After 15 days, the soil is loosened, which aids the growth of the seedlings.

Diseases and Pests

Damping-off occasionally occurs to seedlings in the seedbed. Seedlings decay, appear bulky and limp, and die. It arises more easily in warm and moist conditions. Plant density should be controlled, with appropriate thinning of seedlings. Seedbed ventilation should be improved and drainage maintained. A 58% metalaxyl mancozeb wettable powder liquid spray (1000× dilution) may be used to control the disease. During the seedling stage, a certain amount of damage from midges may occur, with small numbers of gall midges (cecidomyiids) appearing on the plants. These do not harm seedling growth in general but should be prevented and controlled to reduce the degree of damage later in the fields. Either 10% imidacloprid or 48% chlorpyrifos liquid sprays (both 1000× dilution) can be used. To comprehensively prevent and treat diseases and insect pests, seedlings should be treated with a spray before transplanting, such as a 50% carbendazim wettable powder with 48% chlorpyrifos (both 1,000× dilution).

Transplanting Phase

Land Preparation

The area should have good light and temperature conditions, high amounts of organic matter, and no flooding during the rainy season. It should be far from rubbish dumps and landfills. Tall weeds, plant residue, stones, and other foreign objects should be removed first and 500 kg/mu (equivalent to 0.75 kg/m^2) of manure evenly spread over the field before plowing to a depth of 20 cm. Furrows should then be laid out according to the direction of the summer wind, with a width of 1.2 m. Raised soil on furrow top should be slopped on both sides.

Seedling Selection

Seedlings are transplanted when they are 20 cm tall. Those with robust main stems, short internode spaces, and strong signs of growth are selected. The seedbed is watered with a suitable amount of pure water, and the roots of the seedlings are lifted up with the attached soil to minimize the damage. Seedlings cannot be pulled up by the stems or leaves. Deformed or overly thin seedlings, which can be caused by a variety of reasons, grow poorly after transplanting and have low yields. These should be discarded.

Transplantation

This should take place before light rain, after heavy rain, or on a cloudy day. Seedlings should be transplanted with the attached soil at a density of 70 cm between specimens and 80 cm between rows on level ground or 60 cm between specimens and 80 cm between rows on a slope. Thereafter, the roots of each plant should be watered once and again for several consecutive days in sunny weather.

Replacements of Transplanted Seedlings

Seven to ten days after transplanting, dead, weak, and diseased seedlings should be promptly removed. They should be replaced with other seedlings on cloudy or rainy days, to ensure the basic number of 1200–1400 plants per mu (about 1790 plants per hectare or 7500 per acre.)

Field Management

Weeding and Ridging

One and a half months after transplanting, the soil should be tilled and hoed. Tilling should be shallow; hoeing should be small-scale and neat. This avoids damage to the roots of the seedlings. After two and a half months, before the crop closure stage, further hoeing and earthing up takes place. Fertilizer should be applied before earthing up, with the earth being used as cover. The additional earth should be 15 cm away from the main stem.

Fertilizer

The first application should be 20 days after transplanting, when 5 kg/mu (equivalent to 7.5 kg/ha or 31 kg/acre) of compound fertilizer should be spread on the field when the soil is moist or before light rain. The second application takes place after tilling and hoeing, with 10 kg/mu (about 15 kg/ha or 62 kg/acre) used under the same conditions. For the third application, 25 kg/mu (about 37 kg/ha or 156 kg/acre) is added before earthing up. Care must be taken to maintain a certain distance between the fertilizer and the plants to avoid damage to the branches. Depending on soil fertility and growth conditions, the number and frequency of fertilizer applications may be altered. If required, manure can also be added.

Irrigation and Drainage

A. annua cannot tolerate waterlogging. After storms or several days of rain, the field must be promptly drained to prevent flooding. Because the plant is mostly cultivated in hilly areas, irrigation reservoirs may be established based on the terrain. These can also be used to collect water from surrounding fields.

Diseases and Pests

Root and stalk rot arise because of poor drainage and hardening of the soil. In the later growth stage, some of the leaves begin to wilt, which gradually spreads to the entire plant within a few days. The leaves do not fall and maintain their green color, but the roots or stems turn black. This is dispersed, with no clear locus. A 50% carbendazim wettable powder in 600 times water volume can be applied to the affected plants or further diluted 1000 times in water for a spray. Proper crop rotation and interplanting can reduce incidence.

Powdery mildew and rust mainly occur in the later growth period. Signs of powdery mildew include white, powdery filaments on the surfaces of the lower leaves. Rust appears as rust brown specks on the backs of the lower leaves. A 20% triadimefon miscible oil spray, diluted 800 times, can be used as treatment.

Aphids gather in the tips of stems and new-growth areas. In large groups, their consumption of sap leads to yellowing of the top part of the plant, which then turns black and withers. A 10% imidacloprid spray, diluted 1000 times, can control the aphids.

Cabbageworms hide on the undersides of leaves, on which they feed. In a severe infestation, they can consume all the leaves on a plant in 3 days. They can be controlled using a 40% phoxim miscible oil spray diluted 1000 times.

Chrysanthemum gall midges are common in wet, low-lying environments. They lay eggs, either diffusely or in clusters, on leaf joints and growth zones. One day after hatching, the larvae can enter plant tissues. Galls are formed after 5 days, gradually expanding with the growth of the larvae. The galls are green or purple–green, with a sharp apex and round base, and a diameter of 0.6 cm. Severely infested plants are covered in galls; growth slows, the plant becomes dwarfed and deformed, and the leaves atrophy. A 10% imidacloprid (diluted 1000 times) or 48% chlorpyrifos (diluted 1000 times) spray may be used against them. If a small number of galls are found, they should be promptly removed. In cases of severe infection and gall proliferation during the later growth stages, the harvest can be brought forward.

Toxic pesticides should be avoided. Use of chemical pesticides should cease 15 days before leaves are harvested.

2. Harvesting

A. annua leaves can be harvested at least 120 days after transplanting from the peak of the vegetative growth phase to the onset of budding. In 1997, Zhong Fenglin et al. were the first to report that the optimal time for harvesting was during this window, when artemisinin and vegetative content were high.[52] These findings were subsequently confirmed by many researchers.

Different schools of thought have emerged regarding the methods of drying *A. annua* leaves. A 1997 experiment by Zhong Fenglin et al. found that sun-dried leaves had the highest artemisinin content, with an average of 1.0%. This was 23.76% higher than in air-dried and 70.94% higher than in oven-dried samples. Hence, production facilities should opt for sun drying, followed by air drying.[52] In 2001, Ding Derong confirmed that sun drying was best, producing artemisinin content 101.7% higher than in oven-dried and 34.48% higher than in air-dried leaves (unpublished work, mentioned in Yang Shuiping's thesis).[53] Then, in 2009, Feng Shixin et al. air-dried entire *A. annua* specimens for different periods, followed by sun drying. It was found that the relationship of artemisinin content to air drying time followed a parabolic function, peaking significantly after 4–5 days of air drying before declining gradually. Artemisinin content fell as drying temperature rose. At 40°C, the content was higher.[54]

Hu Sihua et al. studied the best method for harvesting *A. annua* leaves in 2007. Many of the findings showed that leaves left on the branch and air-dried naturally had 28.87% higher artemisinin content than those that were sun-dried within a day, which is currently the most widespread practice. In large-scale production, air drying of leaves on the branch consumes large amounts of space and time, requiring better facilities. For small-scale farmers, however, it is easy to air dry branches with leaves after they have been collected. This method produces leaves with markedly higher artemisinin content than sun drying at the same day.

Harvesting methods should depend on weather and production conditions. First, one must ensure that the leaves are not exposed to rain, which causes them to rot and significantly reduces yield and artemisinin content. Experiments showed that drought stress could increase artemisinin content by 20.25%. A convenient and economical method—either physical or chemical—could be explored to produce a water-scarce environment before the harvest.

In practice, the following method of harvesting is used. The entire plant is cut down, bundled, and air-dried in vertical stacks in a cool, dry place, out of the rain. Where air-drying facilities are not available, moving directly to sun drying is possible. After air drying, the plants are spread out and sun-dried until the leaves become prickly. The leaves are then beaten off with a bamboo stick, whereupon twigs and other impurities are sifted out. During sifting, the leaves and twigs are stirred by hand in the sieve to separate them. Leaves should be dry and clean, with no signs of rot. They are packed in cloth bags and placed on a frame off the floor in a shaded, airy location. Dampness should be prevented.

2.2 CELLULAR ENGINEERING OF *ARTEMISIA ANNUA*

Wang Hong[1], *Ye Hechun*[2]
[1]*University of Chinese Academy of Sciences, Beijing, China;* [2]*Institute of Botany, Chinese Academy of Sciences, Beijing, China*

A. OVERVIEW OF TISSUE AND ORGAN CULTURING

Producing artemisinin via cell culturing preserves natural resources and is free of normal restrictions on cultivation. Various cellular and genetic engineering methods can also be used to produce new *A. annua* strains with high artemisinin yield. Moreover, natural artemisinin is nontoxic to humans and cannot be replaced by synthetics.[55]

The first reports on callus induction of *A. annua* appeared in the early 1980s. In 1983, Zeng Meiyi of the Institute of Chinese Materia Medica, China Academy of Chinese Medical Sciences, and Li Guofeng of the Institute of Botany, Chinese Academy of Sciences collaborated to induce the callus formation (clumps of undifferentiated plant cells), its differentiation, and to study the changes in artemisinin content. They succeeded in the induction process and in regenerating plantlets.[25] Thin-layer chromatography with its color reaction and UV spectrophotometry after alkali treatment were used to detect artemisinin in extracts from calluses, calluses with shoots, and regenerated plantlets. The results showed that calluses with shoots had basically the same chromatographic characteristics as the R_f value of the artemisinin standard sample, as well as the same absorption peak. Colorimetric reactions were not seen in calluses. It showed that, in Hainan specimens, formation of shoots in calluses was accompanied by production of artemisinin, with a content of 0.008% of dry weight. When differentiated seedlings further developed into flowering plants, a similar analysis showed that artemisinin content could reach 0.92% of dry weight, much higher than in natural field plants (0.56%).

Biosynthesis of artemisinin was thus related to the degree of differentiation in plant tissues. This was consistent with the results obtained by Butcher in 1977, namely, that it was more difficult to produce terpenoids from calluses because such compounds were only produced when calluses differentiated into tissue-bearing structures.[56] In 1993, however, Woerdenbag et al. adjusted the microelements in an MS medium, rotating culturing regenerated shoots in media containing NAA 0.05 mg/L and 6-BA 0.2 mg/L rotating culture.[57] Even without roots, the shoots could synthesize artemisinin. Jha et al. in 1988[58] and Tawfiq et al. in 1989[59] also cultured tissues and shoots, in which artemisinic acid was detected.

Different combinations of cytokinins and auxins in a culture medium readily result in callus formation.[60–63] However, these calluses are usually more compact.[64] A 1 mg/L 6-benzylaminopurine (6-BA) and 1 mg/L 2,4-D medium can produce large numbers of looser calluses. Even then, induction is successful in only 10% of the explants.[63] Many works have shown that undifferentiated calluses or cell cultures contain only trace amounts[64–66] of or no artemisinin.[25,59,62,63,67] Hence, partial differentiation is a precondition for biosynthesis of artemisinin.[60,62,67] Artemisinin was not detected in most culture media,[63,67] except in a 1986 study by Nair et al. where a small amount of artemisinin was detected.[64]

The results are mixed on whether tissues cultured in vitro contain artemisinin. Some studies suggested that, in seedlings with a certain degree of differentiation, the formation of roots is important for the synthesis of artemisinin.[60,67,68] Seedlings without roots contained only trace amounts of artemisinin.[58,60,63,67,69,70] Most experiments found no artemisinin in roots,[59–61,63,67] apart from Nair et al. in 1986[64] and Jha et al. in 1988.[58] Hairy root culture systems had been developed by the end of the 20th century,[68,71–74] and Weathers et al. reported in 1994 that hairy roots had relatively high artemisinin content (0.4%).[68] However, this was extremely inconsistent, and in 1995 Jaziri et al. could not verify the presence of artemisinin in roots.[74] Because artemisinin is relatively toxic to plants,[75] it can only be concentrated in the glandular secretary trichomes on the plant's surface.[75–77] It is therefore difficult to explain the presence of artemisinin in hairy roots. Nevertheless, if roots can indeed produce artemisinin, it would significantly clarify the synthetic pathways of artemisinin.

Many studies have reported that growth regulators have a great impact on the synthesis of artemisinin in seedling cultures. Artemisinin content increased sevenfold in shoot clusters treated with 100 μg/mL miconazole.[65] In 1993, Woerdenbag et al.[57] discovered that when 0.2 mg/L 6-BA, 0.05 mg/L NAA, and 1% sucrose were added to an MS culture medium, the resulting shoot clusters had an artemisinin content of 0.16%. Supplements of 10 mg/L gibberellic acid (54%), 0.5 g/L casein hydrolyzate, and 10 mg/L or 20 mg/L naftifine (40%) could increase artemisinin production to different degrees. Other regulators, such as miconazole; inducers, such as cellulose; precursors, such as mevalonic acid (MVA); and gene expression regulators, such as colchicine, had negative or no effects on artemisinin synthesis in shoot cultures. In 1994, Weathers et al. found that 0.1 mg/L BA and 10 mg/L KT caused artemisinin production in shoot clusters to increase by 30%.[68] This was due to an increase in biomass, rather than in artemisinin content (mg/g dry weight).

B. HAIRY ROOT CULTURES

Some secondary metabolites exist only in highly differentiated plant tissues and are only present in trace amounts or cannot be synthesized in in vitro cultured cells. Therefore, scientists have turned their attention to root induction by *Agrobacterium*. These roots grow quickly in hormone-free media, have stable heritability, and are highly differentiated. Such results cannot be obtained by normal, chemical in vitro cell or tissue or organ cultures. The secondary metabolism and metabolic engineering laboratory at the Institute of Botany, Chinese Academy of Sciences initiated research into root cultures in the early 1990s. This is a summary of the results.

1. Hairy Roots, Untransformed Roots, and Callus Tissues

Agrobacterium rhizogenes ATCC15834 was used to induce hairy roots in *A. annua* line 025.[78] Hairy roots, untransformed roots, and callus tissues were cultured in the same medium to compare their biomass. Biomass of hairy roots (gram per vial dry weight) was significantly higher than those of untransformed roots and calluses. There was no marked difference between the latter two cultures.

2. Selection of High-Yield Hairy Root Lines

A total of 747 roots were clipped and cultured antiseptically. Of these, 53 fast-growing lines were selected. After three generations, a further seven lines were chosen.[78] These had all been induced by *A. rhizogenes* ATCC15834 from *A. annua* line 025. Biomass and artemisinin content were significantly different among the seven lines. The HR-9 line had the highest artemisinin yield, reaching 33.25 mg/L per month.

3. Dynamics of Hairy Root Growth and Artemisinin Biosynthesis

To determine a suitable culturing time, the dynamics of growth and artemisinin synthesis were measured in the HR-9 line.[78] Artemisinin content had strong "correlation to growth." Artemisinin content decreased gradually during the exponential growth stage and increased when the growth rate slowed. After growth ceased, artemisinin content remained stable despite prolonging the culture period. Artemisinin production reached 33.25 mg/L per month after 21 days of culturing. Therefore, the optimal culture period was 21 days.

C. INDUCTION OF SHOOT CLUSTERS

Because artemisinin content is relatively low in *A. annua*, researchers attempted to use tissue culturing to resolve the problem of production. However, biosynthesis of artemisinin is closely related to tissue differentiation. The ability of calluses and cell suspensions to synthesize artemisinin is low. Although synthesis takes place in roots, content is extremely low. Some reports have even said that roots do not synthesize artemisinin. Certain studies showed that seedlings without roots could synthesize artemisinin, and the content was stable. Focus has therefore shifted to the culturing of shoot clusters. The secondary metabolism and metabolic engineering laboratory at the Institute of Botany, Chinese Academy of Sciences investigated the factors influencing shoot cluster induction. An induction system was established as a foundation for future industrial production of artemisinin.

1. Effects of Medium Composition on Induction Rate

MS was used as a basic medium. Different combinations of hormones, such as 6-BA; KT; zeatin; 2,4-D; NAA; and indolebutyric acid, were used to induce shoot clusters on different explants, including leaf discs, leaves, stem segments, and leaves with stalks.[79] Addition of 2 mg/L 6-BA with 0.05, 0.1, or 0.15 mg/L NAA could produce an induction rate of 98.7%, 97%, and 96.8%, respectively, in leaf discs. The resulting shoot clusters grew well. The different NAA concentrations produced widely varying rates of vitrification in shoot clusters at 15.4%, 10.5%, and 5.4%, respectively. The ideal combination for an induction medium was 2.0 mg/L 6-BA and 0.15 mg/L NAA. Other combinations produced shoot clusters but with different degrees of callus formation, hyperhydricity, and browning. These affected the induction rate.

Different basic media had significant effects on cell growth and formation of secondary metabolites. These resulting variations can be traced back to the composition of the medium. From there, the chemical factors influencing growth and secondary metabolism can be inferred.

Six basic media (MS, N6, DCR—a formulated basal culture medium for fir and pine, B5, Wolter and Skoog medium (WS), and White) were compared.[79] Statistical analysis showed that MS, WS, and White had different degrees of inhibition on the induction of shoot clusters. High-performance liquid chromatography (HPLC) indicated that shoot clusters cultured in DCR had the highest artemisinin content, followed by those cultured in MS. When MS and DCR were compared, K^+ and nitrogen sources were four to five times higher in MS than in DCR, whereas Ca^{2+} was one-third higher in DCR than in MS. Hence, K^+ and nitrogen sources may help in the induction of shoot clusters, whereas Ca^{2+} enhances the synthesis of artemisinin.

Leaf disc explants were cultured in media with different pH and oven-dried after the induction rate was calculated. Artemisinin content was measured using HPLC.[79] The results indicated that pH 5.5–5.8 was suitable for inducing shoot clusters and pH 5.8 for artemisinin synthesis.

2. Effects of Light, Temperature, and Photoperiod

Leaf disc explants were cultured under light intensities of 0–6000 lux to investigate the effects of light intensity on the induction rate of shoot clusters.[79] It was found that the induction rate was not affected by light conditions. Only in complete darkness was the rate lowered, but it could still be as high as 80%. Shoot clusters cultured in the dark grew faster than those exposed to light, but they were thinner, weaker, and had fewer shoots. An experiment with different temperatures (15–40°C) found that overlying high or low temperatures significantly affected the induction rate.[79] At the optimum temperature of 25°C, the induction rate could reach over 97%. A change of ±3°C had little effect on the induction rate.

Seedlings were cultured aseptically at the same temperature and exposed to photoperiods of 8, 12, 16, 20, and 24 h for 4 weeks. Leaf discs were obtained from them and cultured using the same five photoperiods. Shoot clusters were not induced under continuous illumination, even if the concentration of cytokinin was increased to 4.0 or 8.0 mg/L. A photoperiod of 16 h, applied for 2 weeks, could restore the capacity to induce shoot clusters. This period had the highest induction rate for seedlings and leaf discs, reaching 97.6%.

D. HAIRY ROOTS AND ADVENTITIOUS SHOOTS CULTURED IN BIOREACTORS

The most successful research into artemisinin production in bioreactors was conducted by the Institute of Chemical Metallurgy (now the Institute of Process Engineering) and the Institute of Botany, both at the Chinese Academy of Sciences. A new type of mist bioreactor was used to culture hairy roots and adventitious shoots. To make full use of the space in the bioreactor, reduce nutrient solution loss and prevent bacterial contamination, the original ultrasonic mist bioreactor was improved on. The new ultrasonic mist airlift loop bioreactor was used in preliminary studies of artemisinin production by adventitious shoots[80] and hairy root cultures.[81] Good results were obtained. Growth rates for adventitious shoots were significantly higher in the bioreactor than in flasks: 2.4 times higher than in solid cultures and 2.1 times higher than in shake flasks. Artemisinin content was 1.5 times higher than in solid cultures and 1.8 times higher than in shake flasks. Yield was higher by 2.9 and 3.2 times, respectively. The mist bioreactor provided suitable conditions for adventitious shoots, promoting growth and the synthesis of artemisinin.

In 1986, Nair et al.[64] experimented with callus and cell suspension cultures. They found that the former did not produce artemisinin but the latter did. Tawfiq et al. did not find artemisinin in cell or tissue suspension cultures in a 1989 study.[59] Based on their previous experiment with *Rauvolfia serpentina* adventitious shoots, Fulzele et al. formulated a theoretical framework by which cultured *A. annua* shoots could produce useful secondary metabolites. In 1991, they successfully cultured these shoots in a bioreactor, but there was no increase in artemisinin content.[67]

All these biotechnological studies provided a strong theoretical basis for the industrial production of artemisinin. However, a key problem remained. The artemisinin contents of both the *A. annua* plant material used and the resulting biotechnological cultures were still relatively low. This was the biggest obstacle to reducing the cost of artemisinin production.

E. ESTABLISHMENT OF GENETIC TRANSFORMATION SYSTEM

Following the discovery of artemisinin's unique antimalarial properties, it became critical to expand the supply of this compound. Since the 1980s, scholars in China and abroad have tried to produce artemisinin via plant biotechnology. A root-culture system mediated by root-inducing (Ri) plasmids and a shoot-cluster culture system mediated by tumor-inducing (Ti) plasmids have been established.

1. Root-Inducing Plasmid Mediated Transformation

Ri plasmids are independent, double-stranded, circular DNA molecules separated from the chromosomal DNA of *Agrobacterium rhizogenes*. It is 180–250 kb in size. When plant tissues are infected by *A. rhizogenes*, T-DNA on Ri plasmids can enter plant cells, combining with their genomes and causing these cells to transform. Therefore, Ri plasmids are ideal vectors for genetically engineering the plant.

Ri plasmids can induce hairy roots at the site of infection. In the 1980s, attention turned to using these plasmids to produce hairy root cultures via genetic transformation and thus generate useful secondary metabolites. One potential method for bioengineering artemisinin was the culturing of *A. annua* hairy roots with Ri plasmids from *A. rhizogenes*. The Institute of Botany, Chinese Academy of Sciences was the first to study Ri plasmid induction of *A. annua* hairy root cultures in 1994. Subsequently, more detailed research on the factors influencing Ri-mediated transformation was conducted.[71–73]

Effects of Different Agrobacterium rhizogenes *Lines*

A. rhizogenes lines ATCC15834, R1000, and A4 were used. The plant expression vector was pBI121, and leaves from *A. annua* line 001 were used as explants.[71,72] It was found that different *A. rhizogenes* lines bearing expression vectors produced widely varying rates of induction. The highest induction rate, 71.8%, came from ATCC15834, followed by A4 and R1000. The morphology of the hairy roots also showed large differences. Those induced by ATCC15834 were stout, with many branches and extremely dense hairs. They were markedly different from normal roots, exhibiting the typical morphology of hairy roots. Because these hairy roots grew rapidly and had relatively stable inheritance, they could be selectively cultivated easily. Most of the hairy roots induced by A4 and R1000 were similar to normal roots, with fewer hairs and slower growth on average.

Effects of Artemisia annua *Genotype*

Sterile seedlings from high-yielding lines from Sichuan, 001 and 025, were used together with *A. rhizogenes* ATCC15834.[71] The induction rate varied greatly with different *A. annua* genotypes, with the rate for 001 at twice that of 025. Hairy roots induced from 001 were thicker, with denser hair and branches. The above two experiments showed that induction rate was primarily determined by the interaction between the type of *A. rhizogenes* and genotype of *A. annua* that were used.

Effects of Seedling Age and Type of Explant

Sterile seedlings from the 001 line, aged 5–50 days, were used together with *A. rhizogenes* ATCC15834. No obvious differences were observed in the induction rates (60.02%–67.5%) in seedlings aged 5–20 days. For seedlings above 30 days old, the induction rate dropped significantly (43.9%–55.3%). Given that the biomass of 15- to 20-day-old seedlings was higher, they were ideal for bioengineering.

In addition, different explants were employed, including whole leaves, leaf discs, stem segments, and main stems. The induction rates of different explants varied widely, being highest in whole leaves (57.1%) and leaf discs (57.4%). It was lowest in stem segments (5.1%). The induction rate for sterile main stems could reach 80%, but supply was limited. The morphology of induced hairy roots differed: Those from the bases of whole leaves were thick and fast growing, whereas those from wound sites on leaf discs were weaker. Considering the above factors, whole leaves could serve as ideal explants.

Effects of Preculture and Culture Period and Concentration of Agrobacterium rhizogenes

Leaf explants were used to investigate the effects of preculturing on induction rate.[71] The rate was around 70% with or without preculturing. However, explants precultured for 2 days experienced hairy root induction in 7 days, as opposed to 12–14 days for explants without preculturing. Hence, preculturing brought forward the time of induction.

Culture period had a significant impact on the induction rate. At $OD_{660}=0.75$, the induction rate was highest (94.2%), showing that a suitable culture period was important for raising the rate of transformation. In a test on plasmid concentration, *A. rhizogenes* ATCC15834 ($OD_{660}=0.75$) was diluted 5–20 times. The effect on induction rate was small but over- or underdilution did cause the rate to decrease.

Effects of Phenols

Phenolic compounds such as acetosyringone, potato extract, dinitrophenol, and dichlorophenol were added to the *A. rhizogenes* culture medium to investigate the effects.[71] The induction rates for the aforementioned phenolic compounds were 86.4%, 90.5%, 87.1%, and 86.3%, respectively, as opposed to 96.3% for the control. Hence, these phenolic compounds were not suitable for hairy root induction in *A. annua*.

2. Tumor-Inducing Plasmid Mediated Transformation

In 1996, Vergauwe et al. established a Ti plasmid–mediated transformation system by infecting *A. annua* leaves with *Agrobacterium tumefaciens*.[82] They subsequently studied the effects of various factors—such as explant type and age, strain of *A. tumefaciens*, and type of binary vector—on the transformation.[83] In 1999, the Institute of Botany, Chinese Academy of Sciences also developed a Ti plasmid–mediated transformation system using GFP as a reporter gene. Shoot clusters were induced using explants such as leaves, flower buds, and floral reproductive organs at different developmental stages. Artemisinin was detected in the shoot clusters. Various factors affecting induction, growth, and artemisinin biosynthesis were also studied. This confirmed an optimal medium for growth and biosynthesis.[79,84] Han et al. also conducted detailed research on the factors influencing the transformation rate in 2005, further refining the plasmid transformation system.[85]

Effects of Agrobacterium tumefaciens *Strain and* Artemisia annua *Genotype*

The study used *A. annua* line 001, from Sichuan Province, and seeds from farm strain, named NJ, from Hunan Province. *A. tumefaciens* strains LBA4404 and EHA105 were employed.[85] Results indicated that *A. tumefaciens* strain and plant genotype played decisive roles in determining the transformation rate. EHA105 was superior to LBA4404, and 001 had a higher transformation rate than NJ. Because 001 was more amenable to transformation, had higher artemisinin content, produced a higher shoot-cluster induction rate, and was easier to manipulate because of its smaller leaves, it was used in further experiments.

The best explant for this system was leaves with petioles. Compared with root and stem segments, leaves had the highest shoot-cluster induction rate and were more easily infected with *A. tumefaciens*. Transformation rate was very low with 8-day-old cotyledons or hypocotyls. Seeds could germinate post-transformation but were unable to form roots. Transgenic plants could not be obtained from them.

Effects of Preculture Period

Preculturing is not necessary for transformation; on the contrary, prolonged preculturing hampers induction in resistant shoots. Shoot induction occurs at an early stage; therefore, lengthy preculturing and an additional coculture period of 2–3 days could cause cells in the leaf's wound site to miss this induction window. After 4 days of preculturing and 3 days of coculturing, some leaves began to form roots at the petiole on the selection medium.[85]

Effects of Composition of Infection Medium

A. tumefaciens grows better in an Luria-Bertani medium (LB) than in an MS medium. However, the LB medium was not suitable for kanamycin-resistant shoot-cluster induction.[85] Some of its components—such as tryptone, yeast extract, and NaCl—may inhibit shoot induction.

Effects of Coculturing

Coculturing is one of the most important steps in the genetic transformation of plants because T-DNA is incorporated into the plant's genomic DNA during this process. Leaves were cocultured in a liquid medium before induction and selection. The induction rate of kanamycin-resistant shoot clusters was 24.4% in the 001 strain, as opposed to 32.4% in the control.[85] Control leaves were cultured for 2 days in MS medium and not infected with *A. tumefaciens*. Hence, it was deduced that liquid coculturing could lead to a large number of false-positive shoot clusters. With the MS medium, the shoots' sensitivity to kanamycin fell. This phenomenon should be further studied.

Some studies have reported that adding hormones to the coculture medium can promote cell division in the explant and maintain cell activity, increasing the growth of transformed cells and raising the transformation rate. However, this study showed that hormones could result in false positives. Because shoot-cluster induction occurs at an early stage and can be triggered in a short time, the addition of hormones could result in induction during the coculture stage. These shoots continue to grow in the induction-selection medium, producing false positives.

Acetosyringone as an inducer for the *VirA* gene in the Ti plasmid has a significant effect on transformation in certain plants. This is especially important for plants that cannot produce phenolic compounds that act as effective inducers for *VirA*, such as monocotyledons and some dicotyledons. However, adding 10 mg/L acetosyringone to the coculture medium did little to increase the transformation rate of *A. annua*.

The coculture period has some influence on transformation rate. *A. tumefaciens* cannot enact transformation at the wound site immediately, only forming tumors after 16 h at the site. Therefore, too short a coculture period is not suitable for transformation. If the period is too long, *A. tumefaciens* may grow too intensively, becoming toxic to plant cells. Therefore, this period has a great impact on transformation rate, with a 2–3-day period producing higher rates. During this time, *A. tumefaciens* will have just colonized the medium. Its rapid growth allows the colonies to encircle the leaf in 4 days, influencing the development of plant cells.

Based on the above results, optimal conditions for the genetic transformation of *A. annua* include the use of leaves with petioles as explants and a coculture period of 2–3 days in a solid MS medium.

ACKNOWLEDGMENTS

The authors would like to thank Chen Dahua, Geng Sa, Zhang Long, Cai Guoqin, Qin Mingbo, Liu Benye, Liu Chunzhao, Han Junli, and others for their contributions to this work.

2.3 BIOSYNTHETIC PATHWAYS AND BIOENGINEERING OF ARTEMISININ

Wang Hong[1], *Ye Hechun*[2]
[1]*University of Chinese Academy of Sciences, Beijing, China;* [2]*Institute of Botany, Chinese Academy of Sciences, Beijing, China*

A. BIOSYNTHETIC PATHWAYS

Artemisinin is a sesquiterpene. Its biosynthetic pathway belongs to the sesquiterpene branch of the isoprenoid metabolic pathway and has not been fully ascertained until now. With farnesyl diphosphate (FPP) as a boundary, artemisinin's biosynthetic pathway can be divided into the upstream and downstream stages. The former begins with acetyl-CoA (or glyceraldehyde 3-phosphate and phosphoenolpyruvic acid) and ends with the synthesis of FPP. The latter starts with FPP and ends with artemisinin. The pathway of FPP synthesis, a common sesquiterpene precursor, is relatively well established, but the pathway from FPP to artemisinin is still not fully elucidated.

1. Upstream Pathways

In plants, the biosynthesis of the FPP precursors, isopentenyl pyrophosphate (IPP) and dimethylallyl pyrophosphate (DMAPP), follows two pathways: the MVA pathway and the 1-deoxyxylulose 5-phosphate (DXP) pathway. The MVA pathway occurs in the cytoplasm. First, three molecules of acetyl-CoA undergo an intermolecular double condensation reaction to form 3-hydroxy-3-methylglutaryl-CoA (HMG-CoA). This is catalyzed by acetyl-CoA thiolase and HMG-CoA synthase. Then, HMG-CoA forms MVA in the presence of 3-hydroxy-3-methylglutaryl-CoA reductase (HMGR). MVA is phosphorylated to form MVA-5-pyrophosphate, which is then decarboxylated and dehydrated to produce IPP. Isomerization of IPP results in DMAPP. Catalyzed by farnesyl diphosphate synthase (FPS), these two "activated" isoprene units form geranyl diphosphate (GPP) via an electrophilic reaction mechanism and head-to-tail condensation. GPP reacts with another IPP unit to form FPP (Fig. 2.1).

The DXP pathway, also known as the non-MVA pathway or 2-C-methyl-D-erythritol-4-phosphate (MEP) pathway, occurs in plastids. This pathway only gradually came to light in the 1990s. Now that all the genes that encode the enzymes involved in the DXP pathway have been cloned, the enzyme catalysis pathways are basically clear. The pathway begins with two intermediates from carbohydrate metabolism, pyruvic acid, and glyceraldehyde 3-phosphate. A condensation reaction occurs in the presence of deoxyxylulose-5-phosphate synthase to form DXP. With DXP reductoisomerase, DXP undergoes intramolecular rearrangement and reduction to form MEP. A reaction with 2-C-methyl-D-erythritol-4-phosphate cytidyl transferase and activation with CTP (cytidine triphosphate) cause MEP to become 4-(cytidine-diphospho)-2-C-methyl-D-erythritol (CDP-ME). CDP-ME is phosphorylated by 4-(cytidine-diphospho)-2-C-methyl-D-erythritolkinase to form

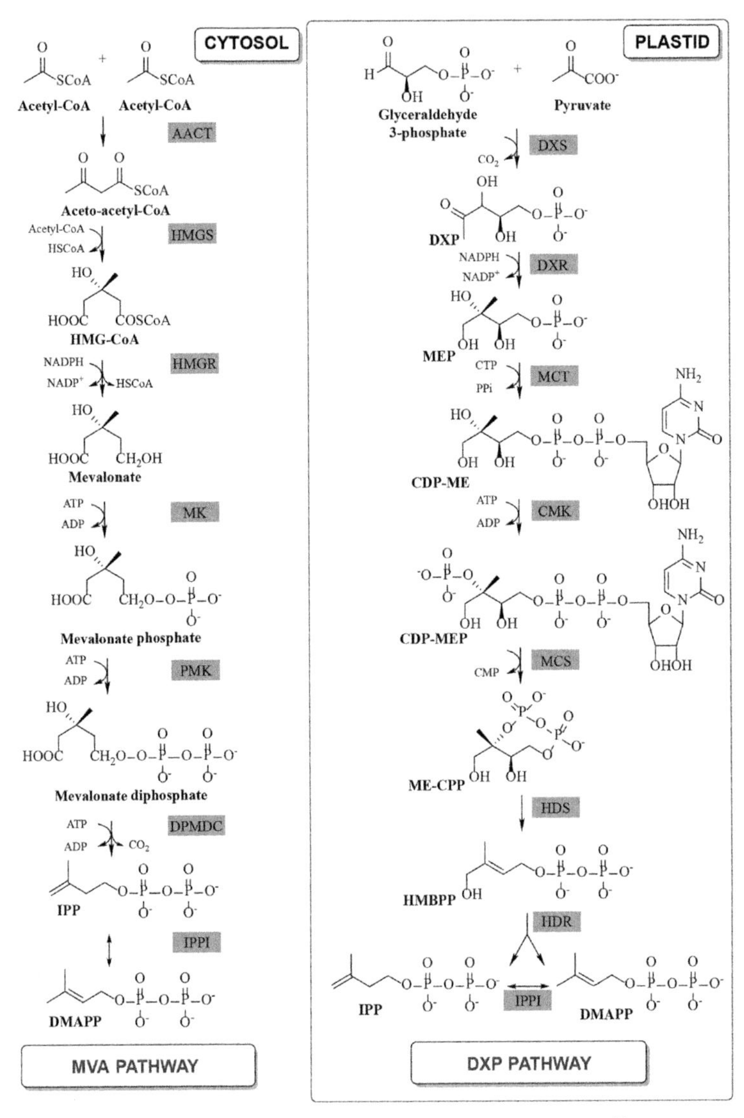

FIGURE 2.1 Biosynthetic pathways of isopentenyl pyrophosphate in plants.[86] *CMK*, 4-(cytidine-diphospho)-2-C-methyl-D-erythritolkinase; *DXR*, DXP reductoisomerase; *DXS*, deoxyxylulose-5-phosphate synthase; *HDR*, 1-hydroxy-2-methyl-2-(E)-butenyl-4-diphosphate reductase; *HDS*, 1-hydroxy-2-methyl-2-(E)-butenyl-4-diphosphate synthase; *HMGR*, 3-hydroxy-3-methylglutaryl-CoA reductase; *IPP*, isopentenyl pyrophosphate; *MCS*, 2-C-methyl-D-erythritol-2,4-cyclodiphosphate synthase; *MCT*, 2-C-methyl-D-erythritol-4-phosphate cytidyl transferase.

CDP-ME. CMP is freed from the molecule via a reaction with 2-C-methyl-D-erythritol-2,4-cyclodiphosphate synthase, producing 2-C-methyl-D-erythritol-2,4-cyclodiphosphate. Reaction with 1-hydroxy-2-methyl-2-(E)-butenyl-4-diphosphate synthase forms 1-hydroxy-2-methyl-2-(E)-butenyl-4-phosphate. Finally, this is catalyzed by 1-hydroxy-2-methyl-2-(E)-butenyl-4-diphosphate reductase to form IPP and DMAPP (Fig. 2.1).

The isoprenoids formed via the MVA pathway are mainly sterols, sesquiterpenes, triterpenes, ubiquinone, and polyterpenes. The DXP/MEP pathway produces mainly monoterpenes, diterpenes, carotenoids, chlorophyll, and plastoquinones. Some controversy still exists on the origin and presence of IPP and DMAPP in artemisinin synthesis. It is generally believed that IPP and DMAPP are synthesized in the cytoplasm. Recent studies indicate that the artemisinin precursor IPP originates both in the traditional MVA pathway and the newly discovered MEP pathway. When GPP is synthesized in the plastids and transported to the cytoplasm, it could be combined with another IPP to form FPP.[87]

2. Downstream Pathways

The biosynthetic pathway from FPP to artemisinin has not yet been fully determined. In recent years, with the cloning and identification of some of the key genes involved, the preliminary steps in this pathway have been basically established. In 1999, Bouwmeester et al.[88] were the first to isolate a sesquiterpene intermediate, amorpha-4,11-diene, with a structure very similar to that of the artemisinin precursor, artemisinic acid. They speculated that it might be involved in artemisinin biosynthesis. Then, Mercke et al.,[89] Chang et al.,[90] and Wallaart et al.[91] cloned the genes for amorpha-4,11-diene synthase (ADS). Subsequent catalytic mechanism studies showed that ADS formed amorpha-4,11-diene with FPP as a substrate.[92]

Bertea et al.[93] analyzed terpenoids from the leaves and glandular secretory trichomes of *A. annua* in 2005 and found that amorpha-4,11-diene was hydroxylated to form artemisinic alcohol, which could then be oxidized to yield artemisinic aldehyde. After that, the C11–C13 double bond was reduced to produce dihydroartemisinic aldehyde. This was then oxidized to form dihydroartemisinic acid. Hence, Bertea et al. proposed the following biosynthetic pathway: amorpha-4,11-diene to artemisinic alcohol, then to artemisinic aldehyde, followed by dihydroartemisinic aldehyde, dihydroartemisinic acid, and finally artemisinin.[93] In 2006, Teoh et al.[94] reported that cytochrome P450 monooxygenase (CYP71AV1) and its reductase might catalyze amorpha-4,11-diene to form artemisinic alcohol, then artemisinic aldehyde, and finally artemisinin. Direct evidence was provided for this three-step pathway. The C11–C13 double bond in artemisinic aldehyde could be reduced by artemisinic aldehyde reductase (DBR2) to yield dihydroartemisinic aldehyde.[95] When catalyzed by aldehyde dehydrogenase (ALDH1), this was then oxidized to form dihydroartemisinic acid.[96] The biosynthetic pathway

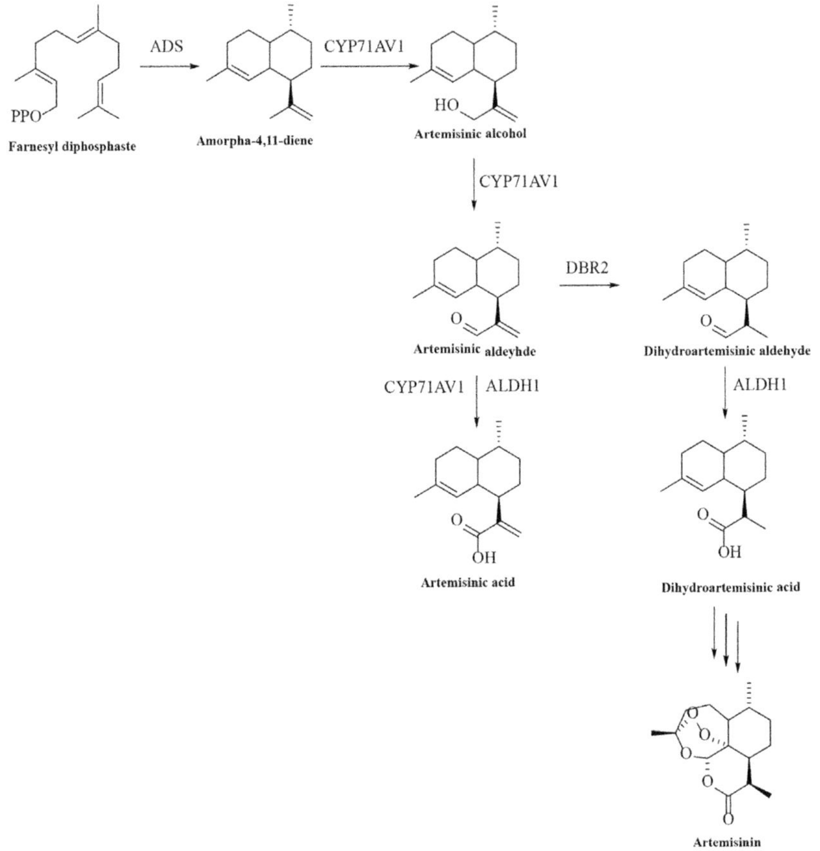

FIGURE 2.2 Proposed biosynthetic pathway of artemisinin.[96] *ADS*, amorpha-4,11-diene synthase; *ALDH1*, aldehyde dehydrogenase; *CYP71AV1*, cytochrome P450 monooxygenase; *DBR2*, double bond reductase.

from FPP to artemisinin is summed up in Fig. 2.2. The steps from FPP to artemisinic and dihydroartemisinic acid are currently well understood, but subsequent transformations are still unclear.

B. GENES INVOLVED IN BIOSYNTHESIS

1. 3-Hydroxy-3-Methylglutaryl-CoA Reductase Gene

HMGR catalyzes HMG-CoA to form MVA, an irreversible reaction. Hence, this is considered the first rate-limiting step in the MVA pathway. In plants, *HMGR* is a multigene family. Different *HMGR* genes could individually or jointly control carbon flow in the MVA pathway in the cytoplasm. Many studies have shown that *HMGR* gene expression and enzyme activity are positively correlated with the amount of isoprenoids produced.

Currently, three *HMGR* genes have been cloned from *A. annua* plants: Kang et al., GenBank accession Nos. U14624 and U14625; Chen et al., GenBank accession No. AF142473. This has laid the foundation for studying the role of *HMGR* genes in the isoprenoid metabolic pathway.

2. Farnesyl Diphosphate Synthase Gene

FPS (FDS) is a prenyl transferase that catalyzes the condensation of IPP and DMAPP to yield GPP. GPP is further condensed with IPP to generate FPP. FPP is located at the multibranch site of the isoprenoid pathway and is a sesquiterpene, sterol, and ubiquinone precursor. FPP synthesis is also strictly regulated, suggesting that FPS may play an important role in the isoprenoid pathway in plants.

In 1996, Matsushita et al.[97] cloned the *FPS* gene from *A. annua*. Subsequently, Chen et al. at the Institute of Botany cloned two *FPS* cDNAs from a high-yielding line of *A. annua*, 025, from Sichuan (GenBank accession Nos. AF136602, AF112881). The function of one of the cDNAs, *AaFPS1*, was further identified, and its enzymatic properties were analyzed.[98] The protein coded by this cDNA had FPS activity, i.e., *AaFPS1* coded for FPS in *A. annua*.

3. Amorpha-4,11-Diene Synthase Gene

In 1999, Bouwmeester et al.[88] reported that amorpha-4,11-diene could be a sesquiterpene intermediate in the artemisinin biosynthetic pathway. They isolated and purified ADS. In *A. annua*, the amount of amorpha-4,11-diene was very low, but ADS activity was relatively high. It indicated that the formation of amorpha-4,11-diene via FPP cyclization was a rate-limiting reaction. Therefore, cloning of the *ADS* gene and increasing its expression in *A. annua* could be one of the most promising ways to increase artemisinin content. From 2000 to 2001, several research groups successively reported on the cloning and functional identification of *ADS* genes.[89–91] In 2006, Li et al. at the Institute of Botany isolated cDNA and a genomic DNA sequence of the *ADS* gene from a high-yielding line of *A. annua*, 001. The corresponding genomic organization and tissue-specific expression were also analyzed.[99]

4. Squalene Synthase Gene

In the isoprenoid biosynthetic pathway, sesquiterpene synthase and squalene synthase (SQS) are key enzymes located at a branching point. They are responsible for the synthesis of sesquiterpenes and sterols, respectively. Thus, cloning of the SQS gene and inhibiting its expression via genetic manipulation could block SQS synthesis. Carbon flows directed to the synthesis of sterols could be diverted to artemisinin, instead increasing the content of the latter.

In 2003, at the Institute of Botany Li Zhengqiu et al. used rapid amplification of cDNA ends (RACE) to clone a full-length cDNA (AaSQS) and a genomic

DNA (gAaSQS) sequence for SQS from *A. annua*. The expression of recombinant SQS was induced in *Escherichia coli*. *E. coli* with full-length AaSQS cDNA produced no detectable expression-specific protein bands. By contrast, overexpression of a protein—molecular weight ~44.5 kDa—was seen in *E. coli* with c-terminal truncation corresponding to the coding sequence of 30 amino acids. In vitro enzyme catalysis established that recombinant SQS promoted the conversion of FPP into squalene.[100] A comparison between the coding sequence of AaSQS cDNA and gAaSQS showed that the latter had 14 exons and 13 introns.[101]

5. Cytochrome P450 Monooxygenase Gene

After discovering ADS in 1999, Bouwmeester et al.[88] speculated on a pathway from amorpha-4,11-diene to artemisinic acid. They suggested that the enzyme involved in this process hydroxylated amorpha-4,11-diene at position C12, forming artemisinic alcohol. Two subsequent oxidation reactions produced artemisinic acid. Using expression sequence tagging, Teoh et al.[94] isolated a cytochrome P450 gene, *CYP71AV1*, from the cDNA library of *A. annua* glandular trichomes in 2006. In vitro experiments showed that *CYP71AV1* not only catalyzed the hydroxylation of amorpha-4,11-diene at C12 to form artemisinic alcohol but also catalyzed the next two oxidation reactions that formed artemisinic acid. Similarly, Ro et al.[102] cloned the *CYP71AV1* gene and introduced it into engineered yeast, successfully using yeast to synthesize artemisinic acid. The content of artemisinic acid could reach 100 mg/L.

6. Artemisinic Aldehyde Delta11(13) Double Bond Reductase Gene

In 2008, Zhang et al.[95] found that *A. annua* bud extracts could reduce the C11–C13 double bond in artemisinic aldehyde. Aldehyde Delta11(13) reductase was isolated from the buds. Protein sequencing was used to identify the gene that coded for aldehyde Delta11(13) reductase, *DBR2*, from the cDNA library of glandular trichomes. The coding region of *DBR2* was 1331 bp long, encoding a protein with 415 amino acids. Its relative molecular weight was 45.6 kDa.

7. Aldehyde Dehydrogenase Gene

In 2009, Teoh et al.[96] cloned an *ALDH1* gene from *A. annua*. Its cDNA encoded a 499 amino acid protein with a molecular weight of 53.8 kDa. *ALDH1* was only expressed in the above-ground part of *A. annua*, most notably in the glandular trichomes but not in the roots. The expression specificity of *ALDH1* was similar to those of other enzymatic genes involved in artemisinin biosynthesis, suggesting that it might also be a part of the process. Tests on prokaryotic expression and enzyme function found that ALDH1 oxidized both artemisinic and dihydroartemisinic aldehyde to form artemisinic and dihydroartemisinic acid, respectively.

8. Dihydroartemisinic Aldehyde Reductase Gene

In 2010, Rydén et al.[103] cloned *Red1*, a dihydroartemisinic aldehyde reductase gene, from *A. annua* flowers. Red1 is an oxidation–reduction reaction enzyme, which can act on a wide variety of substrates, such as ketones, menthol, and neomenthol. In addition, it has a higher affinity to dihydroartemisinic aldehydes and monoterpenes. By reducing dihydroartemisinic aldehyde to dihydroartemisinic alcohol, Red1 may act as a repressor in the biosynthetic pathway.

9. Cinnamyl Alcohol Dehydrogenase Gene

In 2012, Li et al. of the Institute of Botany, Chinese Academy of Sciences identified the expression sequence tag (EST) of alcohol dehydrogenase from the cDNA library of *A. annua* glandular trichomes.[104] RACE technique was used to obtain the full-length cDNA of the cinnamyl alcohol dehydrogenase gene, *AaCAD* (GenBank accession No. EU417964). Its length was 1364 bp; open reading frame encoded a 361 amino acid protein (ACB54931). Its predicted theoretical isoelectric point (pI) was 5.54. In vitro function tests showed that AaCAD had strong bona fide CAD activity. At pH 7.5, AaCAD could effectively reduce cinnamic aldehyde, sinapyl aldehyde, and coniferyl aldehyde to their corresponding alcohols with NADPH as cofactor. The reverse reaction also occurred with high efficiency at pH 9.5, with $NADP^+$ as cofactor. Further analyses showed that AaCAD could reduce the artemisinin biosynthesis intermediate, artemisinic aldehyde, to artemisinic alcohol at pH 7.5. The reverse reaction was weaker under alkaline conditions (pH 9.5). AaCAD's catalytic activity was much higher in the former reaction.

10. Amorpha-4,11-Diene Synthase Gene Promoter

In 2004, Zhang Yansheng from the Institute of Botany was the first to clone the *ADS* promoter sequence (GenBank accession no. AY528931). Other laboratories subsequently reported on this sequence.[105] Pu Gaobin, also of the Institute,[106] cloned a 2850 bp *ADS* promoter regulatory region from the high-yielding 001 strain in 2007. By comparing the RACE results and the promoter sequence, the transcription start site was established at 44 bp upstream from the translation start site and 27 bp downstream from the TATA box. The promoter's cis-acting elements included plant hormone response elements such as response to abscisic acid, ethene, and auxin, as well as environmental stress response elements such as low temperature and heat shock. W-box elements related to fungal elicitors were also present.

11. WRKY Transcription Factor

ADS is a key enzyme in the biosynthetic pathway of artemisinin. The *ADS* gene promoter contains W-box cis-elements (binding site of WRKY transcription factors). To identify the regulatory genes involved in artemisinin biosynthesis,

Ma et al.[107] from the Institute of Botany constructed a cDNA library using *A. annua* glandular trichomes, which are the main artemisinin biosynthesis and accumulation sites. Two EST fragments for the WRKY transcription factor were then established. RACE was used to clone *AaWRKY1* and *AaWRKY2*. *AaWRKY1* encodes a protein with 311 amino acids, containing a conserved WRKY domain and a (C-X7-CX23-H-X1-C)-type zinc finger domain. This was characteristic of group III WRKY proteins.

AaWRKY1 and *ADS* genes are strongly expressed in the trichomes, induced by methyl jasmonate (MeJA) and the fungal elicitor chitosan. Transient expression analysis with GFP as a reporter gene showed that *AaWRKY1* was located in the nuclei. Electromobility shift assay (EMSA) and yeast one-hybrid tests found that AaWRKY1 could bind to the W-box cis-elements in *ADS* promoters. Analysis of *AaWRKY1* overexpression in transgenic tobacco and transient overexpression in agroinfiltrated tobacco leaves indicated that *AaWRKY1* could activate ADS promoters. In sum, *ADS* is the target gene of the *AaWRKY1* transcription factor. It was transiently expressed in agroinfiltrated *A. annua* leaves and could activate and intensify the expression of several key enzymes in the artemisinin biosynthetic pathway, such as HMGR, ADS, CYP71AV1, and DBR2.[107]

12. AaEFR Transcription Factors

Recently, Yu et al.[108] cloned two jasmonate-responsive AP-type transcription factor genes, *AaEFR1* and *AaEFR2*, from *A. annua*. These factors were highly expressed in the flowers. Yeast one-hybrid and EMSA tests showed that *AaEFR1* and *AaEFR2* could bind with the CBF2 motif in the *ADS* gene promoter and the RAA motif in the *CYP71AV1* gene promoter. Overexpression of these two genes could enhance the expression of two key enzymatic genes involved in biosynthesis, *CYP71AV1* and *ADS*, thereby promoting the synthesis of artemisinin and artemisinic acid.

C. REGULATION OF ARTEMISININ BIOSYNTHESIS

Since the discovery of artemisinin's medicinal properties, the regulation of artemisinin biosynthesis has been an important research question, with new studies constantly appearing. In recent years, genes involved in biosynthesis have been cloned and the pathway has been further investigated. Methods of increasing artemisinin content by the introduction of heterologous genes have also been gradually developed, with fairly good results.

1. Molecular Regulation

Overexpression of Farnesyl Diphosphate Synthase Gene

As we have seen, the artemisinin biosynthetic pathway belongs to the sesquiterpene branch of the isoprenoid pathway (Fig. 2.3). FPP is located at the multibranch site of this pathway, and it is generally believed that the synthesis of FPP is tightly

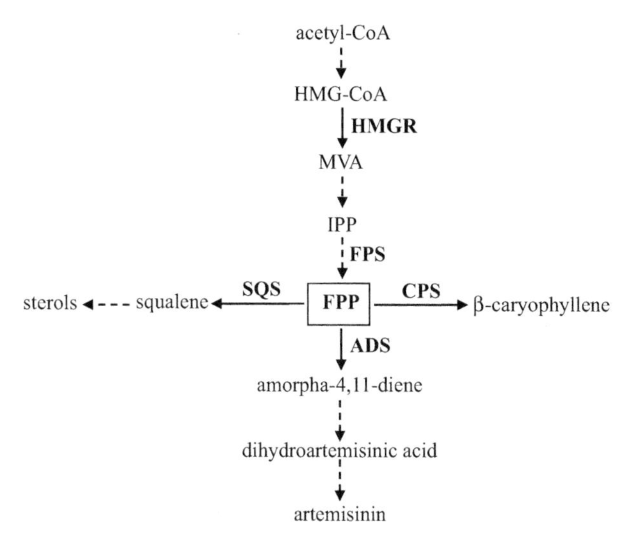

FIGURE 2.3 Mevalonic acid pathway yielding artemisinin and other isoprenoid products. *APS*, amorpha-4,11-diene synthase; *CPS*, β-caryophyllene synthase; *FPP*, farnesyl diphosphate; *FPS*, farnesyl diphosphate synthase; *HMGR*, 3-hydroxy-3-methylglutaryl-CoA reductase; *IPP*, isopentenyl pyrophosphate; *MVA*, mevalonic acid; *SQS*, squalene synthase.

regulated. Hence, FPS, which catalyzes the synthesis of FPP, may play a regulatory role. FPS is a 1,4-isopentenyl transferase, which catalyzes IPP and DMAPP condensation to form GPP. GPP then condenses with IPP to form FPP (Fig. 2.3). FPP is cyclized by ADS to yield the first cyclic compound, amorpha-4,11-diene. Hence, FPP content will directly influence artemisinin production. The Institute of Botany overexpressed homologous *FPS* from cotton and endogenous *FPS* in two different *A. annua* lines to study the regulatory effects of this gene.

In 2000, Chen et al. used *A. tumefasciens* LBA4404 to introduce cotton *FPS* cDNA into *A. annua* strain 025 (artemisinin content 3 mg/g dry weight).[109] Polymerase chain reaction (PCR) and Southern blot analyses showed that the gene was integrated into the *A. annua* genome; Northern blot analysis indicated that it was expressed at the transcriptional level in at least five transgenic lines. Artemisinin content in these transgenic plants was around 10 mg/g dry weight, an increase of one to two times compared to the control.

Han et al. overexpressed the *FPS* gene, cloned in the Institute of Botany, in the high-yielding 001 strain in 2006.[110] FPS enzyme activity was two to three times higher in transgenic plants, with artemisinin content reaching 0.9% dry weight or 1.34 times that of the control. It suggested that FPS may play a regulatory role in artemisinin biosynthesis.[110]

Overexpression of Amorpha-4,11-Diene Synthase Gene

As discussed above, ADS catalyzes FPP to form amorpha-4,11-diene. Because the carbon skeletons of amorpha-4,11-diene and artemisinic acid are very similar, the former is considered a sesquiterpene precursor. ADS catalyzes the first step in the sesquiterpene branch. Based on some of its metabolic regulatory characteristics, it was proposed that this reaction was a rate-limiting step in the biosynthetic pathway. Hence, overexpression of ADS may increase carbon flow to artemisinin synthesis, thereby increasing artemisinin content. At the same time, intermediate downstream products can be measured. Changes in their concentration can be used to deduce the functions of each step in the pathway.

In 2007, Li Zhengqiu of the Institute of Botany[111] overexpressed the *ADS* gene in *A. annua*; artemisinin content increased significantly in some of the transgenic plants, with the highest being 41% more than the control. The content of artemisinic and dihydroartemisinic acid in transgenic plants rose to 49% and 79% higher than the control, respectively. Hence, the synthesis of amorpha-4,11-diene is a rate-limiting step in the biosynthetic pathway. Furthermore, conversion of artemisinic or dihydroartemisinic acid could also be a downstream rate-limiting step.

Overexpression of 3-Hydroxy-3-Methylglutaryl-CoA Reductase Gene

HMGR is thought to be a key enzyme in the isoprenoid pathway, controlling carbon flow in its early stages. In 2000, Aquil et al.[112] successfully transferred periwinkle *HMGR* genes to *A. annua*. Transgenic plants with overexpressed *HMGR* had higher enzyme activity than in the control. Of the seven transgenic lines, four had similar artemisinin content to the control. The others saw some increase in content, by 17%, 18%, and 23%, respectively. The results showed that HMGR had a regulatory role in the biosynthetic pathway.

In 2001, Alam et al. introduced a periwinkle *HMGR* gene and an *A. annua ADS* gene into *A. annua* using the same vector, controlled by ubiquitin and CaMV 35S promoters, respectively.[113] In all eight transgenic lines, HMGR and ADS enzyme activity increased by a maximum of 71.83% and 61.76%, respectively, compared with the control. Two of the eight plants had no significant change in artemisinin content. The rest had higher artemisinin content, with the three highest lines showing an increase of 7.95, 7.60, and 7.65 times, respectively. The joint overexpression of two key enzymes affected the regulation of artemisinin biosynthesis more than the sum of their individual effects. Such overexpression could have a great influence on artemisinin biosynthesis.

Inhibition of Squalene Synthase Gene

SQS catalyzes two FPP molecules to produce squalene. In *A. annua*, FPP acts as a precursor for the biosynthesis of sterols and artemisinin (Fig. 2.3). Thus, by blocking or limiting the flow of FPP toward the synthesis of sterols, FPP may accumulate and be channeled toward the production of sesquiterpenes. This would increase artemisinin content and is one of the methods by which genetic engineering could regulate artemisinin biosynthesis.

Recently, Wang et al. from the Institute of Botany[114] successfully transferred an antisense *AaSQS* cDNA fragment into *A. annua*. In the transgenic lines ASQ3 and ASQ5, the *SQS* gene was partially suppressed on mRNA level, but line ASQ2 was not significantly different from the control. Examination of the metabolites indicated that the artemisinin contents of ASQ3 and ASQ5 were 23.2% and 21.5% higher than the control, respectively. No notable changes were seen in ASQ2 when compared with the 001 control strain. Squalene content was 19.4% lower in ASQ3 and 21.6% lower in ASQ5. The suppression of SQS genes via antisense technology not only inhibits the production of squalene but also promotes carbon flow toward the synthesis of artemisinin. It is an effective way of increasing artemisinin yield.

Zhang et al.[115] used RNA interference technology to inhibit the expression of the *SQS* gene in 2009. The transgenic plants had 3.14 times more artemisinin content than the control, a large increase. This further showed that SQS had an important regulatory function in the artemisinin biosynthetic pathway.

Inhibition of β-Caryophyllene Synthase Gene

β-Caryophyllene synthase (CPS) catalyzes FPP to produce β-caryophyllene. It is a sesquiterpene cyclase that acts in the sesquiterpene branch and competes with ADS (in the artemisinin synthesis path) for the FPP precursor (Fig. 2.3). Antisense technology can effectively block the expression of the *CPS* gene. In 2011, Chen et al. inserted a 750 bp antisense fragment of the *CPS* gene into the plant expression vector pBI121, introducing it into *A. annua* via agroinfiltration.[116] Real time PCR (RT-PCR) showed that the expression of endogenous *CPS* genes was suppressed in the transgenic plants. At the same time, the expression of the key enzymatic genes involved in artemisinin biosynthesis, such as *ADS*, *ALDH1*, *DBR2*, and *CYP71AV1*, was enhanced. It could be that inhibition of the *CPS* gene allowed more FPP to enter the artemisinin synthesis pathway, and this increase in substrate induced higher expression in the corresponding enzymatic genes.

All transgenic plants had significantly lower levels of *CPS* gene expression than the control, with a corresponding sharp drop in β-caryophyllene content. The transgenic lines T2, T3, and T9 saw a reduction in β-caryophyllene content of 54.1%, 62.7%, and 50%, respectively. RT-PCR analysis showed that the inhibition of the endogenous *CPS* gene in these three lines was stronger than in other transgenic lines, indicating that the degree of suppression correlated positively to β-caryophyllene content. The T4 line was not significantly different in terms of artemisinin content, but other lines had much higher content in general, with the increase reaching 54.9% in the T10 strain. Hence, suppressing metabolic pathways that compete with artemisinin synthesis for the same substrate is a feasible way to raise artemisinin production.[116]

Overexpression of Transcription Factors Related to Artemisinin Biosynthesis

In 2002, Yu et al.[108] cloned two transcription factors, *AaEFR1* and *AaEFR2*, from *A. annua*. They studied the effects of overexpression and RNAi inhibition of these two factors on artemisinin biosynthesis. Lines where *AaERF* was overexpressed had a roughly twofold increase in *AaERF1* and *AaERF2* transcription levels. ADS transcription increased 2–8 times, and CYP71AV1 transcription increased 1.2 to 5 times. Artemisinin content rose by 19%–67% and artemisinic acid by 11%–76% in lines with *AaERF1* overexpression. In lines where *AaERF2* was overexpressed, artemisinin and artemisinic acid content rose by 24%–51% and 17%–121%, respectively.

RNAi inhibition of *AaERF1* allowed the expression of *AaERF1* to fall to 27%–62% of the control. *ADS* and *CYP71AV1* expression decreased to 13%–35% and 7%–38% of the control, respectively. Artemisinin and artemisinic acid content were only 58%–76% and 30%–55% of the control, respectively. RNAi inhibition of *AaERF2* caused the expression of *AaERF2* to diminish to 19%–55% of the control, with *ADS* and *CYP71AV1* expression falling to 4%–9% and 3%–12% of the control, respectively. The contents of artemisinin and artemisinic acid dropped to 51%–70% and 37%–69% of the control, respectively. No obvious morphological differences were seen in the *AaERF* overexpression and RNAi inhibition lines. *AaERFs* were positive regulators in artemisinin biosynthesis.[108]

2. Elicitor Regulation

Salicylic Acid

Salicylic acid (SA) is a simple phenolic compound produced in plants. As a phytohormone, it acts as a signaling molecule regulating several important metabolic processes. In recent years, research into SA has mainly focused on the induction of pathogenesis-related gene expression and signal transduction of secondary metabolism regulators. SA's regulatory function depends on the generation of reactive oxygen species, which act as secondary messengers in vivo. This could be related to the inhibition of oxidase activity and increase in reactive oxygen. If SA can raise the level of reactive oxygen and induce the synthesis of secondary metabolites, theoretically it could also promote the formation of artemisinin. Pu et al. of the Institute of Botany studied the effects of exogenous SA on artemisinin synthesis and its mechanism of action.[117]

The results showed that exogenous SA induced artemisinin biosynthesis in at least two ways: by increasing the production of reactive oxygen, which promoted the conversion of dihydroartemisinic acid into artemisinin and by upregulating the expression of some genes involved in artemisinin biosynthesis. Therefore, in an agricultural setting, SA can be sprayed on *A. annua* plants before harvesting to increase artemisinin content quickly, effectively, and cheaply.[117]

Chitosan

Chitosan is a natural, nontoxic, biodegradable polymer derived from the cell walls of fungi or marine animals. It can induce resistance to pathogens in plants. As a biological elicitor, chitosan can produce a series of defense responses, such as stomatal closure, and promote the synthesis of phenols and terpenoids. Currently, chitosan has gained attention as a potential regulator of secondary metabolites. Reports have emerged on the impact of chitosan on artemisinin biosynthesis in *A. annua* roots and in vitro cultured cells. However, the effects of spraying chitosan on *A. annua* plants have not been reported. In 2011, Lei et al. of the Institute of Botany studied the effects of a chitosan spray, applied to *A. annua* leaves, on artemisinin synthesis. They also conducted a preliminary investigation into the action mechanism of chitosan on artemisinin production.

Chitosan induces artemisinin biosynthesis in two ways: by activating the expression of enzymatic genes involved in artemisinin biosynthesis, thereby increasing the formation of artemisinin precursors and by promoting the accumulation of reactive oxygen, which enhances the conversion of dihydroartemisinic acid into artemisinin. Chitosan did not inhibit the growth of *A. annua*. A chitosan spray applied 2 days before harvesting could be a rapid and effective way of increasing artemisinin yield.[118]

D. ARTEMISININ AND PRECURSORS SYNTHESIS IN PLANT BIOREACTORS

The precursor in artemisinin biosynthesis, FPP, is widespread in plants. Now, as key enzymatic genes have been cloned, attention has shifted to the use of transgenic plants as bioreactors to effectively produce artemisinin and its precursors. Tobacco has long been used as a heterologous host to express important nucleoproteins and chloroplast proteins. It is also a model system for transient gene expression. Because the leaves of the tobacco plant have relatively large biomass and genetic transformation techniques for tobacco are relatively well developed, the plant is highly suitable for use as a bioreactor in the production of medically important compounds.

1. Precursors

Shen Haiyan from the Institute of Botany[119] found that when the *ADS* gene in *A. annua* was inserted into tobacco, the *ADS* gene in the resulting transgenic tobacco plants could turn their own FPP into the artemisinin precursor amorpha-4,11-diene. Lei Caiyan, also from the Institute,[120] transferred an *ADS–FPS* fusion gene, as well as an *ADS–FPS* fusion gene with an *Arabidopsis thaliana* plastid signal peptide, into tobacco. The addition of the signal peptide increased the content of amorpha-4,11-diene in transgenic tobacco around 50 times. This was 300 times as high as the content in transgenic tobacco with only the *ADS* gene.

Recent studies have shown that metabolic compartmentation plays an important role in regulating terpenoid metabolism.[121] The *A. annua ADS* gene and avian *FPS* gene were simultaneously introduced to tobacco plastids, resulting in overexpression. The content of amorpha-4,11-diene in these tobacco plants increased more than 4000 times when compared with ordinary transgenic plants. This demonstrated the importance of metabolic compartmentation.[121]

In 2010, van Herpen et al.[122] inserted the *ADS*, *FPS*, and *HMGR* genes, all containing mitochondrial signal peptides, into tobacco. Amorpha-4,11-diene was detected in the transgenic tobacco. When the above three genes were inserted together with the *CYP71AV1* gene, artemisinic acid and other oxidative products of amorpha-4,11-diene could not be detected in the transgenic plants. However, glycosylated products of artemisinic acid could be detected.

Zhang et al.[123] also used tobacco as a platform for the production of artemisinin precursors in 2011. When *ADS* and *FPS* genes were cooverexpressed in transgenic tobacco, amorpha-4,11-diene and artemisinic alcohol accumulated in the transgenic plants. If the *DBR2* gene was overexpressed at the same time, mainly dihydroartemisinic alcohol was accumulated. It was speculated that the cellular microenvironment in transgenic tobacco was more conducive to the formation of artemisinic alcohol but was not conducive to the oxidation of this alcohol to form artemisinic acid.

2. Artemisinin

In 2011, Farhi et al.[124] inserted five genes involved in artemisinin biosynthesis—truncated *tHMGR*, *ADS/mtADS* (with and without mitochondrial signal peptide), *CPY71AV1*, *CPR*, and *DBR2*—into tobacco, under the control of different promoters and terminators. Artemisinin could be detected in transgenic tobacco containing *ADS* or *mtADS*. This was the first report showing that a transgenic tobacco bioreactor could produce artemisinin and not its precursors. Although artemisinin content was far lower in the transgenic tobacco than in *A. annua* plants, this was an unprecedented result. It was believed that further refinement could allow artemisinin content to increase. Farhi's study also demonstrated the potential of tobacco as a bioreactor for the combinatorial biosynthesis of therapeutic compounds.

E. COMBINATORIAL BIOSYNTHESIS OF ARTEMISININ AND PRECURSORS IN MICROORGANISMS

As research into artemisinin's biosynthetic pathway has advanced, it is now possible to use heterologous microorganisms as bioreactors to produce artemisinin and its precursors. This method involves the use of metabolic engineering to produce special secondary metabolites and seems very promising. Many terpenoid and other secondary metabolic pathways in plants have been "transferred" to a series of genetically modified microorganisms, mainly *E. coli* and yeast.

1. *Escherichia coli*

It is possible to use microorganisms as a platform for terpenoid biosynthesis because many of the catalytic steps involved in this process are highly conservative in, and shared by, higher plants and prokaryotes. *E. coli* is an ideal host for this because of its rapid growth rate and ease of genetic manipulation.

E. coli uses the DXP pathway to produce IPP. Because modification and regulation of the DXP pathway are relatively complex, Martin et al.[125] assembled eight key genes in the yeast MVA pathway into two operons and inserted them into *E. coli*. One of the two operons, *MevT*, converted acetyl-CoA into MVA. The other, *MBIS*, converted MVA into FPP. It was hoped that these two operons could bypass the DXP pathway to effectively produce FPP. The results showed that all the genes in this metabolic pathway were expressed, and a great deal of FPP was produced and accumulated in *E. coli*. However, excessive accumulation of intermediates such as FPP or IPP and DMAPP seriously inhibited growth. Hence, a site-directed mutagenic form of the *A. annua ADS* gene was coexpressed, which corresponded to the preferred codons in *E. coli*. This successfully allowed FPP to be converted to nontoxic amorpha-4,11-diene.

In addition, Pitera et al.[126] found that *E. coli* with only *MevT* operons accumulated HMGR-CoA, which was toxic to *E. coli*. However, when the *HMGR* gene was coexpressed, HMGR-CoA was not accumulated and the content of MVA increased twofold. Successful overexpression of genes involved in either the common or specific terpenoid pathways could catalyze conversion of the corresponding substrates. Based on these findings, Newman et al.[127] optimized the fermentation process in 2006, allowing the content of amorpha-4,11-diene to rise to 280–480 mg/L. Yuan et al.[128] discovered that incorporating operons into the plasmid structure could increase the content of MVA. This could also be achieved by integrating the MVA pathway into the bacterial chromosome.

Expression of normal-functioning cytochrome P450 genes is the biggest challenge to using engineered *E. coli* to produce terpenoids. Deficiencies exist in posttranslational modification and correct folding of proteins, the integration of proteins in the membrane system, and the binding of some cofactors. Hence, the ability of *E. coli* to functionally express plant *CYP* genes is a major concern. In 2007, Chang et al. engineered *E. coli* to express the *CYP71AV1* gene,[129] but both in vivo and in vitro tests could not determine if it was active. Low concentrations of artemisinic alcohol were only observed after optimization of codons and modification of the N-terminal transmembrane domain. When the *CPR* gene from *A. annua* was also inserted into *E. coli* to match the *CYP71AV1* gene, the content of artemisinic alcohol content increased to 5.6 mg/L. This was 12 times the amount before the addition of the *CPR* gene but only 1/20 of the amount produced by high-yielding engineered yeast.

2. Yeast

Compared with *E. coli*, there are clear advantages to using engineered yeast. Yeast is a eukaryote and is genetically closer to plants. The fermentation period is also relatively short (4–5 days), and the production of desired products in yeast is more than two orders of magnitude higher than in plants.[130] Yeast also has internal membranes and can better express proteins that are integrated with or connected to membranes. Desired products are secreted out of the cell, simplifying the extraction and purification process. However, yeast has a high level of terpenoid metabolism, making it necessary to regulate this metabolism and channel it toward the synthesis of FPP instead of sterols.[131] Glycosylation is also relatively common in yeast, and the glycosylation of proteins and desired products must be addressed.

In 2003, Jackson et al.[132] expressed the *A. annua epi*-cedrol synthase gene in engineered yeast to produce the corresponding sesquiterpene. Transgenic yeast could convert its FPP into *epi*-cedrol. After refinement, the yield could be six times that of *E. coli*.

Lindahl et al.[133] compared two different methods of engineering *A. annua ADS* gene expression in yeast in 2006. The gene was introduced via plasmids or inserted into the yeast genome by homologous recombination. Both methods resulted in successful gene expression, yielding 600 and 100 μg/L amorpha-4,11-diene, respectively. The difference seemed to be related to the level of heterologous gene expression, which was lower in the genome-integrated gene than in the plasmid-inserted gene.

Keasling's laboratory was a center of research into the production of artemisinin precursors using engineered yeast. The team included Ro et al.[102] who introduced multiple heterologous genes and genetically engineered many homologous genes to successfully produce engineered yeast that yielded high amounts of amorpha-4,11-diene and artemisinic acid. They found that inserting the *ADS* gene alone produced yeast, which yielded very small amounts of amorpha-4,11-diene. Coexpression of the *HMGR* gene improved the amount of FPP and increased the yield of amorpha-4,11-diene approximately fivefold. This suggested that increasing FPP provided more "raw material" for artemisinin biosynthesis.

Downregulation of the key gene *ERG9*, involved in sterol biosynthesis, and overexpression of mutagenic *ERG20*, a transcription factor gene involved in sterol metabolism, suppressed the sterol synthetic pathway. Thereafter, amorpha-4,11-diene content rose to 105 mg/L. Cutting off the FPP pathway allowed more carbon to flow toward artemisinin biosynthesis. By integrating an additional copy of the *HMGR* gene into the genome and overexpressing the *FPS* gene at the same time, amorpha-4,11-diene content rose further to 153 mg/L, 30 times higher than with the introduction of a single *ADS* gene. This showed that yeast was a prime target for metabolic engineering. To obtain the artemisinin precursor, artemisinic acid, two additional genes from *A. annua*—*CYP71AV1* and *CPR*—were introduced into the yeast with high amorpha-4,11-diene content. Artemisinic alcohol and artemisinic aldehyde were present in very low amounts in the culture medium and in yeast cell debris. After extraction was refined, artemisinic acid content could reach 100 mg/L.[102]

In 2012, Westfall et al.[134] further modified the yeast with high amorpha-4,11-diene content (S228C) produced by Keasling's laboratory. All the enzymes involved in the MVA pathway were integrated into S228C, allowing artemisinic acid content to increase twofold. After optimizing fermentation conditions, the content of amorpha-4,11-diene reached more than 40 g/L, 10 times higher than that of artemisinic acid. This provided a convenient source for the subsequent chemical synthesis of the dihydroartemisinic acid precursor. These results demonstrated the potential of using engineered yeast to efficiently produce artemisinin precursors.

After *DBR2* gene was isolated, Zhang et al. integrated the *FPS*, *ADS*, *CYP71AV1*, *CPR*, and *DBR2* genes together into yeast.[95] Dihydroartemisinic acid was detected in the engineered yeast. They also found that overexpression of the *DBR2* gene in plants resulted in a decrease in artemisinic acid content and an increase of dihydroartemisinic acid. This implied that artemisinic aldehyde was diverted away from the artemisinic acid to artemisinin pathway and toward the dihydroartemisinin pathway.

As research into the artemisinin biosynthetic pathway progresses, using yeast as a platform for direct, large-scale, industrial production of artemisinin is proving extremely promising.

F. OTHER STUDIES RELATED TO ARTEMISININ BIOSYNTHESIS

1. Effects of *ipt* Gene on Physiological and Biochemical Characteristics of *Artemisia annua*

Research has shown that the glandular secretory trichomes found on the surfaces of *A. annua* leaves and flower buds could be important sites of artemisinin synthesis and/or storage. The trichomes' specialized chloroplasts could play crucial roles in artemisinin biosynthesis. To study the effects of cytokinins on growth, development, and artemisinin biosynthesis, Geng et al. from the Institute of Botany transferred the isopentenyl transferase gene (*ipt*) in *A. tumefaciens* into *A. annua*, under the control of the CaMV 35S promoter. The relationships between chlorophyll, cytokinin, and artemisinin levels were investigated.[135]

The *A. annua* plants used were of the 025 strain kept at the Institute of Botany. One of the most important characteristics of the *ipt* transgenic plants was the loss of apical dominance. Lateral buds grew significantly faster than in the control plants. This was a common feature in *ipt* transgenic plants. Cytokinin content rose to a level that inhibited root growth, producing significantly lower root biomass than in the control. Levels of cytokinin were measured and found to be two to three times higher than in control plants. This promoted chlorophyll biosynthesis, resulting in chlorophyll content 20%–60% higher than in control plants. Artemisinin content rose by 30%–70% compared to the control. Hence, cytokinin, chlorophyll, and artemisinin levels were positively correlated.

2. Effects of *FPF1* and *CO* Genes on Flowering Time and Artemisinin Biosynthesis

In *A. annua*, artemisinin content varies according to the plant's developmental stage. Content is low in young plants, gradually increasing as the plant grows and develops and reaching a peak before or during flowering. Hence, there could be some connection between flowering and artemisinin synthesis. Wang et al. of the Institute of Botany used *Agrobacterium tumefaciens* to insert two early flowering genes—flowering promoting factor 1 (*FPF1*) and *CONSTANS* (*CO*)—from *Arabidopsis thaliana* into *A. annua*. Artemesinin content in the transgenic plants was measured and compared to the flowering period. This would provide a theoretical basis for the regulation of artemisinin biosynthesis.[136,137]

In short-day conditions, the *FPF1* transgenic plants flowered around 20 days earlier than the nontransgenic control. For the *CO* transgenic plants, this was around 15 days earlier. These results demonstrated that heterologous *FPF1* and *CO* could indeed promote early flowering in *A. annua* as well as in *A. thaliana*. Artemisinin content was measured at different growth stages (vegetative growth, before flowering, full flowering, and postflowering) of the *FPF1* transgenic plants. Before flowering, artemisinin content increased with each growth stage in both transgenic and control plants, regardless of whether they experienced short-day conditions. At the first appearance of flower buds, artemisinin content peaked in the transgenic plants. Control plants in short-day conditions also had relatively high artemisinin content during this time. Transgenic and control plants without short-day conditions remained in the vegetative stage, with no significant difference in artemisinin content when compared with their corresponding short-day plants.

After flowering, artemisinin content in *FPF1* transgenic plants decreased markedly. *FPF1* transgenic plants and control plants that had not experienced short-day conditions were still in the vegetative growth stage, with artemisinin contents maintained at a high level. Similar results were seen with the *CO* transgenic plants. Hence, the introduction of heterologous early flowering genes *FPF1* and *CO* could cause *A. annua* to flower earlier, but artemisinin content in the transgenic plants was not significantly different when compared with control plants of the same age, which remained at the vegetative stage. Flowering itself did not affect the peak in artemisinin content, which occurred just before flowering.[136,137]

3. Analysis of Terpenoid Profiles

Artemisinin's biosynthetic pathway belongs to the sesquiterpene branch of the isoprenoid metabolic pathway. The actual artemisinin biosynthetic pathway and its regulatory mechanisms are still not entirely clear. Metabolomics is a promising way of addressing this problem. This is a newly established discipline, which studies the metabolic products of a particular biological system and their interrelation. It takes metabolic activity as a comprehensive, systematic whole and is an important tool for the analysis of genetic function and metabolic pathways.

Recently, the Institute of Botany adopted metabolomics techniques such as gas chromatography–mass spectrometry and comprehensive two dimensional gas chromatography–time of flight mass spectrometry (GC×GC–TOFMS). It comparatively analyzed the terpenoid metabolic spectrums of different transgenic *A. annua* at varying growth stages, with different genomes and with MeJA treatment. These data were subject to multivariate and bioinformatics analyses to determine the compounds closely associated with artemisinin biosynthesis. This clarified the relationship between their synthetic pathways and artemisinin pathway and hence elucidated the artemisinin pathway and its regulatory mechanisms. Metabolic engineering could then be used to increase artemisinin yield and provide a scientific means for meeting market demand.

Terpenoid Profiles at Different Developmental Stages

In 2008, Ma et al.[138] used metabolomics to analyze the terpenoid compounds present in two *A. annua* strains at different developmental stages. These stages included the vegetative growth stage, namely 30, 35, and 45 days after planting (T1, T2, and T3, respectively), the preflower budding stage, at 55 (T4, photoperiod thereafter changed to 8/16 h) and 65 days (T5) after planting, and the flower budding stage, at 75 (T6) and 85 days (T7) after planting. The biosynthesis and accumulation of terpenoids underwent dynamic changes at different developmental stages. Terpenoid compounds increased in variety and content as the vegetative growth stage progressed, peaking at the end of this stage and at the preflowering budding stage. This decreased over time during the reproductive growth stage. Multidimensional partial least-square discriminant analysis confirmed that the 001 strain had 17 compounds with significant changes in concentration at different growth stages. Of these, 15 were terpenoids.

In the SP-18 strain, 18 compounds changed significantly across the developmental stages, including 16 terpenoids. Artemisinin, artemisinic acid, dihydroartemisinic acid, and arteannuin B were all marker compounds that underwent marked changes in concentration. Artemisinic and dihydroartemisinic acid peaked at the end of the vegetative growth stage, falling rapidly during the reproductive growth stage. Changes in artemisinin and arteannuin B content were less drastic. They were at a relatively high level during the vegetative growth stage, with a slight increase at the prebudding stage. A decrease was seen during the budding stage itself. Therefore, the early part of this stage was ideal for obtaining metabolic samples, which could guide harvesting techniques.

Terpenoid Profiles of Different Genotypes

Although *A. annua* is a single species, different plants can vary widely in terms of the concentration of artemisinin and its precursors. In 2001, Delabays et al.[20] reported that, if environmental influence was excluded, the main factor causing variations in artemisinin content was the plant's genotype. SP-18 and 001 were high-yielding lines maintained by the metabolic engineering laboratory of the Institute of Botany of Chinese Academy of Sciences. They have obvious differences in morphology, and SP-18 has higher artemisinin content than that of 001. Wang et al.[139] of the metabolic engineering laboratory used metabolomics to analyze the accumulation of terpenoids in 001 and SP-18.

Differences in metabolite profile were seen between SP-18 and 001. Multidimensional partial least-square determinant analysis found 22 compounds that differed significantly between the two genotypes. Twelve were sesquiterpenoids, three were monoterpenes, four were triterpenoids, and three were unknown compounds. Artemisinin and its related precursors—dihydroartemisinic acid, artemisinic acid, and arteannuin B—were all marker compounds, which indicated that they varied relatively widely between the two genotypes.

Compounds specific to SP-18 were camphor and two unidentified sesquiterpenoids; for 001, these were borneol and β-farnesene. Both genotypes also had different accumulation patterns for artemisinin and its precursors. Dihydroartemisinic acid and artemisinin were higher in SP-18, whereas artemisinic acid and arteannuin B were very low. The 001 strain, by contrast, had lower dihydroartemisinic acid and artemisinin and higher artemisinic acid and arteannuin B. Therefore, in high-yielding strains, artemisinin and dihydroartemisinic acid content were positively correlated. Taken together with Brown's and Sy's[140] in vivo labeling experiments, it indicated that the conversion of dihydroartemisinic acid into artemisinin could be a rate-limiting step.

In 2000, using different *A. annua* chemotypes, Wallaart et al.[141] studied the relationship between the accumulation patterns of artemisinin precursors and artemisinin concentration. Strains with higher dihydroartemisinic acid content also had higher artemisinin. Those with higher artemisinic acid content had lower artemisinin. This experiment used both high- and low-yielding chemotypes, whereas 2009 experiment of Wang et al.[139] used only high-yielding genotype, both with higher dihydroartemisinic acid content. However, artemisinic acid and arteannuin B content in 001 were five times those of SP-18, which had very low concentrations of both compounds. These results suggested that artemisinin and dihydroartemisinic acid content were closely related. Recently, Bertea et al.[93] and Brown and Sy[140] reported that artemisinin was most probably derived from dihydroartemisinic acid, which was in line with the findings from Wang et al.[139]

Terpenoid Profiles of Transgenic Lines

As mentioned in a previous section, the biosynthetic pathway from FPP to dihydroartemisinic acid is well known, and the genes involved have been cloned and identified.[95,96,142] The subsequent steps from dihydroartemisinic acid to artemisinin are still not fully known. The cloning of genes involved in artemisinin biosynthesis has made it possible to genetically engineer plants with higher artemisinin content. Mention has been made of an experiment involving transgenic *A. annua* with *FPS* cDNA from cotton (Section 2.3,B,2). Artemisinin content increased one to two times when compared to the control and could reach 1.0% dry weight.[109] Further experiments produced other transgenic lines of *A. annua* with overexpressed endogenous *FPS* (F4, F6 and F18)[110] and *ADS* (A4, A6, A8 and A9).[111] A line with *ADS* RNAi, Ami, was also obtained.[111] All these studies showed that artemisinin content was related to gene expression in transgenic plants, but further changes in the terpenoid metabolic profile were not examined.

In 2009, Ma et al.[143] used metabolomics to analyze the terpenoid metabolic profiles of transgenic *A. annua*, measuring changes in metabolic products and their effects on artemisinin synthesis and terpenoid metabolic networks. The analysis was conducted using comprehensive GC×GC–TOFMS. Partial least squares discriminant and orthogonal signal correction-partial least squares analysis was conducted on 200 peaks. After the introduction of heterologous genes, significant changes occurred in the terpenoid metabolic profile. Compared to the controls, overexpression of the *ADS* gene caused the largest alterations in artemisinin and its related precursors. Such changes were less drastic in lines where the *FPS* gene was overexpressed. Of the compounds, which experienced changes—and which could therefore serve as markers—70% were sesquiterpenes.

Effects of Methyl Jasmonate

MeJA is a signaling molecule belonging to the jasmonate family of plant hormones. It has multiple physiological functions, acting as a cellular messenger in metabolic processes to activate specific defense-related genes. It elicits the production of compounds related to the plant's defense response. Thus, heterologous MeJA is a good elicitor for the induction of secondary metabolites in plants and plant cell cultures. Many studies have shown that MeJA can induce terpenoid synthesis. To increase artemisinin production, therefore, Wang et al.[144] investigated the effects of heterologous MeJA on the biosynthesis of artemisinin and several other terpenoid compounds.

The results showed that 300 μmol/L of MeJA could raise artemisinin content by 38% in 8 days. Terpenoid metabolic profiling showed that MeJA promoted the biosynthesis of not only artemisinin but also many other compounds, especially sesquiterpenes and triterpenes. Orthogonal signal correction-partial least squares analysis found nine marker compounds that increased significantly after MeJA. Of these, six were sesquiterpenes and three were triterpenes. Squalene, one of the marker compounds, increased in content by 67%; another unidentified sesquiterpene increased by 60%. These compounds could have a regulatory effect on artemisinin. Spraying of exogenous MeJA could be used as an effective means of raising artemisinin production.

ACKNOWLEDGMENTS

The authors would like to thank Professors Liu Benye, Chen Dahua, Geng Sa, Liu Yan, Zhao Yujun, Zhang Yansheng, Han Junli, Li Zhenqiu, Wang Huahong, Ma Dongming, Pu Gaobin, and Lei Caiyan, as well as Song Yugang, Shen Haiyan, Chen Jianlin, and Li Xing for their contributions to this work.

REFERENCES

1. World Health Organization. *WHO monograph on good agricultural and collection practices (GACP) for Artemisia annua L.* Geneva: WHO; 2006.
2. Shishkin BK, Bobrov EG, editors. *Flora of the USSR, vol. XXVI, Compositae Giseke* [Translated by Doon Scientific Translation Co.]. Koenigstein: Koeltz Scientific Books; 1995.
3. Hu SL. Correcting the nomenclature of *Artemisia apiacea* Hance, *Artemisia annua* Linnaeus, and the *Seseli* genus. *Prim J Chin Materia Medica* 1993;**7**(3):4–6. [胡世林. 青蒿、黄花蒿与邪蒿的订正. 基层中药杂志, 1993, 7(3): 4–6.]
4. Hu SL. Examining *Artemisia apiacea* in the materia medica. *Asia Pac Tradit Med* 2006;**1**:28–30. [胡世林. 青蒿的本草考证. 亚太传统医药, 2006, 1: 28–30.]
5. State Pharmacopoeia Commission. *Pharmacopoeia of the People's Republic of China*. 2010 edition. Beijing: China Medical Science Press; 2010. p. 184. [国家药典委员会. 中华人民共和国药典. 2010年版. 一部. 北京: 中国医药科技出版社, 2010: 184.]
6. Flora Republicae Popularis Sinicae Editorial Board, Chinese Academy of Sciences, editor. *Flora Republicae Popularis Sinicae*, vol. 76, Part II. Beijing: Science Press; 1991. p. 62–4. [中国科学院中国植物志编辑委员会. 中国植物志. 第七十六卷. 第二分册. 北京: 科学出版社, 1991: 62–64.]
7. Zhang J, editor. *A detailed chronological report of Project 523 and the discovery and development of Qinghaosu (artemisinin)*. English edition. Houston: Strategic Book Publishing and Rights Co.; 2013. p. 60–1.
8. Zhang JQ, Chen Q, Liu ZH, et al. Determining artemisinin content in *A. annua*. *Shaanxi Chem Ind* 1996;**3**:34–5. [张积强, 陈强, 刘宗怀, 等. 青蒿中青蒿素含量的测定. 陕西化工, 1996, 3: 34–35.]
9. Huang HB, Chen JM, Feng JF. RP-HPLC determination of artemisinin content in *A. annua*. *J Guangxi Univ* 1994;**19**(2):194–6 (Natural Science Edition). [黄海滨, 岑家铭, 奉建芳. RP-HPLC测定青蒿中青蒿素的含量. 广西大学学报(自然科学版), 1994, 19(2): 194–196.]
10. Li DP, Liang XY, Chen XZ, et al. Thin layer chromatography-ultraviolet spectrophotometry in determining artemisinin content in *A. annua* from different regions of Guangxi. *Guihaia* 1995;**15**(3):254–5. [李典鹏, 梁小艳, 陈秀珍, 等. 采用薄层层析—紫外分光光度法测定广西不同产地黄花蒿中青蒿素含量. 广西植物, 1995, 15(3): 254–255.]
11. Zhong GY, Zhou HR, Ling Y, et al. Research into high-quality *A. annua* germplasm resources. *Chin Tradit Herb Drugs* 1998;**29**(4):264–6. [钟国跃, 周华蓉, 凌云, 等. 黄花蒿优质种质资源研究. 中草药, 1998, 29(4): 264–266.]
12. Li JH, Zhang HL, Li XL, et al. SFE-HPLC determination of artemisinin content in *A. annua* from Inner Mongolia. *J Chin Med Mater* 2000;**23**(12):756–8. [李吉和, 张惠灵, 李小玲, 等. 内蒙古地区黄花蒿中青蒿素的SFE-HPLC测定. 中药材, 2000, 23(12): 756–758.]
13. Qi XJ, Li SY, He YJ. Determining artemisinin content in *A. annua* from the Tianjin area. *Chin Tradit Herb Drugs* 2006;**37**(3):449–50. [齐向娟, 李士雨, 何玉娟. 天津地区黄花蒿中青蒿素测定. 中草药, 2006, 37(3): 449–450.]
14. Wang YL, Zhang L, Zhang HM. Determination of artemisinin content in *A. annua* from Changde, Hunan. *Pharm Care Res* 2007;**7**(5):381–2. [王玉亮, 张磊, 张汉明. 湖南省常德地区青蒿中青蒿素的含量分析. 药学服务与研究, 2007, 7(5): 381–382.]
15. Zhou YD, Zhao M, Zhang RM, et al. Biomass and artemisinin content of wild *A. annua* in Heilongjiang. *J Northeast For Univ* 2009;**37**(5):70–1. [周亚丹, 赵敏, 张荣沭, 等. 黑龙江野生黄花蒿生物量及青蒿素含量. 东北林业大学学报, 2009, 37(5): 70–71.]

16. Woerdenbag HJ, Pras N, Chan NG, et al. Artemisinin, related sesquiterpenes, and essential oil in *Artemisia annua* during a vegetation period in Vietnam. *Planta Med* 1994;**60**(3):272–5.
17. Chan KL, Teo CK, Jinadasa S, et al. Selection of high artemisinin yielding *Artemisia annua*. *Planta Med* 1995;**61**(3):285–7.
18. Ferreira JFS, Janick J. Production and detection of artemisinin from *Artemisia annua*. *Acta Hortic* 1995;**390**:41–9.
19. Ferreira JF, Simon JE, Janick J. Developmental studies of *Artemisia annua*: flowering and artemisinin production under greenhouse and field conditions. *Planta Med* 1995;**61**(2):167–70.
20. Delabays N, Simonnet X, Gaudin M. The genetics of artemisinin content in *Artemisia annua* L. and the breeding of high yielding cultivars. *Curr Med Chem* 2001;**8**(15):1795–801.
21. Zhu WP. *Research into the domestication of wild A. annua and the selection of cultivar phenotypes with high artemisinin content* [Ph.D. thesis]. Hunan Agricultural University; 2003. [朱卫平. 野生黄花蒿的引种驯化和高青蒿素含量栽培品种选育目标性状的研究. 长沙：湖南农业大学博士学位论文, 2003.]
22. Zhang L. *Analysis of factors influencing artemisinin content and the reproductive structure of A. annua* [Ph.D. thesis]. Institute of Botany, Chinese Academy of Sciences; 2003. [张龙. 青蒿素含量影响因子分析及青蒿生殖结构的研究. 北京：中国科学院植物研究所博士学位论文, 2003.]
23. Tang Q. *Comparative test on A. annua strains and a study on the structure of the reproductive system in seeds* [Master's thesis]. Hunan Agricultural University; 2007. [唐其. 黄花蒿品系比较试验及种子繁殖体系构建的研究. 长沙：湖南农业大学硕士学位论文, 2007.]
24. Xu CQ, Wei X, Li F, et al. Research into propagation techniques for *A. annua*. *Guihaia* 1998;**18**(3):271–4. [许成琼，韦霄，李锋，等. 黄花蒿繁殖技术研究. 广西植物, 1998, 18(3): 271–274.]
25. He XC, Zeng MY, Li GF, et al. Induced differentiation of calluses in *A. annua* and changes in artemisinin content. *Acta Bot Sin* 1983;**25**:87–90. [贺锡纯，曾美怡，李国凤，等. 青蒿愈伤组织的诱导分化及青蒿素含量的变化. 植物学报, 1983, 25: 87–90.]
26. Yang YW, Li BJ, Zhang YX. A preliminary study on *A. annua* tissue cultures. *J Yunnan Univ Tradit Chin Med* 2001;**24**(2):8. [杨耀文，李保军，张廷襄. 黄花蒿组织培养的初步研究. 云南中医学院学报, 2001, 24(2): 8.]
27. Wang MQ. *A. annua* tissue cultures and plant regeneration. *J Beijing Univ Tradit Chin Med* 2004;**27**(2):74–5. [王梦琼. 青蒿的组织培养及植株再生. 北京中医药大学学报, 2004, 27(2): 74–75.]
28. Wu XL, Liu F, Li LY, et al. Induction and regeneration of *A. annua* leaf calluses. *Lishizhen Med Materia Medica Res* 2007;**18**(5):1106–7. [伍晓丽，刘飞，李隆云，等. 青蒿叶片愈伤组织的诱导和植株再生. 时珍国医国药, 2007, 18(5): 1106–1107.]
29. Li XJ. *Study on A. annua tissue culture and induction of chromosome mutation* [Master's theses]. Guangxi University; 2008. [李晓娟. 黄花蒿组织培养及染色体变异诱导的研究. 南宁：广西大学硕士学位论文, 2008.]
30. Yang SP, Yang X, Huang JG, et al. Progress in research into artemisinin production. *J Trop Subtrop Bot* 2004;**12**(2):189–94. [杨水平，杨宪，黄建国，等. 青蒿素生产研究进展. 热带亚热带植物学报, 2004, 12(2): 189–194.]
31. Ma XJ, Wei SG, Feng SX, et al. A new *A. annua* cultivar, Guihao No. 3. *Acta Hortic Sin* 2010;**37**(1):169–70. [马小军，韦树根，冯世鑫，等. 黄花蒿新品种“桂蒿3号”. 园艺学报, 2010, 37(1): 169–170.]

32. Xiang JX, Tan DJ, Yang YK, et al. Selective propagation of the Hubei No. 1 cultivar of *A. annua*. *J Anhui Agric Sci* 2008;**36**(32):14050–1. [向极钎，覃大吉，杨永康，等. 鄂青蒿1号品种选育. 安徽农业科学, 2008, 36(32): 14050–14051.]

33. Xiang JX, Tan DJ, Yang YK, et al. Selective propagation of a high-yield, high-quality *A. annua* strain and the creation of a seed production base. *J Hubei Inst Natl* 2010;**28**(4):387–90 (Natural Science Edition). [向极钎，覃大吉，杨永康，等. 高产优质黄花蒿品种选育及种子生产基地建设. 湖北民族学院学报(自然科学版), 2010, 28(4): 387–390.]

34. Li LY, Wu YK, Ma P, et al. Selective propagation and promotion of a new *A. annua* cultivar, Yu Qing No. 1. *China J Chin Materia Medica* 2010;**35**(19):2516–22. [李隆云，吴叶宽，马鹏 等. 青蒿新品种“渝青1号”的选育及其示范推广. 中国中药杂志, 2010, 35(19): 2516–2522.]

35. Delabays N, Collet G, Benakis A. Selection and breeding for high artemisinin (Qinghaosu) yielding strains of *Artemisinin annua*. *Acta Hortic* 1993;**330**:203–7.

36. Delabays N, Obrist F. Test of the viability and conservation of the pollen of *Artemisia annua* L. *Acta Hortic* 1998;**457**:115–8.

37. Debrunner N, Dvorak V, Magalhaes P, et al. Selection of genotypes of *Artemisia annua* L. for the agricultural production of artemisinin. In: Pank F, editor. *Proceedings of an international symposium on breeding research in medicinal and aromatic plants, Quedlinburg, Germany, 30 June–4 July 1996*. 1996. p. 321–7.

38. De Magalhaes PM, Pereira B, Sartoratto A, et al. New hybrid lines of the antimalarial species *Artemisia annua* L. *Acta Hortic* 1999;**502**:377–81.

39. Chen HR. Selective propagation of the new *A. annua* cultivar, Jingxia No. 1. *Compend Chin Technol Res* 1992;**1**:186–7. [陈和荣. 黄花蒿新品系“京厦 I 号”的选育. 中国技术成果大全, 1992, 1: 186–187.]

40. Eelco Wallaart T, Pras N, Quax WJ. Seasonal variation of artemisinin and its biosynthetic precursors in tetraploid *Artemisia annua* plants compared with the diploid wild-type. *Planta Med* 1999;**65**(8):723–8.

41. Xun XH, Jiang TW, Peng XY, et al. Regeneration pathways of test-tube *A. annua* seedlings and induction of polyploidy. *J Hunan Agric Univ* 2003;**29**(2):115–9 (Natural Science Edition). [寻晓红，蒋泰文，彭晓英，等. 黄花蒿试管苗再生途径及多倍体诱发的研究. 湖南农业大学学报(自然科学版), 2003, 29(2): 115–119.]

42. Mu SY. *A. annua tissue culture and propagation of tetraploid plants* [Master's thesis]. Southwest University; 2007. [穆胜玉. 青蒿的组织培养及其四倍体育种. 重庆：西南大学硕士学位论文. 2007.]

43. Dong QS, Ma XJ, Feng SX, et al. Experimental research into *A. annua* seed germination. *China Seed Ind* 2008;**8**:47–8. [董青松，马小军，冯世鑫，等. 黄花蒿种子发芽试验研究. 中国种业, 2008, 8: 47–48.]

44. Li HL, Xu YM, Li LY, et al. A study on quality control and standards for *A. annua* seeds. *Seed* 2008;**27**(11):1–4. [李红莉，徐有明，李隆云，等. 青蒿种子品质检验及质量标准的研究. 种子, 2008, 27(11): 1–4.]

45. Li HL, Li LY, Xu YM. Effects of different storage methods on *A. annua* seed germination. *China J Chin Materia Medica* 2009;**34**(12):1585–7. [李红莉，李隆云，徐有明. 不同贮藏方式对青蒿种子发芽的影响. 中国中药杂志, 2009, 34(12): 1585–1587.]

46. Chen HR, Chen M, Zhong FL, et al. Several factors influencing the active component of the traditional medicine *A. annua*. *Bull Chin Materia Medica* 1986;**11**(7):9–11. [陈和荣，陈敏，钟凤林，等. 影响中药青蒿有效成分的几个因子. 中药通报, 1986, 11(7): 9–11.]

47. Cao YL, Chen XB, Wang XS, et al. Growth and artemisinin content of cultivated *A. annua*. *J Chin Med Mater* 1996;**19**(8):379–80. [曹有龙，陈晓斌，王新坤，等. 栽培的青蒿生长量及青蒿素含量. 中药材, 1996, 19(8): 379–380.]

48. Lu HS. Potential of *A. annua* development and utilization, and techniques for its cultivation. *For By Prod Spec China* 2002;**1**:6. [路红顺. 黄花蒿的开发利用价值与栽培技术. 中国林副特产, 2002, 1:6.]
49. Wang SG, Liang Y. Physiological ecology of *A. annua* and its comprehensive utilization. *Chin Wild Plant Resour* 2003;**8**:47–9. [王三根，梁颖. 中药青蒿的生态生理及其综合利用. 中国野生植物资源, 2003, 8: 47–49.]
50. Liang HL, Wei X, Tang H, et al. An investigation into the main pests of *A. annua*, and preventive and treatment measures. *J Chin Med Mater* 2007;**30**(11):1349–52. [梁惠凌，韦霄，唐辉，等. 黄花蒿主要病虫害调查及防治措施. 中药材, 2007, 30(11): 1349–1352.]
51. Jiang YS, Wei JQ, Liang HL, et al. Standard operating procedures (SOPs) for *A. annua* production. *Guihaia* 2008;**28**(3):363–6. [蒋运生，韦记青，梁惠凌，等. 黄花蒿规范化生产标准操作规程(SOP). 广西植物, 2008, 28(3): 363–366.]
52. Zhong FL, Chen HR, Chen M. Experimental research into the optimal harvest time, part, and drying method for *A. annua*. *China J Chin Materia Medica* 1997;**22**(7):405–6. [钟凤林，陈和荣，陈敏. 青蒿最佳采收时期、采收部位和干燥方式的实验研究. 中国中药杂志, 1997, 22(7): 405–406.]
53. Yang SP. *Resource investigation and germplasm screening of Artemisia annua L, and optimization of nitrogen, phosphorus, potassium and cultivation density* [Ph.D. thesis]. Southwest University; 2011. [杨水平. 重庆青蒿资源调查、种质筛选及栽培中合理的氮磷钾肥和密度. 重庆：西南大学博士学位论文, 2011.]
54. Feng SX, Ma XJ, Yan ZG, et al. Effects of processing method on artemisinin content extracted from *A. annua*. *Guihaia* 2009;**29**(6):857–9. [冯世鑫，马小军，闫志刚，等. 加工方法对黄花蒿提取青蒿素含量的影响. 广西植物, 2009, 29(6): 857–859.]
55. Klayman DL. Qinghaosu (artemisinin): an antimalarial drug from China. *Science* 1985;**228**(4703):1049–55.
56. Butcher DN. Secondary products in tissue culture. In: Reinert J, Bajaj YPS, editors. *Plant cell, tissue and organ culture*. Berlin: Springer-Verlag; 1977. p. 688–93.
57. Woerdenbag HJ, Luers JFJ, van Uden W, et al. Production of the new antimalarial drug artemisinin in shoot cultures of *Artemisia annua* L. *Plant Cell Tiss Organ Cult* 1993;**32**(2):247–57.
58. Jha S, Jha TB, Mahato SB. Tissue culture of *Artemisia annua* L.: a potential source of an antimalarial drug. *Curr Sci* 1988;**57**(6):344–6.
59. Tawfiq NK, Anderson LA, Roberts MF, et al. Antiplasmodial activity of *Artemisia annua* plant cell cultures. *Plant Cell Rep* 1989;**8**(7):425–8.
60. Martinez BC, Staba J. The production of artemisinin in *Artemisia annua* L. tissue cultures. *Adv Cell Cult* 1988;**6**:69–87.
61. Kim NC, Kim SU. Biosynthesis of artemisinin from 11,12-dihydroarteannuic acid. *J Korean Agric Chem Soc* 1992;**35**(2):106–9.
62. Brown GD. Production of anti-malarial and anti-migraine drugs in tissue culture of *Artemisia annua* and *Tanacetum parthenium*. *Acta Hortic* 1993;**330**:269–76.
63. Ferreira JFS, Janick J. Roots as an enhancing factor for the production of artemisinin in shoot cultures of *Artemisia annua*. *Plant Cell Tiss Organ Cult* 1996;**44**(3):211–7.
64. Nair MS, Acton N, Klayman DL, et al. Production of artemisinin in tissue cultures of *Artemisia annua*. *J Nat Prod* 1986;**49**(3):504–7.
65. Kudakasseril GJ, Lam L, Staba EJ. Effect of sterol inhibitors on the incorporation of 14C-isopentenyl pyrophosphate into artemisinin by a cell-free system from *Artemisia annua* tissue cultures and plants. *Planta Med* 1987;**53**(3):280–4.

66. Woerdenbag HJ, Pras N, van Uden W, et al. High peroxidase activity in cell cultures of *Artemisia annua* with minute artemisinin contents. *Nat Prod Lett* 1992;**1**(2):121–8.
67. Fulzele DP, Sipahimalani AT, Heble MR. Tissue cultures of *Artemisia annua*: organogenesis and artemisinin production. *Phytother Res* 1991;**5**(4):149–53.
68. Weathers PJ, Cheetham RD, Follansbee E, et al. Artemisinin production by transformed roots of *Artemisia annua*. *Biotechnol Lett* 1994;**16**(12):1281–6.
69. Woerdenbag HJ, Moskal TA, Pras N, et al. Cytotoxicity of artemisinin-related endoperoxides to Ehrlich ascites tumor cells. *J Nat Prod* 1993;**56**(6):849–56.
70. Paniego NB, Giulietti AM. *Artemisia annua* L.: dedifferentiated and differentiated cultures. *Plant Cell Tiss Organ Cult* 1994;**36**(2):163–8.
71. Liu BY, Ye HC, Li GF, et al. Factors affecting the transformation of *Artemisia annua* with *Agrobacterium rhizogenes*. *Chin J Appl Environ Biol* 1998;**4**:349–53. [刘本叶，叶和春，李国凤，等. 发根农杆菌转化青蒿影响因素的研究. 应用与环境生物学报, 1998a, 4: 394–353.]
72. Liu BY, Ye HC, Li GF, et al. Growth of *A. annua* hairy roots and the dynamics of artemisinin biosynthesis. *Chin J Biotechnol* 1998;**14**:401–4. [刘本叶，叶和春，李国凤，等. 青蒿发根生长及青蒿素生物合成动态的研究. 生物工程学报, 1998b, 14: 401–404.]
73. Qin MB, Li GZ, Yun Y, et al. Induction of hairy roots in *A. annua* with *Agrobacterium rhizogenes* and its *in vitro* culture. *Acta Bot Sin* 1994;**36**:165–70. [秦明波，李国珍，云月，等. 发根农杆菌诱导青蒿发根产生及离体培养. 植物学报, 1994, 36: 165–170.]
74. Jaziri M, Shimomura K, Yoshimatsu K, et al. Establishment of normal and transformed root cultures of *Artemisia annua* L. for artemisinin production. *J Plant Physiol* 1995;**145**(1–2):175–7.
75. Duke SO, Paul RN. Development and fine structure of the glandular trichomes of *Artemisia annua* L. *Int J Plant Sci* 1993;**154**(1):107–18.
76. Duke MV, Paul RN, Elsohly HN, et al. Localization of artemisinin and artemisitene in foliar tissues of glanded and glandless biotypes of *Artemisia annua* L. *Int J Plant Sci* 1994;**155**(3):365–72.
77. Ferreira JFS, Janick J. Floral morphology of *Artemisia annua* with special reference to trichomes. *Int J Plant Sci* 1995;**156**:807–15.
78. Liu BY, Ye HC, Li GF, et al. Studies on dynamics of growth and biosynthesis of artemisinin in hairy roots of *Artemisia annua* L. *Chin J Biotechnol* 1999;**14**(4):249–54.
79. Geng S, Ji SD, Yuan JY, et al. Factors affecting induction of shoot cluster of *Artemisia annua*. *Chin Tradit Herbal Drugs* 2004;**35**:566–72.
80. Liu CZ, Wang YC, Kang XZ, et al. Artemisinin production by adventitious shoots of *Artemisia annua* in a novel mist bioreactor. *Acta Bot Sin* 1999;**41**(5):524–7. [刘春朝，王玉春，康学真，欧阳藩，叶和春，李国凤. 利用新型雾化生物反应器培养青蒿不定芽生产青蒿素. 植物学报, 1999, 41(5): 524–527.]
81. Liu CZ, Wang YC, Ouyang F, et al. Cultivation of *A. annua* hairy roots and production of artemisinin with an airlift loop bioreactor. *Acta Bot Sin* 1999;**41**:181–3. [刘春朝，王玉春，欧阳藩，等. 利用气升式内环流生物反应器培养青蒿毛状根生产青蒿素. 植物学报, 1999, 41: 181–183.]
82. Vergauwe A, Cammaert R, Vandenberghe D, et al. *Agrobacterium tumefaciens*-mediated transformation of *Artemisia annua* L. and regeneration of transgenic plants. *Plant Cell Rep* 1996;**15**(12):929–33.
83. Vergauwe A, van Geldre E, Inze D, et al. Factors influencing *Agrobacterium tumefaciens*-Mediated transformation of *Artemisia annua* L. *Plant Cell Rep* 1998;**18**(1):105–10.
84. Chen DH, Ye HC, Li GF, et al. The expression of green fluorescent protein genes in transgenic *A. annua* seedlings. *Acta Bot Sin* 1999;**41**:490–3. [陈大华，叶和春，李国凤，等. 绿色荧光蛋白基因在青蒿转基因芽中的表达. 植物学报, 1999, 41: 490–493.]

85. Han JL, Wang H, Ye HC, et al. High efficiency of genetic transformation and regeneration of *Artemisia annua* L. via *Agrobacterium tumefaciens*-mediated procedure. *Plant Sci* 2005;**168**(1):73–80.
86. Bouvier F, Rahier A, Camara B. Biogenesis, Molecular Regulation and Function of Plant Isoprenoids. *Prog Lipid Res* 2005;**44**(6):357–429.
87. Schramek N, Wang H, Römisch-Margl W, et al. Artemisinin biosynthesis in growing plants of *Artemisia annua*. A $^{13}CO_2$ study. *Phytochemistry* 2010;**71**(2–3):179–87.
88. Bouwmeester HJ, Wallaart TE, Janssen MH, et al. Amorpha-4,11-diene synthase catalyses the first probable step in artemisinin biosynthesis. *Phytochemistry* 1999;**52**(5):843–54.
89. Mercke P, Bengtsson M, Bouwmeester HJ, et al. Molecular cloning, expression, and characterization of amorpha-4,11-diene synthase, a key enzyme of artemisinin biosynthesis in *Artemisia annua* L. *Arch Biochem Biophys* 2000;**381**(2):173–80.
90. Chang YJ, Song SH, Park SH, et al. Amorpha-4,11-diene synthase of *Artemisia annua*: cDNA isolation and bacterial expression of a terpene synthase involved in artemisinin biosynthesis. *Arch Biochem Biophys* 2000;**383**(2):178–84.
91. Wallaart TE, Bouwmeester HJ, Hille J, et al. Amorpha-4,11-diene synthase: cloning and functional expression of a key enzyme in the biosynthetic pathway of the novel antimalarial drug artemisinin. *Planta* 2001;**212**(3):460–5.
92. Picaud S, Mercke P, He XF, et al. Amorpha-4,11-diene synthase: mechanism and stereochemistry of the enzymatic cyclization of farnesyl diphosphate. *Arch Biochem Biophys* 2006;**448**(1–2):150–5.
93. Bertea CM, Freije JR, van der Woude H, et al. Identification of intermediates and enzymes involved in the early steps of artemisinin biosynthesis in *Artemisia annua*. *Planta Med* 2005;**71**(1):40–7.
94. Teoh KH, Polichuk DR, Reed DW, et al. *Artemisia annua* L. (*Asteraceae*) trichome-specific cDNAs reveal CYP71AV1, a cytochrome P450 with a key role in the biosynthesis of the antimalarial sesquiterpene lactone artemisinin. *FEBS Lett* 2006;**580**(5):1411–6.
95. Zhang Y, Teoh KH, Reed DW, et al. The molecular cloning of artemisinic aldehyde Delta11(13) reductase and its role in glandular trichome-dependent biosynthesis of artemisinin in *Artemisia annua*. *J Biol Chem* 2008;**283**(31):21501–8.
96. Teoh KH, Polichuk DR, Reed DW, et al. Molecular cloning of an aldehyde dehydrogenase implicated in artemisinin biosynthesis in *Artemisia annua*. *Botany* 2009;**87**(6):635–42.
97. Matsushita Y, Kang W, Charlwood BV. Cloning and analysis of a cDNA encoding farnesyl diphosphate synthase from *Artemisia annua*. *Gene* 1996;**172**(2):207–9.
98. Zhao YJ, Ye HC, Li GF, et al. Cloning and enzymology analysis of farnesyl pyrophosphate synthase gene from a superior strain of *Artemisia annua* L. *Chin Sci Bull* 2003;**48**(1):63–7.
99. Li ZQ, Liu Y, Liu BY, et al. The cloning, *E. coli* expression and molecular analysis of amorpha-4,11-diene synthase from a high-yield strain of *Artemisia annua* L. *J Integr Plant Biol* 2006;**48**(12):1486–92.
100. Li ZQ, Wang HH, Wang H, et al. *E. coli* expression, purification, and functional identification of squalene synthase from *Artemisia annua* L. *Chin J Appl Environ Biol* 2007;**13**(3):309–12. [李振秋，王花红，王红，等. 中药青蒿鲨烯合酶的大肠杆菌表达、纯化与功能鉴定. 应用与环境生物学报, 2007, 13: 309–312.]
101. Liu Y, Ye HC, Li GF. Molecular cloning, *E. coli* expression and genomic organization of squalene synthase from *Artemisia annua*. *Acta Bot Sin* 2003;**45**(5):608–13.
102. Ro DK, Paradise EM, Ouellet M, et al. Production of the antimalarial drug precursor artemisinic acid in engineered yeast. *Nature* 2006;**440**(7086):940–3.

103. Rydén AM, Ruyter-Spira C, Quax W, et al. The molecular cloning of dihydroartemisinic aldehyde reductase and its implication in artemisinin biosynthesis in *Artemisia annua*. *Planta Med* 2010;**76**(15):1778–83.
104. Li X, Ma DM, Chen JL, et al. Biochemical characterization and identification of a cinnamyl alcohol dehydrogenase from *Artemisia annua*. *Plant Sci* 2012;**193–194**:85–95.
105. Kim SH, Chang YJ, Kim SU. Tissue specificity and developmental pattern of amorpha-4,11-diene synthase (ADS) proved by ADS promoter-driven GUS expression in the heterologous plant, *Arabidopsis thaliana*. *Planta Med* 2008;**74**(2):188–93.
106. Pu GB. *Expression and regulation of amorpha-4,11-diene synthase gene in Artemisia annua L* [Ph.D. thesis]. Institute of Botany, Chinese Academy of Sciences; 2007. p. 39–53. [蒲高斌. 中药青蒿紫穗槐二烯合酶基因表达及调控的研究. 北京: 中国科学院植物研究所博士学位论文, 2007: 39–53.]
107. Ma D, Pu G, Lei C, et al. Isolation and characterization of AaWRKY1, an *Artemisia annua* transcription factor that regulates the amorpha-4,11-diene synthase gene, a key gene of artemisinin biosynthesis. *Plant Cell Physiol* 2009;**50**(12):2146–61.
108. Yu ZX, Li JX, Yang CQ, et al. The jasmonate-responsive AP2/ERF transcription factors AaERF1 and AaERF2 positively regulate artemisinin biosynthesis in *Artemisia annua* L. *Mol Plant* 2012;**5**(2):353–65.
109. Chen DH, Ye HC, Li GF. Expression of a chimeric farnesyl diphophate synthase gene in *Artemisia annua* L. transgenic plants via *Agrobacterium tumefaciens*-mediated transformation. *Plant Sci* 2000;**155**(2):179–85.
110. Han JL, Liu BY, Ye HC, et al. Effects of overexpression of *Artemisia annua* farnesyl diphosphate synthase on the artemisinin content in *Artemisia annua*. *J Integr Plant Biol* 2006;**48**(4):482–7.
111. Li ZQ. *Functional analysis and transformation of genes involved in artemisinin biosynthesis* [Ph.D. thesis]. Institute of Botany, Chinese Academy of Sciences; 2007. p. 81–91. [李振秋. 青蒿素生物合成相关基因的功能分析及遗传转化研究. 北京: 中国科学院植物研究所博士学位论文, 2007: 81–91.]
112. Aquil S, Husaini AM, Abdin MZ, et al. Overexpression of the HMG-CoA reductase gene leads to enhanced artemisinin biosynthesis in transgenic *Artemisia annua* plants. *Planta Med* 2009;**75**(13):1453–8.
113. Alam P, Abdin MZ. Over-expression of HMG-CoA reductase and amorpha-4,11-diene synthase genes in *Artemisia annua* L. and its influence on artemisinin content. *Plant Cell Rep* 2011;**30**(10):1919–28.
114. Wang H, Song YG, Shen HY, et al. Effect of antisense squalene synthase gene expression on the increase of artemisinin content in *Artemisia annua*. In: Ciftci YO, editor. *Transgenic plants – advances and limitations*. InTech; 2012. p. 397–406.
115. Zhang L, Jing F, Li F, et al. Development of transgenic *Artemisia annua* (Chinese wormwood) plants with an enhanced content of artemisinin, an effective anti-malarial drug, by hairpin-RNA-mediated gene silencing. *Biotechnol Appl Biochem* 2009;**52**(Pt 3):199–207.
116. Chen JL, Fang HM, Ji PY, et al. Artemisinin biosynthesis enhancement in transgenic *Artemisia annua* plants by down-regulation of β-caryophyllene synthase gene. *Planta Med* 2011;**77**(15):1759–65.
117. Pu GB, Ma DM, Chen JL, et al. Salicylic acid activates artemisinin biosynthesis in *Artemisia annua* L. *Plant Cell Rep* 2009;**28**(7):1127–35.
118. Lei CY, Ma DM, Pu GB, et al. Foliar application of chitosan activates artemisinin biosynthesis in *Artemisia annua* L. *Ind Crops Prod* 2011;**33**(1):176–82.

119. Shen HY. *Genetic transformation of tobacco with genes involved in artemisinin biosynthesis*. [Master's thesis]. Institute of Botany, Chinese Academy of Sciences; 2007. p. 40–52. [申海燕. 青蒿素生物合成相关基因对烟草遗传转化的研究. 北京：中国科学院植物研究所硕士学位论文, 2007: 40–52.]
120. Lei CY. *Regulation of artemisinin biosynthesis by FPS, ADS fusion genes and chitosan* [Ph.D. thesis]. Institute of Botany, Chinese Academy of Sciences; 2010. p. 23–66. [雷彩燕. FPS、ADS 融合基因及壳聚糖对青蒿素生物合成的调控. 北京：中国科学院植物研究所博士学位论文, 2010: 23–66.]
121. Wu S, Schalk M, Clark A, et al. Redirection of cytosolic or plastidic isoprenoid precursors elevates terpene production in plants. *Nat Biotechnol* 2006;**24**(11):1441–7.
122. van Herpen TWJM, Cankar K, Nogueira M, et al. *Nicotiana benthamiana* as a production platform for artemisinin precursors. *PLoS One* 2010;**5**(12):e14222.
123. Zhang Y, Nowak G, Reed DW, et al. The production of artemisinin precursors in tobacco. *Plant Biotechnol J* 2011;**9**(4):445–54.
124. Farhi M, Marhevka E, Ben-Ari J, et al. Generation of the potent anti-malarial drug artemisinin in tobacco. *Nat Biotechnol* 2011;**29**(12):1072–4.
125. Martin VJJ, Pitera DJ, Withers ST, et al. Engineering a mevalonate pathway in *Escherichia coli* for production of terpenoids. *Nat Biotechnol* 2003;**21**(7):796–802.
126. Pitera DJ, Paddon CJ, Newman JD, et al. Balancing a heterologous mevalonate pathway for improved isoprenoid production in *Escherichia coli*. *Metab Eng* 2007;**9**(2):193–207.
127. Newman JD, Marshall J, Chang M, et al. High-level production of amorpha-4,11-diene in a two-phase partitioning bioreactor of metabolically engineered *Escherichia coli*. *Biotechnol Bioeng* 2006;**95**(4):684–91.
128. Yuan LZ, Rouviere PE, LaRossa RA, et al. Chromosomal promoter replacement of the isoprenoid pathway for enhancing carotenoid production in *E. coli*. *Metab Eng* 2006;**8**(1):79–90.
129. Chang MC, Eachus RA, Trieu W, et al. Engineering *Escherichia coli* for production of functionalized terpenoids using plant P450s. *Nat Chem Biol* 2007;**3**(5):274–7.
130. Arsenault PR, Wobbe KK, Weathers PJ. Recent advances in artemisinin production through heterologous expression. *Curr Med Chem* 2008;**15**(27):2886–96.
131. Zeng Q, Qiu F, Yuan L. Production of artemisinin by genetically-modified microbes. *Biotechnol Lett* 2007;**30**(4):581–92.
132. Jackson BE, Hart-Wells EA, Matsuda SP. Metabolic engineering to produce sesquiterpenes in yeast. *Org Lett* 2003;**5**(10):1629–32.
133. Lindahl AL, Olsson ME, Mercke P, et al. Production of the artemisinin precursor amorpha-4,11-diene by engineered *Saccharomyces cerevisiae*. *Biotechnol Lett* 2006;**28**(8):571–80.
134. Westfall PJ, Pitera DJ, Lenihan JR, et al. Production of amorphadiene in yeast, and its conversion to dihydroartemisinic acid, precursor to the antimalarial agent artemisinin. *Proc Natl Acad Sci USA* 2012;**109**(3):E111–8.
135. Geng S, Ma M, Ye HC, et al. Effects of *ipt* gene expression on the physiological and chemical characteristics of *Artemisia annua* L. *Plant Sci* 2001;**160**(4):691–8.
136. Wang H, Ge L, Ye HC, et al. Studies on the effects of *fpf1* gene on *Artemisia annua* flowering time and on the linkage between flowering and artemisinin biosynthesis. *Planta Med* 2004;**70**(4):347–52.
137. Wang H, Liu Y, Chong K, et al. Earlier flowering induced by over-expression of *CO* gene does not accompany increase of artemisinin biosynthesis in *Artemisia annua*. *Plant Biol (Stuttg.)* 2007;**9**(3):442–6.

138. Ma CF, Wang HH, Lu X, et al. Metabolic fingerprinting investigation of *Artemisia annua* L. in different stages of development by gas chromatography and gas chromatography-mass spectrometry. *J Chromatogr A* 2008;**1186**(1–2):412–9.
139. Wang HH, Ma CF, Ma LQ, et al. Secondary metabolic profiling and artemisinin biosynthesis of two genotypes of *Artemisia annua* L. *Planta Med* 2009;**75**(15):1625–33.
140. Brown GD, Sy LK. *In vivo* transformations of artemisinic acid in *Artemisia annua* plants. *Tetrahedron* 2007;**63**(38):9548–66.
141. Wallaart TE, Pras N, Beekman AC, et al. Seasonal variation of artemisinin and its biosynthetic precursors in plants of *Artemisia annua* of different geographical origin: proof for the existence of chemotypes. *Planta Med* 2000;**66**(1):57–62.
142. Covello PS, Teoh KH, Polichuk DR, et al. Functional genomics and the biosynthesis of artemisinin. *Phytochemistry* 2007;**68**(14):1864–71.
143. Ma CF, Wang HH, Lu X, et al. Terpenoid metabolic profiling analysis of transgenic *Artemisia annua* L. by comprehensive two-dimensional gas chromatography time-of-flight mass spectrometry. *Metabolomics* 2009;**5**(4):497–506.
144. Wang HH, Ma CF, Li ZQ, et al. Effects of exogenous methyl jasmonate on artemisinin biosynthesis and secondary metabolites in *Artemisia annua* L. *Ind Crops Prod* 2010;**31**(2):214–8.

Chapter 3

Artemisinin Chemical Research

Chapter Outline

Artemisinin-Based and Other Antimalarials. https://doi.org/10.1016/B978-0-12-813133-6.00003-2

3.1 EXTRACTION AND ISOLATION

Luo Zeyuan[1,2]
[1]*Yunnan Institute of Materia Medica, Kunming, Yunnan Province, China;* [2]*Sichuan Institute of Materia Medica, Chengdu, Sichuan Province, China*

Once an active antimalarial component in *Artemisia annua* Linnaeus had been identified at the Institute of Chinese Materia Medica (ICMM) of the China Academy of Chinese Medical Sciences, research began there on the extraction and isolation of this component. Qinghaosu II (artemisinin) was thereby obtained, which could suppress rodent malaria. At the same time, studies were undertaken at the Shandong Institute of Chinese Medicine (SICM), currently Shandong Institute of Chinese Medical Sciences, and Yunnan Institutes of Materia Medica (YIMM) on the active component of *A. annua* L.

At the beginning of 1973, Luo Zeyuan from the YIMM investigated the antimalarial activity of *A. annua* L. f. *macrocephala* Pamp,[1] producing first a crude ether extract, which was purified via silica gel chromatography to obtain active crystals of *huanghaosu* (artemisinin). This active compound was also isolated from *A. annua* L. purchased in a Chinese medicinal material cooperation in Youyang, Chongqing. Further details on the discovery, testing, and nomenclature of this component can be found in Chapter 1. In 1974, to meet clinical needs and the demands of industrial production, Zhan Eryi of the YIMM began comparing different methods of extraction and isolation, eventually settling on #120 gasoline as the extraction solvent. This method became the foundation of large- and industrial-scale extraction in the Kunming Pharmaceutical Co. Ltd. and the Wulingshan Pharmaceutical Factory, which was built under the guidance of Wei Zhengxing from the SICM, yielding artemisinin by the ton. The ether extraction method developed at the ICMM was not suitable for mass production because of safety concerns and hence did not become part of the production technology. In November 1973, the SICM had also isolated artemisinin from local *A. annua* L. but its acetone extraction method also was not applied to large-scale production.

A. THE YUNNAN INSTITUTE OF MATERIA MEDICA'S #120 GASOLINE[a] EXTRACTION METHOD[2]

Early research into the extraction and isolation of active antimalarial compounds took place in tandem with pharmacological screening against rodent malaria. Dry crushed *A. annua* L. f. *macrocephala* Pamp leaves were placed in a Soxhlet extractor together with each of petroleum ether, ether, ethyl acetate, or methanol. Alternatively, the crushed leaves were decocted in water, yielding five extracts for screening. This work demonstrated that

a. Note: #120 Gasoline is a refinery fraction, also called ligroin, or high boiling petroleum ether, b.p. range 90–120°C, comprising mainly aliphatic, including cycloaliphatic hydrocarbons, and smaller amounts of aromatic hydrocarbons.

the active antimalarial compounds were mainly in the ether extract. This was then mixed with a suitable amount of diatomaceous earth and placed at the top of a silica gel column filled with a petroleum ether solvent (boiling point 60–90°C). Elution with petroleum ether was followed by gradient elution with petroleum ether–ethyl acetate mixture, commencing from the mix proportion of petroleum ether–ethyl acetate 95:5. The eluted fractions were collected, concentrated, and analyzed using thin layer chromatography. Fractions with similar components were combined and submitted for pharmacological screening.

The fraction eluted with 85:15 petroleum ether–ethyl acetate had the highest activity against rodent malaria. Therefore, this fraction was submitted to column chromatography with silica gel for a second time. When an 88:12 mixture of petroleum ether–ethyl acetate solvent was used for elution, the resulting fraction left a crystalline residue consisting of clusters of needle-like white crystals after the solvent had been removed by evaporation. These were washed in a small amount of petroleum ether, then mixed with 60% aqueous ethanol (prepared by diluting 95% medical ethanol with water), and heated until they dissolved. A small amount of activated carbon was added to decolorize the solution. Once it had cooled, colorless needle-like crystals formed as precipitate. Pharmacological tests confirmed that the crystals were the effective component.

In June 1973, researchers at the Yunnan Institute discovered that the *A. annua* L. growing in Youyang, Chongqing, contained far more of the antimalarial compound (0.3%) than the *A. annua* L. f. *macrocephala* Pamp from Kunming (0.03%). In 1974, they began to improve the extraction methods. After comparing various techniques, a #120 gasoline solvent extraction method was singled out as practical and suitable for industrial production. Crushed leaves of *A. annua* L. or *A. annua* L. f. *macrocephala* Pamp were saturated in #120 gasoline and were submitted to three extractions, each one taking 24 h. The extracts were each filtered, concentrated under vacuum, and the concentrates left overnight. Crude crystals of artemisinin precipitated from the concentrates. They were collected by filtration and recrystallized via heating with 50% aqueous ethanol and activated carbon, forming white needle-like crystals of artemisinin.

From instructions by Project 523 Provincial Office in Yunnan, the province's Chemical Industries Department and pharmaceutical companies, and senior work units, the Yunnan Institute undertook initial research into the production technology of artemisinin and its expansion from November 1975 to early 1976 (Table 3.1).

1. Extraction Process

Cold Maceration

A few layers of hessian fabric were laid on the nozzle at the bottom of an extraction tank. Leaves and young stems of *A. annua* were thoroughly dried,

TABLE 3.1 Experimental Results of Scaled-Up Gasoline Extraction of Artemisinin

Batch No.	Raw Material (kg)	Gasoline (kg)[a]	50% Ethanol (kg)	Activated Carbon (kg)	Artemisinin Produced (kg)	Yield (%)
1[b]	460	350	66	0.25	1.47	0.32
2[b]	588	420	84	0.25	1.88	0.32
3[c]	1168	998	100	0.4	2.55	0.22

[a] *#120 Gasoline, ethanol, etc., represent the amount used. The 50% ethanol was prepared by diluting 95% medical ethanol with water.*
[b] *Batch nos. 1 and 2 were extracted separately in November 1975. The raw material was* huanghuahao *purchased in Youyang, Chongqing, in May 1975, and identified as* A. annua *L.*
[c] *Batch no. 3 represents the combined results of three extractions carried out in September 1976. The raw material was purchased in July 1976 and identified as* A. annua *L.*

roughly crushed, wet with a suitable amount of #120 gasoline, and added to the tank. The mixture was compacted and more gasoline added from an elevated tank. The bottom nozzle of the extraction tank was opened until the solvent flowed freely out of it, whereupon it was closed. This allowed the plant matter inside to become suitably saturated with extraction solvent. The effluent was added to the elevated tank and poured into the extraction tank again. Once the level of solvent was 50 cm above that of the plant material, the mixture was left for 24 h. The effluent solvent was retrieved from the bottom opening and stored in a tank. This process was carried out three times.

Concentration

The extract from each run was concentrated at 60–70°C, under a reduced pressure of 300–450 mmHg. Once the volume had been reduced by a certain amount, the concentrate was collected while hot because the crystals precipitated on cooling. It was then left for 24 h, whereupon formation of crystals took place. The crystals were collected by filtration, washed once or twice in a small amount of #120 gasoline, filtered again, and dried. The crystals were spread out on a pallet and put in a ventilated drying room at constant temperature to remove remaining traces of the solvent.

Recrystallization

The crystals were then dissolved in the amount of 50% aqueous ethanol corresponding to 60 times the weight of the crystals. 5%–10% of their weight of activated carbon was added. The mixture was boiled for 5 min and filtered while hot. The filtrate was reheated to dissolve the precipitated crystals and then left for 24 h. Colorless needle-like crystals were formed. These were filtered and then dried and ventilated at below 60°C to remove the ethanol. The resulting crystals could be used to make tablets. For injections, they had to be recrystallized again with 50 times their weight of 50% ethanol.

2. Industrial Production Trial

In February 1978, the expanded gasoline extraction method was used in a production trial under the supervision of senior engineers She Deyi, Wu Zuomin, and others of the State Administration of Medicine (now the General Administration of Food and Drugs). The production process was run on 21 batches of plant material weighing 300 kg each. Of these, three batches were used to improve the expanded production techniques, six were used to test the production technology itself, and the rest to manufacture crystals. A total of 19.46 kg of artemisinin crystals were obtained, with a yield of 0.31%. These crystals were used to prepare clinical samples or supplied to partner units for their research.

3. Issues

The experience of the Guangzhou University of Chinese Medicine in the 1990s showed that repeated use of recycled #120 gasoline resulted in an increased proportion of essential oil from *A. annua* in the extract. This decreased the solubility of artemisinin in the recycled gasoline and reduced the yield. Hence, the recycled gasoline had to be fractionally distilled after several uses to remove the essential oil and restore the yield. The disposal of wax and other waste products also had to be dealt with after the isolation of artemisinin.

Initially, researchers had tried to extract artemisinin via reflux of leaf material in gasoline. This raised yield and shortened extraction time but greatly increased the amount of wax, pigments, oils, and other impurities. Hence, it was necessary to use silica gel column chromatography to isolate the artemisinin. Because of limitations in equipment, this method could not be scaled up. Further research is required. Considering the costs involved, the cold maceration method is best for raw material with a high artemisinin content.

B. THE INSTITUTE OF CHINESE MATERIA MEDICA'S ETHER EXTRACTION METHOD[3]

In the second half of 1971, Tu Youyou and others at the ICMM produced a series of extracts from the *A. annua* L. plant, which they had purchased in Beijing. Methods included decoction in water, dissolution in 95% ethanol, production of essential oils, cold maceration or reflux in ethanol, reflux in benzene, and cold maceration in gasoline. The resulting extracts were screened against rodent malaria, which showed that the ether extract had the highest efficacy and lowest toxicity. This extract was further divided into a neutral part and acid part. The latter was found to be inactive and toxic. However, a 1 g/kg oral dose of the neutral part, taken for 3 days, had an inhibition rate of 100% in rodent malaria and low toxicity. Clinical trials confirmed its efficacy against vivax and falciparum malaria.

The neutral part of the ether extract therefore contained the active component. It was mixed with polyamide, percolated in 47% ethanol, and concentrated under vacuum. Ether extraction was carried out on the concentrate, which was then submitted to silica gel chromatography. The eluents were petroleum ether and 10% and 15% ethyl acetate–petroleum ether. Crystals I, II, and III were isolated using this method and again tested against rodent malaria. Only crystal II (*qinghaosu* II, later called artemisinin) achieved full parasite clearance at a dose of 50–100 mg/kg. Nevertheless, this method used a considerable amount of ether and was not suitable for large-scale production. A diluted ethanol extraction method was tried, in which crushed *A. annua* L. leaves were percolated in 47% ethanol. This produced a liquid 8–10 times the volume of the plant material. It was concentrated under vacuum at 70°C to 1/3 of the original volume, extracted with ether, and then concentrated again. The acid portion of the concentrate was removed using 2% sodium hydroxide. Silica gel column chromatography was run on the remaining neutral portion, producing *qinghaosu* II crystals. However, both methods were not scaled up.

C. THE SHANDONG INSTITUTE OF CHINESE MEDICINE'S ACETONE EXTRACTION METHOD[4]

Wei Zhenxing of the Shandong Institute produced an extract of *A. annua* L. from the province's Taian region in November 1973. Dried leaves weighing 100 kg were placed in an immersion tank. Acetone was added and the mixture was cold-macerated three times. The amount of acetone used each time was four to five times the volume of the plant material. Each immersion lasted 24 h. The acetone was removed by evaporation after the first and second rounds of immersion to obtain the acetone extract. The solution used in the third immersion was reserved for next batch of leaves. The acetone extract was mixed with 30 L of 95% ethanol, dissolved at 50°C, and left overnight at 10°C. Once the substances had precipitated out, they were filtered out with cloth, washed twice in ethanol, and combined with the original solution. This was concentrated under vacuum at 60°C to retrieve the ethanol and reduce the volume to 5 L.

The concentrate was evenly mixed with 7 kg silica gel and air-dried. It was then spread into a 1 cm thick layer and dried for a further 2 h at 60°C. A chromatography column was packed with 6 kg of active silica gel and #120 gasoline. The dried mixture, weighing 12 kg, was placed on the top of the column and eluted with gasoline at a flow rate of 200 mL/min, until the eluate was almost colorless (around 10 L). Then the column was eluted with #120 gasoline–ethyl acetate (95:5 *V/V*), and deep yellow (around 15 L) and light yellow (around 90 L) eluates were collected separately. The light yellow eluate was concentrated to 3 L, whereupon crystals began to form once it was left standing. They were collected by filtration and washed twice with #120 gasoline, yielding needle-like crystals of artemisinin. Raw materials from Shandong could provide 140 g and above of crystals, a yield of 0.14%.

This acetone extraction method was not tested in expanded form or used in industrial production.

D. THE GUANGXI INSTITUTE OF BOTANY'S AND GUILIN AROMATICS FACTORY'S #120 GASOLINE EXTRACTION METHOD[5]

Initially, this extraction and isolation method was essentially the same as the ICMM's diluted ethanol process, although #120 gasoline was used instead. Dried *A. annua* L. were purchased in a suburb of Guilin and herbal medicine shops in Yulin. The leaves and stems (length 5–20 mm) were cut up, and 80 kg of the fragments were placed in around six times the amount of gasoline (490 kg, boiling point range 88–115°C, 99% first cut). The mixture was placed in a rotating scraper-type extractor, which was run for 4 h. The liquid extract was filtered out and around five times the amount of gasoline added to the remainder (420 kg). Further extraction was carried out for 2 h using the extractor. This second-round extract was filtered and added to the first-round extract. The mixture was concentrated at normal pressure to retrieve the solvent until the extract reached around 20% of the original volume. This was further concentrated under vacuum until the volume reached 4% of the original material. The residual solution was placed in a crystallization tank and left for 24 h. The crystals were collected, purified, and the purity measured.

The same amount of material and solvent were put in an extraction tank under static conditions to compare the yields obtained by the rotating process described above. It was established that the rotating extraction cut the processing time sevenfold and raised the ultimate yield by 13%–18%. Petroleum ether was also compared to #120 gasoline. The former produced higher yields, but the latter was significantly less costly and could greatly reduce outlays. Hence, it was chosen as the solvent. Rotating extraction and gasoline were also used to test the relationship between yield and the number of extraction rounds and extraction time. Three rounds were found to be the ideal. A "5–3–2" extraction time (first round for 5 h, second for 3 h, third for 2 h, total 10 h) was better than a "4–2–2" (total 8 h) extraction time.

3.2 RESEARCH INTO OTHER COMPONENTS OF *ARTEMISIA ANNUA* LINNAEUS

Luo Zeyuan[1,2]

[1]*Yunnan Institute of Materia Medica, Kunming, Yunnan Province, China;* [2]*Sichuan Institute of Materia Medica, Chengdu, Sichuan Province, China*

In the 1950s, Takemoto, Nakajima et al.[6] studied the composition of *A. annua* essential oils.[7] In 1973, Jeramic et al.[8] reported that a new sesquiterpene, arteannuin B, had been isolated from *A. annua*.[9] Ever since the Chinese discovery of artemisinin, much attention has been devoted to this plant. In more than 30 years, detailed and comprehensive phytochemical studies have taken place worldwide on its various parts, including the leaves, stems, flowers, roots, seeds, glandular trichomes, and even parasitic fungi. More than 220 compounds have been isolated and identified from this plant.[10]

What follows is a review of the research, both in China and abroad, into a few key chemical compounds found in *A. annua*: terpenoids, flavonoids, and coumarins.

A. TERPENOIDS

1. Sesquiterpenes

Beginning in the 1980s, Chinese researchers isolated artemisinin **1** and various other sesquiterpenes from indigenous *A. annua* L. (Fig. 3.1). This included artemisinin A **2**,[11] arteannuin B **3**,[11] deoxyartemisinin **4**,[11] 3-hydroxyartemisinin, or

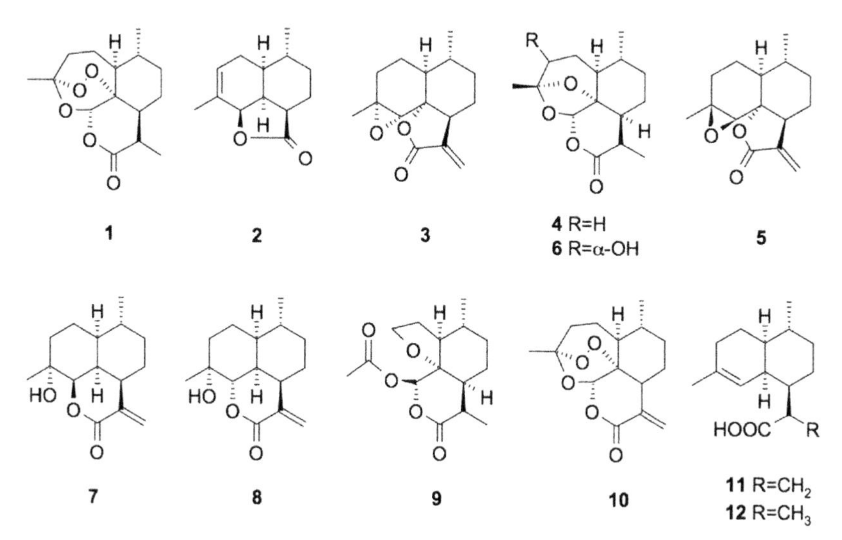

FIGURE 3.1 Structures of sesquiterpenes found in *Artemisia annua*.

artemisinin D **6**,[12] artemisinin E **7**,[12] artemisinin F **8**,[13] artemisinin G **9**,[14] artemisinic acid,[15,16] **11**,[17] artemisinic acid methyl ester **13**,[18] and epoxyarteannuic acid **20**.[19] Arteannuin B **3**,[8] arteannuin C **5**,[20] and artemisitene **10**[21] were first discovered by overseas researchers. In 1985, Acton et al. isolated artemisitene **10** from *A. annua* indigenous to the United States. Its chemical structure was the same as that of artemisinin, and the R_f values were extremely close, making it hard to isolate. The content of artemisitene in *A. annua* was lower, and its antimalarial activity was inferior to that of artemisinin.[21]

Chinese scientists isolated artemisinin G **9**, which was identified via NMR spectroscopy. The chemical structure of this compound was the same as a decomposition product of artemisinin when the latter was heated at 190°C for 10 min[22] or refluxed in xylene for 22 h.[23] Further research showed that artemisinin G, deoxyartemisinin **4**, and 3-hydroxyartemisinin **6**[24] were similar to artemisinin metabolites formed in vivo[25] or to the products of reactions with ferrous ions.[10]

After isolating artemisinic acid **11**, Chinese researchers also obtained its analogue, 11,13-dihydroartemisinic acid **12**, which was present in trace amounts.[26] From a biogenesis perspective, these compounds and amorpha-4,11-diene **22** could be considered precursors[27] in the biosynthesis of artemisinin.[28]

Some compounds similar to artemisinic acid have been identified in the glandular trichomes and leaves of *A. annua* (Fig. 3.2), including artemisinic alcohol 14,[29] dihydroartemisinic alcohol **15**,[30] artemisinic aldehyde **16**, and dihydroartemisinic aldehyde **17**.[30] The leaves also yielded cadin-4,7-(11)-dien-12-al **18**,[31] 6,7-dehydroartemisinic acid **19**,[32] epoxyarteannuic acid **20**,[19] and dihydroartemisinic acid hydroperoxide **21**.[33]

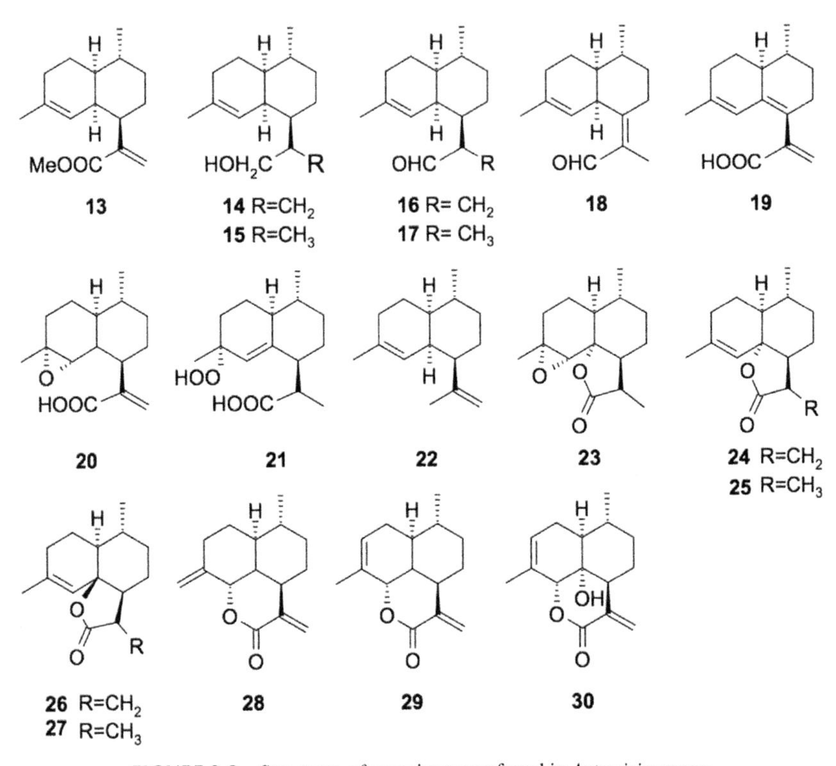

FIGURE 3.2 Structures of sesquiterpenes found in *Artemisia annua*.

The content of arteannuin B **3** in *A. annua* was relatively high. Six compounds have been found with similar structures, including arteannuin C **5**, dihydroarteannuin B **23**,[34] deoxyarteannuin B **24**,[35] dihydro-deoxyarteannuin B **25**,[35] epi-deoxyarteannuin B **26**,[36] and dihydro-epi-deoxyarteannuin B **27**.[37]

Two unusual types of candinanolide sesquiterpenes, annulide **28**[38] and isoannulide **29**,[35] have also been identified. A lactone, 6a-hydroxyisoannulide **30**, has been reported.[39]

2. Essential Oils and Other Mono-, Di-, and Triterpenes

A. annua L. has a unique fragrance, which originates in the essential oil in its leaves. These form another group of important compounds and can comprise 0.02%–0.49% of the fresh leaves.[40] Liquid chromatography (LC)/mass spectrometry (MS) was used to analyze and identify more than 70 chemical components in the essential oil. The major components (Fig. 3.3) included artemisia ketone **31**, isoartemisia ketone **32**, 1,8-cineole **33**, and camphor **34**. More details on the various mono-, di-, and triterpenes[41] can be found in other publications.[42]

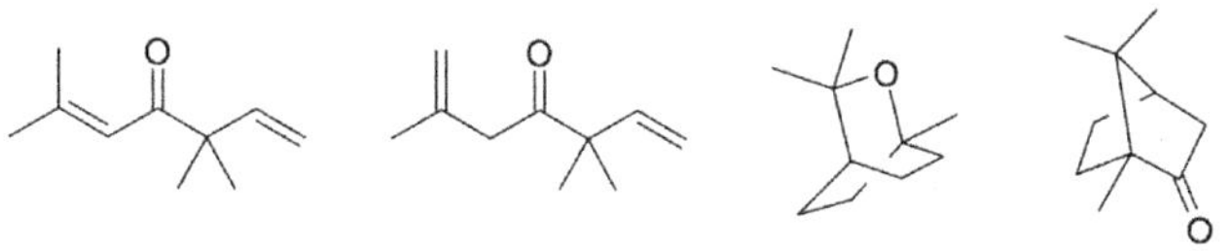

31 32 33 34

FIGURE 3.3 Main components of the monoterpenes in *Artemisia annua* essential oil.

B. FLAVONOIDS AND COUMARINS

As of 2010, 46 flavonoids have been isolated from *A. annua*, including apigenin, artemetin, astragalin, axillarin, casticin, chrysoeriol, chrysosplenetin, chrysosplenol, chrysosplenol D, cirsilineol, cirsiliol, and so on. Seven common coumarins have also been identified, such as scopoletin, scopolin, esculetin, 6,8-dimethoxy-7-hydroxy coumarin, 5,6-dimethoxy-7-hydroxy-coumarin, 5-hydroxy-6,7-dimethoxy-coumarin or tomentin, and coumarin.

3.3 PHYSICAL AND SPECTRAL CHARACTERISTICS OF ARTEMISININ[43,44]

Wu Yulin
Shanghai Institute of Organic Chemistry, Chinese Academy of Sciences, Shanghai, China

Artemisinin forms colorless, needle-shaped crystals with a melting point of 156–157°C, $[\alpha]_D^{17}=+66.3$ (c = 1.64, $CHCl_3$). It is readily soluble in chloroform, acetone, ethyl acetate, benzene, and acetic acid, soluble in ethanol, methanol, and diethyl ether, slightly soluble in petroleum ether, and seemingly insoluble in water.

Its elemental analysis—calculated C = 63.81%, H = 7.85%; found C = 63.72%, H = 7.86%—and high-resolution mass spectrometry (m/e 282.1467 M+) indicated a molecular formula of $C_{15}H_{22}O_5$. Infrared spectroscopy (KBr) shows the absorption peaks of a δ-lactone group at $1745\,cm^{-1}$ and peroxide group at 831, 881, and $1115\,cm^{-1}$.

Its ^1H NRM spectrum (100 MHz, CCl_4, ppm) shows the presence of three methyl groups at δ0.93 (3H, doublet, $J=6.0\,Hz$, C_{14}—CH_3), 1.06 (3H, doublet, $J=6.5\,Hz$, C_{13}—CH_3), 1.36 (3H, singlet, C_{15}—CH_3), and a multiplet at δ3.08–3.44 (1H, multiplet, C_{11}—H). A sharp proton singlet at 5.68 (1H, singlet, C_5—H) is assigned to an acetal hydrogen. Irradiation of the proton at δ3.08–3.44 caused the μ three-proton doublet at 1.06 to collapse into a singlet. Conversely, irradiation of the proton at δ1.06 caused the multiplet at δ3.08–3.44 to turn into a doublet ($J=4.5\,Hz$). ^{13}C NRM spectroscopy (22.63 MHz, CCl_4) showed 15 carbon resonances at δ12, 19, 23, 25, 25.1, 32.5, 33, 35.5, 37, 45, 50, 79.5, 93.5, 105, and 172.

As analytical instruments improve and research into artemisinin progresses both in China and abroad, data on its physical and spectral characteristics will be continually refined. The emergence of 2D NMR, for example, has allowed NMR properties of artemisinin to be more thoroughly understood.

3.4 DETERMINING THE STRUCTURE OF ARTEMISININ

A. EXPLORATION OF STRUCTURE VIA DEGRADATION REACTIONS AND SPECTROSCOPIC ANALYSIS

Wu Yulin
Shanghai Institute of Organic Chemistry, Chinese Academy of Sciences, Shanghai, China

Based on the data from elemental analysis, MS, infrared spectroscopy, and NMR spectroscopy, scientists proposed the following molecular formula for artemisinin: $C_{15}H_{22}O_5$. However, further determination of the structure proved more difficult. The main problem was how the 5 oxygen atoms were arranged in the skeleton of 15 carbon atoms. Back then, it was thought that artemisinin could be a peroxide. Nevertheless, it was a relatively stable compound, which contradicted the general idea that peroxides decomposed easily.

In April 1975, researchers from the ICMM had discovered another natural antimalarial, yingzhaosu A, which contained a peroxide group (see Chapter 1, Section 1.1, B, 2). This inspired the scientists in the Collaborating Group who were researching the structure of artemisinin. Qualitative and quantitative analysis confirmed that artemisinin also had such a group. The reactions of this peroxide group, MS, and existing data from its reactions (hydrogenation, $NaBH_4$ reduction, acid and alkali reactions) allowed substructural aspects of the artemisinin molecule to be elucidated (Fig. 3.4).[44,45]

Later, referring to the structure of arteannuin B isolated from *A. annua*[9] by Yugoslavian chemists and some fragments above, a possible structure of artemisinin was proposed (Fig. 3.4). This proposed structure was reported in the Qinghaosu Research Conference organized by Project 523 Head Office in November 1975. Compared to the ultimate artemisinin structure determined by

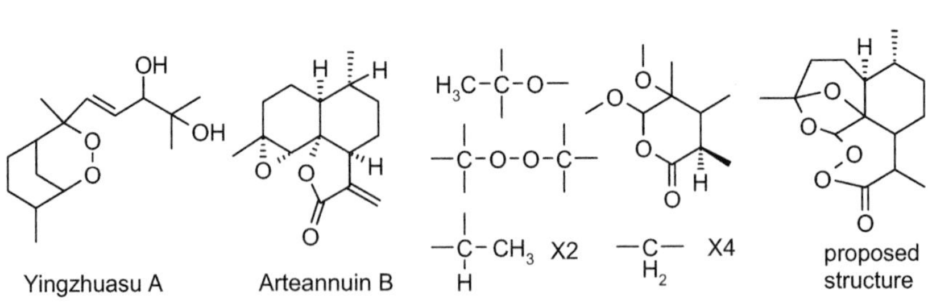

Yingzhuasu A   Arteannuin B   proposed structure

FIGURE 3.4 Structure of yingzhaosu A, arteannuin B, some fragments of the structure of artemisinin, and the proposed structure.

X-ray crystal analysis (see below) at the Institute of Biophysics at the Chinese Academy of Sciences (IBP of CAS), the carbon skeleton of the proposed structure was correct, but the peroxide group was improperly placed in the lactone ring (Fig. 3.4).

B. ARTEMISININ STRUCTURE DETERMINED BY X-RAY CRYSTALLOGRAPHY

Liang Li
Institute of Biophysics, Chinese Academy of Sciences, Beijing, China

The structure of artemisinin[43] was unambiguously determined via X-ray diffraction, and its absolute configuration[46,47] was ascertained by X-ray anomalous scattering. These articles were published in 1977 and 1979 in Chinese and 1980 in English, respectively. The scientific results were given a high commendation by Klayman[48] in a peer-reviewed article in *Science* in 1985.

1. Preliminary Determination of Crystal and Molecular Structure

In 1973, researchers in the ICMM began chemical studies on artemisinin. Thereafter, ICMM collaborated with Shanghai Institute of Organic Chemistry (SIOC) of CAS to use spectroscopy and chemical reactions to probe the molecular structure and found some of the molecular segments. This effort resulted in multiple tentative structures of artemisinin, but none proved to be correct.

In March 1975, Li Pengfei's group at the IBP of CAS began to determine the structure of artemisinin by X-ray crystallography. For that purpose, ICMM provided single crystals and the molecular formula of $C_{15}H_{22}O_5$. The space group and the crystal unit cell were first obtained from Weissenberg photographs. Artemisinin was crystallized in an orthorhombic system, space group $D_2^4-P2_12_12_1$. There was one molecule in an asymmetric unit. Based on this initial information, the cell parameter $a=24.077(6)$ Å, $b=9.443(3)$ Å, $c=6.356(1)$ Å, and the intensities of 1553 independent reflections were routinely measured with a PW-1100 four-circle diffractometer using $CuK_\alpha=1.5418$ Å.

Because the artemisinin molecule contained no heavy atoms, newly developed direct methods[49] had to be utilized for resolving the phase problem of crystallography. However, the methods had never been used in China before. It was also impossible to obtain any crystallographic software in China in 1975. Therefore, the first challenge was to understand the philosophy in the new methods and mathematical algorithms. The researchers of Li's group started from a symbolic addition procedure[50] with specification of origin/enantiomorph, defining phases carefully. They then used the triple phase relationships.[51] In practice, a breakthrough occurred with successfully coded "tangent formulation,"[52] which was programmed by Li Pengfei. Then, it was debugged by Liang Li, who had also made a mandatory program of Fourier syntheses and other crystallographic programs in Bianyi Chengxu Yuyan

(BCY), a type of ALGOL compiler. With thorough analysis and assessment, they first found the 17 nonhydrogen atoms of the molecule on a set of E-maps. Moreover, they computed a series of successive Fourier synthesis based on the 1553 independent reflections, the details of which can be found in the Refs. [46,47]. In November 1975, they confirmed the 15 carbon atoms and 5 oxygen atoms by analyzing the last set of electron density maps, which brought an accountability having an R value down to 0.22.

The R value was defined by the following equation:

$$R = \frac{\sum || F_{\text{obs}} | - | F_{\text{calc}} ||}{\sum | F_{\text{obs}} |}$$

where F is the so-called structure factor and the sum extends over all the reflections measured and their calculated counterparts, respectively.

So far, this molecular structure of artemisinin was revealed by the conventional X-ray diffraction method, exclusively. In comparison to the standard organic chemical bonds, the distances and angles between atoms were chemically reasonable in this structure. This structure allowed interpretation of what had been observed by the spectral analysis and chemical reactions. It also showed that the structure of antimalarial artemisinin quite differed from the structure of arteannuin B, which did not possess any antimalarial potency. The molecular structure of artemisinin is shown in Fig. 3.5. The 22 hydrogen atoms were derived from a theoretical calculation on the crystal structure. The structure in Fig. 3.5 depicted a special structure feature, which was rather unique among natural products in its unusual endoperoxide of a trioxane ring.

It was a milestone in artemisinin research that Liang Li, on behalf of the IBP of CAS, reported their findings at the Project 523 conference on November 30, 1975. This chemical structure of artemisinin determined by X-ray diffraction method was unanimously confirmed by researchers from the ICMM, SIOC of CAS, and IBP of CAS in a special meeting chaired by Professor Liang Xiaotian. The quotation from Nobel Prize Laureate Dorothy Crowfoot truly applies to the

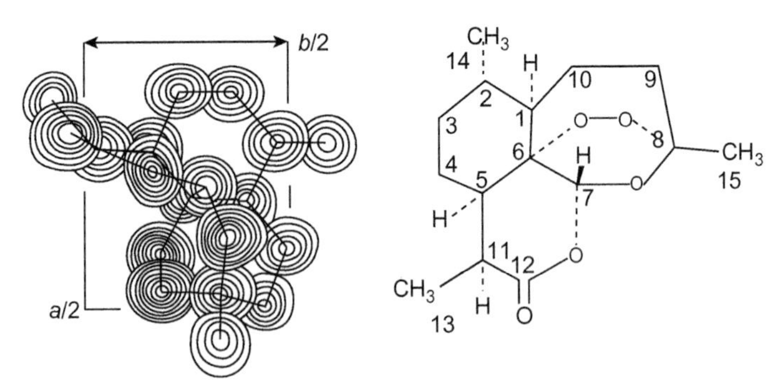

FIGURE 3.5 Overlap of three-dimensional electron-density maps and molecular structure of artemisinin.

determination of artemisinin structure: "A great advantage of X-ray analysis as a method of chemical structure analysis is its power to show some totally unexpected and surprising structure with, at the same time, complete certainty."[53]

With Professor Zhu Xiuchang's help in searching available scientific documentations at the time, an unprecedented sesquiterpene lactone was disclosed in this chemical structure of artemisinin. The original article on crystal and molecular structure of artemisinin was written by Liang Li and planned for submission to *Kexue Tongbao* (*Chinese Science Bulletin*) under a collective authorship in January 1976. Unfortunately, spectral and chemical methods were inserted into the manuscript[54] and mismatched the methods used for structural determination without acknowledging the author. The article was finally published in 1977.[43] This new type of molecule received much attention and was immediately recorded in *Chemical Abstracts*.[43] Realizing the errors in the published article in *Kexue Tongbao*[43] (*Chinese Science Bulletin*), *Chemical Abstracts* corrected the erroneous notion of a "relative configuration" to the original term "structure." *Chemical Abstracts* also differentiated the impractical methods from X-ray analysis. Since then, this crystal structure determined by IBP of CAS has been confirmed and cited widely by others.[48,56]

2. Least-Squares Refinement of the Structure

An accurate structural model was a prerequisite to identify the configuration of artemisinin. Liang Li therefore introduced a full-matrix least-squares refinement procedure to China, also programmed the procedure with BCY, and compiled it on the TQ-16 computer.[57] After several iteration cycles, the discrepancy R-value was brought down from 0.22 to 0.1070. At this point, the spatial distribution of the 22 hydrogen atoms was found from a set of different electron-density maps. Further least-squares refinements were carried out including the 22 hydrogen atoms, until all maximum shift/error ratio parameters were less than 0.05 at convergence, the details of which can be found in the Refs. [46,47]. These efforts resulted in the first accurate crystal structure in China with an R-value of 0.07, which had reached the international standard at that time.

3. Determination of the Absolute Configuration

The absolute configuration was identified with the X-ray anomalous scattering method, which was pioneered by J. M. Bijvoet on the absolute structure of (+)-sodium rubidium tartrate in 1951.[58] While determining an absolute configuration of compounds having only light atoms was still a developing field back then, Liang Li successfully tackled the problem.

First, the $|F_{hkl}|_{theor}$ and $|F_{-h-k-l}|_{theor}$ were calculated by utilizing the refined atomic coordinates, and the anomalous factors, $\Delta f'$ and $\Delta f''$, of both oxygen and carbon atoms were included. Then the experimental $|F_{hkl}|_{exp}$ and $|F_{-h-k-l}|_{exp}$

were precisely measured by specially designed scanning tactics, which enabled the accumulation of the weak anomalous scatterings emitted by light atoms. The redundant gathering of X-ray anomalous dispersion was carried on the 15 Bijvoet pairs having neighbor pairs with appropriate intensities, $B \geq 0.006$ for calibrating absorption BA, respectively.

$$B = (Q_H - 1) \Big/ \frac{1}{2} (Q_H + 1)$$

$$BA = \left(\frac{Q_{H_1}}{Q_{H_2}} - 1 \right) \Big/ \frac{1}{2} \left(\frac{Q_{H_1}}{Q_{H_2}} + 1 \right)$$

where $Q_H = |F_{hkl}| / |F_{-h-k-l}|$.

In comparing the sign of B_{theor} with that of B_{exp}, the absolute configuration should be exactly the same as the crystal structure model if the signs agreed with each other; otherwise, the absolute configuration would be the mirror image of this model. At the end of 1977, the absolute configuration of artemisinin had been determined sufficiently well by direct detection of Bijvoet differences on the theoretical and the experimental B and BA. It was the mirror image of the preliminary crystal structure, i.e., the enantiomer of the sketched crystal structure.[43] Regardless of the rectifying absorption or not, both B and BA indicated the truth. The meaningful absolute structure was described in the previously cited Refs. [46,47] and is shown in Fig. 3.6.

The absolute configuration determination was completed on time in accordance with the National Project 523's plan.[59] The knowledge of the absolute configuration shed some light on the studies of artemisinin by chemical reactions and synthesis.[44]

In April 1978 Liang Li completed the manuscript on "Crystal Structure and Absolute Configuration of Qinghaosu"[46,47] signed by a collective author as Qinghaosu Research Group, IBP of CAS to honor Li Pengfei (1938–76), who

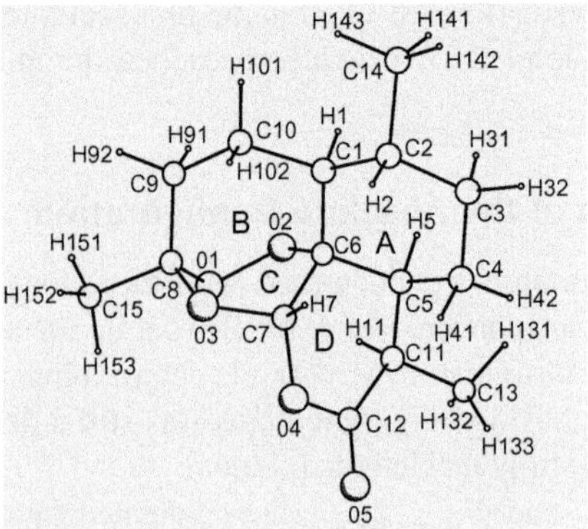

H143 H141
H101 C14 H142
H1
H91 C10 C1 C2 H31
H92 H102 H32
C9 H2 C3
B 02 H5
H151 C6 A
01 C4
C
H152 C15 C8 H7 C5 H42
03 C7 D H11 H41 H131
H153 C11
C13
04 C12 H132
H133
05

FIGURE 3.6A Absolute configuration of artemisinin shown on a ball-and-stick model.

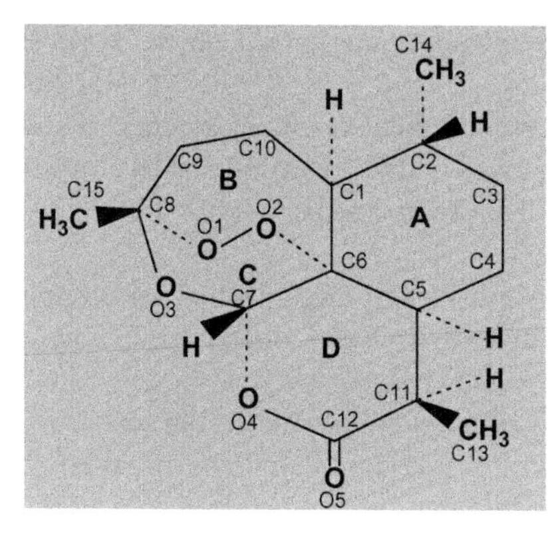

FIGURE 3.6B Stereostructure of artemisinin displayed in molecular formula.

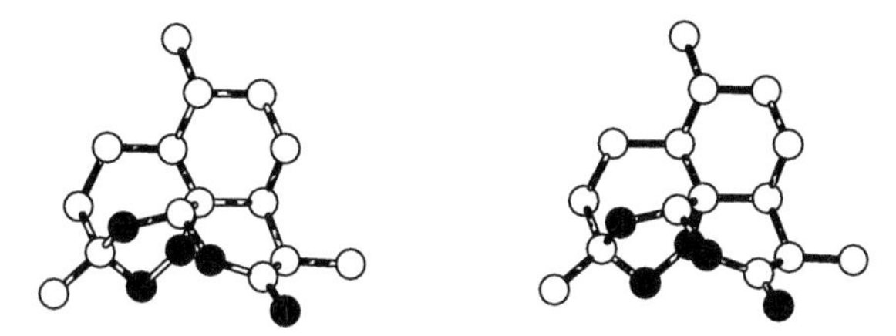

FIGURE 3.6C Stereo-molecular model of artemisinin. *(Cited from Klayman DL. Qinghaosu (artemisinin): an antimalarial drug from China.* Science *1985;***228***(4703):1049–55. https://doi.org/10.1126/science.3887571.)*

had devoted his life, and the work of others to determine the stereostructure of artemisinin in the IBP of CAS. The article was published in Chinese[46] in 1979 and in English[47] in 1980. The latter had been collected in *Chemical Abstracts* in 1980.[47]

The plant *A. annua* L. has been extensively used as an effective antimalarial for a long time. The structures of artemisinin and its derivatives, such as artesunate,[60,61] all possess an unusual endoperoxide linkage of 1,2,4-trioxane ring. This structural feature has been discussed in the "Crystal Structure and Absolute Configuration of Qinghaosu."[46,47] This unique structure has been proven to be critical to antimalarial activity.[48,62,63] "...it is generally believed that the drug acts through a carbon-centered free radical mechanism where the free radicals are generated by iron and/or heme and alkylate essential parasite proteins or heme."[64] Furthermore, a recent research article[65] discussed that "...hem, rather than free ferrous iron, is predominantly responsible for artemisinin activation. The hem derives primarily from the parasite's hem biosynthesis pathway at the early ring stage and from hemoglobin digestion at the latter stages." In addition to the bioactivity, the artemisinin crystalline thermal stability has been experimentally proved.[66] It can also be traced back to its structural discussion section[46,47]: "All five oxygen atoms crowd on the same side of the molecule. A carbon-oxygen chain of O=C—O—C—O—C—O—O—C is formed, and the carbon-oxygen bond distances are in a sequence of short, long, short, long, short…, but all lying well within range of that of a normal single bond or a partial double bond." Those are consistent with a principle: material properties rely on their structure.[67]

Since 1978, the World Health Organization has commented highly on the chemical structure illustration. X-ray diffraction technology has made great progress, and artemisinin research is growing vigorously, including exploring drug–target interactions at the atomic resolution. During the past 40 years, countless scientists have referenced the stereostructure[43,46,47] in their publications. This fact shows that the research achievement of the X-ray crystallography group in the IBP of CAS reached the world-leading edge at the end of the 1970s.

4. Nomenclature

The chemical structure of artemisinin was completely different from those of known antimalarial drugs. It is a sesquiterpene with a peroxide group, with seven chiral centers, a 1,2,4-trioxane unit, and a unique carbon–oxygen chain. Its molecular model is shown in Fig. 3.7.

The nomenclature of artemisinin has been described in Section 1. Several versions of its English name have appeared in Chinese works, such as qinghaosu, artemisinin, and arteannuin. In Chinese reports, it was numbered according to the system used for the carbon skeleton of cadinane sesquiterpenes (Fig. 3.8). The CAS used the name "artemisinin" during registration and retained its Chinese Pinyin Romanization, "qinghaosu." Its CAS number is 63968-64-9. The IUPAC systematic name was used, namely (3*R*,5a*S*,6*R*,8a*S*,9*R*,12*S*,12a*R*)-octahydro-3,6,9-trimethyl-3,12-epoxy-12*H*-pyrano[4,3-*j*]-1,2-benzodioxepin-10(3*H*)-one.

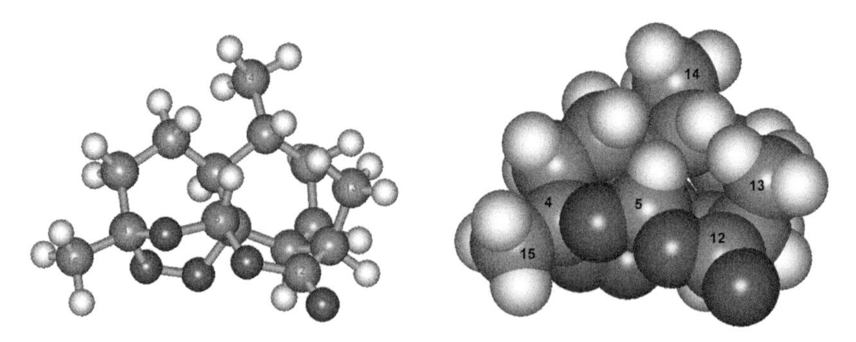

FIGURE 3.7 Molecular model of artemisinin. (Red (dark gray in print versions), oxygen; green (light gray in print versions), carbon; white, hydrogen) left: ball and cylinder model, right: overlapping spheres model.

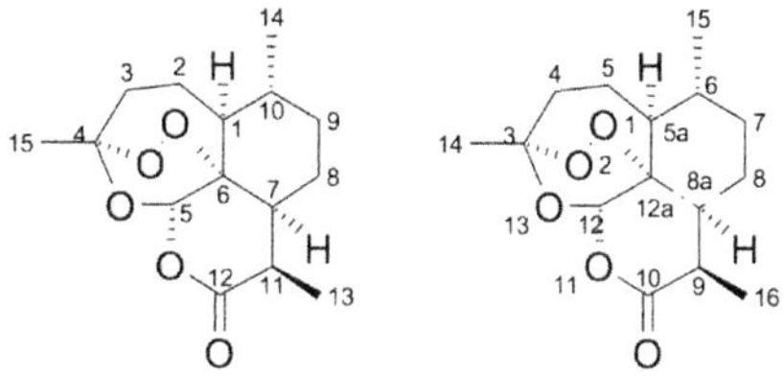

FIGURE 3.8 Two different numbering systems for artemisinin.

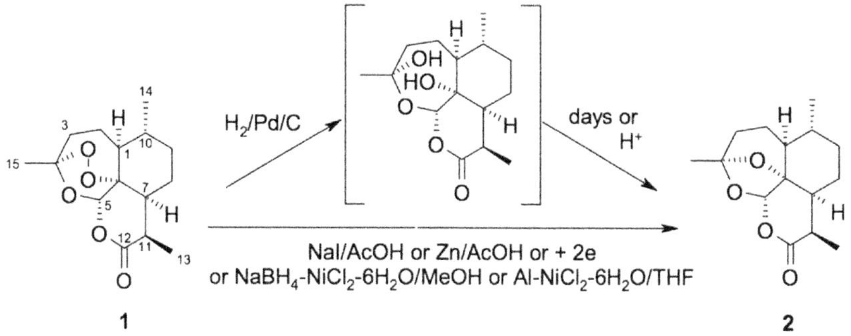

FIGURE 3.9 Reduction of the peroxide group.

3.5 CHEMICAL REACTIONS OF ARTEMISININ

Wu Yulin
Shanghai Institute of Organic Chemistry, Chinese Academy of Sciences, Shanghai, China

A. REDUCTION OF THE PEROXIDE GROUP

Artemisinin is a highly oxidized compound. Its structure contains linked peroxide, ketal, acetal, and lactone groups (Fig. 3.6 above), which are all reducible under different conditions. However, the peroxide and lactone groups are reduced first. In the presence of palladium on charcoal, the peroxide group could be reduced by hydrogenation to form a dihydroxy intermediate. When left standing or in the presence of a catalytic amount of acid, this intermediate became deoxyartemisinin **2**, a stable compound.[44] This compound could also be obtained via reduction with zinc dust–acetic acid (Fig. 3.9).[68]

Recently, several laboratories[69] have studied the electrochemical reduction[70] of artemisinins.[71] This is an irreversible, two-electron reduction.[72] When this author repeated the reaction, the only product isolated was deoxyartemisinin **2**, but the reaction itself was very slow (Fig. 3.9).[73]

Another important reaction involving the peroxide group is single-electron reduction with ferrous or copper (I) ions. In the early 1990s, the author began studying the reactions of artemisinin in an acetonitrile–water (1:1) solution with equimolar ferrous sulfate at room temperature. The two major products were tetrahydrofuran **3** and 3α-hydroxy deoxyartemisinin **4**.[74] These were identified as naturally occurring[12] in *A. annua*[14] and as metabolites formed in vivo[24] from artemisinin.[25] Thereafter, the reactions of a number of artemisinin derivatives with ferrous sulfate were investigated. These also yielded the aforementioned two compounds.[74]

Later, the electron spin resonance of the primary and secondary carbon-centered free radicals, formed in the reaction of artemisinin and ferrous sulfate,[75] was detected via use of radical trapping agents in conjunction with electron paramagnetic resonance.[76] Based on this, a reaction mechanism was proposed: Ferrous ions induce homolysis of the peroxide chain, producing two types of oxygen-centered free radical that rapidly rearranges to form carbon-centered free radicals (Fig. 3.10). Compounds **3** and **4** were produced by the primary and secondary carbon-centered free radicals, respectively. It was presumed that these reactions serve as the basis of mechanism of action of artemisinin. Recently these proposals have been criticized on the basis that such

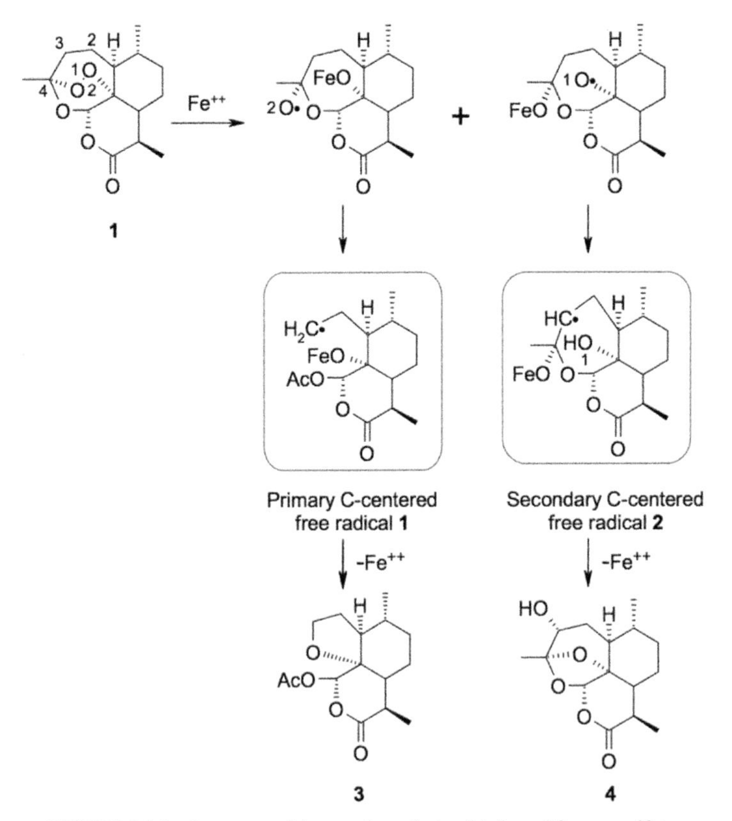

FIGURE 3.10 Summary of the reaction of artemisinin and ferrous sulfate.

radicals can only be intercepted in very small amounts and are too short-lived and incapable of alkylating protein in the aerobic environment of the malaria parasite.[77]

B. REDUCTION OF THE LACTONE GROUP

The lactone group of artemisinin can be reduced by sodium, potassium, and zinc borohydride to form the lactol dihydroartemisinin **5**. This is significant because lactones cannot ordinarily be reduced by sodium borohydride under these conditions (0–5°C, methanol solvent). Notably, the reagent reduces the lactone group but does not affect the peroxide group. If deoxyartemisinin **2** is used as the substrate, the lactone group is not reduced by sodium borohydride. Stronger reducing agents, such as diisobutylaluminum hydride, were needed to produce the lactol deoxy-dihydroartemisinin **6** (Fig. 3.11). It can also be produced if the hydrogenation of compound **5** is catalyzed with palladium on carbon; therefore, it can be said that the peroxide group promotes the reduction of the lactone group, but the exact process is still not completely understood.

Derivatives are not easily obtained directly from artemisinin itself. However, as the reduction reaction that provides dihydroartemisinin proceeds in high yields (>90%), this in turn enables derivatives to be prepared from dihydroartemisinin. Importantly, dihydroartemisinin and its derivatives have higher antimalarial activity than artemisinin itself.

Artemisinin can be further reduced with sodium borohydride in the presence of boron trifluoride diethyl etherate to form deoxoartemisinin **7**.[78] Alternatively, this compound can be obtained by reducing dihydroartemisinin **5** with $BH_3{\cdot}Net_3$ and Me_3SiCl in dimethyl ether.[79]

Lithium aluminum hydride, a more powerful reducing agent, reduces not only the lactone and peroxide groups but also the acetal and ketal groups.[80] This yields completely reduced (e.g., 8) and partially reduced products (Fig. 3.12).[81]

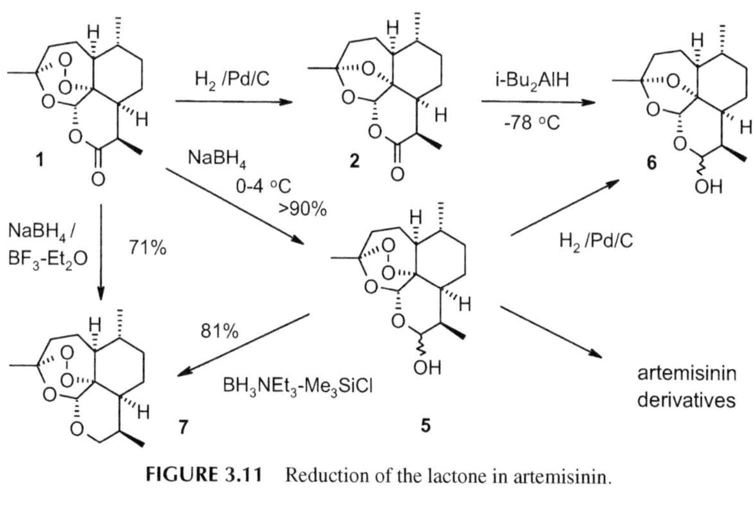

FIGURE 3.11 Reduction of the lactone in artemisinin.

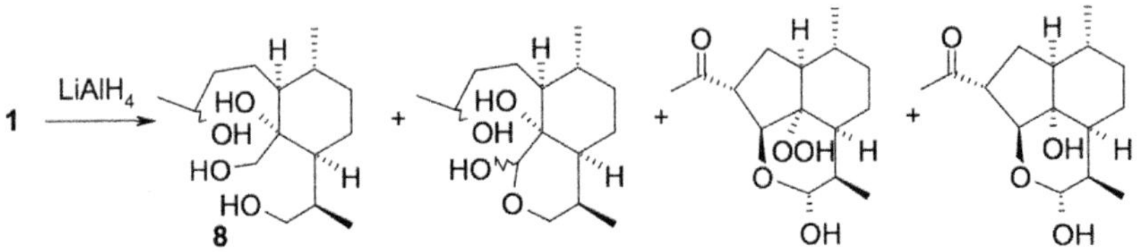

FIGURE 3.12 Products formed from reduction of artemisinin with the powerful reducing agent lithium aluminum hydride.

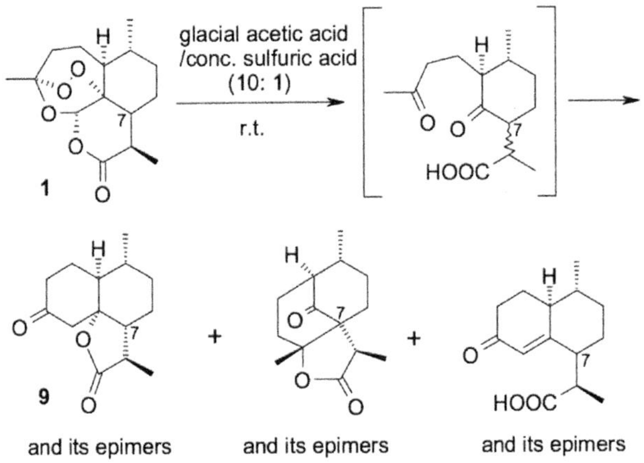

FIGURE 3.13 Reaction of artemisinin in glacial acetic acid-concentrated sulfuric acid.

C. ACID DEGRADATION OF ARTEMISININ

At room temperature, artemisinin and a mixture of glacial acetic acid and concentrated sulfuric acid (10:1) yield several products that have lost carbon atoms, including a range of ketolides and α,β-unsaturated ketones.[82] The main product is compound **9**. X-ray crystal analysis confirmed that configuration at C7 is inverted to that at C7 of artemisinin.[83] It was deduced that these products were the result of further reaction of the acidic diketone intermediate (Fig. 3.13).

When artemisinin is heated under reflux in methanol and an acid catalyst, the acetal and ketal groups break apart to form a mixture of methyl esters **10**. These on treatment with glacial acetic acid and concentrated sulfuric acid (10:1) at 0–5°C yield the diketone ester **11**, which retains the configuration at C7 of artemisinin. A small amount of artemisinin could be recovered. The yield of this two-step process can exceed 90% once the allowance is made for unreacted material. The intermediate **10** can be separated and reduced to produce either deoxyartemisinin **2** or compound **12**. In the latter case, peroxide group is retained and the lactone group is reformed.[84] The diketone ester **11** (Fig. 3.14)[85]

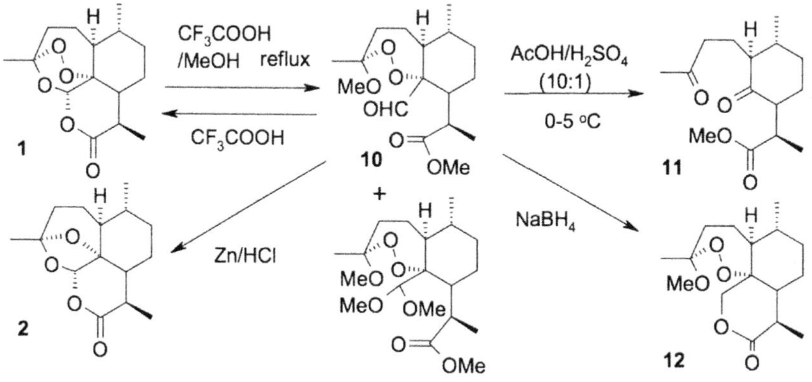

FIGURE 3.14 Acid degradation of artemisinin.

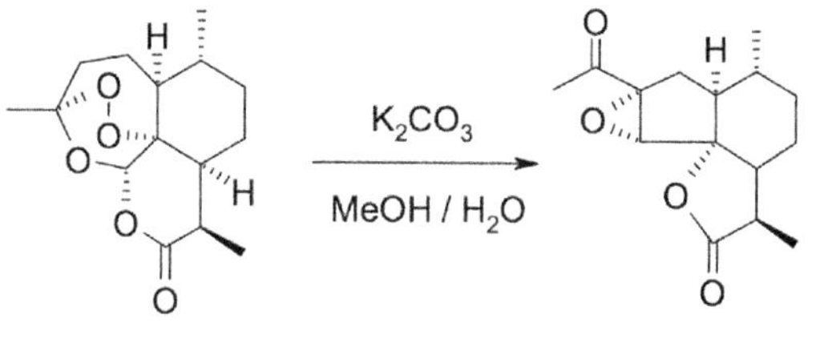

FIGURE 3.15 Alkali degradation of artemisinin.

is a very useful relay intermediate[86] and has been extensively used[87] in the synthesis of artemisinin[88] and its analogues.[89]

D. ALKALINE DEGRADATION OF ARTEMISININ

Artemisinin is highly unstable in alkaline media. At room temperature, it rapidly forms several complex products[44] in a solution of potassium carbonate, methanol, and water. Of these, octahydroindene **13** (Fig. 3.15) comprises 10% of the yield. This tendency must be borne in mind when working with artemisinin.

Quantitative UV analysis of artemisinin makes use of the compound's alkali degradation reactions. (See UV Spectrophotometry in Section 3.7.).

E. PYROLYSIS OF ARTEMISININ

Compared to other common peroxides, artemisinin is a relatively stable compound. Even at its melting point of 156–157°C, no clear signs of decomposition are seen.[22] However, decomposition takes place after 10 min at 190°C.[23] Compounds **14** (4%), **3** (12%), and **4** (10%) can be isolated from the products (Fig. 3.16).

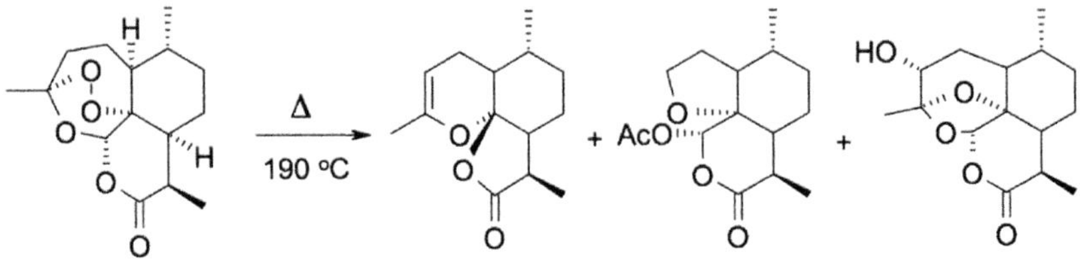

1 → (Δ, 190 °C) → 14 + 3 + 4

FIGURE 3.16 Pyrolysis of artemisinin.

3.6 SYNTHESIS OF ARTEMISININ

Wu Yulin
Shanghai Institute of Organic Chemistry, Chinese Academy of Sciences, Shanghai, China

Artemisinin has been an attractive target for organic synthesis because of its exceptional antimalarial activity and unique chemical structure. As of 2003, over 10 synthetic routes to artemisinin using optically active monoterpenes or sesquiterpenes as starting materials have been published. (Schmid and Hoheinz in 1983 carried out the total synthesis of qinghaosu [artemisinin].).[b]

The introduction of the peroxy group was achieved either by photooxidation through the [2+2] addition of singlet oxygen with an enol ether or by so-called biomimetic approaches. The synthetic routes are outlined in Fig. 3.17. Compound **21** (an enol methyl ether) was the common substrate of the key photooxygenation step in the two earliest synthetic Routes **A**[90–92] and **B**.[93] (−)-Isopulegol **16** and artemisinic acid **19** were taken as starting material in routes **B** and **A**, respectively. Artemisinic acid **19** was converted into **21** via dihydroartemisinic acid **20** to complete the partial synthesis. Then the compounds **19** and **20** themselves were synthesized from the commercially available citronellal **15** to complete the formal total syntheses. In 1985, the degradation product **11** was used as a relay intermediate in a convenient preparation of the key compound **22** in Route **C**.[85]

Routes **D**[94] and **E**[95] were developed in 1990 and 2003, respectively. Both used (+)-isolimonene **18** as the starting material. The intramolecular Diels–Alder reaction or iodolactonization followed by Michael addition and Wittig reaction, respectively, produced the key intermediates **21** and **22**. In Route **F**,[96,97] also developed in 1990, dihydroartemisinic acid **20**, was converted into the cyclic enol ether **23** that was submitted to photooxygenation to give artemisinin in better overall yield than that obtained via Route **A**. This route proceeded via deoxoartemisinin **7**, which has superior antimalarial activity to that of artemisinin.

b. Schmid et al. was the first group outside China to synthesize artemisinin (G. Schmid, W. Hofheinz, Total Synthesis of Qinghaosu. *J Am Chem Soc* 1983;**105**(3):624–25.)

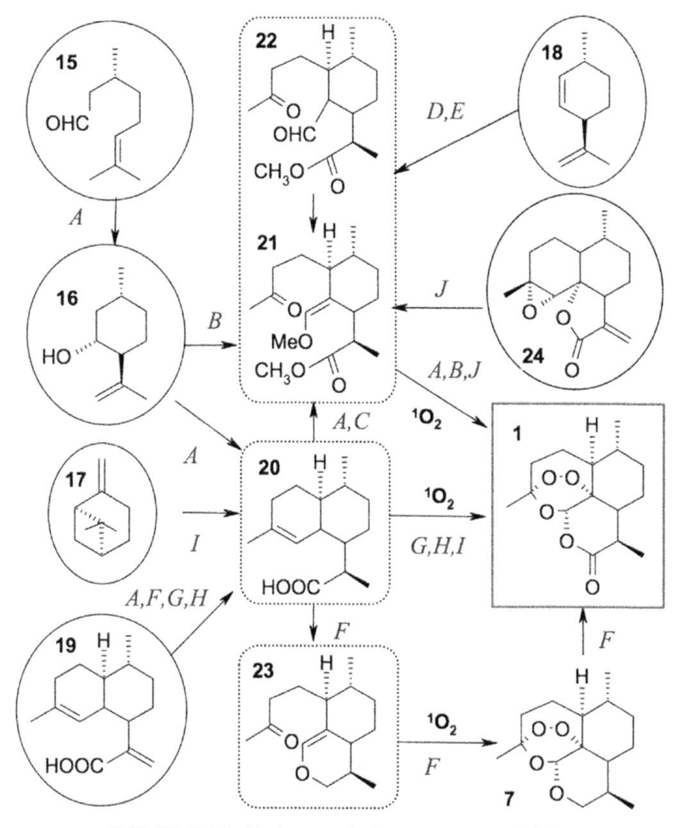

FIGURE 3.17 Various synthetic routes to artemisinin.

Route **G**[98,99] and later Route **H**[100–102] represent biomimetic synthesis involved direct photooxygenation of dihydroartemisinic acid **20**, another component of *A. annua.* In 1993, Route **I**[103] was used to synthesize compound **20** from β-pinene **17**; this then followed Route **G** to produce artemisinin. Route J[104,105] described in 1992 and 1998 commenced with artemisinin B **24**, another component found in relatively large amounts in *A. annua.* Its key steps were the deoxygenation followed by ozonization and elimination to give the photooxygenation substrate **21**.

One exception to the photooxygenation protocol involved synthesis of artemisinin using ozonolysis of vinyl silane to introduce the peroxy group (Fig. 3.18).[106,107]

After a hiatus of some years, a new wave of syntheses was reported in 2010–13.

In 2010 the Yadav group synthesized 6-epi-methyl dihydroartemisinic acid from citronellal **15** without a protecting group and then converted this into artemisinin according to Route **G**. However, this route was unsuitable for preparation of artemisinin in larger scale and there was no improvement in the key step of incorporating the peroxy group.[108]

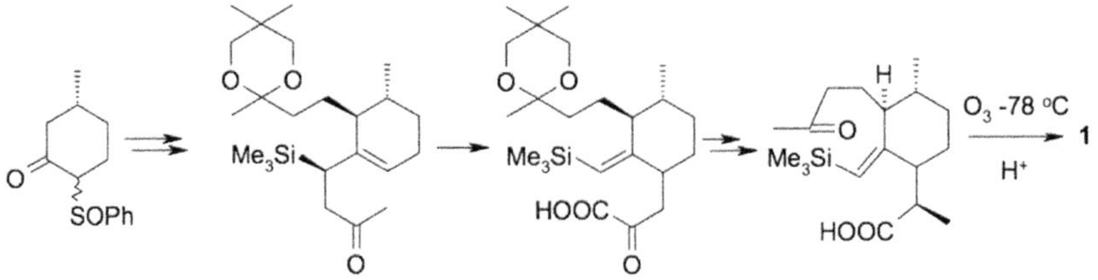

FIGURE 3.18 Peroxide group produced via ozonolysis of vinyl silane.

In 2012, Seeberger et al. in Max Planck Institute for Colloids and Interfaces described a variation[109] based on previously recorded routes[99,100] involving dihydroartemisinic acid. Interestingly in the Seeberger report, enol intermediates were claimed without reference to the much earlier work by Haynes et al., which described in detail the preparation and identification of the enol intermediates and their conversion and the biosynthetic significance thereof into artemisinin.[100–102] The technique involved photooxygenation carried out in a continuous flow reactor using tetraphenylporphyrin as a photosensitizer. The reaction product was then rearranged to artemisinin by adding trifluoroacetic acid as previously reported by Acton and coworkers in earlier publications unfortunately overlooked. The overall yield for artemisinin was up to 39%. The report claimed that the flow technique could be used to produce large amounts of artemisinin at "low cost." Practically speaking, however, synthetic artemisinin manufactured in this way can only be competitive if the price of artemisinic acid is less than one-fourth that of artemisinin. However, it is noted that the extraction process yielding artemisinin can be modified to provide artemisinic acid as well; if it turns out that artemisinins may find application in dealing with diseases other than malaria, it may become profitable to evaluate new extraction and synthesis procedures.[110,111]

One important advance in terms of academic significance was a new method of inducing the peroxide group without photooxygenation. Wu et al. of the SIOC reported in 2011 that hexavalent molybdenum catalyzes an epoxide ring-opening reaction with hydrogen peroxide resulting in direct introduction of a hydroperoxide group on the tertiary carbon atom. From there, 12-deoxoartemisinin **7** could be synthesized. Ye Bin's method[97] was then used to produce artemisinin.[112] At a press conference on 4 July 2012, Zhang Wenbin of Shanghai Jiao Tong University announced another new synthetic route from dihydroartemisinic acid to artemisinin. It involved hydrogen peroxide and a catalyst, required one step, and provided artemisinin in 60% yield (Fig. 3.19).[113] Nevertheless, this will only have practical significance when the cost of artemisinic acid becomes less than one-third that of artemisinin.

In August 2012, Zhu et al. of Indiana University described a new total synthesis route. It used the readily available starting material cyclohexanone and

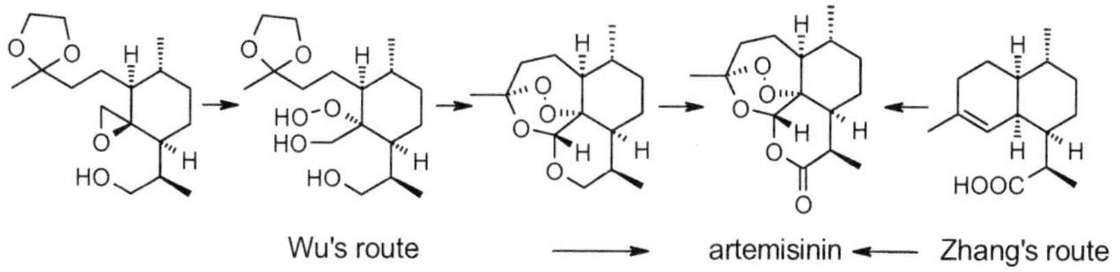

FIGURE 3.19 Two peroxide induction routes utilizing hydrogen peroxide.

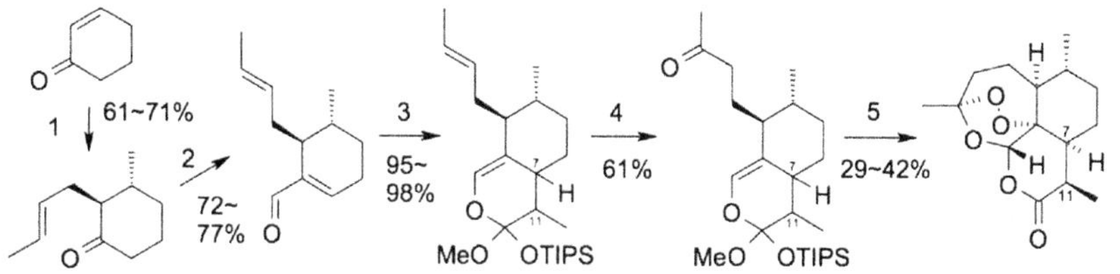

FIGURE 3.20 Synthesis of artemisinin in five steps.

provided artemisinin in five steps (Fig. 3.20).[114] The final step, which involved introduction of the peroxide group, still employed earlier methods involving singlet oxygen and an enol ether intermediate. Instead of photooxygenation, however, singlet oxygen was produced using hydrogen peroxide and ammonium heptamolybdate under thermal conditions. The yield was not high, being comparable to that of photooxygenation. The first and third steps were unusual and the fourth had a remarkable and almost unbelievable degree of selectivity over the chiral centers at C-7 and C-11 generated in the third step. The report stated that this route had been followed at gram scale. From a laboratory perspective, this represents a considerable achievement. More work is needed to translate this into large-scale production.

Again, in November 2012, Wu et al. of the SIOC detailed a new synthesis route in the online version of *Tetrahedron* (published in print form in January 2013).[115] This also used dihydroartemisinic acid as a substrate, which was first treated with hydrogen peroxide and ammonium molybdate to generate singlet oxygen under thermal conditions. The crude reaction products were treated with acid at room temperature in the presence of oxygen for 2 days. This produced artemisinin in 41% overall yield.[115] This entailed reactions similar to those described 20 years ago by Roth et al.,[116] Acton et al.,[99] and Haynes et al.[100–102] with the only difference being the absence of photooxygenation for generation of singlet oxygen. It was also similar to Zhang Wenbin's unpublished method, for which a patent application has been made.[117] As with the other methods above, this route will be of practical significance if artemisinic acid was cheaper and more readily available.

A systematic introduction and review of pre-2010 research into the total synthesis of artemisinin can also be found in Chapter 2 of *Compendium of the Total Synthesis of Natural Products—Terpenes*.[118]

3.7 QUANTITATIVE ANALYSIS OF ARTEMISININS COMPOUNDS

Zeng Meiyi
Institute of Chinese Materia Medica, China Academy of Chinese Medical Sciences, Beijing, China

This section introduces the quantitative analysis of common artemisinins, such as artemisinin (Qinghaosu, QHS) itself, artemether, artesunate, and dihydroartemisinin. It is divided into two parts: First, the analysis of pharmaceutical substances and preparations as laid out in the *Pharmacopoeia of the People's Republic of China* (hereafter: *Chinese Pharmacopoeia*) and the WHO's *International Pharmacopoeia* and second, some examples of new developments in the assay of artemisinins in body fluids.

A. METHODS IN THE CHINESE AND INTERNATIONAL PHARMACOPOEIAS

By 1976, the antimalarial activity and chemical structures of artemisinins had been confirmed. To ensure their safe use in a clinical setting, it was necessary to establish quantity and quality standards first. These would be recorded in the *Chinese Pharmacopoeia* as a reference. Thus, developing an assay method for artemisinin was a priority. The techniques listed in the *Pharmacopoeia* deal primarily with pharmaceutical substances and preparations.

Artemisinin and its derivatives possess a peroxide group, but assay methods specific to this group, such as iodometric titration[119] or other colorimetric techniques, are easily affected by ambient air. They are also inconvenient. The UV end absorption of artemisinins is also at 210 nm or below, rendering them unsuitable for UV spectrophotometry. Hence, there was some difficulty in developing assay techniques.

In 1977, Project 523 Head Office organized two exchange workshops on artemisinin assay techniques, combining the expertise of all the researchers involved to produce an accurate and practical method in a short time. All the techniques brought to the workshop had to be presented on the spot and discussed. Through this engagement, it was discovered that alkali reaction of artemisinin could quantitatively produce a new product. Its absorption wavelengths shifted to 292 nm, and it had strong UV absorbance. Hence, a UV spectrophotometric method was developed based on the alkali reaction of artemisinin. It was the first artemisinin assay listed in the *Chinese Pharmacopoeia* (1995) and is still one of the two methods recorded in the latest edition of the WHO's *International Pharmacopoeia* (2013).

1. UV Spectrophotometry

Workers at the Nanjing Medical College collaborating with Shen Xuankun of the Guangzhou University of Chinese Medicine had developed a UV spectrophotometric assay method prior to 1976. It was based on the reaction of artemisinin with dilute alkali solutions, which quantitatively produced a compound with strong UV absorbance at a wavelength of 292 nm. This method drew the attention of the 1977 workshops. Sample preparation and reaction criteria were refined under the guidance of Yan Kedong of the National Institute for the Control of Pharmaceutical and Biological Product. Demonstrations were run, and instruments from different manufacturers and models were compared. As a result, the UV spectrophotometry method was considered accurate, stable, and easy to follow. It was better than other techniques, had a sensitivity of around 10^{-6} g/mL, and could be used for pharmaceutical substances. Artemisinin (QHS) was dissolved in a solution of 95% ethanol–0.2% sodium hydroxide (1:4) in a volumetric flask, warmed in a water bath at 50°C for 30 min, and cooled immediately. Optical density was then assayed at 292 nm.[120]

Later, to clarify the reaction mechanisms, Zeng Meiyi acidified the alkali solution with 10% hydrochloric acid. Using chloroform extraction, colorless granular crystals were obtained with a melting point of 155.5–157.5°C (uncorrected). In a solution of 20% ethanol, the crystal had strong UV absorbance at 260 nm. After being restored to an alkaline state at room temperature, the compound again displayed strong absorption at 292 nm. Hence, the artemisinin–alkaline reaction product was termed Q292 and the crystal was named Q260. The conversion between Q292 and Q260 is reversible and is pH dependent under room temperature. The critical pH value for complete conversion of Q292 to Q260 is at 5.584. In a 20% ethanol solution, Q260 had a UV λ_{max} of 260 nm ($\varepsilon = 11{,}200$). In 95% ethanol–0.2% sodium hydroxide (1:4), the λ_{max} was 292 nm ($\varepsilon = 19{,}400$). The optical density concentration of Q292 and Q260 both showed good linearity and were qualified for quantitative analysis.

Based on its mass spectrometric, proton NMR [$(CD_3)_2CO$, 250 MHz] and infrared characteristics, Liang Xiaotian and Li Lanna interpreted the condensation reaction of artemisinin in alkali solution and the chemical transformation of Q292–Q260. They showed that different alkali concentrations yielded completely different reaction products with artemisinin. Q292 was only quantitatively produced by reaction of artemisinin with the 95% ethanol–0.2% sodium hydroxide (1:4) solution. At higher alkali concentrations, the products would change. This explained why, in quantitative reactions, the sodium hydroxide concentration had to be kept within $0.2 \pm 0.02\%$. The condensation reaction and chemical conversion process is shown in Fig. 3.21.[121,122] Q260 as a new compound was listed in CAS as 88104-60-3.

The alkali reaction mechanism of artemisinin and the keto–enol tautomerism between Q292 (enol) and Q260 (α,β-unsaturated ketone) affected by pH change became a key reference for the precolumn high-performance liquid chromatography (HPLC) assay methods later developed for artemisinin determination in body fluids.

Wang Zhongshan, Zhu Yaohua, Zhang Shuliang et al. developed a UV spectrophotometric assay method for artemether in 1982. An anhydrous ethanol solution of artemether was mixed with 1 mol/L hydrochloric acid–anhydrous

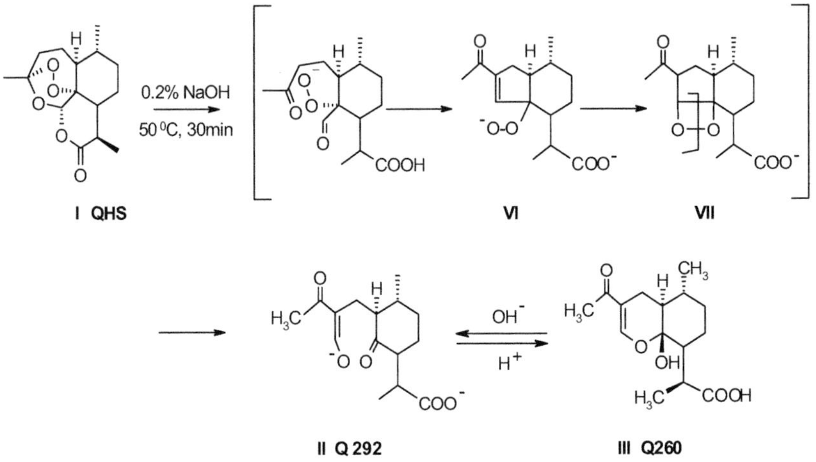

FIGURE 3.21 Quantitative reaction of artemisinin and dilute alkali.

ethanol (1:9) and reacted in a water bath for 5 h at 55 ± 1°C. After cooling, the absorbance was measured at a wavelength of 254 ± 1 nm and the coefficient calculated. The sample was assayed at the same time as the reference, and the ratio between the two reflects the artemether content in the sample.[123] An increase in the acidity of the solution or in temperature could speed up the reaction, but the instability of the products rose with increase of acidity or temperature. Hence, lower acidity and temperature were used to extend the stable period to 3 h. However, 5 h was needed for the reaction.

In 1985, Zeng Meiyi and Zhao Shishan established an alkali reaction-UV spectrophotometric assay for dihydroartemisinin.[124] A solution of dihydroartemisinin in ethanol and 2.0% sodium hydroxide (1:4) was prepared in a volumetric flask and warmed in a water bath at 60°C for 30 min. Once cooled, optical density was measured at 238 nm. At concentrations of 0.05–0.4 mg/10 mL, the relative coefficient $r = .998$ and the coefficient of variation were less than 1.4%. It is well known that dihydroartemisinin is very unstable and readily decomposes[125] to form various products, each in small amounts.[126] Hence, the reference sample must be strictly kept in dark and cool conditions. Before use, it must be checked for decomposition products.

The 2000[127] and 2005[128] editions of the *Chinese Pharmacopoeia* listed the UV spectrophotometric assays for artemisinin, dihydroartemisinin, and dihydroartemisinin tablets. However, UV spectrophotometric techniques were removed from the 2010 edition. Instead, HPLC assays were recorded for artemisinin, artemether, artesunate, and dihydroartemisinin, including their preparations. The UV spectrometric assay was retained only for dihydroartemisinin tablets.[129]

The third and fourth editions of the *International Pharmacopoeia*, published in 2003[130] and 2013, respectively, include the UV spectrophotometric assays for artemisinin, artemether, and dihydroartemisinin, as well as their preparations. It was applied to artemisinin and artemether tablets. However, the same criteria (e.g., alkali concentration, temperature, time, and wavelength) were listed for both the artemisinin and dihydroartemisinin assays. This was in fact an error, as each set of criteria for each individual compound cannot be mixed up.[131]

2. HPLC With UV Detection

Beginning with its 2000 edition (seventh edition),[127] the *Chinese Pharmacopoeia* listed HPLC with UV detection for artemether and artemether capsules. Octadecylsilane bonded to silica gel was used as the stationary phase and acetonitrile–water (55:45) as the mobile phase. The detection wavelength was 210 nm. All criteria were the same as in its 2005 edition.[128] In the 2010 edition (ninth edition), HPLC method was applied to the pharmaceutical substances and preparations of all artemisinins. The aforementioned stationary phase was used for artemisinin and dihydroartemisinin, whereas the mobile phase was a 60:50 mixture of acetonitrile–water. The same detection wavelength was used. For artemether, the mobile phase was 62:38 acetonitrile–water. For artesunate, a 42:58 acetonitrile–pH 3.0 phosphate buffer mobile phase was used. Column temperature was 30°C. Its other criteria were the same as for artemisinin.[129] Most of the criteria recorded in the 2010 edition were the same as those in the 2003 edition of the *International Pharmacopoeia*, except that an assay wavelength of 210 nm was recommended and the proportion of the mobile phase for artesunate was slightly different.[130]

In the 2003 and 2013 editions of the *International Pharmacopoeia*, HPLC with UV detection was listed as the first assay method for artemisinin, dihydroartemisinin, artemether, and artesunate, with UV spectrophotometry or titration as second methods. Either of these techniques could be selected. The HPLC criteria in both editions were essentially the same apart from the artesunate assay, where the ratio of the acetonitrile–pH 3.0 buffer mobile phase was changed from 50:50 to 44:56. The assay wavelength of 216 nm was recommended for all artemisinins.[131]

3. Artesunate Titration

This titration method employed sodium hydroxide with phenolphthalein as an indicator. The artesunate content was calculated from the amount of alkali solution consumed in the test. This titration method was recorded as an assay for artesunate tablets in the 2000[127] and 2005[128] editions of the *Chinese Pharmacopoeia*. In the 2010 edition, this was changed to HPLC for artesunate (newly added) and artesunate injections and tablets.[129] HPLC was also the first assay listed for these three drugs in the third[130] and fourth editions of the *International Pharmacopoeia*.[131]

4. HPLC With UV Detection for Artemisinin-Based Combination Therapies

Artemisinin-based combination therapies (ACTs) comprise two or more compounds. When conducting HPLC analysis, therefore, the component drugs must also be assayed. Their characteristics are often very different, making assay techniques more complicated than for artemisinins alone.

In 1989, Zeng Meiyi, Zhao Shishan, and Wang Yunling developed an HPLC-UV detection assay for artemether and lumefantrine in artemether–lumefantrine tablets. Back then, the tablets had a fixed dose of 20mg artemether and 100mg lumefantrine.[132] Lumefantrine does not readily dissolve in methanol and other organic solvents. Hence, it was ion-paired by dissolving lumefantrine (former names as fluorenemethanol or benflumetol) in a 1%–2% glacial acetic acid–methanol solution. Exploiting the chromatographic characteristics for lumefantrine, methanol–water–glacial acetic acid–diethylamine (93:6:1:0.01) was used as the solvent for both lumefantrine and artemether and as the mobile phase on a Zorbax ODS column to separate these two compounds. By accident, it was also discovered that artemether formed another compound in the mobile phase, which had UV absorbance at 230nm. UV spectrophotometry determined that, in a mobile phase without glacial acetic acid, λ_{max} for artemether was 205nm. It was shifted to 230nm in solution with glacial acetic acid. The absorbance increased, and the new compound was also stable and suitable for quantitative analysis.

Because lumefantrine has UV peak absorbance at 234–235nm, 234nm was chosen as the assay wavelength. Retention time was 4.5min for artemether and 10.5min for lumefantrine. Average sample recovery was 99.5% for artemether and 107.5% for lumefantrine. Based on peak area response factor, the linearity of artemether was good between 0.2 and 1.0mg/mL, $r=0.9996$. The same could be said of lumefantrine from 0.04 to 0.2mg/mL, $r=0.9992$. This HPLC assay allowed both drug components to be analyzed using the same mobile phase and detection wavelength. In 1996 Zeng Meiyi, Lu Zhiliang, and Yang Songcheng et al. adopted this method to determine the concentration of lumefantrine in human plasma (see below).[133]

The 2013 *International Pharmacopoeia* included an HPLC-UV detection assay for artemether–lumefantrine tablets, with an acetonitrile–ion-pair reagent (700:300 and 300:700) as the mobile phase for gradient elution. The ion-pair reagent was prepared by dissolving 5.65g of sodium hexanesulfonate and 2.75g of sodium dihydrogen phosphate in around 900mL of water. The pH was adjusted to 2.3 using phosphoric acid. It was diluted with water to 1000mL and filtered. The assay solvent for the tablets and both component drugs used 200mL of this ion-pair reagent, 60mL water, and 200mL 1-propanol. The mixture was diluted to 1000mL with acetonitrile. Twenty tablets were crushed and a sample equivalent to one tablet was accurately weighed out (~20mg artemether and 120mg lumefantrine). This was placed in a 100mL volumetric flask, dissolved via sonication in 85 mL of the aforementioned solvent, and diluted to volume. The same procedure was carried out with reference samples of 20 mg artemether and 120 mg lumefantrine. The flow rate of the mobile phase was 1.3 mL/min, and 20 μL of the testing and reference samples were introduced alternately. Analysis was based on peak area. For the first 28 min, the assay wavelength was set at 210 nm and then changed to 380 nm. Retention time was 19 min for artemether and 34 min for lumefantrine.[131] This assay employed two ratios of mobile phase and two wavelengths to isolate and measure the component drugs.

In 2006, Li Meiqin, Fan Qi, Zhang Xiaosong et al. established an HPLC–evaporative light scattering detection assay method for organic compounds without UV absorbance. Li et al. used it to measure the dihydroartemisinin content in dihydroartemisinin ACT tablets. However, the other two components of the tablet were not analyzed at the same time.[134]

B. DEVELOPMENTS IN THE ASSAY OF ARTEMISININS IN BODY FLUIDS

As research into artemisinins and ACTs progressed, pharmacokinetic studies on these drugs, their metabolites, and combination components were needed to provide reliable data for clinical regimens. Their concentrations in body fluids were relatively low, and the UV end absorption sensitivity of artemisinins themselves was too low. Hence, this was a critical issue for pharmacokinetic research, to establish a sensitive assay method for artemisinins and their in vivo metabolites in body fluids.

Through the Ministry of Health, the China Qinghaosu Steering Committee invited the WHO to send experts for a workshop on assay methods for artemisinin and its derivatives in body fluids, held at the ICMM in October 1983. The workshop introduced a few practical assay methods and demonstrated the extraction of drugs from body fluids. Participants also spoke about their areas of expertise. Beginning in 1985, Chinese scientists actively developed radioimmunoassay, HPLC–polarographic detection, HPLC-UV detection, liquid chromatography tandem mass spectrometry (LC-MS/MS), and other assay methods in succession, increasing the sensitivity of analysis to a 10^{-9} or 10^{-10} g/mL level. Pharmacokinetic research was strongly supported. At the same time, new body fluid assays were devised for other antimalarial drugs invented in China, such as lumefantrine and naphthoquine phosphate.

1. Radioimmunoassay

The radioimmunoassay method, established by Song Zhenyu, Zhao Kaicun, Liang Xiaotian et al. in 1985, was the first assay for artemisinin and artesunate in plasma to be published in China.[135] It exploited the specificity of antigen–antibody binding and high level of sensitivity of the radioisotope.

This method involved first a reaction of the hydroxide group of dihydroartemisinin at position 12. Then, an artemisinin-12-O-acetic acid–bovine serum albumin (BSA) complex was prepared. Each BSA molecule was connected to around 15 artemisinin molecules. The complex was dissolved in saline and mixed with an equivalent volume of Freund's adjuvant to form a 2 mg/mL antigen complex emulsion. Sheep were given 4 mg intramuscular inoculations of this emulsion. A booster was administered after 3 weeks and once a month thereafter. Plasma antibodies were assayed after 7–10 days. The titer was around 1,800, which did not increase significantly after the boosters. When various amounts of ^{3}H-dihydroartemisinin were added to a fixed amount of antiserum (1:500 dilution), the antigen–antibody binding reaction appeared saturated.

Various amounts of n-labeled artesunate were added to the antiserum at 1:1000 dilution and 7000 DPM ^{3}H-dihydroartemisinin. This produced a standard competition curve, and assay sensitivity was 2–3 ng. The competition curves for dihydroartemisinin, artemisinin, and artemether were similar, with sensitivities at 2–10 ng. The cross-reactions were complete. However, cross-reaction between deoxyartemisinin and artesunate was only 1%. This indicated the method was specific to the peroxide group in artemisinins. The experiment also indicated that incubation temperature and time did not clearly affect antigen–antibody binding. This assay had relatively good reproducibility, with intragroup and intergroup variation coefficient at 4%–11% and 3%–7%, respectively. Because it targeted the peroxide group, however, it could not differentiate between artesunate, artemether, and their active metabolite, dihydroartemisinin. The original drug and the metabolite had to be measured as a whole, representing a weakness in the method.

The drug time-concentration of intravenous artesunate (4 or 14 mg/kg) in dogs was measured using this method. It was found to adhere to a linear, single-compartment model. Average parameters were K=0.028/min, $T_{1/2}$=26.5 min, Vd=1.55 L/kg, and Cl_T=2.33 L/h/kg. This assay was also employed to compare the pharmacokinetics of the same dose of oral artemisinin or dihydroartemisinin in humans.[136]

2. HPLC with Polarographic Detection

For compounds such as artemisinins with peroxide groups, HPLC with polarographic detection was the most direct and relatively most sensitive assay method. Because it involved redox reactions, its difficulty lay in removing oxygen from the ambient air, preventing it from affecting the procedure. If this was not carried out properly, sensitivity would be lost. Demands on equipment and facilities were therefore higher.

In 1985, Yang Shude, Ma Jianmin, Sun Juanhua et al. were the first to use this assay, measuring artesunate and its dihydroartemisinin metabolite in human plasma.[137] For the HPLC, a YWG-C_{18} H_{37} column was used, with a methanol–0.03 mol/L ammonium sulfate–glacial acetic acid (55:45:0.1) mobile phase. Trace amounts of oxygen were removed by pure nitrogen gas with vanadium(III) chloride. The nitrogen was then bubbled through the mobile phase to deoxidize it. The detector was controlled by a polarographic analyzer utilizing sampled direct current. The potential of the static mercury working electrode was −0.24V (reference electrode Ag/AgCl) and drop time was 2s. From 20 to 800ng/mL, the plasma concentrations of artesunate and dihydroartemisinin showed a good linear relationship with peak current, $r=1.000$. Absolute recovery rates for both drugs were above 90% and the coefficients of variation were less than 9%. The sensitivity of the assay may have been affected by the continued presence of oxygen, which influenced the background current.

Zhou Zhongming, Huang Yuexiang, Guo Xingbo et al. also employed this method to measure artemisinin, artesunate, artemether, and dihydroartemisinin content in 1988.[138] A YWG-C_{18} H_{37} column, stainless steel precolumn, and isolation column were used. The mobile phase was methanol–0.02mol/L ammonium sulfate solution (75:25). It was poured into a three-necked, round-bottom flask. Pure nitrogen gas, which had been deoxidized with vanadium(III) chloride, was bubbled through. The liquid was refluxed to boiling and left for 1–2h. Before being introduced, the sample was deoxidized in pure nitrogen for 10min. Also using nitrogen gas, the sample solution was injected into the tube via a valve. Flow rate was 1mL/min. A PAR 310 polarographic detector was controlled by a model 364 polarographic analyzer, with sampled direct current. For artemisinin, dihydroartemisinin, artemether, and artesunate, the volt–ampere curves of the working electrode showed that potential was saturated at −0.90V (reference electrode Ag/AgCl). This was taken as the potential for the assay. A 303 static mercury drop electrode was the working electrode.

The composition of the mobile phase changes to its pH, and the effects on retention time were monitored. A standard solution was tested 10 times to establish the linearity and precision of the calibration curve. The relative standard deviations did not exceed 7% for artemisinin and α- and β-dihydroartemisinin and 8% for artemether and artesunate. The curves for artesunate and dihydroartemisinin were derived via external standard calibration. Good linearity was seen between 10–160ng/mL and 100–1600ng/mL, respectively. The artemether curve was calibrated with artemisinin as the internal standard, producing good linearity at 10–120ng/mL. Linear regression was run using the least-squares method, with peak height as the dependent and drug concentration as the independent variable. The relative coefficients of the seven calibration curves were $r=0.997$–0.999. This method was applied to study the pharmacokinetics of single intramuscular injections of artemether in rats and healthy male volunteers.

Van Agtmael, Butter, Portier et al.[139] introduced some important improvements to this technique in 1998. The nitrogen deoxidizing process was carried out in a tightly closed system, and samples were injected automatically. Artemether and dihydroartemisinin were assayed at 5–220ng/mL to validate the accuracy, precision, and reproducibility of these modifications. For artemether, the interday accuracy was +1% to +6.3%; intraday accuracy was −1% to +4.5%; and intra- and interday precision was less than 6%. For dihydroartemisinin, the interday accuracy was −0.6% to +2.6%; intraday accuracy was −9.5% to −0.3%; and intra- and interday precision was less than 9%. Within 3 months, 70 reproducible samples were produced. The interday accuracy for both drugs was −0.6% to +7.6% and precision was less than 18%. The results were cross-checked with those yielded by two other laboratories, which used comparable techniques. The correlation coefficient was more than 0.98%. This technique was also used successfully in testing samples from uncomplicated falciparum malaria patients in China, who received artemether or an artemether ACT. It examined the pharmacokinetics of artemether and dihydroartemisinin during a 4-day regimen of both drugs.[140] This assay represented the HPLC–polarographic method at its most sensitive.

3. HPLC With UV Detection

In 1986, Zhao Shishan and Zeng Meiyi established an HPLC method with precolumn reaction assay for artemisinin in plasma. This was the first time such a method had been used on artemisinin in body fluids.[141] The precolumn reaction was based on the alkali reaction of artemisinin, which was then acidified to form Q260 (see above). A LiChrosorb-RP_{18} column was used, with a 0.01 mol/L NaH_2PO_4–Na_2HPO_4 in water–methanol (55:45 *V/V*) buffer solution as the mobile phase. Samples were dissolved in a 0.01 mol/L acetic acid sodium acetate in water–methanol (8:2 *V/V*) buffer solution. Because the methanol content in this buffer was lower than that of the mobile phase, the sample was enriched for a short time in the column. Together with an increase in the sample size to 200 μL, this maximized assay sensitivity. The detection limit was 3 ng/mL, and a level of 10^{-9} g/mL could be reached. At a concentration range of 5–500 ng/mL,[141] the recovery rate was >95% and relative standard deviation was 1.2%–18%.[142]

Zhou Zhongming, Zhao Shishan, Nie Shuqin et al.[143] adopted this technique to test healthy volunteers given oral artemisinin and found that the concentrations of the drug in plasma and saliva were closely correlated.[142] It was also employed by Zhou Zhongming, Zengl Meiyi, Nie Shuqin et al.[144] to study the pharmacokinetics and pharmacodynamics of artemisinin suppositories in falciparum malaria patients.[145] This assay required many steps but the demands on equipment were not high, and it was easy to perform. Sensitivity was also satisfactory.

Already discussed is the HPLC-UV detection assay developed by Zeng et al. in 1989 for artemether–lumefantrine tablets. This technique was modified by Zeng Meiyi, Lu Zhiliang, Yang Songcheng et al. to determine lumefantrine in human plasma in 1996 (see above).[133] Their assay used a Spherisorb C_{18} column with a methanol–water–glacial acetic acid–diethylamine (93:6:1:0.03) mobile phase. From the UV to the visible spectrum, lumefantrine had four absorbance peaks, which could be used as detection wavelengths. To minimize interference from endogenous substances, a detection wavelength of 335 nm was chosen. The solubility of lumefantrine was enhanced using the same methods described above; additionally, glacial acetic acid–ethyl acetate (1:100) was used to increase the extraction of the drug from plasma. At concentrations of 5–4000 ng/mL in plasma, the absolute recovery rate of lumefantrine was 92.91%. Because the range of concentrations in body fluids was very wide, it was divided into two sections to enable more accurate analysis. Between 5 and 200 ng/mL, $r=0.9962$; between 200 and 4000 ng/mL, $r=0.9982$. The detection limit was 11.8 ng/mL. The substance used as the internal standard, “8212,” had a recovery rate of 84.85% at 300 ng/mL. Interday (5–4000 ng/mL) precision and accuracy were 2.46%–17.85% and 0.12%–21.20% for each range, respectively. Intraday (10, 50, 500 and 4000 ng/mL) precision and accuracy were 1.17%–8.19% and 0.56%–7.50%, respectively.

In 1999, Lefèvre and Thomsen used this method to investigate the multiple-dose pharmacokinetics of lumefantrine after oral administration of artemether–lumefantrine tablets and the component drugs in falciparum malaria patients, including 36 Chinese, 52 Thais, and 60 Gambian children. They also examined the influence of a high-fat breakfast on the bioavailability of these tablets in healthy volunteers.[146] The method was also adopted by Ezzet, van Vugt, Nosten et al. to study the pharmacokinetics and pharmacodynamics of lumefantrine in malaria patients receiving three different artemether–lumefantrine regimens.[147]

4. HPLC–MS/MS

Huang Jianming, Yu Yunqiu, Weng Weiyu et al. devised an HPLC–MS/MS assay for artemether and its active metabolite dihydroartemisinin in human plasma in 2005.[148] It utilized a C_{18} column and a 0.2% acetic acid–methanol mobile phase for gradient elution. The flow speed was 1.0 mL/min. Positive ions were collected via atmospheric pressure chemical ionization, which produced stronger ionization and increased injection volume, enhancing sensitivity. Selected ion monitoring used *m/z* 221 for artemether and dihydroartemisinin and *m/z* 283 for the artemisinin internal standard. Plasma samples were extracted with methyl tert-butyl ether. At a range of 5–300 ng/mL, artemether and dihydroartemisinin showed good linearity, $r=0.9990$. The detection limit was 2 ng/mL. The intra- and interday precisions were less than 9.3%, and the accuracy (recovery rates) was 92%–105%. When both drugs were at concentrations of 5, 50, and 200 ng/mL, the extraction recovery rates were 80%–96%. This method was used to test the plasma concentrations of artemether and dihydroartemisinin in six healthy male volunteers after a single 200mg artemether tablet. The results were used to generate time–concentration curves for both drugs.

The HPLC–MS/MS method was also employed by Liu Zeyuan et al.[149] to study changes in artemisinin and naphthoquine phosphate concentrations in the plasma and urine of healthy subjects given artemisinin–naphthoquine phosphate tablets.[150,151] It involved a Zorbax Extend-C_{18} column, column temperature 40°C, and a methanol–10mmol/L ammonium acetate solution (80:20) mobile as phase for gradient elution with autoinjection. Multiple reaction monitoring (MRM) was used, alongside electrospray ionization and a Q-Trap triple quadrupole tandem mass spectrometer. The reaction values were *m/z* 283.3→*m/z* 209.1 for artemisinin, *m/z* 410.3→*m/z* 337.1 for naphthoquine, and *m/z* 515.2→*m/z* 276.1 for the internal standard, telmisartan. At a plasma concentration range of 4.00–1000ng/mL, artemisinin had a correlation coefficient of 0.9990; the coefficient for the same range in urine was 0.9975. For naphthoquine, a plasma concentration range of 0.50–500ng/mL had a correlation coefficient of 0.9956; in urine, the correlation coefficient was 0.9991. The detection limits for artemisinin and naphthoquine were 4.00 and 0.50ng/mL, respectively, which was not affected by endogenous substances in plasma and urine. At concentrations of 10, 100, and 800ng/mL, the extraction recovery rates ($n=3$) of artemisinin were 93.3±9.81%, 86.1±5.30%, and 89.2±3.51%, respectively. For naphthoquine at 5, 50, and 400ng/mL, the rates ($n=3$) were 78.3±5.01%, 86.0±2.74%, and 87.9±1.83%, respectively. The rate for telmisartan at 40ng/mL was 87.2±5.10%.

The above concentrations were used to calculate intra- and interday precisions ($n=6$). For artemisinin, they were 4.07%–8.21% and 4.58%–9.91%, respectively; for naphthoquine, they were 7.77%–9.07% and 9.57%–13.60%, respectively. The values were all below 15%. Accuracy for artemisinin was −4.6% to 8.7% and that of naphthoquine was −1.7% to 4.0%. This method was applied to a study of the pharmacokinetics of artemisinin and naphthoquine in the plasma of 10 healthy male volunteers given a single dose of eight ACT tablets (125mg artemisinin and 50mg naphthoquine per tablet). Single-dose safety and the influence of food on bioavailability were also investigated.[152]

In 2006, Liu Ping, Hou Yunan, Shan Chengqi et al. employed HPLC–MS/MS to measure artemether and dihydroartemisinin in the plasma of beagles.[153] An acetonitrile–0.1% acetic acid (66:34) mobile phase was used with a C_{18} precolumn. Analysis was conducted via MRM, with electrospray ionization and a triple quadrupole tandem mass spectrometer. The detection values were *m/z* 321.1→*m/z* 275.1 for artemether, *m/z* 284.3→*m/z* 267.1 for dihydroartemisinin, and *m/z* 283.1→*m/z* 209.2 for the internal standard, artemisinin. At a linear range of 1–400ng/mL, artemether and dihydroartemisinin had intra- and interday precisions of less than 9.0%.

The assays introduced above represent the more sensitive and easily adopted methods for measuring artemisinins in body fluids. They are an indicator of the advances made in this field.

ACKNOWLEGMENT

We would like to thank Richard K. Haynes for his review and suggestions to this chapter.

REFERENCES

1. Luo KJ. Study on the plant containing a novel antimalarial, huanghuahaosu. *Acta Bot Yunnanica* 1980;**2**(1):33–41. [罗开均. 抗疟新型药物黄花蒿素原植物的研究. 云南植物研究，1980，2(1): 33–41.]
2. *Collected data on huanghuahao research*. Held at the Yunnan Institute of Materia Medica; 1977. p. 2. [云南省药物研究所. 黄蒿素资料汇编(内部资料). 1977, 2.]
3. *Antimalarial research into A. annua*. Held at the Institute of Chinese Materia Medica, Academy of Traditional Chinese Medicine; 1978. p. 35–40. [北京中医研究院中药研究所化学室. 青蒿抗疟研究，1978:35–40.]
4. *Study on the extraction techniques of artemisinin* (internal documents 1976). Held at the Shandong Institute of Materia Medica. [山东省中医药研究所. 黄花蒿素提取工艺的研究(内部资料), 1976.]
5. Li SY, Huang J, Wu ZX, et al. Study of the production technology of artemisinin. *Bull Chin Tradit Herb Drugs* 1979;**12**:555–9. [李舒养，黄剑，吴祖祥，等. 黄花蒿素生产工艺的研究. 中草药通讯，1979，12: 555–559.]
6. Takemoto T, Nakajima T. Studies on the essential oil of *Artemisia annua* L. I. Isolation of a new ester compound. *Yakugaku Zasshi* 1957;**77**:1307–9.
7. Takemoto T, Nakajima T. Studies on the essential oil of *Artemisia annua* L. II. Structure of *l*-β-Artemisia alcohol. *Yakugaku Zasshi* 1957;**77**:1310–3.
8. Jeramic D, Joki A, Behbud A, et al. A new type of sesquiterpene lactone isolated from *Artemisia annua* L. Arteannuin B. *Tetrahedron Lett* 1973;**14**(32):3039–42.
9. Uskokovié MR, Williams TH, Blount JF. The structure and absolute configuration of arteannuin B. *Helv Chim Acta* 1974;**57**(3):600–2.
10. Li Y, Wu YL. A golden phoenix arising from the herbal nest - a review and reflection on the study of antimalarial drug qinghaosu. *Front Chem China* 2010;**5**(4):357–422.
11. Tu YY, Ni MY, Zhong YR, et al. Studies on the chemical constituents of the Chinese traditional medicine, artemisia annua L. I. *Acta Pharm Sin* 1981;**16**(5):366–70. [屠呦呦，倪慕云，钟裕蓉，等. 中药青蒿化学成分的研究I. 药学学报，1981，16(5): 366–370.]
12. Tu YY, Ni MY, Zhong YR, et al. Studies on the constituents of *Artemisia annua* L. Part II. *Planta Med* 1982;**44**(3):143–5.
13. Zhu DY, Deng DA, Zhang SG, et al. Structure of artemisilactone. *Acta Chim Sin* 1984;**42**(9):937–9. [朱大元，邓定安，张顺贵，等. 青蒿内酯的结构. 化学学报，1984, 42(9): 937–939.]
14. Wei ZX, Pan JP, Li Y. Artemisinin G, a sesquiterpene from *Artemisia annua*. *Planta Med* 1992;**58**(3):300.
15. Deng DA, Zhu DY, Gao YL, et al. Structure of artemisinic acid. *Acta Chim Sin* 1981;**26**(19):1209–11. [邓定安，朱大元，高耀良，等. 青蒿酸的结构研究. 化学通报，1981，26(19): 1209–1211.]
16. Tu YY, Ni MY, Zhong YR, et al. Chemical constituents in *Artemisia annua* L. and research into artemisinin derivatives (summary). *Bull Chin Mater Med* 1981;**6**(2):31. [屠呦呦，倪慕云，钟裕蓉，等. 中药青蒿化学成分和青蒿素衍生物的研究(简报). 中药通报，1981，6(2): 31.]
17. Tian Y, Wei ZX, Wu ZH. Study on the chemical constituents of *Artemisia annua*, a traditional Chinese herb. *Chin Tradit Herb Drugs* 1982;**13**(6):249–51. [田樱，魏振兴，吴照华. 中药青蒿化学成分的研究. 中草药，1982，13(6): 249–251.]

18. Zhu DY, Zhang S.G, Liu BN, et al. Study on the antibacterial constitutents of *Artemisia annua*. *Chin Tradit Herb Drugs* 1982;**13**(2):54. [朱大元，张顺贵，刘伯年，等. 青蒿抗菌有效成分. 中草药，1982，13(2): 54.]
19. Wu ZH, Wang YY. Studies on the structure and synthesis of artemisinin and its related compounds XI. The identification of Epoxyarteannuinic acid. *Acta Chim Sin* 1984;**42**(6):596–8. [吴照华，王燕燕. 青蒿素及其类似物的结构和合成XI. 环氧青蒿酸的分离和鉴定. 化学学报，1984，42(6): 596–598.]
20. Misra LN. Arteannuin C, a sesquiterpene lactone from *Artemisia annua*. *Phytochemistry* 1986;**25**(12):2892–3.
21. Acton N, Klayman DL. Artemisitene, a new sesquiterpene lactone endoperoxide from artemisia annua. *Planta Med* 1985;**51**(5):441–2.
22. Lin AJ, Klayman DL, Hoch JM, et al. Thermal rearrangement and decomposition products of artemisinin (qinghaosu). *J Org Chem* 1985;**50**(23):4504–8.
23. Luo XD, Yeh HJC, Brossi A. The chemistry of drugs. VI. Thermal decomposition of qinghaosu. *Heterocycles* 1985;**23**(4):881–7.
24. Zhu DY, Huang BS, Chen ZL, et al. Isolation and identification of artemisinin conversion products in human metabolism. *Acta Pharmacol Sin* 1983;**4**(3):194–7. [朱大元，黄宝山，陈仲良，等. 人体代谢青蒿素转化物的分离和鉴定. 中国药理学报，1983，4(3): 194–197.]
25. Lee IS, Hufford CD. Metabolism of antimalarial sesquiterpene lactones. *Pharmacol Ther* 1990;**48**(3):345–55.
26. Huang JJ, Xia ZQ, Wu LF. Study on the constituents of artemisia annua. L. I. Isolation and identification of 11-r(-)-Dihydroarteannuic acid. *Acta Chim Sin* 1987;**45**(6):609–12. [黄敬坚，夏志强，吴莲芬. 青蒿化学成分的研究I. 11-R(-)-双氢青蒿酸的分离和结构鉴定. 化学学报，1987，45(6): 609–612.]
27. Anand A, Kumkum R, Raghunath ST. Biosynthesis of artemisinic acid in *Artemisia annua*. *Phytochemistry* 1990;**29**(7):2129–32.
28. Bouwmeester HJ, Wallaart TE, Janssen MH, et al. Amorpha-4,11-diene synthase catalyses the first probable step in artemisinin biosynthesis. *Phytochemistry* 1999;**52**(5):834–54.
29. Woerdenbag HJ, Bos R, Salomons MC, et al. Volatile constituents of *Artemisia annua* L. (Asteraceae). *Flavour Frag J* 1993;**8**(3):131–7.
30. Bertea CM, Freije JR, van der Woude H, et al. Identification of intermediates and enzymes involved in the early steps of artemisinin biosynthesis in *Artemisia annua*. *Planta Med* 2005;**71**(1):40–7.
31. Ahmad A, Misra LN. Terpenoids from *Artemisia annua* and constituents of its essential oil. *Phytochemistry* 1994;**37**(1):183–6.
32. El-Feraly FS, Al-Meshal IA, Khalifa SI. Epi-deoxyarteannuin B and 6,7-dehydroartemisinic acid from *Artemisia annua*. *J Nat Prod* 1989;**52**(1):196–8.
33. Wallaart TE, Pras N, Quax WJ. Isolation and identification of dihydroartemisinic acid hydroperoxide from *Artemisia annua*: a novel biosynthetic precursor of artemisinin. *J Nat Prod* 1999;**62**(8):1160–2.
34. Elmarakby SA, el-Feraly FS, Elsohly HN, et al. Microbial transformation studies on arteannuin B. *J Nat Prod* 1987;**50**(5):903–9.
35. Sy LK, Brown GD. Deoxyarteannuin-B, Dihydro-deoxyarteannuin-B and *trans*-5-hydroxy-2-isopropenyl-5-methylhex-3-en-1-ol from *Artemisia anuua*. *Phytochemistry* 2001;**58**(8):1159–66.
36. Roth RJ, Acton N. Isolation of epi-deoxyartemisinin B from *Artemisia annua*. *Planta Med* 1987;**53**(6):576.
37. Brown GD. Two new compounds from *Artemisia annua*. *J Nat Prod* 1992;**55**(12):1756–60.

38. Brown GD. Annulide, a sesquiterpene lactone from *Artemisia annua*. *Phytochemistry* 1993;**32**(2):391–2.
39. Pathak A, Jain DC, Bhakuni RS, et al. Deepoxidation of Arteannuin-B with chlorotrimethylsilane and sodium-iodide. *J Nat Prod* 1994;**57**(12):1708–10.
40. Bagchi GD, Haider F, Dwivedi PD, et al. Essential oil constituents of *Artemisia annua* during different growth periods at monsoon conditions of subtropical North Indian plains. *J Essent Oil Res* 2003;**15**(4):248–50.
41. Brown GD, Liang GY, Sy LK. Terpenoids from the seeds of *Artemisia annua*. *Phytochemistry* 2003;**64**(1):303–23.
42. Zheng GQ. Cytotoxic terpenoids and flavonoids from *Artemisia annua*. *Planta Med* 1994;**60**(1):54–7.
43. Collaborating Group for Research into the Structure of Artemisinin. Artemisinin, a novel sesquiterpene lactone. *Chin Sci Bull* 1977;**22**(3):142. [青蒿素结构研究协作组. 一种新型的倍半萜内酯——青蒿素. 科学通报，1977，22(3): 142.] Chemical Abstracts 87, 98788g (1977).
44. Liu JM, Ni MY, Fan JF, et al. The structure and reactions of artemisinin. *Acta Chim Sin* 1979;**37**(2):129–40. [刘静明，倪慕云，樊菊芬，等. 青蒿素的结构和反应. 化学学报，1979，37(2): 129–140.]
45. Li Y, Wu YL. An over four millennium story behind qinghaosu (artemisinin) – a fantastic antimalarial drug from a traditional Chinese herb. *Curr Med Chem* 2003;**10**(21):2197–230.
46. Qinghaosu Research Group, Institute of Biophysics, Chinese Academy of Sciences. *Crystal structure and absolute configuration of qinghaosu*. 1979. p. 1114–28.
47. Qinghaosu Research Group, Institute of Biophysics, Chinese Academy of Sciences. Crystal structure and absolute configuration of qinghaosu. *Sci Sin* 1980;**23**(8):380–96. Chemical Abstracts 93, 71991e (1980).
48. Klayman DL. Qinghaosu (artemisinin): an antimalarial drug from China. *Science* 1985;**228**(4703):1049–55. http://dx.doi.org/10.1126/science.3887571.
49. Zhu NJ. Introduction to the Nobel prize in chemistry 1985- direct methods for the determination of crystal structure. *Knowl Power* 1986;**1**:5. [竺迺珏 1986, "创建测定分子结构的直接法-浅谈1985年诺贝尔化学奖", 知识就是力量，1，5.]
50. Karle IL, Karle J. An application of the symbolic addition method to the structure of l-arginine dihydrate. *Acta Cryst* 1964;**17**:835–41.
51. Cochran W. Relations between the phases of structure factors. *Acta Cryst* 1955;**8**:473–8. http://dx.doi.org/10.1107/S0365110X55001485.
52. Hauptman H, Karle J. Structure invariants and seminvariants for noncentrosymmetric space groups. *Acta Cryst* 1956;**9**:45–55. http://dx.doi.org/10.1107/S0365110X56000097.
53. Crowfoot D. X-ray analysis of complicated molecules. Nobel Lecture. In: *Nobel lectures: chemistry 1942–1962*. December 11, 1964. p. 83.
54. Wang M. Publication process involving the discovery of artemisinin (qinghaosu) before 1985. *Asian Pac J Trop Biomed* 2016;**6**(6):461–7.
55. See reference 48.
56. Chan K, Yuen K, Hiroaki T, et al. Polymorphism of artemisinin from *Artemisia Annua*. *Phytochemistry* 1997;**46**(7):pp.1209–1214.
57. Liang L, Dong YC, Lin ZC. The least-squares methods in crystal structure refinement. *Acta Chim Sin* 1979;**3**:215–7. [梁丽，董贻诚，林政炯 1979 "晶体结构修正的最小二乘方法", 化学通报 3:23–25.]
58. Bijvoet JM, Peerdeman AF, van Bomme AJ. Determination of the absolute configuration of optically active compounds by means of X-rays. *Nature* 1951;**168**:271–2. http://dx.doi.org/10.1038/168271a0.

59. Project 523 Head Office. *Artemisinin research plan*. Held at the Center for the History of Medicine, Peking University; 1976–1977. [青蒿协作组（1976年7月7日）一九七六至一九七七年青蒿素研究計划,国务院523办公室1975年12月会议制订的一九七六至一九七七年青蒿素研究計划(草案).]
60. Liang L, Dong Y, Zhu N. Stereochemistry of a derivative of artemisinin – artesunate $C_{19}H_{28}O_8$. *J Struct Chem* 1986;**5**(2):73–7.
61. Liang L, Zhu N. In: *Stereochemistry of artemisinin series. Molecular structure: chemical reactivity and biological activity. International symposium, Beijing, China*. September 15–21, 1986. p. 103–5.
62. Kepler JA, Phillip A, Lee YW, et al. Endoperoxides as potential antimalarial agents. *J Med Chem* 1987;**30**:1505–9.
63. Gu J, Chen K, Jiang H, et al. A DFT study of artemisinin and 1,2,4-trioxane. *J Mol Struct (Theochem)* 1999;**459**:103–11.
64. Bhattacharjee AK, Karle JM. Stereoelectronic properties of anti-malarial artemisinin analogues in relation to neurotoxicity. *Chem Res Toxicol* 1999;**12**:422–8.
65. Wang J, Zhang C, Chia W, et al. Haem-activated promiscuous targeting of artemisinin in *Plasmodium falciparum*. *Nat Commun* 2015;**6**:10111. http://dx.doi.org/10.1038/ncomms10111.
66. Dong J, Dan Y, Tan Z, et al. Low temperature molar heat capacities and thermal stability of crystalline artemsinin. *Thermochim Acta* 2007;**463**:2–5.
67. Zhou GD. *Structure and properties: the applications of chemical principles*. 3rd ed. Beijing: Higher Education Press; 2009. ISBN: 978-7-04-026586-6. p. 147–8. [周公度(2009). “第四章碳和氮的化学”《结构和物性-化学原理的应用》第三版，147–148页，高等教育出版社[ISBN 978-7-04-026586-6.]
68. Li Y, Yu PL, Chen YX, et al. Studies on analogs of qinghaosu - some acidic degradations of qinghaosu. *Sci Bull* 1986;**31**(35):1038–40.
69. Chen Y, He CX, Zhu SM, et al. Electrocatalytic reduction of artemether by Hemin. *J Electrochem Soc* 1997;**144**(6):1891–4.
70. Chen Y, Zhu SM, Chen HY, et al. Study on the electrochemical characteristics of artemisinin and its derivatives I. Electrochemical reduction of artemisinin at the Hg electrode. *Acta Chim Sin* 1997;**55**(9):921–5. [陈扬，朱世民，陈洪渊，等. 青蒿素及其衍生物电化学性质的研究 I. 青蒿素在汞电极上的电化学还原. 化学学报，1997，55(9): 921–925.]
71. Chen Y, Zheng JM, Zhu SM, et al. Evidence for hemin inducing the cleavage of peroxide bond of artemisinin (qinghaosu): cyclic voltammetry and in situ FT IR spectroelectrochemical studies on the reduction mechanism of artemisinin in the presence of hemin. *Electrochimica Acta* 1999;**44**(14):2345–50.
72. Jiang HL, Chen KX, Tang Y, et al. Theoretical and cyclic voltammetry studies on antimalarial mechanism of artemisinin (qinghaosu) derivatives. *Indian J Chem* 1997;**36B**:154–60.
73. Wu WM, Wu YL. Chemical and electro-chemical reduction of qinghaosu (artemisinin). *J Chem Soc Perkin Trans 1* 2000;**32**(24):4279–83.
74. Wu WM, Yao ZJ, Wu YL, et al. Ferrous ion induced cleavage of the peroxy bond in qinghaosu and its derivatives and the DNA damage associated with this process. *Chem Commun* 1996;**18**:2213–4.
75. Wu WM, Wu YK, Wu YL, et al. Unified mechanism framework for the Fe(II)-Induced cleavage of qinghaosu and derivatives/analogues. The first spin-trapping evidence for the earlier postulated secondary C-4 radical. *J Am Chem Soc* 1998;**120**(14):3316–25.
76. Butler AR, Gilbert BC, Hulme P, et al. EPR evidence for the involvement of free radicals in the iron-catalysed decomposition of qinghaosu (artemisinin) and some derivatives: antimalarial action of some polycyclic endoperoxides. *Free Radic Res* 1998;**28**(5):471–6.

77. Haynes RK, Cheu K-W, N'Da D, Coghi P, Monti D. Considerations on the mechanism of action of artemisinin antimalarials: Part 1-The 'carbon radical' and 'heme' hypotheses. *Infect Disord Drug Targets* 2013;**13**:217–77.
78. Jung M, Li X, Bustos DA, et al. A short and stereospecific synthesis of (+)-deoxoartemisinin and (−)-deoxodesoxyartemisinin. *Tetrahedron Lett* 1989;**30**(44):5973–6.
79. Rong YJ, Ye B, Zhang C, et al. An efficient synthesis of deoxoqinghaosu from dihydroqinghaosu. *Chin Chem Lett* 1993;**4**(10):859–60.
80. Wu YL, Zhang JL. Reduction of artemisinin with lithium aluminium hydride. *Chin J Org Chem* 1986;**6**(2):153–6. [吴毓林，张景丽. 青蒿素的锂铝氢还原. 有机化学，1986，6(2): 153–156.]
81. Sy LK, Hui SM, Cheung KK, et al. A rearranged hydroperoxide from the reduction of artemisinin. *Tetrahedron* 1997;**53**(22):7493–500.
82. Zhou WS, Wen YC. The structures and synthesis of artemisinin and related compounds VI. The structures of the degradation products of artemisinin. *Acta Chim Sin* 1984;**42**(5):455–9. [周维善，温业淳. 青蒿素及其一类物的结构和合成 VI. 青蒿素降解产物的结构. 化学学报，1984，42(5): 455–459.]
83. Gu YX. Determination of the crystal structure of a sesquiterpene lactone with lost carbon. *Acta Phys Sin* 1982;**31**(7):963–8. [古元新. 失碳倍半萜内酯晶体结构测定. 物理学报，1982，31(7): 963–968.]
84. Li Y, Yu PL, Chen YX, et al. Studies on artemisinin analogs – some acidic degradations of artemisinin. *Chin Sci Bull* 1985;**17**:1313–5. [李英，虞佩琳，陈一心，等. 青蒿素类似物的研究——青蒿素的一些酸性降解反应. 科学通报，1985，(17): 1313–1315.]
85. Wu YL, Zhang JL, Li JC. Studies on the synthesis of artemisinin and its analogs – reconstruction of artemisinin from its degradation product. *Acta Chim Sin* 1985;**43**(9):901–3. [吴毓林，张景丽，李金翠. 青蒿素及其类似物的合成研究——由青蒿素降解产物重组青蒿素. 化学学报，1985，43(9): 901–903.]
86. Zhang JL, Li JC, Wu YL. Ozonization synthesis of artemisinin analogues. *Acta Chim Sin* 1988;**23**(6):452–5. [张景丽，李金翠，吴毓林. 臭氧化合成青蒿素类似物. 化学学报，1988，23(6): 452–455.]
87. Rong YJ, Yulin W. Synthesis of C-4-substituted qinghaosu analogues. *J Chem Soc Perkin Trans* 1993;**1**(18):2147–8.
88. Ye B, Zhang C, Wu YL. Synthesis studies on 15-nor-qinghaosu. *Chin Chem Lett* 1993;**4**(7):569–72.
89. Sy LK, Cheung KK, Zhu NY, et al. Structure elucidation of Arteannuin O, a novel cadinane diol from *Artemisia annua*, and the synthesis of Arteannuins K, L, M and O. *Tetrahedron* 2001;**57**(40):8481–93.
90. Xu XX, Zhu J, Huang DZ, et al. Studies on the structure and synthesis of artemisinin and its related compounds X. The stereocontrolled synthesis of artemisinin and deoxyartemisinin from artemisinic acid. *Acta Chim Sin* 1983;**41**(6):574–6. [许杏祥，朱杰，黄大中，等. 青蒿素及其一类物的结构和合成 X. 从青蒿酸立体控制合成青蒿素和脱氧青蒿素. 化学学报，1983，41(6): 574–576.]
91. Xu XX, Zhu J, Huang DZ, et al. Total synthesis of arteannuin and deoxyarteannuin. *Tetrahedron* 1986;**42**(3):819–28.
92. Xu XX, Zhu J, Huang DZ, et al. Studies on the structure and synthesis of artemisinin and its related compounds. XVII. Stereocontrolled total synthesis of methyl dihydroartemisinic acid – total synthesis of artemisinin. *Acta Chim Sin* 1984;**42**(9):940–2. [许杏祥，朱杰，黄大中，等. 青蒿素及其一类物的结构和合成XVII. 双氢青蒿酸甲酯的立体控制性全合成——青蒿素全合成. 化学学报，1984，42(9): 940–942.]

93. Schmid G, Hofheinz W. Total synthesis of qinghaosu. *J Am Chem Soc* 1983;**105**(3):624–5.
94. Ravindranathan T, Kumar MA, Menon RB, et al. Stereoselective synthesis of artemisinin. *Tetrahedron Lett* 1990;**31**(5):755–8.
95. Yadav JS, Babu RS, Sabitha G. Stereoselective total synthesis of (+)-artemisinin. *Tetrahedron Lett* 2003;**44**(2):387–9.
96. Ye B, Wu YL. Syntheses of carba-analogues of qinghaosu. *Tetrahedron* 1989;**45**(23):7287–90.
97. Ye B, Wu YL. An efficient synthesis of qinghaosu and deoxoqinghaosu from arteannuic acid. *J Chem Soc Chem Commun* 1990:726–7.
98. Roth RJ, Acton N. A simple conversion of artemisinic acid into artemisinin. *J Nat Prod* 1989;**52**(5):1183–5.
99. Acton N, Roth RJ. On the conversion of dihydroartemisinic acid into artemisinin. *J Org Chem* 1992;**57**(13):3610–4.
100. Haynes RK, Vonwiller SC. Catalysed oxygenation of allylic hydroperoxides derived from qinghao (artemisinic) acid. conversion of qinghao acid into dehydroqinghaosu (artemisitene) and qinghaosu (artemisinin). *J Chem Soc Chem Commun* 1990:451–3.
101. Haynes RK, Vonwiller SC, Warner JA, et al. Copper(II)-induced cleavage-oxygenation of allylic hydroperoxides derived from qinghao acid in the synthesis of qinghaosu derivatives – evidence for the intermediacy of enols. *J Am Chem Soc* 1995;**117**:11098–105.
102. Haynes RK, Vonwiller SC. From qinghao, marvellous herb of antiquity to the antimalarial trioxane qinghaosu and some remarkable new chemistry. *Acc Chem Res* 1997;**30**:73–9.
103. Liu HJ, Yeh WL, Chew SY. A total synthesis of the antimalarial natural product (+)-Qinghaosu. *Tetrahedron Lett* 1993;**34**(28):4435–8.
104. Lansbury PT, Nowak DM. An efficient partial synthesis of (+)-artemisinin and (+)-deoxoartemisinin. *Tetrahedron Lett* 1992;**33**(8):1029–32.
105. Nowak DM, Lansbury PT. Synthesis of (+)-artemisinin and (+)-deoxoartemisinin from Arteannuin B and Arteannuin acid. *Tetrahedron* 1998;**54**(3/4):319–36.
106. Avery MA, Jennings-White C, Chong WKM. The total synthesis of (+)-artemisinin and (+)-9-desmethylartemisinin. *Tetrahedron Lett* 1987;**28**(40):4629–32.
107. Avery MA, Chong WKM, Jennings-White C. Stereoselective total synthesis of (+)-artemisinin, the antimalarial constituent of *artemisia annua* L. *J Am Chem Soc* 1992;**114**(3):974–9.
108. Yadav JS, Thirupathaiah B, Srihari P. A concise stereoselective total synthesis of (+)-artemisinin. *Tetrahedron* 2010;**66**(11):2005–9.
109. Levesque F, Seeberger PH. Continuous-flow synthesis of the anti-malaria drug artemisinin. *Angew Chem Int Ed* 2012;**51**(7):1706–9.
110. Vonwiller SC, Haynes RK, King G, Wang HJ. An improved method for the isolation of qinghao (artemisinic) acid from artemisia annua. *Planta Med* 1993;**59**:562–3. 66.
111. Haynes RK, Vonwiller SC. Extraction of qinghaosu (artemisinin) and qinghao (artemisinic) acid: preparation of artemether and new analogues. *Trans R Soc Trop Hyg Med* 1994;**88**:S23–6.
112. Hao HD, Li Y, Han WB, et al. A hydrogen peroxide based access to qinghaosu (artemisinin). *Org Lett* 2011;**13**(16):4212–5.
113. Yi R, Gu WM, Cao J. Effective artificial synthesis of artemisinin. *Xinmin Evening News* July 4, 2012. [易蓉，顾伟民，曹杰. 青蒿素实现"高效人工合成".新民晚报，2012年7月4日.]
114. Zhu CY, Cook SP. A concise synthesis of (+)-artemisinin. *J Am Chem Soc* 2012;**134**(33):13577–9.
115. Chen HJ, Han WB, Hao HD, et al. A facile and scalable synthesis of qinghaosu (artemisinin). *Tetrahedron* 2013;**69**(3):3112–4.
116. Roth RJ, Acton N. A simple conversion of artemisininic acid into artemisinin. *J Nat Prod* 1989;**52**(5):1183–5.

117. Zhang WB, Liu DL, Yuan JJ. *A method of preparing artemisinin from artemisinic acid* China Patent CN 102718773 B. May 8, 2013. [张万斌，刘德龙，袁乾家. 一种由青蒿酸制备青蒿素的方法. 中国专利公开号CN102718773A公开日2012.10.10.]

118. Wu YL. *Compendium of the total synthesis of natural products*. Beijing: Science Press; 2010. p. 39–65. [吴毓林. 天然产物全合成荟萃——萜类. 北京：科学出版社，2010：39–65.]

119. Zeng MY. A modified iodometric method in determination of organic bridged peroxides-iodometric determination of qinghaosu. *Chin J Pharm Anal* 1984;**4**(6):327–9. [曾美怡. 桥式有机过氧化物碘量法的改进——碘量法测定青蒿素. 药物分析杂志，1984，4(6): 327–329.]

120. Sheng XK, Yan KD, Luo ZY, et al. UV analysis in the determination of artemisinin content. *Chin J Pharm Anal* 1983;**3**(1):24–6. [沈璇坤，严克东，罗泽渊，等.紫外分光光度法测定青蒿素含量. 药物分析杂志，1983，3(1): 24–26.]

121. Zeng MY, Li LN, Liang XT. Mechanism basis for the quantitative UV analysis of qinghaosu via alkali treatment. *Chin J Pharm Anal* 1985;**5**(5):268–71. [曾美怡，李兰娜，梁晓天. 紫外分光光度法测定青蒿素含量的化学反应机理. 药物分析杂志，1985，5(5): 268–271.]

122. Zeng MY, Li LN, Chen SF, et al. Chemical transformation of qinghaosu, a peroxidic antimalarial. *Tetrahedron* 1983;**39**(18):2941–6.

123. Wang ZS, Zhu YH, Zhang SL, et al. UV assay method for artemether. *Chin J Pharm Anal* 1982;**2**(1):65–8. [王仲山，朱耀华，张叔良，等. 蒿甲醚的紫外分光光度测定法. 药物分析杂志，1982，2(2): 65–68.]

124. Zeng MY, Zhao SS. Quantitative determination of dihydroartemisinin by UV spectrophotometry via alkaline reaction. *Chin J Pharm Anal* 1986;**6**(3):135–8. [曾美怡，赵世善. 紫外分光光度法测定双氢青蒿素含量. 药物分析杂志，1986，6(3): 135–138.]

125. Jansen FH. The pharmaceutical death-ride of dihydroartemisinin. *Malar J* 2010;**9**:212–6.

126. Haynes RK, Chan HW, Lung CM, et al. Artesunate and dihydroartemisinin (DHA): unusual decomposition products formed under mild conditions and comments on the fitness of DHA as an antimalarial drug. *ChemMedChem* 2007;**2**(10):1448–63.

127. Editorial Commission of Pharmacopoeia of China. *Pharmacopoeia of the People's Republic of China (7th edition), Part II*. Beijing: Chemical Industry Press; 2000. [国家药典委员会.中华人民共和国药典. (第7版)二部. 北京：化学工业出版社，2000.]

128. Editorial Commission of Pharmacopoeia of China. *Pharmacopoeia of the People's Republic of China (8th edition), Part II*. Beijing: Chemical Industry Press; 2005. [国家药典委员会.中华人民共和国药典. (第8 版)二部. 北京：化学工业出版社，2005.]

129. Editorial Commission of Pharmacopoeia of China. *Pharmacopoeia of the People's Republic of China (9th edition), Part II*. Beijing: Chemical Industry Press; 2010. [国家药典委员会.中华人民共和国药典. (第9版)二部. 北京：化学工业出版社，2010.]

130. World Health Organization. 3rd ed. *The international pharmacopeia*, vol. 5. Geneva: World Health Organization Department of Essential Medicines and Pharmaceutical Policies; 2003.

131. World Health Organization. *The international pharmacopeia, 4th edition, including first, second, and third Supplements*. Geneva: World Health Organization Department of Essential Medicines and Pharmaceutical Policies; 2013.

132. Zeng MY, Zhao SS, Wang YL. Determination of artemether and fluorenemethanol in artemether combination tablets by HPLC and a study on chemical conversion of artemether in mobile phase. In: *WHO/swg-chemal conference, 24–26 April 1989*. April 1989.

133. Zeng MY, Lu ZL, Yang SC, et al. Determination of benflumetol in human plasma by reversed-phase high-performance liquid chromatography with ultraviolet detection. *J Chromatogr B Biomed Appl* 1996;**681**(2):299–306.

134. Li MQ, Fan Q, Zhang XS. HPLC-ELSD determination of dihydroartemisinin in compound dihydroartemisinin tablets. *Chin J Pharm Anal Engl Ed* 2006;**26**(8):1163–5.
135. Song ZY, Zhao KC, Liang XT, et al. Radioimmunoassay of qinghaosu and artesunate. *Acta Pharm Sin* 1985;**20**(8):610–4. [宋振玉，赵凯存，梁晓天，等. 青蒿酯和青蒿素的放射免疫测定法. 药学学报，1985，20(8): 610–614.]
136. Zhao KC, Song ZY. Pharmacokinetics of dihydroqinghaosu in hunman volunteers and comparison with qinghaosu. *Acta Pharm Sin* 1993;**28**(5):342–5. [赵凯存，宋振玉，双氢青蒿素在人的药代动力学及与青蒿素的比较，药学学报，1993，28(5)：342–345.]
137. Yang SD, Ma JM, Sun JH, et al. Determination of artesunate and dihydroqinghaosu in human plasma by high performance liquid chromatography with reductive mode electrochemical detection. *Acta Pharm Sin* 1985;**20**(6):457–62. [杨树德，马建民，孙娟华，等. 还原型电化学极谱检测高效液相色谱法测定人血浆中青蒿酯和双氢青蒿素. 药学学报，1985, 20(6): 457–462.]
138. Zhou ZM, Huang YX, Guo XB, et al. HPLC with polarographic detection of artemisinin and its derivatives and application of the method to the pharmacokinetic study of artemether. *J Liq Chromatogr* 1988;**11**(5):1117–37.
139. van Agtmael MA, Butter JJ, Portier EJ, et al. Validation of an improved reversed-phase high-performance liquid chromatography assay with reductive electrochemical detection for the determination of artemisinin derivatives in man. *Ther Drug Monit* 1998;**20**(1):109–16.
140. van Agtmael MA, Shan CQ, Jiao XQ, et al. Multiple dose pharmacokinetics of artemether in Chinese patients with uncomplicated falciparum malaria. *Int J Ant A* 1999;**12**(2):151–8.
141. Zhao SS, Zeng MY. Application of precolumn reaction to high-performance liquid chromatography of qinghaosu in animal plasma. *Anal Chem* 1986;**58**(2):289–92.
142. Zhao SS. High-performance liquid chromatographic determination of artemisinine (qinghaosu) in human plasma and saliva. *Analyst* 1987;**112**(5):661–4.
143. Zhou ZM, Nie SQ, Sun XM, et al. Study on the correlation between artemisinin concentrations in saliva and blood. *Chin Pharmacol* 1990;**7**(2):49. [周钟鸣，聂淑琴，孙晓淼，等，青蒿素的唾液浓度和血药浓度的相关性研究，中国药理通讯，1990, 7(2):49.]
144. Zhou ZM, Zeng MY, Nie SQ, et al. Pharmacokinetics and pharmacodynamics of artemisinin in malaria patients. *Chin Pharmacol* 1991;**8**(2):38. [周钟鸣，曾美怡，聂淑琴，等. 青蒿素在恶性疟病人的药代动力学和药效动力学，中国药理通讯, 1991, 8(2):38.]
145. Zhou ZM. Pharmacokinetic studies of qinghaosu and its derivatives by HPLC method. In: *WHO/swg-chemal conference, 24–26 April 1989*. April 1989.
146. Lefèvre G, Thomsen MS. Clinical pharmacokinetics of artemether and lumefantrine (Riamet®). *Clin Drug Investig* 1999;**18**(6):467–80.
147. Ezzet F, van Vugt M, Nosten F, et al. Pharmacokinetics and pharmacodynamics of lumefantrine (benflumetol) in acute falciparum malaria. *Antimicrob Agents Chemother* 2000;**44**(3):697–704.
148. Huang JM, Yu YQ, Weng WY, et al. Determination of Artemether and its Metabolite dihydroartemisinin in Human Plasma byHPLC with atmospheric pressure chemical ionization mass spectrometry. *Chin J Clin Pharm* 2005;**14**(5):274–8. [黄建明，郁韵秋，翁伟宇，等. 高效液相质谱联用法测定人血浆中蒿甲醚及其活性代谢物双氢青蒿素的浓度. 中国临床药学杂志，2005，14(5): 274–278.]
149. Liu ZY, Li HY, Gao HZ, et al. Study on the pharmacokinetics of naphthoquine phosphate compound tablets in humans. Clinical data for new drug registration 1992 (held at the Institute of Microbiology and Epidemiology, Academy of Military Medical Sciences). [刘泽源，李海燕，高洪志，等. 复方磷酸萘酚喹片人体药代动力学研究. 新药注册：临床研究资料, 1992. (资料存在军事医学科学院微生物流行病研究所档案室).]

150. Gao HZ, Li HY, Xu SG, et al. Determination the concentration of artemisinin in healthy human plasma and study on its pharmacokinetics by LC-MS-MS. *Chin J Clin Pharmacol* 2009;**25**(2):138–40. [高洪志，李海燕，徐树光，等. 高效液相色谱-串联质谱法测定健康人体血浆中青蒿素浓度及其药代动力学研究.中国临床药理学杂志，2009, 25(2):138–140.]
151. Fang Y, Wang J, Pan ZH, et al. LC-MS/MS simultaneous determination of artemisinin and naphthoquine phoshate in human plasma. *Chin J Pharm Anal* 2009;**29**(8):1264–9. [方翼，王静，潘志恒，等. LC-MS/MS法测定人血浆中青蒿素和磷酸萘酚喹的浓度，药物分析杂志，2009，29(8):1264–1269.]
152. Qu HY, Gao HZ, Hao GT, et al. Single-dose safety, pharmacokinetics, and food effects studies of compound naphthoquine phosphate tablets in healthy volunteers. *J Clin Pharmacol* 2010;**50**(11):1310–8.
153. Liu P, Hou YN, Shan CQ, et al. HPLC/ms/ms in the determination of artemether and its metabolite dihydroartemisinin in the plasma of beagles. *Chin J Pharm Analysis* 2006;**26**(5):58. [刘萍，侯禹男，单成启，等. HPLC/MS/MS 法测定比格犬血浆中蒿甲醚及其代谢物双氢青蒿素的浓度. 药物分析杂志，2006，26(5): 58.]

Chapter 4

Artemisinin Derivatives and Analogues

Li Ying
Shanghai Institute of Materia Medica, Chinese Academy of Sciences, Shanghai, China

Chapter Outline

Beginning in 1974, clinical trials involving many different artemisinin preparations—tablets, oil solutions, and oil or water suspensions—were carried out in malaria-endemic areas across China. They produced a large volume of data on the drug's rapid action, low toxicity, and, most outstandingly, efficacy against drug-resistant falciparum malaria. The cure rate for tablets representing a total dose of 5 g of artemisinin administered over 3 days reached 90% against falciparum malaria. However, the 30-day recrudescence rate was around 50%.[1] In contrast, this rate was around 10% for oil or water suspensions, but it was found that the artemisinin crystals became larger in the injection media during standing, rendering it impossible to use. In February 1976, the Project 523 Leading Group instructed the Shanghai Institute of Materia Medica to modify the structure of artemisinin to address the problems of high recrudescence rates and the difficulties of formulating injections for critically ill patients because of the poor solubility of artemisinin. Therefore, new artemisinins were sought with higher efficacy, greater solubility, and a recrudescence rate of less than 10%.

Based on knowledge already at hand on the chemical structure of artemisinin and its reactions, studies examining the relationship between structure and antimalarial activity, as well as the synthesis and selection of new derivatives, were

Artemisinin-Based and Other Antimalarials. https://doi.org/10.1016/B978-0-12-813133-6.00004-4

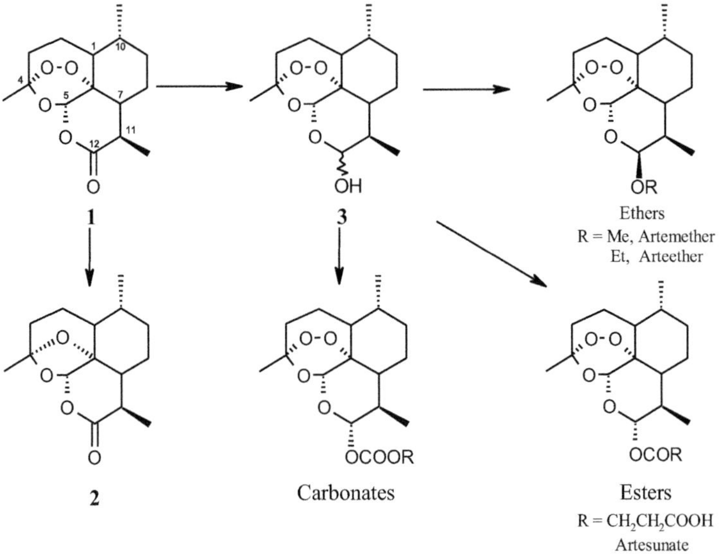

FIGURE 4.1 Reduction of artemisinin to produce the inert deoxy compound **2**, reduction of artemisinin to yield dihydroartemisinin (**3**), and conversion of (**3**) into ether, carboxylic, and carbonic acid ester derivatives.

carried out at the Institute. Artemisinin (compound **1** in Fig. 4.1) is a sesquiterpene lactone containing a peroxide group. Its five oxygen atoms are connected in an unusual way, namely —C—O—O—C—O—C—O—C=O or in a 1,2,4-trioxane. At the outset, a key focus of research was on the peroxide functionality. According to published methods,[2] artemisinin was shown to undergo catalytic hydrogenation to form deoxy artemisinin (**2**). Tests on rodent malaria indicated that compound **2**, which had lost the peroxide ring, had no antimalarial activity, indicating that the peroxide is essential for antimalarial activity. A few other simple peroxides were then screened, including the monoterpene ascaridole, which all displayed poor antimalarial activities. Therefore, the peroxide group and other features of the core structure of artemisinin had to be preserved.[3]

At the Shanghai Institute of Organic Chemistry, the chemical reactions of artemisinin were studied; it was found that many reactions led to products that had lost the peroxide group or other structural fragments. The only exception was observed when artemisinin was reduced with sodium borohydride at 5°C to generate dihydroartemisinin (**3**); the key peroxide group and core structure were retained. It was more effective against rodent malaria than artemisinin, but solubility was not improved and the stability was poor. Dihydroartemisinin has a relatively active hemiacetal structure, which was exploited for preparation of other new derivatives. It was found that dihydroartemisinin reacts with alcohol in acidic conditions to generate more stable acetals. Ester derivatives were also synthesized (Fig. 4.1).

Thus, in the second half of 1976, at the Shanghai Institute of Materia Medica, the synthesis of three types of artemisinin derivatives, ethers, carboxylic, and carbonic acid esters was carried out.[3] These compounds were screened against chloroquine-resistant *Plasmodium berghei* with artemisinin as a control, to determine their 90% suppressive dose (SD_{90}). The first 25 compounds tested were dissolved in vegetable oils and administered via intramuscular injection. Most showed higher efficacy against rodent malaria than artemisinin. The SD_{90} of artemisinin was 6.20 mg/kg; for dihydroartemisinin, artemether, and arteether, the SD_{90} values were 3.65, 1.02, and 1.95 mg/kg, respectively. This indicated that the efficacy of artemether was six times that of artemisinin and three times that of dihydroartemisinin. Many of the carboxylic and carbonic acid ester derivatives had 10 times the antimalarial activity of artemisinin.[4] Because of its high solubility in oil and its chemical stability, artemether was selected as a candidate for further research.

In 1977, artesunate was synthesized at the Guilin Pharmaceutical Factory.[5] Its intravenous ED_{50} and ED_{90} against sensitive *P. berghei* were 0.3 and 1.14 mg/kg, respectively. Following extensive research into their pharmacology, preparations, and clinical use, intravenous artesunate and intramuscular artemether injections were approved for production as new antimalarial drugs in 1987. They were the first to be evaluated according to China's newly established Provisions for New Drug Approval.

Thereafter, studies on artemisinin derivatives continued at the Shanghai Institute of Materia Medica; the focus was on chemical structure and its relationship to efficacy. From the early 1980s, many foreign research institutes also became involved in the investigation of artemisinin derivatives and analogues. The early work carried out in China was confirmed by the later foreign studies and was used as a benchmark for these studies. In the past 30 years, thousands of derivatives and analogues (including almost 1000 in China alone) have been synthesized and tested. Unfortunately, none of these new-generation derivatives can replace artemether or artesunate. Many reports[6] have appeared in the course[7] of this prolonged search.[8] The following is a brief introduction to the key findings from China and abroad, with a focus on recent research.

4.1 ARTEMISININ OIL-SOLUBLE DERIVATIVES

Early research in China centered mainly on these compounds and most of the types of derivatives mentioned above fell into this category. In initial studies at the Shanghai Institute of Materia Medica, it was found that carbonic and carboxylic acid ester derivatives had higher antimalarial efficacy than artemether, but they were less soluble in oil and less stable than the ether. Therefore, these derivatives did not pass the preliminary selection process.

During the visit to China in April 1985 by representatives of the Scientific Working Group of CHEMAL, the high oil solubility of artemether and its far greater stability compared to artesunate were demonstrated. These WHO experts thought that alcohol would be a by-product of arteether metabolism in vivo and that artemether metabolism would produce methanol. Arteether would thus be safer, and it was chosen for further development.[9] Although subsequent research showed that the neurotoxicity of arteether was higher than that of artemether, clinical evidence of neurotoxicity in artemisinins had not emerged in the 1980s. Therefore, arteether was introduced in 2000.[10] The clinical dose of arteether-oil injections was double that of artemether-oil injections, confirming a 1980 report by Gu Haoming that the SD_{90} of arteether was double that of artemether against *P. berghei*.[4]

Li Ying's group at the Shanghai Institute of Materia Medica subsequently extended its focus from aliphatic to aromatic ethers. Acetyl or trifluoroacetylated derivatives of dihydroartemisinin were used as starting material. In the presence of trifluoroacetic acid or $BF_3 \cdot Et_2O$, reaction with various phenols produced 12-β aryl esters of dihydroartemisinin 4 (Fig. 4.2). The efficacy of most of these compounds against rodent malaria was higher than that of artemisinin but was lower than that of artemether.[11]

Likewise, acetyl and trifluoroacetyl dihydroartemisinin derivatives were found to react with aromatic amines or heterocyclic compounds such as triazole and benzotriazole[12] in the presence of an acid catalyst to yield derivatives containing nitrogen atoms (Fig. 4.3).[13] These also had higher antimalarial activities than artemisinin.[14]

Haynes et al. reported the synthesis of artemisone (BAY 44 9585) employing a new route to prepare alkylamino artemisinins (Fig. 4.4).[15] Artemisone was shown to possess appreciably higher antimalarial activity than artesunate both in vitro and in vivo and, unlike artesunate, is hydrolytically stable at neutral pH.[15] It is the first artemisinin prepared outside China, which has gone into Phase I and Phase II clinical trials.[16] When used with a final dose of mefloquine, it was curative at one-third of the total dose of the comparator drug pair

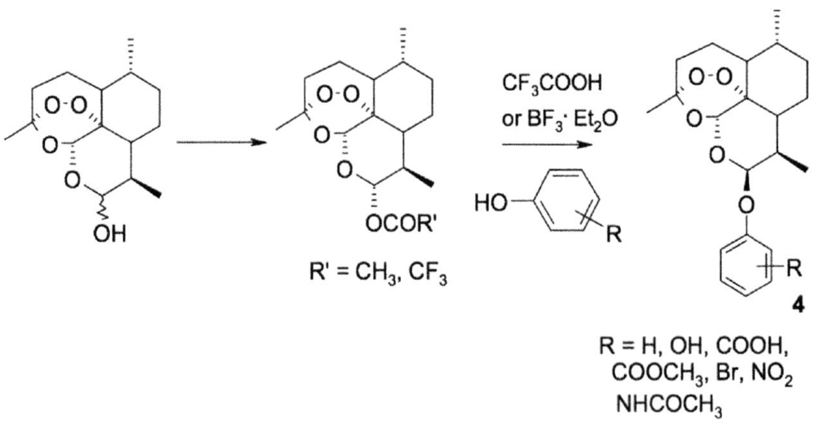

OH

OCOR'

R' = CH_3, CF_3

CF_3COOH or $BF_3 \cdot Et_2O$

HO

R

4

R = H, OH, COOH, $COOCH_3$, Br, NO_2 $NHCOCH_3$

FIGURE 4.2 Synthesis of 12-β aryl esters of dihydroartemisinin.

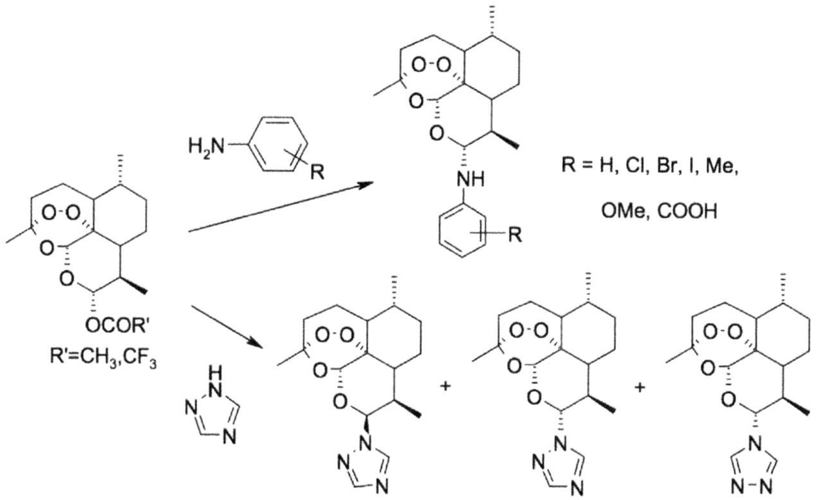

FIGURE 4.3 Synthesis of artemisinin derivatives with nitrogen atoms.

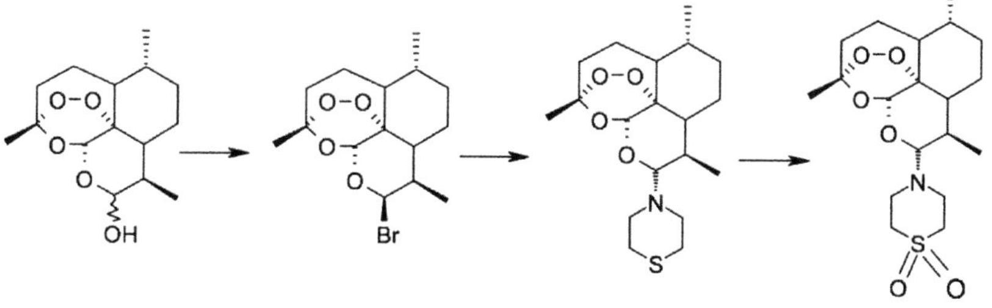

FIGURE 4.4 Synthesis of artemisone.

artesunate–mefloquine and elicited appreciably low parasite clearance times.[17] Significantly, when used in combination with a low dose of chloroquine, it prevented recrudescence and improved cure rates against cerebral malaria in a rodent model[18] and therefore shows promise for treatment of cerebral malaria.

In the 1990s, Li Ying's group commenced preparation of 12α-aryldeoxoartemisinins (**5**) by reacting acetyl dihydroartemisinin and aromatic hydrocarbons with an electron-donating group in the presence of boron trifluoride etherate $BF_3 \cdot Et_2O$.[19] Wu Yulin et al. at the Shanghai Institute of Organic Chemistry then discovered that this reaction also produced a by-product 11α-epimers (**6**) (Fig. 4.5). The antimalarial activities of the 12α-aryldeoxoartemisinins were higher than that of artemisinin. Some, such as 12α-(-2′-hydroxy) naphthyldeoxoartemisinin, were more effective than artemether, but their 11α-epimers performed poorly.[20]

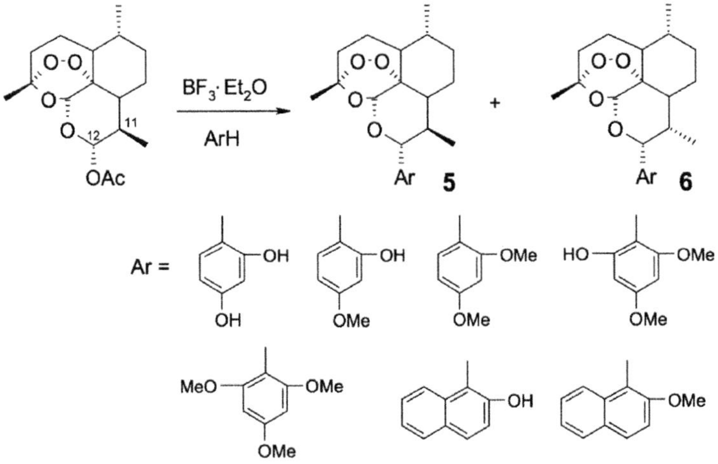

FIGURE 4.5 Synthesis of 12α-aryldeoxoartemisinins.

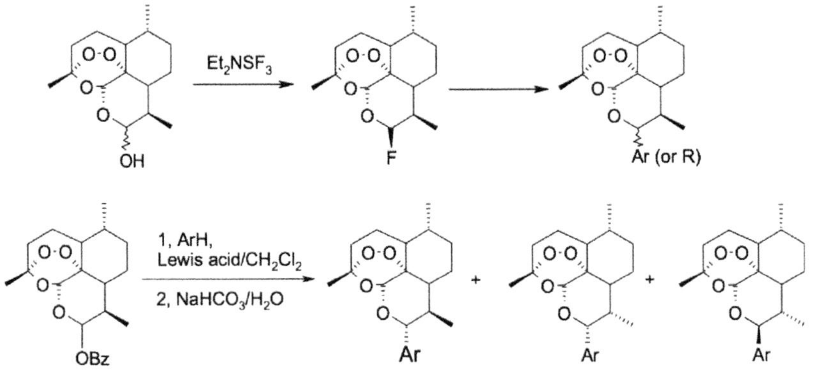

FIGURE 4.6 Synthesis of 12-aryldeoxoartemisinins.

Later, overseas researchers[21,22] also prepared 12-aryldeoxoartemisinins using different processes bases on C-glycosidation technology (Fig. 4.6).[23] The group led by Haynes first converted dihydroartemisinin into a fluoro compound, which was then easily transformed into arylated derivatives.[24,25]

4.2 ARTEMISININ WATER-SOLUBLE DERIVATIVES

Sodium artesunate (**7**), synthesized at the Guilin Pharmaceutical Factory,[5] was the first water-soluble artemisinin derivative, which could be used for intravenous or intramuscular injections to treat severely ill or unconscious malaria patients.[26] However, its aqueous solution was unstable. If left standing for just a few moments, a precipitate of dihydroartemisinin would form (Fig. 4.7). Therefore, the sodium bicarbonate aqueous solution had to be mixed with artesunate to form aqueous sodium artesunate immediately before injection.

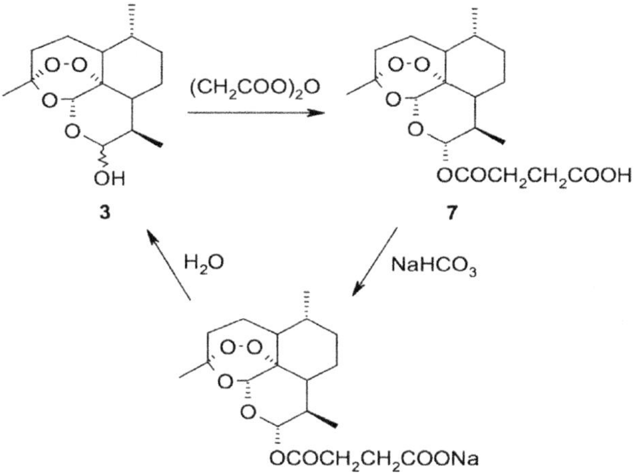

FIGURE 4.7 Preparation and hydrolysis of sodium artesunate.

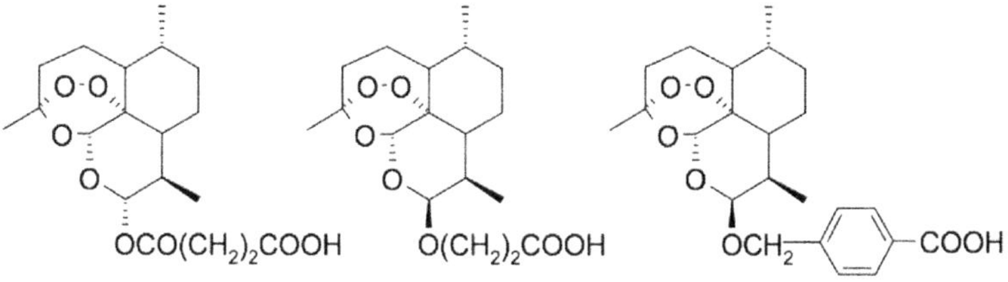

FIGURE 4.8 Artesunate and two artemisinin derivatives with a carboxylic acid group.

To find more stable and effective water-soluble artemisinin derivatives, scientists in China and abroad synthesized a wide variety of new compounds. The hydrolytic instability of artesunate was because of the lability of its ester linkage attached to C-12. Hence, it was proposed that replacing it with an ether or C—C linkage would resolve this problem (Fig. 4.8). Li's group synthesized compound **8**, but its efficacy was lower than that of artesunate.[27] Studies on artelinic acid (**9**) had previously been prioritized by the Walter Reed Army Institute of Research.[28] It was stable and had a far longer half-life (1.5–3 h) than artesunate (20–30 min).[29] In animal tests, however, its efficacy was lower than that of artesunate.[30]

Haynes et al.[31] and Jung et al.[32] reported on analogues of artelinic acid (**10**) and artesunate (**11**) with C—C linkages at 12-C. Both had been synthesized from artemisinic acid with the goal of removing the easily hydrolyzed ester linkage (Fig. 4.9). Preliminary tests showed that compound **10** had greater antimalarial activity than arteether.

Most antimalarial drugs contain basic nitrogen atoms either embedded in an aromatic core or within alkyl side chains. Thus, salts are easily prepared from

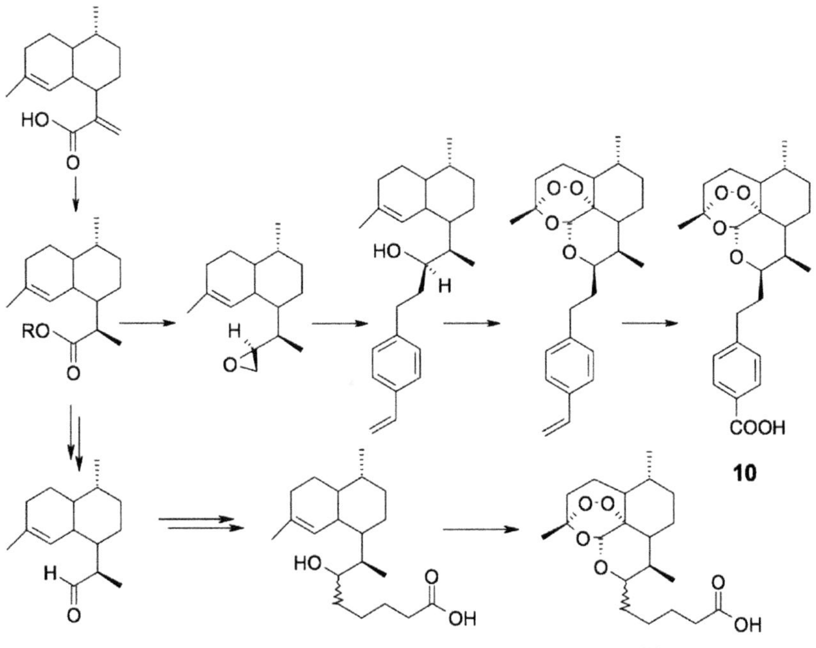

10

11

FIGURE 4.9 Synthesis of analogues of artesunate and artelinic acid.

an inorganic acid and are frequently used in injections or oral formulations in a clinical setting. Artemisinin is a neutral sesquiterpene lactone. If basic groups are to be introduced, its stability and antimalarial activities could be affected. Li's group, together with the antimalarial laboratory at the Academy of Military Medical Sciences' Institute of Microbiology and Epidemiology, prepared and studied four types of artemisinin derivatives (**12–15**) bearing amino groups (Fig. 4.10).[27,33]

Salts obtained from these compounds and organic acids (succinic acid or oxalic acid) were water-soluble and had relatively better stability. Salts of compounds **13** and **14** were less effective against rodent malaria than artesunate, but the salt of compound **12** was four to five times more active. Thus, it was used in a 7-day test against simian malaria. Two oral doses, 10 and 3.16 mg/kg, were compared. At the lower dose, compound **12** did not achieve the same degree of parasite clearance as with artesunate, and recrudescence was high. It was therefore discarded. It was thought that these water-soluble compounds were rapidly excreted, whereas the artesunate metabolite, dihydroartemisinin, could remain in vivo and produce sustained antimalarial effects.[27]

Mannich bases are commonly associated with antimalarial drugs, as exemplified by amodiaquine. Thus, compound **15** was prepared, and like compound **12**, it was more effective against rodent malaria but not as good as artesunate against simian malaria.[33]

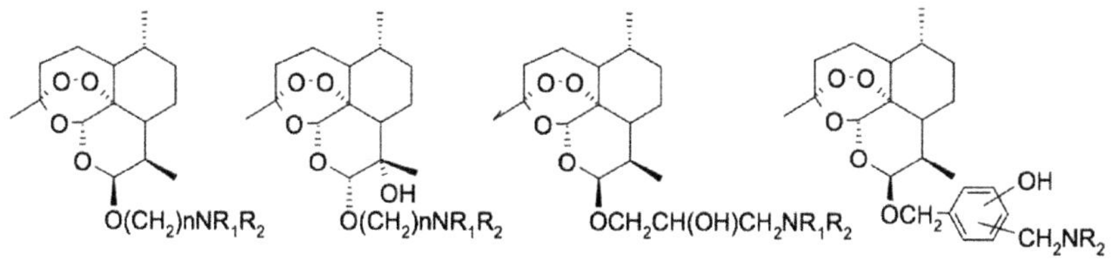

FIGURE 4.10 Artemisinin derivatives with amino groups.

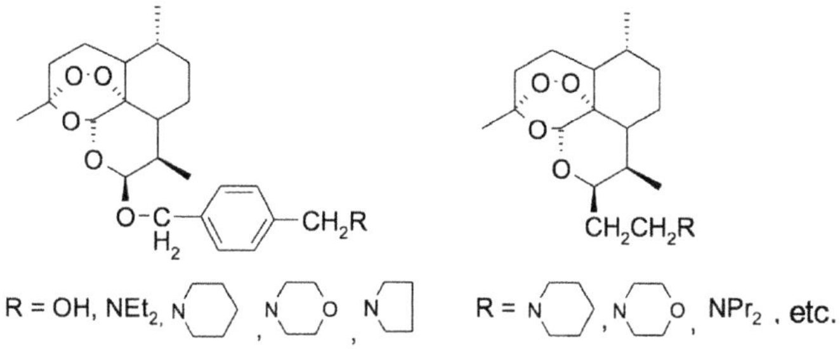

FIGURE 4.11 Other alkaline artemisinin derivatives.

Recent research has discovered that some water-soluble artemisinins have relatively strong immunosuppressive effects. Their efficacy was confirmed in in vitro and in vivo tests in mice with dinitrofluorobenzene-induced delayed hypersensitivity, sheep red blood cell-induced antibody response, experimental autoimmune encephalomyelitis, polyarthritis, and systemic lupus erythematosus (SLE). One such compound, SM934 (**12**, $n=2$, $NR_1R_2=NH_2$, maleate), was easy to prepare, readily water-soluble, had high bioavailability and low toxicity, and could raise the survival rate of MRL-1pr/1pr mice. Therefore, it was selected as a candidate for further study as a treatment for SLE.[34] Preclinical studies on SM934 have been completed and it was approved for clinic trial.

Research on basic artemisinin derivatives was also conducted[35] outside China (Fig. 4.11).[36] O'Neill et al. proposed that a basic group could increase the content of the artemisinin derivative in the acidic parasitic food vacuole, resulting in higher efficacy.[37] Compounds **16** ($R=NEt_2$) and nine other derivatives in **17** showed higher in vitro activities compared to artemisinin.

4.3 ARTEMISININ DIMERIC AND TRIMERIC DERIVATIVES

In acidic conditions or with a dehydrating agent, dihydroartemisinin forms the anhydro product **18** and products **19** and **20**, which arise via intermolecular condensation reactions.[38] Compound **18** has lower antimalarial activity, but

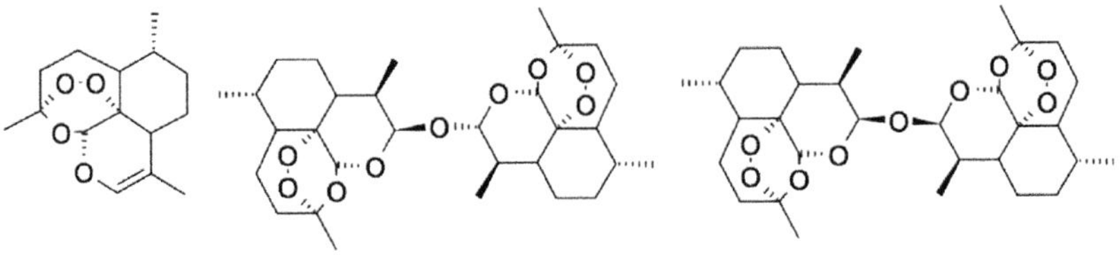

18 **19** **20**

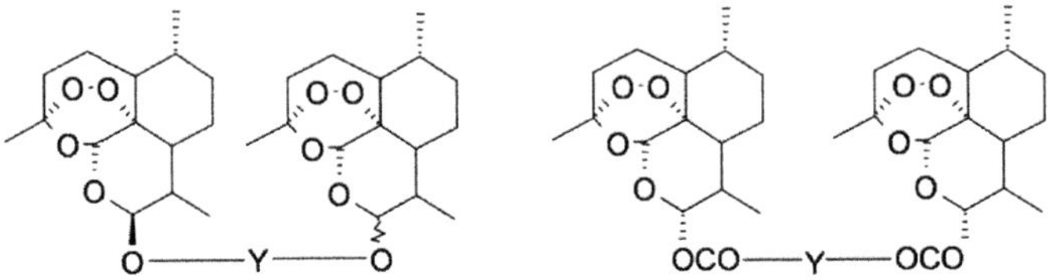

21
Y = $(CH_2)n$, n = 2 ~ 6

22
Y = $(CH_2)n$, n = 3 ~ 5, CH=CH, CH=C(CH3)

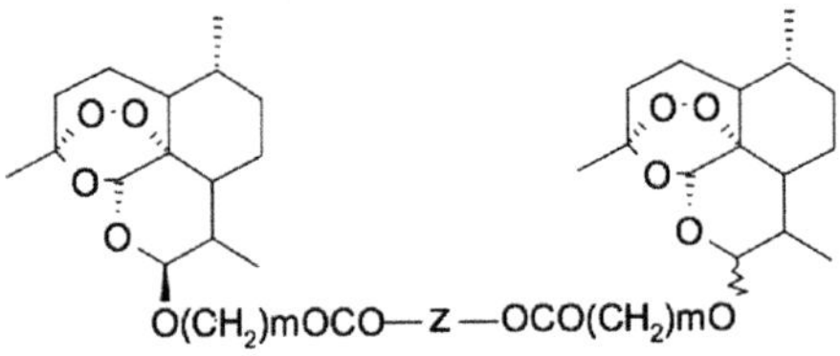

$O(CH_2)mOCO$—Z—$OCO(CH_2)mO$

23
m = 2 ~ 3, Z = $(CH_2)n$, n = 3 ~ 5, CH=CH, phenyl etc.

FIGURE 4.12 Anhydro product **18** and products arising by intermolecular condensation reactions of dihydroartemisinin.

the activity of compounds **19** and **20** are superior to dihydroartemisinin.[39] Li's group synthesized a series of ether and ester artemisinin dimers (**21–23**) using diprotic acid or glycol (Fig. 4.12).[40] These compounds were generally more effective than artemisinin itself, although not better than artemether.[41]

Recently, Li's group prepared a series of artemisinin dimers containing nitrogen atoms (Fig. 4.13).[42] Some of them were found to have antineoplastic activity in in vitro and animal tests.[43] At Ruijin Hospital, which is affiliated to Shanghai Jiao Tong University's School of Medicine, it was discovered that one compound,[44] SM 1044,[45] had good suppressive effects on leukemia[46] in several animal models.[47]

In the past 10 years, many artemisinin dimers[48] comprising different linking groups have been produced by several overseas research groups (Fig. 4.14). Several of these showed higher antimalarial activity than artemisinin in vitro. Some[49] also have strong[50] and selective[51] suppressive[52] effects on the growth of human cancer cells.[53–56]

Artemisinin trimers and tetramers also have been synthesized.[57] Some show greater in vitro efficacy than artemisinin and have antineoplastic[50] effects.[56]

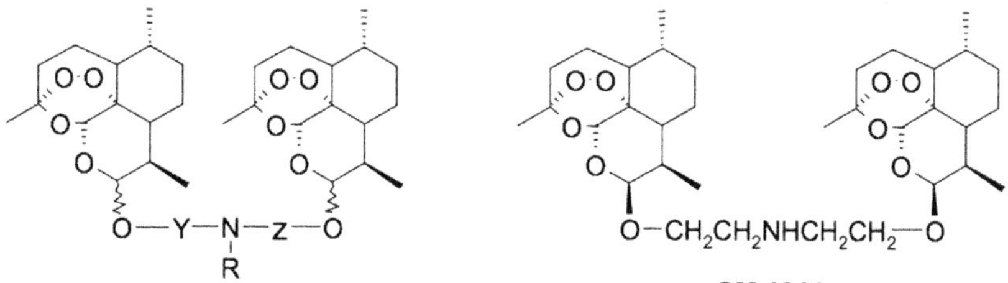

SM 1044

FIGURE 4.13 Artemisinin dimers containing nitrogen atoms.

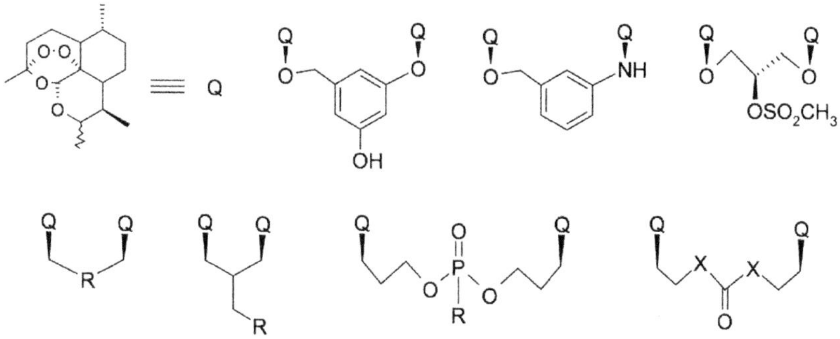

FIGURE 4.14 New dimeric and trimeric derivatives of artemisinin.

4.4 ARTEMISININ SIMPLIFIED ANALOGUES

It is clear that the peroxide within the 1,2,4-trioxane moiety in the artemisinin molecule is responsible for its activity. Posner et al. synthesized a series of 1,2,4-trioxanes, in which compounds **24** and **25** (Fig. 4.15) had higher efficacy than artemisinin.[55] Jefford et al. prepared Fenozan B07, which was also superior to artemisinin but somewhat inferior to artemether.[58] Recently, Wu Yikang et al. at the Shanghai Institute of Organic Chemistry also prepared several such compounds **26–28**. Some which contained two trioxane components, for example, compound **28**, showed higher antimalarial activity in vitro than artesunate.[59]

In the 1970s, Chinese chemists discovered yingzhaosu A, also derived from a plant, which is a sesquiterpene with a peroxide group (Fig. 4.16).[60] It proved to be active against malaria in animal tests but was unstable. Its plant source was also rare. Hence, research on this compound was halted. Inspired by this, however, chemists from the pharmaceutical company Roche synthesized yingzhaosu A analogues, of which arteflene (Ro 42-1611)[61] was selected for further study. However, because of the difficulty of synthesizing this compound, it could not be produced in large quantities and research stalled.[62]

Although the peroxide group is essential for the antimalarial activity of artemisinin, peroxides with simplified structures in many cases tend to show feebler

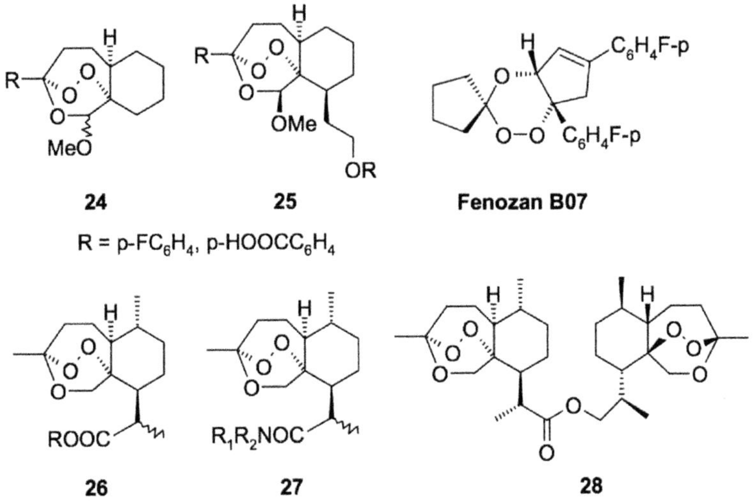

24 25 Fenozan B07

R = p-FC_6H_4, p-$HOOCC_6H_4$

26 27 28

FIGURE 4.15 Fully synthetic 1,2,4-trioxanes.

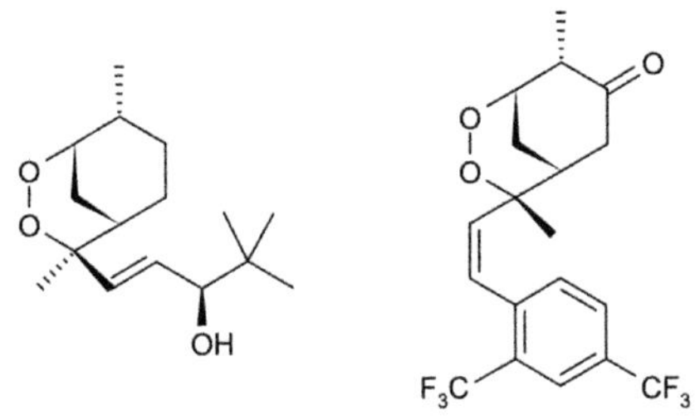

Yinzhaosu A Arteflene

FIGURE 4.16 Structures of yingzhaosu A and arteflene.

antimalarial activities effects. However, 1,2,4,5-tetraoxanes, with two peroxide groups, are relatively easy to prepare and display good antimalarial activities. In general, they form when a ketone is treated with hydrogen peroxide under acidic conditions. Several reports have confirmed the antimalarial activity of these compounds.[48] Some examples are presented in Fig. 4.17.

Compound WR 14899929 (**29**, R-Me)[63] was synthesized in a single-step process from p-methylcyclohexanone and hydrogen peroxide. Costs were low. It showed antimalarial activity in vitro and possessed activity comparable with that of artemisinin against rodent malaria.[62]

It was found that tetraoxanes could be replaced by ozonides. These are also peroxides—this was a fortuitous discovery by Vennerstrom, who observed that an ozonide formed as a by-product in his earlier tetraoxane work, and it was found to be active. Vennerstrom et al. synthesized a large number of dispiro ozonides.[64] Among these, OZ277 and OZ439[65] were selected for further study (Fig. 4.18).

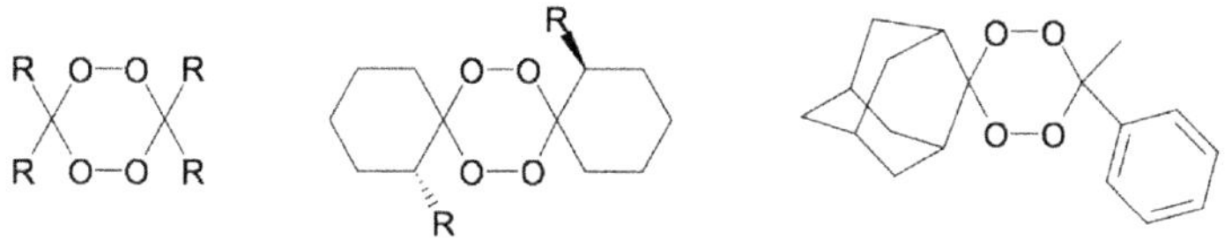

29 (R=Me, WR 148999)

FIGURE 4.17 Examples of 1,2,4,5-tetraoxanes.

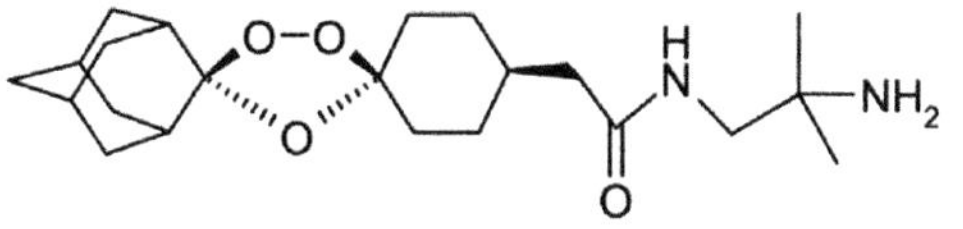

OZ277

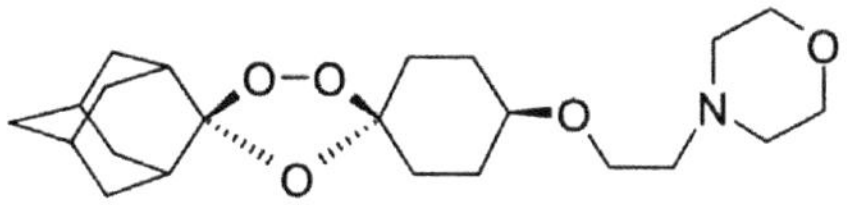

OZ439

FIGURE 4.18 Structures of OZ277 and OZ439.

Clinical trials on OZ277, which has an adamantyl group, had been conducted from 2003 to 2007, with unsatisfactory results.[66] Ranbaxy suggested that because the cost of producing OZ277 was much lower than that of artemisinin, it could be used in developing countries. Thus, they developed a fixed-combination therapy, Synriam, comprising OZ277 (arterolane) and piperaquine phosphate. Clinical trials of a 3-day, three-tablet regimen achieved a 95% cure rate.[67,68]

In 2008, Vennerstrom et al. chose to work on OZ439. A single 20 mg/kg dose could cure normal-strain rodent malaria. Phase I clinical trials (single administration, highest dose 1600 mg) are complete; however, Phase II trials[65] at this stage seem to indicate that a single dose did not effect a cure.

4.5 TRIOXAQUINES

Dechy-Cabaret et al. synthesized various compounds they called trioxaquines, a combination of quinoline and 1,2,4-trioxanes (Fig. 4.19). The idea was that such compounds would exercise dual mechanisms of action in vivo and achieve high levels of antimalarial efficacy.[69]

In 2008, they reported that one such compound, PA 1103/SAR 116242, had been selected out of over 100 others as a candidate for development[70] but was not further developed.

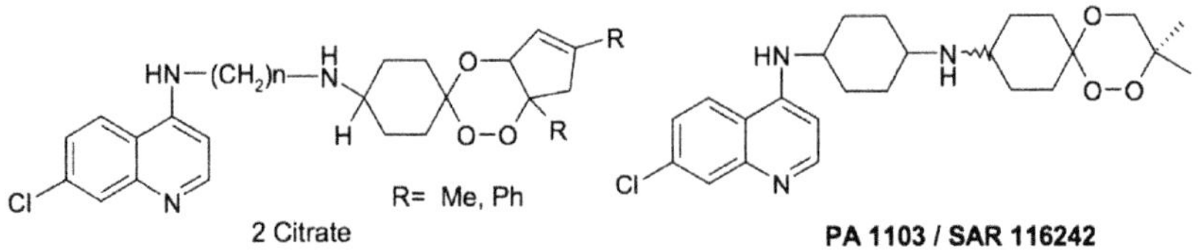

FIGURE 4.19 Structures of some trioxaquines.

4.6 ARTEMISININS THERMAL STABILITY AND PURITY ANALYSIS

Compared to other peroxides, artemisinin is relatively stable; even so, the peroxide is unstable relative to other nonperoxidic antimalarial drugs. Therefore, artemisinins must be kept away from reducing agents, acids, and bases and stored in dark and cold conditions.

In Section 3.5, we discussed the decomposition of artemisinin in acidic or alkaline conditions, its reactions with reducing agents, and its interactions with ferrous ions, which are related to its mechanism of action. Here, research into the thermal stability of artemisinin, dihydroartemisinin, and artesunate will be examined.

In 1985, Lin et al.[71] and Luo et al.[72] reported that when artemisinin was exposed to a neutral solvent (e.g., ethanol, isopropyl alcohol, or toluene), heated, and refluxed for several days, no changes were observed. However, after 5 h of reflux with tetralin at 200°C or during 22 h reflux with xylene at 180–190°C, many decomposition products were formed. The structures of four of these have been determined (Fig. 4.20).

In 1986, Lin et al. reported that after being heated at 190°C for 3 min, dihydroartemisinin decomposed to give the products (Fig. 4.21) including **31** and **33**.[73]

In 2007, Haynes et al. heated dihydroartemisinin at 100°C for 14 h under nitrogen gas. This yielded a mixture of dihydroartemisinin and its decomposition products. If dihydroartemisinin was dissolved in toluene and heated at 100°C for 24 h, it decomposed completely, the main products of which were **30** and **31**.[74]

This study also included the thermal decomposition of artesunate. Artesunate was heated under nitrogen gas at 100°C for 39 h. Of its decomposition products, eight were isolated and identified (Fig. 4.22). These include formyl dihydroartemisinin **34** (13%), 12β-artesunate **35** (10.2%), and the dimers **36** (4.0%), **37** (4.4%), and **38** (1.7%).[74]

Artesunate in aqueous solution at pH 1.2 has a half-life of 26 min and hydrolyzes relatively quickly to form dihydroartemisinin. At pH 7.4 and 23°C, the half-life is 10 h.[74] When reacted with 5 mol/L hydrochloric acid–ethanol (1:1) at room temperature for 1.5 h, artesunate decomposes to form dihydroartemisinin (**3**, 12%), (**31**, 30%), (**33**, 1.2%), 12β-arteether (34%), and 12α-arteether (15%) (Fig. 4.23).[74]

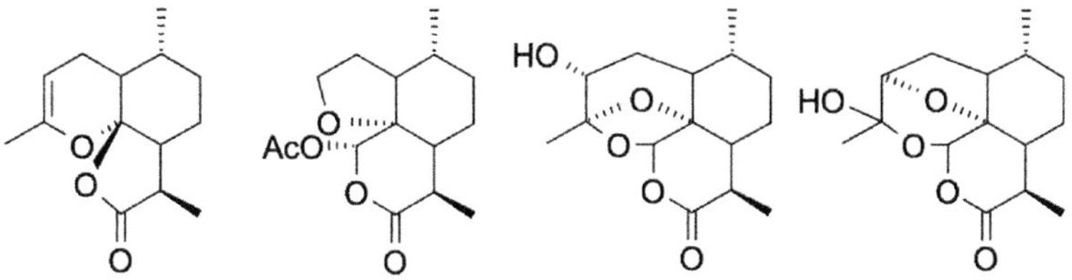

FIGURE 4.20 Thermal decomposition products of artemisinin.

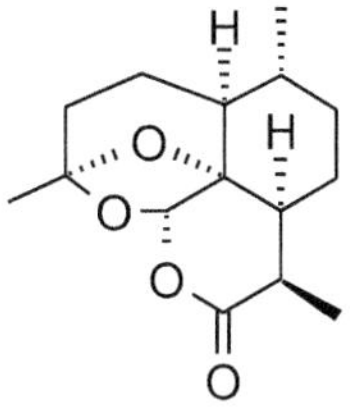

2 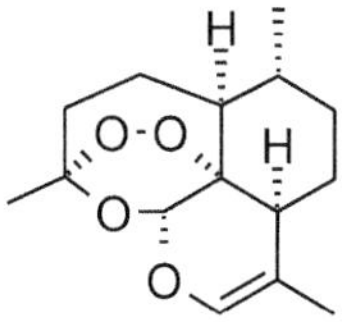 18 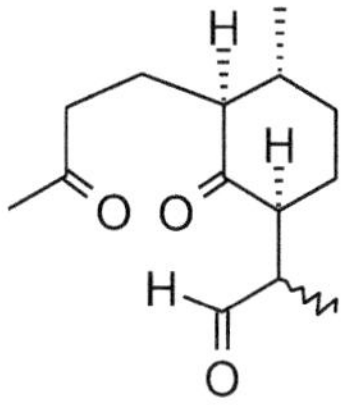 30

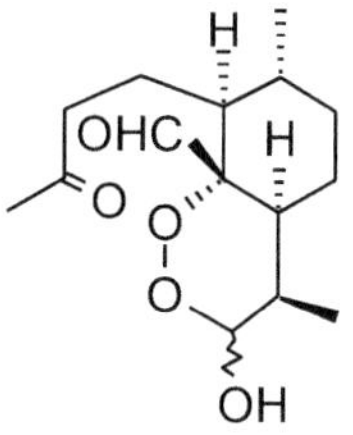

31 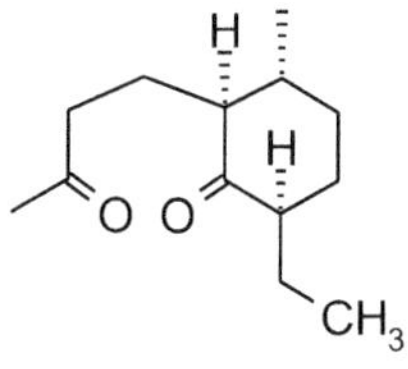 32 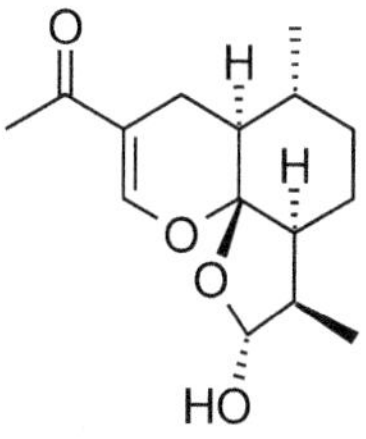 33

FIGURE 4.21 Thermal decomposition products of dihydroartemisinin.

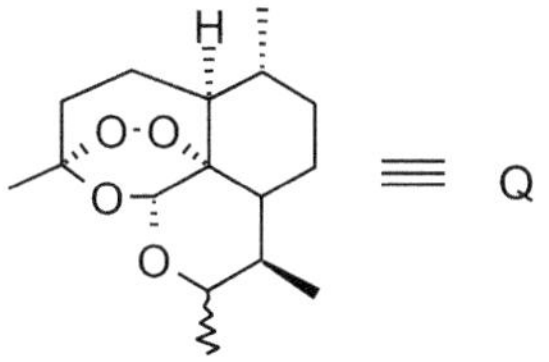

≡ Q

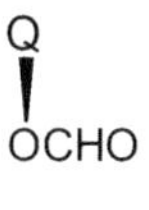

Q–OCHO
34

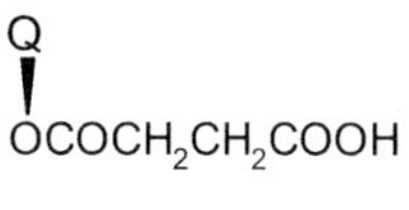

Q–$OCOCH_2CH_2COOH$
35

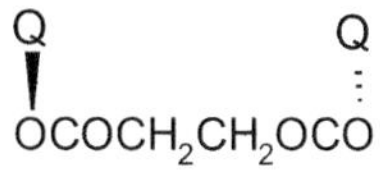

Q–$OCOCH_2CH_2OCO$–Q
36

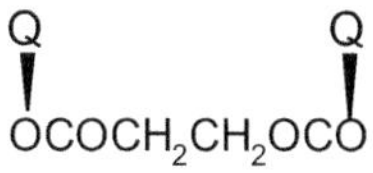

Q–$OCOCH_2CH_2OCO$–Q
37

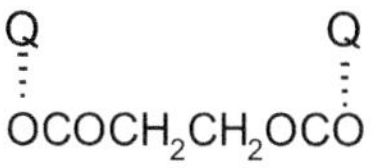

Q–$OCOCH_2CH_2OCO$–Q
38

FIGURE 4.22 Thermal decomposition products of artesunate.

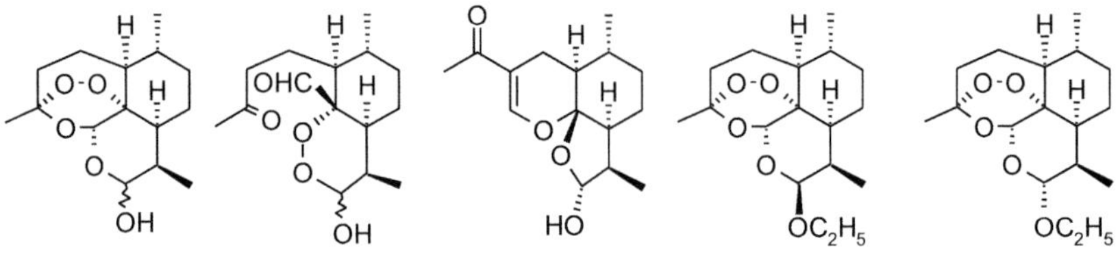

FIGURE 4.23 Decomposition products of artesunate formed in hydrochloric acid–ethanol solution.

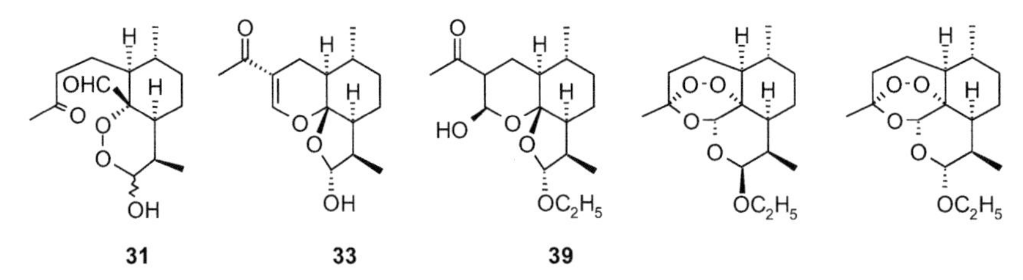

FIGURE 4.24 Decomposition products of dihydroartemisinin formed in hydrochloric acid–ethanol solution.

Dihydroartemisinin in hydrochloric acid–ethanol solution at room temperature during 2h gave a variety of products including **31** (48%), **33** (9%), **39** (4.4%), 12β-arteether (9.4%), and 12α-arteether (16%) (Fig. 4.24).[74]

As the overall knowledge of the stability of artemisinins grows, production facilities are being developed to store these drugs in cold and dark conditions to prolong their shelf lives. At the Kunming Pharmaceutical Corporation, for example, the active pharmaceutical ingredient artemether and the artemether formulations it manufactures are stored at 5°C. This can extend shelf life to 4–5 years.

ACKNOWLEGMENT

We would like to thank Richard K. Haynes for his review and suggestions to this chapter.

REFERENCES

1. China Cooperative Research Group on Qinghaosu and its Derivatives as Antimalarials. Clinical studies on the treatment of malaria with qinghaosu and its derivatives. *J Tradit Chin Med* 1982;**2**(1):45–50.
2. Liu JM, Ni MY, Fan JF, et al. Structure and reactions of arteannuin. *Acta Chim Sin* 1979;**37**(2):129–40. [刘静明，倪慕云，樊菊芬，等. 青蒿素的结构和反应. 化学学报, 1979, 37(2): 129–140.]
3. Li Y, Yu PL, Chen YX, et al. Studies on artemisinin analogs I. The synthesis of ethers, and carboxylic and carbonic acid esters of dihydroartemisinin. *Acta Pharm Sin* 1981;**16**(6):429–39. [李英，虞佩琳，陈一心，等. 青蒿素类似物的研究 I. 还原青蒿素的醚类、羧酸酯类及碳酸酯类的合成.药学学报, 1981, 16(6): 429–439.]

4. Gu HM, Lu BF, Qu ZQ. Antimalarial activity of artemsinin derivatives against chloroquine-resistant *Plasmodium berghei*. *Acta Pharmacol Sin* 1980;**1**(1):48–50. [顾浩明，吕宝芬，瞿志强. 青蒿素衍生物对伯氏疟原虫抗氯喹株的抗疟活性. 中国药理学报，1980，1(1): 48–50.]
5. Liu X. Study on artemisinin derivatives. *Chin Pharm Bull* 1980;**15**:183. [刘旭. 青蒿素衍生物的研究. 药学通报, 1980, 15: 183.]
6. O'Neill PM, Posner GH. A medicinal chemistry perspective on artemisinin and related endoperoxides. *J Med Chem* 2004;**47**(12):2945–64.
7. Li Y, Huang H, Wu YL. Qinghaosu (artemisinin) – a fantastic antimalarial drug from a traditional Chinese medicine. In: Liang XT, Fang WS, editors. *Medicinal chemistry of bioactive natural products*. New Jersey: Wiley; 2006. p. 183–256.
8. Li Y, Wu YL. A golden phoenix arising from the herbal nest – a review and reflection on the study of antimalarial drug qinghaosu. *Front Chem China* 2010;**5**(4):375–422.
9. Buchs P, Brossi A. *Synthesis of artemisininelactol derivatives* US Patent 5011951A, published. April 30, 1991.
10. Special Programme for Research, Training in Tropical Diseases (TDR). *Drugs brought to registration*. 2016. http://www.who.int/tdr/about/products/drugs_brought_registration/en/.
11. Liang J, Li Y. Synthesis of aryl ether derivatives of artemisinin. *Chin J Med Chem* 1996;**6**(1): 22–5. [梁洁，李英. 青蒿素芳香醚类衍生物的合成. 中国药物化学杂志, 1996, 6(1): 22–25.]
12. Yang YH, Li Y, Shi YL, et al. Artemisinin derivatives with 12-aniline substitution: synthesis and antimalarial activity. *Bioorg Med Chem Lett* 1995;**5**(16):1791–4.
13. Li Y, Liao XB. *Artemisinin derivatives containing nitrogen heterocyclic groups and their preparation* China Patent CN 1296009 A, published. May 23, 2001. [李英，廖细斌. 含氮杂环基青蒿素衍生物及其制备方法：中国，公开号CN1296009A.公开日2001-5-23.]
14. WHO screening report. (unpublished data). [WHO抗疟筛选报告.]
15. Haynes RK, Fugmann B, Stetter J, et al. Artemisone – a highly active antimalarial drug of the artemisinin class. *Angew Chem Int Ed Engl* 2006;**45**(13):2082–8.
16. Nagelschmitz J, Voith B, Wensing G, et al. First assessment in humans of the safety, tolerability, pharmacokinetics, and ex vivo pharmacodynamic antimalarial activity of the new artemisinin derivative artemisone. *Antimicrob Agents Chemother* 2008;**52**(9):3085–91.
17. Krudsood S, Wilairatana P, Chalermrut K, et al. Artemifone, a new anti-malarial artemisinin derivative: open pilot trial to investigate the antiparasitic activity of bay 44-9585 in patients with uncomplicated *P. Falciparum* malaria. In: *Abstracts from the XVI international congress for tropical medicine and malaria: medicine and health in the tropics. Marseille, France*. September 11–15, 2005. p. 142. P054.
18. Waknine-Grinberg JH, Hunt N, Bentura-Marciano A, et al. Artemisone effective against murine cerebral malaria". *Malar J* 2010;**9**:227.
19. Li Y, Yang YH, Liang J, et al. *Artemisinin derivatives containing phenyl and heterocyclic groups and their preparation* China Patent CN 1049435 C, published. February 16, 2000. [李英，杨永华，梁洁，等. 含苯基和杂环基的青蒿素衍生物及其制备方法：中国，授权公告号CN 1049435C.]
20. Wang DY, Wu YK, Wu YL, et al. Synthesis, Iron(II)-induced cleavage and *in vivo a*ntimalarial efficacy of 10-(2-hydroxy-1-naphthyl)-deoxoqinghaosu (-deoxoartemisinin). *J Chem Soc Perkin Trans* 1999;**1**(13):1827–31.
21. Posner GH, Parker MH, Northrop J, et al. Orally active, hydrolytically stable, semisynthetic, antimalarial trioxanes in the artemisinin family. *J Med Chem* 1999;**42**(2):300–4.
22. Woo SH, Parker MH, Ploypradith P, et al. Direct conversion of pyranose anomeric OH→F→R in the artemisinin family of antimalarial trioxanes. *Tetradedron Lett* 1998;**39**(12):1533–6.

23. Haynes RK, Chan HW, Cheung MK, et al. Stereoselective preparation of 10α- and 10β-aryl derivatives of dihydroartemisinin. *Eur J Org Chem* 2003;**11**:2098–114.
24. Haynes RK, Lam WL, Chan HW, et al. *C-10 halogen-, amino- and carbon-substituted derivatives of artemisinin for treatment of malaria, coccidiosis, and neosporosis* European Patent PCT/GB99/02267, published. July 1999.
25. Haynes RK, Lam WL, Hsiao WWL, et al. *Artemisinin derivatives bearing DNA intercalating and binding groups displaying anticancer activity* European Patent PCT/GB99/02276, published. July 1999.
26. Liu X. *The preparation of an antimalarial drug, artesunate* China Patent CN 85100781 B, published. July 6, 1988. [刘旭. 抗疟药物青蒿琥酯的制备方法：中国，授权公告号CN85100781B.]
27. Li Y, Zhu YM, Jiang HJ, et al. Synthesis and antimalarial activity of artemisinin derivatives containing an amino group. *J Med Chem* 2000;**43**(8):1635–40.
28. Lin AJ, Klayman DL, Milhous WK. Antimalarial activity of new water-soluble dihydroartemisinin derivatives. *J Med Chem* 1987;**30**(11):2147–50.
29. Lin AJ, Miller RE. Antimalarial activity of new dihydroartemisinin derivatives. 6. Alpha-alkylbenzylic ethers. *J Med Chem* 1995;**38**(5):764–70.
30. Li Y, Zhu YM, Jiang HJ, et al. Antimalarial activity of a new type of water-soluble artemisinin derivative. In: Shen JX, editor. *Antimalarial drug development in China*. Beijing: National Institutes of Pharmaceutical Research and Development; 1990. p. 35.
31. Haynes RK, Vonwiller SC. Efficient preparation of novel qinghaosu (artemisinin) derivatives: conversion of qinghaosu (artemisinic) acid into deoxoqinghaosu derivatives and 5-Carba-4-Deoxoartesunic acid. *Synlett* 1992;**6**:481–3.
32. Jung M, Lee K, Kendrick H, et al. Synthesis, stability, and antimalarial activity of new hydrolytically stable and water-soluble (+)-Deoxoartelinic acid. *J Med Chem* 2002;**45**(22):4940–4.
33. Li Y, Yang ZS, Zhang H, et al. Artemisinin derivatives bearing mannich base group: synthesis and antimalarial activity. *Bioorg Med Chem* 2003;**11**(2):4363–8.
34. Li Y, Zuo JP, Yang ZS, et al. *Water-soluble artemisinin derivatives, their preparation, pharmaceutical formulations and uses* China Patent CN 101223177 B, published. November 16, 2011. [李英，左建平，杨忠顺，等. 水溶性青蒿素衍生物、制法、药物组合物及用途：中国，授权公告号CN101223177B.]
35. Hindley S, Ward SA, Storr RC, et al. Mechanism-based design of parasite-targeted artemisinin derivatives: synthesis and antimalarial activity of new diamine containing analogues. *J Med Chem* 2002;**45**(5):1052–63.
36. Jung M, Bae J. An efficient synthesis of novel analogs of water-soluble and hydrolytically stable deoxoartemisinin. *Heterocycles* 2000;**53**(2):261–4.
37. O'Neill PM, Bishop LP, Storr RC, et al. Mechanism-based design of parasite-targeted artemisinin derivatives: synthesis and antimalarial activity of benzylamino and alkylamino ether analogues of artemisinin. *J Med Chem* 1996;**39**(22):4511–4.
38. China Cooperative Research Group on Qinghaosu and its Derivatives as Antimalarials. The chemistry and synthesis of qinghaosu derivatives. *J Tradit Chin Med* 1982;**2**(1):9–16.
39. Galal AM, Ahmad MS, El-Feraly FS, et al. Preparation and characterization of a new artemisinin-derived dimer. *J Nat Prod* 1996;**59**(10):917–20.
40. Chen YX, Yu PL, Li Y, et al. Studies on artemisinin analogues III. The synthesis of diprotic acid diesters and monoesters of dihydroartemisinin. *Acta Pharm Sin* 1985;**20**(2):105–11. [陈一心，虞佩琳，李英，等. 青蒿素类似物的研究III. 二氢青蒿素二元酸双酯和单酯类衍生物的合成. 药学学报, 1985, 20(2): 105–111.]
41. Chen YX, Yu PL, Li Y, et al. Studies on artemisinin analogues VII. The synthesis of bis(dihydroartemisinin) and bis(dihydrodeoxoartemisinin) ethers. *Acta Pharm Sin*

1985;**20**(6):470–d. [陈一心，虞佩琳，李英，等. 青蒿素类似物的研究VII. 双(二氢青蒿素)醚和双(二氢脱氧青蒿素)醚类衍生物的合成. 药学学报, 1985, 20(6): 470–473.]

42. Li Y, Zhu Y, Zhang Y, et al. *An artemisinin dimer containing nitrogen atoms, its preparation and uses* China Patent CN 102153564 A, published. August 17, 2011. [李英，朱焰，张瑜，等. 含氮原子的青蒿素二聚体、其制备方法及用途：中国，公开号CN102153564A.]
43. Liu JM, Fei AM, Ni RM, et al. Study on Kasumi-1 cell apoptosis induced by SM 1044, a new type of artemisinin derivative, and its mechanism. *J Exp Hematol* 2011;**19**(3):607–10. [刘静静，费爱梅，聂瑞敏，等. 新型青蒿素衍生物SM 1044 诱导Kasumi-1 细胞凋亡及机制的研究. 中国实验血液学杂志, 2011, 19(3): 607–610.]
44. Mi JQ, Li Y, Liu JJ, et al. *The uses of an artemisinin dimer* China Patent CN 102614168 A, published. August 1, 2012. [糜坚青，李英，刘静静，等. 一种青蒿素二聚体的用途：中国，公开号: CN102614168A.]
45. Mi JQ, Li Y, Ni RM, et al. *Artemisinin derivatives and their pharmaceutical salts in the preparation of drugs to treat acute leukemia* China Patent CN 103202835 A, published. May 13, 2015. [糜坚青，李英，聂瑞敏，等. 青蒿素衍生物及其药用盐用于制备治疗急性白血病的药物：中国, CN 103202835A.]
46. Mi JQ, Li Y, Peng Y, et al. *Artemisinin derivatives and their pharmaceutical salts in the preparation of drugs to treat acute myeloid leukemia* Chinese Patent CN 103202836 A, published. July 17, 2013. [糜坚青，李英，彭宇，等. 青蒿素衍生物及其药用盐用于制备治疗急性髓细胞性白血病的药物：中国, CN 103202836A.]
47. Mi JQ, Li Y, Ni RM, et al. *Artemisinin derivatives and their pharmaceutical salts as a new drug for leukemia treatment* China Patent CN 103202837 A, published. July 17, 2013. [糜坚青，李英，聂瑞敏，等. 青蒿素衍生物及其药用盐用于治疗白血病的药物：中国, CN 103202837A.]
48. Tang Y, Dong Y, Vennerstrom JL. Synthetic peroxides as antimalarials. *Med Res Rev* 2004;**24**(4):425–48.
49. Posner GH, Ploypradith P, Parker MH, et al. Antimalarial, antiproliferative, and antitumor activities of artemisinin-derived, chemically robust, trioxane dimers. *J Med Chem* 1999;**42**(21):4275–80.
50. Jung M, Lee S, Ham J, et al. Antitumor activity of novel deoxoartemisinin monomers, dimers, and trimers. *J Med Chem* 2003;**46**(6):987–94.
51. Jeyadevan JP, Bray PG, Chawick J, et al. Antimalarial and antitumor evaluation of novel C-10 non-acetal dimers of 10-beta-(2-hydroxyethyl)deoxoartemisinin. *J Med Chem* 2004;**47**(5):1290–8.
52. Posner GH, McRiner AJ, Paik IH, et al. Anticancer and antimalarial efficacy and safety of artemisinin-derived trioxane dimers in rodents. *J Med Chem* 2004;**47**(5):1299–301.
53. Slade D, Galal AM, Gul W, et al. Antiprotozoal, anticancer and antimicrobial activities of dihydroartemisinin acetal dimers and monomers. *Bioorg Med Chem* 2009;**17**(23):7949–57.
54. Posner GH, Paik IH, Sur S, et al. Orally active, antimalarial, anticancer, artemisinin-derived trioxane dimers with high stability and efficacy. *J Med Chem* 2003;**46**(6):1060–5.
55. Posner GH, Jeon HB, Parker MH, et al. Antimalarial simplified 3-aryltrioxanes: synthesis and preclinical efficacy/toxicity testing in rodents. *J Med Chem* 2001;**44**(19):3054–8.
56. Ekthawatchai S, Kamchonwongpaisan S, Kongsaeree P, et al. C-16 artemisinin derivatives and their antimalarial and cytotoxic activities: syntheses of artemisinin monomers, dimers, trimers, and tetramers by nucleophilic additions to artemisitene. *J Med Chem* 2001;**44**(26):4688–95.
57. Kumar N, Sharma M, Rawat DS. Medicinal chemistry perspectives of trioxanes and tetraoxanes. *Curr Med Chem* 2011;**18**(25):3889–928.
58. Jefford CW, Rossier JC, Mihous WK. The structure and antimalarial activity of some 1,2,4-trioxanes, 1,2,4,5-tetroxanes, and bicyclic endoperoxides. Implications for the mode of action. *Heterocycles* 2000;**52**(3):1345–50.

59. Hao HD, Wittlin S, Wu Y. Potent antimalarial 1,2,4-trioxanes through perhydrolysis of epoxides. *Chem Eur J* 2013;**19**(23):7605–19.
60. Liang XT, Yu DQ, Wu WL, et al. The chemical structure of yingzhaosu a. *Acta Chim Sin* 1979;**37**(3):215–30. [梁晓天，于德泉，吴伟良，等. 鹰爪素A的化学结构. 化学学报, 1979, 37(3): 215–230.]
61. Hofheinz W, Burgin H, Gocke E, et al. Ro 42-1611 (arteflene), a new effective antimalarial: chemical structure and biological activity. *Trop Med Parasitol* 1994;**45**(3):261–5.
62. Vennerstrom JL, Dong Y, Andersen SL, et al. Synthesis and antimalarial activity of sixteen dispiro-1,2,4,5-tetraoxanes: alkyl-substituted 7,8,15,16-tetraoxadispiro[5.2.5.2]hexadecanes. *J Med Chem* 2000;**43**(14):2753–8.
63. Vennerstrom JL, Ager Jr AL, Andersen SL, et al. Assessment of the antimalarial potential of tetraoxane WR 148999. *Am J Trop Med Hyg* 2000;**62**(5):573–8.
64. Vennerstrom JL, Arbe-Barnes S, Brun R, et al. Identification of an antimalarial synthetic trioxolane drug development candidate. *Nature* 2004;**430**(7002):900–4.
65. Charman SA, Arbe-Barnes S, Bathurst IC, et al. Synthetic ozonide drug candidate OZ439 offers new hope for a single-dose cure of uncomplicated malaria. *Proc Natl Acad Sci USA* 2011;**108**(11):4400.
66. Valecha N, Looareesuwan S, Martensson A, et al. Arterolane, a new synthetic trioxolane for treatment of uncomplicated *Plasmodium falciparum* malaria: a phase II, multicenter, randomized, dose-finding clinical trial. *Clin Infect Dis* 2010;**51**:684–91.
67. Rathi A. Ranbaxy launches new antimalarial. *Chem World* 2012;**9**(6):7.
68. Valecha N, Savargaonkar D, Srivastava B, et al. Comparison of the safety and efficacy of fixed-dose combination of arterolane maleate and piperaquine phosphate with chloroquine in acute, uncomplicated *Plasmodium vivax* malaria: a phase III, multi-centric, open-label study. *Malar J* January 27, 2016;**15**:42.
69. Dechy-Cabaret O, Benoit-Vical F, Loup C, et al. Synthesis and antimalarial activity of trioxaquine derivatives. *Chem Eur J* 2004;**10**(7):1625–36.
70. Coslédan F, Fraisse L, Pellet A, et al. Selection of a trioxaquine as an antimalarial drug candidate. *Proc Natl Acad Sci USA* 2008;**105**(45):17579–84.
71. Lin AJ, Klayman DL, Hoch JM, et al. Thermal rearrangement and decomposition products of artemisinin (qinghaosu). *J Org Chem* 1985;**50**(23):4504–8.
72. Luo XD, Yeh HJC, Brossi A. The chemistry of drugs VI. Thermal decomposition of qinghaosu. *Heterocycles* 1985;**23**(4):881–7.
73. Lin AJ, Theoharides AD, Klayman DL. Thermal-decomposition products of dihydroartemisinin (dihydroqinghaosu). *Tetrahedron* 1986;**42**:2181–4.
74. Haynes RK, Chan HW, Lung CM, et al. Artesunate and dihydroartemisinin (DHA): unusual decomposition products formed under mild conditions and comments on the fitness of DHA as an antimalarial drug. *ChemMedChem* 2007;**2**(10):1448–63.

Chapter 5

Artemisinin and Derivatives: Pharmacodynamics, Toxicology, Pharmacokinetics, Mechanism of Action, Resistance, and Immune Regulation

Chapter Outline

Artemisinin-Based and Other Antimalarials. https://doi.org/10.1016/B978-0-12-813133-6.00005-6

5.1 ANTIMALARIAL PHARMACODYNAMICS

Ye Zuguang, Li Zelin†
Institute of Chinese Materia Medica, China Academy of Chinese Medical Sciences, Beijing, China

From the late 1970s to the early 1980s, the following organizations were involved in the antimalarial pharmacology of artemisinin and derivatives as individual workers or in collaboration teams between them: Institute of Chinese Materia Medica (ICMM) at the China Academy of Chinese Medical Sciences (CACMS), Institute of Microbiology and Epidemiology (IME) at the Academy of Military Medical Sciences (AMMS), Chinese Academy of Sciences' Shanghai Institute of Materia Medica (SIMM of CAS), Guangxi Medical University, Guangxi University of Chinese Medicine (Guangxi UCM), and Guangzhou University of Chinese Medicine (Guangzhou UCM). Included were studies on the effects of artemisinins on the parasite's erythrocytic stage in rodents and monkeys, on rodent and human falciparum malaria in vitro, and on the tissue stage of chicken, rodent, and monkey malaria, as well as research into drug resistance. In 1981 World Health Organization (WHO) held the fourth conference of its scientific working group on the chemotherapy of malaria in Beijing, centering on the theme of artemisinin. China's conference preparatory group organized units across the country to speak on their findings, according to their various specialties. Under the name "Cooperative Research Group on *Qinghaosu* and its Derivatives,"[1] they published research undertaken in China over many years. This section focuses on pharmacology studies,[2] supplemented by relevant post-1980s material.

A. EFFECTS ON THE MALARIA PARASITE'S ERYTHROCYTIC STAGE

1. Antimalarial Effects of Artemisinins on *Plasmodium berghei* In Vivo

Antimalarial Potency of Artemisinin

Healthy mice weighing 18–22 g were randomly divided into groups. Blood was drawn from the orbit of mice infected with *Plasmodium berghei* and diluted in saline containing an anticoagulant. Each healthy mouse was injected intraperitoneally with 1×10^7 parasites. The mice were medicated after 24 h for three consecutive days. Caudal blood smears were prepared 24 h after the last administration of artemisinin to determine the parasite clearance rates in separate dose groups. The Burn–Karber method[3,4] was used to calculate the ED_{50}.

P. berghei was also used in early studies by the ICMM on the suppressive effects of artemisinin. Mice were medicated orally once a day for 3 days, starting 24 h after infection. Caudal blood smears were done 24 h after the last administration to determine parasite clearance and calculate the ED_{50}. The ED_{50} of oral artemisinin was

†Deceased.

46.3 ± 6.9 mg/kg. When the 3-day dose was at least four times that of the effective dose (i.e., a total dose of 1200 mg/kg), no parasites were seen 1 month after parasite clearance. At this dose, no signs of infection appeared among healthy mice 1 month after receiving whole blood inoculations from the above infected mice.[2]

In 1980, the pharmacology laboratory at the Sichuan Institute of Chinese Materia Medica (now Sichuan Institute of Chinese Medical Sciences, Sichuan ICMS; or Sichuan Academy of Chinese Medical Sciences SACMS) gave mice intraperitoneal injections of *P. berghei*.[5] Once a day for 3 days beginning 24 h after infection, the mice were given an artemisinin suspension orally. Blood smears were made on the fifth day and parasite clearance observed. According to the Burn–Karber formula, the ED_{50} was 60.7 ± 7.8 mg/kg. The median lethal doses (LD_{50}) of a single oral dose of artemisinin was 4530 ± 672.7 mg/kg, and the therapeutic index was 75. It showed that artemisinin was an ideal antimalarial with high efficacy and low toxicity.

Mice with mild infections were given different intragastric doses of an artemisinin suspension orally 24 h after they were injected with the parasite. They were medicated twice a day for 9 days. Blood smears were prepared on Days 5, 10, 15, 20, and 30, and parasite clearance and recrudescence were observed. The results indicated that 300 mg/kg oral artemisinin per dose could produce parasite clearance in 5 days, with no recrudescence in 30 days. Mice with severe infections were given the same artemisinin suspension 72 h after being injected with the parasite. They were given daily doses of 100, 200, or 500 mg/kg, in 3- or 5-day treatment groups. Blood smears were examined on Days 4, 6, 8, 12, 14, and 17 days after infection. The recrudescence and mortality rates were higher in the 3-day treatment groups. This was a first sign that a longer treatment course was necessary.

The effects of different artemisinin doses and routes of administration on parasitemia were also investigated. Five days after infection, the animals were divided into different drug groups: single subcutaneous injections of 100 or 450 mg/kg or single oral doses of 100, 250, or 500 mg/kg. At fixed intervals after medication, blood smears were examined to observe parasitemia. Subcutaneous administration resulted in slower parasite clearance than oral medication at the same dose. However, parasitemia took longer to recrudesce. This could be due to slower absorption of the injected suspension. A single subcutaneous injection of 450 mg/kg artemisinin could prevent recrudescence, which showed that, absorption although slow, the effects were long lasting. It could overcome the short-term recrudescence seen in artemisinin treatment. A 1% artemisinin and carboxymethylcellulose suspension injected subcutaneously either once or three times at a total dose of 300 mg/kg could stop recrudescence before 30 days.[5]

In 1981, Peters, Li Zelin et al.[6] gave mice infected with chloroquine-sensitive *P. berghei* subcutaneous injections of artemisinin once a day for 4 days, based on Peters' 4-day suppressive test.[7] The resulting ED_{90} was 2.3 mg/kg. Similar treatment was also given to mice infected with *P. berghei* with strains resistant to primaquine, cycloguanil, pyrimethamine, sulfaphenazole, mefloquine, and menoctone. The ED_{90} values against these resistant strains were similar to the value against chloroquine-sensitive strain. This indicated that artemisinin was not only effective against chloroquine-sensitive *P. berghei* but also against several other drug-resistant strains, and there was no significant cross-resistance between artemisinin and other antimalarial drugs.

Efficacy of Artemisinins and Chloroquine on Chloroquine-Sensitive Plasmodium berghei[1]

On Day 0, Shanghai-outbred mice weighing 18–22 g were given intraperitoneal injections of 5×10^6 erythrocytes infected with *P. berghei*. The drugs were given once daily on Days 1–3. Blood smears were prepared on Day 4, stained, and examined. The ED_{50} and ED_{90} were calculated using a simplified probit method. Chloroquine was the standard control (see Table 5.1).

As can be seen from Table 5.1, intramuscular injections of an artemether-oil solution were the most effective out of all the drugs, including the chloroquine control. Next was an intramuscular injection of aqueous solution of sodium artesunate, followed by intramuscular injections of an artemisinin-oil suspension.

2. Efficacy of Artemisinins on Simian Malaria In Vivo

Artemisinin: Oral and Suppositories

Monkeys (rhesus) were infected with *Plasmodium cynomolgi* sporozoites and medicated when parasites were present in the blood. Chloroquine or chloroquine–primaquine was the control. Once parasitemia had reached a certain level, seven such monkeys were given 200 mg/kg oral artemisinin once daily for 3 days. On Days 2–3 after medication, the parasites were completely cleared from the blood. This was a similar level of efficacy to chloroquine. As shown in Table 5.2, artemisinin could kill the erythrocytic stage of simian malaria, but it was not necessarily a complete cure.[2]

TABLE 5.1 Efficacy of Artemisinin, Artemether, Artesunate, and Chloroquine on Chloroquine-Sensitive *Plasmodium berghei*

Drug	Preparation	Administration	ED_{50} (mg/kg/day)	ED_{90} (mg/kg/day)
Artemisinin	Aqueous suspension	p.o.	10.8	28.3
	Aqueous suspension	im	4.90	8.01
	Oil suspension	im	0.77	2.15
Artemether	Oil solution	im	0.37	0.53
Artesunate	Solution	im	0.54	1.77
	Solution	iv	0.94	3.10
Chloroquine	Solution	p.o.	1.85	2.60
	Solution	im	0.60	1.12
	Solution	iv	0.67	1.25

TABLE 5.2 Effects of Artemisinin on the Erythrocytic Stage of *Plasmodium cynomolgi*

Drug/Total Dose (mg/kg)	Subject No.	No. of Days After Infection	Parasite Rate Before Treatment (%)	Clearance Time (Days)	Time of Recrudescence (Days)
Artemisinin 600	58	31	1.00	3	11
Artemisinin 600	50	31	6.00	2	–[a]
Artemisinin 600	59	11	0.01	3	–[a]
Primaquine + chloroquine 9 + 60	43	47	2.00	2	–
Chloroquine 60	45	47	16.00	2	–
Artemisinin 600	22	18	6.00	2	23
Artemisinin 600	23	18	2.00	2	11
Artemisinin 600	67	18	0.10	2	11
Artemisinin 600	66	18	13.00	2	8
Primaquine + chloroquine 9 + 60	70	18	26.00	3	–
Primaquine + chloroquine 9 + 60	218	18	0.30	2	–
Chloroquine 20	64	18	1.00	2	–

[a]*No reappearance of the parasite in half a year.*

In 1986, Li Zelin et al. commenced preclinical, in vivo pharmacodynamic (PD) studies of artemisinin suppositories in preparation for new-drug registration. Monkeys infected with *Plasmodium knowlesi* were divided into high-, medium-, and low-dose groups, with three animals in each. They were given 25, 50, or 100 mg/kg artemisinin suppositories rectally, once daily for 4 days. Changes in parasitemia and clearance times and rates were observed 24 and 48 h after treatment. Artemisinin suppositories could reliably kill the parasites, with a significant dose–effect relationship. The results are in Tables 5.3 and 5.4.[8]

TABLE 5.3 Changes in *Plasmodium knowlesi* Parasitemia After Treatment With Artemisinin Suppositories at Various Dosages

Time After Medication (h)	Posttreatment Parasitemia Ratio (Parasitemia After Medication/Parasitemia Before Medication)			
	Control Group	25 mg/kg Group	50 mg/kg Group	100 mg/kg Group
24	5.847	0.643	0.090	0.083
48	19.737	1.430	0.017	0.044

TABLE 5.4 Effects of Artemisinin Suppositories on *Plasmodium knowlesi*

Group (mg/kg)	Subject No.	Clearance Time (h)	Average Clearance Time (h)	Elimination Rate (%)
100	3	36	46	100
	10	36		
	109	66		
50	2	–	–	33.3
	5	–		
	110	66		
25	4	–	–	0
	7	–		
	9	–		
0	1	–	–	0
	6	–		
	8	–		

The speed of parasite clearance and rate of reduction in parasitemia were compared across different doses. For all three dose groups, parasitemia fell by 90% or more 24h after administration. After 66h, the average maximum decline in parasitemia for the high-, medium- and low-dose groups was 100%, 99.80%, and 99.15% (two animals), respectively, as shown in Table 5.5.

The results showed that artemisinin suppositories had a rapid therapeutic effect on monkeys infected with *P. knowlesi*.

Artemisinin and Artemether Intramuscular Injections[1]

In 1982, the Cooperative Research Group on Artemisinin and its Derivatives infected rhesus monkeys intravenously with erythrocytes containing *P. cynomolgi*. When parasitemia reached a certain level after an incubation period, the monkeys were given intramuscular injections of different doses of an artemisinin-oil suspension or artemether-oil solution once daily for 3days. At fixed intervals, blood smears were obtained to observe parasite clearance and recrudescence times. The results are shown in Table 5.6.

Table 5.6 indicates that artemisinin and artemether were both very effective against asexual *P. cynomolgi*, but artemether was better.

Efficacy of Intravenous Sodium Artesunate on Plasmodium knowlesi[1]

Using the above protocol, rhesus monkeys infected with *P. knowlesi* were given 6mg/kg intravenous sodium artesunate once daily for 3days. Full parasite

TABLE 5.5 Parasite Clearance Speeds of Artemisinin Suppositories on *Plasmodium knowlesi*

	Average Decline in Parasitemia		
Time of Medication (h)	100mg/kg Group	50mg/kg Group	25mg/kg Group
6	34.93	24.97	67.90
12	70.83	76.30	87.60
24	91.76	90.97	92.40[a]
36	99.06	98.83	98.85[a]
42	99.70	99.13	99.10[a]
60	99.80	99.60	99.35[a]
66	100.00	99.80	99.15[a]

[a]*For two animals (one died). Other figures represent three animals.*

TABLE 5.6 Efficacy of Intramuscular Artemisinin and Artemether Injections on the Erythrocytic Stage *Plasmodium cynomolgi*

Drug	Preparation	Dose (mg/kg/day)	Parasite Clearance Time (Days)	Time of Recrudescence (Days)
Artemisinin	Oil suspension	20	2	No recrudes.
		20	2	No recrudes.
		10	2	20
		10	2	23
		4	4	8
		4	2	24
		1	3	9
Artemether	Oil solution	8	2	No recrudes.
		8	2	No recrudes.
		4	2	No recrudes.
		4	1	No recrudes.
		4	2	20
		2	1	No recrudes.
		2	2	No recrudes.
		1	2	16

clearance occurred 16–20h after the first administration, and there was no recrudescence within 31 days.

3. Effects of Artemisinins on *Plasmodium falciparum* In Vitro

The strains used in these in vitro tests included the *Wellcome/Liverpool/West African* strain, commonly used in Britain, or strains isolated from Africa and Thailand. The Hainan strain *FCC1/HN* was used in domestic (China) experiments.

Effective Concentrations of Artemisinin, Artemether, and Sodium Artesunate on FCC1/HN-Strain Plasmodium falciparum[1]

In 1982, the Cooperative Research Group on Artemisinin determined the effective inhibitory concentrations of artemisinin, artemether, and sodium artesunate on the erythrocytic stage of human *Plasmodium falciparum* cultured in vitro.

TABLE 5.7 Effects of Artemisinin and Its Derivatives on In Vitro *FCCl/HN*-Strain *Plasmodium falciparum*

Drug	EC_{50} (ng/mL)	EC_{90} (ng/mL)	Efficacy Ratio Relative to Chloroquine	
			EC_{50}	EC_{90}
Chloroquine	2.24	7.95	1	1
Artemisinin	1.99	4.52	1.13	1.76
Artemether	2.19	4.12	1.02	1.92
Sodium artesunate	0.14	1.18	16.12	6.74

A 180 μL suspension containing 2.5% (*V/V*) red blood cells (RBCs) with *FCC1/HN*-strain *P. falciparum* (infection rate around 1%) was added to each well of the culture plate, along with 20 μL dextrose saline with different concentrations of the drugs. Chloroquine or wells with no drug added were used as the controls. After culturing for 48 h, blood smears were obtained. The EC_{50} and EC_{90} of artemisinin, artemether, and artesunate sodium were calculated using Finney's method[9] (Table 5.7). The EC_{50} and EC_{90} of all three drugs were significantly lower than those of chloroquine, with sodium artesunate having the lowest values.

In 1988, Gu Haoming et al.[10] of the SIMM used fluorescence spectroscopy to evaluate the antimalarial effects of artemisinin and artemether on *FCC1/HN*-strain *P. falciparum* cultured in vitro. *P. falciparum* was cultured for 40 h with artemisinin, artemether, or chloroquine. Ethidium bromide was used as a fluorescent tag. Fluorescence was measured after solubilization with sodium dodecyl sulfate. The inhibitory concentration (IC_{50}) of artemisinin, artemether, and chloroquine were measured at 75.2, 29.4, and 43.2 nmol/L, respectively. The in vitro tests showed that artemether was more effective than artemisinin and chloroquine.

Effects of Artemisinin, Artemether, Dihydroartemisinin, and α-Propoxycarbonyl Dihydroartemisinin on Wellcome/Liverpool/ West African-Strain Plasmodium falciparum

In 1983, Li Zelin et al. observed the antimalarial effects of artemisinin, artemether, dihydroartemisinin, and α-propoxycarbonyl dihydroartemisinin (SM242) on *Wellcome/Liverpool/West African*-strain *P. falciparum*, using changes in the incorporation of [G-^{3}H]hypoxanthine.[11] These were compared to a no-drug control. The IC_{50} of artemisinin, dihydroartemisinin, SM242, and artemether were measured at 126, 0.585, 1.1, and 1.2 nmol/L, respectively. The relative activity of the latter three drugs was 215, 115, and 105 times that of artemisinin, respectively.

Therefore, the authors suggested that reduction of the lactone group in the artemisinin molecule increased the in vitro antimalarial activity of dihydroartemisinin, artemether, and SM242 by more than 100 times. Research into suppression times (see Section 5.4 of this chapter on the mechanism of action) also showed no suppressive effects within 80 min but significant inhibition after 160 min.

Effects of Artemisinin and Artemether on Plasmodium falciparum *Isolates From Thailand*

You Jiqing studied the effects of artemisinin and artemether on *P. falciparum* isolates from Thailand in 1986.[12] Two nonclones (K_1 and K_{31}) and nine clones (T_9) were isolated from Thailand. In addition, a nonclone (G_1) isolated from Gambia was used as a control. Before the test, all isolates had been cultured in petri dishes using the Trager and Jensen candle-jar method.[13] The culture medium was RPMI 1640 with 10% serum. The concentrations of artemisinin and artemether in contact with the isolates were 10^{-7} to 10^{-12} mol/L. Isolates were cultured in the wells of microtest plates for 36–72 h at 37°C. After 24 and 48 h, the culture mediums containing drugs were replaced. Samples were collected and blood smears examined to measure the minimum inhibitory concentration (MIC).

The results showed that at 36, 45, or 72 h of exposure to the drug, the MIC values were roughly similar. Hence, a 36-h exposure period was taken as a standard. The MIC was determined based on the onset of parasitemia in the blood, which was 0.5%–1% in the initial experiment. The results indicated that the MIC of artemisinin and artemether were 10^{-7}–10^{-8} mol/L and 10^{-8} mol/L, respectively. Although the parasite's sensitivity to the two drugs was somewhat different, this was not large. These studies confirmed that artemisinin and artemether were very effective against 11 *P. falciparum* strains with varying resistance to pyrimethamine, Fansidar, and chloroquine.

Conclusion: The above experiments, described in detail, showed that artemisinin had rapid and significant suppressive effects on rodent malaria in vivo, with an ED_{50} of 46.3 ± 6.9 mg/kg for oral administration. In vivo the drug's main drawback was a high rate of recrudescence after a short, 3- to 4-day course of treatment. This rate could be reduced by administering the effective oral dose via intramuscular injection. Artemisinin, artemether, and sodium artesunate could effectively kill *P. berghei* and *P. cynomolgi* in vivo in the erythrocytic stage. The ED_{90} of the three drugs on rodent malaria was 2.15, 0.53, and 1.77 mg/kg/day, respectively, when delivered via injection.

B. EFFECTS OF ARTEMISININ ON MALARIA PARASITE'S EXOERYTHROCYTIC STAGE

1. Experiments in Chickens

In a joint experiment with the Shanghai Institute of Parasitic Diseases, the ICMM of CACMS infected chicks with avian malaria sporozoites.[2] 9–11 days later, when a large number of gametocytes were seen in the blood, the chickens were exposed to *Aedes aegypti* or *Aedes albopictus* mosquitoes. After feeding, the mosquitoes were kept for around 9 days at $27 \pm 2°C$ and a relative humidity of about 80%. Dissection of the salivary glands confirmed the presence of sporozoites. The mosquitoes were then lightly anesthetized with chloroform. Females were ground up in saline, resulting in a suspension of 0.05–0.1 carrier mosquitoes per 0.2 mL saline.

Chicks were then injected with 0.2 mL of this suspension intramuscularly (into chest muscle). Oral artemisinin was given 30 min after infection, once a day for 6 days. From Day 7, blood smears were prepared once a day. In the control group, parasitemia appeared 7–8 days after infection and the chicks died in 9–14 days. Secondary exoerythrocytic parasites could be seen in brain tissue slides. Doses of 100–300 mg/kg artemisinin were used, with primaquine, pyrimethamine, and unmedicated groups as controls. Efficacy was evaluated based on the absence of or delay in parasitemia, prolonged survival of the chick, absence of secondary exoerythrocytic parasites, and loss of infectivity. Artemisinin had no effect on the preerythrocytic stage of chicken malaria. When the dose was increased to 400 mg/kg, the animals all died in 7 days because of drug toxicity.

In 1982, the Collaborative Research Group on Artemisinin infected chickens with 1.2×10^3 sporozoites isolated from *Anopheles stephensi* via intramuscular injection.[1] After 30 min, the chickens were given an aqueous artemisinin suspension orally, once daily for 6 days. Blood smears were obtained from the second day to observe parasitemia. Mortality was recorded. Pyrimethamine and an unmedicated group were the controls. It showed that 200 mg/kg/day artemisinin had no effect on chickens infected with *Plasmodium gallinaceum*. All animals died when the dose was raised to 400 mg/kg/day.

2. Effects of Artemisinin on Exoerythrocytic *Plasmodium yoelii nigeriensis*

Peters, Li Zelin et al. in 1981 infected mice with *Plasmodium yoelii nigeriensis* sporozoites.[6] The mice received a 300 mg/kg subcutaneous injection of artemisinin before infection. It had no prophylactic effect.

3. Effects of Artemisinin, Chlorquine, Primaquine, Pyrimethamine on Exoerythrocytic *Plasmodium yoelii*[1]

C57 inbred black mice aged 6–8 weeks were intravenously infected with 2.5×10^4 sporozoites isolated from *A. stephensi*. Three to four hours later, 100 mg/kg artemisinin was administered subcutaneously. Blood smears were prepared on Days 7 and 14 to determine whether artemisinin could be used as prophylaxis. Both 100 mg/kg artemisinin and chloroquine were ineffective against exoerythrocytic *P. yoelii*, whereas 20–50 mg/kg primaquine, 5–10 mg/kg robenidine, and 0.1–0.3 mg/kg pyrimethamine were effective.

TABLE 5.8 Effects of Oral Artemisinin on Exoerythrocytic *Plasmodium cynomolgi*

Drug	Dose (mg/kg) × Number of Days Administered	First and Last Days of Medication	Parasitemia	Day After Medication When Parasitemia Noted
Artemisinin	100 × 6	D_0–D_5	+	8
Artemisinin	100 × 6		+	8
Primaquine[a]	3 × 6		+	22
Chloroquine[a]	10 × 6		+	12
Artemisinin	100 × 6	D_5–D_{10}	+	8
Artemisinin	100 × 6		+	7
Primaquine[a]	3 × 6		–	–
Chloroquine[a]	10 × 6		+	8
Control			+	8

[a]*Dose calculated according to the drug base.*

4. In Vivo Experiments With Simian Malaria[1]

On Day 0, normal cynomolgus monkeys were intravenously infected with 8.7×10^6 *P. cynomolgi* sporozoites isolated from *A. stephensi*. Thereafter, on Days 0–5 or Days 5–10, the monkeys were given 100 mg/kg oral artemisinin in aqueous suspension for 6 days. Blood smears were examined on Days 7–12. The results (Table 5.8) showed the appearance of parasitemia on Days 7–8, indicating that 100 mg/kg oral artemisinin did not prevent parasitemia regardless of whether it was given on the day of infection or on the fifth day. The drug was ineffective against exoerythrocytic *P. cynomolgi*.

C. EFFECT OF ARTEMISININS ON CHLOROQUINE-RESISTANT STRAIN OF *PLASMODIUM BERGHEI* IN VIVO

1. Efficacy of Artemisinin on Chloroquine-Sensitive and Chloroquine-Resistant Strains of *P. berghei*

In 1979, the pharmacology laboratory[2] of the ICMM of CACMS infected mice with normal and chloroquine-resistant *P. berghei* via intraperitoneal injection with infected blood diluted to 5×10^6 parasitized RBCs per 0.2 mL. Artemisinin was administered orally 2 days after infection, for 3 days. On the fourth day, blood smears were prepared, stained, and examined to calculate the ED_{50}. The drug's efficacy against both strains of rodent malaria was compared, to determine if there was cross-resistance between artemisinin and chloroquine. The results (Tables 5.9 and 5.10) showed that the ED_{50} of artemisinin against normal *P. berghei* was 46.3±6.9 mg/kg. Against the chloroquine-resistant strain, the ED_{50} was 167.6±20.9 mg/kg. The resistance ratio was only 3.6, based on the normal strain. It suggested a slight degree of cross-resistance between artemisinin and chloroquine.

The pharmacology laboratory at the Sichuan Institute of Chinese Materia Medica (now Sichuan ICMS)[5] also carried out research into the effects of artemisinin on normal and chloroquine-resistant strains of rodent malaria in

TABLE 5.9 Efficacy of Artemisinin on Normal-Strain *Plasmodium berghei*

Drug and Dose (mg/kg × 3d)	Number of Animals	Number of Animals With Full Clearance	Elimination Rate (%)	ED_{50} (M ± SD)
Artemisinin, 31.7	10	2	20	46.3 ± 6.9
Artemisinin, 42.2	10	4	40	
Artemisinin, 56.3	10	8	80	
Artemisinin, 75.0	10	9	90	
Artemisinin, 100	10	10	100	
Control	10	0	0	

TABLE 5.10 Efficacy of Artemisinin on Chloroquine-Resistant *Plasmodium berghei*

Drug and Dose (mg/kg × 3d)	Number of Animals	Number of Animals With Full Clearance	Elimination Rate (%)	ED_{50} (M ± SD)
Artemisinin, 79.1	10	0	0	167.6±20.9
Artemisinin, 105.5	10	0	0	
Artemisinin, 140.6	10	3	30	
Artemisinin, 187.5	10	6	60	
Artemisinin, 250	10	10	100	

1980. Mice infected via intraperitoneal injections were given an artemisinin suspension orally. The parasite clearance rate was observed. The ED_{50} for the normal strain was 60.7±7.8 mg/kg, as opposed to 238.8±26.5 mg/kg for the chloroquine-resistant strain. The resistance index was 3.9, similar to the results obtained by the ICMM.

In 1981, Peters, Li Zelin et al.[6] also investigated drug resistance to artemisinin. The study included both normal strain of malaria and strains resistant to seven common antimalarial drugs, including chloroquine, quinine, and primaquine. The ED_{50} values for each strain were measured. The ED_{50} of artemisinin against a highly chloroquine-resistant strain was 2.5 times its ED_{50} against a sensitive strain, and it did not differ significantly from the ED_{50} against mildly chloroquine-resistant and other drug-resistant strains.

Mice infected with chloroquine-sensitive *P. berghei* were given subcutaneous injections of artemisinin once a day for 4 days. The ED_{90} was measured at 2.3 mg/kg. The drug's ED_{90} against moderately chloroquine-resistant *P. berghei* was 3.9 times this figure. Against highly chloroquine-resistant *P. berghei*, the ED_{90} was 34 times that of the normal strain.

2. Effects of Artemisinins on Chloroquine-Sensitive and Chloroquine-Resistant Strains of *Plasmodium berghei*[1]

On Day 0, mice were infected with normal or highly drug-resistant *P. berghei*. From Days 1–3, two groups of mice were given either an aqueous suspension of artemisinin orally or an intravenous injection of aqueous sodium artesunate solution. Blood smears were obtained on Day 4. A third group of mice was given intramuscular injections of an artemether-oil solution from Days 2–4, with blood smears prepared on Day 6. The results (Table 5.11) showed a slight degree of cross-resistance between chloroquine and the other three drugs.

In 1986, Li Chengshao et al. from the Shandong Institute of Chinese Medicine (now Shandong Institute of Chinese Medical Sciences, Shandong ICMS; or

TABLE 5.11 Efficacy of Artemisinin, Artemether, and Artesunate Against Chloroquine-Resistant *Plasmodium berghei*

Drug	Administration Route	Normal Strain ED_{50} (mg/kg)	Resistant Strain ED_{50} (mg/kg)	Resistance Index
Artemisinin	Oral	46.3	167.6	3.6
Artemether	Intramuscular	0.72	1.2	1.7
Sodium artesunate	Intravenous	0.3	0.5	1.7

Shandong Academy of Chinese Medicine SACM) used Peters' dose escalation method[7] on the ANKA and N strains of *P. berghei* to conduct resistance studies of artemisinin.[14] Following the Peters protocol, the average number of days' delay before parasitemia reached 2% of erythrocytes was taken as the qualitative indicator. Changes in parasitic resistance were observed over successive passages. Quantitative indicators of resistance included the ED_{50} of artemisinin, the resistance index (I_{50}), the 50% experimental curative dose (CD_{50}), and the ratio of the CD_{50} of the resistant strain to that of its parent strain. These were measured using the 4-day suppressive test. After cultivation for more than a year, the ED_{50} of artemisinin against *RQ/ANKA* and *RQ/N P. berghei* were more than 50 times the ED_{50} against the parent strains. The *RQ/ANKA* strain showed significant cross-resistance to sodium artesunate and artemether, with resistance indices of 13.1 and 11.7, respectively. This strain also had mild cross-resistance to primaquine—resistance index 2.9—and no significant cross-resistance to chloroquine.

Li Guofu et al. of the IME of AMMS used the dose escalation method in vivo to cultivate artemisinin- and artemether-resistant strains of *P. berghei* in 1990.[15] Resistance appeared from the 10th passage. Artemisinin-resistant (*RQ*) and artemether-resistant (*RAr*) strains showed no cross-resistance to chloroquine, primaquine, and amodiaquine.

5.2 TOXICOLOGY

Yuan Shoujun
Institute of Radiology and Radiation Medicine, Academy of Military Medical Sciences, Beijing, China

As discussed in previous chapters, artemisinin was extracted from *Artemisia annua* as an effective antimalarial monomer. The drug and its derivatives have been used to treat various forms of malaria worldwide, saving thousands of lives. Nevertheless, attention was drawn not only to the efficacy of artemisinin drugs but also to their toxicity and safety in the human body and in animal experiments. These drugs showed relatively mild side effects in a clinical setting and few adverse ones. The therapeutic window was large. In the laboratory, toxicological research often used higher doses and longer courses of medication, which produced more toxic effects. However, such toxicity was reversible.

Since the 1970s, Chinese pharmaceutical researchers have undertaken in-depth work into the safety of artemisinin and its derivatives, and many reports have been published. Three stages of toxicological research are presented here based on these publications, the reports of the various work units and institutes involved, and interviews with the scientists.

A. STUDIES IN THE 1970S

Systematic reports on toxicology research into artemisinins were published by the ICMM of CACMS[2] and Guangxi UCM in 1979.[16] These are summarized below.

1. Acute Toxicity of Single-Dose

The ICMM evaluated the toxicity of single-dose oral administration of artemisinin and its extracts in mice. The LD_{50} of the active component of artemisinin, dilute alcohol extract of *A. annua*, artemisinin itself, and reduced artemisinin were 7425, 4162, 5105, and 465 mg/kg, respectively. Comparison of the LD_{50} and ED_{50} (46.3 mg/kg) of artemisinin yielded a therapeutic index of 110, showing that the drug was safe and had a large therapeutic window.

The acute toxicity studies of single-dose artemisinin conducted by the Guangxi UCM were more comprehensive, evaluating oral, intramuscular, and intravenous routes of administration. It showed that the LD_{50} of oral artemisinin in mice was 3800 mg/kg, and the safety index was 120 ($LD_5/ED_{95}=3000/25$). The maximum tolerated dose in rats (MTD) was greater than 12,000 mg/kg. The MTD for other animals such as pigeons, guinea pigs, rabbits, cats, dogs, and monkeys were all greater than 1000 mg/kg.

For a single intramuscular injection of artemisinin, the LD_{50} in mice was 838 mg/kg. In rats, it was 2645 mg/kg. The MTD for pigeons was 125 mg/kg, lethal dose 250 mg/kg; for guinea pigs, the MTD was 250 mg/kg, lethal dose 500 mg/kg. For rabbits, the MTD was 500 mg/kg, lethal dose 1000 mg/kg; for cats, the MTD was 250 mg/kg, lethal dose 500 mg/kg. The MTD for dogs was 250 mg/kg.

For a single intravenous injection of artemisinin, the LD_{50} in mice was 631 mg/kg. It was 1100 mg/kg in rats. For pigeons, the MTD was 50 mg/kg, lethal dose 150 mg/kg. For rabbits, the MTD was 300 mg/kg, lethal dose 500 mg/kg. The MTD for dogs was 50 mg/kg. Signs of intoxication appeared at 100 mg/kg and lethal dose was 250 mg/kg.

The signs of acute toxicity in animals were mainly respiratory depression, absence of nervous reflexes, and convulsions, culminating in respiratory and cardiac arrest. Some species (e.g., dogs) displayed a vomiting response.

2. Subacute Toxicity of Multidoses Over 3–7 Days

The ICMM of CACMS reported that, at the therapeutic dose of artemisinin, cats experienced a decrease in heart rate. When the dose was raised to 2000 mg/kg, this also appeared in mice. When a dose of 100–800 mg/kg was administered for 3 days, there was a transient increase in transaminase in mice. Pathology examination showed cloudy swelling of liver cells and petechiae in brain tissue.

The Chinese medicinal herb laboratory of Guangxi UCM reported that 100 mg/kg oral artemisinin administered for 5 days produced no toxic reactions in rats, rabbits, cats, dogs, and other animals.

3. Toxicity of Multidoses Over 10–14 Days

Rats and rabbits were given oral doses once daily for 14 days. Individual rabbits displayed lower leukocyte counts and higher transaminase levels, but no other signs of toxicity were seen. When the dose was increased to 5000 mg/kg for rats, the animals showed a slight decrease in food intake and body weight and lung and liver congestion. No reproductive toxicity was observed in mice given 200 mg/kg oral artemisinin for 10 days.

4. Toxicity in Humans

Three volunteers were given artemisinin once a day for 3 days, total doses 3500, 5000, and 5000 mg, respectively. One person experienced transient numbness in the limbs and another had an elevated heart rate. In another experiment, eight people received an active extract of *A. annua*. Three were medicated once a day for 7 days with increased dose each day. The dose was 350 mg on Day 1, 500 mg on Day 2, 1000 mg on Day 3, 2000 mg on Day 4, 3000 mg on Day 5, 4000 mg on Day 6, and 5000 mg on Day 7. Five volunteers were given 3000 mg twice daily for 3 days. There were two cases of mild abdominal pain, two cases of fecal occult blood, and one case of elevated transaminase.

In sum, the studies undertaken during this period provided adequate data on the toxicity of a single and multiple doses of artemisinin. The main indicator, LD_{50}, in animals, ranged relatively widely because of the varying laboratory conditions. However, the experiments all showed a significant gap between the effective dose or exposure concentration and the LD_{50}, and that the safety parameters and therapeutic index of artemisinin were relatively high. The test in humans also showed that the drug was relatively safe. Nevertheless, because the principles of drug safety evaluation were not understood at the time, the multidose toxicity studies were inadequate in terms of their experimental design and indicators observed. They fell short of the basic requirements of current drug safety tests. Often, it is only *after* toxicity studies of short-term regimens in large animals that human trials are conducted.

B. STUDIES IN THE 1980S

1. Artemisinin

In the early 1980s, the ability of artemisinin-based drugs to control malaria gained the attention of the WHO. Therefore, in 1981, the WHO sent experts to China to learn more. In response, China's Ministry of Health requested all relevant work units to assemble their research material on artemisinin. Regarding the incomplete pharmacology and toxicology data, the Ministry's office in charge of artemisinin tasked various units with conducting supplementary studies. For example, the Institute of Radiation Medicine of the Academy of Military Medical Sciences was assigned to perform multidose toxicology studies on large animals. The collected reports, including those on toxicology, were inspected by the group of WHO experts. Subsequently, the National Cooperative Group on Qinghaosu Research put together the toxicology data on artemisinin-based drugs, such as artemisinin, artemether, and sodium artesunate, and published them in English in the *Journal of Traditional Chinese Medicine*.[17] The research units involved included the ICMM of CACMS, AMMS, Shandong Institute of Chinese Medicine (now Shandong ICMS; or Shandong Academy of Chinese Medicine SACM), Yunnan Institute of Materia Medica, Sichuan Institute of Chinese Materia Medica (now Sichuan ICMS; or Sichuan Academy of Chinese Medical Sciences SACMS), SIMM of CAS, Guangzhou UCM, Guangxi Medical University, Guangxi UCM. The article reflected the state of toxicity studies on artemisinin and its derivatives at the time. Its main findings and reports from other Chinese research units from that period are summarized below.

Single-Dose[17]

The LD_{50} of oral artemisinin in mice was 4228 mg/kg; for intramuscular oil injections, the LD_{50} was 3840 mg/kg. The LD_{50} of oral artemisinin in rats was 5576 mg/kg; for intramuscular oil injections, the LD_{50} was 2571 mg/kg.

Two dogs were given intramuscular injections of 800 and 400 mg/kg artemisinin. After 15 min, the dog given the higher dose displayed tonic rigidity, clonic convulsions, and even opisthotonus. The dog given the low dose became excitable and barked wildly. These symptoms disappeared within 30 min. Forty hours after the injection, reticulocyte count fell and serum glutamic pyruvic transaminase (SGPT) and serum alkaline phosphatase (ALP) activity increased slightly. This was more obvious in the dog given the large dose. The indicators returned to normal after 10 days, and no pathological changes were seen in the internal organs and bone marrow.

These symptoms relating to the nervous system were also seen in pigeons, guinea pigs, rabbits, cats, and other animals given relatively large doses of artemisinin. However, sensitivity varied across species. Other symptoms included respiratory depression, sensory delay, and heart rhythm abnormalities.

Short-Term Multidose

Forty rats weighing 200–220 g were divided into four groups of ten. The medicated groups were given intramuscular oil injections of artemisinin, at doses of 600, 400, and 200 mg/kg, respectively. The rats were injected once a day for 7 days. There were no deaths and no abnormalities in body weight, food intake, and serum transaminase levels. Pathology examinations revealed mild hyperemia and degeneration in the hearts, livers, spleens, lungs, kidneys, and other organs of the animals in the large- and middle-dose groups. Dogs given 100 mg/kg oral artemisinin for 5 days showed no signs of toxicity.

Fourteen-Day Multidose

Rats received 250, 500, or 1000 mg/kg oral artemisinin for 14 days. No abnormalities were seen in their weights, ECGs, routine blood tests, and pathology examinations of the heart, liver, spleen, lung, kidney, brain, stomach, and other organs.

In another test, rhesus monkeys were divided into dose groups of four to six, with an equal number of males and females, as below:

- 192 mg/kg intramuscular artemisinin-oil injections once daily for 14 days;
- 96 mg/kg intramuscular artemisinin-oil injections once daily for 14 days;
- 48 mg/kg intramuscular artemisinin-oil injections once daily for 14 days;
- 24 mg/kg intramuscular artemisinin-oil injections once daily for 14 days;
- Control group, with intramuscular injections of an equal volume of solvent.

The items for preclinical evaluation included general observations, such as behavior, appetite, body temperature, body weight, gastrointestinal reactions, and reactions at the injection site; routine urine and hematology tests; bone marrow smears; blood biochemical tests, including protein electrophoresis, blood urea nitrogen (BUN), creatinine, creatine phosphokinase (CPK), lactate dehydrogenase (LDH), glutamic oxaloacetic transaminase (GOT), glutamic pyruvic transaminase (GPT), serum alkaline phosphate (ALP), total bilirubin, glucose, triglyceride, cholesterol, calcium (Ca), phosphate (P), potassium (K^+), and sodium (Na^+); and in vivo corticosteroid levels. Pathology examinations were carried out on the tissues of the major organs. This involved hematoxylin and eosin (H&E) staining; fat staining; hemosiderin, collagen, and bilirubin special staining; and electron microscopy of the heart, liver, kidney, and bone marrow. The main results were as follows:

- 192 mg/kg group: Three animals died within 3 days of medication. The main signs of toxicity were decreased appetite, lethargic responses, and lowered heart rate; a fall in peripheral red blood cell count (RBCs), hematocrit, and hemoglobin levels; an elevated erythrocyte sedimentation rate and absence of reticulocytes; and a reduction in leukocyte count and neutrophil percentage. Bone marrow smears and biopsies showed reduced hematopoietic function, especially for erythroid cells, and the myeloid to erythroid ratio increased. The blood biochemical panel showed that the level of in vivo corticosteroids was significantly reduced. CPK, GOT, BUN, triglyceride, cholesterol, P, and K^+ showed an upward trend, and glucose and Na^+ tended to decrease. Light microscopy and electron microscopy found cytoplasmic coagulation and mitochondrial swelling in the heart muscle, cloudy swelling of the renal tubular epithelial cells, and glycogen deposition and vacuolar degeneration in the liver cells. Detailed histopathological[18] and electron microscopic[19] results are reported in other publications.
- 96 mg/kg group: Symptoms of toxicity were similar to those in the 192 mg/kg group but only two animals died.
- 48 mg/kg group: No animals died. Behavioral, biochemical, and pathology indicators showed no abnormalities. Laboratory tests found an absence of reticulocytes; reduction in red cell count, hemoglobin, and hematocrit levels; and an elevated erythrocyte sedimentation rate.
- 24 mg/kg group: Changes in reticulocytes only.

The monkeys were observed over a recovery period. All pathological signs returned to normal 22 days after medication ceased.

In sum, a 14-day course of artemisinin was safe for monkeys at a dose of 24 mg/kg and below. Toxic reactions appeared at doses greater than 48 mg/kg, with death occurring above 96 mg/kg. The main sign of toxicity was damage to the hematopoietic function of the bone marrow, especially for erythroid cells. The heart was also affected. These abnormalities were reversible.

The above tests began to approach and comprehend the guiding principles of modern, international drug safety research.[20,21] In particular, the experimental design of the rhesus monkey multiple-doses study was in line with the country's current regulations and guidelines of toxicology evaluation. It reflected the highest standard of artemisinin toxicological research at the time and provided important indicators for the safe use of artemisinin in a clinical setting. For instance, it emphasized that reticulocyte count should be monitored. The study also set the upper limit of a safe dose of artemisinin.[22]

Special Toxicity

Muscle Irritation

Japanese long-eared rabbits were given 1 mL oil preparations or aqueous suspensions of artemisinin, containing 50 mg artemisinin. No local damage was seen by the naked eye and by microscopy.

Micronucleus

Kunming mice were given artemisinin at 1/80, 1/40, 1/20, 1/10, and 1/5 of the LD_{50} dose. The test was negative.

Ames Test

Artemisinin exposure concentrations of 300, 30, 3, 0.3, and 0.03 μg/0.1 mL were used. The results were negative.

Teratogenicity

Pregnant Wistar rats were given oral artemisinin at 1/400, 1/200, and 1/25 the LD_{50} dose within the first 6 days of pregnancy. All fetuses developed normally, with no deformities. When the rats were medicated in the middle (7th–12th day) to late (13th–19th day) stages of pregnancy, none of the fetuses survived. When a dose of 1/200 or 1/25 the LD_{50} of artemisinin was administered on the 6th–15th day of pregnancy, all the rats miscarried; at 1/400 of the LD_{50}, half the rats miscarried.

The gestation period was divided into the early (6th–8th day), middle (9th–11th day), and late (12th–14th day) stages of organ formation. The drug was administered at different stages of organogenesis. A dose of 1/25 of the LD_{50}, given during the early stage, resulted in 6.1% of the fetal rats developing umbilical hernia deformities. At the middle and late stages, all the rats miscarried. The results were similar in mice.

Special toxicity tests were also conducted by other work units during this time, including Ames, micronucleus, and chromosomal aberration tests (V79).[23,24] These were all negative. The toxicity of oral artemisinin on fetal rats mainly affected the middle and late stages of organogenesis. It had no effects on the form and function of the male sex organs.

The *Journal of Traditional Chinese Medicine* summarized the toxicology studies on artemisinins.[17] Apart from artemisinin itself, the report also included its derivatives, artemether and sodium artesunate. During this period, toxicology research findings on artemether and sodium artesunate were also published by the Guangxi Medical University, the ICMM, and the IME of AMMS. These are presented below.

2. Artemether[17]

Acute Toxicity

Mice received intramuscular injections of artemether and were observed for 7 days. The LD_{50} was 263 mg/kg and the therapeutic index was 447. Signs of toxicity were lethargy, limp, and deathlike posturing, anorexia, piloerection, and a lowered heart rate. During this time, another experiment was reported in which white mice were given oral artemether. The LD_{50} was 567 mg/kg, the LD_5 was 229.83 mg/kg, the therapeutic index (LD_{50}/ED_{50}) was 12.9, and the safety index (LD_5/ED_{95}) was 3.4.[25]

A female dog was given 130 mg/kg oral artemether, with a second dose 2 days later. No signs of toxicity were observed. In another experiment, three dogs received incremental oral doses of 10, 20, and 40 mg/kg once a day for 3 days. One dog displayed a one-time vomiting response.[25]

No signs of toxicity were seen in a male monkey given an intramuscular injection of 141 mg/kg artemether.

Several rabbits received intramuscular injections of 160 mg/kg artemether. It was well tolerated by all the animals.

Multidose

Rats received intramuscular injections of artemether over 9–14 days, total dose 40–360 mg/kg. The weights of the rats in the high-dose group decreased and histology examination showed slight change in the liver cells.

Six rabbits were divided into three groups of two. The medicated groups were given 10 or 40 mg/kg oral artemether once a day for 6 days. No abnormalities were seen in their behavior, body weights, food intake, and excretion, as well as in hematology, liver and kidney function tests, ECGs, and pathology examinations of major organs.

Monkeys were given intramuscular injections of artemether totaling 97 and 292 mg/kg over 1–3 months. No abnormal signs appeared.

In an experiment on dogs, one group received intramuscular injections of artemether once a day for 3 days. This was repeated after an interval of 7 days. A second group was given the same regimen for a total of 3 months. The total doses of the two groups were 24 and 972 mg/kg, respectively. Apart from weight loss and a slight morphological change in the liver cells, the various indicators in the high-dose group were normal. This included routine urine tests, blood biochemistry panels, ECGs, and histopathology examination of the heart, spleen, lung, kidney, brain, adrenal glands, pituitary gland, gonads, and other organs. Another experiment on dogs used 4 and 16 mg/kg doses of oral artemether, given once a day for 6 days. Behavior, liver and kidney function, and hematology tests showed no abnormalities. Histopathology examination revealed only one case with liver congestion and interstitial inflammation in the kidneys.[25]

Wu Boan et al. published a detailed report on a toxicology study with beagles given multiple intramuscular injections of artemether.[26] The dogs were randomly divided into five groups of four. The medicated groups received 6, 19, or 32 mg/kg artemether. The fourth group received the same volume of solvent, and the fifth was a negative control. The drug was given once a day for 15 days. Clear signs of toxicity appeared in the medium- and high-dose groups, such as loss of appetite and weight, salivation, trembling of the muscles, and fever. Three animals in the high-dose group died, with coma and respiratory depression preceding death. ECGs showed prolonged QT and PR intervals, with first-degree atrioventricular block appearing in some animals. Hematology tests found a reduction in reticulocyte and red cell count in the middle- and high-dose groups. Total white cell count and neutrophil count also decreased in the high-dose group. Bone marrow smears revealed significant damage to erythroid cells. Creatine kinase (CK) and triglyceride levels were elevated in the middle- and high-dose groups. The high-dose group tested positive for urinary protein and blood. The histopathology results showed varying degrees of hyperemia, hemorrhage, and edema in the middle- and high-dose groups, visible to the naked eye. Microscopic examination revealed degeneration and necrosis of the gastrointestinal mucosa, vacuolar degeneration of the liver cells, and inflammatory cell infiltration. Other signs included subendocardial hemorrhage; subacute inflammation and hemorrhage of the lungs; partial atrophy or hemorrhage of the glomeruli and oozing of protein and red cells into the glomerular capsule; and proliferation of reticular cells in the thymus, lymph nodes, and spleen, with phagocytic activity. These indicators returned to normal 14 days after medication ceased. The main target organ of artemether toxicity is the hematopoietic tissue of the bone marrow. This was the first domestic experiment to use standard laboratory animals in a multidose study of the toxicity of artemisinins.

Special Toxicity

Local Irritation

Histological examination of biopsies of intramuscular injection sites was performed with mice, rats, rabbits, dogs, and monkeys. Muscle fibers were normal and blood CPK was not elevated.

Teratogenicity

Artemether had significant toxic effects on mouse and rat embryos. In the first 6–15 days of gestation, intramuscular injections of 10.72 mg/kg artemether over 10 days caused a 30.1% abortion rate. No malformations were found in the surviving fetuses. A dose of 21.44 mg/kg, administered on the 9th–11th day of pregnancy, caused miscarriage and death in all the pregnant rats.

Toxicity in Humans

Three volunteers were given 100 mg/kg AMT once daily for 3 days. No signs of toxicity were seen in ECGs, liver and kidney function, and routine blood and urine tests.[25]

3. Artesunate

Acute Toxicity

The LD_{50} of intravenous sodium artesunate in mice was 520 mg/kg; for intramuscular injection, it was 475 mg/kg.[17] The therapeutic index (LD_{50}/ED_{50}) of intravenous sodium artesunate on a sensitive strain of *P. berghei* was 1733. On a chloroquine-resistant strain, the index was 1040.[27] Another experiment found that the LD_{50} of intravenous sodium artesunate was 552.82 mg/kg. The therapeutic index (LD_{50}/ED_{50}) was 15.3, and the safety index was 4.8. The LD_{50} of oral sodium artesunate was 1002.57 mg/kg, with a therapeutic index of 13.04 and safety index of 4.19.[28]

Three dogs received intravenous sodium artesunate.[17] The dose increased daily, starting at 37.5 mg/kg on the first day, 75 mg/kg on the second, and 100 mg/kg on the third. The dogs' food intake decreased. They were supine and lethargic, refused to eat, and appeared fatigued but recovered 2 days after medication ceased.[27] In another experiment, four dogs were given 20, 40, or 80 mg/kg oral sodium artesunate once daily for 3 days, and no signs of toxicity were seen.[28]

Ning Dianxi et al. were the first to use the approximate lethal dose method to evaluate the toxicity of a single dose of intravenous sodium artesunate in dogs.[29] The results indicated that the maximum dose with no toxic side effects was 70 mg/kg, and the approximate lethal dose was 240 mg/kg. The main signs of toxicity were gastrointestinal and neurological symptoms, and blood was visible in the urine. Reticulocyte count was reduced and ALP levels elevated. Pathology examination showed suppression of erythropoiesis in the bone marrow, blood congestion in the organs, bile stasis, necrosis of the gastrointestinal mucosa, intestinal bleeding, and degeneration of the liver and kidney parenchymal cells.

Anesthetized rabbits were given intravenous sodium artesunate. When the cumulative dose reached 300 mg/kg, the heart rate slowed, ECGs showed a prolonged PP interval, and blood pressure decreased. At 600 mg/kg, the sinus rhythm was significantly delayed, heart rate decreased by half, and blood pressure fell to 40 mmHg. At 800 mg/kg, the T wave became abnormal, flattened, bidirectional, or inverted. Finally, at 1200 mg/kg, cardiac arrest and death occurred.

Multidose

Six monkeys were divided into three groups. The medicated groups were given 10 or 32 mg/kg intravenous sodium artesunate once daily for 14 days. No abnormalities were seen on histopathology examination.

Six dogs were divided into three groups, two of which received 10 or 40 mg/kg intravenous sodium artesunate once daily for 14 days, and a control group. Their weights, appetite, feces, urine, behavior, hematological tests, SGPT, and nonprotein nitrogen were the same as before medication. Histopathology examination found only a small amount of blood congestion and hyperemia in part of the liver. Another experiment divided six dogs into three groups. The two medicated groups were given 5 or 20 mg/kg intravenous sodium artesunate once daily for 6 days. One animal in the high-dose group displayed elevated transaminase and a small amount of bleeding or hyperemia in the liver. No other abnormalities were seen.[28]

Ning Dianxi et al. also undertook more systematic toxicity studies of multiple doses of intravenous sodium artesunate.[30] Thirty mongrel dogs, half male and half female, were divided into five groups of six. One of the groups was the control. The four medicated groups were given 12.25, 22.5, 45, or 90 mg/kg intravenous sodium artesunate, administered at a rate of 2 mL/min, once daily for 14 days. All dogs in the 90 mg/kg group were severely intoxicated and died on Days 9–11 of medication. Two dogs in the 45 mg/kg group died on Days 12–13. No clear signs of toxicity were seen in the 22.5 mg/kg group, but some indicators were abnormal. Apart from a slight disruption to the maturation of erythroid cells in the bone marrow, no abnormalities were seen in the 11.25 mg/kg group.

Symptoms of toxicity included vomiting, salivation, reduced food intake or refusal to feed, bloody mucus in the stool, fever, listlessness, lethargy, weakness, and weight loss. ECGs showed prolonged QT intervals at doses greater than 22.5 mg/kg and prolonged PR intervals and changes to the T wave above 45 mg/kg. A significant decrease in peripheral reticulocytes could be seen across all groups and nucleated erythrocytes. There was a drop in leukocytes, neutrophils, and eosinophils in the two higher-dose groups. Bone marrow smears mainly revealed disruption to erythropoiesis. In the 12.25 mg/kg group, erythroid hyperplasia, a fall in the myelo–erythroid ratio, and inhibition of the maturation of intermediate erythroblasts were seen. The erythroid percentage declined, and maturation of early erythroblasts inhibited in the 22.5 mg/kg group. Additionally, in the 45 mg/kg group, some animals displayed complete suppression of erythropoiesis. Nucleated erythrocytes were absent in bone marrow smears from the 90 mg/kg group. Myeloid hematopoietic disorders occurred in the 45 and 90 mg/kg groups, manifesting as decreased myeloid maturation and mitotic indices. In the two lower-dose groups, bone marrow biopsies showed that the hematopoietic tissue had filled the intertrabecular spaces. Scattered foci of early erythroblasts could be seen. Reduction and suppression of the various types of hematopoietic cell were found in the other two groups. ALP, cholesterol, triglyceride, SGOT, LDH, CK, creatinine, and urea levels were elevated in the two higher-dose groups, while blood glucose declined. Some dogs tested positive for protein in the urine.

Pathology abnormalities were mainly seen in the two higher-dose groups, with multiorgan congestion visible to the naked eye. Microscopic examination revealed degeneration, loss, and necrosis of the gastrointestinal mucosa; liver sinusoidal dilatation, congestion, liver cell swelling, vacuolar degeneration and necrosis, and inflammatory cell infiltration; congestion and interstitial vasodilation in part of the glomeruli; and varying degrees of bleeding in the epicardium. There was an inflammatory response at the injection site. Positive indicators gradually returned to or approached normal levels 28 days after medication ceased.

Special Toxicity

Local Irritation

This test involved rabbits, dogs, and monkeys given intramuscular or intravenous sodium artesunate. Gross and histopathology examination of muscles and blood vessels at the injection site showed no injury due to irritation.[17]

Mutagenicity

Li Zelin et al. reported that sodium artesunate had no mutagenic effects in an Ames test with histidine-deficient strains of *Salmonella typhimurium*, such as TA98, TA100, and TA102. Micronucleus tests were negative.[31] When V79 cells were exposed to sodium artesunate, the cultured cells showed no chromosomal aberrations with or without S9.

Embryotoxicity

Clear embryotoxic effects were seen in pregnant rats given 0.54, 0.81, 3.25, or 26 mg/kg sodium artesunate during organogenesis (Days 6–15 of gestation). Between 32.8% and 100% of the embryos were absorbed, and the maximum dose that produced no effects was 0.41 mg/kg. When 26 mg/kg sodium artesunate was administered in early organogenesis, 78.9% of the embryos were absorbed, 17.1% survived, and 3.5% had umbilical deformities. In middle to late organogenesis, all embryos were absorbed.[32] Sodium artesunate administered in early pregnancy (Days 1–6) resulted in 82.3% live births, with individual cases of umbilical malformation and delay in skeletal development.[33]

C. TRENDS IN TOXICOLOGY RESEARCH FROM THE 1990s ONWARD

1. Emphasis on Artemisinin Combination Therapy (ACT)

Beginning in the 1990s, more attention was devoted to artemisinin-based combination therapies (ACTs) to accelerate the fast action, prolong the antimalarial efficacy of new drug combinations, and avoid the emergence of drug resistance.

Therefore, many toxicological studies of this period focused on ACTs, such as artemether–lumefantrine, artemisinin–piperaquine (Artequick), dihydroartemisinin–piperaquine phosphate, artemisinin–naphthoquine, and artesunate–pyronaridine phosphate tablets. The Artequick studies met basic, modern, new-drug safety evaluation requirements[34] and were conducted in laboratories according to the Good Laboratory Practice standard (GLP).[35]

2. Relationship Between in Vivo Drug Exposure (Toxicokinetics) and Toxicity

Modern drug safety evaluation demanded greater emphasis on the relationship between in vivo exposure and toxicity. During this period, a few overseas reports were published about in vivo exposure from repeated administrations of artemisinin and the resulting toxicity. They indicated that such regimens had a tendency toward autoinduction, in which in vivo exposure concentrations decreased as the doses increased. Long-term in vivo exposure could cause neurological and reproductive toxicity.[36] Neurological toxicity appeared in monkeys after 179.5 h of exposure and in rats after 67.1 h.[37]

3. Studies Based on the Mechanism of Action of Artemisinin

As artemisinin's mechanism of action[38] gradually became understood, it was established that the drug's ability to kill the malaria parasite came from its peroxide group, which generated free radicals with Fe^{2+}. These free radicals damaged parasitic macromolecules. Such findings were a springboard for toxicity studies. For instance, Lou Xiaoe et al. discovered the mechanism by which artesunate damaged the tissues during pregnancy.[39] The drug severely reduced glutathione (GSH) peroxidase activity in rat embryos, while malondialdehyde (MDA) levels significantly increased and led to oxidative damage in the embryo.

The mechanism by which artemisinins inhibited erythropoiesis was also studied; Yao Lang et al. explored how artesunate suppressed the maturation of erythroid precursor cells in the bone marrow and their effects on heme biosynthesis.[40] Fluorescence spectroscopy revealed that peripheral erythrocyte protoporphyrin levels were elevated in beagles given oral artesunate, and sideroblasts appeared in the peripheral blood and bone marrow. It showed that inhibition of heme biosynthesis could be one pathway by which erythropoiesis was suppressed.

4. Focus on Neurological Toxicity of Artemisinins

The characteristics[41] and mechanisms[42] of neurotoxicity were studied. From the 1990s onward, attention focused on the neurotoxicity of artemisinins even though such effects had not been reported in humans. The first report came from the Walter Reed Army Institute of Research, which found that neurological symptoms such as ataxia and tremors appeared in dogs given large doses of intramuscular artemether, leading to death. Other studies confirmed this and determined that the drug caused dose-dependent damage to the brainstem. Artemisinin-type drugs produced this damage in animals such as rats, dogs, and rhesus monkeys. Dogs and rats were more sensitive to this than monkeys were. Some studies ranked the neurotoxicity of artemisinins, with dihydroartemisinin being the most toxic, followed by arteether, artemether, acidic benzyl artemisia ether, artemisinin, and artesunate as the least toxic.

The drugs' neurotoxic and antimalarial effects could be created by the same mechanism of action, namely disruption of the endoperoxide bridge catalyzed by ferrous ions or heme, resulting in an unstable organic free radical or electrophilic substances; these intermediates alkalized specific proteins in both parasitic cells and neurons, causing cell death. Studies on neuroblastoma and glioma tumor cell lines exposed to artemisinin revealed that artemisinins inhibited the growth of neuronal axons. Electron microscopy showed that dihydroartemisinin targeted the mitochondrial membrane and endoplasmic reticulum in neurons, creating similar morphological changes as in *P. falciparum*. With erythrocytes as a model and MDA as a lipid peroxidation indicator, it was observed that artemisinins could increase heme peroxidation by two to three times and heme-catalyzed lipid peroxidation by six times.

Chinese laboratories also studied the neurotoxic mechanisms of artemisinins.[43] Dihydroartemisinin and artemether could cause swelling in the mitochondria of the neurons; breakage, reduction, or absence of the mitochondrial cristae; and inhibition of mitochondrial respiratory chain synthase I and IV, reducing the ability to oxidize NADH and reduce cytochrome C.

5. Evaluation on Toxicity and Safety in Humans

In the 1970s and 1980s, many drugs studied in China were used in human trials immediately after preliminary safety evaluations in large animals. The test subjects were usually the researchers themselves. Beginning in the 1990s, according to the regulations in modern drug research and development, human trials had to conform to appropriate specifications and demands. Artemisinin-type drug trials began to be published.

For instance, Kongpatanakul et al. ran safety studies on dihydroartemisinin and artesunate involving healthy volunteers in Thailand. This trial was designed using randomized, single-blind, and crossover methods.[44] The subjects received 300 mg of the drug daily for 2 days. Adverse reactions were observed and laboratory tests conducted. Drug tolerance was good. Five days after medication, in vivo reticulocyte count reached its lowest level. On Day 7, hemoglobin levels fell significantly. The indicators were lower for artesunate than for dihydroartemisinin.

Price et al. studied the toxicity of artemisinins on falciparum malaria patients in the western border regions of Thailand.[45] A total of 836 patients received oral artesunate or artemether; 2826 received mefloquine in combination and 1303 received mefloquine alone. Adverse reactions included nausea, vomiting, anorexia, and dizziness, and the incidence was significantly higher in the mefloquine combination group. The artemisinins were well tolerated. No allergic, neurotoxic, behavioral, cardiovascular, and skin toxicity reactions were seen. In addition, a study by McGready et al. showed no clear toxic effects of artemisinins in pregnant women.[46] Artesunate and artemether were used to treat 83 pregnant women with falciparum malaria in Thailand. The drugs were well tolerated with no drug-related side effects. By the end of the follow-up, 73 women (88%) had given birth to live infants, 3 (4%) miscarried, and 2 (3%) were still pregnant. Contact was lost with five cases. The newborns had no congenital abnormalities; all 46 infants followed up for more than a year developed normally.

Li et al. investigated the severe embryotoxic effects of artemisinins in rodents versus their relative safety for humans. This could be due to differences between rodents and primates or humans in the erythroid cell and blood vessel formation stage during early gestation.[47] Embryotoxicity in rodents occurred during this stage. Given the short duration of this stage of only 1 day in rodents, single or multiple doses of the drug could result in the death of a high proportion of early erythroid cells and severe anemia in the embryo. Therefore, embryotoxicity was high in rodents. This developmental stage was longer in primates, and 12 days of drug exposure were needed to cause the same degree of embryonic cell death. Although human data are lacking, it seems certain than the same longer exposure time would apply, and the 2- to 3-day course required for antimalarial treatment would not result in embryotoxic effects. Moreover, most pregnant women currently take oral ACTs, in which the peak concentration of artemisinins is lower and the exposure time shorter. The probability of incurring embryotoxic effects is therefore low.

D. CONCLUSIONS

Artemisinins have become some of the most important antimalarials. They are effective and do not easily lead to resistance. Various signs of toxicity were seen in animal experiments but were seldom and relatively absent in humans.[48] Hence, they are truly a valuable contribution to humanity from the Chinese scientists. At the same time, it is noted that the treatment doses of these drugs are small and regimens are short. As the biological efficacy of artemisinins is studied further, other uses may be found in the future and dose and exposure time may increase. Therefore, attention must be paid to the toxicity of high doses administered over a long period. Especially research into genotoxicity, embryotoxicity, and cytotoxicity must be strengthened.

ACKNOWLEDGMENTS

Deepest thanks to Prof. Song Shuyuan of the Pharmacology and Toxicology Laboratory of the AMMS for data and valuable advice and to the AMMS archives for providing original records.

5.3 PHARMACOKINETICS OF ARTEMISININ AND ITS DERIVATIVES

Zhou Zhongming, Li Tao, Yang Weipeng
Institute of Chinese Materia Medica, China Academy of Chinese Medical Sciences, Beijing, China

The first pharmacokinetic (PK) study on artemisinins appeared in 1977, titled "A Preliminary Study of the Absorption, Distribution, Excretion and Metabolism of ^{3}H-Artemisinin and ^{3}H-Reduced Artemisinin in Mice,"[49] published by the National Radioisotope Technical Exchange Group. Since then, 38 years have passed. In October 1981, the first international paper on pharmacokinetics, "Metabolism and Pharmacokinetics of Qinghaosu and its Derivatives," was delivered in Beijing at the fourth WHO–SWG-CHEMAL meeting. It was presented by the Collaborative Research Group on Qinghaosu and its Derivatives as Antimalarials and signed by seven major research units: the ICMM of CACMS; Institute of Materia Medica, Chinese Academy of Medical Sciences; SIMM of CAS; Guangzhou UCM; Sichuan Institute of Chinese Materia Medica (now Sichuan ICMS; or Sichuan Academy of Chinese Medical Sciences SACMS); Guangxi Medical University; and Beijing Normal University.[50]

Other units also took part in PK research into artemisinins, including the AMMS, Shandong Institute of Chinese Medicine (now Shandong ICMS), China Institute of Atomic Energy of CAS, and Shanghai Institute of Organic Chemistry (SIOC of CAS). Many studies were conducted by two or more collaborating units, reflecting the spirit of cooperation. Because of artemisinins' unique effects against falciparum and cerebral malaria and their more extensive pharmacological activity, they attracted the attention and active participation of international medical and pharmaceutical centers. The rapid advance in methods of analysis and new technologies allowed the study of artemisinins in vivo to become more systematic and in-depth. This played an important role in drug development and clinical practice. This section focuses on PK studies of artemisinin, dihydroartemisinin (or reduced artemisinin), artemether, and artesunate and their ACTs.

A. ARTEMISININ

1. Artemisinin: Nonclinical Pharmacokinetics

The chemical structure of artemisinin had no specific or sensitive spectroscopic characteristics to aid analysis. To begin with, in vivo pathways were traced using random tritium (^{3}H) isotopic labeling.[49] After mice were fed ^{3}H-artemisinin, radioactivity soon appeared in the blood, reaching a peak after 1 h. This was followed by a rapid decline, dropping to half the peak value 4 h later. Thereafter, radioactivity subsided slowly. Radioactivity in the tissues, organs, and secretions was highest after 1 h and then gradually declined. It was concentrated in the bile and kidneys, followed by the liver, and could cross the blood–brain barrier. A whole-body autoradiography confirmed this (Fig. 5.1).

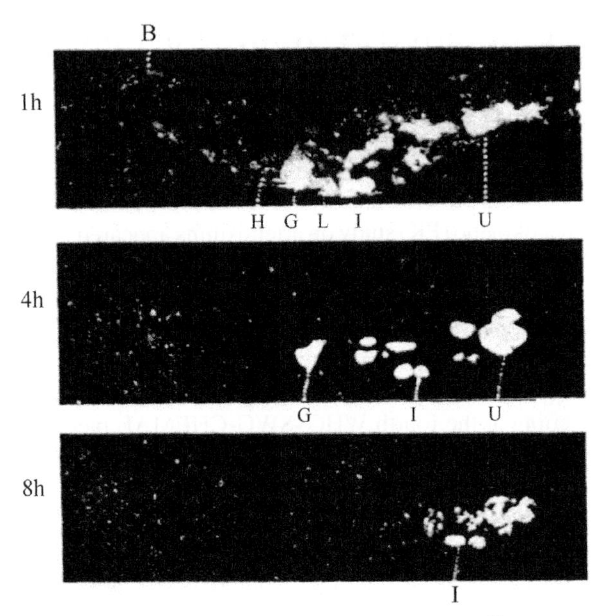

FIGURE 5.1 Whole-body autoradiogram of ^{3}H-artemisinin in mice.[50] *B*, brain; *G*, gallbladder; *H*, heart; *I*, intestine; *L*, liver; *U*, urinary bladder.

Excretion was rapid, primarily in the urine. Within 24h, 56.3 ± 9.8% was excreted in the urine and 22.2 ± 2.3% in feces. From 0 to 24h, 80% of radioactivity was excreted via urine and feces. Another experiment on ^{3}H-artemisinin in mice produced similar findings.[51]

If a sample preparation was not combined with radiation thin layer chromatography (TLC), the radiation intensity of the test results could include ^{3}H-artemisinin and radioactive residues and decomposition products. Because the radionuclide tracer was very sensitive and sample preparation was simple, it was suitable for large-scale experiments. Radionuclide tracers were used to study the in vivo pathways—mainly absorption, distribution, and excretion—of artemisinin, several of its derivatives, and many medicinal compounds. Such tracers had certain limitations. Therefore, comparative tests had to be run using other methods. The low sensitivity of early assay methods for artemisinin meant that only in vitro studies could be conducted. Alternatively, the dose needed to be increased in animals to obtain the PK parameters.

In 1982, Shu Hanlin, Li Wanhai et al. established a vanillin colorimetric assay method for artemisinin,[52] using it initially to study plasma protein binding.[53] The sensitivity was 1.4 g/mL, and the binding rate in human serum protein was 64.5 ± 6.8%. This method was easy to execute but because of its low specificity and sensitivity it could be applied to biological samples but not to analyze plasma concentrations in animals.

Based on a color reaction of artemisinin, Niu et al. used TLC to study the in vivo pathways of oral, intramuscular, and intravenous artemisinin in rats in 1985.[54] The biological sample processing method yielded an average recovery rate of 86.4%, and the sensitivity was 0.2 g/mL. Absorption was fast and complete through the gastrointestinal tract, but metabolism was rapid. After an hour, only around 8.3% of the original drug was seen in the liver, resulting in low bioavailability. One hour after rats were given 900 mg/kg oral artemisinin, the drug was concentrated in the liver, where the majority of it was metabolized. Metabolites were excreted in urine. The drug was widely distributed in the body and could cross the blood–brain and placental barriers. An hour after a dose of 150 mg/kg intravenous artemisinin, most of the drug was found in the lungs and kidneys. Because the first-pass effect did not take place in the liver, blood concentration was higher and was maintained for longer. The drug–time curve corresponded to an open two-compartment model. The parameters were $T_{1/2}\alpha=2.66$ min, $T_{1/2}\beta=30.13$ min, $k_{21}=0.057$/min, $k_{12}=0.121$/min, $k_{10}=0.105$/min, $V_C=0.90$ L/kg, and $V_B=4.1$ L/kg.

The rats were also given intramuscular injections of an artemisinin-oil suspension. This was absorbed relatively rapidly and blood concentration was high. Bioavailability was twice that of a water suspension. The amount of the original drug excreted in urine after 48 h was only 2.6% (intravenous, 150 mg/kg), 0.7% (intramuscular, 150 mg/kg), and 0.2% (oral, 300 mg/kg).

In 1987, based on the principle that artemisinin produced fluorescence after heating with sulfuric and acetic acids, Wang Dunrui et al. used fluorescence TLC to examine the PKs of an intraperitoneal injection of 500 mg/kg artemisinin-oil in mice.[55] The lower detection limit was 60 ng/mL. From 60 to 1000 ng/mL, the relationship between artemisinin concentration and intensity of fluorescence was linear. The drug–time curve in serum followed a two-compartment model. The parameters were $T_{1/2}\alpha=0.95$ h and $T_{1/2}\beta=1.87$ h. Serum concentration peaked at 0.5 h. After around an hour, the concentration of artemisinin fell, while the concentration of reduced artemisinin increased gradually. Because artemisinin and reduced artemisinin had different fluorescence spectrum characteristics, this method could distinguish between them. However, it was uncertain if the method could pinpoint other metabolites. Moreover, only two animals were used at any point in time and the number of samples was limited. The parameters thus represented only preliminary findings.

Song Zhenyu, Zhao Kaicun et al. established a radioimmunoassay (RIA) method[56] based on artemisinin's active peroxide group to study the in vivo PKs of a 10 mg/kg intramuscular injection of the drug in dogs.[57] Artemisinin was rapidly absorbed and plasma concentration peaked at 2 h. Peak concentration was 0.2 μg/mL. Elimination was also fast, with a half-life of 1.6 h and a mean retention time (MRT) of 3.3 h. For dogs given 50 mg/kg artemisinin orally or 20 mg/kg rectally, the drug was not detected in the blood, indicating that the bioavailability of intramuscular injections was higher. The sensitivity was 2–3 ng/mL, with less blood used. This method was simple, with no cross-reaction to metabolites without the dioxide bridge. It could thus be used only to measure the PKs of artemisinin or dihydroartemisinin. With artemether and artesunate, however, the primary metabolite was dihydroartemisinin. Like the original drugs, it also had a peroxide group. Hence, the measured result would include both the original drug and dihydroartemisinin. Nevertheless, because the drugs' antimalarial activity stems from this peroxide group, such results would still be clinically significant.

In 1986, Zhao Shishan and Zeng Meiyi established an HPLC method with pre-column reaction assay for artemisinin in animal and human plasma. It was based on the alkali reaction of artemisinin to form a product, which was then acidified to form another product named Q260 with strong UV absorption.[58] The assay sensitivity was optimized by increasing the sample size to 200 μL and decreasing the methanol content in the sample solution in comparison with the mobile phase, to enrich the sample in a short time in the column. The detection limit was 3 ng/mL. The linearity was good at a concentration range of 5–500 ng/mL. Sample preparation procedures were straightforward and could be assayed with ordinary HPLC and UV detection. Sensitivity and specificity were relatively good, and this method is still used today.

Compounds with peroxide groups such as artemisinins possess the characteristic of electrochemical redox reaction. In 1988 Zhou Zhongming et al. of the ICMM of CACMS worked with the Institute of Chemistry of CAS and the Guangzhou UCM to develop the HPLC with polarographic detector for determination of artemisinins in biological samples. The sensitivity of this method is qualified for its use in PK studies.[59]

Beginning in 2006, liquid chromatography–tandem mass spectrometry (LC–MS/MS) has been used to investigate the PKs of artemisinin.[60] This technology combined the advantages of chromatographic separation and mass spectrometric detection. It overcame artemisinin's lack of spectrometric characteristics. Sensitivity and specificity were high and analysis time was short. Hence, this method has now been widely adopted.

Xing et al. and Yan et al. applied LC–MS/MS to study the PKs of 40 mg/kg oral artemisinin in rat plasma.[60,61] The linear range was 1.0–200 ng/mL, the lower quantification limit was 1.0 ng/mL, and the recovery rate was 95.4 ± 4.5%. The parameters of a single administration were $C_{max} = 84.24 \pm 21.25$ ng/mL, $T_{max} = 1.20 \pm 0.27$ h, $AUC_{0\text{-}\infty} = 126.53 \pm 37.25$ ng·h/mL, $CL/F = 375.54 \pm 162.62$ L/min/kg, and $T_{1/2} = 1.01 \pm 0.12$ h. The parameters of a 5-day course were $C_{max} = 18.03 \pm 10.21$ ng/mL, $T_{max} = 1.20 \pm 0.27$ h, $AUC_{0\text{-}\infty} = 39.91 \pm 20.27$ ng·h/mL, $CL/F = 1365.84 \pm 795.80$ L/min/kg, and $T_{1/2} = 1.08 \pm 0.12$ h. This indicated that the activity of artemisinin in vivo was time dependent. In 2011, Xing, Yan et al. applied this method to the PKs of 1000 mg oral artemisinin in dogs.[62] The linear range was 2.0–500 ng/mL, the lower quantification limit was 2.0 ng/mL, and the recovery rate was 78%–90%. Only trace amounts of the drug were detected in plasma, indicating that it was poorly absorbed and rapidly metabolized in dogs. Further investigation was needed to clarify the specific mechanisms.

In 2012, Du et al. used HPLC–HRMS, LTQ Orbitrap mass spectrometer to qualitatively and quantitatively analyze artemisinin and its metabolites in rat plasma.[63] The linear range was 5.0–200 ng/mL, the lower quantification limit was 5.0 ng/mL, and the recovery rate was 95.4 ± 4.5%. Dose and route of administration were similar to the publication of Xing et al.[60] However, it yielded markedly different drug–time charts and results. The C_{max} value was above 600 ng/mL, a difference of about 10 times. The artemisinin was prepared using soybean oil, which may have affected the results.

Lie et al. employed ultra performance liquid chromatography–MS/MS in their study of the PKs of 8 mg/kg intravenously and 100 mg/kg orally of artemisinin in rat serum.[64] The lower quantification limit was 4.0 ng/mL. A high degree of linearity was seen at concentration range of 4.0–10,000 ng/mL, and the sample preparation method resulted in a recovery rate of 80.0%–107.3%. Only 100 µL of serum was used and the analysis took 3 min. Three rats were used in the preliminary test on intravenous artemisinin. The noncompartmental parameters were $C_{max} = 51.13 \pm 28.05$ µg/mL, $T_{1/2} = 101.23 \pm 36.71$ min, $V_d = 1898.31 \pm 635.15$ mL/kg, $CL = 15.59 \pm 11.61$ mL/min/kg, and $AUC_{0\text{-}6h} = 10.80 \pm 5.78$ µg·h/mL.

Because artemisinin was rapidly absorbed and metabolized and had a short half-life, some research units began to develop transdermal and effective sustained-release preparations in the 1990s. This would reduce the rate of recrudescence and improve therapeutic efficacy as a whole. Zhou Zhongming et al.[65] studied the in vivo PKs of single-dose, 50 mg/kg ordinary and controlled-release transdermal patches (artemisinin transdermal therapeutic system or QHS-TTS) in rabbits.[66] The drug-time curve of ordinary patches followed a linear, open, bell-shaped, single-compartment model. The parameters were $AUC_{0\text{–}36h} = 12{,}375$ ng·h/mL, $T_{max} = 8.25$ h, $C_{max} = 614.15$ ng/mL, $T_{1/2(ka)} = 3.50$ h, $T_{1/2(ke)} = 7.14$ h, and $T_L = 1.18$ h. These patches could eliminate rodent malaria without recrudescence. For the matrix-type controlled-release patches, plasma concentration began to rise after 10 h, stabilized at 36 h, and could be maintained for more than 5 days after a single dose. However, the long retention time was a problem possibly causing resistance to develop.

Li Guodong et al. investigated a dose of 20 mg/kg oral artemisinin, both normal and sustained-release, in New Zealand rabbits.[67] The main parameters of the normal oral preparation were $AUC_{0\text{-}12h} = 7.88 \pm 1.56$ ng·h/mL, $MRT = 5.84 \pm 1.09$ h, and $T_{1/2} = 0.99 \pm 0.17$ h. For the sustained-release oral preparation, they were $AUC_{0\text{-}12h} = 56.58 \pm 6.84$ ng·h/mL, $MRT = 8.77 \pm 1.56$ h, and $T_{1/2} = 1.01 \pm 0.53$ h. The peak concentration of the latter preparation was relatively higher than that of the former; the elimination rate was slow and the in vivo retention time was long. Relative bioavailability was six times higher than for normal artemisinin. This was an initial sign that efficient sustained-release oral preparations were effective in delaying drug release in the body. HPLC method with precolumn reaction assay and UV detection was used in these studies. As polymer materials, solvents, and penetration enhancers are developed, more efficient, safer, and convenient transdermal artemisinin and other sustained-release preparations have been produced, such as sustained-release transdermal patches,[68] self-emulsifying formulations,[69] artemisinin-based nanoliposomes, etc.[70] The in vivo PKs is being evaluated.

In the 1980s, Chinese scholars turned to bioeffectiveness (mainly toxicological and pharmacological) to determine the PKs of Chinese medicinal herbs. Li Chengshao et al. used pharmacological effectiveness[71] to study the PKs of an intramuscular injection of artemisinin-oil on rodent malaria.[72] The half-life was measured based on the decrease in drug effectiveness over time, starting at 100% at 0h. Time was set as the horizontal axis and the probit rate of decrease in drug efficacy as the vertical axis. Linear regression yielded a half-life of 10.5h. Extrapolation using the SD_{50} produced a PK half-life of 11h. Hence, the two methods came to basically the same results. The elimination rate constant, $K=0.21$/h, and the half-life of the effective dose, $T_{1/2ED}=3.34$h, were inferred based on the dose–response and time–effectiveness curves. Such methods bypassed artemisinin's lack of spectral characteristics. Examining relative concentrations in body fluids was a more direct form of evaluation, which aided in the study of complex drugs for which no specific assay methods existed. However, they yielded widely varying results with a large margin of error. It was also difficult to select sensitive indicators. Therefore, these methods are rarely used today.

2. Artemisinin: Clinical Pharmacokinetics

Oral Administration

Initially, because of artemisinin's relative insolubility, oral administration was commonly and effectively used in a clinical setting. In 1988, Zhao employed HPLC-UV method to determine the concentration–time curve of oral artemisinin capsules in the plasma and saliva of healthy Chinese volunteers.[73] The study was conducted as part of a wider test comparing these capsules to artemisinin suppositories.[74] Recently, Li Tao et al. reevaluated the data using Phoenix WinNonlin PK modeling software and a noncompartmental method. The concentration–time curve and parameters are shown in Fig. 5.2 and Table 5.12. The peak time for the capsules was 2.5h, after which drug concentration decreased rapidly. The in vivo MRT was 4.8h, AUC_{0-24h} was 4.2µg·h/mL, the clearance rate (CL_F_obs) was 2.4L/h/kg, and apparent volume of distribution (Vz_F_obs) was 14.5L/kg.

Titulaer et al. improved on this method's sample preparation process[71] and studied the PKs of 400mg oral artemisinin in 10 healthy volunteers.[75] The peak concentration was 260±94µg/L, $AUC_{0-\infty}$ was 819±190µg·h/L, peak time was 1.0±0.5h, absorption half-life was 0.54±0.29h, elimination half-life was 1.9±0.6h, and MRT was 3.4±0.7h.

Using RIA,[56] Zhao Kaicun et al. conducted a similar study on three healthy Chinese volunteers given 15mg/kg artemisinin tablets.[76] Plasma concentration

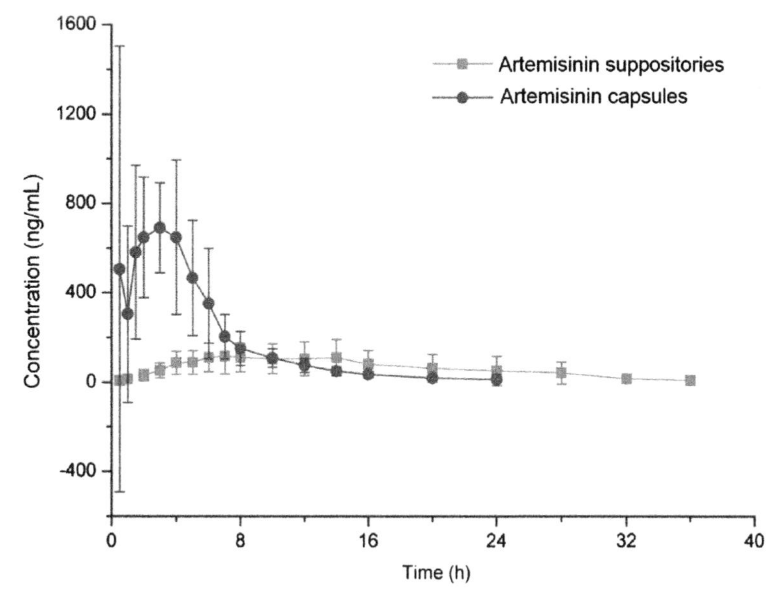

FIGURE 5.2 Drug plasma concentration–time curve in healthy volunteers after a single dose of 10 mg/kg artemisinin capsules ($n=6$) or suppositories ($n=9$).

peaked 1.5±0.32 h after medication, at 0.09±0.01 μg/mL. Serum concentration half-life was 2.77±0.22 h, apparent volume of distribution (V_{ss}) was 152.34±35.90 L/kg, clearance rate was 46.55±9.52 L/kg/h, $AUC_{0\text{-}36h}$ was 0.33±0.07 μg·h/mL, and in vivo MRT was 3.27±0.32 h.

Duc et al. tested 12 healthy Vietnamese volunteers given 500 mg artemisinin capsules orally.[77] The parameters are listed in Table 5.13. Absorption was rapid, with a peak time of 1.8 h. Absorption half-life was 0.58 h and peak concentration was 391 μg/L. Distribution was relatively broad, with an apparent volume of 19.4 L/kg. Clearance rate was high at 256 L/h and elimination half-life was 2.59 h.

Sidhu and Ashton studied four healthy Caucasians given 500 mg artemisinin capsules orally.[78] The elimination half-life, peak concentration, and clearance rate were 2.3±0.8 h, 150±92 ng/mL, and 16±7 L/h/kg, respectively, and the plasma protein-binding rate was around 88% in venous blood. Later, Ashton et al. used modified HPLC-UV[79] to test eight healthy Vietnamese volunteers who received 250, 500, or 1000 mg artemisinin capsules.[80] The parameters for each dose are shown in Table 5.14. Plasma concentration peaked in less than 2.8 h, the mean half-life of elimination was consistently less than 2.84 h, and average in vivo MRT was less than 5.49 h. As the dose increased, C_{max} and AUC doubled, the clearance rate fell, and the half-life tended to rise.

Benakis et al. adopted the HPLC method[81] in a study with seven healthy Swiss volunteers given 500 mg artemisinin tablets.[82] The results conformed to

TABLE 5.12 Pharmacokinetic Parameters (Mean ± SD) in Healthy Volunteers and Malaria Patients After a Single Dose of 10 mg/kg Artemisinin Capsules or Suppositories

	Healthy Volunteers			Malaria Patients
	Suppositories	Capsules		Suppositories
Parameters	Plasma[b]	Plasma[c]	Saliva[d]	Plasma[e]
HL_Lambda_z (h)	4.3 ± 1.2	4.2 ± 2.2	2.2 ± 0.9	3.3 ± 0.7
T_{max} (h)	11.1 ± 6.1	2.5 ± 1.4	2.5 ± 1.8	9.3 ± 3.5
C_{max} (ng/mL)	168 ± 70	1145 ± 457	179 ± 46	176 ± 93
$AUC_{0\text{-}24h}$ (ng·h/mL)[a]	2116 ± 1197	4202 ± 1091	565 ± 160	2360 ± 1310
Vz_F_obs (mL/kg)	41915 ± 32141	14504 ± 9259	57519 ± 30257	27575 ± 19631
Cl_F_obs (mL/h/kg)	7573 ± 7383	2400 ± 495	18195 ± 4203	5520 ± 3369
$AUMC_{last}$ (h·h·ng/mL)	28676 ± 21373	20994 ± 9406	2264 ± 1253	28484 ± 18856
MRT_{last} (h)	12 ± 4.2	4.8 ± 1.3	3.8 ± 1.4	12 ± 1.9

[a] *Artemisinin suppositories* $AUC_{0\text{-}36h}$.
[b] n = 9.
[c] n = 6.
[d] n = 3–4.
[e] n = 6.

TABLE 5.13 Pharmacokinetic Parameters of a Single Oral Dose of 500 mg of Artemisinin in 12 Healthy Vietnamese Subjects

	(M ± SD)	Min	Max
V_d/F (L/kg)	19.4 ± 6.9	9.9	30.1
AUC (μg·h/L)	2054 ± 455	1246	2938
$T_{1/2obs}$ (h)	0.58 ± 0.54	0.06	2.05
T_{max} (h)	1.81 ± 0.73	0.87	3.6
C_{max} (μg/L)	391 ± 147	174	683
$T_{1/2el}$ (h)	2.59 ± 0.55	1.58	3.42
CL/F (L/h)	256 ± 63	170	401

From Duc DD, De Vries PJ, Nguyen X, et al. The pharmacokinetics of a single dose of artemisinin in healthy Vietnamese subjects. *Am J Trop Med Hyg* 1994;**51**(6):785.

TABLE 5.14 Artemisinin Pharmacokinetic Parameters (Mean ± SD) After Oral Administration of 250, 500, and 1000 mg Artemisinin as Single Doses in Randomized Order in Eight Healthy Male Subjects

	Dose (mg)		
	250	500	1000
$AUC_{0-\infty}$ (ng·h/mL)	674 ± 386	1560 ± 755	3758 ± 2170
$AUC_{0-\infty}$/Dose (ng·h/(mL·mg))	2.70 ± 1.54	3.12 ± 1.51	3.76 ± 2.17
CL_{oral}/BW (L/(h·kg))	8.90 ± 4.00	7.83 ± 5.22	6.19 ± 2.66
C_{max} (ng/mL)	205 ± 127	450 ± 324	792 ± 498
T_{max} (h)	2.8 ± 1.9	2.3 ± 0.9	2.8 ± 1.6
T_{lag} (h)	1.5 ± 1.0	1.1 ± 0.4	1.0 ± 0.4
Half-life (h)	1.38 ± 0.40[a]	2.00 ± 0.60	2.84 ± 1.08
MRT (h)	4.18 ± 0.99[a]	4.49 ± 0.60	5.49 ± 1.52
V_{dxx}/BW (L/kg)	38.4 ± 18.9[a]	35.5 ± 24.8	33.7 ± 16.1

[a]n = 7.
From Ashton M, Gordi T, Hai TN, et al. Artemisinin pharmacokinetics in healthy adults after 250, 500 and 1000 mg single oral doses. *Biopharm Drug Dispos* 1998;**19**(4);245–50.

an open, one-compartment model. Plasma peak time was 100 min, peak concentration was 0.36 μg/mL, time of half-maximal concentration was 0.62 h, distribution half-life was 2.61 h, elimination half-life was 4.34 h, and $AUC_{0\text{-}\infty}$ was 1.19 μg·h/mL.

Gao Hongzhi et al. examined 10 healthy Chinese volunteers after taking 100 mg/kg artemisinin tablets.[83] C_{max} was 466.50 ± 120.45 μg/L, $AUC_{0\text{-}\infty}$ was 2.82 ± 0.87 mg·h/L, $T_{1/2}$ was 3.58 ± 0.86 h, and T_{max} was 2.15 ± 0.91 h.

Several reports deal with the in vivo pathways of oral artemisinin in malaria patients. Hassan Alin et al. tested 18 falciparum malaria patients in Tanzania who had received 500 mg artemisinin capsules.[84] The parameters were C_{max} = 615.4 ± 387.0 ng/mL, T_{max} = 2.5 h, elimination half-life = 2.2 ± 0.6 h, $AUC_{0\text{-}\infty}$ = 2234 ± 1502 ng·h/mL, and clearance rate = 314.3 ± 189.4 L/h.

In Vietnam, De Vries et al. studied 11 malaria patients treated with 500 mg artemisinin capsules.[85] The parameters were: V/F = 22.8 ± 16.6 L/kg, $AUC_{0\text{-}\infty}$ = 2302 ± 2023 μg·h/L, T_{max} = 2.88 ± 1.71 h, C_{max} = 364 ± 250 μg/L, $T_{1/2el}$ = 2.72 ± 1.76 h, MRT = 4.95 ± 1.96 h, and CL = 428 ± 342 L/h.

Ashton studied the PKs in 15 Vietnamese malaria patients who were given 500 mg artemisinin capsules once and after completion of a 5-day treatment course[86] (500 mg once per day on Days 1 and 5; 250 mg twice per day on Days 2–4.) The results are shown in Table 5.15. The parameters after a single dose (Day 1) were $AUC_{0\text{-}\infty}$ = 2780 ± 1717 ng·h/mL, CL = 299 ± 326 L/h, $T_{1/2}$ = 2.0 ± 0.5 h, C_{max} = 706 ± 414 ng/mL, median T_{max} = 2.5 h, and MRT = 4.5 ± 1.3 h. After 5 days, the AUC, CL, and C_{max} were markedly different. Hence, the parameters had a certain degree of time dependence.

Gordi et al. examined 77 cases of uncomplicated malaria in Vietnam who had received artemisinin capsules in different regimens.[87] The standard regimen, as above, was 500 mg on Days 1 and 5 and one 250 mg dose in the morning and evening of Days 2–4. Another regimen with gradually increasing doses was 100 mg on Day 1, 50 mg in the morning and evening of Day 2, 125 mg in the morning and evening of Days 3–4, and 500 mg on Day 5. The average parasite elimination time of the second regimen was 50 ± 23 h, longer than that of the standard regimen (34 ± 14 h). The parameters are shown in Table 5.16. Blood concentration was not high in the second regimen, and the shorter parasite elimination time did not appear either. Its parameters were not significantly different from those of the standard regimen. Therefore, it was not recommended for clinical use.

Low bioavailability may be one of the reasons behind artemisinin's high recrudescence rate. Augustijns et al. used a Caco-2 intestinal epithelial cell model to investigate intestinal absorption.[88] The apparent permeability coefficient of artemisinin, P_{app}, was (30.4 ± 1.7) × 10^{-6} cm/s, indicating high intestinal absorption. Svensson et al.[89] proved that artemisinin was not a substrate of P-glycoprotein. Hence, the small intestine should not be the cause of the reduction in the bioavailability of oral artemisinin. Now, it is known that artemisinin bioavailability is mainly affected by the drug's solubility and the first-pass effect in the liver.

TABLE 5.15 Artemisinin Pharmacokinetic Parameters (Mean ± SD) Determined on Days 1 and 5 of a Five-Day Oral or Rectal Regimen of 500 mg/day Artemisinin in Vietnamese Men With Uncomplicated Falciparum Malaria

	Oral ($n = 15$)		Rectal ($n = 15$)	
	Day 1	Day 5	Day 1	Day 5
$AUC_{0\text{-}10h}$ (ng·h/mL)	2434 ± 1372	468 ± 724[a]	698 ± 446[c]	181 ± 257[a]
$AUC_{0\text{-}\infty}$ (ng·h/mL)	2780 ± 1717[d]	686 ± 854[b,e]	865 ± 518[c,f]	395 ± 305[g]
CL_{oral} (L/h)	299 ± 326[d]	1618 ± 1188[b,e]		
$T_{1/2}$ (h)	2.0 ± 0.5[d]	1.9 ± 0.9[e]	2.0 ± 1.4[f]	3.5 ± 2.6[g]
C_{max} (ng/mL)	706 ± 414	134 ± 136[a]	185 ± 92[c]	41 ± 45[a,c]
T_{max} (h) (Median)	2.5	2	4	2.5
Range	0.5–5	1–6	2–10	1–8
T_{lag}(h) (Median)	1	1	1	1
Range	0.5–1.5	0–4	0.5–2.5	0.5–2.5
MRT (h)	4.5 ± 1.3[d]	4.8 ± 1.4[e]	4.8 ± 2.2[f]	7.3 ± 4.3[g]

[a]*P ≤ .002.*
[b]*P ≤ .01, comparison between Days 1 and 5.*
[c]*P ≤ .002, comparison between oral and rectal administration (Mann–Whitney U test) on the same day.*
Number of cases: [d]n = 13; [e]n = 11; [f]n = 9; [g]n = 4
From Ashton M, Sy ND, Huong NV, et al. Artemisinin kinetics and dynamics during oral and rectal treatment of uncomplicated malaria. *Clin Pharmacol Ther* 1998;**63**(4);482–93.

Rectal Administration

In clinical practice, patients with severe malaria often cannot be given drugs orally or vomit easily. Compliance with oral medication in children is also relatively poor. Hence, researchers had focused on rectal administration as an alternative delivery method. In 1982, the ICMM successfully formulated artemisinin suppositories. Clinical trials conducted by the malaria laboratory of the Guangzhou UCM proved that the suppositories were effective in treating falciparum,[90] vivax,[91] and severe malaria.[92]

In 1986, Zhou Zhongming et al. conducted a comparative PKs study of artemisinin suppositories and capsules in nine and six healthy volunteers, respectively.[93] The concentration–time curves are shown in Fig. 5.2. The same authors presented the parameters for artemisinin capsules in Table 5.12. Compared to the capsules, the suppositories had a later peak time. Peak concentration was low, clearance rate was slow, and the MRT was prolonged to 12h, 2.5 times of

TABLE 5.16 Pharmacokinetic Parameters for Vietnamese Male and Female Adults With Uncomplicated Falciparum Malaria Orally Treated With Artemisinin by a Standard- or Escalating-Dose Regimen

Regimen (mg)	CL_s/F (L/h)		$T_{1/2}$ (h)		C_{max}/D[a]		T_{max} (h)		MRT (h)	
	Mean ± SD	95% CI	Mean ± SD	95% CI	Mean ± SD	95% CI[b]	Mean ± SD	95% CI	Mean ± SD	95% CI
Standard										
Day 1, 500	1019 ± 633[c,d]	783, 1255	1.3 ± 1.1	0.9, 1.7	244 ± 202	162, 311	2.9 ± 1.9	2.3, 3.8	4.2 ± 1.4	3.7, 4.8
Day 5, 500	8140 ± 7837	5917, 12895	1.4 ± 2.4	0.8, 1.7	53 ± 47	37, 71	2.6 ± 1.2	2.1, 3.0	4.4 ± 3.9	2.8, 5.8
Escalating										
Day 1, 100	2019 ± 1277[c]	1503, 2535	1.1 ± 0.7	0.8, 1.4	185 ± 194	93, 232	2.8 ± 1.5	2.4, 3.7	3.5 ± 1.1	3.2, 4.0
Day 5, 500	7160 ± 5614	4983, 9336	1.1 ± 0.8	0.8, 1.4	57 ± 43	37, 70	2.5 ± 1.0	2.1, 2.9	3.7 ± 1.3	3.3, 4.2

[a] C_{max}/D, maximum plasma concentration adjusted for the dose.
[b] 95% CI, 95% confidence interval.
[c] $P < .01$ for within-group differences on the first and last treatment days ($n = 26–31$)
[d] $P < .01$ for between-group differences on the first treatment day ($n = 26–31$).
From Gordi T, Huong DX, Hai TN, et al. Artemisinin pharmacokinetics and efficacy in uncomplicated-malaria patients treated with two different dosage regimens. *Antimicrob Agents Chemother* 2002;**46**(4):1026–31.

that of the capsules. This allowed the drug to act in the body for a longer time. Relative bioavailability, calculated based on *AUC*, was 50% that of the capsules. This was basically consistent with the clinical results of Titulaer et al.[75]

The ICMM and Guangzhou UCM ran studies of the in vivo PKs and PDs of a single dose of the suppositories in malaria patients.[94] After medication, patients' body temperature tended to fall, and parasite clearance had basically occurred in 24 h. The plasma concentration effect–time curve is shown in Fig. 5.3, and the parameters were similar to those in healthy people.

These results were essentially similar to those of Koopmans et al.,[95] whose study involved 600 mg artemisinin suppositories with eight Vietnamese malaria patients. Both studies agreed in terms of parasite clearance, low plasma concentration, and that the suppositories had only 30% of the bioavailability of the oral formulations. However, the peak time was faster (7 h), which may be due to a different suppository matrix. The results also corresponded to the studies by Ashton et al.[86] and Titulaer et al.,[75] but the differences between individuals were large.

In short, after malaria patients were given 10 mg/kg or 600 mg artemisinin suppositories, blood concentration was low but could still reach effective treatment levels. The MRT of the suppositories was extended, which allowed the drug to maintain efficacy for longer. Relative to oral preparations, the suppositories had a certain clinical significance. These studies indicated that selecting preparations and routes of administration based on relative bioavailability alone was one sided. Confirmation had to be sought through pharmacokinetic–pharmacodynamic (PK–PD) tests.

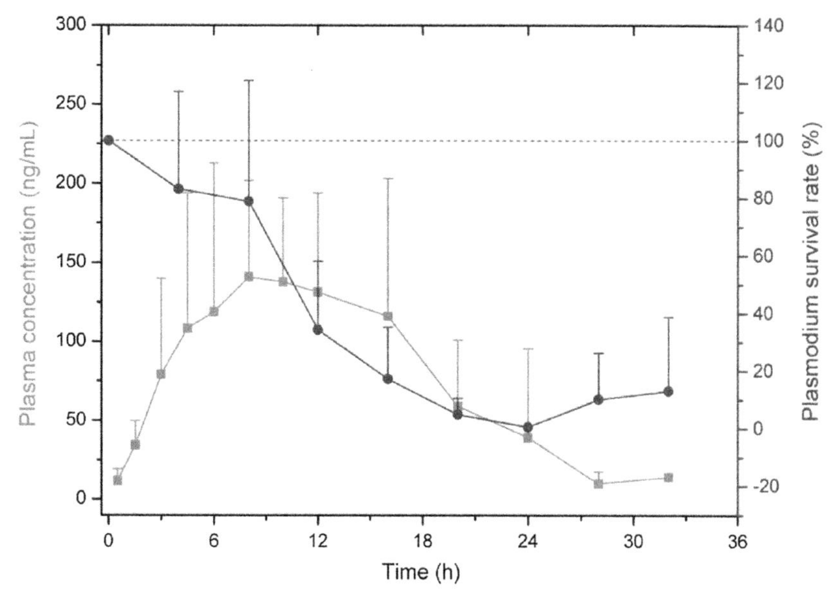

FIGURE 5.3 Plasma concentration effect–time curve of malaria patients after a single dose of artemisinin suppositories (—■— artemisinin plasma concentration–time; —●— parasite survival rate–time).

Intramuscular Administration

The study by Titulaer et al. on oral artemisinin also included 10 healthy Dutch volunteers given an injection of 400 mg artemisinin-oil suspensions.[75] Peak concentration was 209 ± 97 μg/L, $AUC_{0\text{-}\infty}$ was 2499 ± 1055 μg·h/L, peak time was 3.4 ± 2.0 h, absorption half-life was 1.59 ± 0.51 h, elimination half-life was 7.44 ± 3.83 h, and MRT was 10.6 ± 5.8 h. Compared with oral administration, intramuscular injections were absorbed slowly but had a longer elimination half-life and MRT and higher bioavailability. Similar results were obtained in animal experiments.[54]

Relationship Between Concentrations in Saliva, Capillaries, and Blood

In an early study, Zhou Zhongming et al. explored the relationship between drug concentrations in saliva and plasma[73] after artemisinin capsules were administered.[74] Typical concentration–time curves for saliva and plasma are shown in Fig. 5.4.

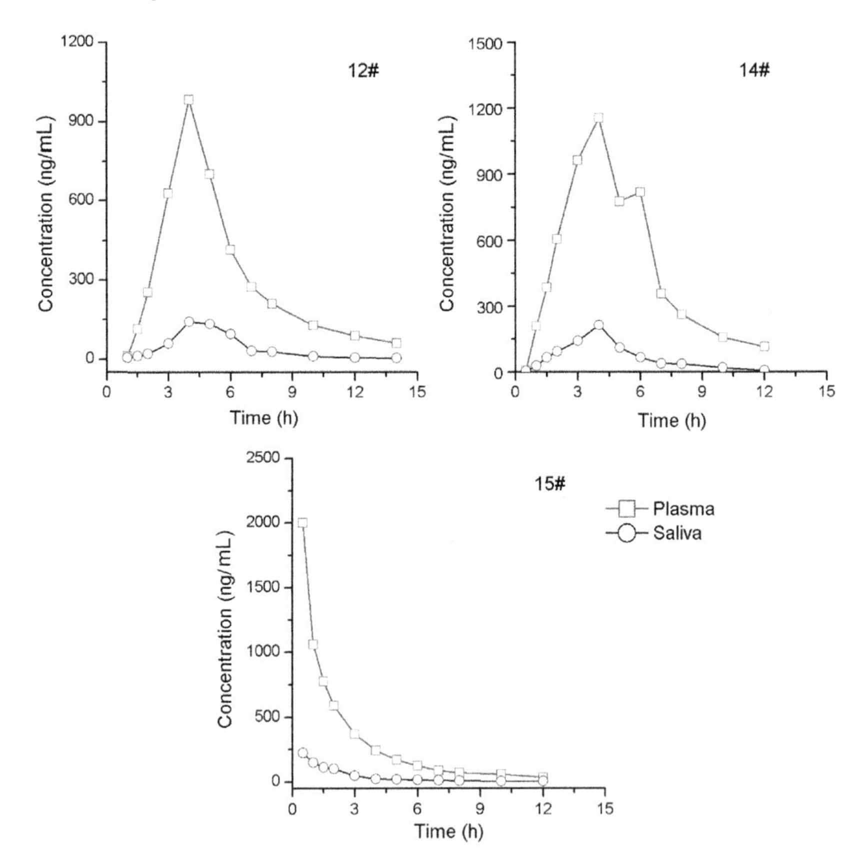

FIGURE 5.4 Drug concentration–time curves in the blood and saliva of three volunteers given 10 mg/kg artemisinin capsules.

Changes in concentration were consistent in both saliva and plasma, and the in vivo peak time was basically the same. Linear regression was run on the paired data. It showed a linear relationship between concentration in saliva (S) and in plasma (P), $P = .039 + 6.66S$ ($r = .949$, $P < .001$). Hence, plasma concentration was around 6.66 times saliva concentration.

Sidhu and Ashton also investigated the relationship between saliva, capillary, and venous blood concentration in four healthy male volunteers.[78] After a 3-h period, the ratio of drug concentration in saliva to free drug concentration in venous blood began to stabilize at 0.65 ± 0.05; concentration in the latter was about 1.54 times the concentration in the former. Concentration in saliva was closer to that in the capillaries. Gordi et al.[96] found similar results in 18 malaria patients. The experiments run by Sidhu and Gordi et al. also determined the protein-binding rate. After the 3-h stabilization period, drug concentration in saliva and the capillaries and free-drug concentration in plasma basically equalized, with ratios of around 1.18 and 0.99. This provided a basis for PK studies and drug formulations in which saliva or capillary blood replaced plasma.

Drug Metabolism

In the 1980s, Zhu Dayuan et al. studied the metabolic conversion of oral artemisinin[97] and found that, in humans,[98] a portion of the drug was excreted in its original form in feces.[99] Four metabolites were detected in urine (Fig. 5.5): deoxyartemisinin (1), deoxydihydroartemisinin (2), 9,10-dihydroxydeoxyartemisinin (3), and a derivative of hydrogenated indene (4).[99] Their structural features all showed that the peroxide bridge had been lost and they had no antimalarial activity.

It indicated that the metabolic process also deactivated the drug, and that the peroxide bridge was important for antimalarial efficacy. Recently, Liu et al. used LC–HRMS (LTQ Orbitrap) to identify in vivo first- and second-phase

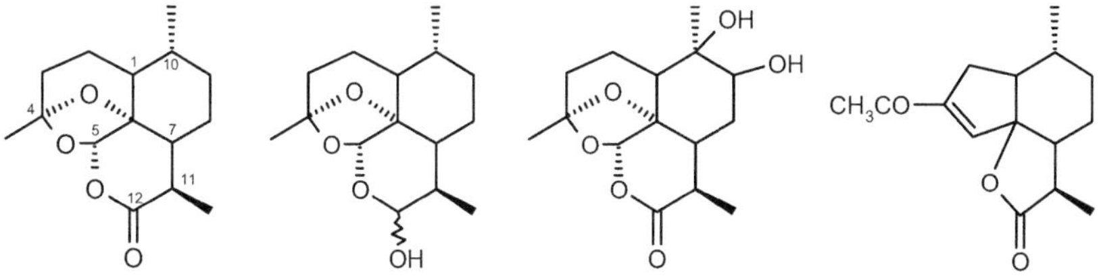

FIGURE 5.5 Metabolites of artemisinin in human urine. *(From Zhu D, Huang B, Chen Z, et al. Isolation and identification of biological metabolites and conversion products of artemisinin I. Isolation and identification in humans.* Acta Pharm Sin *1980;**15**(7);509–12. [朱大元，黄宝山，陈仲良，等. 青蒿素生物代谢转化物的分离鉴定 I. 人体代谢转化物的分离和鉴定. 药学学报，1980，15(7)：509–512.]; Zhu D, Huang B, Chen Z. Isolation and identification of conversion products of artemisinin metabolism in humans.* Acta Pharmacol Sin *1983;**4**(3):194–7. [朱大元，黄宝山，陈仲良. 人体代谢青蒿素转化物的分离鉴定. 中国药理学报，1983，4(3)：194–197.])*

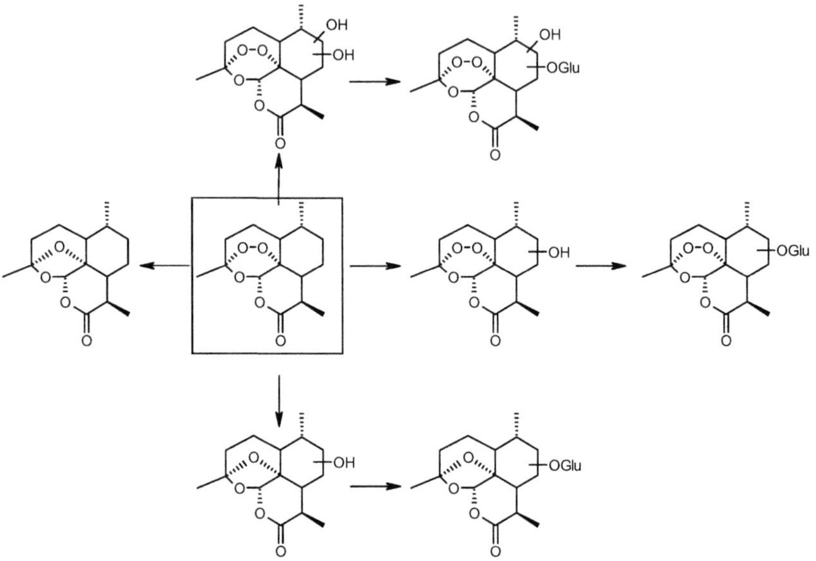

FIGURE 5.6 Proposed metabolic pathway of artemisinin. *(From Liu T, Du F, Wan Y, et al. Rapid identification of phase I and II metabolites of artemisinin antimalarials using LTQ-orbitrap hybrid mass spectrometer in combination with online hydrogen/deuterium exchange technique.* J Mass Spectrom *2011;**46**(8):725–33.)*

metabolites in rats.[100] The study suggested a possible metabolic pathway (Fig. 5.6) and found that deactivation took place mainly in the liver.

Svensson and Ashton proved that artemisinin metabolism in the liver microsomes was primarily related to CYP2B6, followed by CYP3A4. CYP2A6 was relatively uninvolved.[101] Mukanganyama et al. proposed that the metabolic process was also related to the function of UDP-glucuronosyltransferases.[102]

3. Factors Affecting Artemisinin's Metabolism and Pharmacokinetics

Formulation Factor

Different artemisinin preparations, such as suppositories, injections, and capsules, have different in vivo PKs. Koopmans et al. studied two different types of artemisinin suppository—in fat and in polyethylene glycol—in male Vietnamese volunteers.[103] The parameters of the two kinds of suppository showed no clear differences. Ngo et al. studied the relationship between the in vitro solubility of three types of artemisinin tablets and their bioavailability in rabbits.[104] The correlation between solubility and in vivo bioavailability was good.

Ashton et al.[105] found that the in vivo PKs of artemisinin was time dependent.[106] As Table 5.17 shows the *AUC* and C_{max} decreased after continuous administration and the clearance rate increased.

TABLE 5.17 Mean ± SD Artemisinin Pharmacokinetic Parameters on Days 1, 4, and 7 of a 7-Day Oral Regimen of 500 mg of Artemisinin Daily to 10 Healthy Male Vietnamese Adult Subjects, of Which 7 Were Restudied on Day 21 After a 2-Week Washout Period

Day	*AUC* (ng·h/mL)	C_{max} (ng/mL)	$T_{1/2}$ (h)	*CL/F* (L/h)
1	1373 ± 912 (754, 1779)	311 ± 232 (165, 395)	3.0 ± 1.2[a] (2.0, 3.8)	502 ± 283 (281, 664)
4	402 ± 251 (147, 605)	148 ± 93 (74, 197)	3.8 ± 2.0[a] (2.3, 5.2)	3033 ± 4636 (826, 3398)
7	346 ± 398 (61, 485)	110 ± 104 (34, 149)	4.8 ± 5.0[b] (1.2, 9.8)	8048 ± 12895 (1031, 8190)
21	862 ± 503[c] (137, 2180)	195 ± 126[c] (59, 359)	2.7 ± 0.9[c] (1.8, 3.6)	553 ± 186[c] (364, 767)

Values in parentheses represent the 95% confidence intervals around the (geometric) average based on log-normal distributions.
[a] $n = 8–9$.
[b] $n = 5$.
[c] $n = 6–7$.
From Ashton M, Hai TN, Sy ND, et al. Artemisinin pharmacokinetics is time-dependent during repeated oral administration in healthy male adults. *Drug Metab Dispos* 1998;**26**(1):25–7.

Gupta et al.[107] and Xing et al.[108] confirmed that autoinduction appeared in the in vivo metabolism of artemisinin, which could be related to the effects of CYP2A6[109] and CYP2B6.[110] Intestinal absorption experiments of Augustijns et al. indicated that P-gp and other intestinal transporters were uninvolved.[88]

Organism Factor

Ashton et al. found sex-dependent differences in the PKs of artemisinin in rats (Table 5.18).[111] As described above, there were no clear differences in PKs between healthy individuals and malaria patients. The influence of age will be discussed in the section on population pharmacokinetics (in Section 4 below).

Interactions With Other Drugs or Food

Hassan Alin et al. studied the PKs of a 500 mg dose of artemisinin, taken after an artemisinin-only regimen or an artemisinin-mefloquine regimen, in malaria patients.[84] The C_{max} and elimination half-life in plasma were roughly the same, but the *AUC* increased significantly and clearance rate decreased (Table 5.19). Svensson et al. found that when artemisinin was combined with a mefloquine enantiomer,[112] the latter had no effect on artemisinin PKs.

TABLE 5.18 Artemisinin Pharmacokinetic Parameters After Short-Term Constant Rate I.V. Infusion (20 mg/kg) or I.P. Injection (50 mg/kg) of Artemisinin Emulsion in the Male and Female Sprague-Dawley Rat

	Intravenous Infusion (20 mg/kg)		Intraperitoneal Injection (50 mg/kg)	
	Female	Male	Female	Male
n	5	5	6	6
$AUC_{0-\infty}$ (ng·h/mL)	1894 (1605, 2234)	1813 (1502, 2188)	5562 (4688, 6598)	2490 (1746, 3550)[a]
CL (L/h/kg)	10.6 (7.5, 15.0)	12.0 (10.4, 13.0)	8.71 (7.33, 10.3)	21.7 (12.1, 38.8)[a]
CL_u (L/h/kg)	90.6 (64.2, 128)	132 (118, 147)[b]	74.4 (62.7, 88.4)	246 (137, 441)[a]
V_{ss} (L/kg)	8.77 (5.99, 12.8)	8.75 (7.62, 10.1)		
$V_{u,ss}$ (L/kg)	74.9 (51.2, 110)	99.5 (86.6, 114)		
Terminal half-life (h)	0.52 (0.35, 0.76)	0.57 (0.44, 0.73)	0.59 (0.47, 0.73)	0.84 (0.70, 1.00)

Values in parentheses represent the 95% confidence intervals of log-transformed data; CL_u, unbound clearance; V_{ss}, steady-state volume of distribution; $V_{u,ss}$, steady-state volume of distribution of unbound drug.
[a] *P<.001 when compared with females.*
[b] *P<.05 when compared with females.*
From Ashton M. Quantitative in vivo and in vitro sex differences in artemisinin metabolism in rat. *Xenobiotica* 1999;**29**(2):195–204.

Zhou Zhongming et al. looked into the effects of food in rats given 80 mg/kg oral artemisinin.[113] Absorption was delayed, peak time prolonged, peak concentration reduced, and elimination time extended in the group given food. However, the *AUC* and bioavailability were higher. The parameters for this group were C_{max} = 1.55 μg/mL, T_{max} = 3 h, *AUC* = 12.5 μg·h/mL, $T_{1/2(ka)}$ = 1.36 h, and $T_{1/2(ke)}$ = 2.9 h. It suggested that, in a clinical setting, the first dose of artemisinin might be taken on an empty stomach, so that a higher blood concentration may be reached rapidly. Subsequent doses may be taken after meals to maintain blood concentration for a longer time. However, Dien et al. conducted a similar study with healthy, male Vietnamese volunteers given 500 mg oral artemisinin capsules.[114] It showed that the impact of food on the PKs of artemisinin was not statistically significant. Food was not a main factor influencing the behavior of artemisinin. The differences between these studies could be species-specific or due to varying drug preparations and types of food.

TABLE 5.19 Mean ± SD Artemisinin Pharmacokinetic Parameters After the First 500 mg Oral Administration to Falciparum Malaria Patients Who Received Artemisinin Alone (Treatment A) or in Combination With Mefloquine (Treatment A + M)

	Regimen A ($n=18$)	Regimen A+M ($n=20$)	P Value
$AUC_{0\text{-}8h}$ (ng·h/mL)	2014 ± 1359 (505.7, 5502)	2786 ± 1608 (1096, 6894)	<.05
$AUC_{0\text{-}\infty}$ (ng·h/mL)	2234 ± 1502[a] (675.9, 6412)	3252 ± 1873[b] (1337, 7409)	<.05
$T_{1/2}$ (h)	2.2 ± 0.6[a] (1.1, 3.4)	2.5 ± 0.7[b] (1.5, 4.0)	Not significant
CL/F (L/h)	314.3 ± 189.4[a] (78.00, 740.0)	195.4 ± 86.91[b] (67.00, 374.0)	<.05
V_{ss}/F (L)	1578 ± 986.7[a] (403.0, 3822)	976.8 ± 464.9[b] (292.0, 2095)	Not significant
C_{max} (ng/mL)	587.4 ± 385 (99.4, 1602)	818.3 ± 493.1 (262.1, 2073)	Not significant
T_{max} (h)	2.5 (1.0, 4.5)	2 (1.0, 6.0)	Not significant

[a]$n = 15$.
[b]$n = 17$.
Regimen A, Artemisinin; *Regimen A+M*, artemisinin + mefloquine.
From Hassan Alin M, Ashton M, Kihamia C, et al. Clinical efficacy and pharmacokinetics of artemisinin monotherapy and in combination with mefloquine in patients with falciparum malaria. *Br J Clin Pharmacol* 1996;**41**(6):587–92.

4. Population Pharmacokinetics

Pregnant women, the elderly, and children have very different physiological indicators from normal adults. In clinical practice, one cannot automatically apply adult doses to these groups, and it is very important to establish individual regimens for them. Population PKs is a powerful tool in this respect. In practice, artemisinin is often used in combination therapies, the population PKs of which can be found in later chapters.

Here, one key study on PKs in children will be discussed.[115] This study chiefly analyzed the impact of weight and dose on the PK parameters of artemisinin. It involved 23 child patients (aged 2–12 years) and 31 adult patients (aged 16–45) with uncomplicated malaria. The dose was 10 mg/kg/day oral artemisinin for 5 days. On Day 1, 107 capillary blood samples were taken; on Day 5, 33 such samples were obtained. Plasma drug concentration was determined using HPLC. The parameters were calculated with NONMEM software.

Parasite clearance and defervescence times were similar in children and adults. The data on plasma concentration were plotted separately for children and adults, and the analysis yielded a one-compartment model. The *CL/F*, *V/F*, and K_a values were based on the Day 1 samples.

Clearance and volume of distribution in adults were 432 L/h (RSD% = 19%) and 1600 V/F(L) (RSD% = 28%); in children, they were 14.4 L/h/kg (RSD% = 24%) and 37.9 L/kg (RSD% = 33%). Interindividual variation in *CL/F*, *V/F*, and K_a was 45% (RSD% = 44%), 104% (RSD% = 36%), and 576% (RSD% = 21%), respectively. Intraindividual variation in *CL/F* and *V/F* was 53% (RSD% = 32%) and 86% (RSD% = 36%). Residual variation σ was 47% (RSD% = 41%). Oral bioavailability gradually decreased by a factor of 6.9 from Days 1–5; this was similar in children and adults. A final evaluation was performed reanalyzing the 53 datasets with case deletion diagnostic statistics. Apart from K_a, the variation coefficients of the other parameters were less than 10, indicating that the model was stable. Average clearance rates in adults and children were 402 and 13.2 L/h/kg, respectively, and mean volumes of distribution were 1504 and 36.7 L/kg, respectively.

Body weight clearly affected *CL/F*, *V/F*, and other parameters. Therefore, it was recommended that the clinical dose of artemisinin for children should be based on weight, whereas adults could receive the standard dose. Moreover, artemisinin concentrations showed a certain degree of time dependence, which affected recrudescence rates and drug regimens.

5. Pharmacokinetic–Pharmacodynamic Models

Previous studies have shown that in vivo artemisinin metabolism was characterized by autoinduction and first-pass saturation in the liver. To better explore artemisinin's time-dependent PKs, Gordi et al.[116] investigated two dose regimens. A total of 24 healthy, male Vietnamese adults were randomly divided into two groups. One group was given 500 mg oral artemisinin a day for 5 days; the other received increasing doses of 100, 100, 250, 250, and 500 mg over 5 days. Besides this, two people from each group took 500 mg of the drug on Days 7, 10, 13, 16, 20, and 24. Saliva samples were collected on Days 1, 3, 5, and the last day of medication and drug concentrations measured via HPLC. The data was analyzed using a semiphysiologic PK model, which included the autoinduction effects on metabolizing enzyme precursors and a two-compartment model, with the addition of a liver compartment to account for hepatic enzyme autoinduction and high hepatic extraction. The PK model is shown in Fig. 5.7.

Artemisinin's autoinduction occurred at around 1.9 h, and the enzymatic elimination half-life was about 37.9 h. Hepatic extraction ratio was about 0.93, reaching 0.99 after autoinduction. The model showed that autoinduction primarily affected artemisinin's bioavailability and not the systemic clearance. Hepatic elimination saturation could explain why the dose–*AUC*

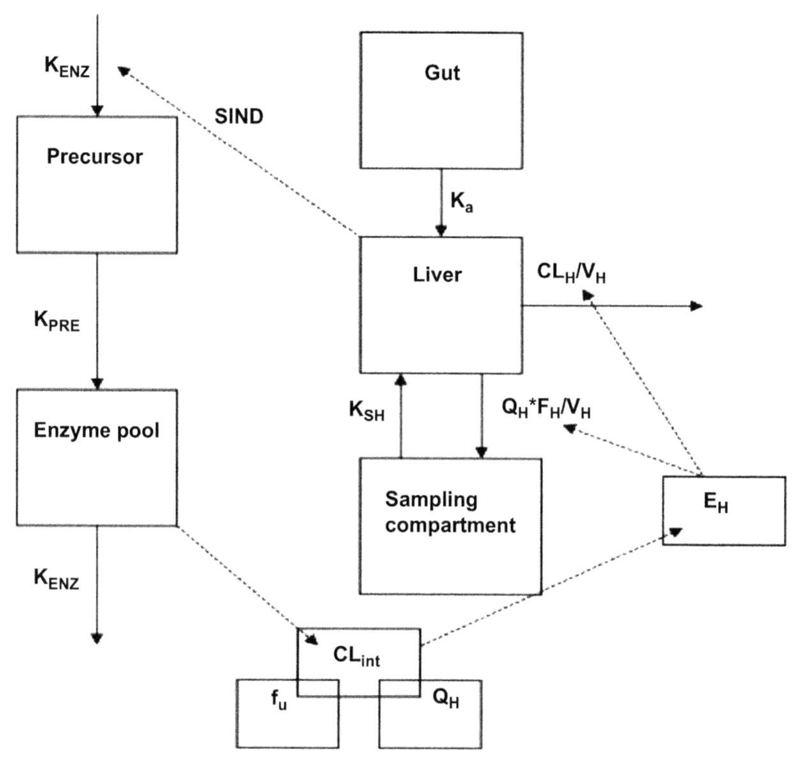

FIGURE 5.7 Schematic diagram of time–drug concentration in saliva. CL_H, hepatic clearance; CL_{int}, intrinsic clearance; E_H, extraction ratio; F_H, bioavailability from the liver compartment to the sampling compartment; f_u, plasma unbound fraction; k_a, absorption constant rate; k_{ENZ}, zero-order production rate constants for the enzyme precursor or first-order elimination rate of the metabolizing enzymes; k_{PRE}, first-order production rate constant for the metabolizing enzymes; k_{SH}, transfer rate constant of artemisinin from the sampling compartment to the hepatic compartment (set equal to Q_H/V_S, V_S being the volume of distribution of the sampling compartment); Q_H, hepatic plasma flow; S_{IND}, slope of the inducing effect of artemisinin hepatic concentration on the production rate of enzyme precursor; V_H, volume of the liver compartment (set equal to 1). *(From Gordi T, Xie R, Huong NV, et al. A semiphysiological pharmacokinetic model for artemisinin in healthy subjects incorporating autoinduction of metabolism and saturable first-pass hepatic extraction.* Br J Clin Pharmacol *2005;**59**(2):189–98.)*

relationship was nonlinear. In short, the experiment confirmed artemisinin's rapid autoinduction and reduction in bioavailability over time. The different dose regimens could effectively model the drug's in vivo PKs, including autoinduction and first-pass elimination. PK parameters are shown in Table 5.20.

In a later experiment, Gordi et al.[117] adopted the same protocol with 77 Vietnamese malaria patients. Using drug concentration in saliva and parasitemia at different points in time, a semimechanistic, combined PK–PD model was

TABLE 5.20 Typical Pharmacokinetic Parameter Values for Artemisinin and Associated Interoccasional (IOV) and Interindividual (IIV) Variability in 23 Healthy Vietnamese Subjects

Parameter	Estimate (RSE%)	IOV (RSE%)	IIV (RSE%)
$T_{1/2ENZ}$ (h)	37.9 (22)	–	–
S_{IND} (L/ng)	0.018 (27)	–	–
CL_{int} (L/h)	2880 (27)	–	0.32 (44)
V_s (L)	48.8 (20)	–	–
Lag time (h)	0.50 (Fixed)	2.5 (39)	–
k_a (1/h)	0.18 (30)	0.55 (32)	–
f_u	0.14 (Fixed)	–	–
K_m (ng/mL)	1370 (70)	–	–
MIT (h)	1.85 (40)	–	–
Residual error	0.5 (13)	–	–

CL_{int}, intrinsic clearance; f_u, plasma unbound fraction; k_a, absorption constant rate; K_m, hepatic artemisinin concentration resulting in 50% of maximum intrinsic clearance; *Lag time*, absorption lag time; *MIT*, mean induction time; *RSE%*, relative standard error in percent; S_{IND}, slope of the inducing effect of artemisinin hepatic concentration on the production rate of enzyme precursor; $T_{1/2,ENZ}$, enzyme elimination half-life; V_s, volume of sampling compartment.
From Gordi T, Xie R, Huong NV, et al. A semiphysiological pharmacokinetic model for artemisinin in healthy subjects incorporating autoinduction of metabolism and saturable first-pass hepatic extraction. *Br J Clin Pharmacol* 2005;**59**(2):189–98.

produced (Fig. 5.8). PK and PD parameters are listed in Tables 5.21 and 5.22, respectively. The semiphysiologic PK model incorporated the autoinduction of drug eliminating enzymes. The PD model included the visible and invisible compartments reflecting different stages of parasite life cycle and also divided the parasites into sensitive, insensitive, and injured stages. Individual differences in PK and PD data were large. Before autoinduction, hepatic extraction ratio of artemisinin was 0.87, the distribution volume was 27L, and drug half-life was 0.7h. The PD model was able to take the numerical process of parasite elimination into account. The longest mean transit time taken for the parasites to pass from the sensitive visible to invisible to insensitive visible stages was 34.5h in a life cycle, and the injured parasite half-life was 2.7h.

B. DIHYDROARTEMISININ

Li Tao, Zhou Zhongming, Yang Weipeng
Institute of Chinese Materia Medica, China Academy of Chinese Medical Sciences, Beijing, China

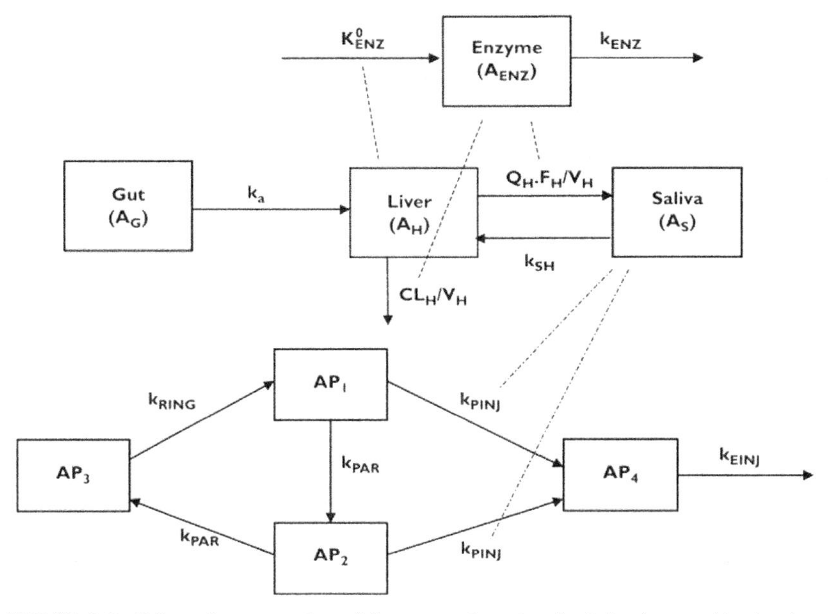

FIGURE 5.8 Schematic presentation of the proposed semimechanistic pharmacokinetic–pharmacodynamic (PK–PD) model. CL_H, hepatic clearance; F_H, hepatic bioavailability; K^0_{ENZ} k_{ENZ}, zero-order production rate and first-order elimination rate of induced enzymes; k_a, absorption rate constant; k_{EINJ}, parasite removal rate constant from blood by spleen; k_{PAR}, rate constant for transfer of mature parasites (trophozoites) to invisible parasites (schizonts); k_{PINJ}, the second-order transfer rate constant from trophozoite and schizont forms to injured parasites caused by artemisinin; k_{RING}, rate constant for transfer of schizonts to early forms(rings); k_{SH}, transfer rate of artemisinin from the sampling to the hepatic compartment; Q_H, hepatic plasma flow; V_H, hepatic volume of distribution. *(From Gordi T, Xie R, Jusko WJ. Semi-mechanistic pharmacokinetic/pharmacodynamic modelling of the antimalarial effect of artemisinin.* Br J Clin Pharmacol *2005;**60**(6):594–604.)*

1. Nonclinical Pharmacokinetics

The isotope research group at the ICMM of CACMS and the Department of Biology at Beijing Normal University studied the absorption, distribution, excretion, and metabolism of dihydroartemisinin in mice using tritium labeling (^{3}H label on C12 hydroxyl),[118] liquid scintillation counting, and radio TLC.[49] Absorption of oral ^{3}H-dihydroartemisinin was rapid, with radioactivity in the blood peaking 1 h after medication and falling by at least half in 4 h. As for distribution in the organs and tissues, radioactivity peaked in the liver and heart 0.5 h after medication, as opposed to after 1 h in the other tissues. Radioactivity after 1 h was highest in the kidneys, followed in descending order by the lungs, liver, spleen, bone and marrow, heart, muscles, and brain. It decreased as plasma concentration fell. Excretion of radioactivity via bile was at its highest after 1 h. Within 24 h, 82.7% of radioactivity was excreted in urine and feces, mainly through urine (67.1 ± 1.7%). Radio TLC showed that ^{3}H-dihydroartemisinin was relatively stable in vitro and swiftly metabolized in vivo. The original drug could not be detected in urine.

TABLE 5.21 Artemisinin Pharmacokinetic Parameters and Associated Interoccasional (IOV) and Interindividual (IIV) Variability in Vietnamese Malaria Patients

Parameter	Estimates (RSE%)	IIV (RSE%)	IOV (RSE%)
$T_{1/2ENZ}$ (h)	37.9 (Fixed)	–	–
IND_{max} (L/ng)	13.4 (25)	–	–
IND_{50} (ng/mL)	140 (62)	–	–
CL_{int} (L/h)	1550 (13)	–	–
Female effect on Cl_{int} (L/h)	438 (61)	–	–
V_s (L)	27.0 (14)	1.1 (40)	–
Lag time (h)	0.50 (Fixed)	–	1.5 (31)
k_a (1/h)	0.3 (23)	–	0.53 (43)
K_m (ng/mL)	1960 (53)	0.67 (54)	–
Residuals	0.65 (12)	–	–

CL_{int}, intrinsic clearance; *IND_{50}*, liver artemisinin concentrations that produce 50% of the maximum induction; *IND_{MAX}*, maximum induction rate of artemisinin hepatic concentration on the production rate of enzymes; *k_a*, absorption constant rate; *K_m*, hepatic artemisinin concentration resulting in 50% of maximum intrinsic clearance; *Lag time*, absorption lag time; *RSE%*, relative standard error in percent; *$T_{1/2,ENZ}$*, enzyme elimination half-life; *V_S*, volume of sampling compartment.
From Gordi T, Xie R, Jusko WJ. Semi-mechanistic pharmacokinetic/pharmacodynamic modelling of the antimalarial effect of artemisinin. *Br J Clin Pharmacol* 2005;**60**(6):594–604.

TABLE 5.22 Artemisinin Pharmacodynamic Parameters and Associated Interindividual (IIV) Variability in Vietnamese Malaria Patients

Parameter	Estimates (RSE%)	IIV (RSE%)
T_{PAR} (h)	34.4 (5.5)	0.26 (70)
T_{RING} (h)	13.6	–
K_{PINJ} (ng/mL/h)	1.58 (59)	–
R_F	15 (FIXED)	–
K_{EINJ} (1/h)	0.26 (9.1)	–
INV	26200 (28)	2.1 (20)
Proportional residual error	0.88 (6.3)	–

INV, initial value of invisible parasites; *K_{EINJ}*, elimination rate of injured parasites; *K_{PINJ}*, injury rate of trophozoites and schizonts; *R_F*, parasite replication factor per life cycle; *RSE%*, relative standard error in percent; *T_{PAR}*, transfer time of parasites from trophozoites to schizonts and further to ring forms; *T_{RING}*, transfer time of parasites from ring forms to schizonts, equal to 48-T_{PAR}.
From Gordi T, Xie R, Jusko WJ. Semi-mechanistic pharmacokinetic/pharmacodynamic modelling of the antimalarial effect of artemisinin. *Br J Clin Pharmacol* 2005;**60**(6):594–604.

Yao Qianyuan et al. employed TLC to investigate the PKs of dihydroartemisinin emulsions and oil preparations in mice and rabbits. The mice were given intraperitoneal injections and oral doses; the rabbits received only oral doses.[119] In mice, injections were absorbed more rapidly than oral preparations, but there was no clear difference in elimination time. Both emulsions and oil preparations conformed to a one-compartment model. Oil preparations were absorbed more slowly and in smaller amounts. In rabbits given 800 mg/kg oral emulsion, peak time was around 4 h longer than in mice. Elimination did not take place after 10 h, and plasma concentration was lower in male rabbits. Distribution in mice given intraperitoneal injections of 300 mg/kg emulsion was highest in the liver, followed by the spleen, kidneys, muscles, heart, and brain. Concentration peaked in the heart and muscles after around 10 min, in the kidneys after 15 min, and in the liver after 5 min. Elimination time was around 1 h.

Using RIA, Song Zhenyu et al.[56] examined the PKs of dihydroartemisinin tablets in rabbits and dogs.[120] An equal number of male and female rabbits were given 10, 20, or 30 mg/kg tablets. Male dogs received 20 mg/kg tablets. The drug was well absorbed, with plasma concentration at 0.03–0.13 μg/mL in rabbits and 0.13 μg/mL in dogs. MRT was 1.36–1.73 h in rabbits and 3.04 h in dogs. Elimination half-life was 1.00–1.19 h in rabbits and 2.10 h in dogs.

Li Rui et al.[121] and Theoharides et al.[122] conducted a study in dogs using gas chromatography–mass spectrometry (GC–MS). The drug-time curve showed that dihydroartemisinin had a first-order reaction rate in dogs, in line with a one-compartment model. MRT was short and the amount excreted via urine and bile was small. The drug was further metabolized in vivo.

Chen Laishun et al.[123] and Huo Jiceng et al.[124] also adopted the TLC method to examine the PKs of artesunate and its dihydroartemisinin metabolite in rats, milk cows, and cattle. In rats, the results conformed to a one-compartment open model, parameters $V = 0.2394 \pm 0.06555$ L/kg, $K_{10} = 0.2777 \pm 0.04632$/min, $T_{1/2} = 11.50 \pm 7.64$ min, $T_{lag} = 0.867 \pm 0.76$ min, $T_{max} = 5.47 \pm 1.51$ min, and $C_{max} = 7.58 \pm 1.46$ μg/mL. There were no significant differences in the MRTs and elimination half-lives of dihydroartemisinin in milk cows and cattle, showing that dihydroartemisinin metabolism did not vary across species.

Using bioefficacy, Li Chenshao et al.[125] demonstrated that dihydroartemisinin was absorbed more rapidly than artemisinin. Its dose–effect curve shifted to the left, showing that it had higher antimalarial activity.

Lee et al.[126] performed in vitro research on rodents indicated that initial metabolism took place mainly in the liver. A study involving rat liver microsomes, conducted by Leskovac et al.,[127] showed that dihydroartemisinin was metabolized primarily into four monohydroxylated metabolites, but the position of the hydroxyl group was undetermined. Ilett et al.[128] found that the main metabolite in bile was α-dihydroartemisinin-β-glucuronide. The complete metabolic pathways are still being investigated.

2. Clinical Pharmacokinetics

Yang Shude et al.[129] established an HPLC polarographic assay for dihydroartemisinin in human plasma. However, no further reports have appeared.

Song Zhenyu et al.[56] used RIA[130] to compare the PKs of dihydroartemisinin tablets and suppositories with artemisinin in humans.[76] The dihydroartemisinin tablets were rapidly absorbed, with serum drug concentration peaking 1.33h after medication. Peak concentrations were 0.13 μg/mL for a 1.1 mg/kg dose and 0.71 μg/mL for a 2.2 mg/kg dose. MRTs were 2.36 and 2.26h, and plasma elimination half-lives were 1.63 and 1.57h, respectively. For 8 mg/kg dihydroartemisinin suppositories, concentration peaked at 0.11 μg/mL 4.7h after medication. MRT was 6.96h. Bioavailability was higher than in the corresponding artemisinin preparations. The suppositories had a longer MRT, which could extend drug efficacy.

A study was conducted in Vietnam on five healthy male and three healthy female volunteers by Le et al. using HPLC with electrochemical detection. The volunteers were given single oral doses of 4.7 or 5.58 mg/kg.[131] The drug–time data adhered to a one-compartment model, with a lag time (Table 5.23). There were no significant differences between males and females. Interindividual variation was relatively large, reaching 47.8% and 45.3% for *AUC* and C_{max}, respectively. Metabolism was rapid, with an elimination half-life of around 2h. From a clinical perspective, doses should be given at least twice daily or combined with antimalarials with long half-lives.

Liu Yiming et al. tested Chinese volunteers[132] given single intravenous dihydroartemisinin drips.[133] Thirty healthy volunteers were randomly divided into three groups, each receiving an intravenous drip of 40, 80, or 160 mg dihydroartemisinin. An HPLC–MS/MS assay was used, with PK parameters calculated on WinNonLin software based on a noncompartmental model. The parameters of the three groups were in ascending order of dose: C_{max} 561.5±127.4 ng/mL, 1080±210 ng/mL, and 2533±503 ng/mL; $T_{1/2}$ 1.69±0.52h, 1.88±0.66h, and 1.92±0.53h; and $AUC_{0-24.5h}$ 575.6±98.7 ng·h/mL, 1370±289 ng·h/mL, and 2893±649 ng·h/mL. There was strong linearity between C_{max}, $AUC_{0\text{-}24.5h}$, and the dose, with the former two increasing in proportion to the dose. The study also analyzed sex differences. In the 40 mg group, no significant differences were found in C_{max}/D, $AUC_{0\text{-}24.5h}$/D, and $AUC_{0\text{-}\infty}$/D ($P>.05$). In the 80 mg dose group, however, there was no clear difference in C_{max}/D but strong variation in $AUC_{0\text{-}24.5h}$/D and $AUC_{0\text{-}\infty}$/D ($P<.05$). The two parameters were higher in females than in males. For the 160 mg group, C_{max}/D, $AUC_{0\text{-}24.5h}$/D, and $AUC_{0\text{-}\infty}$/D all showed statistically significant differences ($P<.05$), being higher in females than in males. It suggested that for doses higher than 80 mg, adjustments needed to be made based on sex. The differences could be due to the significantly lower body weights of the female subjects or to variations in elimination. Further studies are required. No serious adverse effects were seen. Therefore, a single intravenous drip of 40–160 mg dihydroartemisinin was safe in healthy Chinese subjects.

TABLE 5.23 Pharmacokinetics of Single Dose of Oral Dihydroartemisinin in Vietnamese Healthy Volunteers; Data Presented as Median (Range)

	Male Volunteers						Female Volunteers				Median (Range) for Pool Data
Pharmacokinetics	1	2	3	4	5	Median (range)	1	2	3	Median (Range)	
T_{lag} (h)	0.40	0.09	0.35	0.45	0.25	0.35 (0.09-0.45)	0.44	0.59	0.78	0.59 (0.44-0.78)	0.41 (0.09-0.78)
$T_{1/2a}$ (h)	0.71	1.43	0.78	0.17	0.27	0.71 (0.17-1.43)	0.25	0.61	0.54	0.54 (0.25-0.61)	0.58 (0.17-1.43)
$T_{1/2e}$ (h)	1.67	2.16	1.51	3.43	1.77	1.77 (1.51-3.43)	2.39	2.24	1.69	2.24 (1.69-2.39)	1.97 (1.51-3.43)
Tmax (h)	1.63	2.15	1.61	1.21	1.07	1.61 (1.07-2.15)	1.30	2.04	1.87	1.87 (1.3-2.04)	1.62 (1.07-2.15)
Cmax(ng/ml)	401	464	754	311	128	401 (128-754)	537	787	467	537 (467-787)	466* (420-3.535)
Cmax/dosage (ng/ml/dosage)	78.5	100.5	116.2	67.4	27.2	78.5 (27.2-116.2)	96.2	124.6	99.2	99.2 (96.2-124.6)	97.7 (27.2-124.6)
AUC(ng h/ml)	1.426	2.546	2.610	1.594	420	1.594 (420-2.610)	2.139	3.535	1.570	2.139 (1.570-3.535)	1.867** (420-3.535)
AUC/dosage (ng h/ml/dosage)	279.3	551.6	402.4	345.4	89.3	345.4 (89.3-551.6)	383.2	559.7	333.6	383.2 (333.6-559.7)	364.3 (89.3-559.7)
Cl/f(ml/min/kg)	59.7	30.2	41.4	48.2	186.6	48.2 (30.2-186.6)	43.5	29.8	50	43.5 (29.8-50)	45.9 (29.8-186.6)
Vz/f	8.6	5.7	5.4	14.3	28.7	8.6 (5.4-28.7)	9.0	5.8	7.3	7.3 (5.8-9.0)	7.95 (5.4-28.7)

*mean [SD]:481 [218] ng/ml,coefficient of variation (CV):45.3%
**mean [SD] 1980 [942] ngxh/ml, coefficient of variation (CV):47.6%
From Le NH, Na-Bangchang K, Le TD et al. Pharmacokinetics of a single oral dose of dihydroartemisinin in Vietnamese healthy volunteers. *SE Asian J Trop Med Public Health* 1999;**30**(1):11–6.

C. ARTEMETHER

Zhou Zhongming, Wang Yiwei, Li Tao
Institute of Chinese Materia Medica, China Academy of Chinese Medical Sciences, Beijing, China

1. Nonclinical Pharmacokinetics

In 1981, Yang Qichao et al. used radioisotope tracers to investigate the absorption, distribution, metabolism, and excretion of ^{3}H-artemether (tritium gas exposure, random labeling) in mice.[134] After intragastric administration, radioactivity rose swiftly in the blood and peaked after 0.5h, whereupon it declined rapidly. It was widely distributed in the body. In descending order, radioactivity was highest in bile, followed by the kidneys, intestines, liver, lungs, bone and marrow, heart, spleen, and brain. This was consistent with the overall autoradiography results. An hour after medication, radioactivity was highest in the stomach and intestinal contents and the bladder, followed by the salivary glands, liver, esophagus, heart, and blood vessels (dorsal vein). Radioactivity was lower in the other organs, either undetectable or in trace amounts. After 4h, it was still strongest in the intestines, bladder, and salivary glands and decreased in the heart and liver. From 8 to 24h, radioactivity was concentrated mainly in the feces. Between 0 and 24h, excretion in urine and feces accounted for around 85% of radioactivity. Radioactive TLC of blood and urine showed that 1h after medication, only 7% of the original drug was found in blood and 1.8% in urine. It indicated that artemether was rarely excreted in its original form. Most of the original drug had been rapidly metabolized.

Thereafter, Jiang Jirong et al. studied the absorption, distribution, and excretion of ^{3}H-artemether and ^{14}C-artemether (^{14}C label on 12-methyl ether) in mice and rats, also employing tracers.[135] Mice were given intravenous injections of 1mCi/kg ^{3}H-artemether. After 30min, the distribution of radioactivity in the organs was highest in the liver, followed in descending order by the lungs, adrenal glands, kidneys, spleen, stomach, heart, muscles, colon, small intestine, femur, testes, and brain. Urinary excretion accounted for 41.3±3.0% of the total radioactivity in 24h and 56.0±13% in 72h. Fecal excretion accounted for 26±10% in 24h and 39±16% in 72h. Within 24h, excretion in urine and feces made up 68±10% of the total dose. Bile excretion made up 32.5% of the total dose after 3h. Distribution was largely the same in rats injected with 0.25mCi/kg intravenous ^{3}H-artemether, although radioactivity was significantly higher in the lungs after 3h. In mice given intravenous and intramuscular injections of ^{14}C-artemether, the $^{14}CO_2$ excreted via respiration accounted for 31±3.0% and 14.9±2.0% of the dose, respectively. This rose to 50±4.0% in mice that received phenobarbital and the intravenous injection, significantly higher than in the injection-only group. It indicated that artemether metabolism involved demethylation and was related to hepatic enzyme activity. Radioisotope tracers were used to determine total radioactivity in blood, tissues, and excreta.

In 1982, Zhang Yindi et al.[136] and Zeng Yanlin et al.[137] adopted a TLC assay for artemether in rabbit plasma. Artemether showed good linearity from 0.2 to 2μg;

specificity was high, the detection limit was 0.1 μg, and the plasma recovery rate was 83%. However, the variation coefficient was large, between 9.6% and 13.0%. The PKs of an intravenous artemether lipid emulsion (80 mg/kg) in rabbits was consistent with a linear two-compartment model. The parameters were $t_{1/2}$(α and β) = 0.144 ± 0.077 h and 0.896 ± 0.371 h; K_{21}, K_{10}, and K_{12} = 1.235 ± 0.705/h, 4.143 ± 1.370/h, and 1.140 ± 0.951/h, respectively; V_c, V_d (area), and $V_{d(ss)}$ were 0.609 ± 0.119 L/kg, 2.985 ± 0.787 L/kg, and 1.054 ± 0.202 L/kg, respectively; and clearance rate was 2.401 ± 0.339 L·h/kg. Intramuscular injections of 250 or 125 mg/kg artemether-oil were absorbed slowly, with an absorption rate constant of 0.0377 ± 0.0119/h. Individual variation in absorption rate was large, between 19% and 74%, average 36.14 ± 18.39%.

Zou Chongda improved this method using thin-layer chromatographic densitometry with artemisinin as an internal standard. It was applied to a study on the distribution of an artemether lipid emulsion in rats.[138] After 15, 30, and 60 min, concentration was highest in plasma, followed by the brain, heart, and muscles. It was lowest in the liver. The concentration decreased most rapidly in plasma. Only a small amount of the drug could be detected in plasma after 120 min, and it was completely undetectable in the other organs. After 15 min, very high concentrations were seen in the brain and muscle, suggesting that artemether was swiftly transported and could pass the blood–brain barrier. A comparison of the concentration–time curves for the brain and plasma showed that the drug was eliminated from the brain more slowly than from plasma, indicating that it stayed in the brain longer. Hence, artemether may be suited for the treatment of cerebral malaria. The drug's retention time in the metabolically active liver, kidney, and other organs was short, and it was present in smaller amounts than in the brain and muscles. It was rapidly metabolized in the liver and kidney, showing that both organs were the main sites of metabolism. This was in line with the results of the isotopic assays.

Huang et al. assayed artemether in plasma and whole blood with HPLC-polarographic detection.[139] Artemisinin was the internal standard, and the detection limit was 10 ng. A good linear relationship was seen at 10–1000 ng. The sample recovery rate was 71%–100%. A 10 mg/kg intravenous injection could be used in whole-blood assays in rats and plasma assays in humans. It was found that impurity peaks were higher in whole blood than in plasma. Moreover, because the speed of decay in the whole-blood samples was very high, a precolumn component was added. Later, HPLC-UV detection[140] emerged as a method to further isolate artemether and its metabolites.[141]

Zhang Zhirong et al. established a capillary chromatography assay for artemether in dog plasma.[142] The minimum detectable concentration was 0.02 μg/mL; linear range was 0.05–12.5 μg/mL, r = .994; and recovery rate was 103.3%. Mongrels were given 1200 mg/kg conventional tablets and delayed-release tablets of artemether. There were significant differences in AUC, T_{max}, MRT, and $T_{1/2}$ between both types of tablet (Table 5.24). It showed that the sustained-release tablets had a longer effective time.

TABLE 5.24 Parameters of Delayed-Release (DT) and Conventional (CT) Artemether Tablets

	AUC (μg·h/mL)		T_{max} (h)		C_{max} (μg/mL)		*MRT* (h)		$T_{1/2}$ (h)	
No.	**CT**	**DT**	**CT**	**DT**	**CT**	**DT**	**CT**	**DT**	**CT**	**DT**
1	18.974	24.056	2.065	2.454	5.416	4.539	3.426	4.281	1.011	1.647
2	10.925	27.012	2.173	2.805	2.899	4.644	3.755	5.104	1.068	2.045
3	9.625	23.562	2.563	2.699	2.597	4.432	3.529	4.751	1.198	1.733
4	15.587	20.019	2.150	2.451	4.662	4.151	3.418	4.095	1.168	1.464
5	11.868	25.241	1.388	3.152	4.416	3.823	2.616	5.390		
Mean	13.396	23.978	2.068	2.712	3.998	4.318	3.348	4.726	1.111	1.722
SD	3.827	2.581	0.426	0.290	1.204	0.332	0.432	0.549	0.078	0.301
T value	–5.13		–2.80		–0.57		–4.43		–5.03	
t Value					3.355					
	$P<.01$		$P<.05$		$P<.05$		$P<.01$		$P<.01$	

From Zhang Z, Hu H, Hong Z, et al. In vivo pharmacokinetics of delayed-release artemether tablets in dogs. *West China J Pharm Sci* 2000;**15**(5):335–7. [张志荣，胡海燕，洪诤，等. 蒿甲醚缓释片在狗体内的药代动力学. 华西药学杂志, 2000, 15(5): 335–337.]

Bai Kehua et al.[143] and Xing et al.[144] compared the PKs of single- and multidose artemether and its metabolite dihydroartemisinin in rats via LC–MS. For the rats medicated for 5 days, the $AUC_{0\text{-}24h}$ of artemether in rats was 50.3 ± 13.3 ng·h/mL, significantly lower than the $AUC_{0\text{-}24h}$ of the single-dose group (23.4 ± 15.4 ng·h/mL, $P<.05$). The $AUC_{0\text{-}24h}$ of dihydroartemisinin was 16.4 ± 8.0 ng·h/mL, also significantly lower than that of the single-dose group (42.1 ± 8.2 ng·h/mL). However, the oral clearance rate markedly increased after multiple doses ($P<.05$): 27.2 ± 11.6 L/(min·kg) for artemether in the multidose group, as opposed to 10.5 ± 2.7 L/(min·kg) in the single-dose group. The values for dihydroartemisinin were 11.7 ± 2.7 L/(min·kg) in the single-dose group and 33.4 ± 12.4 L/(min·kg) in the multiple-dose group. For artemether and dihydroartemisinin, the reduction in *AUC* and increase in *CL/F* both showed time dependence. There were no clear changes in MRT and T_{max}.

The main artemether metabolites are dihydroartemisinin, deoxydihydroartemisinin, and artemether furan esters. Mohamed et al. applied solid phase extraction and GC–MS to measure artemether and dihydroartemisinin in plasma.[145] Artemisinin was the internal standard, and selected ion monitoring was used. The recovery rates of artemether, dihydroartemisinin, and artemisinin were 94.9%, 92.2%, and 81.3%, respectively. The parameters of artemether and dihydroartemisinin were C_{max} = 245.2 and 35.6 ng/mL and $AUC_{0\text{-}8h}$ = 2463.6 and 111.8 ng·h/mL, respectively.

Liu Ping et al. developed a rapid and sensitive LC–MS/MS assay[146] with multiple-reaction monitoring for artemether and dihydroartemisinin in plasma. The values were m/z321.1→m/z275.1 for artemether, m/z284.3→m/z267.1 for dihydroartemisinin, and m/z283.1→m/z209.2 for artemisinin (internal standard). The linear range for artemether and dihydroartemisinin was 1–400 ng/mL, and intra- and interday precisions were less than 9.0%. Methodology and stability evaluations were conducted. This method was used to test the plasma of six dogs given single intravenous or intramuscular injections of 2 mg/kg artemether. The results are shown in Fig. 5.9.

Yang Huasheng et al. compared the plasma concentrations of artemether and dihydroartemisinin in mice given transdermal or oral artemether.[147] The C_{max}, T_{max}, *AUC*, and MRT of the artemether patch were significantly higher than those of the oral preparations ($P<.05$, Table 5.25). It demonstrated the sustained-release capability of the patches. In vitro experiments with the skin of rats were conducted using TK-12A transdermal diffusion devices.[148] Penetration enhancers at different concentrations could increase the steady permeation rate, cumulative permeation dose, and skin retention to varying degrees. A linear correlation existed between skin retention and cumulative dose. Artemether was an ideal candidate for transdermal preparations.

Hu Haiyan and Zhang Zhirong refined the in vitro experimental conditions of sustained-release artemether tablets.[149] Artificial gastric and intestinal fluid containing 0.4% sodium dodecyl sulfate were chosen as the release medium. The conditions for acid digestion were optimized. It was confirmed that, among

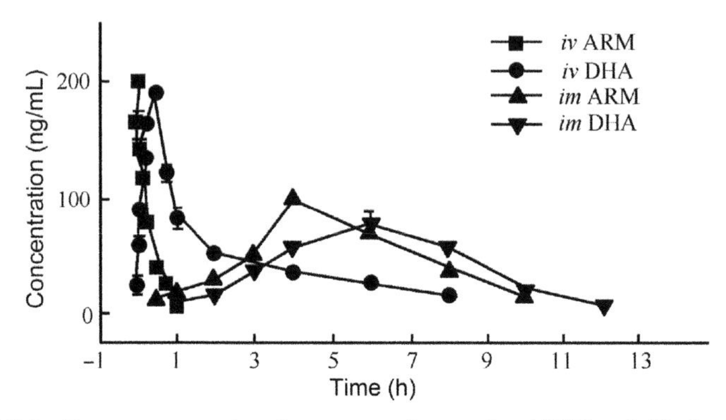

FIGURE 5.9 Plasma concentration–time curves of artemether (ARM) and dihydroartemisinin (DHA). *(From Liu P, Hou Y, Shan C, et al. HPLC-MS-MS in the determination of artemether and its metabolite dihydroartemisinin in the plasma of beagles.* Chin J Pharm Anal *2006;**26**(5):585–8. [刘萍，侯禹男，单成启，等. HPLC-MS-MS法测定比格犬血浆中蒿甲醚及其代谢物双氢青蒿素的浓度. 药物分析杂志, 2006, 26(5): 585–588.])*

various other factors, release of the drug was more sensitive to ionic strength. The experiment showed that sustained-release tablets had clear differences when compared to normal tablets (Fig. 5.10). Moreover, rapid-release tablets had improved taste and disintegrated more quickly.[150] Self-microemulsifying delivery systems[151] and microemulsion injections were more stable than commercially available artemether-oil preparations and had 1.5 times the antimalarial activity.[152]

Li Wanhai et al. studied artemether and plasma protein binding.[53] The binding rates in different species were 57.5% (mice), 61.1% (monkeys), and 76.9% (humans).

Chen Laishun et al. examined the impact of phenobarbital induction on the metabolism of artemether by rat liver microsomal enzymes[153] and the effects of deuterium isotopes.[154] There was a strong link between metabolism and the NADPH-cytochrome P450 system. O-demethylation was the main metabolic pathway. Three artemether metabolites—dihydroartemisinin, deoxy-dihydroartemisinin, and artemether furan ester—were examined using the above method. It confirmed that apart from demethylation and deoxidation, artemether metabolism involved hydroxylation at position 9. The hydroxylation products were stereoselective (9α-hydroxyartemether, 9β-hydroxyartemether). Preliminary in vitro tests showed that two of these metabolites had antimalarial effects. Under simulated in vivo acid and alkali conditions, Chen Yougen et al.[155] isolated a series of artemether degradation products, with dihydroartemisinin and artemether furan esters being the main compounds. It was also concluded that, in mammals, the metabolism of artemether to form dihydroartemisinin was not necessarily produced by hepatic drug-metabolizing enzymes.

With linear ion trap-mass spectrometry, Liu et al. detected 77 Phase I metabolites—including artemisinin and dihydroartemisinin—in rat liver microsomes

TABLE 5.25 Pharmacokinetic Parameters of Transdermal and Oral Artemether in Rats

	Transdermal Preparations		Oral Preparations	
Parameter	Artemether	Dihydroartemisinin	Artemether	Dihydroartemisinin
$T_{1/2}$ (h)	16.45 ± 7.28	35.48 ± 19.45	1.10 ± 0.33	2.63 ± 3.85
T_{max} (h)	19.00 ± 24.72	12.40 ± 19.92	0.18 ± 0.09	0.50 ± 0.46
C_{max} (μg/L)	16.42 ± 1.96	5.60 ± 1.65	466.36 ± 262.82	330.80 ± 192.46
AUC (μg·h/L)	707.57 ± 117.41	172.53 ± 58.33	537.07 ± 145.51	436.61 ± 72.39
MRT (h)	43.04 ± 4.21	34.63 ± 7.89	1.73 ± 0.13	1.47 ± 0.20

From Yang H, Ye Z, Liang B, et al. Pharmacokinetics of artemether patches in rats. *China J Chin Mater Med* 2008;**33**(12):1459–62. [杨华生，叶祖光，梁秉文，等. 蒿甲醚贴剂小鼠体内药代动力学研究. 中国中药杂志, 2008, 33(12): 1459–1462.]

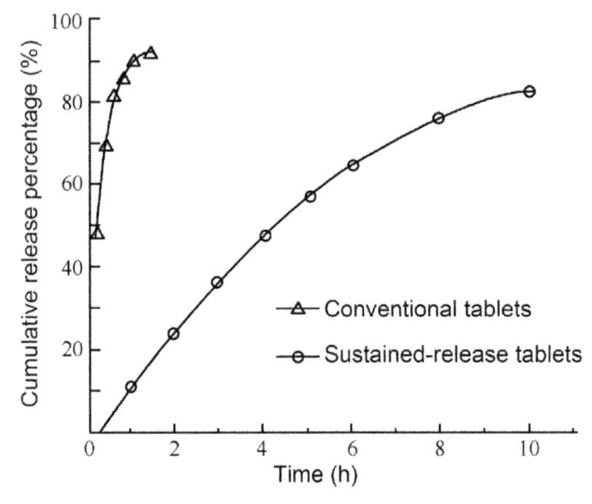

FIGURE 5.10 In vitro release profiles of conventional and sustained-release artemether tablets. *(From Hu H, Zhang Z. Study on the in vitro release of artemether sustained-release tablets.* Chin J Synth Chem *1999;7(4):334–9. [*胡海燕，张志荣. 蒿甲醚缓释片体外释放研究. 合成化学*, 1999, 7(4): 334–339.])*

and urine[156] and 12 Phase II metabolites in bile. It suggested that Phase I metabolism primarily involved hydroxylation, dehydrogenation, demethylation, and deoxygenation, whereas glucuronidation reactions occurred in Phase II.

Maggs et al. used LC–MS/MS and radiometric HPLC to study the biliary metabolites of [13–^{14}C]-artemether in anesthetized Wistar rats.[157] Within 0–3 h after intravenous drug administration, the major metabolites found were the glucuronides of 9α-hydroxyartemether (33.4 ± 6.8%) and α-dihydroartemisinin (22.5 ± 4.4%); four stereochemically unassigned monohydroxy-artemether glucuronides (**II**, 3.1 ± 0.9%; **IV**, 4.4 ± 1.7%; **V**, 21.4 ± 3.0%; **VI**, 3.0 ± 1.1%); and a dihydroxyartemether glucuronide (6.0 ± 2.1%). A sixth monohydroxyartemether glucuronide (**VIIa**) and deoxydihydroartemisinin were detected in trace amounts. The furano acetate isomer of DHA glucuronide, indicative of the formation of a radical intermediate, was also found in trace amounts. The metabolic scheme of artemether deduced by the biliary metabolites is shown in Fig. 5.11.

2. Clinical Pharmacokinetics

Monotherapy

Lin Bingliu et al.[158] ran an RIA on the PKs of artemether in normal humans (Table 5.26). After intramuscular injections of 3.2, 6.0, and 10.0 mg/kg artemether-oil preparation, peak concentration C_{max} and peak time T_{max} increased with the dose. The plasma concentration half-life also rose. The data obtained were of practical clinical significance.

Zhou et al. measured the effects of voltage, mobile phase ratio, and pH on HPLC-polarographic analysis of artemisinin and its derivatives

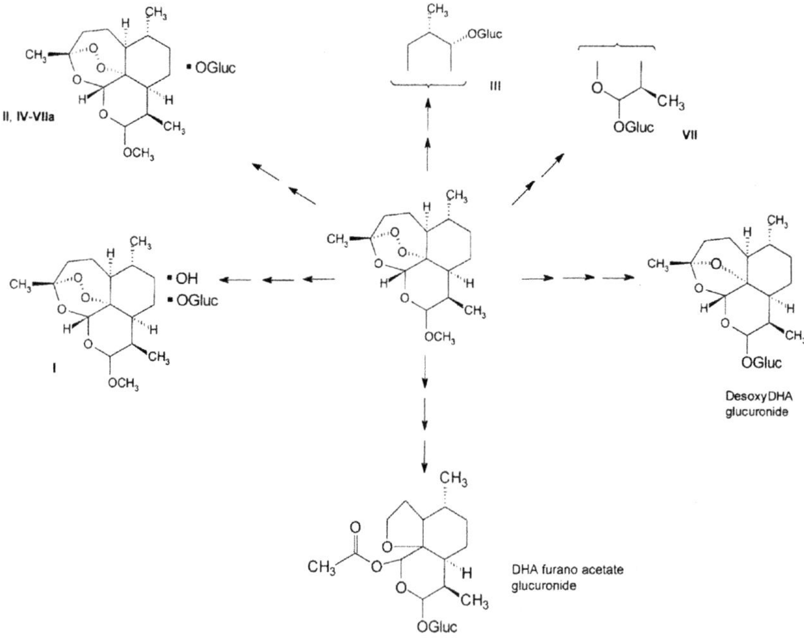

FIGURE 5.11 Metabolic schema for the proposed biliary metabolites of [14C-] artemether in male rats. *(From Maggs JL, Bishop LP, Edwards G, et al. Biliary metabolites of beta-artemether in rats: biotransformations of an antimalarial endoperoxide.* Drug Metab Dispos *2000;**28**(2):209–17.)*

TABLE 5.26 Pharmacokinetic Parameters of Intramuscular Artemether Injections at Different Doses

	Dose (mg/kg)		
	3.2	**6.0**	**10.0**
K_a (1/h)	0.63 ± 0.16	0.21 ± 0.12	0.18 ± 0.06
K (1/h)	0.19 ± 0.03	0.13 ± 0.04	0.13 ± 0.04
$T_{1/2Ka}$ (h)	1.10	3.96	3.92
$T_{1/2K}$ (h)	3.58	5.54	6.12
T_{max} (h)	3.05	7.04	6.73
C_{max} (μg/mL)	0.25	0.59	1.53
AUC (μg·h/mL)	2.19	9.64	29.79
Vd (L/kg)	7.62	4.69	3.17

From Lin B, Chen Z, Li Z, et al. Phase I clinical pharmacokinetics trial on artemether. *J Guangzhou Coll Tradit Chin Med* 1987;**4**(3):48–9. [林炳鎏，陈芝喜，李志强，等. 蒿甲醚I期临床药代动力学的研究. 广州中医学院学报, 1987, 4(3): 48–49.]

(dihydroartemisinin, artemether, and artesunate).[59] It allowed artesunate to be more effectively isolated from dihydroartemisinin, artemisinin, and artemether. The plasma concentration and PK parameters of intramuscular artemether in healthy humans were also obtained (Fig. 5.12, Table 5.27). Compared with

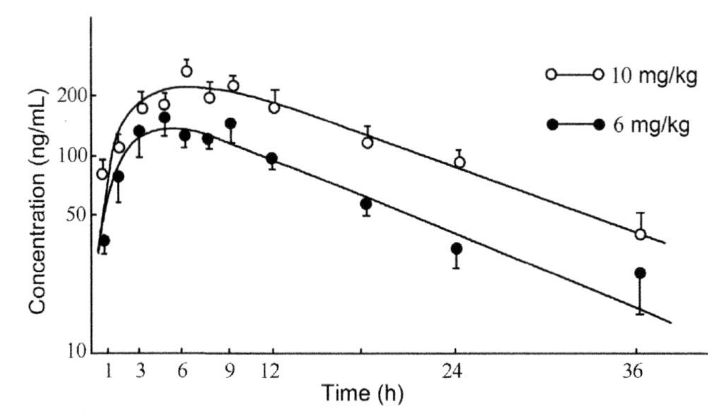

FIGURE 5.12 Drug plasma concentration–time curves of 6 and 10 mg/kg intramuscular artemether in healthy male subjects. *(From Zhou ZM, Huang YX, Xie GH, et al. HPLC with polarographic detection of artemisinin and its derivatives and application of the method to the pharmacokinetic study of artemether.* J Liq Chromatogr *1988;**11**(5):1117–37.)*

TABLE 5.27 Pharmacokinetic Parameters of 6 and 10 mg/kg Intramuscular Artemether in Healthy Male Subjects

Parameter	Mean ± SD(n = 3)	
Dose (mg/kg)	6	10
C_o (μg/mL)	0.2903 ± 0.1250	0.3260 ± 0.0322[a]
K_a (/h)	0.3547 ± 0.2290	0.4268 ± 0.2541[a]
K (/h)	0.1190 ± 0.0624	0.0646 ± 0.0157[a]
$t_{1/2}K_a$ (h)	2.6351 ± 1.7021	2.0805 ± 1.2225
$t_{1/2}K$ (h)	7.7215 ± 5.6269	11.1163 ± 2.4552
$AUC_{0\text{-}36h}$ (μg·h/mL)	2.5690 ± 0.4377	5.1580 ± 0.7512[c]
T_{max} (h)	5.2358 ± 1.2988	6.2727 ± 2.5687[a]
C_{max} (μg/mL)	0.1448 ± 0.0391	0.2237 ± 0.0262[b]

Compared with 6 mg/kg group.
[a]$P > .05$.
[b]$P < .05$.
[c]$P < .01$.
From Zhou ZM, Huang YX, Xie GH, et al. HPLC with polarographic detection of artemisinin and its derivatives and application of the method to the pharmacokinetic study of artemether. *J Liq Chromatogr* 1988;**11**(5):1117–37.

the above results (Table 5.26), the *AUC* was significantly lower. This could be because the concentration measured via RIA included both artemether and the dihydroartemisinin metabolite.

Another report by Hanpithakpong et al. assayed artemether in human plasma with LC–MS.[159] The linear range was 1.43–500 ng/mL, and the detection limit was 0.36 ng/mL. The amount of plasma needed for a sample was reduced to only 50 μL. Huang et al.[160] added ammonium formate to the mobile phase, which improved artemether's low-ionization capacity and achieved closer compliance to previous studies. The recovery rate was 73%–81% for AMT and 90%–99% for dihydroartemisinin, and the linear range was 2–200 ng/mL.

Mordi et al.[161] investigated the 72-h PKs of artemether and dihydroartemisinin in healthy Malaysians after a single 200 mg dose of artemether. HPLC-electrochemical detection was used. Artemether peaked at 310 ± 153 μg/L after 1.88 ± 0.21 h. Elimination half-life was 2.00 ± 0.59 h and *AUC* was 671 ± 271 μg·h/L. Dihydroartemisinin peaked at 273 ± 64 μg/L at 1.92 ± 0.13 h; *AUC* was 753 ± 233 μg·h/L. It suggested that artemether's short half-life contributed to the high rate of recrudescence. In addition, dihydroartemisinin degenerated rapidly at room temperature. Hence laboratory specimens should be stored at −20°C as soon as possible after sampling.

Six healthy Thai males were given 200 mg oral artemether. In addition, eight Thai male malaria patients received 100 mg oral artemether twice daily for 4 days.[162] The C_{max} of the malaria patients (231 ng/mL) was higher than with the healthy volunteers (118 ng/mL), and the steady-state plasma concentration in the patients was 36–60 ng/mL after multiple doses. This provided a basis for the clinical regimen. In another test, 19 Vietnamese patients with severe malaria were given intramuscular injections of artemether or artesunate to compare the drugs' pathways.[163] The C_{max} of artesunate was 5710 nmol/L and $T_{1/2}$ was 30 min; the C_{max} of artemether was 574 nmol/L and *MRT* was 10 h. It indicated that artesunate was rapidly absorbed, and its plasma concentration was significantly higher than that of artemether. Artesunate also yielded more of the active dihydroartemisinin metabolite than artemether did. Therefore, artesunate should be the drug of choice for patients with severe malaria.

Karbwang et al.[164] injected malaria patients experiencing acute renal failure with intramuscular artemether. Their C_{max}, *AUC*, and $T_{1/2}$ (2.38 ng/mL, 35.4 ng·h/mL, and 7 h, respectively) were higher than in patients with uncomplicated malaria (1.56 ng/mL, 25.2 ng·h/mL, and 5.7 h, respectively). However, their *Vz/F* and *CL/F* (5.5 L/kg, 7.4 mL/min·kg) were lower than in uncomplicated malaria cases (8.6 L/kg, 19.1 mL/min·kg). Renal failure markedly altered the drug's pathways, raising bioavailability, reducing systemic clearance, and changing plasma protein binding.

Manning et al.[165] tested the plasma and cerebrospinal fluid of 32 children with malaria in Papua New Guinea (average age 39 months). Artemether concentrations were higher in children with meningitis, around 4.6 times the level in normal individuals. It suggested that the blood–brain barrier was weakened in malaria patients.

Plasma antimalarial activity was higher after an oral dose than after an intramuscular injection. However, there was a significant first-pass effect. Bioavailability was similar for rectal and intramuscular administration, but some factors in the absorption process remained unclear.[166] Compared to the intramuscular injection, oral artemether was rapidly but not completely absorbed, retention time was short, and the recrudescence rate was high.[167]

Van Agtmael[168] examined the effects of CYP2D6 and CYP2C19 on the metabolism of artemether and dihydroartemisinin in Caucasians. No significant differences were found when artemether was combined with quinidine, a CYP2D6 antagonist, or omeprazole, a CYP2C19 antagonist. It suggested that artemether and dihydroartemisinin were not substrates of CYP2D6 and CYP2C19.

Mithwani et al. studied the population PKs of intramuscular artemether in 100 African children with severe malaria.[169] After a 3.2mg/kg injection, artemether in plasma conformed to a one-compartment model with first-order absorption and elimination. A large degree of individual variation was seen. The *CL/F* of AMT was 14.3L/h (interindividual variation 53%) and its terminal half-life was 18.5h. The data for dihydroartemisinin fit a one-compartment linear model. Assuming complete conversion of artemether into dihydroartemisinin, the *CL/F* was 93.5L/h (interindividual variation 90.2%). Its clearance and formation rate were linked. If artemether possessed flip-flop kinetics, the elimination half-life was 21min and the corresponding volume of distribution was 8.44L (interindividual variation 104%). Breathing difficulty, metabolic acidosis, demographics, and disease factors did not affect the in vivo fate of artemether. In children experiencing malaria and acidosis, the PK parameters were more susceptible to absorption kinetics relative to the body's physiological and pathological characteristics. Drug formulation may be a major contributing factor.

The PK data for oral artemether in pregnant women indicated that plasma concentration and exposure were lower than in normal women, which would affect clinical practice.[170]

Drug Interactions in Joint Administration

A combination of mefloquine and artemether[171] significantly reduced the bioavailability of mefloquine. The AUC and C_{max} increased, T_{max} was prolonged, while no clear differences were seen in $T_{1/2}$. Joint administration of oral artemether and pyrimethamine[172] showed that both drugs were rapidly absorbed, but pyrimethamine was eliminated more slowly, with a longer elimination half-life than that of artemether. The bioavailability of pyrimethamine also rose. C_{max} increased and the apparent volume of distribution decreased. When artemether was used with carbamazepine,[173] the bioavailability of the latter grew markedly and its $AUC_{0\text{-}\infty}$, C_{max}, and T_{max} rose. Combination regimens would therefore need to be adjusted.

Food Interactions

Van Agtmael et al. found that taking artemether with grapefruit juice could significantly improve the oral bioavailability of the drug. The C_{max} increased from 42 to 107 ng/mL; T_{max} was prolonged from 2.1 to 3.6 h; and *AUC* rose from 177 to 336 ng·h/mL. The elimination half-life was not greatly affected. Trace amounts of artemether were detected in the saliva of volunteers, but no dihydroartemisinin was found. Intestinal CYP3A4 was involved in artemether metabolism,[174] resulting in time-dependent degeneration.[175]

D. ARTESUNATE

Zhou Zhongming, Zhang Huihui, Yang Weipeng
Institute of Chinese Materia Medica, China Academy of Chinese Medical Sciences, Beijing, China

1. Nonclinical Pharmacokinetics

Artesunate is an active derivative of artemisinin. It is usually dissolved in a 5% sodium bicarbonate solution to produce intravenous injections. Its kinetics varies across animal species.

Li Rui et al. studied the absorption, distribution, metabolism, and excretion of intravenous sodium artesunate in rats.[176] After a 200 mg/kg injection, the drug was rapidly converted into dihydroartemisinin, which peaked in the various organs after 10 min. Concentration was greatest in the heart, followed in descending order by the muscles, lungs, spleen, kidneys, brain, blood, and liver. After 80 min, only the heart, liver, brain, kidney, muscles, and blood still contained traces of dihydroartemisinin. The drug had completely disappeared after 120 min. The PKs of intravenous artesunate conformed to a one-compartment model: $T_{1/2}$ was 15.6 min and apparent volume of distribution V_d was 1.13 L/kg.

To further ascertain the fate of sodium artesunate in vivo, Li Rui et al. used GC, HPLC, and TLC in a dog model.[177] It showed that, both in vivo and in vitro, artesunate could be converted into two dihydroartemisinin isomers, α (RT 5.2 min) and β (RT 6.4 min). The β isomer was more effective. In dogs, dihydroartemisinin had a first-order rate reaction and adhered to a one-compartment model. The $T_{1/2}$ was 10 min and apparent volume of distribution was 0.36 L/kg (Table 5.28). Artesunate could induce hepatic drug-metabolizing enzymes, with a clear first-pass effect. These results indicated that artesunate was widely distributed in the body and rapidly metabolized. Its active metabolite, dihydroartemisinin, was swiftly eliminated from the blood, and its concentration in tissues was much greater than in blood. It was also more concentrated in the brain than in plasma, providing a good means for treating cerebral falciparum malaria and underscoring artesunate's therapeutic value.

TABLE 5.28 Dihydroartemisinin Plasma Concentrations in Dogs Given Intravenous Artesunate (Mean ± SD)

Time (min)	Quantitative Relative Concentrations C of TLCS Peak Area (μg/mL)	logC	t·logC	t^2
5	39.81 ± 0.7	1.5999	7.9999	25
10	28.30 ± 2.0	1.4518	14.5180	100
15	20.50 ± 0.9	1.3118	19.6770	228
20	14.70 ± 0.9	1.1673	23.3460	400
25	9.60 ± 0	0.9823	24.5580	628
30	–			
60	–			

From Li R, Zhou L, Li X, et al. GC-MS in the study of the in vivo fate of sodium artesunate. *Acta Pharm Sin* 1985;**20**(7):485–90. [李锐，周莉玲，李迅，等. 应用气一质联用等法对青蒿酯钠体内命运的研究. 药学学报, 1985, 20(7): 485–490.]

Song Zhenyu et al.[56] and Zhao Kaicun et al.[57] established an RIA method for artesunate in dogs. The assay required only small samples, which could be tested directly without further treatment. Sensitivity was high (2–3 ng) and reproducibility was good. An intravenous injection of 6 mg/kg artesunate, delivered in 1 min, produced a drug–time curve with a correlation coefficient of 0.993, r=.996. Blood concentration conformed to a linear, one-compartment kinetic model. Average PK parameters: $T_{1/2}=0.45$ h, $V_d=0.15$ L/kg, and $CL_T=0.2$ L/h/kg.

Topical artesunate and RIA were also used to study PKs in mice and rabbits.[178] Some hair was removed from rabbits, and a 25 mg/kg topical preparation was applied. Plasma concentration peaked at an average of 2 h, peak concentration 1.80 μg/mL. MRT in rabbits was 3.54 h, and the elimination half-life was around 2.46 h. Mice were given 6.7, 31.3, and 71.4 mg/kg topical artesunate. Plasma concentrations peaked 0.5–4 h after medication, with a peak concentration of 0.82, 2.05, and 7.11 μg/mL, respectively. MRT was 3.39, 2.79, and 3.54 h, and the elimination half-life was 2.35, 1.93, and 2.45 h, respectively. Topical artesunate was well absorbed and plasma concentrations were maintained for a relatively long time. The rapid absorption preserved the advantages of artesunate while resolving the issue of its short half-life.

Chen Zhixi et al. improved the RIA method in an experiment on oral artesunate in dogs.[179] The kinetics adhered to a one-compartment model. The parameters of a 20 mg/kg dose were $T_{1/2}(K_a)=38.74 \pm 9.54$ min, peak time 80.28 ± 7.19 min, and $AUC_{0\text{-}9h}=1747 \pm 70$ mg·min/L. For a 40 mg/kg dose, the parameters were $T_{1/2}(K_a)=25.93\pm9.29$ min, peak time 86.64 ± 25.71 min, and $AUC_{0\text{-}9h}=2822\pm414$ mg·min/L. The peak concentrations were similar, showing that the tablets were rapidly absorbed.

Huo Jiceng et al. developed a rapid-scan TLC assay for artesunate in milk cows. Cows were injected with 5 mg/kg intravenous artesunate.[180] The plasma drug–time data was consistent with a two-compartment model.[181] The distribution half-life was 2.43 ± 0.88 min, indicating that artesunate was swiftly distributed. The volume in the central compartment was small, showing that distribution was also broad. Elimination was also rapid, as shown in the elimination half-life of 31.45 ± 12.93 min.

Another assay method adopted by Edlund et al. involved HPLC with post-column derivatization and UV detection. It was used to measure artesunate and dihydroartemisinin.[79] The detection limit for both drugs was 50 nmol/L. Zhou et al. also used HPLC with electrochemical reduction to analyze artesunate and dihydroartemisinin in blood.[182] Rabbits were given 100 mg/kg intravenous artesunate. The drug disappeared from plasma swiftly, with a measured $T_{1/2}$ of only 1.7 min. It was rapidly metabolized to produce dihydroartemisinin, with an elimination half-life $T_{1/2}\beta$ of 29 ± 2 min. Culturing in blood from different species showed that artesunate was hydrolyzed more rapidly in rabbit blood than in rat and human blood. This hydrolytic reaction was catalyzed by esterases in plasma and erythrocytes. The results suggested that the PK studies of artesunate should focus on dihydroartemisinin and its metabolites.

Shen Xuesong et al. prepared artesunate nanoliposomes via film dispersion and used UV spectrophotometry to determine artesunate concentration.[70] Rats were randomly divided into two groups, one given the nanoliposomes and the other 50 mg/kg oral artesunate as active pharmaceutical ingredient. Blood samples of 1 mL were then collected at different times. The drug–time curve is shown in Fig. 5.13 and PK parameters in Table 5.29. The liposome curves and the data from the first 6 h of the artesunate active pharmaceutical ingredient group were consistent with the Higuchi equation: $Q=17.63T_{1/2}+16.09$ ($r=.99$). Concentration–time data were in line with a one-compartment model. Compared to the active pharmaceutical ingredient group, the liposomes showed significant delayed release in vitro. Rats given oral liposomes also displayed a longer elimination half-life, lower total elimination rate, and smaller apparent distribution volume. Therefore, such formulations had long-lasting effectiveness and sustained release.

2. Clinical Pharmacokinetics

Yang Shude et al.[129] used HPLC with polarographic detection to study artesunate in human plasma.[183] Four volunteers received 3.3–4.4 mg/kg intravenous artesunate and the levels of artesunate and its active metabolite, dihydroartemisinin, were measured. Artesunate concentration declined rapidly, with the drug–time curve conforming to a biexponential function. It was mainly converted

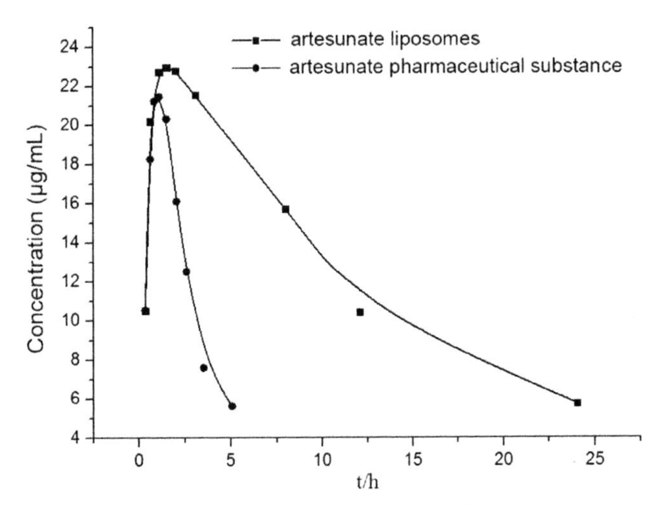

FIGURE 5.13 Drug concentration–time curves of artesunate liposomes and its pharmaceutical substance (50 mg/kg) in rats. *(From Shen X, Xu W, Jin M, et al. The in vivo pharmacodynamics of artemisinin nanoliposomes in rats.* Lishizhen Med Mater Med Res *2011;**22**(2):384–6. [沈雪松，徐伟，金美华，等. 青蒿素类纳米脂质体在大鼠体内的药动学研究. 时珍国医国药，2011, 22(002)：384-386.])*

TABLE 5.29 Pharmacokinetic Parameters of Artesunate Liposomes and Active Pharmaceutical Ingredient (API) in Rats

Group	C_{max} (μg/mL)	T_{max} (h)	AUC (μg·h/mL)	$T_{1/2a}$ (h)	$T_{1/2e}$ (h)
Liposomes	86.33	4.21	1708.40	1.19	10.34
API	128.84	0.7	446.703	0.19	1.85

From Shen X, Xu W, Jin M, et al. The in vivo pharmacodynamics of artemisinin nanoliposomes in rats. *Lishizhen Med Mater Med Res* 2011;**22**(2):384–6. [沈雪松，徐伟，金美华，等. 青蒿素类纳米脂质体在大鼠体内的药动学研究. 时珍国医国药，2011，22(002)：384–386.]

to dihydroartemisinin, which had an average terminal phase half-life of only around 45 min. The plasma concentration of artesunate fell swiftly in the distribution phase. At the elimination phase, it was negligible compared to dihydroartemisinin. Therefore, dihydroartemisinin occupied a dominant position in terms of plasma concentration over time.

Chen Zhixi studied the Phase I clinical PKs of artesunate using RIA.[184] Eight healthy male volunteers were given 2, 3.75, or 4.22 mg/kg intravenous artesunate. The drug–time curve conformed to an open two-compartment model, elimination was swift, and the average $t_{1/2}$ was 34.2 min. Elimination was slower in the high-dose group than in the medium- or low-dose groups, indicating that the elimination rate could be dose related. Hence, the number of doses should be increased in a clinical setting to maintain the effective concentration in blood.

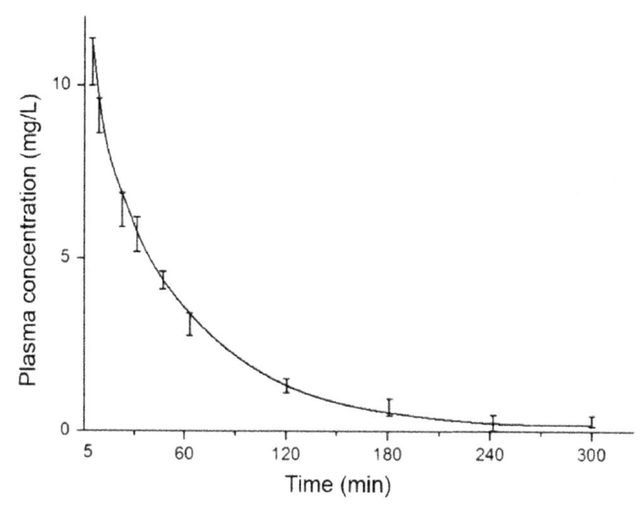

FIGURE 5.14 Plasma concentration–time curves of intravenous artesunate in six healthy male subjects. *(From Lin B, Chen Z, Jia K, et al. Bioavailability of oral artesunate tablets in healthy volunteers.* J Guangzhou Coll Tradit Chin Med *1990;7(3):163–6. [*林炳流，陈芝喜，贾可亮，等. 健康志愿者口服青蒿琥酯片剂的生物利用度. 广州中医学院学报, *1990, 7(3):163–166.])*

Zhao Kaicun et al. also used this method to study the PKs of artesunate in healthy volunteers.[185] Intravenous artesunate, at doses of 2.0 or 3.8 mg/kg, yielded a two-compartment model drug–time curve. The $t_{1/2}\beta$ was 0.5 and 0.6–0.8 h, respectively; steady-state distribution volumes were 0.1–0.6 and 0.3–0.4 L/kg, respectively; and overall clearance rates were 0.4–1.2 and 0.2–0.5 L/kg·h. A small amount of the drug was excreted in urine, a total of only 1.8% after 7 h. Therefore, elimination took place largely through metabolic transformation. The excretion half-life in urine was twice the plasma elimination half-life.

Lin Bingliu et al.[186] adopted this assay for artesunate tablets and injections in healthy volunteers. Oral artesunate conformed to a linear, open, one-compartment model, whereas intravenous artesunate followed a linear, open, two-compartment model (Figs. 5.14 and 5.15). Peak time for the tablets was 53.07 ± 20.58 min, peak concentration was 1.94 ± 1.05 μg/mL, $T_{1/2}\beta$ was 41.35 ± 7.89 min, and absolute bioavailability was 40.39 ± 14.99%. The elimination half-life of intravenous artesunate $T_{1/2}\beta$ was 33.96 ± 4.73 min. The tablets were absorbed more rapidly but to a smaller extent than the injections.

Morris et al. conducted a comprehensive overview of the clinical PKs of intravenous, intramuscular, oral, and rectal artesunate.[187] High-intravenous doses fell rapidly in vivo. The half-life was less than 15 min, clearance rate (*CL/F*) was 2–3 L/kg/h, and distribution volume (*V/F*) was 0.1–0.3 L/kg. Dihydroartemisinin concentration peaked after 25 min; the elimination half-life was 30–60 min, *CL/F* was 0.5–1.5 L/kg/h, and *V/F* was 0.5–1.0 L/kg. Compared to intravenous artesunate, an intramuscular injection had lower

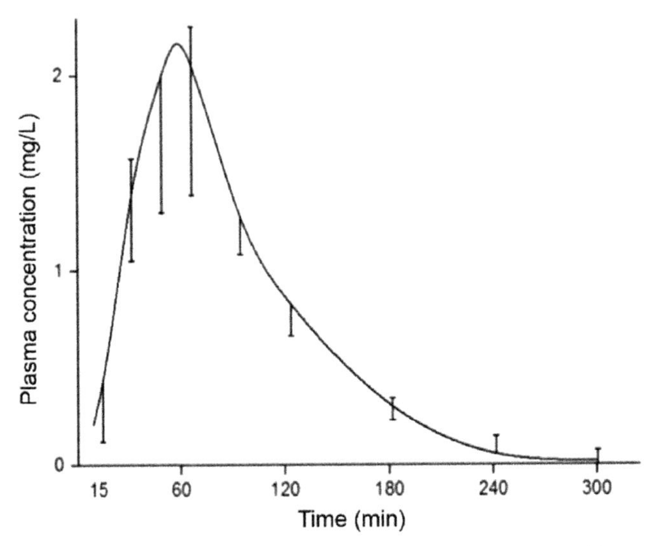

FIGURE 5.15 Plasma concentration–time curves of oral artesunate in six healthy male subjects. *(From Lin B, Chen Z, Jia K, et al. Bioavailability of oral artesunate tablets in healthy volunteers.* J Guangzhou Coll Tradit Chin Med *1990;7(3):163–6.)*

peak concentration, a longer half-life, and a higher distribution volume. Dihydroartemisinin reached a peak later and bioavailability was over 86%. As with intramuscular administration, the bioavailability of oral artesunate was also high (>80%). Its C_{max} was 1 h and elimination half-life was 20–45 min; the C_{max} of dihydroartemisinin was 2 h and its elimination half-life was 0.5–1.5 h. The *AUC* of dihydroartemisinin was more than 10 times that of artesunate. The parameters for rectal administration were similar to those for oral artesunate, except that C_{max} was delayed and the half-life extended. Drug interaction studies found that oral artesunate did not significantly change the PK parameters of atovaquone–proguanil, chlorproguanil–dapsone, sulfadoxine–pyrimethamine (SP), mefloquine–piperaquine, and dihydroartemisinin.

Navaratnam et al. studied the PKs of artesunate and amodiaquine in fixed and loose combination in healthy people.[188] Twenty-four healthy Malaysian volunteers were entered into a 2×2 crossover study. The subjects were randomized and divided into two groups. In the first period, one group was given 200 mg artesunate + 540 mg amodiaquine fixed combination and the other 200 mg artesunate and 612 mg amodiaquine loose combination. After a washout period of 60 days, the two groups switched, each was given the alternative drug combination. The partner drugs artesunate, amodiaquine, and their metabolites dihydroartemisinin and desethylamodiaquine were measured. The *AUCs* of dihydroartemisinin and desethylamodiaquine in the fixed combination were 1522 ± 633 and 30,021 ± 14,211 ng·h/mL, respectively. In the nonfixed combination group, the *AUCs* were 1688 ± 767 and 40,261 ± 19,824 ng·h/mL, respectively. Both artesunate and amodiaquine when used in fixed or nonfixed combination were easily absorbed and had similar PK profiles.

Batty et al. isolated and measured artesunate and dihydroartemisinin via HPLC.[189] In humans, oral artesunate conformed to a one-compartment model and intravenous artesunate to a two-compartment one. The respective $T_{1/2}$ were 41.35 and 33.96 min.

Bethell et al. tested 10 Vietnamese children, aged 6–15 years, who had malaria. They received 3 mg/kg oral artesunate.[190] Plasma concentrations were measured in vitro. Oral artesunate was rapidly absorbed and biological efficacy peaked in 1.7 h, at a concentration equivalent to 664 ng. However, this method yielded large interindividual variation.

Adjei et al. studied the concomitant use of artesunate and amodiaquine in children.[191] A total of 103 uncomplicated falciparum malaria patients in Ghana, aged 1–14 years, were involved. Of these, 15 children received only amodiaquine and 88 were given the combination. The clearance rate of the *N*-desethylamodiaquine metabolite increased with body weight in a nonlinear fashion. Distribution volume of *N*-desethylamodiaquine was higher in the combination group ($P < .001$), and the distribution half-life was longer ($P < .01$). *AUC* and elimination half-life were similar.

Mwesigwa et al. examined an artemether–lumefantrine (AL, abbreviated as ASL in the publication cited) and an artesunate–amodiaquine combination in 21 Ugandan children, aged 5–13 years, who had uncomplicated falciparum malaria.[192] The children received artemether–lumefantrine twice daily or artesunate–amodiaquine once daily for 3 days. Blood samples were collected and tested on the last day. For the artemether–lumefantrine group, the C_{max} and $AUC_{0\text{-}\infty}$ of lumefantrine were 6757 ng/mL and 210 µg·h/mL, respectively. The C_{max} and $AUC_{0\text{-}\infty}$ of artemether were 34 ng/mL and 168 ng·h/mL, respectively, and those of dihydroartemisinin were 119 ng/mL and 382 ng·h/mL, respectively. In the artesunate–amodiaquine group, the C_{max} of artesunate was 51 ng/mL, and its $AUC_{0\text{-}\infty}$ was 113 ng·h/mL. The C_{max} of dihydroartemisinin was 473 ng/mL; its $AUC_{0\text{-}\infty}$ was 1404 ng·h/mL. The C_{max} and $AUC_{0\text{-}\infty}$ of amodiaquine were 5.2 ng/mL and 39.3 ng·h/mL, respectively; the values for *N*-desethylamodiaquine were 235 ng/mL and 148 µg·h/mL, respectively. Compared to previous studies in adults, lumefantrine exposure in children was reduced. Dihydroartemisinin exposure in the artesunate–amodiaquine group was similar in children and adults. Special attention should therefore be paid to the use of combinations involving artemisinin derivatives in children.

Onyamboko et al.[193] studied 26 pregnant women with uncomplicated malaria who were in the 2nd to 26th and 32nd to 36th weeks of pregnancy; 26 female malaria patients who were 3 months postpartum; and 25 nonpregnant women with uncomplicated malaria. The women received 200 mg oral artesunate. The *CL/F* of dihydroartemisinin in the pregnant, postpartum, and nonpregnant patients were 1.39, 1.26, and 1.07 L/kg/h, respectively. The *V/F* was 2.84, 3.00, and 2.45 L/kg, respectively. In pregnant patients, the *AUC* of dihydroartemisinin was 68% that of nonpregnant patients and was similar to that of postpartum patients. The PKs of artesunate and dihydroartemisinin could be influenced by pregnancy.

The population PKs of an artesunate–pyronaridine combination was studied by Tan et al.[194] Single or multiple doses of 2–5 mg/kg oral artesunate were given to 91 healthy Korean volunteers, either alone or with pyronaridine, both before and after food. A more predictive drug-metabolite PK model was constructed based on the plasma concentration data. This included a drug delivery compartment, a central compartment (artesunate), and a peripheral compartment (dihydroartemisinin). Artesunate was rapidly absorbed, with an absorption rate K_a of 3.85/h. The population CL/F was 1190 L/h (interindividual variability (IIV) 28%) and the V_3/F was 1210 L (IIV 57.4%). For dihydroartemisinin, the CL_M/F was 93.7 L/h (IIV 36.2%) and the V_2/F was 97.1 L (IIV 30%). The apparent intercompartmental clearance rate Q/F of dihydroartemisinin was 5.74 L/h, and the peripheral compartment volume of distribution V_4/F was 18.5 L. Before medication, some volunteers consumed high-fat, high-calorie foods, resulting in an 84% decrease in K_a. Weight also affected CL_M/F. Each unit change in body weight produced a 1.9 unit change in CL_M/F.

Combinations and joint regimens involving artemisinins are recommended by the WHO for malaria treatment. They overcome parasitic drug resistance, raise efficacy, and have become commonplace in clinical practice.

ACKNOWLEDGMENTS

Deepest thanks to Professors Li Zelin, Zeng Meiyi, and Li Ying for their information and valuable advice.

5.4 ANTIMALARIAL MECHANISMS OF ARTEMISININ AND ITS DERIVATIVES

Plasmodium is a single-celled, parasitic protozoan, which invades the host's RBCs. It has a complete cellular structure and biochemical metabolic system. Artemisinins are a new class of antimalarial drug: sesquiterpenes whose molecules contain a peroxide group. Because of their unique structure, their mechanisms of antimalarial action are completely different from those of known antimalarials. After the discovery of artemisinin's antimalarial activity, researchers immediately set to work investigating its mechanisms of action.

A. MORPHOLOGICAL EFFECTS ON MALARIA PARASITES

Li Zelin[†]*, Ye Zuguang*

Institute of Chinese Materia Medica, China Academy of Chinese Medical Sciences, Beijing, China

In 1976, Li Zelin of the Pharmacology Laboratory of the ICMM of CACMS studied the effects of 100 and 800 mg/kg oral artemisinin on the ultrastructure of the erythrocytic stage of *P. berghei* using electron microscopy.[2] Then, in

[†]Deceased.

1980, Peters conducted similar observations in the United Kingdom with artemisinin-oil subcutaneous injections. In vitro cultivation of human *P. falciparum* (Wellcome/Liverpool/West African strain) was also undertaken. With electron microscopic autoradiography and tritium labeling, the location of artemisinin and dihydroartemisinin in *P. falciparum* and their relationship to ultrastructural changes were studied. This elucidated the drug's mechanism of action. It suggested that artemisinin affected nucleic acid and protein metabolism. This enabled the design of further biochemical experiments. This work was completed by Li (1981) and Gu Haoming (1982) in the United Kingdom.

Ye Zuguang, also of the Pharmacology Laboratory, was the first in China to cultivate *P. falciparum* in vitro. The FCC/HN strain was used, and the effects of artemisinin on the parasite and erythrocyte membranes of the host were studied via transmission and scanning electron microscopy.

Subsequently, a number of Chinese research units actively investigated the effects of artemisinin, artemether, artesunate, and dihydroartemisinin on the ultrastructures of various species of Plasmodium with different hosts, such as rodents (*P. berghei*, *P. berghei* ANKA, *P. yoelii*), monkeys (*Plasmodium inui*), and humans (*P. falciparum* Hainan strain, *P. falciparum* Liverpool strain).

1. Effects of Artemisinin on the Ultrastructure of Rodent Malaria

Li Zelin's team[2] infected mice intraperitoneally with *P. berghei*. 3–4 days later, when the parasitized rate of erythrocytes reached 10% and above, they were included in the experiment. The mice were divided into four groups: normal control; lowest effective dose (100 mg/kg); erythrocyte parasite clearance dose (800 mg/kg); and chloroquine control (40 mg/kg). 4, 8, 12, 16, 20, and 24 h after oral medication, blood was collected, smeared, fixed, embedded, sectioned, and double-stained with uranyl acetate and lead citrate. The samples were studied under a JEM-100B electron microscope (Tables 5.30 and 5.31) .

The ultrastructural changes were similar in the 800 mg/kg group, except that they occurred more quickly.

The effects of artemisinin on the ultrastructures of erythrocyte phase of rodent malaria parasites primarily involved the membranes. It acted first on the food vacuole membranes, mitochondria, and limiting membranes and then the endoplasmic reticulum. The mitochondria swelled. The food vacuole membranes and limiting membranes had widened intermembrane gaps or took on a whorl-patterned appearance. Next, changes appeared in the endoplasmic reticulum and nuclear membrane. The endoplasmic reticulum displayed a linear arrangement, vacuolization, or disintegrated. The nuclear membrane swelled. Apart from the impact on the membranes, chromatin agglutination was seen in the nucleus. After 12 h, phagocytic vacuoles were evident, some of which were expelled from the parasite. This progressed rapidly until the internal structures of some of the trophozoites disintegrated at 20–24 h, leaving only many phagocytic vacuoles. Higher artemisinin doses resulted only in faster effects, with no differences in the target sites or characteristics.

TABLE 5.30 Effects of 100 mg/kg Artemisinin on the Ultrastructure of Erythrocytes of Rodent Malaria

Time (h)	Changes in Ultrastructure
4	No significant changes compared to normal structure (Fig. 5.16).
8	Whorl-pattern lines seen in food vacuole membranes of some trophozoites. Mitochondria swollen, malaria pigment lightened. No significant changes in majority of trophozoites.
12–14	Whorl-pattern lines appeared in limiting membrane, some in multiple layers. Some ribosomes no longer adhered to the endoplasmic reticulum and became dispersed. Endoplasmic reticulum presented in rows, some with vacuolization, or partial disintegration. Nuclear membrane swollen, with widened intermembrane space; nuclear pigment agglutinated in star-shaped clusters. Progressive changes to food vacuole membranes and mitochondria. A few trophozoites displayed phagocytic vacuoles, some of which were expelled into the erythrocyte. Minimal changes seen in a small number of trophozoites. (Figs. 5.17 and 5.18)
16	Limiting membrane significantly swollen, with widened intermembrane space, some with small vacuoles. Food vacuoles had whorl-lined membranes. Mitochondria swollen. Endoplasmic reticulum vacuolized or disintegrated. Phagocytic vacuoles increased in number, some were expelled from the parasite. Pigment grains increased and enlarged. Nuclear membrane swollen, chromatin agglutinated in star-shaped clusters, some were nucleolus-like. Merozoites had few changes. Schizonts had widening of the nuclear membrane during mitosis. (Fig. 5.19)
20	Changes to food vacuoles, mitochondria, nuclei, and endoplasmic reticulum in trophozoites similar to those at 16 h. Reduction in ribosomes. Pigment presented in rod-shaped forms, some of which were expelled into the cytoplasm. Phagocytic vacuoles enlarged, a large number of which were expelled into the erythrocyte. Some trophozoites completely disintegrated, leaving only phagocytic vacuoles and barely visible residual limiting membranes. Only a few trophozoites experienced mild changes. (Fig. 5.20)
24	Majority of trophozoites disintegrated. Only phagocytic vacuoles remained in host erythrocytes. Few changes to schizonts. (Fig. 5.21)

The sites and mechanisms of action produced by artemisinin and chloroquine were different. The effects of oral chloroquine in mice were mainly manifested in an expansion of the digestive vacuoles and clumping of pigment. These studies showed that artemisinin acted mostly on the membrane structure and mitochondria.

In 1981, working in Peters' laboratory in the United Kingdom, Ellis and Li Zelin conducted electron microscopy on *P. berghei*-infected mice given

TABLE 5.31 Effects of 40 mg/kg Chloroquine Control on the Erythrocyte Ultrastructure of Rodent Malaria

Time (h)	Changes in Ultrastructure
4	Large phagocytic vacuoles formed, pigment agglutinated in large clumps within.
16	Pigment clusters seen, some of which were expelled into the parasitic cytoplasm. Some expansion in the intermembrane space of the nucleus. (Fig. 5.22)
24	Some phagocytic vacuoles containing pigment clusters were expelled into the host erythrocyte. (Fig. 5.23)

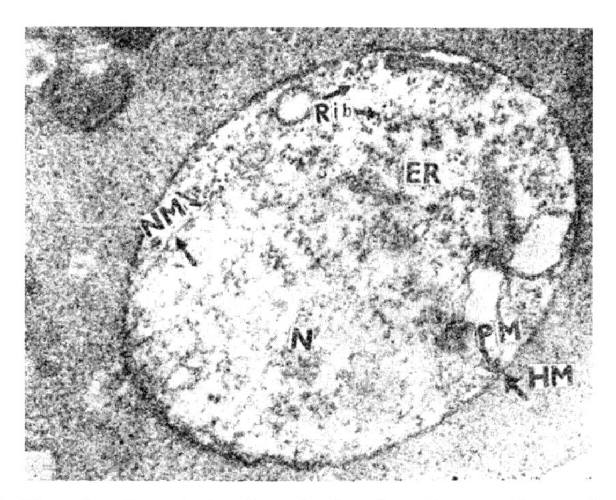

FIGURE 5.16 Normal early rodent malaria trophozoite with parasitic membrane (PM), host erythrocyte membrane (HM), nucleus (N), nuclear membrane (NM), mitochondria (M), ribosome (Rib), and endoplasmic reticulum (ER) (×24,000).

subcutaneous injections of artemisinin-oil suspension. The ultrastructural changes occurred faster than with oral administration and could be seen after 30 min. The number of ribosomes and lines of the endoplasmic reticulum increased significantly. The gap between the parasitic membrane and host membrane widened. The nuclear membrane swelled and became vacuolated. The nucleoplasm agglutinated. Subsequently, the changes in the membrane system progressed, and structural disintegration occurred in 8 h leading to parasitic death. The study also concluded that artemisinin chiefly targeted the membranes. It proposed the use of autoradiography and electron microscopy to further study the drug's action sites. Also, because protein synthesis was implicated in the ribosomes and nucleic acid synthesis was involved in the nucleoplasm, biochemical research should focus on the drug's effects on both synthetic pathways.

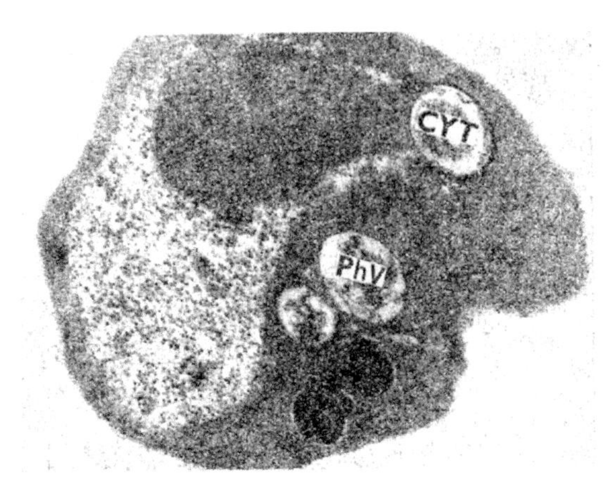

FIGURE 5.17 Trophozoite 12 h after artemisinin 100 mg/kg. Whorls appeared in the cytostome (CYT), and phagocytic vacuoles (Phv) were discharged from the parasite (×15,000).

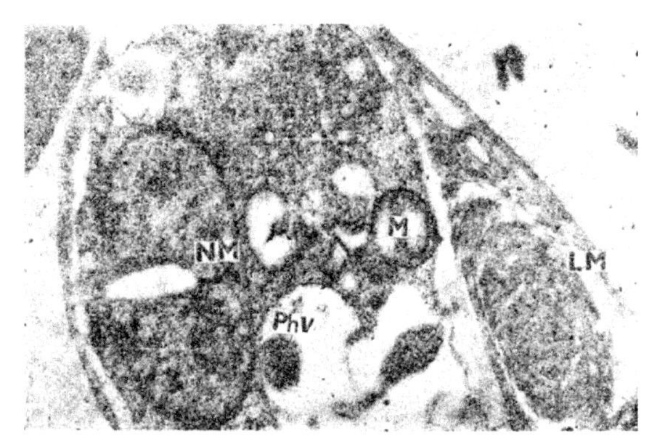

FIGURE 5.18 Trophozoite 14.5 h after artemisinin 100 mg/kg. Whorls appeared in the limiting membrane (LM), the nuclear membrane (NM) was swollen, and nuclear chromatin formed star-shaped clusters. The phagocytic vacuole (PhV) contained parasitic cytoplasm. Mitochondria (M) were swollen (×15,000).

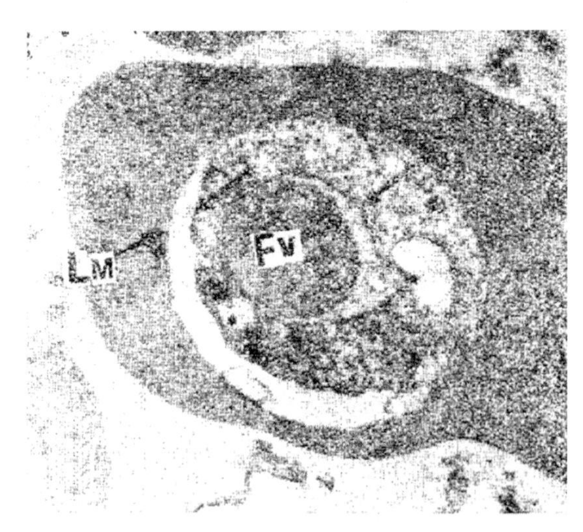

FIGURE 5.19 Trophozoite 16 h after artemisinin 100 mg/kg. Food vacuole (Fv) membranes were swollen, and the gap between the outer and inner membranes widened. The limiting membrane swelled and the membrane gap also widened (×12,000).

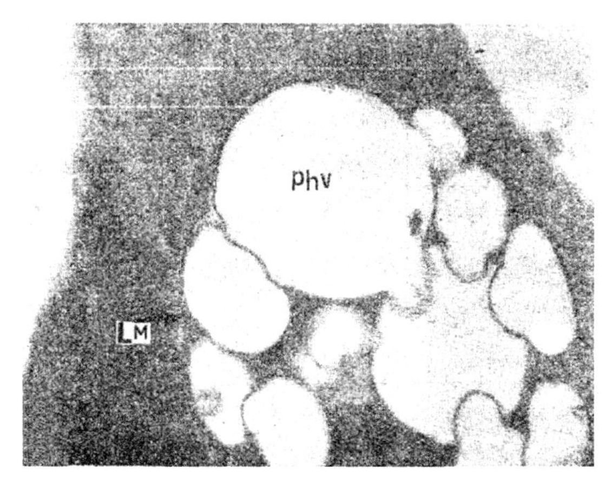

FIGURE 5.20 Trophozoite 20 h after artemisinin 100 mg/kg. All the cytoplasmic structures had disintegrated. Only phagocytic vacuoles (Phv) and indistinct limiting membranes (LM) were seen (×24,000).

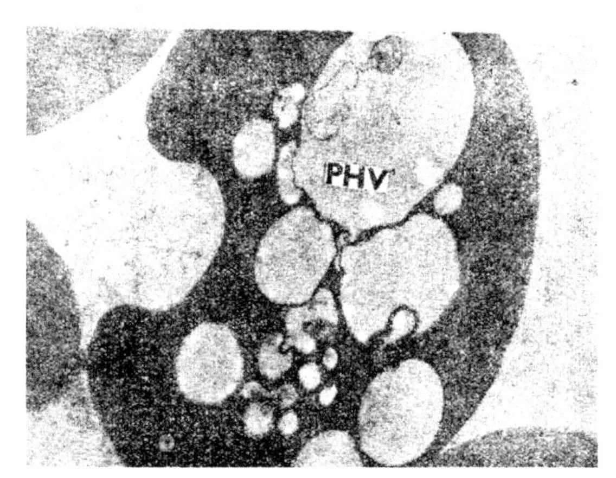

FIGURE 5.21 Trophozoite 24 h after artemisinin 100 mg/kg. All cytoplasmic structures had disintegrated, and residual parasitic phagocytic vacuoles (Phv) were seen in the host erythrocytes (×12,000).

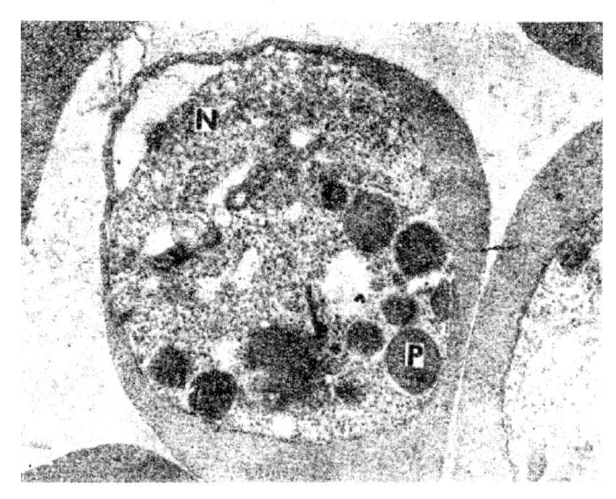

FIGURE 5.22 Trophozoite 16 h after chloroquine 40 mg/kg. Large pigment clusters (P) were discharged into the trophozoite cytoplasm (×18,000).

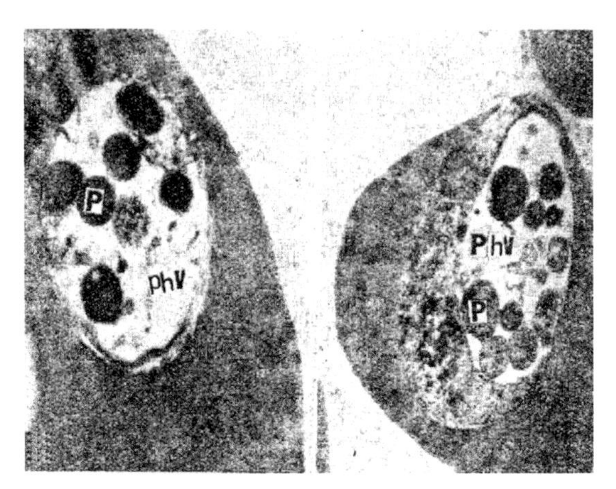

FIGURE 5.23 Trophozoite 24 h after chloroquine 40 mg/kg. The large phagocytic vacuoles (Phv) contained pigment clusters (×145,000).

2. Study on the Localization of Artemisinins (^{3}H-Artemisinin, ^{3}H-Dihydroartemisinin) With Autoradiography and Electron Microscopy

Li Zelin employed autoradiography and electron microscopy to study how the drugs entered the parasite and the relationship between drug localization in the parasitic body and its structural changes. This would enhance understanding of the drugs' mechanism of action and the link between structure and efficacy.[195]

The ^{3}H-dihydroartemisinin experiment used the Wellcome/Liverpool/West African strain of *P. falciparum*. ^{3}H-dihydroartemisinin and ^{3}H-artemisinin were added to the culture medium at concentrations of 10^{-6} and 10^{-7} mol/L, respectively. Samples were obtained after 1 h, fixed with 3% glutaraldehyde, suspended in agar, cut into small pieces, and then fixed with 1% osmium tetroxide. They were dehydrated in methanol, embedded in araldite resin, sectioned, sprayed with a carbon film, and treated with an L4 nuclear emulsion. Isotope exposure took place in a cartridge, whereupon the sample was developed, fixed, stained with lead acetate, and put under an electron microscope.

The results showed that 1 h after ^{3}H-dihydroartemisinin was added to the parasite culture, silver particles of high-electron density appeared in the shape of twisted filaments. More than 95% of these particles appeared in infected erythrocytes; of these, 74.5% were in the parasites. Their distribution in the parasites is shown in Table 5.32. Structural changes included an expansion in the food vacuoles and prominent large agglutinations of pigment crystals in the vacuoles. Changes to the structures of the parasitic ribosomes and endoplasmic reticulum could also be seen.

^{3}H-artemisinin yielded similar results on parasites after 1 h of drug exposure. The vast majority of silver particles appeared in infected erythrocytes, two-thirds

TABLE 5.32 Distribution of Silver Particles Caused by ^{3}H-Dihydroartemisinin as Seen in Autoradiography and Electron Microscopy

	Food Vacuole			
Plasma Membrane	Membrane	Within Vacuole	Pigment	Cytoplasm and Others
49.89%	13.9%	3.6%	11.39%	21.18%

From Li Z. The application of nuclear technology to the study of the action mechanisms of the new antimalarials artemisinin and dihydroartemisinin. *Nucl Sci Tech* 1984;**2**:23–5. [李泽琳. 核技术在抗疟新药青蒿素、二氢青蒿素抗疟作用研究中的应用. 核技术, 1984, 2: 23–25.]

TABLE 5.33 Distribution of Silver Particles Caused by ^{3}H-Artemisinin as Seen in Autoradiography and Electron Microscopy

	Food Vacuole			
Plasma Membrane	Membrane	Within Vacuole	Pigment	Cytoplasm and Others
41%	13%	2%	2%	42%

From Li Z. The application of nuclear technology to the study of the action mechanisms of the new antimalarials artemisinin and dihydroartemisinin. *Nucl Sci Tech* 1984;**2**:23–5. [李泽琳. 核技术在抗疟新药青蒿素、二氢青蒿素抗疟作用研究中的应用. 核技术, 1984, 2: 23–25.]

of which were in the parasites. The distribution of the silver particles in the parasites is shown in Table 5.33. The difference was that the pigment grains were small and formed only 2% of the silver particles. Structural changes included a widening of the membrane gap and formation of vacuoles in the plasma and food vacuole membranes, where 52% of the silver particles were located. Electron density was reduced in areas where the silver particles were present. There was no agglutination, expansion, or increase in pigment grains. After 4h of exposure, there was no change in the distribution of silver particles in the malaria parasites. However, the silver particles present in the areas with membrane widening increased from 52% to 69%. The widening and swelling of the membranes themselves also intensified. After 8h, the silver particles were still mainly located in the plasma membrane, but serious damage had occurred to the membrane by this time. Changes were also seen in the structures of the ribosomes and endoplasmic reticulum.

These findings showed that the effects of artemisinin and dihydroartemisinin were mainly located in the parasitic plasma membrane. A large number of silver particles were present there, associated with structural changes. Some silver particles were also present in the cytoplasm, and alteration of the endoplasmic reticulum structures and nucleoplasm was observed. Dihydroartemisinin could also cause clustering in the malarial pigment, almost all of which displayed the silver particles. Peters and Li Zelin also reported that dihydroartemisinin could cause agglutination of the pigment in rodent malaria.[196,197] Its mode of action and the forms of pigment aggregates were different from those caused by chloroquine and quinine. This could be due to differences in the area where the pigment clustered.

3. Morphological Profile of *Plasmodium falciparum* by Transmission Electron Microscopy and Appearance of Erythrocytic Membrane by Scanning Electron Microscopy

Ye Zuguang used transmission electron microscopy to observe the normal ultrastructures of *P. falciparum* cultured in vitro[198] and the effects of artemisinin and chloroquine on them.[199,200] Pathological changes took place in the parasitic morphology 2 h after exposure to artemisinin at the in vitro effective concentration of 10^{-7} mol/L and after 1 h at concentration 5×10^{-7} mol/L. The changes included a widening of the gaps in the cell, mitochondrial, and nuclear membranes, and swelling of the endoplasmic reticulum and deformation of its membranes. Autophagic vacuoles appeared at the same time (Figs. 5.24 and 5.25), which contained iron-rich pigment clusters. For chloroquine 10^{-6} mol/L, many autophagic vacuoles appeared after 0.5 h, containing a large number of pigment clusters (Fig. 5.26). These were the main morphologic differences between the effects of artemisinin and chloroquine.

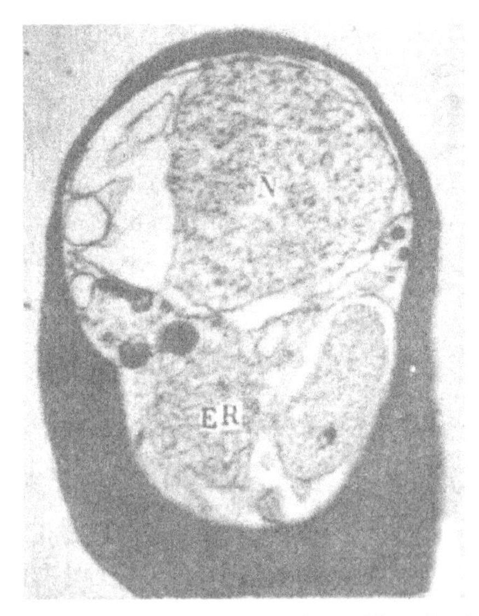

FIGURE 5.24 Widening of membrane gaps and vacuolar swelling of endoplasmic reticulum (ER) in Plasmodium exposed to 5×10^{-7} mol/L artemisinin for 1 h (×21,000). *(From Ye Z, Li Z, Fu X, et al. Effects of artemisinin and chloroquine on the ultrastructure of in vitro cultured erythrocytic* Plasmodium falciparum. J Tradit Chin Med *1982;**4**:65–7. [*叶祖光，李泽琳，傅湘琦，等. 青蒿素和氯喹对体外培养的红内期恶性疟原虫超微结构的影响. 中医杂志, *1982, (4): 65–67.])*

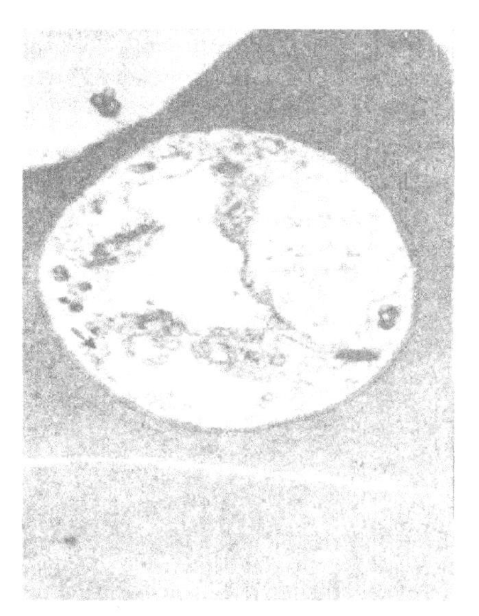

FIGURE 5.25 Swollen endoplasmic reticulum fused to form large autophagic vacuoles in Plasmodium exposed to 5×10^{-7} mol/L artemisinin for 4 h (×15,000). *(From Ye Z, Li Z, Fu X, et al. Effects of artemisinin and chloroquine on the ultrastructure of in vitro cultured erythrocytic* Plasmodium falciparum. J Tradit Chin Med *1982;**4**:65–7. [*叶祖光，李泽琳，傅湘琦，等. 青蒿素和氯喹对体外培养的红内期恶性疟原虫超微结构的影响. 中医杂志, *1982, (4): 65–67.])*

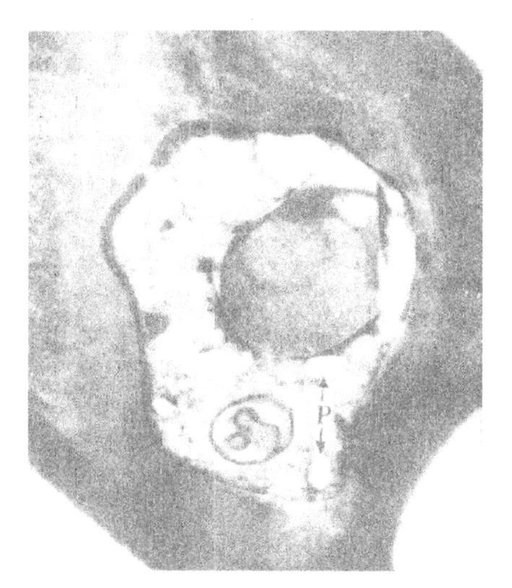

FIGURE 5.26 Large autophagic vacuole with pigment granules in Plasmodium exposed to 10–6 mol/L chloroquine for 4 h (×21,000). *(From Ye Z, Li Z, Fu X, et al. Effects of artemisinin and chloroquine on the ultrastructure of in vitro cultured erythrocytic* Plasmodium falciparum. J Tradit Chin Med *1982;**4**:65–7. [*叶祖光，李泽琳，傅湘琦，等. 青蒿素和氯喹对体外培养的红内期恶性疟原虫超微结构的影响. 中医杂志, *1982, (4): 65–67.])*

In addition to documenting the artemisinin-induced morphological changes on *P. falciparum* by transmission electron microscopy, Ye et al. also employed scanning electron microscopy to observe the surface structure of erythrocytes infected with *P. berghei* and the effects of artimisinin on its morphology.[201] The concave appearance of infected erythrocytes deepened significantly to the extent of forming nest- or bullet hole-shaped cavities. This was caused by the interchange of a large amount of material between the erythrocyte and plasma, brought about by the vigorous parasitic metabolism. This change in erythrocytic surface was termed the "metabolic window." Artemisinin had no significant effect on the morphology and number of the metabolic windows, indicating that the drug did not eliminate parasites through damage to this window. It also showed that artemisinin acted selectively on the parasites and did not affect host cells.

4. Effects of Artemisinins on Ultrastructure of Malaria Parasite

Jiang Jingbo and Aikawa conducted electron microscopy on *P. inui* simian malaria, which had a more obvious mitochondrial structure, and *P. falciparum* cultured in vitro.[202] *P. inui* parasitic mitochondria swelled markedly 2.5 h after artemisinin was administered. This differed from the effects of chloroquine, which was widely used at the time, and corresponded to the findings of Li Zelin et al. It further confirmed that the mitochondria were one of the targets of artemisinin.

Aikawa et al. exposed erythrocytes infected with *P. falciparum* to ^{3}H-dihydroartemisinin and ^{14}C-artemisinin.[202] Both drugs were located in the food vacuoles and the mitochondria. Structural changes first appeared in the mitochondria, endoplasmic reticulum, and nuclear membrane 2 h after exposure; after 4 h, multilayered myelin-like figures appeared in the nuclear and plasma membranes. The ribosomes disappeared and food vacuoles disintegrated. Such changes were seen in around 30% of the parasites, leading to general disorganization of the parasitic structure.

These observations were consistent with the conclusions of 1979 and 1981 publications of Li Zelin. Artemisinin's effects on the parasitic ultrastructures concentrated mainly on the membrane system. It acted first on the food vacuole and surface membranes and the mitochondria. Next, it affected the nuclear membrane and endoplasmic reticulum. It also had an effect on the nuclear chromatin. The drug's mechanism of action lay in disrupting the functions of the surface membrane and mitochondria, blocking the supply of nutrients from the host's erythrocytic cytoplasm. The impact of artemisinin derivatives, such as artemether and artesunate, on the parasitic ultrastructures was basically the same as that of the parent drug artemisinin. Table 5.34 summarizes the findings.

Electron microscopy enables the observation of parasitic organelles and ultrastructures. The timing and extent of damage to these structures can be traced, charting a path for future research into artemisinin's mechanism of action.

TABLE 5.34 Summary of Research Into the Effects of Artemisinins on Parasitic Ultrastructures

Year Published	Drug	Animal/ Plasmodium Strain	Plasmodium Strain Cultured In Vitro	Main Sites of Action	Principal Investigators
1979	Artemisinin	Mouse *Plasmodium berghei*		Cell membrane—mitochondria system	Li Zelin[2], Xue Baoyun
1981	Artemisinin-oil suspension	Mouse *P. berghei*		Cell membrane, nucleoplasm, endoplasmic reticulum	Ellis, Li Zelin
1981	Artesunate	Monkey		Cell membrane, nuclear membrane	Li Rui[203]
1982–83	Artemisinin		*P. falciparum* Hainan strain	Membranes, mitochondria	Ye Zuguang[198,199]
1984–85	^{3}H-Dihydroartemisinin	Mouse *P. berghei*	*P. falciparum* Liverpool strain	Pigment	Ellis, Li Zelin[204]
1985	Artesunate	Monkey *Plasmodium inui*		Mitochondria	Jiang Jingbo[202], Aikawa
1988	Artemether	Mouse *P. berghei*		Mitochondria, membranes	Chen Leyi, Hu Ming[205]
1988	Artesunate	Mouse *P. berghei*		Mitochondria, membranes	Zhao Yi[206]
1993	Artemisinin		*P. falciparum*	Membrane-mitochondria system	Maeno[207], Aikawa
2000	Dihydroartemisinin	Mouse *P. berghei* ANKA strain		Membranes, pigment	Chen Peiquan, Yuan Jie[208]

B. EFFECTS OF ARTEMISININS ON PARASITIC BIOCHEMICAL METABOLISM

Li Zelin†*, Ye Zuguang*
Institute of Chinese Materia Medica, China Academy of Chinese Medical Sciences, Beijing, China

To better understand artemisinin's action mechanism, it is necessary to examine its effects on the parasite's biochemistry and metabolism. This clarifies the aforementioned morphological changes.

1. Interaction of Infected Erythrocytes and ^{3}H-Artemisinin

In 1992, Zhang Kuihan and Li Zelin[209] cocultured *P. falciparum* with tritium-labeled dihydroartemisinin in vitro to observe the drug's concentration effects in infected erythrocytes. Method of Polet et al.[210] was used to determine the concentration effect and its speed. ^{3}H-artemisinin at a concentration of 5×10^{-8} mol/L was added to the *P. falciparum* culture. The specific radioactivity was 74 GBq (2 mCi)/mg, and radiochemical purity was 95.4%. After the drug was introduced, the cells were sampled at 0.5, 1, 2, 3, and 6 h and prepared on a scintillating disk. The dose-response curve was derived using different concentrations of ^{3}H-artemisinin, from 10^{-8} to 10^{-5} mol/L, with samples taken 3 h after drug administration. Samples were washed in saline to remove extracellular radioactive material, underwent a digestion process, and were assayed using a Pack-8000 liquid scintillation counter. The results were as follows.

Drug absorption in infected (around 10.1%) and normal erythrocytes after 6 h can be seen in Fig. 5.27. The drug was rapidly absorbed by infected cells, with

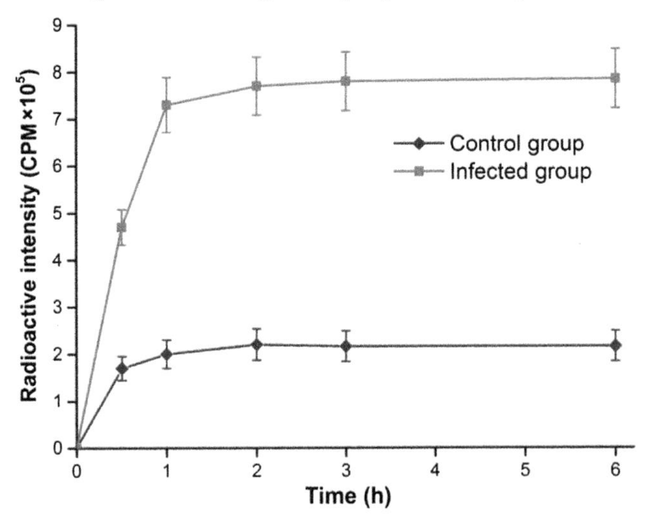

FIGURE 5.27 Absorption of ^{3}H-artemisinin in infected and normal erythrocytes.

†Deceased.

cellular radioactivity reaching 50% of the peak in 30 min. After 1 h, radioactivity was close to the maximal value. Radioactivity in the 10.1% infected erythrocytes was far higher than in the equivalent number of normal cells. The radioactivity of a 100% infection rate was calculated based on the formula below.

^{3}H-artemisinin absorption in 100% infected erythrocytes = A + B

where: A = (B – C) ÷ Rate of infection (ie. 10.1%);
B = Radioactity of 10.1% infected erythrocytes;
C = Radioactity in normal erythrocytes.

Table 5.35 shows the radioactivity in 100% infected cells and an equivalent number of normal cells. The intensity of radioactivity was 18.2 times higher in infected cells after 30 min and 25.4 times higher after 6 h.

The distribution ratios of ^{3}H-artemisinin in erythrocytes and the culture medium are shown in Tables 5.36 and 5.37. Absorption in normal erythrocytes was only slightly higher than in the culture medium at around 1.5 times. By contrast, the concentration effects of infected cells were clear, with radioactivity at 30 times that of the equivalent volume of culture medium.

Fig. 5.28 shows the radioactive absorption dose–response curve of erythrocytes with an infection rate of 10.1%. Between concentration of 10^{-6} and 10^{-8} mol/L, absorption of ^{3}H-artemisinin continually increased in both types of cell.

Artemisinin could swiftly enter erythrocytes infected with *P. falciparum* and with a high degree of concentration. This facilitated the conditions for parasite elimination and was one of the factors behind the drug's high efficacy and rapid action. The differences in ^{3}H-artemisinin concentrations in infected and normal erythrocytes were similar to such variations seen in chloroquine reported in the literature. In this experiment, however, absorption neared its peak 1 h after administration, faster than the 3 h reported study on ^{3}H-chloroquine absorption by Polet et al.[210]

TABLE 5.35 Comparison of Radioactive Absorption From ^{3}H-Artemisinin in 100% Infected and Normal Erythrocytes

Time (h)	Infected Erythrocytes (CPM[a])	Normal Erythrocytes (CPM[a])	Ratio
0.5	32,728	1,800	18.2
1	34,143	2,109	25.7
2	54,725	2,364	23.1
3	55,707	2,287	24.4
6	58,856	2,317	25.4

[a]Counts per Minute

TABLE 5.36 Radioactivity in Equal Volumes of Normal Erythrocytes and Culture Medium

Time (h)	Medium (1 mL)	Normal Erythrocytes (1 mL)	Ratio
0.5	32,100	36,000	1.1
1	35,198	42,180	1.2
2	31,683	47,280	1.5
3	32,206	45,740	1.4
6	30,663	46,340	1.5

TABLE 5.37 Radioactivity in Equal Volumes of Infected Erythrocytes and Culture Medium

Time (h)	Medium (1 mL)	Infected Erythrocytes (1 mL)	Ratio
0.5	27,029	660,800	17.8
1	37,001	1,093,980	29.6
2	36,850	1,105,080	30.0
3	38,439	1,124,940	29.3
6	37,516	1,188,540	31.7

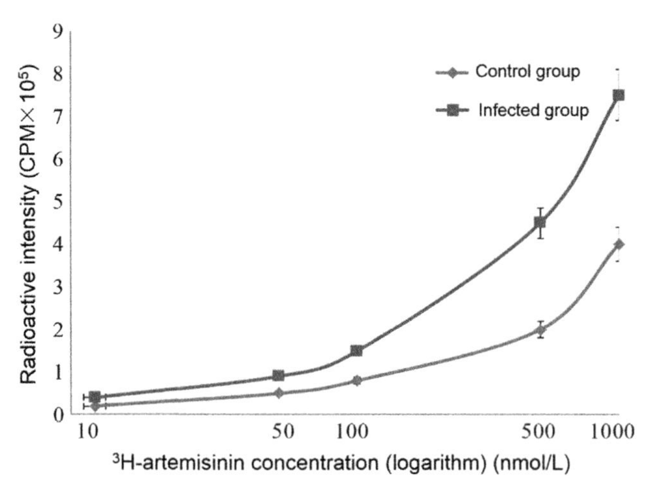

FIGURE 5.28 Dose–response relationship of ^{3}H-artemisinin absorption in infected and normal erythrocytes.

Changes in the biochemistry of infected erythrocytes could be related to their selective absorption and concentration of artemisinin. After infection, a vacuole containing the parasite was formed in the erythrocyte and the parasite itself developed. This significantly increased the lipid volume of the infected cell by up to three to five times that of normal erythrocytes.[211] The parasite also had a high-lipid content, up to five times that of normal cells.[212] It enabled the highly fat-soluble artemisinin molecules to enter the infected cell and hence the parasite, increasing absorption and concentration of the drug.

Like chloroquine, artemisinin was selectively concentrated in infected erythrocytes, while its concentration in normal erythrocytes was very low. This was highly conducive to the drug's antimalarial activity.

2. Effects on Nucleic Acid Metabolism of Malaria Parasite

In 1979, coinciding with the 100th anniversary of the discovery of the malaria parasite, Sherman reviewed the past century of research into parasitic biochemistry. It served as an important guide for studies into the parasite itself and antimalarial drugs. The paper stated that nucleic acid metabolism in Plasmodium involved four aspects: DNA, RNA, the synthesis and nature of pyrimidine, and the mechanisms behind purine transport and salvage.

Plasmodium can synthesize pyrimidine to meet its own nucleic acid synthesis and metabolic demands but is unable to synthesize purine compounds without some exogenous source. Therefore, some purine precursors can be used to observe the drug's effects on the parasitic nucleic acid metabolism. They can also serve as an indicator of the growth and reproduction of the parasite. The quantity of precursor incorporated into the parasitic nucleic acid reflects its growth, reproduction, and living state. Hence, this method was developed as a quantitative assay to evaluate antimalarial efficacy.[213]

In 1983, Li Zelin[11] and Peters added different concentrations of artemisinin, dihydroartemisinin, and artemether to in vitro falciparum malaria culture (Wellcome/Liverpool/West African strain).[195] After being cultured for 24 h, 0.2 μCi ^{3}H-hypoxanthine per 50 nL culture medium was introduced, followed by another 18 h of culturing. Samples were taken, erythrocytes were pyrolyzed, and radioactivity measured to obtain the counts per minute and calculate the 50% inhibitory concentration of hypoxanthine incorporation (ID_{50}). Separately, a fixed concentration of artemisinin or dihydroartemisinin was cultured with the parasites and ^{3}H-hypoxanthine. Samples were prepared at different times and processed using the same procedure. Radioactivity was gaged. The inhibitory effects at different times were compared to those of control groups.

The results showed that artemisinin and its derivatives significantly inhibited ^{3}H-hypoxanthine incorporation. The ID_{50} of artemisinin and dihydroartemisinin was 126 ± 55 and 0.58 ± 0.035 nmol/L, respectively, and the inhibitory ratio of the two drugs was 1:215. It indicated that dihydroartemisinin had a stronger inhibitory effect. Artemether and sodium artesunate were also at least 100 times more effective in this respect than artemisinin. Time–response curves showed that, when compared to the control, 10^{-6} mol/L artemisinin or 10^{-7} mol/L dihydroartemisinin produced significant inhibition after 2 h. This reached dramatic levels after 5 h.

Because the parasite cannot synthesize purine itself, it relies on the purine in the host's erythrocytes, primarily hypoxanthine and adenine. Studies suggested that hypoxanthine[214] was the main precursor of nucleic acid metabolism in Plasmodium.[215] Hypoxanthine entered the malaria parasite to form inosine monophosphate, then adenosine monophosphate (AMP), adenosine diphosphate (ADP), adenosine triphosphate (ATP), and finally nucleic acid. By inhibiting the incorporation of ^{3}H-hypoxanthine, artemisinin and its derivatives suppressed nucleic acid metabolism in the parasite.

Ye Zuguang et al. used tritium-labeled adenosine and hypoxanthine, both purine precursors, to study the effects of artemisinin on nucleic acid metabolism in vitro *P. falciparum*.[216] The antimalarial efficacy of artemisinin was also quantitatively determined (Table 5.38). Artemisinin clearly inhibited the incorporation of both nucleic acid precursors, showing that the drug's effects severely undermined the parasitic nucleic acid metabolism.

Simian malaria tests conducted by Zhao[217] also showed that the action mechanism of artemisinins mainly involved the suppression of nucleic acid metabolism and not protein metabolism.

In their review, Xu Xuxiang and Zhu Shihui wrote that artemisinins did not clearly inhibit carbohydrate metabolism but had significant effects on the incorporation of nucleic acid precursors. The experiments also yielded conflicting data. For example, some tests indicated that artemisinin had no impact on

TABLE 5.38 **Effects of Artemisinin and Chloroquine on the Uptake of ^{3}H-Adenosine and ^{3}H-Hypoxanthine Into Human Erythrocytes Infected With Different Strains of *Plasmodium falciparum***

		IC_{50}		
Drug	**Radioactive Precursor**	**FCMSU$_1$/ Sudan Strain (Chloroquine Sensitive)**	**FCB K$^+$ Strain (Chloroquine Resistant)**	**Resistance Index**
QHS	Hypoxanthine	2.8×10^{-8} mol/L	3.0×10^{-8} mol/L	1.07
	Adenosine	2.8×10^{-8} mol/L	3.2×10^{-8} mol/L	1.14
CQ	Hypoxanthine	5.6 ng/mL	65.4 ng/mL	11.68
	Adenosine	7.6 ng/mL	47.4 ng/mL	6.24

QHS, artemisinin; *CQ*, chloroquine.

adenosine incorporation. Others showed that artemisinin inhibited hypoxanthine uptake. The reasons for this were unclear. The inhibition of precursor uptake into the macromolecule did not necessarily mean that artemisinin's basic action mechanism involved the macromolecule. Instead, artemisinin could be affecting membrane transport processes.[214]

3. Effects on Protein Metabolism of Malaria Parasite

Gu Haoming examined the effects of artemisinin, dihydroartemisinin, and artemether on *P. falciparum* protein synthesis and metabolism, using the incorporation of ^{3}H-isoleucine. This evaluated the drugs' action mechanism from the perspective of biochemical metabolism.[215] Less than 1 h after the drugs were introduced, incorporation of ^{3}H-isoleucine was markedly suppressed. At a concentration of 3×10^{-5} mol/L, the positive dactinomycin control—an inhibitor of eukaryotic protein synthesis—also produced such effects. After 8 h, all three drugs and the positive control had no visible effects on the parasite rate, indicating that no other aspects of the parasite were seriously affected. Isoleucine was the main exogenous amino acid in plasmodium protein synthesis in the infected erythrocyte. Therefore, it could serve as an indicator for protein synthesis and metabolism. At an early stage, artemisinins could clearly inhibit protein synthesis and metabolism.

Ye Zuguang et al. exploited parasitic synchronization technique to control various stages of in vitro *P. falciparum* and investigate the selective effects of the drug on these stages (ring form, trophozoite, and schizont).[218] It was discovered that the trophozoite stage, at which protein synthesis was most active, was also the most sensitive to artemisinin. This was followed by the schizont stage, when nucleic acid synthesis was at its peak.

However, Zhao Yi's simian malaria experiment suggested that artemisinins acted mainly through inhibiting nucleic acid metabolism and not protein metabolism.

Experiment of Gu Haoming et al. using the cell-free protein synthesis system also showed that artemisinins had no effects on protein synthesis and metabolism. Another test conducted by Gu et al. indicated that the drugs did not affect globinase and free amino acid activity in *P. berghei* and hence had no impact on protein metabolism.[219]

These contradictory results demonstrated that one indicator or experimental method alone could not establish the drugs' mechanisms of action. It was unclear if a change in one metabolic indicator was primary or secondary. Only a series of tests could explore and confirm this. Until now, artemisinins' mechanism of action is not entirely clear.

4. Effects on Various Enzymes in Malaria Parasite

In an aerobic and nutrient-rich environment, the energy required by the plasmodium parasite comes mostly from the oxidation of organic food and glucose metabolism. This biological process consumes oxygen and produces carbon dioxide. The process of gaseous metabolism is produced by parasitic respiration.

Current biochemical pharmacology studies showed that artemisinin produced antimalarial effects by undermining parasitic respiration via various enzymes. Electron microscopy also indicated that the drug acted mainly on the parasite's membrane structure, disrupting membrane-mitochondrial function. Howells reported that, in the trophozoite stage, cytochrome oxidase activity in the mitochondria of rodent malaria was high. Artemisinin could be suppressing parasitic respiration by inhibiting cytochrome oxidase and succinate dehydrogenase. Going further, Zhao Yi et al. explored the effects of artemisinin on three enzymes involved in mitochondrial oxidation: cytochrome C oxidase, succinate dehydrogenase, and LDH.

Cytochrome C Oxidase

Adopting Schneider's and Potter's protocol, Zhao Yi's works on rodent malaria studied the effects of dihydroartemisinin on cytochrome C oxidase and compared it with quinine dihydrochloride.[220] Saline, dihydroartemisinin, and a 0.5 mL quinine dihydrochloride solution were poured into the main chambers of reaction flasks. Then, 0.7 mL of 0.1 mol/L phosphate buffer (pH 7.4), 0.3 mL of 4.8×10^{-4} mol/L cytochrome C solution, 0.2 mL of 4.8×10^{-8} mol/L aluminum oxide solution, 0.2 mL of 0.114 mol/L ascorbic acid solution, and 0.1 mL of 2% rat liver homogenate were added to the main chamber of each flask. The central chamber was filled with 0.2 mL of 10% potassium hydroxide solution and a 1.0 cm × 1.0 cm piece of filter paper. This was stabilized for 10 min before observation. Over 40 min of observation, the average oxygen consumption of the saline, 7.5 mmol/L dihydroartemisinin, and 3 mmol/L quinine dihydrochloride groups were 102.40 ± 5.36, 49.16 ± 4.31, and 48.45 ± 10.78 μL, respectively. The dihydroartemisinin and quinine dihydrochloride groups displayed significant inhibition of cytochrome C oxidase activity. The inhibition rates were 51.99% ($P < .01$) and 52.69% ($P < .01$), respectively (Fig. 5.29).

In cellular respiration, cytochrome C oxidase represents the final point of the cytochrome system. These enzymes directly transfer electrons from the respiratory substrate through the cytochrome system to oxygen molecules (i.e., automatic oxidation). Their functions include participating in respiratory oxidation on the mitochondrial surface, catalyzing indophenol reaction with cytochrome C, and providing energy for the whole cell. Artemisinins affected cytochrome C oxidase, inhibiting glucose metabolism and thus parasitic respiration.

Succinate Dehydrogenase

Huang Guojun et al. also used the Schneider and Potter method to determine the effects of dihydroartemisinin on succinate dehydrogenase, comparing it with quinine dihydrochloride.[220] After 40 min of observation, the average oxygen consumption of the saline, 7.5 mmol/L dihydroartemisinin, and 3 mmol/L

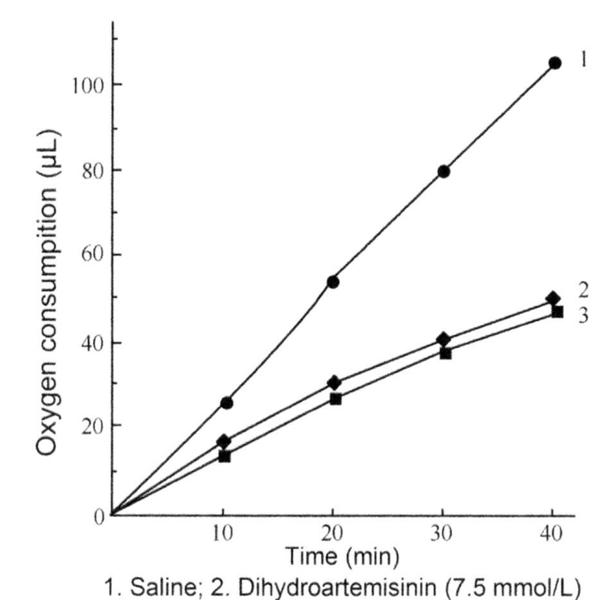

FIGURE 5.29 Effects of different drugs on cytochrome C oxidase. *(From Huang G, Zhao Y. Effect of sodium artesunate on the respiration, glucose metabolism and protein metabolism of* Plasmodium berghei. J Tradit Chin Med *1981;**10**:66–70. [黄国钧, 赵一. 还原青蒿素琥珀酸钠对鼠疟原虫呼吸、糖代谢及蛋白质代谢的影响. 中医杂志, 1981, 10: 66–70.])*

quinine dihydrochloride groups were 37.63 ± 3.11 μL, 24.58 ± 2.33 μL, and 9.30 ± 2.56 μL, respectively. Dihydroartemisinin and quinine dihydrochloride had clear inhibitory effects on the activity of succinate dehydrogenase; the inhibition rates were 34.70% ($P < .01$) and 75.29% ($P < .01$), respectively (Fig. 5.30).

Succinate dehydrogenase is a flavin enzyme, bound to the inner mitochondrial membrane. It is important for oxidative phosphorylation and electron transport, providing electrons for eukaryotic mitochondria and the oxidative and energy-producing respiratory chains in various prokaryotic cells. Hence, it is a mitochondrial marker enzyme. Artemisinins inhibited parasitic respiration by acting on succinate dehydrogenase and disrupting glucose metabolism.

Lactate Dehydrogenase

Using colorimetrics, Huang Guojun et al. observed the impact of dihydroartemisinin on LDH, comparing it with quinine dihydrochloride. Two parallel assay tubes were used for each drug concentration, along with a control tube. Both the assay and control tubes were loaded with 0.1 mL of the drug, 0.5 mL lactic acid matrix solution, and 0.05 mL diluted serum. A 0.1 mL coenzyme solution was

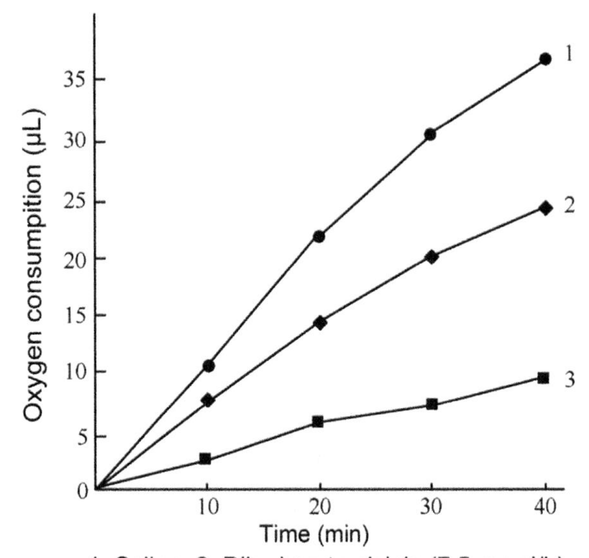

FIGURE 5.30 Effects of different drugs on succinate dehydrogenase. *(From Huang G, Zhao Y. Effect of sodium artesunate on the respiration, glucose metabolism and protein metabolism of* Plasmodium berghei. J Tradit Chin Med *1981;**10**:66–70. [黄国钧, 赵一. 还原青蒿素琥珀酸钠对鼠疟原虫呼吸、糖代谢及蛋白质代谢的影响. 中医杂志, 1981, 10: 66–70.])*

added to the assay tube and 0.1 mL distilled water to the control tube. The final concentration of the drug was calculated based on the volume of liquid at this point. After mixing, the solution was kept warm and the colorimetric test run.

After four tests per group, it was found that LDH activity in the saline group was 525 ± 61 active units. For 3.2, 1.6, and 0.8 mmol/L dihydroartemisinin, this was 358.5 ± 39, 453.8 ± 37, and 499.3 ± 38 active units, respectively. Variation analysis against the saline group showed that 3.2 and 1.6 mmol/L dihydroartemisinin significantly inhibited LDH ($P < .1$ and $P < .05$, respectively).

LDH is a broadly distributed enzyme that catalyzes the interconversion of lactic acid and pyruvic acid. L-LDH acts on L-lactic acid and D-LDH on D-lactic acid. Both use NAD^+ as a hydrogen acceptor. During anaerobic glycolysis, it catalyzes pyruvic acid to form lactic acid by accepting hydrogen from NADH, formed by glyceraldehydes-3-phosphate dehydrogenase. This produces lactic acid. LDH plays key role in the glycolytic process. Experiments have shown that artemisinin and its derivatives can affect this process in malaria parasite by acting on LDH.

Further studies have shown that parasitic metabolism in erythrocytes includes an Embden–Meyerhof pathway, which does not require oxygen and a tricarboxylic acid cycle and pentose phosphate branch, both of which need oxygen. Artemisinin and its derivatives have an inhibitory effect on LDH in glycolysis and succinate dehydrogenase in the tricarboxylic acid cycle. In addition, they can also affect the level of cytochrome oxidase. These drugs inhibit parasitic respiration by acting on various enzyme systems, resulting in their antimalarial effect.

In addition, Li Rui[203] proposed that artemisinins inhibited cytochrome oxidase, interfered with the mitochondrial functions of the membrane system, and impaired the ability of the parasite's digestive enzymes to break the host's hemoglobin down into amino acids. The parasite could not obtain sufficient nutrients and material for protein synthesis, producing autophagocytic vacuoles and leading to parasitic disintegration and death. However, study by Ye Zuguang on the effects of artemisinin on oxidative phosphorylation in yeast and rat mitochondria yielded negative results. Further investigation is needed.

5. Effects on Pigment Clumping in Malaria Parasite

Warhurst and American scholars established a method to measure the impact of different antimalarials on parasitic pigment clumping. Pigment was not a useless product of metabolism as previously thought but acted as intracellular receptors. Unlike quinine, chloroquine caused pigment clumping. Administering quinine first and chloroquine second led to quinine acting as a competitive antagonizer on these receptors. As a result, pigment clumping did not occur.

The first morphological change to appear under the influence of chloroquine was agglutination of the malarial pigment, which in fact was part of the formation of autophagocytic vacuoles. In 10–30 min, small food vacuoles in the cytoplasm of the erythrocytic parasite fuse. The pigments contained in the vacuoles therefore clump together to form larger grains. Finally, a large autophagocytic vacuole is produced, containing clusters of these pigment grains and other cytoplasmic components. Therefore, the mechanism of chloroquine was closely related to the formation of autophagocytic vacuoles. Warhurst et al. not only discovered this chloroquine-induced pigment clumping (CIPC) but also developed it into a method for evaluating antimalarial activity.[221]

Antimalarial drugs can be divided into four main categories depending on whether they cause pigment agglutination and affect CIPC: first, 4-aminoquinoline compounds, such as mepacrine, with similar sites of action to that of chloroquine; second, quinines, which do not cause pigment agglutination themselves but antagonistically inhibit CIPC and compete with chloroquine in terms of action site; third, 8-aminoquinolines, which do not cause agglutination but only inhibit CIPC in larger doses nonantagonistically and have a different initial action site from that of chloroquine; fourth, folate antimetabolites, which neither cause agglutination nor affect CIPC and have a completely different site of action from that of chloroquine. CIPC research therefore considers three factors: whether a drug causes pigment agglutination, whether it inhibits CIPC, and whether an inhibitory effect is competitive or noncompetitive.

Ye Zuguang et al. adopted this methodology to study artemisinin's mechanism of action both in vitro and in vivo and compared it to that of chloroquine.[222] Artemisinin did not induce pigment agglutination in *P. berghei*. No pigment clumping was seen even when the oral dose reached 450 mg/kg in infected mice or at an in vitro concentration of 10^{-3} mol/L (Table 5.39).

However, artemisinin had a certain inhibitory effect on CIPC, but this was not complete. The maximum inhibition rate was 60%, which fell to 50% when the dose was increased. The inhibition rate of the quinine hydrochloride control was 90% and above. Moreover, the competitive antagonistic effects of artemisinin and quinine with chloroquine on pigment clumping CIPC was demonstrated in the in vivo experiment with artemisinin 12.5 mg/kg + chloroquine, quinine hydrochloride 75 mg/kg + chlorquine, with chloroquine alone and quinine hydrochloride alone as controls. The CIPC dose-response curves and regression lines of artemisinin + chlorquine, quinine hydrochloride + chlorquine, and chloroquine alone were shown in Figs. 5.31 and 5.32. Statistical analysis of the data showed that for chloroquine alone, the slopes of the regression lines were 3.49 (in vivo) and 3.38 (in vitro). For the artemisinin–chloroquine combination, the slopes were 1.17 (in vivo) and 1.21 (in vitro). The difference was significant ($P < .05$). Similarly, the dose-response curve of artemisinin was also completely different from that of quinine hydrochloride.

Research into the time-effect relationship also showed that if artemisinin was administered 40 min before chloroquine or if both drugs were simultaneously added to the in vitro culture, the artemisinin's inhibitory effects on CIPC were more marked. This could be because of the damage to parasitic membranes caused by artemisinin, preventing chloroquine from acting on receptors through the membrane system. But if chloroquine entered the parasite first and acted on these receptors, however, it was already too late to administer artemisinin. Hence, when artemisinin was delivered simultaneously with or after chloroquine, no CIPC inhibition was seen. This indicated that artemisinin's site and mechanism of action was different from those of chloroquine and provided a theoretical basis for why artemisinin could be effective in treating patients with chloroquine-resistant *P. falciparum*.

The parasitic pigment (heme) is a product of the digestion and decomposition of human hemoglobin. Warhurst suggested that the nitrogen atom in antimalarial drugs, such as quinoline and chloroquine, combined with parasitic heme to exert antimalarial effects. Therefore, the nitrogen atom was essential for antimalarial activity. However, artemisinins had no nitrogen atom but were able to partially suppress CIPC, which did not support Warhurst's hypothesis. Warhurst et al. had to reconsider these conclusions.

Zhao Yi also reported that artemisinin had no significant effects on folate metabolism in the parasite.[223] The experiment used *P. berghei* and adopted conventional antimalarial-drug screening methods. Each group had six mice given intraperitoneal injections of five million infected erythrocytes. Medication began on the second day after infection, administered twice daily for 3 days.

TABLE 5.39 Pigment Agglutination Rate of Artemisinin In Vivo and In Vitro

	In Vivo Experiments				In Vitro Experiments		
		Agglutination Rate (%)				Agglutination Rate (%)	
Drug	Dose (mg/kg)	Before Drug	80 min After Drug	Adjusted Value[a]	Dose (mol/L)	80 min After Drug	Adjusted Value[b]
Normal saline		6	8	2		6	
Tween		7	6	–1		6	0
Chloroquine	80	6	95	89	10^{-6}	92	86
Artemisinin	50	5	6	1	10^{-5}	9	3
Artemisinin	150	6	6	0	10^{-4}	5	–1
Artemisinin	450	6	8	2	10^{-3}	6	0
Quinine hydrochloride	50	7	6	–1	10^{-6}	8	2
Quinine hydrochloride	150	4	7	3	10^{-5}	6	0

[a]*In vivo adjusted value = Postmedication value – premedication value.*
[b]*In vitro adjusted value = Experimental value – normal saline value.*

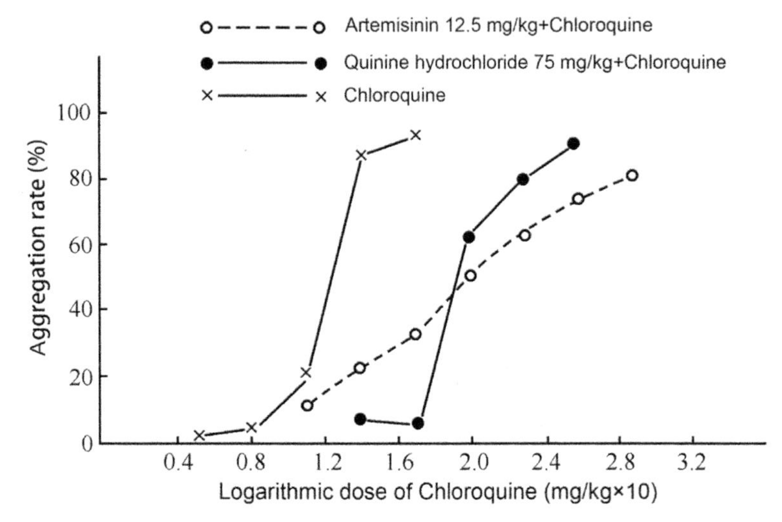

FIGURE 5.31 CIPC (pigment aggregation) dose–response curves of 12.5 mg/kg artemisinin + chloroquine, 75 mg/kg quinine hydrochloride + chloroquine, and chloroquine (in vivo). *(From Ye Z, Li Z. Effects of artemisinin on chloroquine-induced pigment clumping.* J Tradit Chin Med *1981;**4**:62–6. [叶祖光，李泽琳. 青蒿素对氯喹引起的色素凝集的影响. 中医杂志, 1981, 4: 62–66.])*

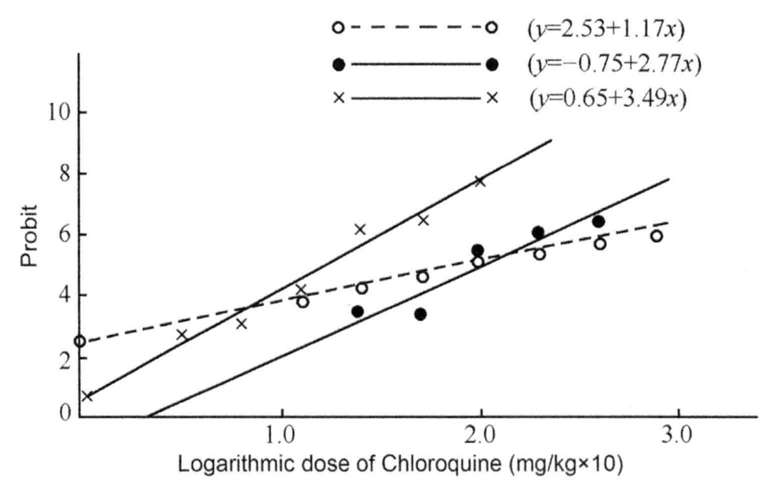

FIGURE 5.32 CIPC dose–response regression lines of 12.5 mg/kg artemisinin + chloroquine, 75 mg/kg quinine hydrochloride + chloroquine, and chloroquine (in vivo) [*identical line symbols* as in Fig. 5.31] *(From Ye Z, Li Z. Effects of artemisinin on chloroquine-induced pigment clumping.* J Tradit Chin Med *1981;**4**:62–6. [叶祖光，李泽琳. 青蒿素对氯喹引起的色素凝集的影响. 中医杂志, 1981, 4: 62–66.])*

TABLE 5.40 Effects of Antimalarial Drugs on *Plasmodium berghei*

Drug Group	Oral Dose Per Administration (mg/kg), Twice Daily, Total Three Days	Inhibition Rate (%)	*P* Value
Artemisinin	50	94.0	
Artemisinin + aminobenzoic acid	50 + 50	81.7	>.05
Artemisinin + folic acid	50 + 5	99.6	>.05
Artemisinin + aminobenzoic acid + folic acid	50 + 50 + 5	95.7	>.05
Artemisinin + sulfadimethoxine	50 + 500	99.9	>.05
Artemisinin + trimethoprim	25 + 50	100	No statistical analysis as inhibition rate reached 100%
Artemisinin + trimethoprim	50 + 100	100	
Artemisinin + sulfadimethoxine + trimethoprim	50 + 500 + 100	100	
Aminobenzoic acid	100	8.6	
Trimethoprim	100	65.4	>.05

Each test group had a corresponding nonmedicated control group. Caudal blood was obtained on the fifth day, smeared, Giemsa stained, and observed under a microscope. The parasite inhibition rate was used as an indicator of the drug's suppressive effects (Table 5.40).

Aminobenzoic acid did not inhibit artemisinin's antimalarial effects. Its mechanism of action was clearly different from those of pyrimethamine, proguanil, and sulfa and sulfone drugs. This indicated that artemisinin's action mechanism did not interfere with folate metabolism. Artemisinins did not interfere with nucleic acid metabolism. For example, artesunate had no impact on DNA synthesis in cell proliferation.[223]

6. Oxidative Damage in Malaria Parasite

The chemical structure of artemisinins included a peroxy group, and their antimalarial activity disappeared if the peroxide bridge was destroyed. This led to the idea that their mechanism of action could be related to their ability to produce free radicals and active oxygen in vivo. A logical inference was that artemisinins were selectively concentrated in parasite cells, releasing active oxygen or generating free radicals. This produced oxidative damage in the membrane system or biological macromolecules. However, there is still insufficient experimental evidence to support this hypothesis, although some scattered evidence has been accumulated.

Many researchers have proven that artemisinins can produce free radicals. Jin Yonggang and Teng Wenghe et al.[224] used electron spin resonance (ESR) spectroscopy to measure the free radicals produced by the metabolic transformation of artesunate both in vivo and in vitro. The results confirmed the above hypothesis. Moreover, the free radicals generated further triggered lipid peroxidation and tissue damage. Vitamin E antioxidants could inhibit free radical production, lipid peroxidation, and tissue damage. Although their article was on the toxicology of artesunate, it also proved that artemisinins generated free radicals, which could cause tissue damage and hence parasite death. Others suggested the drug's ability to create active oxygen led to elimination of the parasite.[225]

Some scholars have proposed that the free radicals generated by artemisinin act on parasitic proteins, deactivating them. The prevailing theory now is that of a "two-step effect." First, the peroxy structure is activated and free radicals are produced. These, in turn, act on the relevant components of parasitic protein to form covalent adducts. It deactivates the protein and produces antimalarial effects. Furthermore, some studies have found the drug's antimalarial activity could be because of its effects on parasitic heme[226] and can be suppressed by oxidants (including catalase) and free radical scavengers (such as vitamin E). The artemisinin–heme complex (QHS-heme) can oxidate sulfhydryl groups on the cell membrane but artemisinin alone cannot. Hence, the drug only exerts antimalarial effects through reduction of Fe^{2+} in heme. The ferrous ions play a crucial role in catalyzing the decomposition of the organic peroxide to generate free radicals, which damage the structure of the membrane system or alkylate biomolecules.

Wu Yulin's group at the SIOC and Li Ying's team at the SIMM[227] studied artemisinins and ferrous ions. They found that these ions reacted with the peroxy group to form oxygen free radicals first, which were quickly rearranged to become carbon-centered free radicals. Using chemical reactions and electron paramagnetic resonance, they confirmed that two types of carbon-centered radical were produced, which could react with cysteine or GSH to generate carbon–sulfur covalent adducts. If present, DNA would be broken. At the same time, it was discovered that artemisinin analogues that produced carbon-centered radicals easily also had strong antimalarial activity. This indicated that such radicals were the key active factors in creating antimalarial effects (see Chapter 3). However, details are still unclear on the subsequent activity of these free radicals and their target sites on the parasite (such as nucleic acids, enzymes containing cysteine residues, etc.), leading to the decomposition and death of the parasite. Further research is needed.

Lin Fubao and Pan Huazhen[228] employed chemiluminescence to demonstrate that, under alkaline conditions (pH 10.2), sodium artesunate (SA) could produce superoxide cations O_2^-, hydrogen peroxide (H_2O_2), and other active oxygens. Luminescence increased along with the concentration of sodium artesunate (Fig. 5.33).

The impact of different pH conditions on the formation of O_2^- from sodium artesunate was measured using cytochrome C reduction. At pH 7.4, sodium artesunate could not generate active oxygen (Fig. 5.34).

Further research into the effects of $FeSO_4$, $FeCl_3$, hemoglobin, and hemin on H_2O_2 generation showed that, at pH 7.4, sodium artesunate could not produce H_2O_2, but this could be induced by $FeSO_4$. The amount of H_2O_2 formed peaked in a few minutes. Hemoglobin also had an inducing effect, but this was significantly weaker than that of $FeSO_4$. $FeCl_3$ and hemin seemed to have no impact on this process (Fig. 5.35).

At pH 7.4, sodium artesunate could increase the concentration of O_2^- in normal and infected erythrocytes. When the drug's concentration was 2.5 μmol/L, the concentration of O_2^- rose 3.1 times in normal erythrocytes and 4.1 times in infected ones (Fig. 5.36).

Sodium artesunate also had significant effect on the formation of H_2O_2 in erythrocytes. Fig. 5.37 shows that at 2.5 μmol/L sodium artesunate, the catalase inhibition rate was 15% in normal erythrocytes and as high as 35% in infected erythrocytes. The production of H_2O_2 was higher in infected than in normal erythrocytes. In addition, measurement of MDA (from lipid peroxidation) confirmed that sodium artesunate could increase levels of MDA in normal and infected erythrocytes. MDA had high reactivity and could cross-link lipids and proteins.

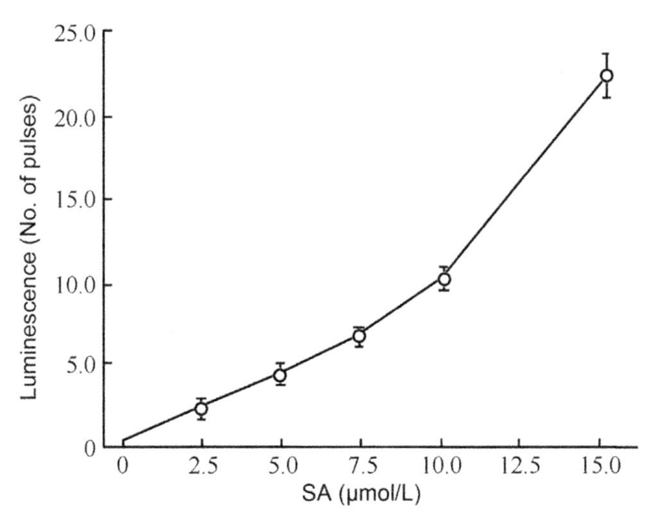

FIGURE 5.33 Relationship of sodium artesunate (SA) concentration to chemiluminescence (pH 10.2, $n=4$). *(From Lin F, Pan H. Antimalarial mechanism of sodium artesunate.* Acta Acad Med Sin *1989;3:180–4. [蔺福宝，潘华珍. 青蒿酯钠的抗疟机理. 中国医学科学院学报, 1989, (3): 180–184.])*

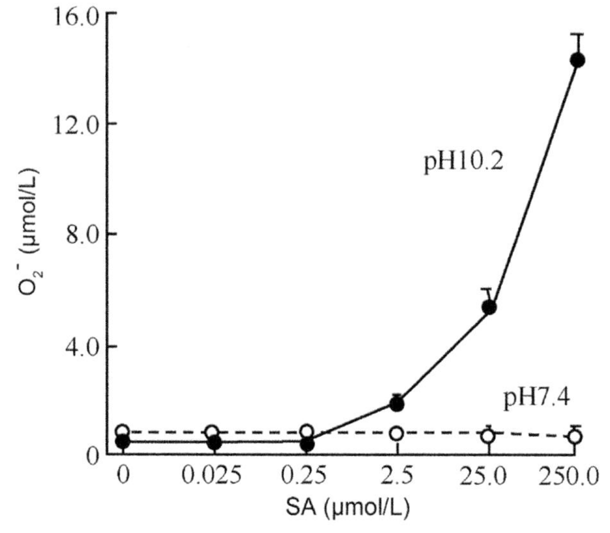

FIGURE 5.34 Formation of O_2^- from sodium artesunate (SA) under different pH conditions (n=4). *(From Lin F, Pan H. Antimalarial mechanism of sodium artesunate.* Acta Acad Med Sin *1989;3:180–4. [蔺福宝，潘华珍. 青蒿酯钠的抗疟机理. 中国医学科学院学报, 1989, (3): 180–184.])*

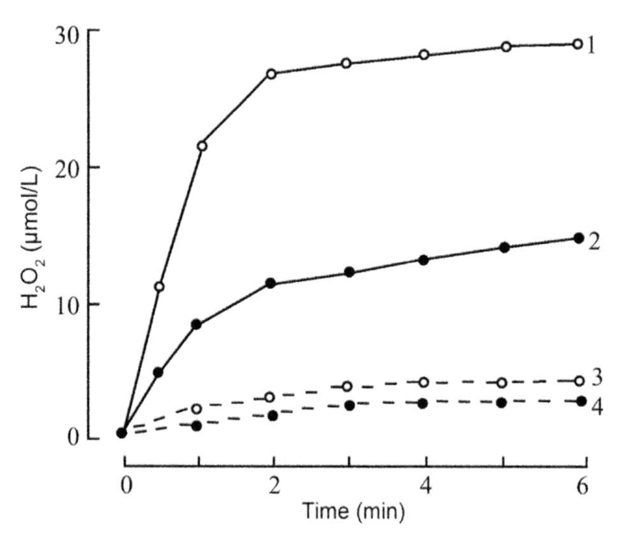

FIGURE 5.35 Effects of $FeSO_4$ (1), $FeCl_3$ (2), hemoglobin (3), and hemin (4) on H_2O_2 generation by sodium artesunate 0.2 mmol/L at pH 7.4. *(From Lin F, Pan H. Antimalarial mechanism of sodium artesunate.* Acta Acad Med Sin *1989;3:180–4. [蔺福宝，潘华珍. 青蒿酯钠的抗疟机理. 中国医学科学院学报, 1989, (3): 180–184.])*

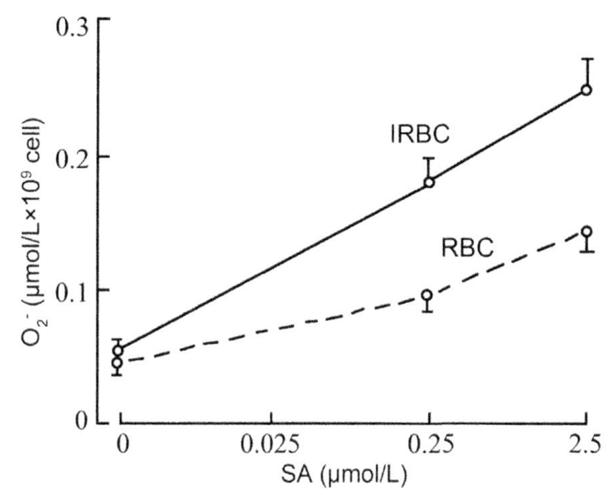

FIGURE 5.36 Effects of sodium artesunate (SA) on O_2^- concentration in normal (RBC) and infected (IRBC) erythrocytes (n=4). *(From Lin F, Pan H. Antimalarial mechanism of sodium artesunate.* Acta Acad Med Sin *1989;3:180–4. [*蔺福宝，潘华珍. 青蒿酯钠的抗疟机理. 中国医学科学院学报*, 1989, (3): 180–184.])*

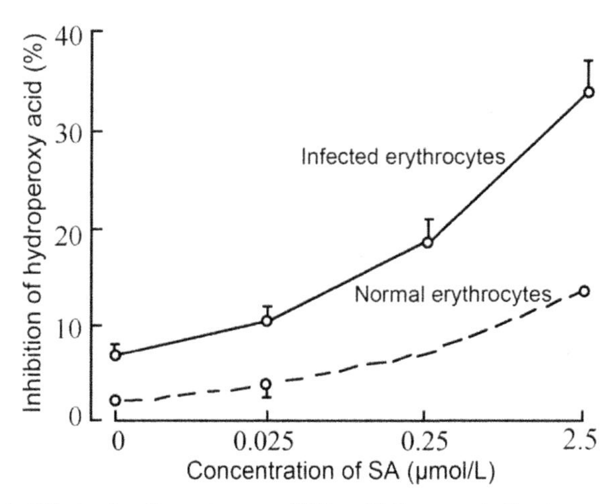

FIGURE 5.37 Effects of sodium artesunate (SA) on H_2O_2 concentration in normal (RBC) and infected (IRBC) erythrocytes (n=4). *(From Lin F, Pan H. Antimalarial mechanism of sodium artesunate.* Acta Acad Med Sin *1989;3:180–4. [*蔺福宝，潘华珍. 青蒿酯钠的抗疟机理. 中国医学科学院学报*, 1989, (3): 180–184.])*

C. FURTHER RESEARCH DEVELOPMENTS IN ANTIMALARIAL MECHANISM OF ACTION OF ARTEMISININ AND ITS DERIVATIVES (ARTEMISININS)

1. Studies on the Antimalarial Mechanism Based on Their Free Radicals From Reaction of Artemisinins With Fe(II)

Wu Yulin
Shanghai Institute of Organic Chemistry, Chinese Academy of Sciences, Shanghai, China

Artemisinin is a sesquiterpene molecule containing carbon, hydrogen, and oxygen but no nitrogen atoms and can be used for the treatment of multidrug-resistant (MDR) strains of *P. falciparum*. It is obvious that its antimalarial mechanism is quite different from that of previous antimalarial drugs such as quinine, chloroquine, etc. Since the discovery of artemisinin, there has been a wide interest on the antimalarial mechanism at the molecular level of artemisinins and other synthetic antimalarial drugs. Research has been on whether and how free radicals from artemisinins' reaction cause damage to malaria parasite either by affecting heme or parasite's DNA.

Formation of Carbon Free Radicals of Artemisinins With Fe(II)

It is known that artemisinin acts on the parasite at its intraerythrocytic asexual stage, in which the parasite digests hemoglobin as its nutritional resource and leaves free heme and small amount of iron. It is also known that the peroxide group of artemisinin and its derivatives is responsible for its antimalarial activity; and that in the Fenton reaction in organisms, hydrogen peroxide reacts with ferrous ions and could produce free radicals. With this background, since the early 1990s,[229–234] several laboratories started studies on the reaction of artemisinins and ferrous ions. At first, the reaction of artemisinin and ferrous sulfate (1:1 in mole) was run in $H_2O—CH_3CN$ (1:1 in volume, pH 4) solution at room temperature. It was found that the two major products tetrahydrofuran compound **3** and 3α-hydroxy-deoxyartemisinin **4** are also the major metabolites of artemisinin in vivo or in humans.[98,126] Based on the analysis of these products, a reaction mechanism of primary C-centered free radical **1** and secondary C-centered free radical **2**, respectively, followed the single-electron rearrangement of oxygen-centered free radicals was suggested (Fig. 5.38).[235,236]

Wu et al. treated artemether and a number of other artemisinin derivatives with ferrous sulfate in the same reaction condition and found similar derivatives of tetrahydrofuran compound **3** and 3α-hydroxy deoxyqinghaosu **4** as the two major products (Fig. 5.39).[236] It can be concluded that the reaction of artemisinins with ferrous ions is definitely a free radical reaction through a short-lived O-centered radical–anion and subsequent primary and secondary C-centered radicals.

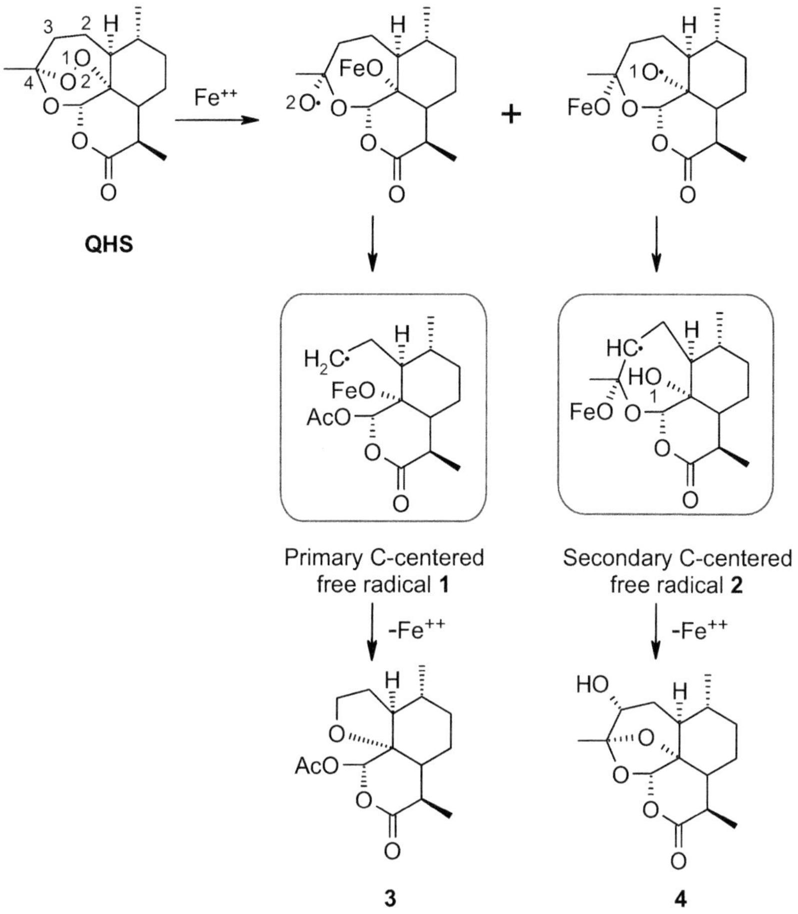

FIGURE 5.38 Formation of carbon-centered free radicals from QHS (artemisinin).

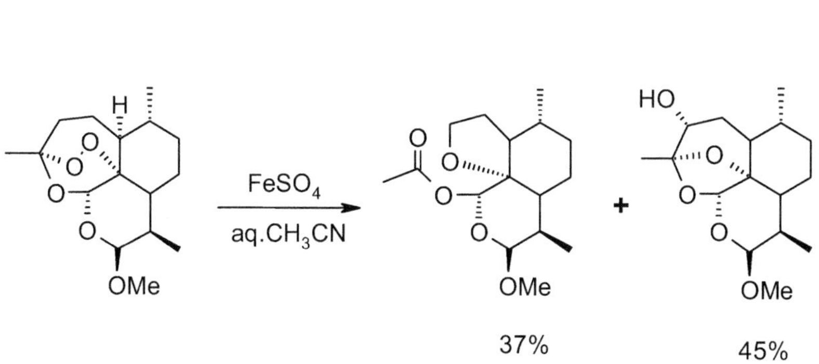

FIGURE 5.39 Reaction products of artemether and ferrous sulfate in $H_2O—CH_3CN$ solution.

In addition, Butler et al. detected the electron spin resonance signals (ESR) of primary and secondary C-centered free radicals.[237]

Antimalarial Activity and the Free Radical Reaction of Artemisinins

To study whether this free radical mechanism is related to artemisinins' antimalarial activity, several stable and UV-detectable C-12 aromatic substituted derivatives of artemisinin were synthesized by Wang et al.[238,239]

Using the usual Lewis acid as the catalyst, the Friedel–Crafts alkylation gave the desired product 2′,4′-dimethoxyphenyl deoxoartemisinin **5** and also 11α-methyl epimer **6** as the by-products (Fig. 5.40). These products were separated and subjected to both bioassay and chemical reaction with ferrous ions. It was found that the derivative with normal configuration at C-11 showed higher antimalarial activity and higher chemical reactivity in the reaction with ferrous ions. On the contrary, the 11α-methyl epimer **6** was less active and almost inert to the reaction (Table 5.41).

In the case of 11α- and11β-epimers of α-hydroxy naphthyl deoxoqinghaosu, similar result was also obtained.[239] Their lower activity may be attributed to the steric hindrance around O-1 atom in 11α-epimer, which blocks the way for Fe^{+2} to attack O-1 (Fig. 5.41).

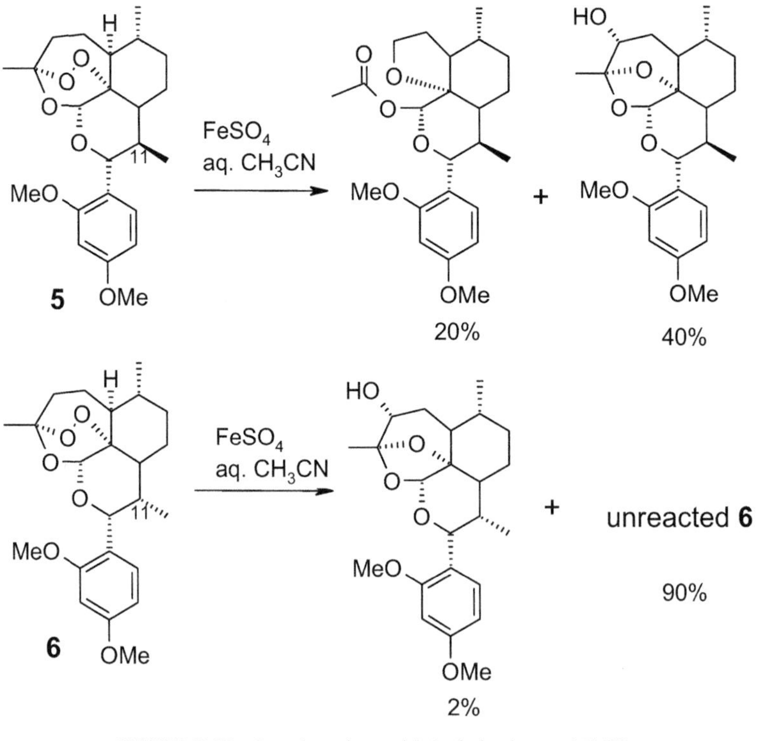

FIGURE 5.40 Reaction of artemisinin derivatives and $FeSO_4$.

TABLE 5.41 ED_{50} and ED_{90} of Artemether, Compounds 5 and 6 in Mice Infected With *Plasmodium berghei* K173 Strain

Compound[a]	ED_{50} (mg/kg)	ED_{90} (mg/kg)
Artemether	1	3.1
Compound **5**	1.27	5.27
Compound **6**	4.18	76.27

[a]*Compounds were given as oral suspensions in Tween 80.*

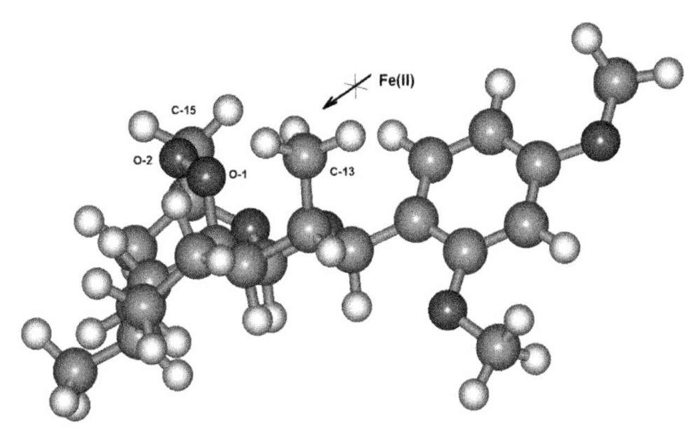

FIGURE 5.41 11α-methyl group blocks the attack of Fe^{+2} on O-1.

Interaction of DNA With Carbon-Centered Free Radicals

In the free radical mechanism of artemisinin, there are two different kinds of free radicals: the oxygen radical from the Fenton reaction and carbon free radicals from the reaction of artemisinin-ferrous ions. Knowing that DNA could be cleaved with Fenton reagent, researchers were curious whether artemisinin–ferrous ions could also cleave DNA. The mature RBC has no nucleus, while the parasite has. If artemisinin can permeate the cell membrane to reach the nucleus of the parasite and reacts with Fe (II), it will possibly interact with DNA. This may explain why artemisinin is only toxic to the parasite but not to the normal RBC.

Given that the pH in blood serum is about 7.35–7.45 and that the ferrous ions will precipitate above pH 7, the DNA damage experiments with artemisinin and stoichiometric ferrous ions were performed in aqueous acetonitrile (1:1) solution, at 37°C and at pH 6.5 adjusted with a phosphate buffer. It was found that the cleavage of calf thymus, salmon, and a supercoiled DNA pUC 18 occurred, yielding a DNA fragment with about 100 base pairs.[235,240]

These experimental results confirmed that these free radicals could cleave the DNA. But whether artemisinin can penetrate cell membrane to react with Fe (II) in the parasite's nucleus to cause DNA damage by free radicals is still uncertain at this stage. Artemisinin's toxicity on tumor cell lines could be the result of tumor cell DNA damage, similar to other antitumor compounds, such as enediyne. Studies showed that these free radicals damaged the deoxygunosine or perhaps the deoxyadenosine of the DNA.

Interaction of Cysteine and Glutathione With Carbon-Centered Free Radicals

It was known that malaria parasite-infected RBCs have a high concentration of reduced GSH and excess GSH might be protecting the parasite from the toxicity of heme. To see the interference of artemisinins might have on GSH, Chinese chemists had artemisinins reacted with cysteine or GSH in the present of a catalytic amount of ferrous ion.[240–243] Isolation and identification of cysteine–artemisinins or GSH–artemisinins adducts (compounds **7–14**) in the result product proved the formation of covalent bonds in these adducts (Figs. 5.42 and 5.43).

As mentioned above, GSH plays an important role in protecting the parasite from oxidative stress. Therefore, these results are significant for understanding the mode of action of artemisinin and its derivatives. The formation of GSH-artemisinins adducts reduces the amount of GSH, and these adducts themselves might also cause inhibition of the GSH reductase and other enzymes in the parasite. On the other hand, it is instructive that those C-centered radicals derived from artemisinin and its derivatives could attack free cysteine and cysteine residue not only in peptides but also probably in proteins. The formation of covalent bond adducts between parasite proteins and artemisinins mentioned in the literature is therefore possible.

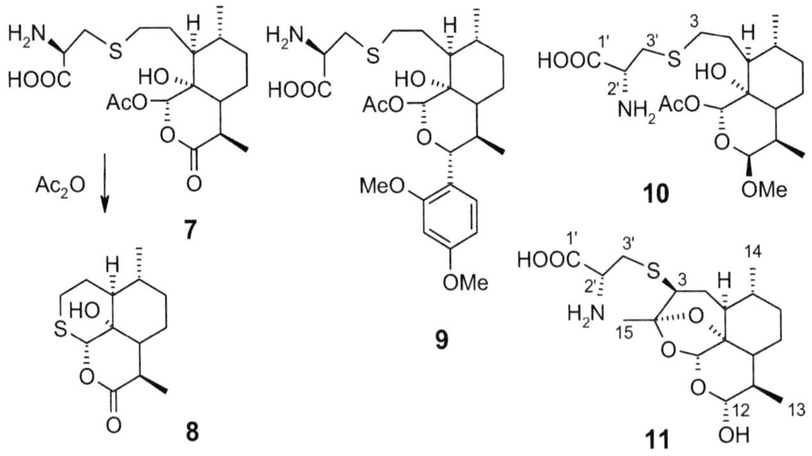

FIGURE 5.42 Cysteine–artemisinins adducts.

Interaction of Heme With Carbon-Centered Free Radicals

Robert and Meunier reviewed their mechanistic study on artemisinin derivatives.[233,234,244,245] They identified some adducts of heme and the primary C-centered free radical **1** through the mesoposition from the reaction of artemisinin with Fe(III)-heme and 2, 3-dimethylhydroquinone in methylenechloride[246] or Fe(III)-heme and GSH in dimethyl sulfoxide.[247]

Fig. 5.44 shows the major artemisinin–heme adduct after demetallation. However, they did not clearly point out how these adducts were related with the action mode of antimalarial or inhibition of the hemozoin formation. In 2003, Haynes et al. pointed out that artemisinin antimalarials did not inhibit hemozoin formation and hence ruled out the role of the artemisinin–heme adduct as an inhibitor.[248] Also, in 2002, a heme-artemisinin named hemart was reported, which stalls all mechanisms of heme polymerization, resulting in the death of the malarial parasite.[249] However, their hemart was synthesized just by mixing equivalent heme and artemisinin in dimethylacetamide at 37°C for 24h. It is not clear whether hemart is a complex or a covalent adduct. The primary C-centered

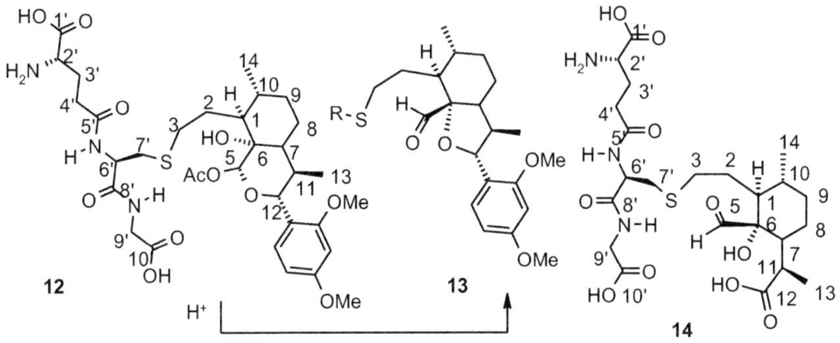

FIGURE 5.43 Glutathione–artemisinins adducts.

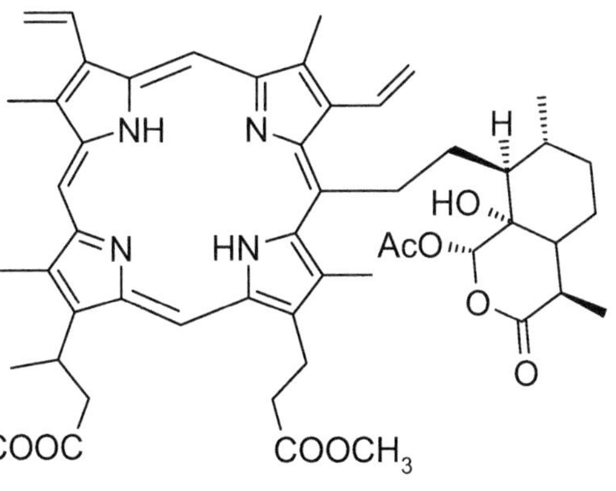

FIGURE 5.44 Major artemisinin–heme adduct after demetallation.

free radical **1** is only formed by reaction of artemisinin and heme in the presence of reducing agents but not in the absence of reductants.

In summary, the mode of antimalarial action of artemisinin has been a problem since its discovery. Several theories have been postulated; however, we are still far from a definite conclusion as mentioned in Yikang Wu's comment.[250] At the moment, the postulation that carbon-centered free radicals as the key active species killing the malaria parasite seems most convincing. Further work is needed to identify the target attacked by these catalytic radicals, resulting in the destruction of the whole biosystem of the parasite.

2. Studies on the Antimalarial Mechanism of Artemisinins Targeting Proteins and Mitochondria

Ma Lina and Ye Zuguang
Institute of Chinese Materia Medica, China Academy of Chinese Medical Sciences, Beijing, China

Hypothesis of Artemisinins Targeting Proteins

Plasmodium falciparum Translationally Controlled Tumor Protein Homolog

A distinct characteristic of artemisinin is that it exerts strong specific action on plasmodium in RBCs without side effects to the human body, which indicates that the carbon free radical of activated artemisinin may react with intraparasitic protein as specific target. Bhisutthibhan et al. reported in 1998 the antimalarial activity of antimalarial endoperoxides was found to alkylate with specific proteins, and one of these target proteins was isolated and identified as the *P. falciparum* translationally controlled tumor protein (TCTP) homolog but no direct evidence demonstrating how it could lead to parasite death because relatively little is known about the physiological roles of TCTPs.[251]

Plasmodium falciparum SERCA Ortholog (PfATP6)

PfATP6 of *P. falciparum* as the protein target of artemisinins was proposed by Eckstein-Ludwig et al. in 2003.[252] PfATP6 is the only sarco/endoplasmic reticulum Ca^{2+}-ATPase (SERCA) sequence in the parasite's genome. The authors expressed PfATP6 in *Xenopus laevis* oocytes membrane preparation. Artemisinins, but not quinine or chloroquine, inhibit PfATP6 in *X. laevis* oocytes with similar potency to thapsigargin, another sesquiterpene lactone, and a highly specific SERCA inhibitor. As predicted, thapsigargin also antagonizes the parasiticidal activity of artemisinin. But deoxyartemisinin lacks an endoperoxide bridge and is ineffective both as an inhibitor of PfATP6 and as an antimalarial. Chelation of iron by desferrioxamine abrogates the antiparasitic activity of artemisinins and correspondingly attenuates inhibition of PfATP6. Futhermore, imaging of parasites with BODIPY-thapsigargin labels the cytosolic compartment and is competed by artemisinin. Fluorescent artemisinin labels parasites similarly and irreversibly in a Fe^{2+}-dependent manner. These data provide evidence that artemisinins act by inhibiting PfATP6 outside the food vacuole after activation by iron. A single amino acid in transmembrane segment 3 of SERCAs can determine susceptibility to artemisinins[253] and when the *S769N PfATPase6* mutation was associated with raised artemether IC_{50}.[254] Since then, much attention has been paid to the hypothesis that PfATP6 is artemisinin-specific target in the malaria parasite.

Nevertheless, in recent years, many opposite conclusions appeared. The 3D structure of PfATP6 was modeled on the basis of the crystal structure of SERCA 1a, the mammalian homolog. AutoDock4 was used to predict the binding affinities of artemisinin (and analogues) and various other antimalarial agents for PfATP6, for which in vitro activity was also reported. No correlation was found between the affinity of the compounds for PfATP6 predicted by AutoDock4 and their antimalarial activities.[255] It is worth noting that some of the experiments used to support PfATP6 doctrine cannot be repeated.[254,256,257] As we all know, the endoperoxide moiety is necessary for antimalarial activity because analogues that lack this group are inactive. Yet, the hypothesis of PfATP6 suggests that artemisinin is similar to the role of thapsigargin, both of which can inhibit SERCA ortholog (PfATP6) of *P. falciparum*, causing the parasite's death. But thapsigargin has no peroxide bridge structure, and recently, the structural similarities between artemisinin and beta-carotene have also been questioned. In a word, series studies provide enough evidence for the lack of association of PfATP6 with the antimalarial activities of artemisinins.[258–260] In 2008, Huang and Zhou, using yeast as a model, found no homolog to be mainly responsible for artemisinin's inhibitory action, and no specific proteins bound potently to dihydroartemisinin.[261] The single-target PfATP6 hypothesis was not looked on favorably. PfATP6 is a potential drug target but unlikely to have played an important role as a target for artemisinin in the parasite *P. falciparum*.[262,263]

Other proteins, including Pfmdr1, Pfcrt, etc., have also been considered to be artemisinin targets, but the evidence is limited and persuasion is weak.[264,265]

Plasmodium falciparum Multiple Protein Targets

In the recent years, Wang et al. in Singapore currently explores the protein target mechanism in *P. falciparum* directly by virtue of unbiased chemical proteomics analysis.[266] Using an alkyne-tagged artemisinin analogue coupled with biotin, 124 protein targets covalently binding with artemisinin were identified, many of which are involved in the essential biological processes of the parasite (Fig. 5.45). Such a broad targeting spectrum disrupts the biochemical landscape of the parasite and causes its death. Furthermore, using alkyne-tagged artemisinin coupled with a fluorescent dye to monitor protein binding, they found that heme, rather than free ferrous iron, is predominantly responsible for artemisinin activation. And the heme derives primarily from the parasite's

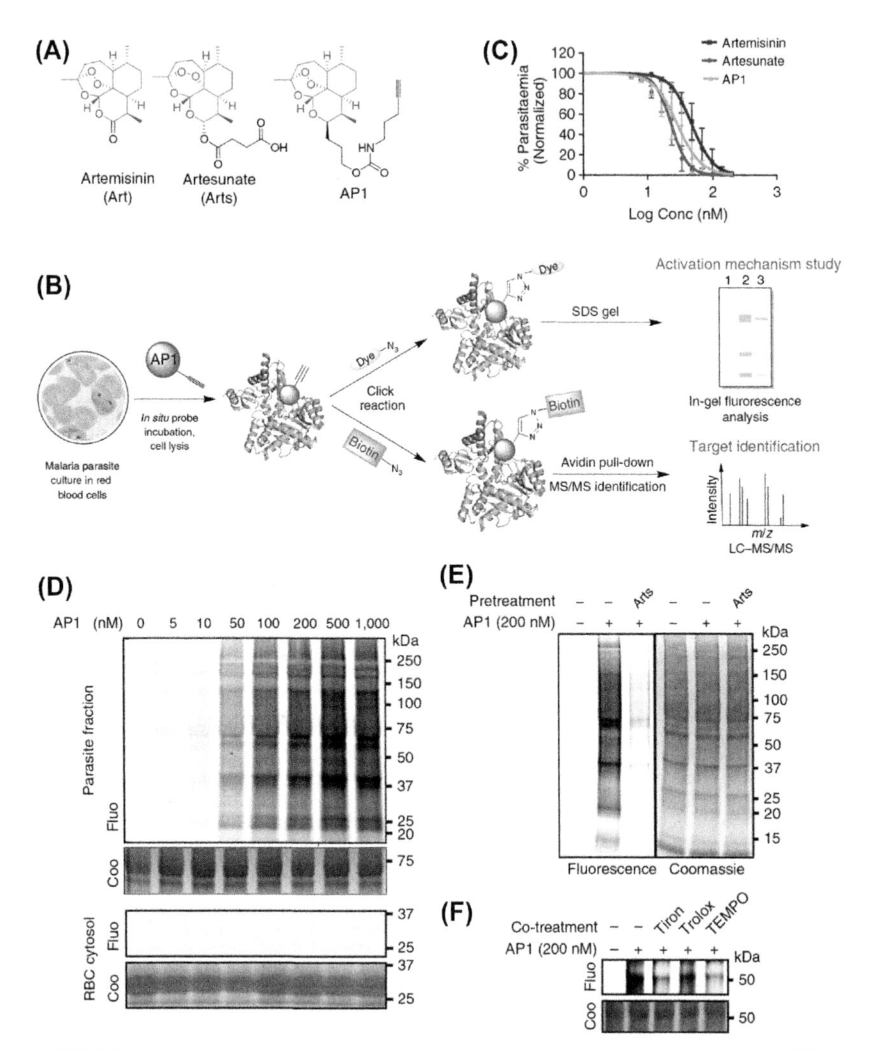

FIGURE 5.45 Chemical proteomics approach to study artemisinin's mechanism of action.[266] (A) Chemical structures of artemisinin (Art), artesunate (Arts), and the alkyne-tagged-clickable probe (AP1). (B) General workflow of the chemical proteomics approach. Fluorescence labeling was used to study the activation mechanism of artemisinin, while biotin pull-downs coupled with liquid chromatography–tandem mass spectrometry (LC–MS/MS) were used to identify protein targets of artemisinin. (C) The killing effect of AP1 is comparable to that of artemisinin and artesunate on *P. falciparum* 3D7. (D) In situ parasite labeling with AP1. The labeling was dose dependent and specific to parasite proteins. Healthy RBC cytosolic proteins were not labeled. (E) The AP1 in situ parasite labeling was artemisinin specific as the excess Arts can largely compete with AP1-target labeling. (F) Free-radical scavenger (Tiron, 1 mM; Trolox, 400 μM; TEMPO, 1 mM) cotreatment reduces the level of parasite protein alkylation by AP1. *Fluo*, fluorescence scanning; *Coo*, coomassie staining. Error bars represent s.d in three independent replicates in (C). *(From Wang JG, Zhang CJ, Chia WN, et al. Haem-activated promiscuous targeting of artemisinin in Plasmodium falciparum.* Nat Commun *2015;6:10111. http://dx.doi.org/10.1038/ncomms 10111.)*

heme biosynthesis pathway at the early ring stage and from hemoglobin digestion at the latter stages. Thus, the "blood-eating" nature of the parasite with the release of extremely high levels of heme confers the high efficacy of artemisinin against the parasites, with minimum side effects toward healthy RBCs. The results support a unifying model to explain the action and specificity of artemisinin in parasite killing. But further biological tests are needed to prove these assumptions.

Hypothesis of Artemisinins Targeting Mitochondria

Srivastava et al. first reported in 1998 that the broad spectrum antimalarial atovaquone could collapse mitochondrial membrane potential in a malaria parasite.[267] To make clear how artemisinins work, Li et al. from Tsinghua University (China) in 2005 developed a yeast (*Saccharomyces cerevisiae*) model and found that artemisinin could inhibit yeast growth through interfering with its mitochondrial functions.[268] However, the mode of action as concluded from the yeast model study lacks direct biochemical supporting evidence and in particular awaits confirmation in malaria parasites. Wang et al. further reported that artemisinin directly acts on mitochondria of *P. berghei* in a similar way as in yeast. Specifically, artemisinin and its homologs exhibit correlated activities against malaria parasite and yeast, with the peroxide bridge playing a key role for their inhibitory action in both organisms. The study showed that artemisinins distributed to malaria parasite mitochondria and directly impaired their functions when isolated mitochondria were tested. The authors also found that large amount of reactive oxygen species (ROS) production was rapidly induced by artemisinin in isolated yeast and malaria parasite mitochondria but not mammalian, and ROS scavengers can improve the effects of artemisinin. Deoxyartemisinin, which lacks an endoperoxide bridge, has no effect on membrane potential or ROS production in parasite mitochondria. OZ209, a distantly related antimalarial endoperoxide, also causes ROS production and depolarization in isolated malaria parasites mitochondria. Finally, interference of mitochondrial electron transport chain (ETC) can alter the sensitivity of the parasite toward artemisinin. Addition of iron chelator desferrioxamine drastically reduces ETC activity and mitigates artemisinin-induced ROS production. The results indicate that mitochondrion is an important direct target, if not the sole one, in the antimalarial action of artemisinins. The authors suggested that fundamental differences among mitochondria from different species delineate the action specificity of this class of drugs, and differing from many other drugs, the action specificity of artemisinins originates from their activation mechanism.[269]

To promote understanding of the antimalarial mechanism of artemisinin and investigate the role of mitochondrial permeability transition pore, a protein complex, which is involved in regulation of mitochondrial volume, in the antimalarial action of artemisinin, Wang et al. in 2009, isolated the mitochondria fraction from malaria parasites and mouse liver. Spectrophotometry was used to check the direct effect of artemisinin on the mitochondrial volume. Isobologram and inhibitors of mitochondrial permeability transition pore were used to check whether mitochondrial permeability transition pore is involved in the antimalarial action of artemisinin. The results showed that artemisinin directly worked on isolated malaria parasite mitochondria and caused mitochondrial swelling but had no effect on the volume of mouse liver mitochondria. In addition, two different mitochondrial permeability transition pore inhibitors antagonized the antimalarial effect of artemisinin. It was concluded artemisinin can work directly on malarial mitochondria to cause mitochondrial swelling, and mitochondrial permeability transition pore is involved in the antimalarial action of artemisinin.[270]

In 2015, Sun et al. used yeast as cellular and genetic model and found that artemisinins are endowed with two major and distinct types of properties: a potent and specific mitochondria-dependent reaction and a more general and less specific heme-mediated reaction. The competitive nature of these two actions could be explained by their shared source of the consumable artemisinins, so that inhibition of the heme-mediated degradation pathway would enable more artemisinin to be available for the mitochondrial action. These properties of artemisinins can be used to interpret the divergent antimalarial and anticancer actions of artemisinins. The heme-mediated nonspecific action and the specific mitochondrial pathway both exist in vivo and while both could be cell inhibitory, these two actions are also antagonistic likely due to their competition for the same source of the consumable artemisinins.[271] This may be why artemisinin can treat different diseases.

Nevertheless, there are also studies questioning the hypothesis of mitochondria. Some researchers found that the digestive vacuole, not the mitochondrial membrane, is an important initial site of endoperoxide antimalarial activity.[272] But Maeno and Kawai observed that the earliest and the most distinctive ultrastructural changes following the administration of artemether were marked swelling of the mitochondria in the parasites.[207,273] No discernible mitochondrial morphologic changes detected on the microscopy do not mean that its function has not changed.

Although tremendous efforts have been devoted to decipher how this class of molecules works, their exact antimalarial mechanism is still an enigma. The biological mode of action of artemisinin has long been controversial. Several hypotheses have been postulated until now. The acknowledged mechanism of artemisinin is to be activated first and then specifically and selectively acts on plasmodium, while the procedure is still unknown. Can we uncover the mysterious veil of artemisinin and push its research to a higher level? And it is our objective to discover its unique therapeutic effects on other diseases. Current research has demonstrated that artemisinin play important roles in antiinfection, autoimmune disease, and cancers in addition to antimalarial.[274] Some drugs even have been licensed for clinical study. We are looking forward to this excellent drug artemisinin bringing more surprises.

5.5 RESISTANCE OF *PLASMODIUM FALCIPARUM* TO ARTEMISININS

A. EARLY WORK IN CHINA AND ABROAD

Li Guoqiao
Institute of Tropical Medicine, Guangzhou University of Chinese Medicine, Guangzhou, Guangdong Province, China

Generally speaking, the emergence of drug resistance means that the efficacy of a drug recommended for the treatment of a disease decreases significantly or disappears. Reports of artemisinin resistance have now been reported in Cambodia,[275,276] Thailand,[277] and some areas in Vietnam.[278] This means that parasite clearance slowed and recrudescence occurred within 28 days. Artemisinin resistance was considered to be the cause when other reasons had been ruled out.

The WHO had earlier defined resistance as continued growth and reproduction of the parasite despite an appropriate dose of antimalarials and level of drug concentration in the blood. This was divided into different levels of resistance—RI, RII, RIII—according to parasite clearance and recrudescence.[279] However, the Worldwide Antimalarial Resistance Network (WWARN) has since made some modifications to this definition, allowing it to more accurately monitor the probability of artemisinin resistance emerging around the world.[280] These new standards included a mathematical model, the parasite clearance estimator (PCE),[280] which could precisely calculate parasite clearance times or clearance rates,[281] taking into account whether clearance occurred in 3 or 7 days and if recrudescence occurred in 28 days.[282]

A December 2005 article published by Jambou et al.[254] stated that to establish a warning system for resistance, blood samples were taken from 530 malaria patients in Cambodia, French Guiana, and Senegal, and the effects of artemisinin on plasmodium from these different regions was compared. There were no signs of resistance in Cambodia. However, sensitivity to artemisinin had decreased in Senegal and French Guiana. In French Guiana, there was evidence to suggest that parasites had developed some SERCA-pfATPase6 gene mutations, which could lead to resistance. The article stated that this was because there were no specifications for the use of artemisinin in Senegal and French Guiana or that artemisinin was administered alone in these areas. In Cambodia, artemisinin use was strictly managed and given only in combination therapies. Nevertheless, this was not entirely accurate because artemisinin as a monotherapy had over 20 years of history in Cambodia.

The article drew attention to the necessity of preventing artemisinin resistance from emerging.[254] Nevertheless, it was based on questionable in vitro test data. It included samples from 289 patients in French Guiana, with a minimum artemether in vitro IC_{50} of 0.2–117 nmol/L. Of the 110 samples from Senegal, the artemether IC_{50} was 0.1–44.7 nmol/L. From the western part of Cambodia near the border with Thailand, 136 samples were obtained with an artesunate

IC_{50} of 0.05–18 nmol/L. In in vitro experiments, artesunate and artemether are not comparable. The article's authors made reference to another paper,[283] which mentioned that artemisinins displayed cross-resistance in in vitro experiments.

In December 2008, Noedl et al.[275] wrote that in vivo, in vitro, molecular labeling, and PK studies suggested the possibility of latent artemisinin resistance on the Cambodia–Thailand border. The overall high cure rates of artesunate showed that only a minuscule number of strains could be considered resistant. It was not a widespread epidemiological phenomenon.

Noedl et al. also published a paper with supplementary clinical, in vitro experimental, and blood concentration data.[284] This was based on 56 patients who received artesunate (4 mg/kg daily, 7 day course of treatment). Four patients showed recrudescence in 28 days. Two were judged to have poor drug absorption and hence inadequate blood concentration levels, rather than experiencing drug resistance. The other two were considered to have drug resistance; their clearance times were 95 and 133 h, as opposed to 52.5 h for patients who were cured. Blood concentration levels were sufficient. The IC_{50} of dihydroartemisinin in these two cases was 10 times that of the reference W2 clone. However, this resistance was not related to pfmdr1 or pfATPase6, which are considered resistant markers.

Dondorp et al.[276] studied two randomly selected locations, one in Pailin, western Cambodia, and the other in Wang Pha, northwestern Thailand. The treatment effectiveness of uncomplicated falciparum malaria was compared. The following regimens were adopted: 2 mg/kg/day oral artesunate over 7 days and 4 mg/kg/day artesunate over 3 days, followed by two doses of mefloquine totaling 25 mg/kg. In vivo and in vitro sensitivity, artesunate PKs, and resistance markers were assessed. A total of 40 patients were involved at each site. The median overall parasite clearance time in Pailin was 84 h (interquartile range 60–96 h); the median in Wang Pha ($P < .001$) was 48 h (interquartile range 36–66 h). Polymerase chain reaction confirmed that, in Pailin, 6 out of 20 cases receiving artesunate monotherapy experienced recrudescence (30%). Only 1 out of 20 cases given artesunate–mefloquine had recrudescence (5%). In Wang Pha, the 20 monotherapy patients saw two cases of recrudescence (10%); the combination treatment group ($P = .31$) had one case of recrudescence (5%). These markedly different parasitic reactions could not be explained by age differences, artesunate or dihydroartemisinin PKs, in vitro isotope susceptibility tests, or pfmdr1 or pfserca-pfATPase6 gene mutations. Compared to northwestern Thailand, the in vivo susceptibility of falciparum malaria to artesunate in western Cambodia had decreased. In vivo resistance manifested in a slower clearance time, but in vitro sensitivity tests showed no changes.

Clearance time was significantly lower in Pailin than in Wang Pha, which was considered a sign of resistance. However, this only showed that in vivo sensitivity to artemisinin was different in the two areas. Moreover, the two groups of patients were of different ages and levels of immunity.

Other authors[277] also found that the parasite clearance half-life in northwestern Thailand had lengthened from 2001 to 2010. The geometric mean of this half-life rose from 2.6h in 2001 to 3.7h in 2010. The proportion of patients experiencing slower clearance (half-life of clearance ≥6.2h) increased from 0.6% in 2001 to 20% in 2010. In western Cambodia, the geometric mean of the clearance half-life was 5.5h, and the proportion with slow clearance increased by 42% between 2007 and 2010. Genetic analysis of the parasites revealed 93 types of single nucleotide polymorphism (SNP). The conclusion was that, based on the genetic evidence, resistance had emerged in *P. falciparum* on the Thai–Myanmar border for at least 8 years and had increased.

In a separate article,[285] it was stated that the appearance of resistance to artemisinin-based combinations was a threat to malaria control. Resistance had been confirmed in western Cambodia and recently in western Thailand. It had not been reported on the Lao border. It caused a drop in the parasite clearance rate, which research suggested was related to latent genome resistance. The genomes of 91 parasitic strains from Cambodia, Thailand, and Laos were studied, with haplotype structure at 6969 polymorphic SNP as a marker. It was found that 33 genetic regions showed evidence of strong selectiveness. An examination of 6969 SNP and microsatellites in the genetic regions of 715 Thai strains revealed a selective area on chromosome 13. This area was strongly correlated with a slower parasite clearance rate ($P = 10^{-12}$–10^{-6}). Mutations in two adjacent regions of chromosome 13 were considered to be the genetic basis of artemisinin resistance.

From 1985 to 1992, Li Guofu et al. at the IME of AMMS carried out important in vivo research on this subject. Strains of *P. berghei* resistant to six antimalarial drugs—artemisinin, artemether, chloroquine, lumefantrine, naphthoquine, and artemether–lumefantrine—were cultivated in mice. Resistance appeared after 20–30 passages, each cultivated for 7 days (Fig. 5.46). After 30 passages, it emerged in artemisinin (QHS) and artemether, which had short half-lives. However, the resistance index (IC_{90}) was not high. After 200 passages, the resistance index was maintained at around 10, namely mild resistance. This was significantly different for drugs with long half-lives. After 40 passages, the resistance index of chloroquine reached 80 and then stabilized at 110. Lumefantrine had the highest resistance index, which peaked at around 400. However, when combined with artemether (A+B), the resistance index stabilized at 20–40.[286] This was an interesting phenomenon. Another drug with a short half-life, quinine, was unfortunately not included. Nevertheless, this was a lengthy and significant experiment.

In the 1980s, Chawira, together with Peters[287] and others,[288] published some important data from in vitro and in vivo experiments.[289] In vivo tests in mice showed that chloroquine-resistant strains of *P. yoelii* rapidly developed resistance to artemisinin (not artesunate), while chloroquine-sensitive strains of *P. berghei* did not. In vitro and in vivo experiments found significant synergistic effects when artemisinin was used together with mefloquine or primaquine. Hence, an artemisinin or artesunate and mefloquine combination (A+M) was recommended. Until recently, when the issue of artemisinin resistance has

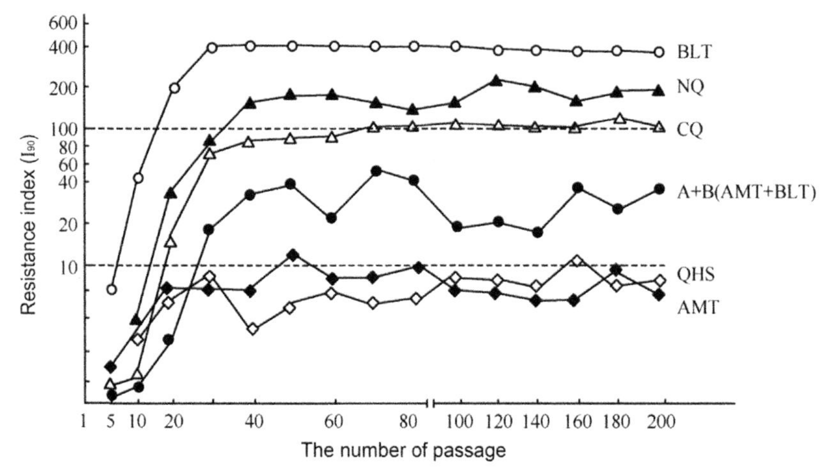

FIGURE 5.46 Cultivation of resistance to six antimalarials in K173 strain *Plasmodium berghei*. *AMT*, artemether; *BLT*, lumefantrine; *CQ*, chloroquine; *NQ*, naphthoquine; *QHS*, artemisinin.

emerged, the artesunate–mefloquine combination has been one of the main recommended ACTs.

Going by clinical practice, areas of China where falciparum malaria is endemic—such as Hainan, Yunnan, Guizhou, Guangxi—swiftly adopted artesunate and artemether as first-line treatments, beginning in 1981. This was because of the drugs' rapid action, high efficacy, low toxicity, and safety. A few areas began using an artemisinin–piperaquine combination only in recent years. In the more than 30 years since, no reports of treatment failure because of drug resistance have appeared, and falciparum malaria has been eradicated in many areas.

Beginning in 1991–92, malaria outbreaks occurred in Vietnam because of the emergence of chloroquine- and Fansidar resistance in *P. falciparum* and a surge in the mobile population following economic development. The annual mortality rate reached 4000–5000. Use of artemisinins spread rapidly, replacing chloroquine, Fansidar, and quinine, and becoming a first-line treatment. Only in 1999 was a dihydroartemisinin–primaquine ACT adopted (CV8). Artemisinins and CV8 were distributed for free by the Vietnamese government. Today, Vietnam has entered the eradication stage, and no signs of treatment failure were seen in the last 20 years (Fig. 5.47). Nevertheless, resistance to drugs with a long half-life such as chloroquine, SP, piperaquine, and mefloquine has emerged after several years. Treatment efficacy fell significantly in 10 years.[278]

More than 190 years have passed since 1820, when French scientists extracted quinine from cinchona bark. It has been widely employed across the world, and use without specifications is commonplace. Given its half-life, quinine resistance should have appeared long ago and caused the drug to lose efficacy. In 1976, Li Guoqiao of the Institute of Tropical Medicine, Guangzhou University of Chinese Medicine traveled to Cambodia to combat malaria. At the

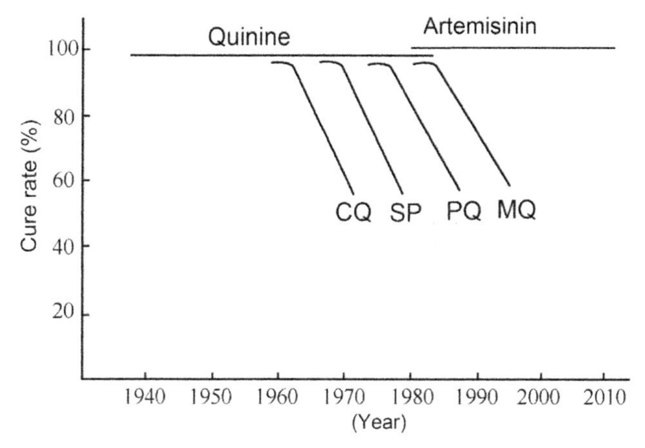

FIGURE 5.47 Changes in the clinical efficacy of antimalarials in Southeast Asia. *CQ*, chloroquine; *MQ*, mefloquine; *PQ*, piperaquine; *SP*, sulfadoxine–pyrimethamine.

time, malaria outbreaks occurred nationwide and there were rumors that quinine was no longer effective. However, when Li used intravenous quinine to treat critical patients, no failures were seen with a regimen of 1.5 g/day at 0, 4, and 8 h (0.5 g/dose, delivered over 12 h). Then, in 1991, Li went to Vietnam to treat cerebral malaria patients with artesunate. There were also rumors of severe quinine resistance, but when the above regimen was used on 30 patients consecutively, there was no treatment failure either.[290] Why has quinine not lost its efficacy after over 190 years? This could be because of its short half-life (6–9 h); after medication ceased, a low blood concentration was not maintained, and the parasite could not develop resistance easily.

Artemisinins also have short half-lives, ranging from 1.5 to 4 h. The half-life of the oil injections is only 7 h. Because of its strong clearance effect and rapid action, only one dose is needed per day and a sustained low blood concentration does not occur (Fig. 5.48). It could be why 200 passages of cultivated resistance did not produce a high resistance index.

One must consider past in vivo and in vitro resistance data. However, they cannot simply predict current and future changes. Quinine and artemisinin do not readily produce drug resistance, unlike drugs with long half-lives.

It is worth considering if artemisinin will go down the path of quinine or chloroquine in future. Artemisinins quickly replaced quinine not because the latter had failed but because it could not act as quickly as artemisinin by far or achieve the same levels of low toxicity, high safety, and ease of use. The half-life of artemisinin is shorter than that of quinine (see above). Hence, the active life of artemisinin may be comparable to quinine's 190 years. Considering that long half-lives readily lead to drug resistance, to prevent the appearance of artemisinin resistance in clinical use, derivatives and transdermal preparations with long half-lives should be prohibited.

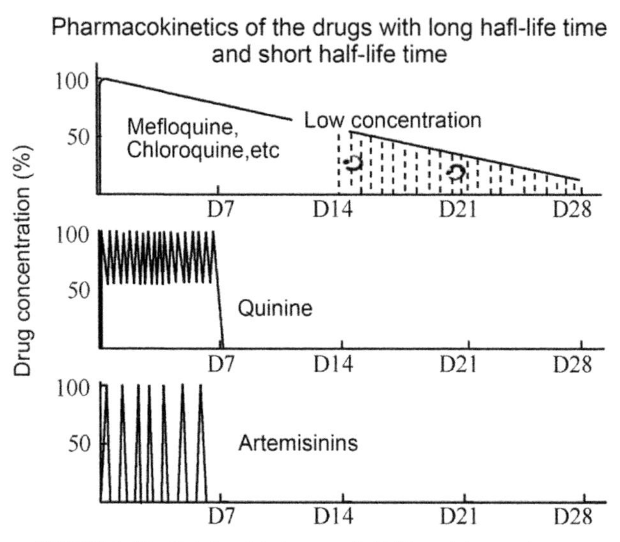

FIGURE 5.48 Relationship between half-life and drug resistance.

Hundreds of millions of ACTs have been used because they have been recommended as first-line treatment drugs worldwide. The possibility of genetic mutations leading to resistance, caused by the drug's selective pressure, is a serious concern. Such reports have appeared in Cambodia and Thailand, and an improved PCE[280] was established by the WWARN (see above).[282] It is an online tool to be used worldwide, enabling early detection of delayed parasite clearance as a marker of possible resistance.

Researchers in areas where it is inconvenient or impossible to use the online PCE can use the rubric designed by the global plan for artemisinin resistance containment[291] to determine drug resistance. This includes two levels. The first, suspected resistance, is defined by increased parasite clearance time and parasites found in over 10% of cases 72h after administering an ACT. The second, confirmed resistance, is defined by treatment failure (no clearance in 7days or continued presence of the parasite after 72h) after oral monotherapy and the presence of adequate blood concentration and recrudescence in 28–42days.

Li believes that resistance should not be determined only by recrudescence because a 7-day regimen of drugs with a short half-life rarely achieves a 100% cure rate. In 1992, when artesunate was brought to southern Vietnam, Li used a 7-day course of oral artesunate, total dose 800mg, to treat 43 cases of falciparum malaria. There were two cases of recrudescence in 28days, but because this was the first time the drug had been used there, it could not be considered drug resistance. From 1975 to 1985, when artemisinins were first used in China, a clearance time of more than 72h was not unheard of. This was strongly influenced by the patient's immune system and also influenced to a great extent by the patience of the staff using the microscope. Therefore, parasite clearance time is not suitable as an indicator of drug resistance. "Suspected resistance" should be defined as a recrudescence rate of more than 15% after a 7-day monotherapy regimen. Other factors, such as immunity, some hemoglobin diseases, and splenectomy, can all affect recrudescence.

B. LATEST DEVELOPMENTS IN RESISTANCE RESEARCH: CLINICAL AND GENETIC

Tran Tinh Hien[1,2]
[1]*Oxford University Clinical Research Unit, Ho Chi Minh City, Vietnam;*
[2]*Nuffield Department of Medicine Oxford University, Oxford, England*

Chloroquine was discovered in 1934 and quickly proved to be one of the most successful and important drugs ever deployed against malaria. However, its therapeutic efficacy was diminished because of the emergence of resistance. Chloroquine resistance emerged in a handful of places, including Southeast Asia (1957), South America (1959), and the Western Pacific region (1960). From there, it spread progressively throughout malaria-endemic areas, including Africa, where surges in malaria mortality were reported.[292] Many years later, studies on chloroquine-sensitive and resistant strains localized a 36 kb segment on chromosome 7 of *P. falciparum*.[293] Further molecular analyses revealed mutations in the *P. falciparum* chloroquine resistance transporter (PfCRT) gene, associated strongly with chloroquine-susceptible and resistant strains in both laboratory and field studies.[294] In addition, mutations in the *Pfmdr1* gene at N86Y, Y184F, S1034C, N1042D, and D1246Y have been reported to be involved in determining drug susceptibility to chloroquine, quinine, mefloquine, halofantrine, lumefantrine, and artemisinin.[295] The failure of chloroquine led to the reemergence of malaria and spread of chloroquine-resistant parasites in Africa,[294] Southeast Asia, and South America.[296]

Chloroquine was replaced with SP as a first-line treatment of malaria. This combination, known as Fansidar, was effective in treating chloroquine-resistant falciparum malaria cases. However, resistance to SP in *P. falciparum* appeared and was conferred by mutations in dihydrofolate reductase (DHFR) and dihydropteroate synthase (DHPS). A combination of five mutations ("quintuple mutant": DHFR-51, -59, and -108 and DHPS-437 and -540) strongly predicts clinical outcome.[297] Mutations in DHFR-164, DHPS-581, and DHPS-613 developed later and are associated with increased SP resistance.[298]

Mefloquine is a quinoline methanol, which is structurally similar to quinine. It is a powerful, long-acting blood schizontocide, effective against all malaria parasites including *P. falciparum* that is resistant to chloroquine, SP, and quinine. The great advantage of mefloquine is that it can be used as a single dose of 15–25 mg mefloquine base/kg. However, there has been a decline in mefloquine efficacy along the Thai–Cambodian border, when used as monotherapy.[299] A combination of mefloquine with artesunate raised effectiveness against resistant malaria to more than 95%.[300]

By the 1990s, *P. falciparum* parasites resistant to multiple drugs (chloroquine, sulfadoxine–antifolate combinations, and mefloquine), also known as MDR strains, were prevalent in Southeast Asia. Since then, artemisinin has been used as the most effective treatment for *P. falciparum*, particularly for patients infected with MDR strains. Unfortunately, evidence has increased over the last decade that artemisinin resistance has emerged and spread within Southeast Asia, first in western Cambodia but now across a widening area of the Greater Mekong Subregion (GMS). Rapid scientific advances in the understanding of this problem have taken place within the last 5 years.[301–303]

A series of clinical efficacy studies was designed to investigate the issue of artemisinin resistance in more detail. These studies confirmed that despite adequate drug levels, parasite clearance rates were twice as slow in the western Cambodian provinces of Battambang,[275] Pailin,[276] and Pursat.[304] Slow clearance was subsequently found to be common across mainland Southeast Asia, both in individual studies[278,305] and in the Tracking Resistance to Artemisinin Collaboration study that employed a common protocol across 10 sites in the region.[306]

In October 2016 the WHO released a report on artemisinin and artemisinin combination therapies[307] in which the following key messages were released:

1. Artemisinin resistance is defined as delayed parasite clearance following treatment with artesunate monotherapy or with an ACT. This represents partial resistance.
2. Delayed parasite clearance does not necessarily lead to treatment failure. In the GMS, high treatment failure rates following treatment with an ACT have almost always been observed in areas where there is concomitant resistance to artemisinin and the partner drug. Outside the GMS, treatment failure with ACTs (artemether–lumefantrine, artesunate–amodiaquine, and artesunate–sufadoxine–pyrimethamine) has occurred in the absence of artemisinin resistance mainly because of partner drug resistance.
3. A molecular marker of artemisinin resistance was identified. Several mutations in the Kelch 13 (K13)-propeller domain were found to be associated with delayed parasite clearance in vitro and in vivo. To date, more than 200 nonsynonymous mutations in the K13 gene have been reported. In Cambodia, Laos, and Vietnam, C580Y, R539T, Y493H, and I543T mutations are frequent or specific. In the western GMS (China, Myanmar, and Thailand), F446L, N458Y, P574L, and R561H mutations are specific. The P553L allele is distributed in these areas. The results of the therapeutic efficacy studies (TES) conducted in 2015 in Cambodia, Laos, and Vietnam, and for which K13 sequencing is available, have shown that C580Y has become the dominant mutation, ranging from 48.8% in Laos to 92.6% in Cambodia. In Africa, nonsynonymous mutations are still rare and highly diverse. Not all nonsynonymous propeller region K13 mutants reported indicate the emergence of artemisinin resistance. Rather, such mutants can represent "passer-by" genotypes in the absence of evidence for the selection of the mutant K13 genotype. In addition, different K13 mutations

have varying effects on the clearance phenotype. The validation of the K13 mutant as a resistance marker will require correlation with slow clearance in clinical studies, reduced drug sensitivity in ex vivo or in vitro assays (e.g., ring-stage assay—RSA 0–3 h), or reduced in vitro sensitivity resulting from the insertion of the K13 mutant in transfection studies.

4. There is no evidence that higher levels of artemisinin resistance (full resistance) have emerged. Nevertheless, partial artemisinin partial resistance could facilitate the selection of partner drug resistance.
5. Piperaquine resistance has emerged in Western Cambodia and, in just a few years, expanded considerably in terms of the proportion of strains and the geographical area affected.
6. The emergence of multidrug resistance, including artemisinin and partner drug resistance causing ACT failure, and the independent emergence of artemisinin resistance in multiple locations in the GMS have led the WHO to recommend the elimination of malaria in this region.

5.6 IMMUNE REGULATION EFFECTS OF ARTEMISININ AND ITS DERIVATIVES

As knowledge of artemisinin's biological activity deepened, it was discovered that the drug had strong regulatory effects on immune function. The following section summarizes its effects on nonspecific and specific immunity.

A. NONSPECIFIC IMMUNITY

Li Zelin[1†], *Li Linna*[2]

[1]Institute of Chinese Materia Medica, China Academy of Chinese Medical Sciences, Beijing, China; [2]Institute of Radiology and Radiation Medicine, Academy of Military Medical Sciences, Beijing, China

Cheng Daoxin[308] studied the in vivo and in vitro effects of artemisinin on the phagocytic function of macrophages in mouse peritoneal cavity and on the enzyme activity of macrophage lysosomal acid phosphatase. The in vivo experiment involved two groups of LACA mice, one given 200 mg/kg intragastric artemisinin and the other an equal volume of excipient. They were dosed once a day for 3 days. Ten hours before the end of the experiment, both groups received intraperitoneal injections of 0.5 mL 5% chicken red blood cell (CRBC) suspension. The mice were killed and an intraperitoneal cavity cell suspension was drawn and incubated at 37°C for 50 min. A large number of macrophages adhered to the glass slides. The loose components were washed off with saline. The slides were dried, fixed with methanol, Giemsa stained, and phagocytic function was observed under a microscope. The average number of macrophages with phagocytized RBCs per 300 macrophages was recorded.

†Deceased.

TABLE 5.42 Effects of Artemisinin on Phagocytic Rate and Index in Mouse Peritoneal Macrophages

Groups	Cases	Phagocytic Rate	Phagocytic Index
Artemisinin	46	46.69±1.80[a]	0.88±0.05[a]
Control	47	38.59±2.36	0.73±0.05

[a]$P<.05$.

TABLE 5.43 Effects of Artemisinin on Phagocytosis of Malaria Pigment or Parasites by Peritoneal Macrophages in Mice

Groups	Cases	Phagocytic Rate
Artemisinin	30	55.26±2.29[a]
Control	30	46.68±1.93

[a]$P<.01$.

The results are shown in Table 5.42. Artemisinin significantly increased the phagocytic rate and index.

For the in vitro test on peritoneal macrophages and *Plasmodium*, two groups of Kunming mice underwent the same treatment. Peritoneal macrophages were harvested into a cell culture medium and adjusted to a dilution of 2×10^6 or 4×10^6 cells/mL, and 1 mL was added to a special dish. This was incubated at 37°C for 2–3 h, whereupon blood from mice infected with *P. berghei* was introduced. The ratio of macrophages to infected erythrocytes was 1:7. The mixture was incubated for 6 h, the glass slides were removed, and the aforementioned staining and fixing process conducted. On each glass slide 300 macrophages were examined. Phagocytic rate was calculated based on phagocytosis of the malarial pigment or of the parasite. The phagocytic rate also increased significantly (Table 5.43).

In further study, the pelleted peritoneal macrophages were disrupted to release acid phosphatase, and its activity expressed as spectrometric intensity of para-nitrophenol when para-nitrophenyl phosphate was added. The study showed that artemisinin also raised lysosomal acid phosphatase activity (Table 5.44).

Other studies also noted the improvement in macrophage function with artemisinin. Zou Zhang et al.[309] examined the antigen-presenting role of macrophages via plaque formation (PFC) and rosette formation (RFC) of splenic lymphocytes of *P. berghei*-infected mice and normal mice after intraperitoneal immunization with 5% sheep red blood cells (SRBC). The infection was induced by intraperitoneal inoculation with 6×10^6 parasitized RBCs. Zou found that PFC and RFC were severely inhibited in infected mice. The extent of inhibition increased

TABLE 5.44 Effects of Artemisinin on Lysosomal Acid Phosphatase Activity in Peritoneal Macrophages in Mice

Groups	Cases	BLB Units/6 × 10^6
Artemisinin	30	0.55 ± 0.037^a
Control	30	0.46 ± 0.027

[a] $P < .05$.

as parasitemia increased. At parasitemia of 30%, the infected mice's PFC and RFC dropped to 0.4% and 7.8% of normal mice. A regimen of 300 mg/kg artemisinin once a day for 7 days eliminated the parasite. Thereafter, PFC and RFC responses gradually recovered. Intraperitoneal injection of 3×10^6 macrophages from normal mice did not improve the PFC and RFC formation in infected mice (Table 5.45).

Zang Qizhong et al.[310] investigated the effects of artemisinin on reticuloendothelial phagocytosis of ^{86}Rb erythrocytes. Artemisinin had no effects on phagocytic function, lymphocytic transformation, and plasma cAMP content in the reticuloendothelial system of normal animals. But in animals with corticoid-induced impaired immunity, artemisinin could raise the compromised lymphocytic transformation rate and lower the elevated plasma cAMP. In instances of high parasitemia, it could also increase the suppressed plasma cAMP.

In an experiment on artemether and spleen weight, Lin Peiying et al.[311] randomly divided 79 mice into high-dose (200 mg/kg/day), low-dose (100 mg/kg/day), chloroquine (200 mg/kg/day), and control groups. Drugs were administered twice a day for 7 days. The mice were killed via orbital bloodletting 24 h after the final dose. Their spleens were obtained and weighed, and the spleen weight per 10 g of body weight was calculated. Artemether increased spleen weight in normal mice ($P < .001$) and in mice given sheep red cell antigens ($P < .001$). Chloroquine had no effect on spleen weight. For mice infected with *P. berghei* spleen weight was significantly lower in those treated with AMT than in those without medication.

Wang Dunrui et al.[312] reported that mouse serum after high doses of artemisinin (500 mg/kg intraperitoneal injection) inhibited ^{3}H-UR incorporation in ConA activated normal allogeneic spleen cells. Deoxyartemisinin also had this effect (Figs. 5.49 and 5.50).

Artemisinin has definite effects on natural killer (NK) cells and macrophages. Some studies indicate that artemisinin had a strong inhibitory effect on NK cells,[313] while increasing the phagocytic activity of macrophages and neutrophils.[314] Artesunate suppressed polymorphonuclear cell infiltration caused by leukotriene B4 and IL-8-induced neutrophil chemotaxis,[315] thus producing nonspecific antiinflammatory effects. NK cells are important immune cells, with the main function of immune regulation. They can act on T and B

TABLE 5.45 Lymphocytic Plaque Formation (PFC) and Rosette Formation (RFC) Reactions in Infected Mice Given Macrophages From Normal Mice

Groups	Cases	Parasite Rate at Immunization	Parasite Rate at Time of Test	PFC/10^6 Spleen Cells	RFC ($\times 10^3$)/10^6 Spleen Cells
Control	7	–	–	828 ± 207	20.7 ± 3.4
Infected (8 day)	14	24.7 (10–50.4)	36.9 (12–55.3)	5 ± 1	0.7 ± 0.4
Infected (8 day) + 3×10^6macrophages	9	22.2 (9.6–33.2)	37.5 (19.7–57.1)	4 ± 1	1.2 ± 1.6

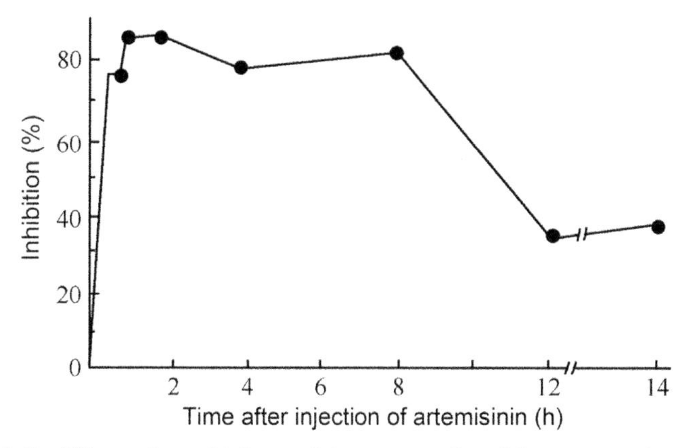

FIGURE 5.49 Effects of artemisinin-containing serum collected from mouse after artemisinin peritoneal injection on ^{3}H-UR incorporation in ConA-activated spleen cells of different mouse.

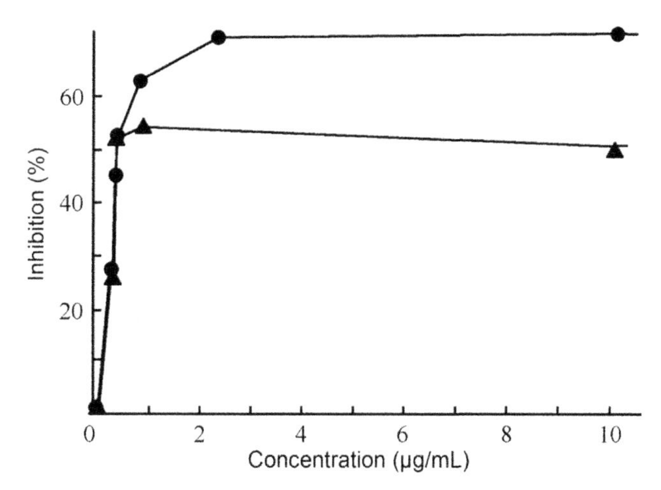

FIGURE 5.50 Effects of artemisinin (–●–) and deoxyartemisinin (–▲–) on ^{3}H-UR incorporation in ConA-activated spleen cells of mouse.

lymphocytes. Artesunate could strongly suppress the antibody-dependent cell-mediated cytotoxicity mechanism mediated by NK cells and its participation in Type II and IV hypersensitivity reactions.[313]

B. HUMORAL IMMUNITY

Li Zelin†
Institute of Chinese Materia Medica, China Academy of Chinese Medical Sciences, Beijing, China

Humoral immunity is mainly manifested in in vivo serum immunoglobulin and the number of B lymphocytes. Many scholars have studied the effects of artemisinins on nonspecific humoral immunity.

†Deceased.

1. Serum IgG

In the following experiments, Lin Peiying et al.[311] studied the effects of oral artemether and chloroquine on IgG levels in ICR mice.

Artemether on Serum IgG of Normal Mice

Seventy-nine mice were randomly assigned to four groups, receiving oral artemether high dose (200 mg/kg/day), low dose (100 mg/kg/day), chloroquine (200 mg/kg/day), and nonmedicated control group. The medicated groups received drug twice daily for 7 days. At completion of the experiment, serum IgG in each mouse was measured using single-radial immunodiffusion. Mean IgG was 12.56 ± 0.72 mg/mL in the control group and 8.82 ± 0.73 mg/mL in the high-dose group. The difference was significant ($P<.001$), indicating that 200 mg/kg/day AMT could significantly reduce serum IgG in normal mice. Chloroquine had no effect.

Artemether on Serum IgG of Mice Immunized With Sheep Red Blood Cells

The same protocol was used, except that each mouse was given an intraperitoneal injection of 0.2 mL antigen (20% SRBC suspension, SRBC 2×10^9/mL) 2 days after the completion of the 7-day medication course. The experiment included 71 mice. Serum IgG was 14.62 ± 0.70 mg/mL in the control group and 12.20 ± 0.74 mg/mL in the artemether low-dose group ($P<.05$). It showed that 100 mg/kg/day artemether could reduce serum IgG in SRBC-immunized mice. Chloroquine had no effect.

Artemether on Serum IgG of Plasmodium berghei*-Infected Mice*

Healthy mice were inoculated with *P. berghei* using standard methods. Then, they were divided into the aforementioned dose groups. Medication was administered 24 h after infection, twice a day for 4 days. Blood smears were prepared 24 h after the last dose, and the parasite clearance and infection rate were observed. Serum was also collected. The inhibition rate was 100% in all medicated groups. IgG levels were no different in the medicated and control groups ($P>.05$).

Different doses of artemether had different effects on serum IgG levels. The same dose also had different effects on different groups of mice. A high dose significantly reduced serum lgG in normal mice, whereas a low dose had no effect. For mice given SRBC, a low dose significantly reduced IgG levels but a high dose did not.

2. B Lymphocytes

Yang Jinhong et al.[316] observed the effects of artemether on the proliferation of spleen B lymphocytes in NIH mice. At 100 and 250 mg/kg doses, artemether had a strong inhibitory effect (Table 5.46).

TABLE 5.46 Effects of Artemether on B Lymphocyte Proliferation in Mouse Spleens

Groups	Cases	Stimulation Index	*P* Value
Control	9	7.18±3.30	
Artemether 50 mg/kg	8	5.80±3.62	$P>.05$
Artemether 100 mg/kg	9	3.30±2.61	$P<.05$
Artemether 250 mg/kg	8	1.64±0.57	$P<.01$

The inhibitory effects of artemether on humoral immunity may be related to its antimalarial mechanism. Some have proposed that the pathogenesis of cerebral malaria was partly because of parasitic mitogens, which could stimulate B lymphocytes to produce a large amount of immunoglobulin M. This led to a marked increase in autoantibodies and immune complexes. Therefore, it was hypothesized that while artemether directly killed the parasite, it inhibited B lymphocyte function at the same time, lowering the production of autoantibodies. This resulted in better efficacy.

C. CELLULAR IMMUNITY

Li Zelin†
Institute of Chinese Materia Medica, China Academy of Chinese Medical Sciences, Beijing, China

In 1981, Qian Ruisheng et al.[317] studied the effects of artemisinin on specific cellular immunity. Sheep red blood cells (SRBC) were used as an antigen in a footpad test in mice. Three groups of mice were given artemisinin-oil suspension (450 mg/kg/day), cyclophosphamide (150 mg/kg/day), or saline for 6 days. On day four of the medication process, half of the mice in each group were sensitized by injecting SRBC into their footpad. Five days later SRBC injection was given to all mice, and the difference in footpad thickness before and 1 day after SRBC challenge established for each mouse. It was found that artemisinin-oil suspension had a significant impact on delayed hypersensitivity reaction. However, it was less marked than with cyclophosphamide ($P<.05$). There was no clear difference in the footpad response of normal, nonsensitized mice, but a strong influence was seen in the normal artemisinin group and the sensitized artemisinin group ($P<.001$). It indicated that artemisinin could act as a cellular immunity adjuvant (Table 5.47).

In the lymphocyte transformation test, heart blood was taken from guinea pigs and 25 U/mL heparin anticoagulant added. An artemisinin-oil suspension

†Deceased.

TABLE 5.47 Effects of Artemisinin-Oil Suspension and Cyclophosphamide in the Footpad Test in Mice

Drug (mg/kg/day)	Difference in Pre- and Post–Challenge Footpad Thickness (mm)	
	Sensitized Group Mean ± SD (Cases)	Normal Group Mean ± SD (Cases)
Artemisinin (450)	0.36 ± 0.05 (8)[a,c]	0.06 ± 0.04 (8)
Cyclophosphamide (150)	0.65 ± 0.12 (5)[b]	0.26 ± 0 (2)
Normal saline	0.18 ± 0.06 (5)	0.20 ± 0.07 (5)

[a] *$P < .05$ compared to the sensitized saline group.*
[b] *$P < .01$ compared to the sensitized saline group.*
[c] *$P < .001$ compared to the normal artemisinin group.*

TABLE 5.48 Effects of Artemisinin and Cyclophosphamide on the Lymphocyte Transformation Rate (%)

Drug	Test			Control	
	LD	MD	HD	Positive	Negative
Artemisinin-oil injection	12	36	50	49	0
Artemisinin-oil injection	14	30	42	43	0
Artemisinin-water injection	13	–	–	49	0
Artemisinin-water injection	12	–	–	43	0
Cyclophosphamide	24	–	–	43	0

"–" indicates hemolysis; LD, MD, and HD indicate artemisinin doses 0.56, 2.80, and 5.60 mg/mL, respectively; cyclophosphamide doses 0.1, 0.5, and 1.0 mg/mL, respectively; positive control group with 50 μg/mL phytohemagglutinin (PHA); negative control group without artemisinin and PHA.

group, artemisinin-water suspension group, and cyclophosphamide group were used, alongside positive and negative controls. The blood was incubated at 37°C for 72 h, centrifuged, smeared, and stained. Microscopic examination of 200 lymphocytes calculates the lymphocyte transformation rate. The results are shown in Table 5.48.

Artemisinin could significantly raise the lymphocyte transformation rate, suggesting that artemisinin not only killed parasites directly but also promoted patients' specific cellular immunity.

TABLE 5.49 Effects of Artemisinin on Lymphocyte Transformation in the Blood of Normal and Prednisone-Treated Rats

Group	Radioactivity (cpm/mL)	*P*
1. Prednisone	113±20	Group 1 vs. 2<.01
2. Prednisone+artemisinin	407±76	
3. Artemisinin	341±87	Group 3 vs. 4>.05
4. Control	224±54	

Zang Qizhong et al.[310] also studied lymphocyte transformation in normal rats and rats treated with corticosteroids (prednisone). It confirmed that artemisinin had no significant stimulus on lymphocyte transformation in normal rats, but it strongly raised T-lymphocyte transformation in the blood of rats with corticoid-induced immunosuppression (Table 5.49).

Shen Ming et al.[318] found that sodium artesunate had strong suppressive effects on ^{3}H-TdR incorporation in mitogen-induced (PHA and ConA) spleen and thymus cells in mice. This was also the case in normal human peripheral lymphocytes. If C57BL or CFW mouse spleen cells were exposed to artemisinin for 30min and then washed, no inhibition appeared. However, if they were washed after 4h, incorporation was still partially suppressed. In vivo tests also showed an inhibitory effect, but this was weak.

Lin Xingyu et al.[319] reported that sodium artesunate significantly inhibited mixed lymphocyte cultures. It also significantly inhibited ^{3}H-TdR incorporation in lipopolysaccharide-stimulated B lymphocytes.

Qian Ruisheng et al.[320] collected serum from mice 2h after they were given artesunate injections. An in vivo-induced antiviral substance was detected, which was trypsin-sensitive, and had the characteristics of a protein. It protected homologous but not heterologous cells. Like β-interferon, it was unaffected at pH 2 but sensitive to heat (see Table 5.50). This suggested that artesunate could be an interferon inducer.

The interferon-inducing effects of artesunate were dose related. The optimal dose was 40–60mg/kg. If the dose was too small or too large, these effects fell or disappeared (Fig. 5.51). This was similar to the dose–effect relationship of levamisole.[321] The induction peaked at 2 and 24h after medication (Fig. 5.52), which was close to the time–effect curve[322] of glycyrrhizin.[312]

Within certain dose and time parameters, the strength of interferon induction was similar to induction by *Astragalus* (Table 5.51). Like other interferon inducers, induction capacity fell with repeated administration and tolerance was produced: a "low reactive period effect" (Fig. 5.53).

Chen Ming et al.[323] assessed the effects of artesunate on specific cellular immunity in rats, based on the migration index (MIT) of peripheral blood

TABLE 5.50 Artesunate-Induced Interferons in Mice

Treatment		Interferon Titers (U/mL)
pH	pH 2	31
	pH 7.2	30
T (°C)	60	0
	4	32
Trypsin (μg/mL)	0	49
	25	48
	100	25
	250	19
L-929 cells		39
Chicken embryo fibroblasts		0
Human embryonic lung cells		0

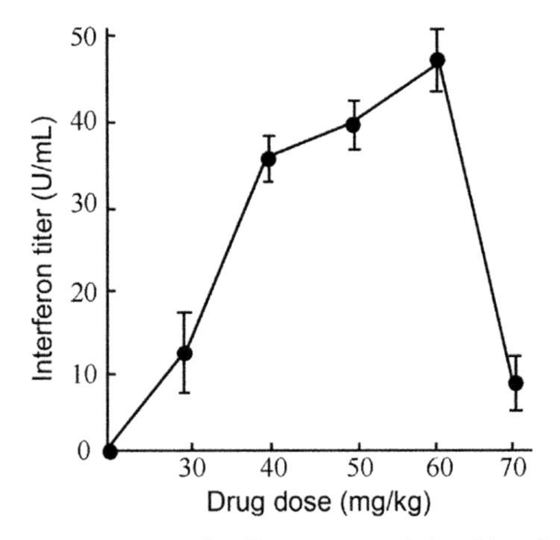

FIGURE 5.51 Dose–response curve of sodium artesunate-induced interferons. *(From Qian R, Li Z, Xie M, et al. Inductive effects of artemisinin on interferons in mice.* Med J Chin People's Lib Army *1985;5:355–8. [钱瑞生，李柱良，谢名荣，等. 青蒿素对小鼠干扰素的诱生作用. 解放军医学杂志, 1985, (5): 355–358.])*

leukocytes and mononuclear leukocyte procoagulant activity (LPCA). The artesunate was injected intraperitoneally. Artesunate could enhance specific cellular immune function (Table 5.52).

Table 5.52 shows that after the injection, peripheral leukocyte MIT was significantly lower than in the control group; LPCA was significantly higher. Chen concluded that the effects of artemisinin on immunity were highly dose

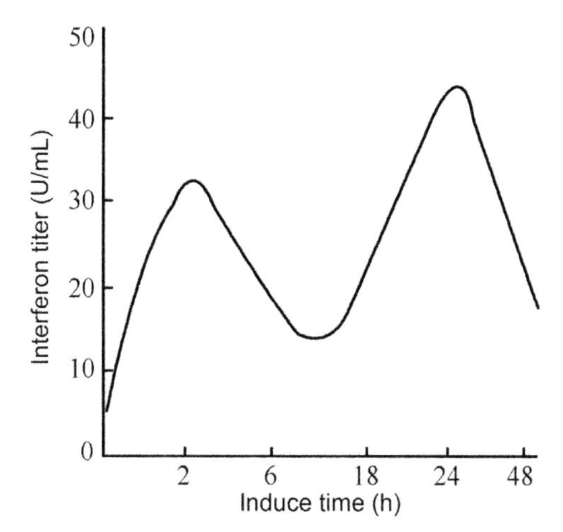

FIGURE 5.52 Time–response curve of sodium artesunate-induced interferons. *(From Qian R, Li Z, Xie M, et al. Inductive effects of artemisinin on interferons in mice.* Med J Chin People's Lib Army *1985;5:355–8. [钱瑞生，李柱良，谢名荣，等. 青蒿素对小鼠干扰素的诱生作用. 解放军医学杂志, 1985, (5): 355–358.])*

TABLE 5.51 Comparison of Induction by Artesunate and *Astragalus*

	Interferon Titers (U/mL)	
Group	**2h**	**24h**
Artesunate	44	34
Astragalus	43	26
Normal saline	0	0

From Qian R, Li Z, Xie M, et al. Inductive effects of artemisinin on interferons in mice. *Med J Chin People's Lib Army* 1985;**5**:355–8. [钱瑞生，李柱良，谢名荣，等. 青蒿素对小鼠干扰素的诱生作用. 解放军医学杂志, 1985, (5): 355–358.]

dependant; in small doses, artemisinin was an immune enhancer, but in high doses, it was an immunosuppressor.

Chen Hua et al.[324] reported that artesunate could raise the inhibitory activity of α1-antitrypsin, promote the proliferation of suppressor T lymphocytes (T_S), curb the production of effector T lymphocytes (T_E), and prevent the release of white blood cell interleukins and other inflammatory mediators. Therefore, it could regulate immunity. Chen studied artesunate's inhibitory effects on allergic contact dermatitis using BALB/C mice. After stimulation, the test group was given 6 days of intramuscular artesunate. Difference in ear thickness and mononuclear cell infiltration of the dermis before and after were taken as indicators and a strong inhibitory effect was demonstrated (Table 5.53). Chen showed that artesunate could promote the activation of suppressor T cells (T_S) and enhance inhibitory

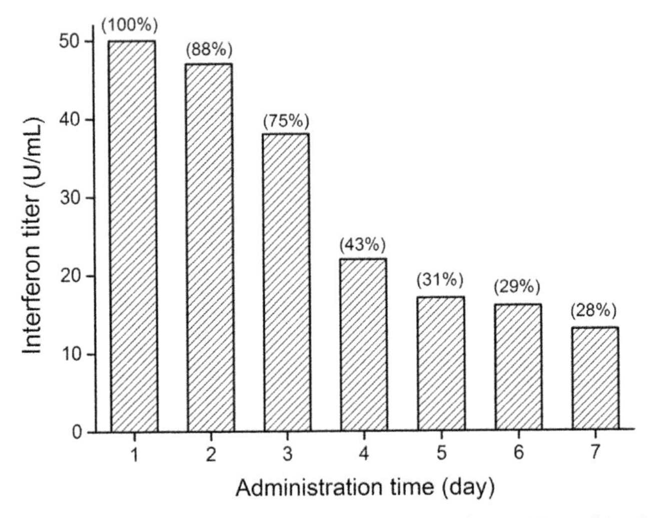

FIGURE 5.53 Low reactive period of artesunate-induced interferons. *(From Qian R, Li Z, Xie M, et al. Inductive effects of artemisinin on interferons in mice.* Med J Chin People's Lib Army *1985;5:355–8. [钱瑞生，李柱良，谢名荣，等. 青蒿素对小鼠干扰素的诱生作用. 解放军医学杂志, 1985, (5): 355–358.])*

TABLE 5.52 Impact of Artesunate on Specific Cellular Immunity in Rats

Group	Cases	Migration Index	*P* Value	Leukocyte Procoagulant Value	*P* Value
Control	15	0.5587 ± 0.1336	<.001	7.67 ± 2.71	<.05
Test	15	0.3119 ± 0.1579		16.22 ± 11.82	

TABLE 5.53 Effects of Intramuscular Artesunate on Allergic Contact Dermatitis in Mice

	Test Group	Control Group	*P* Value
Difference in ear thickness (1×10^{-2} mm) (n)	4.75 ± 0.75 (5)	7.00 ± 0.88 (5)	<.001
Number of mononuclear cell (cells/high power field)	76.2 ± 9.5	165 ± 13.7	<.001

effects on antibody production experiments.[324] In the in vitro test, mice were superimmunized with intraperitoneal injection of SRBC, followed by 14 days of artesunate intramuscular injection. The inhibition rate of antibody production by splenic lymphocytes rose from 51% to 68% ($P<.001$). The in vivo study went further and a new batch of superimmunized mice was given mouse splenic cells created as the in vitro test above. Antibody levels were checked 5 days later. The antibody production inhibition rate was increased from 52% to 67% ($P<.001$). Chen also studied artesunate's effect on α1-antitrypsin.[324] The α1-antitrypsin level established in 20 BALB/C inbred mice was 1.71±0.19 mg/mL. Sixteen BALB/C mice were divided into an SOI (superoptimal immunization) group and an SOI-artesunate group. Orbital blood was collected to determine serum α1-antitrypsin levels. The value was 1.89±0.21 mg/mL for the SOI group and 2.1±0.20 mg/mL for the SOI-artesunate group ($P<.05$). Artesunate increased α1-antitrypsin level.

T_μ lymphocytes are a type of T cells involved in cellular immune responses. Gu Yuanxi et al.[325] studied the effects of three doses of artemisinin and AMT on T, B, T_μ, and T_γ lymphocytes (Figs. 5.54 and 5.55). Artemether had an inhibitory

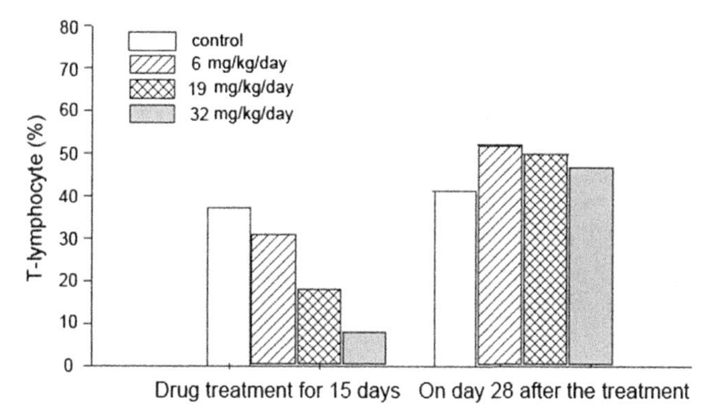

FIGURE 5.54 Effects of artemether on T lymphocyte proliferation in beagles.

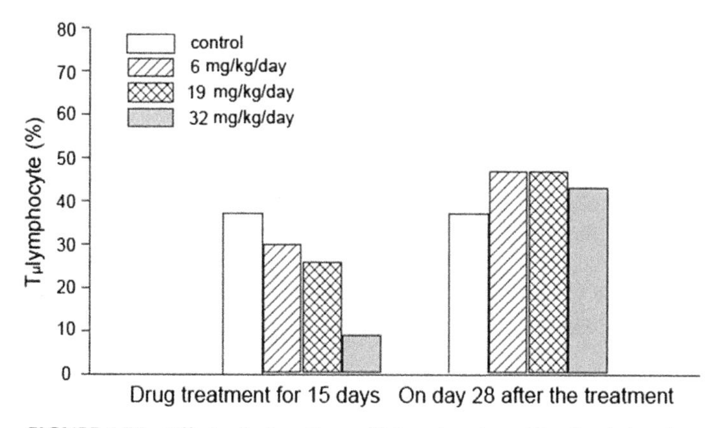

FIGURE 5.55 Effects of artemether on T_μ lymphocyte proliferation in beagles.

TABLE 5.54 Effects of Artemether on Peripheral Blood Lymphocyte Transformation in Phytohemagglutinin-Stimulated Mice

Group	Cases	Lymphocyte Transformation Rate (%)
Artemether	12	60.42 ± 10.22
Control	12	58.08 ± 7.83

effect on T and T_μ cell proliferation. In other words, it inhibited cellular immune function. This returned to normal 28 days after medication stopped.

Zhang Dan et al.[326] reported that artemether could cause the thymus to shrink significantly in mice sensitized with chicken red blood cells (CRBCs). However, it had no effect on peripheral blood lymphocyte transformation in PHA-stimulated mice (Table 5.54).

REFERENCES

1. China Cooperative Research Group on Qinghaosu and its Derivatives as Antimalarials. Antimalarial efficacy and mode of action of *Qinghaosu* and its derivatives in experimental models. *J Tradit Chin Med* 1982;**2**(1):17–24.
2. Pharmacology Laboratory of the Institute of Chinese Materia Medica, Academy of Traditional Chinese Medicine. A pharmacological study of *Artemisia annua* L. *J N Chin Med* 1979;**1**:23–33 [中医研究院中药研究所药理研究室. 青蒿的药理研究.新医药学杂志，1979, 1:23–33.]
3. Kärber G. Beitrag zur kollektiven Behandlung pharmakologischer Reihenversuche. *Naunyn Schmiedebergs Arch für Exp Pathol Pharmakol* 1931;**162**(4):480–3.
4. Sun RY. Practical methods for calculation of LD50. *Acta Pharm Sin* 1963;**10**(2):65–74. [孙瑞元. 简捷实用的半数致死量综合计算法[J] 药学学报, 1963, 10(2): 65–74.]
5. Pharmacology Laboratory of the Sichuan Institute of Chinese Materia Medica. A pharmacological study of artemisinin. *Sichuan Med J* 1980;**1**:33–6 [四川省中药研究所药理室. 青蒿素的药理研究. 四川医学，1980，1：33-36.]
6. Peters W, James DM, Li Z, Robinson BL. Antimalarial activity of arteannuin (artemisinine, *Qinghaosu*), a Chinese plant derivative, against *Plasmodium berghei*: preliminary data. *Trans R Soc Trop Med Hyg* 1981;**75**(4):603–4.
7. Peters W. Drug resistance in *Plasmodium berghei* Vincke and Lips, 1948. I. Chloroquine resistance. *Exp Parasit* 1965;**17**(1):80–9.
8. Institute of Chinese Materia Medica. *Registration dossier for artemisinin suppositories*. 1986 [中医研究院中药研究所. 青蒿素栓申报资料. 1986.]
9. Finney DJ. *Statistical method in biological assay*. 3rd ed. London: Charles Griffin & Co.; 1978.
10. Gu HM, Waki S, Zhu MY, et al. Fluorometric determination of the antimalarial efficacy of artemisinin and artemether against *Plasmodium falciparum in vitro*. *Acta Pharmacol Sin* 1988;**9**(2):160–3.
11. Li ZL, Gu HM, Warhurst DC, et al. Effects of *Qinghaosu* and related compounds on incorporation of [G-^{3}H]hypoxanthine by *Plasmodium falciparum in vitro*. *Trans R Soc Trop Med Hyg* 1983;**77**(4):522–3.

12. You JQ. The sensitivity of *P. falciparum* isolates from Thailand to artemisinin and artemether. *Int J Med Parasit Dis* 1986;**2**:79–80 [尤纪青. 泰国分离的恶性疟原虫对青蒿素和蒿甲醚的敏感性. 国际医学寄生虫病杂志，1986：79-80.]
13. Trager W, Jensen JB. Human malaria parasites in continuous culture. *Science* 1976;**193**(4254):673–5.
14. Li CS, Du YL, Jiang Q. A study of artemisinin resistance in *Plasmodium berghei*. *Acta Pharm Sin* 1986;**11**:811–5 [李成韶，杜以兰，姜齐. 伯氏疟原虫对青蒿素抗药性的研究. 药学学报，1986，11：811-815.]
15. Li GF, Zhou YQ, Jiao XQ. Cultivation of artemisinin- and artemether resistant *Plasmodium berghei*. *J Acad Mil Med Sci* 1990;**2**:97–100 [李国福，周义清，焦岫卿. 青蒿素及蒿甲醚抗药性伯氏疟原虫系的培育. 军事医学科学院院刊，1990，2：97-100.]
16. Zhao Y, Li XJ, Li AY, et al. An experimental report on artemisinin toxicity. *Guangxi J Tradit Chin Med* 1979;**1**:45–6 [赵一，李小娟，李爱媛，等. 青蒿素的毒性试验报告.广西中医药，1979，1：45-46.]
17. China Cooperative Research Group. Studies on the toxicity of Qinghaosu and its derivatives. *J Tradit Chin Med* 1982;**2**(1):31–8.
18. Wang DW, Guan MC, Liu XT. Pathological study of subacute toxicity in rhesus monkeys given intramuscular artemisinin. *Med J Chin People's Lib Army* 1983;**8**(4):257–60 [王德文，关明臣，刘雪桐. 恒河猴肌注青蒿素亚急性毒性病理学研究. 解放军医学杂志，1983，8(4)：257-260.]
19. Wang DW, Liu XT. Effects of artemisinin on simian myocardial ultrastructure. *Bull Acad Mil Med Sci* 1982;**4**(20):459–63 [王德文，刘雪桐. 青蒿素对猴心肌超微结构的影响.军事医学科学院院刊，1982，4(总20)：459-463.]
20. Pharmaceutical Manufacturers Association. *Guidelines for the assessment of drug and medical device safety in animals*. Washington (DC): Pharmaceutical Manufacturers Association; 1977.
21. Association of the British Pharmaceutical Industry. *Guidelines for preclinical and clinical testing of new medicinal products*. London: Association of the British Pharmaceutical Industry; 1977.
22. Song SY, Ding LM. Subacute toxicity of artemisinin-oil suspension in rhesus monkeys. In: *Collected abstracts from the pharmacology conference of the Chinese pharmaceutical association*, vol. II. 1982. p. 123 [宋书元，丁林茂. 青蒿素油悬剂对恒河猴亚急性毒性的研究. 中国药学会全国药理学术会议论文摘要集(下册)，1982：123.]
23. Li GY, Yang LX, Jin YS, et al. Assay report on artemisinin mutagenicity. *J Tradit Chin Med* 1981;**6**(467):67–8 [李广义，杨立新，金玉生，等. 青蒿素诱变性的测定结果报告.中医杂志，1981，6(总467)：67-68.]
24. Yang LX, Wang BS, Wu BF, et al. Special toxicity of artemisinin. *Pharmacol Clin Chin Mater Med* 1985;**6**:208–9 [杨立新，王宾生，邬碧芳，等. 青蒿素特殊毒性研究. 中药药理与临床，1985，6：208-209.]
25. Yang QC, Gan J, Li PS, et al. Pharmacological study of an artemisinin derivative, arteether. *J Guangxi Med Coll* 1981;**2**:5–12 [杨启超，甘俊，李培寿，等. 青蒿素衍生物——青蒿醚的药理研究. 广西医学院学报，1981，2：5-12.]
26. Wu BA, Xu ZH, Shi XC, et al. Long-term toxicity of artemether in beagles. *Chin J Clin Pharmacol* 1987;**3**(3):156–63 [邬伯安，徐在海，石笑春，等. 蒿甲醚对Beagle狗的长期毒性. 中国临床药理学杂志，1987，3(3)：156-163.]
27. Yang QC, Gan J, Li PS, et al. Antimalarial effects and toxicity experiment on sodium artesunate. *J Guangxi Med Coll* 1981;**3**:4–9 [杨启超，甘俊，李培寿，等. 青蒿酯钠的抗疟作用和毒性实验. 广西医学院学报，1981，3：4-9.]

28. Yang QC, Gan J, Li PS, et al. Antimalarial activity and toxicity of an artemisinin derivatives, artesunate. *J Guangxi Med Coll* 1981;**4**:1–6. [杨启超，甘俊，李培寿，等. 青蒿素衍生物——青蒿酯的抗疟活性与毒性. 广西医学院学报，1981，4：1-6.]
29. Ning DX, Xu ZH, Teng XH. Acute toxicity of intravenous artesunate in dogs. *Chin J Pharmacol Toxicol* 1987;**1**(2):129–34. [宁殿玺，徐在海，滕翕和. 青蒿酯钠静脉注射对犬的急性毒性. 中国药理学与毒理学杂志，1987，1(2)：129-134.]
30. Ning DX, Xu ZH, Teng XH. Subacute toxicity of intravenous artesunate in dogs. *Chin J Pharmacol Toxicol* 1987;**1**(2):135–42. [宁殿玺，徐在海，滕翕和. 青蒿酯钠静脉注射对犬的亚急性毒性. 中国药理学与毒理学杂志，1987，1(2)：135-142.]
31. Li ZL, Yang LX, Xue BY, et al. Test on sodium artesunate induced mutation. *Bull Chin Mater Med* 1988;**13**(9):45. [李泽琳，杨立新，薛宝云，等. 青蒿酯钠致突变试验. 中药通报，1988，13(9)：45.]
32. Li ZL, Yang LX, Liu JF, et al. Study on the teratogenicity of sodium artesunate. *Bull Chin Mater Med* 1988;**13**(4):42–4. [李泽琳，杨立新，刘菊福，等. 青蒿酯钠致畸作用的研究. 中药通报，1988，13(4)：42-44.]
33. Li ZL, Yang LX, Liu JF, et al. Embryotoxicity of sodium artesunate. *Chin J Pharmacol Toxicol* 1987;**1**(4):267. [李泽琳，杨立新，刘菊福，等. 青蒿酯钠的胚胎毒性. 中国药理学与毒理学杂志, 1987, 1(4): 267.]
34. Zhang HT, Li XB, Wan HP, et al. Long-term toxicity of artequick in beagles. In: *Collected papers of the national drug toxicology conference*. 2006. p. 21. [张宏涛，利小斌，万红平，等. Artequick对Beagle犬长期毒性研究. 全国药物毒理学术会议论文集，2006：21.]
35. Wan HP, Liang LZ, Cai WM, et al. Toxicity of artequick during the teratogenically sensitive period in rats. In: *Collected papers of the national drug toxicology conference*. 2006. p. 71. [万红平，梁礼珍，蔡五妹，等. Artequick对大鼠的致畸敏感期毒性试验. 全国药物毒理学术会议论文集，2006：71.]
36. Medhi B, Patyar S, Rao RS, et al. Pharmacokinetic and toxicological profile of artemisinin compounds: an update. *Pharmacology* 2009;**84**(6):323–32.
37. Li Q, Hickman M. Toxicokinetic and toxicodynamic (TK/TD) evaluation to determine and predict the neurotoxicity of artemisinin. *Toxicology* 2011;**279**(1–3):1–9.
38. Meshnick SR. Artemisinin: mechanisms of action, resistance and toxicity. *Int J Parasitol* 2002;**32**(13):1655–60.
39. Lou XE, Zhou HJ, Huang HB. Damaging effects of artesunate on the conceptus. *Chin J Pharmacol Toxicol* 2001;**15**(5):398–400. [娄小娥，周慧君，黄泓忭. 青蒿琥酯对孕体组织的损伤.中国药理与毒理学杂志，2001，15(5)：398-400.]
40. Yao L, Teng XH, Xiong CS. Effects of artesunate on the maturation of erythroid cells in the bone marrow of beagles. *Chin J Pharmacol Toxicol* 1993;**7**(3):231–4. [姚朗，滕翕和，熊春生. 青蒿琥酯对比格犬骨髓幼红细胞成熟的影响. 中国药理学与毒理学杂志, 1993, 7(3):231-234.]
41. Zhai HF, Wu BA. An overview of neurotoxicity research into artemisinin-type antimalarials. *Chin J Parasit Dis Control* 1999;**12**(4):303–4. [翟海峰，邬伯安. 青蒿素类抗疟药神经毒性研究概述. 中国寄生虫防治杂志，1999，12(4)：303-304.]
42. H.Zhao Y, Wang JY. Artemisinin neurotoxicity and its mechanisms of action. *Chin J Parasitol Parasit Dis* 2002;**20**(1):49–51. [赵艳红，王京燕. 青蒿素类药物神经毒性及其作用机制. 中国寄生虫学与寄生虫病杂志，2002，20(1)：49-51.]
43. Zhao YH, Wang JY. Effects of artemether on mitochondrial morphology and function in neurons. *Chin J Pharmacol Toxicol* 2003;**17**(3):196–201. [赵艳红，王京燕. 蒿甲醚对神经元细胞线粒体形态和功能的影响. 中国药理学与毒理学杂志，2003，17(3)：196-201.]

44. Kongpatanakul S, Chatsiricharoenkul S, Khuhapinant A, et al. Comparative study of dihydroartemisinin and artesunate safety in healthy Thai volunteers. *Int J Clin Pharmacol Ther* 2009;**47**(9):579–86.
45. Price R, van Vugt M, Phaipun L, et al. Adverse effects in patients with acute falciparum malaria treated with artemisinin derivatives. *Am J Trop Med Hyg* 1999;**60**(4):547–55.
46. McGready R, Cho T, Cho JJ, et al. Artemisinin derivatives in the treatment of falciparum malaria in pregnancy. *Trans R Soc Trop Med Hyg* 1998;**92**(4):430–3.
47. Li Q, Weina PJ. Severe embryotoxicity of artemisinin derivatives in experimental animals, but possibly safe in pregnant women. *Molecules* 2009;**15**(1):40–57.
48. Gordi T, Lepist EI. Artemisinin derivatives: toxic for laboratory animals, safe for humans? *Toxicol Lett* 2004;**147**(2):99–107.
49. A preliminary study of the absorption, distribution, excretion and metabolism of 3H-artemisinin and 3H-reduced artemisinin in mice. In: National Radioisotope Technical Exchange Group, editor. *Application of radioisotopes in preclinical medicine: selected papers from the 1977 conference of the national radioisotope technical exchange group*. Beijing: Atomic Energy Press; 1979. p. 144–9. [^{3}H-青蒿素和^{3}H-还原青蒿素在小鼠体内吸收、分布、排泄和代谢的初步研究.《放射性同位素在基础医学中的应用》编辑组. 放射性同位素在基础医学中的应用：1977年全国放射性同位素技术经验交流会基础医学部分论文选编；北京：原子能出版社，1979：144-149.]
50. China Cooperative Research Group on Qinghaosu and its Derivatives as Antimalarials. Metabolism and pharmacokinetics of Qinghaosu and its derivatives. *J Tradit Chin Med* 1982;**2**(1):25–30.
51. Zang QZ, Qi SB, Wan YD. The *in vivo* absorption, distribution and excretion of ^{3}H-artemisinin in animals. *Sichuan Med J* 1980;**1**(1):30–1. [臧其中，齐尚斌，万尧德. ^{3}H-青蒿素在动物体内的吸收，分布和排泄. 四川医学，1980，1(1)：30-31.]
52. Shu HL, Xu GY, Li WH, et al. A vanillin colorimetric assay method for methyl reduced artemisinin. *Anal Chem* 1982;**10**(11):678–80. [束汉麟，许帼英，李万亥，等. 甲基还原青蒿素的香蓝素比色测定法. 分析化学，1982，10(11)：678-680.]
53. Li WH, Shu HL, Xu GY, et al. Plasma protein binding of artemisinin and its derivatives. *Acta Pharm Sin* 1982;**17**(10):783–6. [李万亥，束汉麟，许国英，等. 青蒿素及其衍生物与血浆蛋白的结合. 药学学报，1982，17(10)：783-786.]
54. Niu XY, Ho LY, Ren ZH, et al. Metabolic fate of Qinghaosu in rats: a new TLC densitometric method for its determination in biological material. *Eur J Drug Metab Pharmacokinet* 1985;**10**(1):55–9.
55. Wang DR, Lin XY, Zhang HZ. A fluorescence TLC assay method for artemisinin and reduced artemisinin. *Acta Pharmacol Sin* 1987;**8**(4):355–8. [王敦瑞，林性玉，张慧珠. 青蒿素和去氧青蒿素荧光薄层扫描测定法. 中国药理学报，1987，8(4)：355-358.]
56. Song ZY, Zhao KC, Liang XT, et al. An RIA assay method for artesunate and artemisinin. *Acta Pharm Sin* 1985;**20**(8):610–4. [宋振玉，赵凯存，梁晓天，等. 青蒿酯和青蒿素的放射免疫测定法. 药学学报，1985，20(8)：610-614.]
57. Zhao KC, Chen QM, Song ZY. A study of the pharmacokinetics of artemisinin and two of its active derivatives in dogs. *Acta Pharm Sin* 1986;**21**(10):736–9. [赵凯存，陈其明，宋振玉. 青蒿素及其两个活性衍生物在狗体内药代动力学的研究. 药学学报，1986，21(10)：736-739.]
58. Zhao SS, Zeng MY. Application of precolumn reaction to high-performance liquid chromatography of Qinghaosu in animal plasma. *Anal Chem* 1986;**58**(2):289–92.
59. Zhou ZM, Huang YX, Xie GH, et al. HPLC with polarographic detection of artemisinin and its derivatives and application of the method to the pharmacokinetic study of artemether. *J Liq Chromatogr* 1988;**11**(5):1117–37.

60. Xing J, Yan HX, Zhang SQ, et al. A high-peformance liquid chromatography/tandem mass spectrometry method for the determination of artemisinin in rat plasma. *Rapid Commun Mass Spectrom* 2006;**20**(9):1463–8.
61. Yan HX, Xing J, Zhang LF, et al. Multiple-dose pharmacokinetics of artemisinin in rats by high-performance liquid chromatography-tandem mass spectrometry. In: *Annual meeting of the Chinese pharmaceutical association, and the eighth Chinese pharmacist week*. 2008.
62. Li C, Wang RL, Zhang LF, et al. A liquid chromatography-mass spectrometry assay method for artemisinin concentration in dog plasma. *Chin Remedies Clin* 2011;**11**(7):748–9. [李超，王锐利，张丽锋，等. 液相色谱-质谱联用法测定犬血浆中青蒿素浓度. 中国药物与临床，2011，11(7)：748-749.]
63. Du FY, Liu T, Shen T, et al. Qualitative-(semi) quantitative data acquisition of artemisinin and its metabolites in rat plasma using an LTQ/Orbitrap mass spectrometer. *J Mass Spectrom* 2012;**47**(2):246–52.
64. Lie L, Pabbisetty D, Carvalho P, et al. Ultra-performance liquid chromatography-tandem mass spectrometric method for the determination of artemisinin in rat serum and its application in pharmacokinetics. *J Chromatogr B* 2008;**867**(1):131–7.
65. Zhou ZM. Studies on the transdermal absorbtion of artemisinin and its antimalarial effect. In: *Combating malaria: proceedings of a UNESCO/WHO meeting of experts*. Paris: UNESCO; 1995. p. 75–83.
66. Zhou ZM, Chen YX, Li ZL, et al. A systematic study on cutaneous absorption of artemisinin. In: *Collected papers from the first national symposium on new drug preparations and the modernization of traditional Chinese medicine*. 2002. p. 212–4. [周钟鸣，陈迎雪，李泽琳，等. 青蒿素透皮吸收的系统研究. 首届全国新型给药技术与中药现代化学术研讨会论文集，2002：212-214.]
67. Li GF, Zhou Q, Zhao CW, et al. A study of the pharmacokinetics high efficiency slow-release artemisinin preparations in rabbits. *Conf Chin Pharm Assoc* 1999:860–1. [李国栋，周全，赵长文，等. 青蒿素高效缓释制剂在兔体内药代动力学研究. 1999中国药学会学术年会，1999：860-861.]
68. Wang NJ, Zhang ZH, Qiu L, et al. Methodological study of HPLC in the determination of the penetration rate of transdermal artemisinin. *China J Chin Mater Med* 2007;**32**(19):2076–7. [王乃婕，张志慧，邱琳，等. HPLC 法测定青蒿素经皮给药体外透皮速率的方法学研究. 中国中药杂志，2007，32(19)：2076-2077.]
69. Yin YJ, Lü XM, Li GD. The preparation of a self-emulsifying artemisinin formulation and its *in vivo* pharmacokinetics in rabbits. *Acad J Second Mil Med Univ* 2008;**29**(7):822–5. [殷玉娟，吕小满，李国栋. 青蒿素自乳化制剂的制备及其在家兔体内的药动学研究. 第二军医大学学报，2008，29(7)：822-825.]
70. Shen XS, Xu W, Jin MH, et al. The *in vivo* pharmacodynamics of artemisinin nanoliposomes in rats. *Lishizhen Med Mater Med Res* 2011;**22**(2):384–6. [沈雪松，徐伟，金美华，等. 青蒿素类纳米脂质体在大鼠体内的药动学研究. 时珍国医国药，2011，22(002)：384-386.]
71. Li CS, Du YL. A study on drug regimens based on the pharmacological half-life of artemisinin. *Acta Pharm Sin* 1984;**19**(6):410–4. [李成韶，杜以兰. 根据青蒿素药效半衰期制订给药方案的探讨. 药学学报，1984，19(6)：410-414.]
72. Li CS, Du YL. The half-life of the effective dose, $t_{1/2\ (ED)}$, and its calculation formula. *Acta Pharm Sin* 1986;**21**(3):165–9. [李成韶，杜以兰. 效量半衰期 t1/2(ED)及其计算公式. 药学学报，1986，21(3)：165-169.]
73. Zhao SS. High-performance liquid chromatographic determination of artemisinine (Qinghaosu) in human plasma and saliva. *Analyst* 1987;**112**(5):661–4.

74. Zhou ZM, Nie SQ, Sun XM, et al. Correlation study of the concentration of artemisinin in saliva and blood. *Chin Pharmacol* 1990;**7**(2):49. [周钟鸣，聂淑琴，孙晓淼，等. 青蒿素的唾液浓度和血液浓度的相关性研究. 中国药理通讯，1990，7(2)：49.]
75. Titulaer HAC, Zuidema J, Kager PA, et al. The pharmacokinetics of artemisinin after oral, intramuscular and rectal administration to volunteers. *J Pharm Pharmacol* 1990;**42**(11):810–3.
76. Zhao KC, Song ZY. Comparison of the pharmacokinetics of dihydroartemisinin and artemisinin in humans. *Acta Pharm Sin* 1993;**28**(5):342–6. [赵凯存，宋振玉. 双氢青蒿素在人的药代动力学及与青蒿素的比较. 药学学报，1993，28(5)：342-346.]
77. Duc DD, De Vries PJ, Nguyen X, et al. The pharmacokinetics of a single dose of artemisinin in healthy Vietnamese subjects. *Am J Trop Med Hyg* 1994;**51**(6):785.
78. Sidhu JS, Ashton M. Single-dose, comparative study of venous, capillary and salivary artemisinin concentrations in healthy, male adults. *Am J Trop Med Hyg* 1997;**56**(1):13–6.
79. Edlund PO, Westerlund D, Carlqvist J, et al. Determination of artesunate and dihydroartemisinine in plasma by liquid chromatography with post-column derivatization and UV-detection. *Acta Pharm Suec* 1984;**21**(4):223–34.
80. Ashton M, Gordi T, Hai TN, et al. Artemisinin pharmacokinetics in healthy adults after 250, 500 and 1000 mg single oral doses. *Biopharm Drug Dispos* 1998;**19**(4):245–50.
81. Benakis A, Schopfer C, Paris M, et al. Pharmacokinetics of arteether in dog. *Eur J Drug Metab Pharmacokinet* 1991;**16**(8):325–8.
82. Benakis A, Paris M, Loutan L, et al. Pharmacokinetics of artemisinin and artesunate after oral administration in healthy volunteers. *Am J Trop Med Hyg* 1997;**56**(1):17.
83. Gao HZ, Li HY, Xu SG, et al. High performance liquid chromatography-tandem mass spectrometry in an assay of artemisinin concentration in healthy human plasma and a study of its pharmacokinetics. *Chin J Clin Pharmacol* 2009;**25**(2):138–40. [高洪志，李海燕，徐树光，等. 高效液相色谱-串联质谱法测定健康人体血浆中青蒿素浓度及其药代动力学研究. 中国临床药理学杂志，2009，25(2)：138-140.]
84. Hassan Alin M, Ashton M, Kihamia C, et al. Clinical efficacy and pharmacokinetics of artemisinin monotherapy and in combination with mefloquine in patients with falciparum malaria. *Br J Clin Pharmacol* 1996;**41**(6):587–92.
85. De Vries PJ, Tran K, Nguyen XK, et al. The pharmacokinetics of a single dose of artemisinin in patients with uncomplicated falciparum malaria. *Am J Trop Med Hyg* 1997;**56**(5):503.
86. Ashton M, Sy ND, Huong NV, et al. Artemisinin kinetics and dynamics during oral and rectal treatment of uncomplicated malaria. *Clin Pharmacol Ther* 1998;**63**(4):482–93.
87. Gordi T, Huong DX, Hai TN, et al. Artemisinin pharmacokinetics and efficacy in uncomplicated-malaria patients treated with two different dosage regimens. *Antimicrob Agents Chemother* 2002;**46**(4):1026–31.
88. Augustijns P, D'Hulst A, Van Daele J, et al. Transport of artemisinin and sodium artesunate in Caco-2 intestinal epithelial cells. *J Pharm Sci* 1996;**85**(6):577–9.
89. Svensson USH, Sandström R, Carlborg Ö, et al. High in situ rat intestinal permeability of artemisinin unaffected by multiple dosing and with no evidence of P-glycoprotein involvement. *Drug Metab Dispos* 1999;**27**(2):227–32.
90. Li GQ, Guo XB, Jian HX, et al. Observations of the therapeutic efficacy of artemisinin suppositories in 100 falciparum malaria cases. *J Tradit Chin Med* 1984;**25**(5):26–8. [李国桥，郭兴伯，简华香，等. 青蒿素栓剂治疗恶性疟 100 例疗效观察. 中医杂志, 1984, 25(5):26-28.]
91. Li YQ, Xie GL, Zhang M. Effects of rectal administration of artemisinin in the treatment of vivax malaria. *Chin J Parasitol Parasit Dis* 1984;**2**(4):279. [李应庆，谢贵林，张明. 青蒿素直肠给药治疗间日疟的效果观察. 中国寄生虫学与寄生虫病杂志，1984，2(4)：279.]

92. Guo XB, Fu LC, Jian HX, et al. Clinical report on artemisinin suppositories in the treatment of 32 cases of severe malaria. *J Guangzhou Coll Tradit Chin Med* 1986;**3**(4):15–7. [郭兴伯，符林春，简华香，等. 青蒿素栓治疗凶险型疟疾 32 例临床报告. 广州中医学院学报，1986，3(4)：15-17.]
93. Zhou ZM, Zeng MY, Li ZL, et al. Pharmacokinetic studies of Qinghaosu and its derivatives by HPLC method. In: *Meeting of the WHO/SWG-CHEMAL*. April 1989. p. 24–6.
94. Zhou ZM, Zeng MY, Nie SQ, et al. Pharmacokinetics and pharmacodynamics of artemisinin in malaria patients. *Chin Pharmacol Bull* 1991;**8**(2):38. [周钟鸣，曾美怡，聂淑琴，等. 青蒿素在疟疾病人的药代动力学和药效动力学. 中国药理通讯，1991，8(2)：38.]
95. Koopmans R, Duc DD, Kager PA, et al. The pharmacokinetics of artemisinin suppositories in Vietnamese patients with malaria. *Trans R Soc Trop Med Hyg* 1998;**92**(4):434–6.
96. Gordi T, Hai TN, Hoai NM, et al. Use of saliva and capillary blood samples as substitutes for venous blood sampling in pharmacokinetic investigations of artemisinin. *Eur J Clin Pharmacol* 2000;**56**(8):561–6.
97. Zhu DY, Huang BS, Chen ZL, et al. Isolation and identification of biological metabolites and conversion products of artemisinin I. Isolation and identification in humans. *Acta Pharm Sin* 1980;**15**(7):509–12. [朱大元，黄宝山，陈仲良，等. 青蒿素生物代谢转化物的分离鉴定Ⅰ. 人体代谢转化物的分离和鉴定. 药学学报，1980，15(7)：509-512.]
98. Zhu DY, Huang BS, Chen ZL. Isolation and identification of conversion products of artemisinin metabolism in humans. *Acta Pharmacol Sin* 1983;**4**(3):194–7. [朱大元，黄宝山，陈仲良. 人体代谢青蒿素转化物的分离鉴定. 中国药理学报，1983，4(3)：194-197.]
99. Yang XW, editor. *The absorption, distribution, metabolism, excretion, toxicity and efficacy of Chinese medicinal ingredients, volume I*. Beijing: China Medical Science Press; 2006. p. 860. [杨秀伟. 中药成分的吸收、分布、代谢、排泄、毒性与药效(上). 北京：中国医药科技出版社，2006：860.]
100. Liu T, Du F, Wan Y, et al. Rapid identification of phase I and II metabolites of artemisinin antimalarials using LTQ-orbitrap hybrid mass spectrometer in combination with online hydrogen/deuterium exchange technique. *J Mass Spectrom* 2011;**46**(8):725–33.
101. Svensson US, Ashton M. Identification of the human cytochrome P450 enzymes involved in the in vitro metabolism of artemisinin. *Br J Clin Pharmacol* 1999;**48**(4):528–35.
102. Mukanganyama S, Naik YS, Widersten M, et al. Proposed reductive metabolism of artemisinin by glutathione transferases in vitro. *Free Radic Res* 2001;**35**(4):427–34.
103. Koopmans R, Ha LD, Duc DD, et al. The pharmacokinetics of artemisinin after administration of two different suppositories to healthy Vietnamese subjects. *Am J Trop Med Hyg* 1999;**60**(2):244–7.
104. Ngo TH, Quintens I, Roets E, et al. Bioavailability of different artemisinin tablet formulations in rabbit plasma — correlation with results obtained by an in vitro dissolution method. *J Pharm Biomed Anal* 1997;**16**(2):185–9.
105. Ashton M, Sy ND, Gordi T, et al. Evidence for time-dependent artemisinin kinetics in adults with uncomplicated malaria. *Pharm Pharmacol Lett* 1996;**6**(3):127–30.
106. Ashton M, Hai TN, Sy ND, et al. Artemisinin pharmacokinetics is time-dependent during repeated oral administration in healthy male adults. *Drug Metab Dispos* 1998;**26**(1):25–7.
107. Gupta S, Svensson USH, Ashton M. In vitro evidence for auto-induction of artemisinin metabolism in the rat. *Eur J Drug Metab Pharmacokinet* 2001;**26**(3):173–8.
108. Xing J, Du F, Liu T, et al. Autoinduction of phase I and phase II metabolism of artemisinin in rats. *Xenobiotica* 2012;**42**(9):929–38.
109. Asimus S, Hai TN, Van Huong N, et al. Artemisinin and CYP2A6 activity in healthy subjects. *Eur J Clin Pharmacol* 2008;**64**(3):283–92.

110. Simonsson USH, Jansson B, Hai TN, et al. Artemisinin autoinduction is caused by involvement of cytochrome P450 2B6 but not 2C9. *Clin Pharmacol Ther* 2003;**74**(1):32–43.
111. Ashton M. Quantitative in vivo and in vitro sex differences in artemisinin metabolism in rat. *Xenobiotica* 1999;**29**(2):195–204.
112. Svensson US, Alin M, Karlsson MO, et al. Population pharmacokinetic and pharmacodynamic modelling of artemisinin and mefloquine enantiomers in patients with falciparum malaria. *Eur J Clin Pharmacol* 2002;**58**(5):339–51.
113. Zhou ZM, Chen YX, Wang YL, et al. Effects of food on artemisinin metabolism. *Chin Pharmacol Bull* 1992;**9**(1):63. [周钟鸣，陈迎雪，王彦礼，等. 食物对青蒿素代谢的影响. 中国药理通讯，1992，9(1)：63.]
114. Dien TK, De Vries PJ, Khanh NX, et al. Effect of food intake on pharmacokinetics of oral artemisinin in healthy Vietnamese subjects. *Antimicrob Agents Chemother* 1997;**41**(5):1069–72.
115. Sidhu J, Ashton M, Huong N, et al. Artemisinin population pharmacokinetics in children and adults with uncomplicated falciparum malaria. *Br J Clin Pharmacol* 1998;**45**(4):347–54.
116. Gordi T, Xie R, Huong NV, et al. A semiphysiological pharmacokinetic model for artemisinin in healthy subjects incorporating autoinduction of metabolism and saturable first-pass hepatic extraction. *Br J Clin Pharmacol* 2005;**59**(2):189–98.
117. Gordi T, Xie R, Jusko WJ. Semi-mechanistic pharmacokinetic/pharmacodynamic modelling of the antimalarial effect of artemisinin. *Br J Clin Pharmacol* 2005;**60**(6):594–604.
118. Jin YT, Liu ZH, Zhou ZM, et al. ^{3}H-Labelled reduced artemisinin. *J Beijing Norm Univ (Nat Sci)* 1979;**2**:73–4. [金显泰,刘正浩,周钟鸣,等. ^{3}H-标记还原青蒿素. 北京师范大学学报(自然科学版), 1979, (2): 73–74.]
119. Yao QY, Yang XD, Zhang DR, et al. Study on the *in vivo* metabolism, absorption and distribution of dihydroartemisinin in animals. *Shandong Pharm Ind* 1985;**1**:10–6. [姚乾元，杨晓东，张典瑞，等. 双氢青蒿素在动物体内代谢和吸收分布的研究. 山东医药工业, 1985, (1): 10–16.]
120. Zhao KC, Song ZY. *In vivo* pharmacokinetics of oral dihydroartemisinin in rabbits and dogs. *Acta Pharm Sin* 1990;**25**(2):147–9. [赵凯存，宋振玉. 口服双氢青蒿素在兔和狗体内的药代动力学研究. 药学学报, 1990, 25(2): 147–149.]
121. Li R, Zhou LL, Li X, et al. The fate of sodium artesunate *in vivo* via GC/MS and other methods. *J Guangzhou Univ Tradit Chin Med* 1984;**1**(2). 12. [李锐，周莉玲，李迅，等. 应用气—质联用等法对青蒿酯钠体内命运的研究. 广州中医药大学学报, 1984, 1(2): 12.]
122. Theoharides AD, Smyth MH, Ashmore RW, et al. Determination of dihydroqinghaosu in blood by pyrolysis gas chromatography/mass spectrometry. *Anal Chem* 1988;**60**(2):115–20.
123. Chen LS, Zeng YL. TLC assay of artemether, artesunate and their dihydroartemisinin metabolite in plasma. *Chin J Pharm* 1989;**20**(2):75–8. [陈来舜，曾衍霖. 蒿甲醚，蒿琥酯及其代谢产物双氢青蒿素在血浆中的薄层扫描定量法. 中国医药工业杂志, 1989, 20(2): 75–78.]
124. Huo JC, Xia WJ, Xue M, et al. The *in vivo* pharmacokinetics of artesunate's active metabolite, dihydroartemisinin, in cattle. *Chin J Vet Sci Technol* 1990;**9**:11–2. [霍继曾，夏文江，薛明，等. 青蒿琥酯的活性代谢物双氢青蒿素在牛体内的药代动力学研究. 中国兽医科技, 1990, (9): 11–12.]
125. Li CS, Du YL, Zhang CL, et al. Pharmacodynamics of the antimalarial efficacy of dihydroartemisinin in mice. *Acta Pharm Sin* 1989;**24**(7):487–9. [李成韶，杜以兰，张翠莲，等. 双氢青蒿素对小鼠抗疟作用的药效动力学. 药学学报, 1989, 24(7): 487–489.]
126. Lee IS, Hufford CD. Metabolism of antimalarial sesquiterpene lactones. *Pharmacol Ther* 1990;**48**(3):345–55.
127. Leskovac V, Theoharides AD. Hepatic metabolism of artemisinin drugs I. Drug metabolism in rat liver microsomes. *Comp Biochem Physiol* 1991;**99**(3):383–90.

128. Ilett KF, Davis TME, Batty KT, et al. Glucuronidation of dihydroartemisinin following administration of artesunate to humans and by human liver microsomes. In: *Proceedings of the 5th international ISSX meeting, Cairns, Australia, October 25–29 No. 13*. 1998. p. 105.
129. Yang SD, Ma JM, Sun JH, et al. Determination of artesunate and dihydroartemisinin in human plasma using HPLC and reductive electrochemical polarography. *Acta Pharm Sin* 1985;**20**(6):457–62. [杨树德，马建民，孙娟华，等. 还原型电化学极谱检测高效液相色谱法测定人血浆中青蒿酯和双氢青蒿素. 药学学报, 1985, 20(6): 457–462.]
130. Zhao KC, Song ZY. Pharmacokinetics and plasma protein binding of dihydroartemisinin. *Chin Pharmacol* 1988;**5**(1):53. [赵凯存，宋振玉. 双氢青蒿素药代动力学及其与血浆蛋白结合的研究. 中国药理通讯, 1988, 5(1): 53.]
131. Le NH, Na-Bangchang K, Le TD, et al. Pharmacokinetics of a single oral dose of dihydroartemisinin in Vietnamese healthy volunteers. *SE Asian J Trop Med Public Health* 1999;**30**(1):11–6.
132. Liu YM, Zeng X, Feng Y, et al. LC-MS in the determination of dihydroartemisinin in human plasma. *Chin J Clin Pharmacol Ther* 2006;**11**(7):814–7. [刘奕明，曾星，冯怡，等.液相色谱－质谱联用法测定人血浆双氢青蒿素浓度. 中国临床药理学与治疗学, 2006, 11(7): 814–817.]
133. Liu YM, Zeng X, Deng YH, et al. Pharmacokinetics and safety of a single intravenous dihydroartemisinin drip in healthy people. *Chin J Clin Pharmacol* 2009;**25**(3):231–4. [刘奕明，曾星，邓远辉，等. 单次静滴双氢青蒿素在健康人体的药代动力学和安全性. 中国临床药理学杂志, 2009, 25(3): 231–234.]
134. Yang QC, Gan J, Liu ZM, et al. Pharmacological study of an artemisinin derivative – arteminin ether. *J Guangxi Med Coll* 1981;**3**:8–14. [杨启超，甘俊，刘忠敏，等. 青蒿素衍生物——青蒿醚的药理研究. 广西医学院学报, 1981, (3): 8–14.]
135. Jiang JY, Yan HY, Zhuang YH, et al. The absorption, distribution and excretion of artemether in mice and rats. *Acta Pharmacol Sin* 1983;**4**(3):197–201. [姜纪荣，严汉英，庄怡华，等. 蒿甲醚在小鼠和大鼠体内的吸收、分布和排泄. 中国药理学报, 1983, 4(3): 197–201.]
136. Zhang YD, Wang CG, Xu GY, et al. TLC quantitative assay for the methyl reduction of artemisinin (artemether) in plasma. *Acta Pharm Sin* 1982;**17**(3):212–7. [张银娣，王长根，许帼英，等. 血浆中甲基还原青蒿素(蒿甲醚)的薄层扫描定量法. 药学学报, 1982, 17(3): 212–217.]
137. Zeng YL, Zhang YD, Xu GY, et al. *In vivo* pharmacokinetics of artemether in rabbits. *Acta Pharm Sin* 1984;**19**(2):31–4. [曾衍霖，张银娣，徐帼英，等. 蒿甲醚在兔体内的药代动力学.药学学报, 1984, 19(2): 31–34.]
138. Zou CD. Rapid absorption of artemether in rats and its *in vivo* distribution. *J Huaqiao Univ (Nat Sci)* 1996;**17**(3):294–6. [邹崇达.大白鼠对蒿甲醚的快速吸收及其在体内的分布.华侨大学学报(自然科学版), 1996, 17(3): 294–296.]
139. Huang YX, Xie GH, Zhou ZM, et al. Determination of artemether in plasma and whole blood using HPLC with flow-through polarographic detection. *Biomed Chromatogr* 1987;**2**(2):53–6.
140. Thomas CG, Ward SA, Edwards G. Selective determination, in plasma, of artemether and its major metabolite, dihydroartemisinin, by high-performance liquid chromatography with ultraviolet detection. *J Chromatogr* 1992;**583**(1):131–6.
141. Muhia DK, Mberu EK, Watkins WM. Differential extraction of artemether and its metabolite dihydroartemisinin from plasma and determination by high-performance liquid chromatography. *J Chromatogr B Biomed Sci Appl* 1994;**660**(1):196–9.
142. Zhang ZR, Hu HY, Hong Z, et al. *In vivo* pharmacokinetics of delayed-release artemether tablets in dogs. *West China J Pharm Sci* 2000;**15**(5):335–7. [张志荣，胡海燕，洪诤，等. 蒿甲醚缓释片在狗体内的药代动力学. 华西药学杂志, 2000, 15(5): 335–337.]

143. Bai KH, Xing J, Wang RL, et al. Time dependence in multiple-dose *in vivo* pharmacokinetics of artemether in rats. In: *Ninth national academic conference on drug and xenobiotics metabolism*. 2009. p. 249. [白克华，邢杰，王锐利，等. 蒿甲醚在大鼠体内多剂量药动学的时间依赖性. 第九届全国药物和化学异物代谢学术会议论文集, 2009: 249.]
144. Xing J, Bai KH, Liu T, et al. The multiple-dosing pharmacokinetics of artemether, artesunate, and their metabolite dihydroartemisinin in rats. *Xenobiotica* 2011;**41**(3):252–8.
145. Mohamed SS, Khalid SA, Ward SA, et al. Simultaneous determination of artemether and its major metabolite dihydroartemisinin in plasma by gas chromatography-mass spectrometry-selected ion monitoring. *J Chromatogr B Biomed Sci Appl* 1999;**731**(2):251–69.
146. Liu P, Hou YN, Shan CQ, et al. HPLC-MS-MS in the determination of artemether and its metabolite dihydroartemisinin in the plasma of beagles. *Chin J Pharm Anal* 2006;**26**(5):585–8. [刘萍，侯禹男，单成启，等. HPLC-MS-MS法测定比格犬血浆中蒿甲醚及其代谢物双氢青蒿素的浓度. 药物分析杂志, 2006, 26(5): 585–588.]
147. Yang HS, Ye ZG, Liang BW, et al. Pharmacokinetics of artemether patches in rats. *China J Chin Mater Med* 2008;**33**(12):1459–62. [杨华生，叶祖光，梁秉文，等. 蒿甲醚贴剂小鼠体内药代动力学研究. 中国中药杂志, 2008, 33(12): 1459–1462.]
148. He JY, Hou SX, Cai Z, et al. Effects of commonly used Chinese medicine penetration enhancers on the transdermal properties of artemether. *China J Chin Mater Med* 2008;**33**(10):1130–2. [何俊瑶，侯世祥，蔡铮，等. 几种常用中药促渗剂对蒿甲醚体外透皮性能的影响. 中国中药杂志, 2008, 33(10): 1130–1132.]
149. Hu HY, Zhang ZR. Study on the *in vitro* release of artemether sustained-release tablets. *Chin J Synth Chem* 1999;**7**(4):334–9. [胡海燕，张志荣. 蒿甲醚缓释片体外释放研究. 合成化学, 1999, 7(4): 334–339.]
150. Shah PP, Mashru RC. Development and evaluation of artemether taste masked rapid disintegrating tablets with improved dissolution using solid dispersion technique. *AAPS PharmSciTech* 2008;**9**(2):494–500.
151. Mandawgade SD, Sharma S, Pathak S, et al. Development of SMEDDS using natural lipophile: application to beta-artemether delivery. *Int J Pharm* 2008;**362**(1–2):179–83.
152. Tayade NG, Nagarsenker MS. Development and evaluation of artemether parenteral microemulsion. *Indian J Pharm Sci* 2010;**72**(5):637–40.
153. Chen LS, Peng N, Yang YM, et al. Biotransformation of artemether in rat liver microsomes. *Chin J Pharmacol Toxicol* 1993;**7**(3):205–9. [陈来舜，彭宁，杨一鸣，等. 蒿甲醚在大鼠肝微粒体中的生物转化. 中国药理学与毒理学杂志, 1993, 7(3): 205–209.]
154. Chen LS, Wang ZJ, Zeng YL. Demethylation metabolism of artemether with rat liver microsome enzymes and effects of deuterium isotopes. *Chin J Pharmacol Toxicol* 1991;**5**(1):53–5. [陈来舜，王再杰，曾衍霖. 蒿甲醚经大鼠肝微粒体酶脱甲基代谢和氘代同位素效应. 中国药理学与毒理学杂志, 1991, 5(1): 53–55.]
155. Chen YG, Yu BY. Pharmacokinetics of artemether in simulated *in vivo* acidic and alkalinic environment. *Chin J Chromatogr* 2002;**20**(1):37–9. [陈有根，余伯阳. 蒿甲醚在模拟体内酸碱环境中的代谢动力学研究. 色谱, 2002, 20(1): 37–39.]
156. Liu T, Du F, Zhu F, et al. Metabolite identification of artemether by data-dependent accurate mass spectrometric analysis using an LTQ-Orbitrap hybrid mass spectrometer in combination with the online hydrogen/deuterium exchange technique. *Rapid Commun Mass Spectrom* 2011;**25**(21):3303–13.
157. Maggs JL, Bishop LP, Edwards G, et al. Biliary metabolites of beta-artemether in rats: biotransformations of an antimalarial endoperoxide. *Drug Metab Dispos* 2000;**28**(2):209–17.
158. Lin BL, Chen ZX, Li ZQ, et al. Phase I clinical pharmacokinetics trial on artemether. *J Guangzhou Coll Tradit Chin Med* 1987;**4**(3):48–9. [林炳鎏，陈芝喜，李志强，等. 蒿甲醚I期临床药代动力学的研究. 广州中医学院学报, 1987, 4(3): 48–49.]

159. Hanpithakpong W, Kamanikom B, Sinqhasivanon P, et al. A liquid chromatographic-tandem mass spectrometric method for determination of artemether and its metabolite dihydroartemisinin in human plasma. *Bioanalysis* 2009;**1**(1):37–46.
160. Huang L, Jayewardene AL, Li X, et al. Development and validation of a high-performance liquid chromatography/tandem mass spectrometry method for the determination of artemether and its active metabolite dihydroartemisinin in human plasma. *J Pharm Biomed Anal* 2009;**50**(5):959–65.
161. Mordi MN, Mansor SM, Navaratnam V, et al. Single dose pharmacokinetics of oral artemether in healthy Malaysian volunteers. *Br J Clin Pharmacol* 1997;**43**(4):363–5.
162. Na-Bangchang K, Karbwang J, Thomas CG, et al. Pharmacokinetics of artemether after oral administration to healthy Thai males and patients with acute, uncomplicated falciparum malaria. *Br J Clin Pharmacol* 1994;**37**(3):249–53.
163. Hien TT, Davis TME, Chuong LV, et al. Comparative pharmacokinetics of intramuscular artesunate and artemether in patients with severe falciparum malaria. *Antimicrob Agents Chemother* 2004;**48**(11):4234–9.
164. Karbwang J, Na-Bangchang K, Tin T, et al. Pharmacokinetics of intramuscular artemether in patients with severe falciparum malaria with or without acute renal failure. *Br J Clin Pharmacol* 1998;**45**(6):597–600.
165. Manning L, Laman M, Paqe-Sharp M, et al. Meningeal inflammation increases artemether concentrations in cerebrospinal fluid in Papua new guinean children treated with intramuscular artemether. *Antimicrob Agents Chemother* 2011;**55**(11):5027–33.
166. Teja-Isavadharm P, Nosten F, Kyle DE, et al. Comparative bioavailability of oral, rectal, and intramuscular artemether in healthy subjects: use of simultaneous measurement by high performance liquid chromatography and bioassay. *Br J Clin Pharmacol* 1996;**42**(5):599–604.
167. Karbwang J, Na-Bangchang K, Congpuong K, et al. Pharmacokinetics and bioavailability of oral and intramuscular artemether. *Eur J Clin Pharmacol* 1997;**52**(4):307–10.
168. Van Agtmael MA, Van Der Graaf CA, Dien TK, et al. The contribution of the enzymes CYP2D6 and CYP2C19 in the demethylation of artemether in healthy subjects. *Eur J Drug Metab Pharmacokinet* 1998;**23**(3):429–36.
169. Mithwani S, Aarons L, Kokwaro G, et al. Population pharmacokinetics of artemether and dihydroartemisinin following single intramuscular dosing of artemether in African children with severe falciparum malaria. *Br J Clin Pharmacol* 1986;**52**(2):146–52.
170. Tarning J, Kloprogge F, Piola P, et al. Population pharmacokinetics of artemether and dihydroartemisinin in pregnant women with uncomplicated plasmodium falciparum malaria in Uganda. *Malar J* 2012;**11**(1):293–305.
171. Na-Bangchang K, Karbwang J, Molunto P, et al. Pharmacokinetics of mefloquine, when given alone and in combination with artemether, in patients with uncomplicated falciparum malaria. *Fundam Clin Pharmacol* 1995;**9**(6):576–82.
172. Tan-ariya P, Na-Bangchang K, Ubalee R, et al. Pharmacokinetic interactions of artemether and pyrimethamine in healthy male Thais. *SE Asian J Trop Med Public Health* 1998;**29**(1):18–23.
173. Sukhija M, Medhi B, Pandhi P. Effects of artemisinin, artemether, arteether on the pharmacokinetics of carbamazepine. *Pharmacology* 2006;**76**(3):110–6.
174. Van Agtmael MA, Gupta V, van der Graaf CA, et al. The effect of grapefruit juice on the time-dependent decline of artemether plasma levels in healthy subjects. *Clin Pharmacol Ther* 1999;**66**(4):408–14.
175. Van Agtmael MA, Gupta V, van der Wösten TH, et al. Grapefruit juice increases the bioavailability of artemether. *Eur J Clin Pharmacol* 1999;**55**(5):405–10.

176. Li R, Liao ZY, Huang GY, et al. *In vivo* pharmacokinetics of artesunate. *Chin Tradit Herb Drugs* 1981;**12**(5):20–2. [李锐，廖灶引，黄桂英，等. 青蒿酯钠体内动力学研究. 中草药, 1981, 12(5): 20–22.]
177. Li R, Zhou LL, Li X, et al. GC-MS in the study of the *in vivo* fate of sodium artesunate. *Acta Pharm Sin* 1985;**20**(7):485–90. [李锐，周莉玲，李迅，等. 应用气一质联用等法对青蒿酯钠体内命运的研究. 药学学报, 1985, 20(7): 485–490.]
178. Zhao KC, Xuan WY, Zhao Y, et al. *In vivo* pharmacokinetics of topical artesunate in mice and rabbits. *Acta Pharm Sin* 1989;**24**(11):813–6. [赵凯存，宣文漪，赵一，等. 青蒿琥酯皮肤擦剂在小鼠和兔体内的药代动力学研究. 药学学报, 1989, 24(11): 813–816.]
179. Chen ZX, Lin BL, Li GQ, et al. *In vivo* pharmacokinetics of artesunate tablets in dogs. *J Radioimmunol* 1996;**9**(2):77–9. [陈芝喜，林炳流，李国桥，等. 青蒿琥酯片剂在狗体内药代动力学. 放射免疫学杂志, 1996, 9(2): 77–79.]
180. Huo JC, Xia WJ, Xue M. Determination of artesunate and dihydroartemisinin using rapid TLC. *J Tradit Chin Vet Med* 1990;**2**:7–8. [霍继曾，夏文江，薛明. 青蒿琥酯和双氢青蒿素的快速薄层扫描测定. 中兽医医药杂志, 1990, 2: 7–8.]
181. Huo JC, Xia WJ, Xue M, et al. *In vivo* pharmacokinetics and metabolism of artesunate in cows. *Acta Pharm Sin* 1991;**26**(3):225–7. [霍继曾，夏文江，薛明，等. 青蒿琥酯在奶牛体内的药代动力学与代谢. 药学学报, 1991, 26(3): 225–227.]
182. Zhou ZM, Anders JC, Chung H, et al. Analysis of artesunic acid and dihydroqinghaosu in blood by high-performance liquid chromatography with reductive electrochemical detection. *J Chromatogr* 1987;**414**(1):77–90.
183. Yang SD, Ma JM, Sun JH, et al. Clinical pharmacokinetics of artesunate. *Chin J Clin Pharmacol* 1985;**1**(2):102–9. [杨树德，马建民，孙娟华，等. 青蒿酯的临床药代动力学. 中国临床药理学杂志, 1985, 1(2): 102–109.]
184. Chen ZX, Lin BL, Guo XB, et al. Phase I clinical study on the pharmacokinetics of sodium artesunate. *J Guangzhou Coll Tradit Chin Med* 1987;**4**(1):40–2. [陈芝喜，林炳流，郭兴伯，等. 青蒿酯钠I期临床药代动力学的研究. 广州中医学院学报, 1987, 4(1): 40–42.]
185. Zhao KC, Chen ZX, Lin BL, et al. Phase I clinical study on the pharmacokinetics of artesunate and artemether. *Chin J Clin Pharmacol* 1988;**4**(2):76–81.
186. Lin BL, Chen ZX, Jia KL, et al. Bioavailability of oral artesunate tablets in healthy volunteers. *J Guangzhou Coll Tradit Chin Med* 1990;**7**(3):163–6. [林炳流，陈芝喜，贾可亮，等. 健康志愿者口服青蒿琥酯片剂的生物利用度. 广州中医学院学报, 1990, 7(3):163–166.]
187. Morris CA, Duparc S, Borghini-Fuhrer I, et al. Review of the clinical pharmacokinetics of artesunate and its active metabolite dihydroartemisinin following intravenous, intramuscular, oral or rectal administration. *Malar J* 2011;**10**:263.
188. Navaratnam V, Ramanathan S, Wahab MSA, et al. Tolerability and pharmacokinetics of non-fixed and fixed combinations of artesunate and amodiaquine in Malaysian healthy normal volunteers. *Eur J Clin Pharmacol* 2009;**65**(8):809–21.
189. Batty KT, Le AT, Ilett KF, et al. A pharmacokinetics and pharmacodynamics study of artesunate for vivax malaria. *Am J Trop Med Hyg* 1998;**59**(5):823–7.
190. Bethell DB, Teja-Isavadharm P, Phuong CXT, et al. Pharmacokinetics of oral artesunate in children with moderately severe plasmodium falciparum malaria. *Trans R Soc Trop Med Hyg* 1997;**91**(2):195–8.
191. Adjei GO, Kristensen K, Goka BQ, et al. Effect of concomitant artesunate administration and cytochrome P4502C8 polymorphisms on the pharmacokinetics of amodiaquine in Ghanaian children with uncomplicated malaria. *Antimicrob Agents Chemother* 2008;**52**(12):4400–6.
192. Mwesigwa J, Parikh S, McGee B, et al. Pharmacokinetics of artemether-lumefantrine and artesunate-amodiaquine in children in Kampala, Uganda. *Antimicrob Agents Chemother* 2010;**54**(1):52–9.

193. Onyamboko MA, Meshnick SR, Fleckenstein L, et al. Pharmacokinetics and pharmacodynamics of artesunate and dihydroartemisinin following oral treatment in pregnant women with asymptomatic plasmodium falciparum infections in Kinshasa DRC. *Malar J* 2011;**10**:49.
194. Tan B, Naik H, Jang IJ, et al. Population pharmacokinetics of artesunate and dihydroartemisinin following single- and multiple-dosing of oral artesunate in healthy subjects. *Malar J* 2009;**8**(1):304.
195. Li ZL. The application of nuclear technology to the study of the action mechanisms of the new antimalarials artemisinin and dihydroartemisinin. *Nucl Sci Tech* 1984;**2**:23–5. [李泽琳. 核技术在抗疟新药青蒿素、二氢青蒿素抗疟作用研究中的应用. 核技术, 1984, 2: 23–25.]
196. Ellis DS, Li ZL, Gu HM, et al. The chemotherapy of rodent malaria, XXXIX. Ultrastructural changes following treatment with artemisinine of *Plasmodium berghei* infection in mice, with observations of the localization of [^{3}H]-dihydroartemisinine in *P. falciparum in vitro*. *Ann Trop Med Parasitol* 1985;**79**(4):367–74.
197. Peters W, Li ZL, Robinson BL, et al. The chemotherapy of rodent malaria, XL. The action of artemisinin and related sesquiterpenes. *Ann Trop Med Parasitol* 1986;**80**(5):483–9.
198. Ye ZG, Li ZL, Fu XQ, et al. The ultrastructure of *in vitro* cultured *Plasmodium falciparum*. *Acta Microbiol Sin* 1983;**23**(20):175–8. [叶祖光，李泽琳，傅湘琦，等. 体外培养的恶性疟原虫的超微结构. 微生物学报, 1983, 23(20): 175–178.]
199. Ye ZG, Li ZL, Fu XQ, et al. Effects of artemisinin and chloroquine on the ultrastructure of *in vitro* cultured erythrocytic *Plasmodium falciparum*. *J Tradit Chin Med* 1982;**4**:65–7. [叶祖光，李泽琳，傅湘琦，等. 青蒿素和氯喹对体外培养的红内期恶性疟原虫超微结构的影响. 中医杂志, 1982, (4): 65–67.]
200. Ye Z, Li Z, Li G, et al. Effects of Qinghaosu and chloroquine on the ultrastructure of the erythrocytic stage of *P. falciparum* in continuous cultivation *in vitro*. *J Tradit Chin Med* 1983;**3**(2):95–102.
201. Ye ZG, Li ZL, Li GQ, et al. Effects of artemisinin on erythrocytes infected with *P. berghei* and the surface structure of free parasites. *Chin J Parasitol Parasit Dis* 1986;**4**(4):260–1. [叶祖光，李泽琳，李桂琴，等. 青蒿素对感染伯氏疟原虫红细胞及游离疟原虫表面结构的影响. 中国寄生虫学与寄生虫病杂志, 1986, 4(4): 260–261.]
202. Jiang J, Jacobs J, Liang D, et al. Qinghaosu-induced changes in the morphology of *Plasmodium inui*. *Am J Trop Med Hyg* 1985;**34**(3):424–8.
203. Li R. *Effects of sodium artesunate on cytochrome oxidase and T lymphocytes*. Collected Data. Guangzhou: Guangzhou College of Traditional Chinese Medicine; 1983. p. 46. [李锐. 青蒿琥酯钠对细胞色素氧化酶和T淋巴红细胞的影响. 科研资料汇编. 广州中医学院, 1983: 46.]
204. Ellis DS, Li ZL, Robinson BL, et al. Preliminary EM studies of the effect of arteannuin on *Plasmodium berghei* in mice. *Trans R Soc Trop Med Hyg* 1981;**75**(4):600.
205. Chen LY, Hu M, Zhang QL. Effects of artemether on the ultrastructure of *in vivo P. berghei* in mice. *Chin J Parasitol Parasit Dis* 1988;**S1**:18–9. [陈乐义，胡明，张起麟. 蒿甲醚对小鼠体内伯氏疟原虫超微结构的影响. 中国寄生虫学与寄生虫病杂志, 1988, S1: 18–19.]
206. Zhao Y. Study on the action mechanism of artesunate's antimalarial effects on *P. berghei*. *Sichuan J Physiol Sci* 1988;**4**(S3):02–3. [赵一. 青蒿琥酯(Artesunate)对伯氏鼠疟(*P.berghei*)抗疟作用机理的研究. 四川生理科学杂志, 1988, 4(S3): 02–03.]
207. Maeno Y, Toyoshima T, Fujioka H, et al. Morphologic effects of artemisinin in *Plasmodium falciparum*. *Am J Trop Med Hyg* 1993;**49**(4):485–91.
208. Chen PQ, Yuan J, Du QY, et al. Effects of dihydroartemisinin on fine structure of erythrocytic stages of *Plasmodium berghei* ANKA strain. *Acta Pharmacol Sin (Engl Ed)* 2000;**21**(3):234–8. [陈沛泉，袁捷，杜巧云，等. 双氢青蒿素对伯氏疟原虫ANKA株超微结构的影响(英文). Acta Pharmacologica Sinica，2000，21(3)：234-238.]

209. Zhang KH, Li ZL. Concentrating Effects of 3H-artemisinin in in vitro cultivated *Plasmodium falciparum*. *Tradit Chin Drug Res Clin Pharmacol* 1992;**3**(4):19–21. [张奎汉，李泽琳. ^{3}H-青蒿素在体外培养人恶性疟原虫内的浓集作用. 中药新药与临床药理, 1992, 3(4): 19–21.]
210. Polet H. Uptake of chloroquine-3H by *Plasmodium knowlesi in vitro*. *J Pharm Exp Ther* 1969;**168**(1):187–92.
211. Sherman IW. Biochemistry of *Plasmodium* (malaria parasites). *Microbiol Rev* 1979;**43**(4):452–95.
212. Lawrence CW. Lipid content of *Plasmodium berghei*-infected rat red blood cells. *Exp Parasitol* 1969;**26**(2):181–6.
213. Desjardins RE, Canfield CJ, Haynes JD, et al. Quantitative assessment of antimalarial activity *in vitro* by a semiautomated microdilution technique. *Antimicrob Agents Chemother* 1979;**16**(6):710–8.
214. Xu XX, Zhu SH, Lin QS. Pharmacology research into artemisinin. *Guangdong Med J* 1986;**7**(4):25–8. [徐旭祥，朱师晦，林全胜. 青蒿素的药理研究. 广东医学, 1986, 7(4): 25–28.]
215. Gu HM, Warhurst DC, Peters W. Rapid action of Qinghaosu and related drugs on incorporation of 3H-isoleucine by *Plasmodium falciparum in vitro*. *Biochem Pharmacol* 1983;**132**(17):2463–6.
216. Ye ZG, Van Dyke K, Wimmer M. Effect of artemisinin and chloroquine on drug-sensitive and drug-resistant strains of *P. falciparum* malaria. *Exp Parasitol* 1987;**64**(3):418–23.
217. Zhao Y. Experimental studies on the effects of artesunate sustained release tablets on *Plasmodium cynomolgi*. *Chem Pharm Bull* 1987;**35**(4):2052–61.
218. Ye ZG, Doak C, Van Dyke K. Selective effect of Qinghaosu on different stages of *Plasmodium falciparum in vitro*. *J Chin Pharm Sci* 1993;**2**(1):64–8.
219. Gu HM, Zhu MY, Xi GL, et al. Effects of 13 compounds on globinase activity and free amino acids in *Plasmodium berghei*. *Acta Pharmacol Sin* 1987;**8**(5):460–4. [顾浩明，朱梅英，奚国良，等. 13个化合物对伯氏疟原虫珠蛋白酶活性及游离氨基酸活性的影响. 中国药理学报, 1987, 8(5): 460–464.]
220. Huang GJ, Zhao Y. Effect of sodium artesunate on the respiration, glucose metabolism and protein metabolism of *Plasmodium berghei*. *J Tradit Chin Med* 1981;**10**:66–70. [黄国钧, 赵一. 还原青蒿素琥珀酸钠对鼠疟原虫呼吸、糖代谢及蛋白质代谢的影响. 中医杂志, 1981, 10: 66–70.]
221. Warhurst DC, Thomas SC. The chemotherapy of rodent malaria, XXXI. The effect of some metabolic inhibitors upon chloroquine-induced pigment clumping (CIPC) in *Plasmodium berghei*. *Ann Trop Med Parasitol* 1978;**72**(3):203–11.
222. Ye ZG, Li ZL. Effects of artemisinin on chloroquine-induced pigment clumping. *J Tradit Chin Med* 1981;**4**:62–6. [叶祖光，李泽琳. 青蒿素对氯喹引起的色素凝集的影响. 中医杂志, 1981, 4: 62–66.]
223. Zhao Y, Li AY, Zhou F. Study on the action mechanism of artemisinin and increasing its efficacy. *J Tradit Chin Med* 1979;**11**:58–60. [赵一，李爱媛，周芳. 青蒿素抗疟作用原理与增效的探讨. 中医杂志, 1979, 11: 58–60.]
224. Jin YG, Teng WH, Sun CP. Radicals and the toxicity mechanism of sodium artesunate. *Chin J Pharmacol Toxicol* 1989;**3**(2):138–42. [靳永刚，滕翕和，孙存普. 自由基与青蒿琥酯毒性机理. 中国药理学与毒理学杂志, 1989, 3(2): 138–142.]
225. Kamchonwongpaisan S, Meshnick SR. The mode of action of the antimalarial artemisinin and its derivatives. *Gen Pharmacol* 1996;**27**(4):587–92.
226. Xu J, Zheng Y, Zhang R, et al. Study on the action mechanism of artemisinin drugs. *Prog Pharm Sci* 2002;**26**(5):274–7. [徐进，郑莹，张睿，等. 青蒿素类药物作用机制的探讨. 药学进展, 2002, 26(5): 274–277.]

227. Li Y, Wu YL. A golden Phoenix arising from the herbal nest - a review and reflection on the study of antimalarial drug Qinghaosu. *Front Chem China* 2010;**5**(4):357–422.
228. Lin FB, Pan HZ. Antimalarial mechanism of sodium artesunate. *Acta Acad Med Sin* 1989;**3**:180–4. [蔺福宝，潘华珍. 青蒿酯钠的抗疟机理. 中国医学科学院学报, 1989, (3): 180–184.]
229. Posner GH, Oh CH. A regiospecifically oxygen-18-labeled 1,2,4-trioxane: a simple chemical model system to probe the mechanism(s) for the antimalarial activity of artemisinin (Qinghaosu). *J Am Chem Soc* 1992;**114**(21):8328–9.
230. Posner GH, Park SB, Gonzalez L, et al. Evidence for the importance of high-valent Fe=O and of a diketone in the molecular mechanism of action of antimalarial trioxane analogs of artemisinin. *J Am Chem Soc* 1996;**118**:3537–8.
231. Cumming JN, Posner GH. Antimalarial activity of artemisinin (Qinghaosu) and related trioanes: mechanism(s) of action. *Adv Pharmacol (San Diago)* 1997;**37**:253–97.
232. Jefford CW, Posner GH. Why artemisinin and certain synthetic peroxides are potent antimalarials—implications for the mode of action. *Curr Med Chem* 2001;**8**(15):1803–26.
233. Olliaro PL, Haynes RK, Meunier B, et al. Possible modes of action of the artemisinin-type compounds. *Trends Parasitol* 2001;**17**:122–6.
234. Robert A, Dechycabaret O, Cazelles J, et al. From mechanistic studies on artemisinin derivatives to new modular antimalarial drugs. *Acc Chem Res* 2002;**35**:167–74.
235. Wu WM, Yao ZJ, Wu YL, et al. Ferrous ion induced cleavage of the peroxy bond in Qinghaosu and its derivatives and the DNA damage associated with this process. *J Chem Soc Chem Commun* 1996;**18**:2213–4.
236. Wu WM, Wu YK, Wu YL, et al. Unified mechanistic framework for the Fe(II)-induced cleavage of Qinghaosu and derivatives/analogues. The first spin-trapping evidence for the previously postulated secondary C-4 radical. *J Am Chem Soc* 1998;**120**(14):3316–25.
237. Butler AR, Gilbert BC, Hulme P, et al. EPR evidence for the involvement of free radicals in the iron-catalysed decomposition of Qinghaosu (artemisinin) and some derivatives: antimalarial action of some polycyclic endoperoxides. *Free Radic Res* 1998;**28**:471–6.
238. Wang DY, Wu YL, Wu YK, et al. Further evidence for the participation of primary carbon-centered free-radicals in the antimalarial action of the Qinghaosu (artemisinin) series of compounds. *J Chem Soc Perkin Trans* 2001;**1**:605–9.
239. Wang DY, Wu YK, Wu YL, et al. Synthesis, iron(II)-induced cleavage and *in vivo* antimalarial efficacy of 10-(2-hydroxy-1-naphthyl) – dexoxqinghaosu (-deoxoartemisinin). *J Chem Soc Perkin Trans* 1999;**1**:1827–31.
240. Wu YL, Chen HB, Jiang K, et al. Interaction of biomolecules with Qinghaosu (artemisinin) and its derivatives in the presence of ferrous ion — an exploration of antimalarial mechanism. *Pure Appl Chem* 1999;**71**:1139–42.
241. Wang DY, Wu YL. A possible antimalarial action mode of Qinghaosu (artemisinin) series compounds. Alkylation of reduced glutathione by C-centered primary radicals produced from antimalarial compound Qinghaosu and 12-(2-4-dimethoxyphenyl)-12-deoxoqinghaosu. *Chem Commun* 2000;**22**:2193–4.
242. Wu YK, Yue ZY, Wu YL. Interaction of Qinghaosu (artemisinin) with cysteine sulfhydryl mediated by traces of non-heme iron. *Angew Chem Int Ed* 1999;**38**:2580–2.
243. Wu WM, Chen YL, Zhai ZL, et al. Study on the mechanism of action of artemether against schistosomes – the identification of cysteine adducts of both carbon-centered free-radicals derived from artemether. *Biorgan Med Chem Lett* 2003;**13**:1645–7.
244. Robert A, Dechycabaret O, Cazelles J, et al. Recent advances in malaria chemotherapy. *J Chin Chem Soc* 2002;**49**(3):301–10.

245. Robert A, Bonduelle C, Laurent SAL, Meunier B. Heme alkylation by artemisinin and trioxaquines. *J Phys Org Chem* 2006;**19**:562–9.
246. Robert A, Cazelles J, Meunier B. Characterization of the alkylation product of heme by the antimalarial drug artemisinin. *Angew Chem Int Ed* 2001;**40**:1954–7.
247. Robert A, Coppel Y, Meunier B. Alkylation of heme by the antimalarial drug artemisinin. *Chem Commun* 2002;**5**:414–5.
248. Haynes RK, Monti D, Taramelli D, et al. Artemisinin antimalarials do not inhibit hemozoin formation. *Antimicrob Agents Chemother* 2003;**47**:1175.
249. Kannan R, Sahal D, Chauhan VS. Heme-artemisinin adducts are crucial mediators of the ability of artemisinin to inhibit heme polymerization. *Chem Biol* 2002;**9**:321–32.
250. Wu YK. How might Qinghaosu (artemisinin) and related-compounds kill the intraerythrocytic malaria parasite — a chemist's view. *Acc Chem Res* 2002;**35**:255–9.
251. Bhisutthibhan J, Pan XQ, Hossler PA, et al. The *Plasmodium falciparum* translationally controlled tumor protein homolog and its reaction with the antimalarial drug artemisinin. *J Biol Chem* 1998;**273**(26):16192–8.
252. Eckstein-Ludwig U, Webb R, van Goethem ID, et al. Artemisinins target the SERCA of *Plasmodium falciparum*. *Nature* 2003;**424**(6951):957–61.
253. Uhlemann AC, Cameron A, Eckstein-Ludwig U, et al. A single amino acid residue can determine the sensitivity of SERCAs to artemisinins. *Nat Struct Mol Biol* 2005;**12**(7):628–9.
254. Jambou R, Legrand E, Niang M, et al. Resistance of *Plasmodium falciparum* field isolates to in-vitro artemether and point mutations of the SERCA-type PfATPase6. *Lancet* 2005;**366**(9501):1960–3.
255. Garah FBE, Stigliani JL, Coslédan F, et al. Docking studies of structurally diverse antimalarial drugs targeting PfATP6: no correlation between in silico binding affinity and *in vitro* antimalarial activity. *ChemMedChem* 2009;**4**(9):1469–79.
256. Cojean S, Hubert V, Le Bras J, et al. Resistance to dihydroartemisinin. *Emerg Infect Dis* 2006;**12**(11):1798.
257. Afonso A, Hunt P, Cheesman S, et al. Malaria parasites can develop stable resistance to artemisinin but lack mutations in candidate genes ATP6 (encoding the sarcoplasmic and endoplasmic reticulum Ca^{2+} ATPase), TCTP, MDR1, and CG10. *Antimicrob Agents Chemother* 2006;**50**(2):480–9.
258. Valderramos SG, Scanfeld D, Uhlemann AC, et al. Investigations into the role of the *Plasmodium falciparum* SERCA (PfATP6) L263E mutation in artemisinin action and resistance. *Antimicrob Agents Chemother* 2010;**54**(9):3842–52.
259. Cui L, Wang Z, Jiang H, et al. Lack of association of the S769N mutation in *Plasmodium falciparum* SERCA (PfATP6) with resistance to artemisinins. *Antimicrob Agents Chemother* 2012;**56**(5):2546–52.
260. Cardi D, Pozza A, Arnou B, et al. Purified E255L mutant SERCA1a and purified PfATP6 are sensitive to SERCA-type inhibitors but insensitive to artemisinins. *J Biol Chem* 2010;**285**(34):26406–16.
261. Huang Q, Zhou B. Exploring artemisinins protein targets in yeast. *J Tsinghua Univ (Sci Tech)* 2008;**48**(3):411–3.
262. Arnou B, Montigny C, Morth JP, et al. The *Plasmodium falciparum* Ca^{2+}-ATPase PfATP6: insensitive to artemisinin, but a potential drug target. *Biochem Soc Trans* 2011;**39**(3):pp.823–831. http://dx.doi.org/10.1042/BST0390823.
263. Kamugisha E, Jing S, Minde M, et al. Efficacy of artemether-lumefantrine in treatment of malaria among under-fives and prevalence of drug resistance markers in Igombe-Mwanza, North-western Tanzania. *Malar J* 2012;**11**(1):58.

264. Duraisingh MT, Roper C, Walliker D, et al. Increased sensitivity to the antimalarials mefloquine and artemisinin is conferred by mutations in the Pfmdr1 gene of *Plasmodium falciparum*. *Mol Microbiol* 2000;**36**(4):955–61.
265. Saha P, Guha SK, Das S, et al. Comparative efficacy of artemisinin combination therapies (ACTs) in *P. falciparum* malaria and polymorphism of PfATPase6, Pfcrt, Pfdhfr and Pfdhps genes in tea gardens of Jalpaiguri district, India. *Antimicrob Agents Chemother* 2012;**56**(5):2511–7. http://dx.doi.org/10.1128/AAC. 05388-11.
266. Wang JG, Zhang CJ, Chia WN, et al. Haem-activated promiscuous targeting of artemisinin in *Plasmodium falciparum*. *Nat Commun* 2015;**6**:10111. http://dx.doi.org/10.1038/ncomms 10111.
267. Srivastava IK, Rottenberg H, Vaidya AB. Atovaquone, a broad spectrum antiparasitic drug, collapses mitochondrial membrane potential in a malarial parasite. *J Biol Chem* 1997;**272**(7):3961–6.
268. Li W, Mo W, Shen D, et al. Yeast model uncovers dual roles of mitochondria in action of artemisinin. *PLoS Genet* 2005;**1**:e36. http://dx.doi.org/10.1371/journal.pgen. 0010036.
269. Wang J, Huang L, Li J, et al. Artemisinin directly targets malarial mitochondria through its specific mitochondrial activation. *PLoS One* 2010;**5**(3):e9582.
270. Wang J, Huang LY, Long YC, et al. Effect of mitochondrial permeability transition pore in antimalarial action of artemisinin. *Prog Mod Biomed* 2009;**9**(21):4006–9.
271. Sun C, Li J, Cao Y, et al. Two distinct and competitive pathways confer the cellcidal actions of artemisinins. *Microb Cell* 2015;**2**(1):pp.14–25.
272. del Pilar Crespo M, Avery TD, Hanssen E, et al. Artemisinin and a series of novel endoperoxide antimalarials exert early effects on digestive vacuole morphology. *Antimicrob Agents Chemother* 2008;**52**(1):98–109.
273. Kawai S, Kano S, Suzuki M. Morphologic effects of artemether on *Plasmodium falciparum* in Aotus Trivirgatus. *Am J Trop Med Hyg* 1993;**49**(6):812.
274. Ho WE, Peh HY, Chan TK, et al. Artemisinins: pharmacological actions beyond anti-malarial. *Pharmacol Ther* 2014;**142**(1):pp.126–139.
275. Noedl H, Se Y, Schaecher K, et al. Evidence of artemisinin-resistant malaria in western Cambodia. *N Engl J Med* 2008;**359**(24):2619–20.
276. Dondorp AM, Nosten F, Yi P, et al. Artemisinin resistance in *Plasmodium falciparum* malaria. *N Engl J Med* 2009;**361**(5):455–67.
277. Phyo AP, Nkhoma S, Stepniewska K, et al. Emergence of artemisinin-resistant malaria on the western border of Thailand: a longitudinal study. *Lancet* 2012;**379**(9830):1960–6.
278. Hien TT, Thuy-Nhien NT, Phu NH, et al. *In vivo* susceptibility of *Plasmodium falciparum* to artesunate in Binh Phuoc province, Vietnam. *Malar J* 2012;**11**(1):355.
279. World Health Organization. *Chemotherapy of malaria and resistance to antimalarials* WHO Technical Report Series 29. Geneva: World Health Organization; 1973.
280. Worldwide Antimalarial Resistance Network. Parasite Clearance Estimator (PCE). http://www.wwarn.org/tools-resources/toolkit/analyse/parasite-clearance-estimator-pce. (The online tool was launched on WWARN's website on May 29, 2013).
281. White NJ. The parasite clearance curve. *Malar J* 2011;**10**:278.
282. Flegg JA, Guerin PJ, White NJ, et al. Standardizing the measurement of parasite clearance in falciparum malaria: the parasite clearance estimator. *Malar J* 2011;**10**:339.
283. Brockman A, Price RN, van Vugt M, et al. *Plasmodium falciparum* antimalarial drug susceptibility on the North-Western border of Thailand during five years of extensive use of artesunate-mefloquine. *Trans R Soc Trop Med Hyg* 2000;**94**(5):537–44.
284. Noedl H, Se Y, Sriwichai S, et al. Artemisinin resistance in Cambodia: a clinical trial designed to address an emerging problem in Southeast Asia. *Clin Infect Dis* 2010;**51**(11):e82–9.

285. Cheeseman IH, Miller BA, Nair S, et al. A major genome region underlying artemisinin resistance in malaria. *Science* 2012;**336**(6077):79–82.
286. Li GF, Shi YL, Jiao XQ, et al. *Establishment of six drug-resistant strain models with K173 strain Plasmodium berghei*. Archives of the Institute of Microbiology and Epidemiology, AMMS; 1993. [李国福，时云林，焦岫卿，等. 伯氏鼠疟原虫K173株6种抗药性虫株模型的建立. 军事医学科学院微生物流行病研究所内部资料, 1993.]
287. Chawira AN, Warhurst DC, Peters W. Qinghaosu resistance in rodent malaria. *Trans R Soc Trop Med Hyg* 1986;**80**(3):477–80.
288. Chawira AN, Warhurst DC, Robinson BL, et al. The effect of combinations of Qinghaosu (artemisinin) with standard antimalarial drugs in the suppressive treatment of malaria in mice. *Trans R Soc Trop Med Hyg* 1987;**81**(4):554–8.
289. Chawira AN, Warhurst DC. The effect of artemisinin combined with standard antimalarials against chloroquine-sensitive and chloroquine-resistant strains of *Plasmodium falciparum in vitro*. *J Trop Med Hyg* 1987;**90**(1):1–8.
290. Li GQ, Guo XB, Fu LC, et al. A randomized comparative study of artesunate and quinine hydrochloride in the treatment of falciparum malaria. *J Guangzhou Univ Tradit Chin Med* 1995;**12**(3):1–5. [李国桥，郭兴伯，苻林春，等. 青蒿琥酯与二盐酸奎宁治疗恶性疟的随机比较. 广州中医药大学学报, 1995, 12(3): 1-5.]
291. World Health Organization. *Global plan for artemisinin resistance containment (GPARC)*. Geneva: World Health Organization; 2011.
292. Antony HA, Parija SC. Antimalarial drug resistance: an overview. *Trop Parasitol* 2016;**6**(1):30–41.
293. Wellems TE, Walker-Jonah A, Panton LJ. Genetic mapping of the chloroquine-resistance locus on *Plasmodium falciparum* chromosome 7. *Proc Natl Acad Sci USA* 1991;**88**:3382–6.
294. Ecker A, Lehane AM, Clain J, et al. PfCRT and its role in antimalarial drug resistance. *Trends Parasitol* 2012;**28**(11):504–14.
295. Reed MB, Saliba KJ, Caruana SR, et al. Pgh1 modulates sensitivity and resistance to multiple antimalarials in *Plasmodium falciparum*. *Nature* 2000;**403**(6772):906–9.
296. Trape JF, Pison G, Spiegel A, et al. Combating malaria in Africa. *Trends Parasitol* 2002;**18**(5):224–30.
297. Kublin JG, Dzinjalamala FK, Kamwendo DD, et al. Molecular markers for failure of sulfadoxine-pyrimethamine and chlorproguanil-dapsone treatment of *Plasmodium falciparum* malaria. *J Infect Dis* 2002;**185**(3):380–8.
298. Plowe CV, Cortese JF, Djimde A, et al. Mutations in *Plasmodium falciparum* dihydrofolate reductase and dihydropteroate synthase and epidemiologic patterns of pyrimethamine-sulfadoxine use and resistance. *J Infect Dis* 1997;**176**(6):1590–6.
299. Smithuis FM, van Woensel JB, Nordlander E, et al. Comparison of two mefloquine regimens for treatment of *Plasmodium falciparum* malaria on the Northeastern Thai-Cambodian border. *Antimicrob Agents Chemother* 1993;**37**(9):1977–81.
300. Nosten F, Luxemburger C, ter Kuile FO, et al. Treatment of multidrug-resistant *Plasmodium falciparum* malaria with 3-day artesunate-mefloquine combination. *J Infect Dis* 1994;**170**(4):971–7.
301. Winzeler EA, Manary MJ. Drug resistance genomics of the antimalarial drug artemisinin. *Genome Biol* 2014;**15**(11):544.
302. Fairhurst RM. Understanding artemisinin-resistant malaria: what a difference a year makes. *Curr Opin Infect Dis* 2015;**28**(5):417–25.
303. Tilley L, Straimer J, Gnädig NF, et al. Artemisinin action and resistance in *Plasmodium falciparum*. *Trends Parasitol* 2016;**32**(9):682–96.

304. Amaratunga C, Sreng S, Suon S, et al. Artemisinin-resistant *Plasmodium falciparum* in Pursat province, western Cambodia: a parasite clearance rate study. *Lancet Infect Dis* 2012;**12**(11):851–8.
305. Kyaw MP, Nyunt MH, Chit K, et al. Reduced susceptibility of *Plasmodium falciparum* to artesunate in southern Myanmar. *PLoS One* 2013;**8**(3):e57689.
306. Ashley EA, Dhorda M, Fairhurst RM, et al. Spread of artemisinin resistance in *Plasmodium falciparum* malaria. *N Engl J Med* 2014;**371**(5):411–23.
307. Global Malaria Program Status Report. *Artemisinin and artemisinin-based combination therapy resistance*. October 2016. http://www.who.int/malaria/publications/atoz/update-artemisinin-resistance- October 2016/en/.
308. Cheng DX. Effects of artemisinin on phagocytosis in peritoneal macrophages in mice. In: *Compendium of research into Chinese Materia Medica (1980–1981*. Beijing: Beijing Medical Institute; 1982. p. 64–8. [程道新. 青蒿素小鼠腹腔巨噬细胞吞噬功能的影响.中医药研究成果汇编(1980-1981), 1982: 64-68.]
309. Zou Z, Cheng DX, Wang YQ. Effects of artemisinin on plaque- and rosette-forming cells in malaria-infected mice. *Chin J Integr Med* 1985;**5**(9):563–7. 517. [邹樟，程道新，王一琴. 青蒿素对感染鼠疟小鼠溶血空斑和玫瑰花形成细胞的影响. 中西医结合杂志, 1985, 5(9): 563-567, 517.]
310. Zang QZ, Zheng ZY, Qi SB, et al. Regulatory effects of artemisinin on immune function. *Nucl Sci Tech* 1984;**2**:55–7. [臧其中，郑振源，齐尚斌，等. 蒿素对免疫功能的调节作用. 核技术, 1984, 2: 55–57.]
311. Lin PY, Pan JQ, Feng ZM. Effects of artemether on serum IgG and spleen weight in mice. *Acta Pharm Sin* 1985;**20**(3):211–3. [林培英，潘竞锵，冯昭明. 蒿甲醚对小鼠血清IgG及脾重的影响. 药学学报, 1985, 20(3): 211-213.]
312. Wang DR, Lin XY, Zhang HZ. Inhibitory effects of artemisinin and deoxy-artemisinin on lymphocyte activation in mice. *Shanghai J Immunol* 1987;**7**(4):199–201. [王敦瑞，林性玉，张惠珠. 青蒿素和去氧青蒿素对小鼠淋巴细胞活化的抑制作用. 上海免疫学杂志, 1987, 7(4): 199-201.]
313. Zhou P, Gao YX. Effects of artesunate on NK cell and ADCC activity. *J Bengbu Med Coll* 1995;**20**(6):363. [周平，高玉祥. 青蒿琥酯对NK细胞活性及ADCC活性的影响. 蚌埠医学院学报, 1995, 20(6): 363.]
314. Gao YX. Study of the effects of artesunate in immunity. *J Bengbu Med Coll* 1998;**23**(5):289. [高玉祥. 青蒿琥酯免疫作用的研究. 蚌埠医学院学报, 1998, 23 (5): 289.]
315. Wang GJ, Gu Q, Zhang HQ. Effects of artesunate on neutrophil chemotaxis. *Chin J Dermatovenerol* 2004;**18**(7):391–3. [王国江，顾青，张海清. 青蒿琥酯对中性粒细胞趋化的影响. 中国皮肤性病学杂志, 2004, 18(7): 391-393.]
316. Yang JH, Li DZ. Effects of artemether injections on immune function in mice. *J Guangzhou Coll Tradit Chin Med* 1990;**7**(2):102–5. [仰锦虹，李道中. 抗疟药蒿甲醚注射液对小鼠免疫功能的影响. 广州中医学院学报, 1990, 7(2): 102-105.]
317. Qian RS, Li ZL, Yu JW, et al. Influence of artemisinin on immune function and its antiviral effects. *J Tradit Chin Med* 1981;**6**:12–8. [钱瑞生，李柱良，余建文，等. 青蒿素的免疫作用和抗病毒作用. 中医杂志, 1981, (6): 63-66.]
318. Shen M, Ge HL, He YX, et al. Immunosuppressive effects of artemisinin. *Shanghai J Immunol* 1982;**6**:7. [沈明，葛海良，何尧祥，等. 青蒿素的免疫抑制作用. 上海免疫学杂志, 1982, (6): 7.]
319. Lin XY, Cheng F, Zhang LY, et al. Comparison of the immunosuppressive effects of artemisinin and two of its derivatives. *Shanghai J Immunol* 1984;**6**:348–51. [林性玉，程枫，张来仪，等. 青蒿素及其二种衍生物的免疫抑制作用比较. 上海免疫学杂志, 1984, 6: 348–351.]

320. Qian RS, Li ZL, Xie MR, et al. Inductive effects of artemisinin on interferons in mice. *Med J Chin People's Lib Army* 1985;**5**:355–8. [钱瑞生，李柱良，谢名荣，等. 青蒿素对小鼠干扰素的诱生作用. 解放军医学杂志, 1985, (5): 355-358.]
321. Matasubara S, Suzuki F, Ishida N. The induction of interferon by levamisole in mice. *Cell Immunol* 1979;**43**:214–9.
322. Abe N, Ebina T, Ishida N, et al. Interferon induction by glycyrrhizin and glycyrrhetinic acid in mice. *Microbiol Immunol* 1982;**26**(6):535–9.
323. Chen M, Zhu ZJ, Wang ZL, et al. Effects of artesunate on immune function in animals. *J Guangxi Med Coll* 1988;**4**:42–5. [陈鸣，朱作金，王朝临，等. 青蒿琥酯对动物免疫功能的影响. 广西医学院学报, 1988, 4: 42–45.]
324. Chen H, Gao YX. Discussion on the immune action mechanisms of artesunate (abstract). *J Bengbu Med Coll* 1988;**3**:195. [陈华，高玉祥. 青蒿琥酯免疫作用机制探讨(摘要). 蚌埠医学院学报, 1988, 3: 195.]
325. Gu YX, Cui YF, Wu BA, et al. Effects of artemether on peripheral T, B, Tμ and Tγ lymphocytes in beagle dog. *J Tradit Chin Med* 1989;**3**:215–9.
326. Zhang D, Lin PY, Pan JQ, et al. Effects of artemether on immune function in mice. *Chin Pharmacol Bull* 1989;**1**:37–9. [张丹，林培英，潘竞锵，等. 蒿甲醚对小鼠免疫功能的影响. 中国药理学通报, 1989, 1: 37–39.]

Chapter 6

Artemisinin and Derivatives: Clinical Studies

Chapter Outline

In 1975, the WHO published "Guidelines for Evaluation of Drugs for Use in Man." Many countries also formulated such guidelines and administrative regulations, which required new drugs to be clinically tested and enhanced the surveillance for adverse reactions. This ensured that the drugs used were safe and effective. China had issued the "Various Regulations on Drug Administration (Draft)" in 1963. A series of regulations and laws were then enacted under various circumstances. On 1 July 1985, the "Drug Administration Law of the People's Republic of China" was in full operation, followed by the publication and implementation of China's "Provisions for New Drug Approval" which included the technical requirements and standards for clinical evaluation of new drugs.

Artemisinin-Based and Other Antimalarials. https://doi.org/10.1016/B978-0-12-813133-6.00006-8

Beginning in 1986, the Ministry of Health invited experts to organize a Drug Evaluation Committee responsible for reviewing new drugs on a regular basis. Besides the early clinical trials on artemisinin and its derivatives artemether, artesunate, and dihydroartemisinin, fresh research was conducted after 1985 in accordance with the new WHO and Chinese guidelines. The drugs were reviewed by the Drug Evaluation Committee and were approved by the Ministry of Health, then marketed as Class A new medications. Currently, they are widely used in areas of Southeast Asia, Africa, South America, and China where malaria is endemic. Arteether had also been invented by Chinese scientists but was rejected because its efficacy was lower than that of artemether. However, it was chosen for further development by WHO but research was subsequently discontinued.

6.1 ARTEMISININ CLINICAL TRIALS

Li Guoqiao, Guo Xingbo
Institute of Tropical Medicine, Guangzhou University of Chinese Medicine, Guangzhou, Guangdong Province, China

A. EARLY CLINICAL RESEARCH ON ORAL AND INJECTABLE FORMULATIONS[1]

In April 1975, after the nationwide collaborative conference on artemisinin as an antimalarial, research units from the Academy of Traditional Chinese Medicine now the China Academy of Chinese Medical Sciences (CACMS), Shandong, Yunnan, Guangdong, Sichuan, Jiangsu, Hubei, Henan, Guangxi, Shanghai, the Chinese Academy of Sciences (CAS), and the People's Liberation Army (PLA) formed the China Collaborative Research Group on Qinghaosu (artemisinin).[2] Artemisinin formulations, including suppositories, were used to treat malaria in Hainan, Yunnan, Sichuan, Shandong, Henan, Jiangsu, Hubei, and Southeast Asia.

1. Formulation, Regimen, and Dose

- Tablet: 3-day course administered 2–4 times a day, each dose 0.3–0.5 g, total dose 2.5–3.2 g. The first dose was doubled in some patients.
- Oil injection: 3-day course, one 0.2–0.4 g intramuscular injection daily, total dose 0.5–0.8 g.
- Oil suspension injection: 3-day course, one 0.2–0.4 g intramuscular injection daily, total dose 0.8–1.2 g.
- Water suspension injection: 3-day course, one 0.4 g intramuscular injection daily, total dose 1.2 g.

2. Evaluation Criteria

- Clinical cure: Body temperature returned to normal 72h after medication, absence of asexual parasites after 120h.
- Effective: Body temperature returned to normal 72h after medication, parasite density decreased significantly in 120h but not eliminated.
- Ineffective: Symptoms persisted 72h after medication, no significant decrease or even increase in parasite density.

3. Vivax Malaria

A total of 1511 cases of vivax malaria were treated with artemisinin. Average defervescence time was 20–30h, average asexual parasite clearance time was 30–40h, and the relapse rate was 10%–30% within a month. Table 6.1 shows results of 738 vivax malaria patients.

The collaborative research group used artemisinin tablets (total dose 3g) to treat 16 cases of vivax malaria in 1975. Another 13 control cases were given chloroquine, total dose 1.5g (base). The parasite clearance time of the artemisinin group was 39.6±13.5h, while that of the chloroquine group was 55.9±16.6h ($P<.01$), indicating that the clearance speed of artemisinin was faster than that of chloroquine. However, the relapse rate of the artemisinin group was 21.4% within a month, while no cases of relapse were seen in the chloroquine group. Similar results were observed in other locations.

4. Falciparum Malaria

Artemisinin was used to treat 527 cases of falciparum malaria in various locations. Fever clearance time was 30–40h, and asexual parasite clearance time was 30–50h. The recrudescence rate was 85% within a month for tablets, and 10%–25% for other formulations (Table 6.2).

Also in 1975, the collaborative research group used 3-day regimens of artemisinin tablets (total dose 2.5g) and chloroquine (total dose 1.5g base) to treat 18 cases of falciparum malaria with each drug. Average asexual parasite clearance time was 37±17.8h for artemisinin and 65.7±29.9h for chloroquine ($P<.01$). Therefore, the clearance speed was faster for artemisinin than for chloroquine. All the cases in the artemisinin group experienced recrudescence in a month, whereas the recrudescence rate for chloroquine was 50%. Although the recrudescence rate for artemisinin was higher than that of chloroquine, the rate dropped to 10%–25% within 1month when the tablet formulation was replaced by the oil, oil-suspension, and water-suspension formulations.

A total of 143 cases which had failed to respond to chloroquine (R_I–R_{III} resistance) were studied in areas where chloroquine-resistant falciparum malaria was prevalent. The 3-day regimen of artemisinin oil, oil suspension,

TABLE 6.1 Different Artemisinin Formulations in the Treatment of Vivax Malaria

Group	No. of Cases	Defervescence Time (h)	Parasite Clearance Time (h)	Relapse in One Month	
				No. of Cases	Relapse Rate (%)
Tablet	128	21.0–33.3	18.0–39.6	128	31.3
Oil injection	318	20.0–30.0	22.0–32.1	213	18.8
Oil suspension injection	132	21.9	28.3–48.0	114	13.2
Water suspension injection	160	35.0	30.0–47.9	160	8.7

TABLE 6.2 Different Artemisinin Formulations in the Treatment of Falciparum Malaria

Group	No. of Cases	Defervescence Time (h)	Parasite Clearance Time (h)	Recrudescence in One Month	
				No. of Cases	Recrudescence Rate (%)
Tablet	83	34.3–45.8	37.0–41.3	60	85.0
Oil injection	223	11.4–67.0	25.3–70.8	140	25.7
Oil suspension injection	40	30.6–31.2	33.9–40.6	38	10.5
Water suspension injection	181	34.7–44.5	46.4–55.1	83	13.3

or water suspension was used and all patients were cured. Therefore, artemisinin was shown to be effective in treating chloroquine-resistant falciparum malaria.

5. Cerebral Malaria

From 1974 to 1978, 141 cases of cerebral malaria were studied in areas where chloroquine-resistant falciparum malaria was endemic. Since injections were not available at the time, nasogastric gavage was used with 36 patients. The cure rate was 91.7%. Intramuscular administration was adopted once injections became available. Of the 141 patients, 131 were cured (92.9%). Ten died, yielding a mortality rate of 7.1%. From statistics across all sites, the average parasite clearance time was 33.3–64.5 h and average defervescence time was 34.1–56.7 h. The average time taken to regain consciousness was 21.5–30.8 h. This did not include a few cases who only regained consciousness after at least 10 days.

6. Side Effects

No obvious side effects were seen in 2099 cases treated with artemisinin. Tests for liver function and recordings of electrocardiograms (ECGs) were conducted on 139 cases before and after treatment, and 75 cases were examined for non-protein nitrogen in the blood. No abnormalities were found. There were no aberrations in patients with heart, liver, or kidney diseases, and in pregnant women. For the water suspension, mild pain was experienced at the injection site, but no other adverse reactions were reported.

B. ARTEMISININ SUPPOSITORIES[3]

1. Phase I Clinical Trial

The trial included 18 healthy male volunteers aged 18–40 years. They were randomly assigned to different dose groups, and a double-blind protocol was used. The regimen for each dose group is shown in Table 6.3. Each volunteer was hospitalized for 6 days (Days −1, 0, and 1–4). Clinical signs were recorded daily. ECGs, hematology tests (including erythrocyte sedimentation rate, hemoglobin, hematocrit, reticulocyte, white cell count and differentials, platelet, and hemoglobin electrophoresis), blood chemistry panels (including serum protein, glucose, urea, creatinine, bilirubin, cholesterol, triglyceride, potassium, sodium, calcium, phosphorus, chlorine, alkaline phosphatase, acid phosphatase, cholinesterase, alanine aminotransferase ALT, aspartate aminotransferase, creatine kinase, and G-6-PD), urine tests (including specific gravity, pH, white blood cell, nitrites, glucose, ketone bodies, protein, bilirubin, red blood cell, hemoglobin), and fecal tests (conventional and for parasite ova) were carried out before treatment and 7, 14, 21, and 28 days thereafter.

TABLE 6.3 Dose Groups in Phase I Clinical Trials for Artemisinin Suppositories

Total Dose (mg)	No. of Cases	Day 1 (mg)		Day 2 (mg)		Day 3 (mg)	
		8 a.m.	12 a.m.	8 a.m.	4 p.m.	8 a.m.	4 p.m.
2800	3	600	600	400	400	400	400
4200	3	900	900	600	600	600	600
5100	3	1200	900	800	700	800	700
5600	3	1200	1200	800	800	800	800
6000	3	1200	1200	900	900	900	900
Control	3	0	0	0	0	0	0

N.B., The number of suppositories given per administration was the same in the dose and control groups.

After the first dose, one case each from the 4200 and 5600 mg groups experienced the urge to defecate and a mild burning sensation around the anus. This did not affect continued administration. One case in the group receiving 4200 mg displayed fever (37.7°C) 3 h after the third dose, which spontaneously eased after 19 h. The group receiving 5600 mg also saw one case of fever (38.8°C) 2 h after the third dose, which lasted for 9 h. All other cases showed no symptoms relating to the central nervous or gastrointestinal systems.

In the group receiving 5100 mg, one case had a slight increase in ALT levels on Days 14 and 21 after medication. This returned to normal on Day 35. ALT levels also rose persistently for one patient in the group receiving 6000 mg on Days 3, 7, 14, and 21. The level was normal on Day 35.

In general, no abnormal changes were seen in ECGs, blood chemistry panels, hematology, routine urine, and routine fecal tests. The results showed that a 3-day course of 4200 mg artemisinin suppositories was well tolerated in healthy individuals.

2. Phase II Clinical Trials

Dose-Finding Experiment With Falciparum Malaria

Suppositories with 100, 200, 300, 400, or 600 mg artemisinin each were used. Table 6.4 shows the dose groups.

The group receiving 1200 mg had seven cases of falciparum malaria. Five of them received medication when the parasite was in the tiny ring-form stage. In all these patients, the parasites developed into thicker, larger ring forms 16 h after medication. One case showed <75% parasite clearance after 48 h; another did not achieve full parasite clearance in 7 days.

The group receiving 1600 mg had 22 cases. In six cases, the parasites developed into thick ring forms 2 days after medication. A minor brood with new tiny ring forms also appeared in one case. Of the 13 cases in the group receiving 2400 mg, 3 did not experience parasite clearance by Day 7.

TABLE 6.4 Artemisinin Suppository Dose Groups

Dose Group (mg)	Day 1 (mg)		Day 2 (mg)		Day 3 (mg)	
	1st Dose	After 4 h	a.m.	p.m.	a.m.	p.m.
1200	400	–	400	–	400	–
1600	400	400	400	–	400	–
2400	400	400	400	400	400	400
2800	600	600	400	400	400	400
3200	800	800	400	400	400	400

TABLE 6.5 Effects of Artemisinin Suppositories on Falciparum Malaria

Dose Group (mg)	No. of Cases	Defervescence Time (h) (M ± SD)	Parasite Clearance Time (h) (M ± SD)
1200	7	21.3 ± 16.3	65.6 ± 17.1[a]
1600	22	23.0 ± 8.9	60.9 ± 21.1[b]
2400	13	25.8 ± 10.7	54.7 ± 11.9[c]
2800	30	15.8 ± 11.9	46.5 ± 16.5
3200	6	16.4 ± 11.2	47.5 ± 16.8

[a]*Parasite clearance in 5/7 cases.*
[b]*Parasite clearance in 15/22 cases.*
[c]*Parasite clearance in 10/13 cases.*

Full parasite clearance was seen in the 30 cases in the group receiving 2800 mg and the six cases in the 3200 mg group.

Clinical symptoms were controlled in all groups. Defervescence and parasite clearance times are shown in Table 6.5.

The results provided the basis for a suppository regimen of 400 mg/dose, administered twice daily, with the initial dose doubled. The total dose over 3 days was 2800 mg. This was recommended for further Phase II trials.

Parasite Clearance Speed

The parasite clearance rate of two artemisinin suppository regimens—1200 mg given in 8 h and 1200 mg given in 4 h on the first day—were compared to those of oral piperaquine phosphate and intravenous quinine hydrochloride. Each drug group had six patients who were observed in identical locations. The groups had three adults and three children aged seven to 10 years, with children receiving half the adult dose. They all came from malaria-endemic areas and had parasite densities of 10,000–100,000 per μL. The first dose was administered when parasites were at the tiny ring-form stage. Parasitemia was observed every 4 h to calculate the time taken to achieve 95% parasite clearance. The regimens and doses were as follows:

- Artemisinin suppositories, total dose 1200 mg on the first day, administered in three doses. Initial dose 400 mg, then 400 mg after 4 and 8 h.
- Artemisinin suppositories, total dose 1200 mg on the first day, administered in two doses. Initial dose 600 mg, then 600 mg after 4 h. This paralleled the doses used on Day 1 of the patients receiving 2800 mg.
- Oral piperaquine phosphate, initial dose 600 mg (piperaquine base), followed by 300 mg after 4 h and 600 mg on Day 2.

- Intravenous quinine hydrochloride drip, 1500 mg daily for 3 days. The dose for the first day was administered within 12 h.

For all four groups blood smears were obtained 24 h after the first dose to calculate parasitemia before proceeding to the doses on Day 2. Average time taken to reach 95% parasite clearance was taken as an indicator. The 1200 mg, three-dose artemisinin suppository group took 24 h; the 1200 mg, two-dose group took 20 h. The piperaquine phosphate group took 28 h and the quinine hydrochloride group took 32 h.

Artemisinin Suppositories and Piperaquine Phosphate With Falciparum Malaria

Both drugs were used to treat 30 cases of falciparum malaria each. A 3-day, 2800 mg regimen of artemisinin suppositories was adopted (initial dose 800 mg, followed by 400 mg after 4 h, and 400 mg in the morning and afternoon of Days 2 and 3). The piperaquine phosphate regimen was the same as in the parasite clearance speed study above. Defervescence and parasite clearance times were 15.8 ± 11.9 h and 46.5 ± 16.5 h, respectively, for the suppositories, and 33 ± 15.3 h and 97.9 ± 26.9 h, respectively, for piperaquine. The recrudescence rate was higher for the suppositories (40%) than for piperaquine (no recrudescence).

3. Expanded Trials

From 1982 to 1984, the 3-day, 2800 mg artemisinin suppository regimen was used to treat 463 patients in Hainan Island, Shenzhen, and Zhaoyang in Hubei. This included 358 cases of falciparum malaria—with 14 cerebral malaria patients and 18 severe cases—and 105 cases of vivax malaria.

Falciparum Malaria

Of the 358 cases, 15 were coinfected with vivax malaria. A total of 246 cases were from malaria-endemic areas (68.7%). Males made up 222 of the cases (62%), and 136 cases were female (38%). In this group, 111 cases were aged less than 16 years (31%), and 247 cases were over 16 (69%). There were 13 pregnant women, two of whom were in the first trimester, six in the second trimester, and five in the third trimester.

After medication, clinical symptoms were rapidly controlled in 355 cases. Defervescence time was 23.8 ± 14.4 h and parasite clearance time was 45.9 ± 17.2 h. For the remaining three cases, one had severe cerebral malaria; parasite clearance was achieved in 22 h, but the patient died of complications after 69 h. One case experienced repeated diarrhea and expelled the suppositories and had to be treated with intravenous quinine hydrochloride. The final case also had diarrhea; body temperature did not return to normal and parasite clearance had not occurred by Day 6. Another course of treatment was given, whereupon defervescence occurred at 148 h and parasite clearance at 160 h.

A total of 83 cases were followed-up for 28 days. There were 38 cases of recrudescence, with a rate of 45.8%. The average time of recrudescence was 18.6±4.6 days. In another 16 cases, *Plasmodium vivax* was found during follow-up and the full 28-day observation was not carried out.

Severe Falciparum Malaria (Including Cerebral Malaria)

Of the 32 severe falciparum malaria cases, 14 had cerebral malaria (nine common, four severe, one very severe; see classification criteria for cerebral malaria at the end of this section). Another 18 cases had other forms of severe malaria, including four cases with hemolytic anemia, two with shock, two with severe jaundice (severe liver damage), seven with severe anemia in the second and third trimesters of pregnancy (Hb < 5 g/dL), and three cases with high parasitemia (parasitized red blood cell rate of 20.6%, 25%, and 26.4%).

Defervescence time for the 32 cases was 36.9±21.3 h, and parasite clearance time was 55.1±14.9 h. Apart from the one severe case who died, the 13 other cerebral malaria patients (92.9%) were cured. Recovery from coma took 26.1±17.7 h, fever clearance time was 42.5±24.9 h, and parasite clearance time was 53.7±16.5 h. The 18 other severe cases were all cured. Defervescence time was 32.9±17.9 h, and parasite clearance time was 59.1±17.8 h.

Vivax Malaria

The suppositories were used to treat 105 cases of vivax malaria; all were clinically cured. Average fever clearance time was 18.5 h, and average parasite clearance time was 37 h. Of the 69 cases followed up for 28 days, 24 cases experienced relapse. The relapse rate was 34.8%; average relapse time was 17 days.

Cerebral Malaria (Post 1985)

After Phase I and II clinical trials for the suppositories were completed and new drug certification was granted, a further 24 cases of cerebral malaria were treated (Collected data from the Asian Primary Healthcare Conference, 1989). There were 15 males and females; nine of the cases were children. Average parasite density was 104,716 per μL. Three of the cases had erythrocytic parasitemia of 31%, 19%, and 18.8%. Onset of coma before treatment was 13.7±12.9 h. Four cases had jaundice, three had hemolytic anemia, and ten had severe anemia (Hb less than 6 g/dL).

Of these, 22 patients were cured (91.7%). Two cases died. Average defervescence time was 35.6±23.5 h, and average parasite clearance time was 57.3±23 h. Time taken to recover from coma was 21.6±13.9 h. Of the two patients who died, one experienced parasite clearance at 36 h. Another regained consciousness 26 h after treatment. Both died due to complications.

The results showed that the suppositories were very effective in treating severe malaria and patients with high levels of parasitemia. The regimen was simple, making it a good emergency treatment in areas with poor medical facilities.

Addendum: Classification Criteria for Cerebral Malaria

- Common. The patient has all the following five criteria: (1) Hospitalization within 12h after onset of coma, with no complications (cerebral edema, respiratory failure, acidosis, shock, renal failure, hemolysis, etc.) or comorbidities (bronchopneumonia, dysentery etc.) on admission. (2) Axillary temperature 41.5°C and below on admission. (3) Absence of convulsions or cessation of convulsions once temperature decreased to 38°C and below, and with appropriate anticonvulsive treatment. (4) Erythrocyte count over 2,500,000 per mm^3, or hemoglobin over 7g/dL. (5) Less than one large ring form (R_3 ring form) observed per field (thick smear, 5×100 magnification), parasitized erythrocytic rate 15% and below.
- Severe. The patient has one of the following six criteria: (1) Axillary temperature at or over 41.5°C; or afebrile; or temperature below 36°C. (2) Temperature drops to 38°C 2–3h after admission, but with persistent convulsions after treatment. (3) Early signs of one or two complications or comorbidities. (4) Jaundice. (5) Erythrocyte count less than 2,500,000 per mm^3, or hemoglobin less than 7g/dL. (6) No antimalarial treatment before admission, with parasites developing normally and more than two late trophozoites or early schizonts per field on thick peripheral blood smears.
- Very Severe. The patient has one of the following eight criteria: (1) Temperature falls to 38°C and below, with no significant reduction in convulsions 3h after treatment with anticonvulsants and dehydrating medication (diuretics), or the upper extremities display inward rotating tonic convulsions. (2) Respiratory or heart failure or shock upon admission, with no improvement 2h after treatment. (3) No urination 24h after admission and no filling of the bladder. (4) Hemoglobinuria, in which the urine is dark brown or has a soy-sauce-like appearance. (5) Disappearance of the pupillary light reflex—ruling out chlorpromazine-induced miosis and insensitivity to light—or pupils of unequal sizes (convulsions may induce pupil dilation, transient unequal size, and insensitivity to light; therefore, observations should take place when convulsions are absent). (6) Late-stage pregnancy. (7) Erythrocyte count less than 1,300,000 per mm^3 or hemoglobin below 4g/dL. Parasitized erythrocytic rate as high as 25%, or late trophozoites over 10 per field on thick peripheral blood smears.

6.2 ARTESUNATE

Guo Xingbo, Li Guoqiao
Institute of Tropical Medicine, Guangzhou University of Chinese Medicine, Guangzhou, Guangdong Province, China

A. INTRAVENOUS ARTESUNATE

Artesunate is an artemisinin derivative. Animal experiments showed that it was three to seven times more effective than artemisinin. In 1985, the Guangzhou University of Chinese Medicine conducted a Phase I clinical trial on intravenous artesunate to observe its toxicity and tolerance in healthy people, to understand its pharmacokinetics, and provide evidence for safe and effective treatment regimens. The trial was approved by the Ministry of Health.

1. Phase I Clinical Trial[4]

The study included 26 volunteers, all male, aged 18–40 years and weighing 50–70 kg. They had no blood, cardiovascular, liver, gastrointestinal, or parasitic diseases; no allergies, no high myopia or retinitis, and no history of alcoholism. They were in good nutritional health; fundus examinations and physical examinations were normal. All the following tests were normal: chest X-rays, ECGs, hematology and blood chemistry panels, G6PD, and urine and fecal tests. The test items were the same as those in the Phase I artemisinin suppository trials. Volunteers had to have normal chest X-rays, erythrocyte sedimentation, hemoglobin electrophoresis, acid phosphatase, and G6PD indicators measured before the trial. Other tests had to be normal on Days −4 and −2.

Artesunate powder injections were supplied by the Guilin No. 2 Pharmaceutical Factory.

Based on a 1981 study on 17 cases of cerebral malaria which showed that artesunate had good efficacy,[5] a predicted therapeutic dose of 2 mg/kg per administration was proposed. Injections were to be given 4, 24, and 48 h after the first dose, namely four injections over 3 days. The total dose was 8 mg/kg. One group was given a lower dose (1.0 mg/kg). Three other groups received 50%, 25%, and 12.5% increments of the predicted therapeutic dose. Table 6.6 shows the dose groups.

TABLE 6.6 Dose Design of Phase I Trial for Intravenous Artesunate Injection

Dose (mg/kg)		% Increase From Predicted Therapeutic Dose	No. of Cases	Control Cases[a]
Total Dose	Single Dose			
4.0	1.0		2	0
8.0[b]	2.0[b]	100.0	6	2
12.0	3.0	50.0	4	2
15.0	3.75	25.0	4	2
16.8	4.2	12.5	2	2

[a]Normal saline is used as control.
[b]Predicted therapeutic dose.

Each package of the drug contained one bottle of 100mg artesunate and 1 mL 5% sodium bicarbonate. Before use, the sodium bicarbonate was injected into the bottle of artesunate. This was shaken for 2 min until the artesunate was completely dissolved. The required amount was drawn into a 1 mL syringe. Saline was added to bring the total volume to 10mL before being injected over 3 min. The drug mixture had to be used within 1 h. Volunteers in the control group received 10mL saline. If any adverse reactions or abnormal indicators appeared in one dose group, the experiment on the next highest dose group was halted. Experiments on dogs had shown that a decrease in reticulocytes and increase in alkaline phosphatase activity were the most sensitive indicators of artesunate toxicity. Therefore, the experiment would stop if reticulocyte count fell by over 50% of the pretreatment levels, or if alkaline phosphatase levels exceeded normal values.

Volunteers were randomly assigned to groups for this study and hospitalized for 9 days (Days −1, 0, 1–7). Body temperature, pulse, and respiratory rate were recorded every 4 h. Blood pressure was measured every morning and evening. Volume of urine, frequency of bowel movements, and stool consistency were recorded every 24 h. On Days 4–7, temperature, pulse, and respiratory rate were checked every morning and afternoon. Vision and fundus examinations were conducted on Days −1, 3, and 7.

Every morning and afternoon, volunteers were questioned for any symptoms relating to their appetite, gastrointestinal system, sleep, central nervous system, cardiovascular system, skin condition etc. Onset and end times were accurately recorded. If discomfort was experienced at any time, this was checked and recorded. All volunteers were given the ECGs and laboratory tests on Days 1, 3, 7, 14, and 28. If any abnormalities were found by Day 28, this was checked weekly until the indicator returned to normal.

Apart from three volunteers—two in the group receiving 4 mg and one in the 8 mg group—the remaining 15 in the artesunate groups experienced a mild bitter taste in the mouth during the injection. This spontaneously disappeared 10 min after medication. Five volunteers—two in the 8 mg group and one each in the 12, 15, and 16.8 mg groups—had mild pain at the injection site, which disappeared after injection. Blood pressure and ECGs measured before and 2 h after medication showed no abnormalities. No symptoms were seen in the cardiovascular, gastrointestinal, central nervous, and other organ systems. Besides reticulocyte count, the hematology, blood chemistry panel, urine, fecal, and ECG indicators were all normal.

Abnormalities appeared in reticulocyte counts. Among the six volunteers in the 8 mg group, two displayed a decrease in reticulocytes to 0.5% on Day 3, the lowest normal found. This was a 29% decline. They returned to premedication levels on Day 7, and no obvious changes were seen in the other four volunteers. The same thing occurred with one out of four volunteers in the group receiving 15 mg. Two volunteers in the 16.8 mg group had subnormal reticulocyte levels. One had levels of 0.5% on Day 3, 0.36% on Day 5, and 0.35% on Day 6. Another had levels of 0.55%, 0.45%, and 0.15% on those days, respectively. Both returned to the premedication state on Day 7 (see Fig. 6.1).

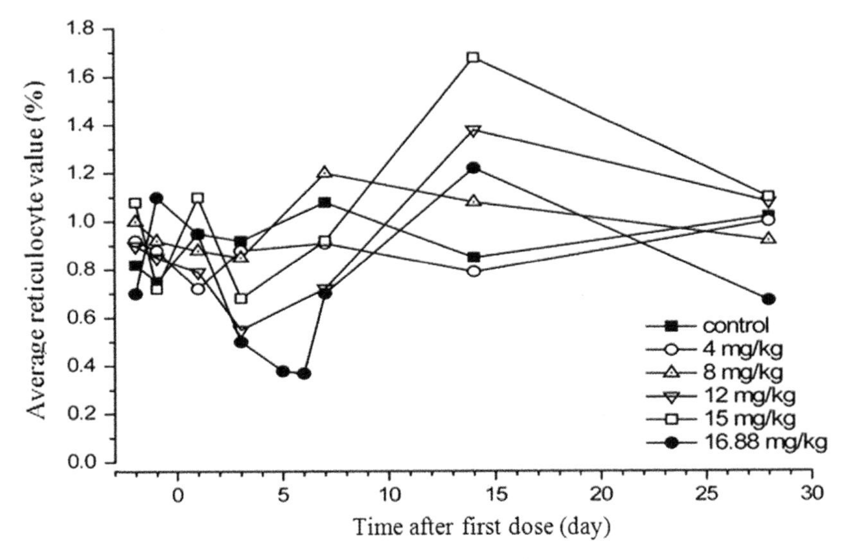

FIGURE 6.1 Reticulocyte levels before and after artesunate treatment.

The Phase I trial showed no adverse reactions apart from the slight bitter taste during injection that disappeared in a few minutes. Besides reticulocyte levels, no abnormalities were seen in the laboratory tests.

Changes in reticulocyte levels were an extremely sensitive indicator of artesunate-induced reactions. At the predicted therapeutic dose of 8 mg, two of the six volunteers experienced a slight decrease in reticulocyte levels but still within normal range. At 16.8 mg, two volunteers had subnormal levels on Days 5 and 6. This returned to premedication levels on Day 7.

Artesunate concentration per 10 mL of the drug solution received by the six volunteers in the 15 and 16.8 mg groups were 191, 203, 221, 224, 260, and 291 mg, respectively. Blood pressure, pulse, and ECGs showed no abnormal changes. This indicated that 10 mL of 2% artesunate solution, infused over 3 min, was safe. Intravenous artesunate had low cardiovascular toxicity, which could not be matched by other, currently available antimalarial injections. From a safety perspective, a 3-day course of 8–12 mg/kg intravenous artesunate (2–3 mg/kg/dose) could be used as a provisional regimen.

2. Phase II Clinical Trials[6]

Dose-Finding Study of Intravenous Artesunate With Falciparum Malaria

Five dose groups were included, with planning for 10 cases in each group and increased if necessary. Failure to control the parasite development or an insufficient dose in three cases in any group was defined as the progression of tiny

TABLE 6.7 Artesunate Dose Group Regimens

Total Dose (mg/kg)	Dose per Administration (mg/kg)			
	First Dose	4h	24h	48h
8.0	2.0	2.0	2.0	2.0
6.0	1.5	1.5	1.5	1.5
4.0	1.0	1.0	1.0	1.0
2.0	0.5	0.5	0.5	0.5
1.0	0.25	0.25	0.25	0.25

ring-form parasites to thick ring forms, or incomplete clearance of the asexual parasite 6 days after treatment. If this occurred, the experiment would be halted for that group. The main efficacy indicator was the average time taken to achieve 95% parasite clearance. Average parasite clearance time was used to determine a suitable therapeutic dose.

Previous studies had shown that 8 mg/kg intramuscular or intravenous artesunate was effective in treating cerebral malaria. However, there was still insufficient evidence for a suitable therapeutic dose. This study therefore adopted decreasing dose groups, starting with 8 mg/kg (Table 6.7).

Five cases of falciparum malaria were treated with a 3-day course of 1 mg/kg. Of these, three failed to achieve parasite clearance by Day 6. Four received the first dose when the parasites were still at the tiny ring-form stage; 16 h after medication, the parasites had developed into thick ring forms in two of these cases. The 2 mg/kg dose was given to eight cases, of whom five did not reach parasite clearance by Day 6. In two cases, the parasites had progressed to the thick ring-form stage after 16 h. Both groups had poor efficacy, showing that the dose was inadequate; therefore, both were withdrawn from the trial.

The 4, 6, and 8 mg/kg groups had 16, 12, and 16 cases, respectively. Average time taken for 95% clearance was 16 h. Six cases in each group were treated when the parasites were still at the tiny ring-form stage. These did not develop further. Defervescence and parasite elimination times are shown in Table 6.8, and parasite clearance speed is shown in Table 6.9 and Fig. 6.2. No adverse reactions were reported during treatment, including the bitter taste and pain at the injection site.

Intravenous artesunate is mainly used as a treatment for patients with severe malaria. Data from 1978 onward indicated that the efficacy of the 3-day course was highly satisfactory. From this dose-finding experiment, a 3-day course of

TABLE 6.8 Different Doses of Artesunate in Treating Falciparum Malaria

Dose Group (mg/kg)	No. of Cases	Average Defervescence Time (h)	Average Parasite Clearance Time (h)
8.0	16	17.4±9.2	60.4±10.1
6.0	12	21.2±13.7	57.6±10.2
4.0	16	19.7±6.4	60.6±14.0
2.0	8	20.2±10.9	a
1.0	5	29.8±21.3	b

[a]*No parasite clearance in 5/8 cases after 6 days.*
[b]*No parasite clearance in 3/5 cases after 6 days.*

240 mg was recommended for Phase II clinical trials. To facilitate the calculation of age-related doses, 60 mg (1.2 mg/kg) doses of intravenous artesunate were recommended for treating falciparum malaria. Subsequent doses should be delivered 4, 24, and 48 h after the first. For patients with severe infections (parasitemia ≥6%), the initial dose should be doubled. In effect, the 3-day course should total 240–300 mg (4.8–6 mg/kg) for adults and 1.5 mg/kg/dose for children below 7 years (Table 6.10).

Intravenous Artesunate Injection Versus Intravenous Quinine Drip in Treating Falciparum Malaria

Thirty adults with falciparum malaria were randomly divided into two groups, receiving either intravenous artesunate or a quinine drip. Artesunate powder injections were supplied by the Guilin No. 2 Pharmaceutical Factory. Each bottle contained 60 mg artesunate and 1 mL of 5% sodium bicarbonate. Before use, the sodium bicarbonate was injected into the artesunate bottle. This was shaken for 2 min until the artesunate had completely dissolved and the mixture was clear. Then, a needle was inserted to remove the air. Finally, 4 mL injectable saline or 5% glucose solution was added to bring the artesunate dose to 12 mg/mL. Corresponding volumes of the mixture were used in the injections for the various dose groups. The dose was injected over 2–3 min.

The first dose of quinine hydrochloride was 20 mg/kg, later reduced to 10 mg/kg for the latter 10 cases due to side effects. Thereafter, 10 mg/kg was administered every 8 h. The total dose for the 3-day course was 90–100 mg/kg.

TABLE 6.9 Parasite Clearance Speeds of Different Artesunate Doses (M ± SD)

Dose Group (mg/kg)	No. of Cases	Parasitemia (%)					
		4h	**8h**	**12h**	**16h**	**20h**	**24h**
8.0	6	85.6 ± 5.1	44.5 ± 6.6	7.6 ± 2.5	1.8 ± 0.6	0.6 ± 0.2	0.2 ± 0.1
6.0	6	79.5 ± 5.4	42.0 ± 7.9	6.5 ± 0.9	1.8 ± 0.9	0.3 ± 0.2	0.1 ± 0.1
4.0	6	81.6 ± 6.1	49.2 ± 4.5	6.4 ± 0.3	2.0 ± 0.4	0.3 ± 0.2	0.1 ± 0.1
2.0	5	78.3 ± 3.8	55.9 ± 6.4	16.3 ± 4.4	4.5 ± 1.3	1.9 ± 0.9	0.8 ± 0.4
1.0	4	78.7 ± 7.3	73.6 ± 8.4	30.2 ± 8.4	8.2 ± 5.5	2.8 ± 1.7	1.0 ± 0.6

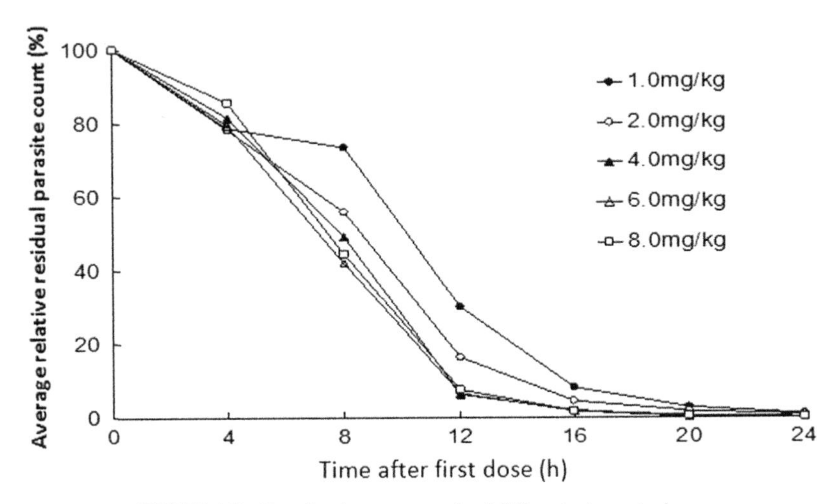

FIGURE 6.2 Parasite clearance speeds of different artesunate doses.

The average time taken for 95% parasite clearance was 16h for the artesunate group, as opposed to 28h for the quinine group. Hence, the artesunate group had significantly faster clearance than the quinine group did (Table 6.11; Fig. 6.3). Average defervescence times were 17.9h for the artesunate group and 29.2h for the quinine group; parasite clearance times were 58.8 and 78.8h, respectively. Both indicators were significantly faster in the artesunate group than in the quinine group ($P<.01$). However, the recrudescence rates were similar (56.7% artesunate, 50% quinine).

No clinical adverse reactions were reported in the artesunate group. The quinine group had 12 cases of tinnitus, eight cases of deafness, five cases of nausea, four cases of vomiting, two cases of dizziness, and one case of headache. All symptoms gradually disappeared after medication ceased. No drug-related abnormalities were seen in the hematology testing, blood chemistry panels, and routine urine tests of both groups. Abnormal findings on ECGs were seen in some cases. Two cases of incomplete right bundle branch block appeared in the quinine group. Three days after medication, three cases in the artesunate and 12 cases in the quinine group presented with sinus bradycardia. After 6days, this persisted in one case in the artesunate group and 11 cases in the quinine group.

3. Phase III Clinical Trials[7]

From 1985 to 1986, Phase III clinical trials of intravenous artesunate were conducted in Hainan and Yunnan's Xishuangbanna area. A total of 346 malaria cases were treated using the Phase II recommended dose (Table 6.10 above). This included 289 cases of falciparum malaria (with 31 cases of cerebral malaria), 56 cases of vivax malaria (one in coma, cause unidentified), and 1 case of *Plasmodium malariae* quartan malaria (in coma, cause unidentified).

TABLE 6.10 Intravenous Artesunate Doses

Age Group (years)	Total Dose (mg)	Dose per Injection (mg)			
		1st Injection	4 h	24 h	48 h
≥16	240	60	60	60	60
11–15	180	45	45	45	45
7–10	120	30	30	30	30
<7	6 mg/kg	1.5 mg/kg	1.5 mg/kg	1.5 mg/kg	1.5 mg/kg

N.B., Initial dose to be doubled for patients with severe infections (parasitemia ≥6%).

TABLE 6.11 Parasite Clearance Speed for Artesunate and Quinine

Drug	No. of Cases	Average Parasitemia (1000 per μL³)	Parasitemia (%, M ± SD)							
			4 h	8 h	12 h	16 h	20 h	24 h	28 h	32 h
Artesunate	6	41.3	81.4 ± 4.6	47.5 ± 10.6	15.2 ± 4.7	4.0 ± 1.6	2.0 ± 1.1	0.8 ± 0.4	0.3 ± 0.1	0.1 ± 0.05
Quinine	6	34.8	98.6 ± 5.8	98.1 ± 8.0	86.4 ± 5.5	80.1 ± 14.0	34.4 ± 9.9	9.3 ± 5.7	1.2 ± 0.3	0.8 ± 0.3

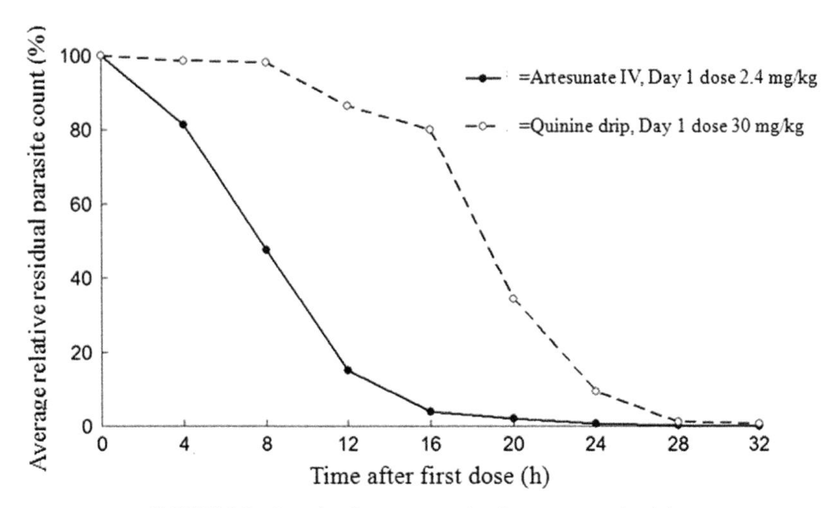

FIGURE 6.3 Parasite clearance speeds of artesunate and quinine.

Uncomplicated Falciparum Malaria

Of the 258 cases of uncomplicated falciparum malaria, 196 cases (75.9%) came from malaria-endemic areas, whereas 62 (24.1%) were from nonendemic areas. Males made up 154 of the cases (59.7%); 104 cases were female (40.3%). The youngest patient was 2 months old, and the oldest was 66 years old. In total, 78 cases were aged less than 16 years (26.2%); 180 cases were over 16 years (73.8%). When admitted to hospital, 14 cases were in the defervescence stage. For the remaining 244 cases, 23 (8.9%) had temperatures below 38°C, 58 (22.5%) had temperatures of 38–38.9°C, and 163 (63.2%) had temperatures of or above 39°C. Hepatomegaly was seen in 124 cases (48.1%), splenomegaly in 89 cases (34.5%), hemolysis in 2 cases, and severe liver damage in 1 case. Asexual parasitemia was 1045–467,677 per μL, mean 23,928 per μL. Erythrocyte count was below 2 million/μL in 15 cases (5.8%), 2–2.99 million/μL in 40 cases (15.5%), and above 3 million/μL for the rest. Artesunate could rapidly control clinical symptoms and eliminate parasites in the 258 cases (Table 6.12).

A total of 89 cases were followed up for 28 days. Recrudescence occurred in 44 cases, a rate of 49.4%. Artesunate injections were primarily developed to eliminate parasites and control symptoms rapidly and are used to treat patients with severe malaria. If oral administration is possible, a treatment course with an antimalarial with a long half-life—such as piperaquine, which is now used in combination with artemisinin—should also be given to prevent recrudescence.

TABLE 6.12 Artesunate in the Treatment of 258 Falciparum Malaria Cases

Research Units	No. of Cases	Defervescence Time (h) (M ± SD)	Parasite Clearance Time (h) (M ± SD)
Guangzhou University of Chinese Medicine[7]	173	16.1 ± 8.3	56.4 ± 14.5
Xishuangbanna Hospital, Yunnan	41	20.8 ± 12.5	37.7 ± 10.8
Changjiang, Dongfang County, and Tongshen Farm Hospital, Hainan	44	14.6–27.0	32.3–53.8

Randomized Comparative Study of Artesunate and Fansimef in Treating Falciparum Malaria in Children[8]

Fansimef tablets—each containing 250 mg mefloquine, 500 mg sulfadoxine, and 25 mg pyrimethamine—have been recommended by the WHO to treat chloroquine-resistant falciparum malaria. It was thus selected as the control. Patients were randomly divided between the two drug groups, each with 40 children. The regimen shown in Table 6.10 was used for artesunate. For Fansimef, those aged 7–10 years were given one tablet, those aged 11–15 received 1.5 tablets, and those aged above 16 received two tablets. Defervescence times were similar for artesunate (15.5 ± 7.5 h) and Fansimef (16.1 ± 11.4 h). Parasite clearance times were significantly faster for artesunate (54.1 ± 15.2 h) than for Fansimef (90.7 ± 38.9 h). No clinical side effects were seen in the artesunate group, whereas 10% of the cases in the Fansimef group experienced nausea and vomiting. A 28-day follow-up showed 40% recrudescence in the artesunate group, and none in the Fansimef group.

Cerebral Malaria[9]

Of the 33 cases of cerebral malaria, 13 were from malaria-endemic areas (39.4%), while 20 (60.6%) were from elsewhere. They were aged 4–55 years: 10 were below 16 years (30.3%) and 23 (69.7%) cases aged 16 and above. Two (6%) did not have fever (temperature on the low side). For the remaining 31 cases, 4 (12.1%) had temperatures of less than 38°C, 8 (24.3%) had temperatures of 38–38.9°C, and 19 (57.6%) had temperatures at or above 39°C. Thirteen (39.4%) had hepatomegaly, and seven (21.2%) had splenomegaly. Average duration of coma before treatment was 12.1 ± 7.9 h. Thirteen patients had one or more complications. Two were in the second or third trimester of pregnancy, and one was in the postpartum period. Falciparum malaria accounted for 31 of the cases, vivax malaria for one, and quartan malaria for one. Parasitemia was 1375–1,425,000 per μL. Seventeen cases (51.5%) were classified as having common cerebral malaria, 13 (39.4%) as severe, and 3 (9.1%) as very severe cerebral malaria.

After treatment, 31 cases were cured (93.9%) and 2 died (6.1%). Defervescence time was 29.9 ± 35.8 h, parasite clearance time was 54.9 ± 30.6 h, and time taken to recover from coma was 28.5 ± 38.9 h. One case presented with symptoms of brainstem damage, shock, renal, respiratory, and heart failure, severe anemia, acidosis, and postpartum infection. Parasitemia was 141,708 per μL, and calculated late trophozoite and schizont counts in peripheral blood smears were as high as 48,441 per μL. Another case had parasitemia of 1,425,000 per μL and an erythrocytic infection rate of 30.9%, combined with shock and acidosis. After treatment with artesunate and other supportive measures, the parasites were rapidly controlled and both cases were cured. No premature births or stillbirths occurred in the two patients who were in the late stages of pregnancy. The patients who died regained basic consciousness within 5 days of treatment, and parasite clearance was achieved in 67 and 77 h. However, they died on Days 11 and 12 due to other complications.

Vivax Malaria

The trial included 55 cases of vivax malaria; 39 (70.9%) were from malaria-endemic areas and 16 (29.1%) from nonendemic areas. Males made up 39 of the cases. The patients were aged 1–66 years; 22 (40%) were below 16 years, and 33 were 16 and above (60%). Nine cases (16.1%) were in the defervescence stage on admission. The other 46 had fever. Three had temperatures below 38°C (5.5%), 8 had temperatures between 38 and 38.9°C (14.5%), and 35 cases had temperatures at or above 39°C (63.6%). There was hepatomegaly in 14 cases (25.4%) and splenomegaly in 13 cases (23.6%). Average parasite density was 7750 per μL.

Clinical symptoms were rapidly controlled in all 55 cases. Defervescence time was 9.9 ± 6.4 h, and parasite clearance time was 44.7 ± 11.9 h. Thirty cases were followed up for 28 days, with 16 cases of relapse (53.3%).

Seven-Day Course of Intravenous Injections in Treating 40 Cases of Falciparum Malaria[10]

The results of the quinine comparison study showed that a 3-day course of 240 mg intravenous artesunate injections—initial dose 60 mg, followed by 60 mg 4, 24, and 48 h later—produced a recrudescence rate of 56.7% (see above). To see if a longer course of treatment could reduce the recrudescence rate, 40 cases of falciparum malaria were given a 7-day regimen. Intravenous artesunate was injected once a day, with 120 mg on the first day and 60 mg/day on Days 2–7, total dose 480 mg. All 40 cases were adults with uncomplicated falciparum malaria. Defervescence time was 14.3 ± 6.1 h, parasite clearance time was 51.5 ± 12.2 h, and recrudescence occurred in 2 out of 36 cases observed for 28 days (5.6%). No adverse reactions were seen. Hematology tests, AST/ALT, bilirubin, creatinine, routine urine tests, and ECGs showed no abnormalities after medication.

Side Effects

In total, 500 cases were treated with intravenous artesunate injections in the Phase II and III clinical trials. One patient with falciparum malaria experienced rashes 3 days after treatment, but it was impossible to confirm if this was due to the concurrent use of diisopropylamine and glucurolactone, or to artesunate itself. None of the other cases had systemic or local adverse reactions. Even when the initial dose was doubled to 120 mg, injected over 2–3 min, the drug was considered relatively safe. Heart rate, blood pressure, and ECGs were measured before and after injection. No cardiovascular adverse reactions were observed.

Hematology tests carried out on 87 patients before and after medication (on Days 3 and 6) showed no abnormalities. No aberrations were also seen in routine urine, serum bilirubin, and creatinine tests conducted at the same time for 86 cases. Fifty cases tested for alkaline phosphatase also yielded normal results.

For ALT, 2 out of the 50 cases in the 240 mg group displayed mild elevation before treatment. On Day 6 after medication, four cases experienced a mild increase in ALT levels, which returned to normal on Day 14. In the 240 mg group, AST levels were raised in seven cases before treatment; three returned to normal on Day 6 and four on Day 14. The 480 mg group had one case displaying elevated AST before treatment. On Day 6, a slight increase was seen in three cases, which returned to normal on Day 14. A link between changes in ALT and AST levels and drug-related side effects has not been proven. Such fluctuations may be due to the disease itself. In 33 cases of cerebral malaria, ALT levels were raised in 7 out of 20 patients. On Day 3, the levels had increased in nine cases; on Day 6, they had increased in seven cases. The elevation was clearly caused by parasite-induced multiple-organ damage in instances of severe infection.

In 82 cases given ECGs, mild sinus bradycardia appeared in only a few cases. One case showed first-degree atrioventricular block on Day 3, which returned to normal on Day 6. Such changes did not increase with a longer course of treatment or higher dose. Of the 40 patients given the 7-day, 480 mg course, 36 were given ECGs, which showed no abnormal changes. In the Phase I study, no ECG abnormalities were seen in the four volunteers given 3.75 mg/kg (equivalent to 750 mg over 3 days) and two given 4.22 mg/kg (844 mg over 3 days). Hence, a 3-day, 240 mg regimen was safe for the heart.

B. INTRAMUSCULAR ARTESUNATE IN TREATING FALCIPARUM MALARIA AND COMPARISON WITH PIPERAQUINE[11]

Intravenous artesunate injections had been shown to be reliably effective. However, they were not convenient to administer. From 1988 to 1989, Hainan's Dongfang County used intramuscular artesunate injections to treat falciparum malaria.

Three regimens were compared in adults: a 3-day, 240 mg course; a 5-day, 360 mg course; and a 7-day, 480 mg course. Their efficacy and occurrence of side effects were studied. Oral piperaquine phosphate was used as the control in this randomized test.

Inclusion criteria: Patients displayed the clinical symptoms of malaria within 5 days of admission; asexual parasitemia was over 1000 per μL. There was no age limit. Some patients were included if the onset of infection had occurred more than 5 days before, and parasitemia was over 5000 per μL. No other antimalarials, or drugs with antimalarial effects, had been used since the onset of symptoms. This included sulfonamides, tetracycline, sulfones, etc.

Artesunate powder injections were supplied by the Guilin No. 2 Pharmaceutical Factory in the form of a 60 mg bottle and a 1 mL ampoule of 5% injectable sodium bicarbonate. The batch number was 8701133. Before use, the contents were mixed to form sodium artesunate. The mixture was shaken for 2–3 min until completely dissolved. A needle was inserted to remove air, and 2 mL sterile water added. The resulting mixture had a concentration of 20 mg artesunate per milliliter. An appropriate amount was loaded into a syringe and injected. One 60 mg intramuscular injection was given daily, with the initial dose doubled. Children's doses were scaled down based on age. The regimen is shown in Table 6.13.

Sugar-coated piperaquine phosphate tablets were used. They were produced by the Shanghai No. 11 Pharmaceutical Factory, batch number 860801. Each tablet contained 150 mg piperaquine base. The total dose was 1500 mg in adults; 600 mg was administered at the first dose, followed by 300 mg after 6 h and 600 mg after 24 h. Children's doses were scaled down based on age (Table 6.14).

Data on the four studied groups is down in Table 6.15.

The 3-, 5-, and 7-day regimens had 30, 100, and 50 cases of falciparum malaria, respectively; all were cured. Average defervescence and parasite clearance times were similar in the three groups ($P > .05$). For the 3-day regimen, 25 cases were followed up for 28 days, resulting in 13 cases (52.0%) of recrudescence. For the

TABLE 6.13 Three-, Five-, and Seven-Day Artesunate Regimens

Age Group (years)	Total Dose per Regimen (mg)			Dose per day (mg)	
	3 day	5 day	7 day	Day 1	Days 2–7
≥16	240	360	480	120	60
11–15	180	270	360	90	45
7–10	120	180	240	60	30
<7	6/kg	9/kg	12/kg	3/kg	1.5/kg

5-day regimen, 9 out of 82 follow-up cases experienced recrudescence (9.8%). For the 7-day regimen, the corresponding number was one case out of 40 (2.5%). The recrudescence rates in the 5- and 7-day groups were not significantly different ($P>.05$), but the difference between the 7- and 3-day groups was $P<.01$.

Piperaquine phosphate was used with 51 patients. Average defervescence and parasite clearance times were significantly slower than with artesunate (Table 6.16). Clinical symptoms were not controlled within 48 h for 13 of the cases; parasitemia did not fall markedly or even continued to increase. Another 11 cases responded to treatment, but the parasites were not eliminated by Day 7. All these cases were treated with other drugs instead. Based on the 4-week observation protocol for chloroquine sensitivity, it was concluded that 17 cases in the piperaquine group were sensitive (cured, 33.3%), 10 cases displayed R_I resistance (19.6%), 11 cases had R_{II} resistance (21.6%), and 13 had R_{III} resistance (25.5%).

During the injection, some patients in the three artesunate groups experienced pain at the injection site, but this was tolerable. There was no local swelling or induration, and no other adverse reactions were reported in the three groups. In the piperaquine phosphate group, one case experienced nausea, and two had headache and dizziness.

After treatment, there were no obvious changes in the hematology tests, blood chemistry panels and urine tests of all four groups. ECGs were conducted on 50 cases in the 5-day regimen group and 40 cases in the 7-day regimen group. Sinus bradycardia (not less than 50 beats/min) was reported in one patient in the 5-day group on Day 6 and in four cases in the 7-day group on Days 3 and 6. In the piperaquine phosphate group, 31 patients were given ECGs. Frequent or occasional premature atrial contraction was seen in one case each on Days 3 and 6. One case presented with first-degree atrioventricular block on Day 3.

TABLE 6.14 Piperaquine Phosphate Regimen

Age Group (years)	Total Dose (mg)	Dose per Administration (mg)		
		First Dose	6 h	24 h
≥16	1500 (10 tablets)	600 (4 tablets)	300 (2 tablets)	600
11–15	1125 (7.5 tablets)	450 (3 tablets)	225 (1.5 tablets)	450
7–10	750 (5 tablets)	300 (2 tablets)	150 (1 tablet)	300
4–9	562.5 (3.75 tablets)	225 (1.5 tablets)	112.5 (0.75 tablets)	225
1–3	375 (2.5 tablets)	150 (1 tablet)	75 (0.5 tablets)	150

TABLE 6.15 Data of Artesunate Intramuscular Injection Groups and Oral Piperaquine Phosphate Group

Group	No. of Cases	Average Age (years)	Average Time of Infection (days)	Average Temperature (°C)	Average Parasitemia ($10^4/\mu L$)	Migrant Population (%)
Artesunate, 3d	30	18.8±9.8	5.9±3.1	38.9±1.0	3.1±3.8	16.7
Artesunate, 5d	100	16.0±10.8	5.7±4.1	39.0±1.5	4.2±7.1	16.0
Artesunate, 7d	50	18.5±9.5	5.0±2.8	38.9±1.0	4.2±9.3	32.0
Piperaquine phosphate, 2d	51	18.5±8.8	5.7±4.3	39.1±0.9	3.0±4.1	23.5

TABLE 6.16 Efficacy Across Four Studied Groups

Group	No. of Cases	Defervescence Time (h)	Parasite Clearance Time (h)	Recrudescence Rate (%)
Artesunate, 3d	30	22.4±12.0	66.4±8.5	52.0
Artesunate, 5d	100	21.6±10.7	68.2±8.7	9.8
Artesunate, 7d	50	19.5±11.8	64.1±8.1	2.5
Piperaquine phosphate	51	37.3±16.8[a]	104.6±16.8[a]	37.0[b]

[a] *$P < .01$ when compared to the artesunate groups.*
[b] *Only 27 sensitive and R_I cases included.*

A total of 180 patients were treated in the artesunate groups, and all were cured. Clinical symptoms were rapidly controlled and parasites swiftly eliminated. Defervescence time was 19–22 h, and parasite clearance time was 64–68 h, comparable to the results obtained with the intravenous injection. The recrudescence rate for the 3-day group was 52%, as opposed to 56.7% for the intravenous injection. It showed that the intramuscular and intravenous injections had similar efficacies against falciparum malaria. The recrudescence rates of the 5- and 7-day regimens were 9.3% and 2.5%, respectively, significantly lower than that of the 3-day group. Lengthening the course of treatment could reduce the recrudescence rate.

In the piperaquine phosphate control group, clinical symptoms were not controlled in 13 of the 51 cases, in which parasitemia did not fall or increased instead. Symptoms were controlled in 11 cases, but the parasites had not been eliminated by Day 7. In 1987, piperaquine had been used to treat 53 cases of falciparum malaria in the same region. The cure rate was 83%; 13.2% had R_I resistance and 3.8% had R_{II} resistance. The latest results demonstrated that falciparum malaria had reduced susceptibility to piperaquine in this area, and that artesunate was effective in treating chloroquine- and piperaquine-resistant falciparum malaria.

As mentioned above, some patients experienced pain at the injection site, but this was tolerable. It did not affect treatment and there were no other adverse reactions. Blood and other tests showed no abnormalities. In the ECGs, four cases (10%) in the 7-day group and three cases (9.7%) in the piperaquine phosphate group displayed sinus bradycardia. Previous trials saw sinus bradycardia in 12 out of 30 cases (40%) given quinine and 4 cases out of 18 (22.2%) given artemisinin–pyronaridine phosphate combination. This could be due to the drugs' stimulating effects on the heart, or to the disease itself.

The efficacies of the intravenous and intramuscular injections were similar, but the latter were easier to use. By increasing the dose and extending the course of treatment, the cure rate could reach 90% and above. Therefore, artesunate was not only an emergency drug for the rapid treatment of severe malaria but also a highly effective therapy.

C. ARTESUNATE TABLETS

1. Phase I Clinical Trial[12]

The tolerability of artesunate tablets and their side effects in healthy volunteers were studied to provide evidence for a safe regimen. A total of 22 volunteers were included in the study; all were male, aged 18–45 years. The average age was 29.7 years, average weight 59.4 kg. Volunteers had no blood, cardiovascular, liver, gastrointestinal, parasitic, or other diseases, as well as no allergies, high myopia, or retinitis. They had no history of alcoholism, were in good nutritional condition, and fundus and physical examinations showed no pathologies. Chest X-rays were normal. ECGs, hematology and blood chemistry panels, and urine and fecal tests were conducted twice before medication; all were normal.

Artesunate tablets, each containing 50 mg artesunate, and placebos (flour tablets) were supplied by the Guilin Pharmaceutical Factory. Four single-dose groups and three 5-day regimen groups were used. The predicted therapeutic dose was 2 mg/kg. For the 5-day regimen groups, drug was administered once a day for 5 days, with the first dose doubled. The doses and regimens are in Table 6.17.

The doses were calculated based on body weight. Volunteers in the control group were given identical placebos (flour tablets).

The study was double-blind and volunteers were randomly assigned. Each volunteer was hospitalized for 9 days, from Day −1 to Day 7. Follow-up checks were performed on Days 21 and 28. Body temperature, pulse, and respiratory rate were recorded every 4 h. Blood pressure and frequency and consistency of urination and defecation were recorded every morning and evening until Day 4. On Days 5–7, body temperature, pulse, and respiratory rate were measured in the morning and afternoon, alongside frequency and consistency of urination and defecation. Vision and fundus tests were run on Days −1, 3, and 7. Every morning and evening, volunteers were surveyed on their appetite, sleep, and symptoms relating to the skin and gastrointestinal, central nervous, and cardiovascular systems. Time of onset and disappearance were accurately recorded. If there was discomfort at any other time, this was checked and recorded. ECGs, hematology tests, blood chemistry panels, and routine urine tests were run on Days 1, 3, 7, 14, and 28. Any abnormalities on Day 28 were reviewed weekly until the indicator returned to normal.

If adverse reactions appeared clinically or in laboratory indicators of any dose group higher than the predicted therapeutic dose, the test was stopped for the next dose group. The test was also halted at the next dose group if reticulocyte count decreased by over 50% of the initial value, or if ALT activity rose above normal and did not recover within 7 days.

No clinical symptoms and laboratory indicators sufficient to stop the test appeared in the 8 mg/kg single-dose and 24 mg/kg 5-day groups. Therefore, the trial proceeded as planned with the 12 mg/kg single-dose and 36 mg/kg 5-day groups.

After medication with artesunate tablets, three cases experienced fever. The first case was in the 12 mg/kg, single-dose group; fever occurred 12 h after treatment, which lasted for 8 h. The second case was in the 12 mg/kg, 5-day group, with onset of fever at 12 h after the first dose and lasting 4 h. The third case was in the 36 mg/kg, 5-day group; the fever began 32 h after the first dose and lasted for 8 h. The highest temperature was 38°C, accompanied by fatigue and malaise. All cases recovered without treatment. No clinical symptoms arose in the other volunteers.

All medicated volunteers had reduced reticulocyte levels on Day 3. Except for the highest dose group, they returned to normal on Day 7. Changes in reticulocyte levels before and after administration are shown in Fig. 6.4 (single-dose groups) and Fig. 6.5 (5-day groups).

TABLE 6.17 Artesunate Tablet Regimens and Doses in Phase I Clinical Trial

Regimen	Total Dose (mg/kg)	Day 1	Day 2	Day 3	Day 4	Day 5	No. of Medicated Cases	No. of Control Cases
Single	2	2	–	–	–	–	2	0
	4	4	–	–	–	–	2	0
	8	8	–	–	–	–	2	1
	12	12	–	–	–	–	2	1
5-day	12	4	2	2	2	2	3	1
	24	8	4	4	4	4	3	1
	36	12	6	6	6	6	3	1
Total no. of cases							17	5

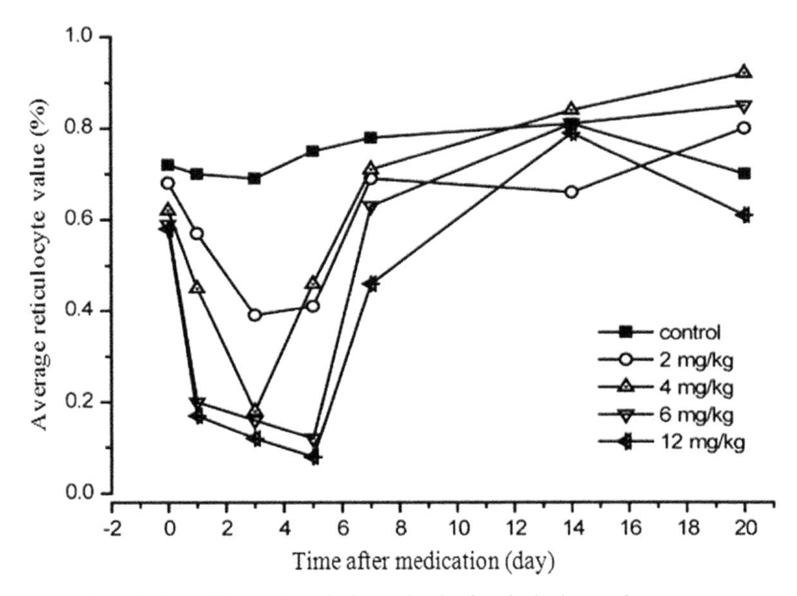

FIGURE 6.4 Changes in reticulocyte levels after single-dose oral artesunate.

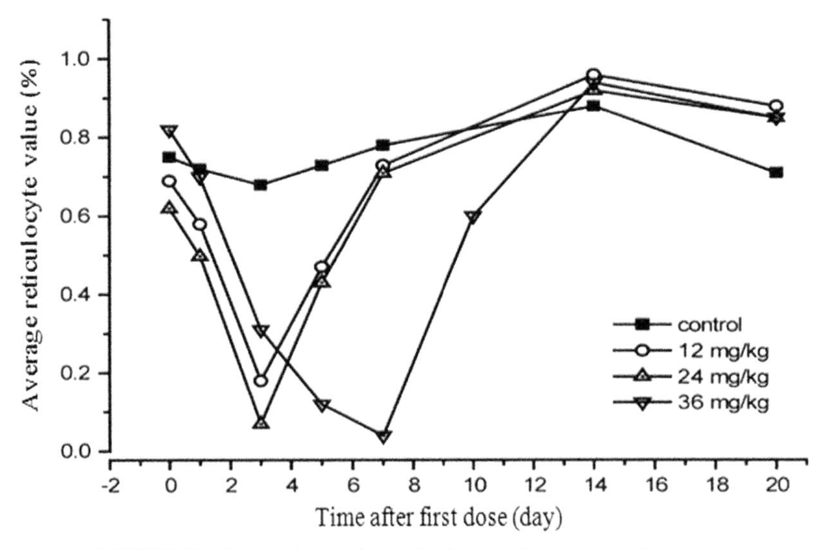

FIGURE 6.5 Changes in reticulocyte levels in the 5-day course of artesunate.

Transaminase levels increased slightly in two cases. On Day 3, a small rise in ALT and AST was seen in one case in the 24mg/kg, 5-day group and one case in the 36mg/kg, 5-day group. They returned to normal on Day 7. This did not occur in other volunteers. Other laboratory tests and ECGs were reviewed several times after treatment. No abnormalities were seen.

No clinical, ECGs, or laboratory abnormalities were detected for the control group.

The Phase I trial yielded no adverse clinical symptoms or signs, except for three cases with low fever of short duration. The incidence of fever was 3 out of 17 (17.6%). A pharmacokinetics study of intravenous artesunate showed that the drug was rapidly transformed into dihydroartemisinin upon entering the body. The $t_{1/2\beta}$ of artesunate was around 30min, and it was difficult to detect in blood after 3h. By contrast, the $t_{1/2\beta}$ of dihydroartemisinin was around 50min, and the drug was almost undetectable in blood after 6h. In this trial, all cases of fever occurred 8h after medication. It indicated that the fever was not caused by artesunate but may be related to its metabolites.

The observations on reticulocyte levels showed that a fall in reticulocytes was a very sensitive indicator of the biological responses to artesunate tablets. Except for a mild decrease in one case in the 2mg/kg, single-dose group, a significant reduction was seen in all other cases on Day 3. The degree and persistence of this decline increased with the dose. In general, this returned to normal on Day 7. Levels in the highest, 36mg/kg dose group only recovered on Day 10. A serum enzyme assay showed that changes in AST and ALT were also sensitive indicators.

The results showed that the maximum tolerated dose of artesunate tablets in healthy people was 12mg/kg in a single dose or 24mg/kg over 5days. It was proposed that a 5-day regimen of 12mg/kg artesunate tablets could be used as a therapeutic dose.

2. Phase II Clinical Trials[13,14]

Artesunate and Piperaquine Phosphate Tablets

Patients with malaria symptoms were selected who had asexual parasitemia of over 1000 per μL, or who had been symptomatic more than 5days before and had parasitemia of over 5000 per μL. They had not used any antimalarials or drugs with antimalarial effects, such as sulfonamides, tetracycline, sulfones, etc.

Three drug groups were used: 3- and 5-day courses of artesunate, and a 2-day course of piperaquine phosphate as control. Artesunate tablets were supplied by the Guilin Pharmaceutical Factory, each containing 20mg artesunate. The first dose was 80mg, followed by 40mg after 6h, then 40mg every morning and afternoon. The total dose was 280mg over 3days or 440mg in 5days. Children aged 11–15years received 3/4 of the adult dose, those aged 7–10 received half the adult dose, those aged 3–6 received 3/8 of the adult dose, and those aged two and below received 1/4 of the adult dose.

Each piperaquine phosphate tablet contained 150mg piperaquine base. Adults received 10 tablets over 2days: four at the first dose, two at 6h, and four at 24h. The total dose was 1500mg. The doses for different age groups were the same as those outlined above. Data on the patients are shown in Table 6.18. The results are shown in Table 6.19.

Defervescence and parasite clearance times for the two artesunate groups were similar ($P>.05$). In the 280mg, 3-day group, 41 cases were followed up for 28days. There were 21 cases of recrudescence, a rate of 51.2%. The average time of recrudescence was 20.5days. The same follow-up was conducted on 45 cases in the 440mg, 5-day group. This had two cases of recrudescence (4.4%). Average time of recrudescence was 27days. The recrudescence rate was lower in the 5-day group than in the 3-day one ($P<.01$). Defervescence and parasite clearance times were longer for the piperaquine phosphate group than for the 5-day group ($P<.01$). The recrudescence rate (17.9%) was also higher than that of the 5-day group ($P<.05$). The average time of recrudescence was 26days. R_{II} level resistance was also seen in two of the 50 cases in the piperaquine phosphate group (no parasite clearance by Day 7).

No adverse reactions were seen in both artesunate groups. The piperaquine phosphate group had three cases of nausea, three cases of vomiting, two cases of dizziness, and two cases of headache. The rate of gastrointestinal side effects was 12%, and the rate of side effects involving the central nervous system was 8%. Before medication and on Days 3 and 7, all three groups underwent hematology testing, serum total bilirubin, ALT, AST, alkaline phosphatase, creatinine, and urine tests, as well as ECGs. No clear abnormalities were seen.

Both artesunate groups had similar defervescence and parasite clearance times, but the recrudescence rate of the 5-day group was obviously lower than that of the 3-day group. A 5-day course of artesunate tablets, total dose 440mg (equivalent to 8.8mg/kg) had no significant clinical side effects and produced no abnormal changes in the tests mentioned above. It indicated that such a regimen was well tolerated.

In both artesunate groups, the average time taken to achieve 95% parasite clearance after the first dose was 20h. This showed that oral artesunate acted rapidly. Tablets were easier to administer than injections, and the recrudescence rate of the 440mg, 5-day course was low. Therefore, it was a good therapeutic regimen. There were no obvious adverse reactions in the artesunate groups, as opposed to a 12% rate of gastrointestinal and 8% incidence of central nervous system side effects in the piperaquine phosphate group. It indicated that artesunate had a lower toxicity than piperaquine phosphate.

3. Phase III Clinical Trials[15]

In the 1990s, a 5-day course of 600mg artesunate was the standard regimen for falciparum malaria. It was used in falciparum malaria patients with partial immunity, yielding a 28-day cure rate of over 90%. This rate could fall

TABLE 6.18 Pretreatment Data on Three Drug Groups

Group	No. of Cases	Average Temperature (°C)	Average Parasitemia (n/μL)	Migrant Population (%)	Incidence of Splenomegaly (%)
Artesunate (3d, 280 mg)	50	39.1 ± 1.1	29,648	20.6	17.5
Artesunate (5d, 440 mg)	50	39.0 ± 1.2	30,284	28.0	16.0
Piperaquine phosphate (2d, 1500 mg)	50	38.9 ± 1.3	29,380	22.0	22.0

TABLE 6.19 Artesunate and Piperaquine Phosphate Tablets in the Treatment of Falciparum Malaria

				28-day Follow-Up		
Group	Cases	Defervescence Time (M ± SD)	Parasite Clearance Time (M ± SD)	No. of Cases	Recrudescence Cases	Recrudescence Rate (%)
Artesunate (3d, 280 mg)	50	17.9 ± 9.3	61.7 ± 14.9	41	21	51.2
Artesunate (5d, 440 mg)	50	18.2 ± 8.6	65.5 ± 16.9	45	2	4.4
Piperaquine phosphate (2d, 1500 mg)	50	27.5 ± 21.6	100.3 ± 20.3[a]	39	7	17.9

[a]*Mean in 48 cases.*

in nonimmune patients. To refine the appropriate dose and regimen of artesunate for patients without immunity, the Institute of Tropical Medicine at the Guangzhou University of Chinese Medicine conducted a randomized trial in southern Vietnam. It compared the standard 5-day, 600 mg regimen with a 7-day, 800 mg course.

Patients had the clinical symptoms of malaria, were aged 7–65 years, and had asexual parasitemia of 1000 per μL and above or large ring-form parasitemia of 100 per μL and above, accompanied by fever. They had not used antimalarials or drugs with antimalarial effects, such as tetracycline or sulfonamides. As far as possible, patients were chosen who had come from elsewhere, had no history of malaria, or had only been infected a few times before. Pregnant women, patients with severe nausea and vomiting, and those aged below 7 or above 65 were excluded. Cases were randomly divided. Both dose groups had 50 patients each.

Artesunate tablets were supplied by the Guilin Pharmaceutical Factory, batch number 940801. Each tablet contained 50 mg artesunate. Drugs were administered once daily, with 100 mg/dose. The initial dose was doubled. The 5-day course had a total dose of 600 mg; the 7-day course totaled 800 mg. The doses were reduced for children related to age (Table 6.20). Each administration was supervised. Another dose was given if the patient vomited within 2 h of medication.

All patients were hospitalized for 28 days to eliminate the possibility of reinfection after the 7-day course. The patients' data are shown in Table 6.21.

Clinical symptoms were rapidly controlled in all cases. Defervescence and parasite clearance times were similar in both groups. The 28-day follow-up was completed for 47 cases in the 5-day group and 48 cases in the 7-day group. The former had nine cases of recrudescence, a rate of 19.1%; the latter had two cases of recrudescence, a rate of 4.2%. This was a significant difference ($P<.01$, Table 6.22).

No adverse reactions were seen in both groups. Hematology tests, blood chemistry panels, ECGs, and routine urine tests showed no abnormalities.

TABLE 6.20 Artesunate Tablet Regimens in Treating Falciparum Malaria

Age (years)	Dose on Day 0 (mg)	Dose on Days 1–6 (mg)	Dose Over 5 days (mg)	Dose Over 7 days (mg)
≥16	200	100	600	800
11–15	150	75	450	600
7–10	100	50	300	400

TABLE 6.21 Pretreatment Conditions of Trial Patients

Regimen	No. of Cases	Age (years)	% Nonimmune	Temperature (°C)	Parasitemia pcs/μL	% Splenomegaly	% Hepatomegaly	WBC 1000/μL	RBC 10,000/μL
5d	50	24.9±9.5	88.0	38.7±0.8	17.229±23.776	30.0	16.0	5.864±1.463	345±64
7d	50	25.4±9.1	80.0	38.8±1.0	13.622±14.101	32.0	30.0	6.151±1.874	324±73

N.B., For nonimmune patients, see case selection.

TABLE 6.22 Two Artesunate Regimens in Treating Falciparum Malaria

Regimen	No. of Cases	Defervescence Time (h) M±SD	Parasite Clearance Time (h) M±SD	Cure Rate (%)	Recrudescence Rate (%)
5d	50	24.0±13.6	61.7±21.1	80.9	19.1
7d	50	20.0±10.1	55.8±15.8	95.8	4.2

Artemisinins have been used for malaria treatment since the early 1980s. Until now, no studies with a large sample size reporting a near-100% cure rate in nonimmune patients have appeared. Bunnag et al.[16] had reported that a 1200 mg, 5-day artesunate regimen achieved a 100% cure rate, but only six cases were treated, making the sample size too limited. Guo Xingbo et al. used a 480 mg, 5-day course of artemether intramuscular injections on 105 falciparum malaria patients with partial immunity. The cure rate was 96.2% but was reduced to 88.2% in 85 nonimmune patients. This was a significant difference. The results of a Phase II clinical trial of dihydroartemisinin showed that a 480 mg, 7-day course cured 205 partially immune patients, a cure rate of 98%. A Phase III trial on the same regimen, which treated 117 cases with partial immunity, yielded a cure rate of 95.7%. For 30 nonimmune patients, however, the cure rate was only 86.7%. An antimalarial drug should be evaluated primarily based on its efficacy in nonimmune patients, making the conclusions more reliable.

A standard regimen and dose of artemisinin-type drugs capable of producing a near-100% cure rate has not been identified. The mentioned study explored regimens for oral artesunate. It suggested that, given a higher proportion of nonimmune patients, the cure rate of a 600 mg, 5-day course of artesunate was 80.9%. An 800 mg, 7-day course had a cure rate as high as 95.8%. The difference was statistically significant. When combined with what is known about artemisinins through clinical practice, the 800 mg artesunate regimen could be used therapeutically and recommended as a standard treatment.

6.3 ARTEMETHER

Guo Xingbo, Li Guoqiao

Institute of Tropical Medicine, Guangzhou University of Chinese Medicine, Guangzhou, Guangdong Province, China

Artemether is an artemisinin derivative developed by the Shanghai Institute of Materia Medica, China Academy of Sciences (SIMM of CAS). From 1978 to 1980, various regimens of artemether-oil injection were used to treat 1088 patients in Hainan, Yunnan, Guangxi, Henan, Hubei, and other sites. This included 829 cases of falciparum malaria[17,18] and 259 cases of vivax malaria. This yielded a clinical cure rate of 100%. A regimen of 600–640 mg was used with 457 cases of falciparum malaria. Of these, 343 cases were followed up for a month; the recrudescence rate was 7.8%. No confirmed, drug-related adverse reactions were reported. To further evaluate the efficacy and toxicity of the drug and explore a suitable treatment dose, the Guangzhou University of Chinese Medicine (Guangzhou UCM) and clinical research units in Yunnan and Hainan conducted Phase I, II, and III clinical trials on the oil injection in 1986 based on China's Provisions for New Drug Approval. The injections were produced by the Kunming Pharmaceutical Factory.

A. PHASE I CLINICAL TRIAL[19]

The double-blind trial included 32 healthy male adult volunteers, randomly grouped. Three- and 5-day courses were used; for the regimens, see Table 6.23. The same volume of drug was administered, with 2 mL injected each time, but at different concentrations. The control group received 2 mL oil injections. Volunteers were hospitalized for 9 days (from 2 days before medication to Day 7). Clinical signs were observed and recorded daily. ECGs, hematology tests (five factors), blood chemistry panels (12 factors), and routine urine tests were conducted on Days −2, −1, 1, 3, 7, 14, and 28. Routine fecal tests were run until Day 7.

Abnormal reactions occurred in the clinical symptoms and laboratory tests in both the 3- and 5-day groups, to different degrees. These manifested in three main ways.

1. Fever

All drug groups had one or a few cases of fever on Day 2 (Table 6.24). These cases also showed a concurrent loss of appetite, to varying degrees. Some had slightly elevated white blood cell count and neutrophils, but this was only observed on Day 1.

2. Decrease in Reticulocytes

Reticulocyte levels fell in each drug group on Day 3, which became more obvious on Day 5. Levels were on the lower end of the normal range in the 7.2 mg/kg, 5-day group; the 9.6 mg/kg, 5-day group; and the 12 mg/kg, 5-day group. In the 12 mg/kg, 3-day group and the 16 mg/kg, 3-day group, levels had decreased to 0.4% and 0.46%, respectively, on Day 3, and 0.47% and 0.36%, respectively, on Day 5. Reticulocyte count began to rise on Day 7 and reached premedication levels on Day 14 (Fig. 6.6).

TABLE 6.23 Artemether Injection Dose Design in the Phase I Clinical Trial

Total Dose (mg/kg)	Dose Per Day (mg/kg)					Medicated Cases	Control Cases
	D1	D2	D3	D4	D5		
7.2, 5d	2.4	1.2	1.2	1.2	1.2	3	1
9.6, 5d	3.2	1.6	1.6	1.6	1.6	5	2
12.0, 5d	4.0	2.0	2.0	2.0	2.0	4	2
12.0, 3d	6.0	3.0	3.0	–	–	6	2
16.0, 3d	8.0	4.0	4.0	–	–	5	2
Total						23	9

TABLE 6.24 Fever in Volunteers Given Artemether Injections

Dose Group	No. of Cases	Cases With Fever	Average Highest Temperature (°C)	Length of Fever (h)
7.2, 5d	3	1	37.8	32
9.6, 5d	5	1	37.8	4
12.0, 5d	4	1	38.5	24
12.0, 3d	6	4	38.5	50.5
16.0, 3d	5	4	39.1	59
Control group	9	1[a]	38.5	4

[a]*Volunteer also presented with upper respiratory tract symptoms.*

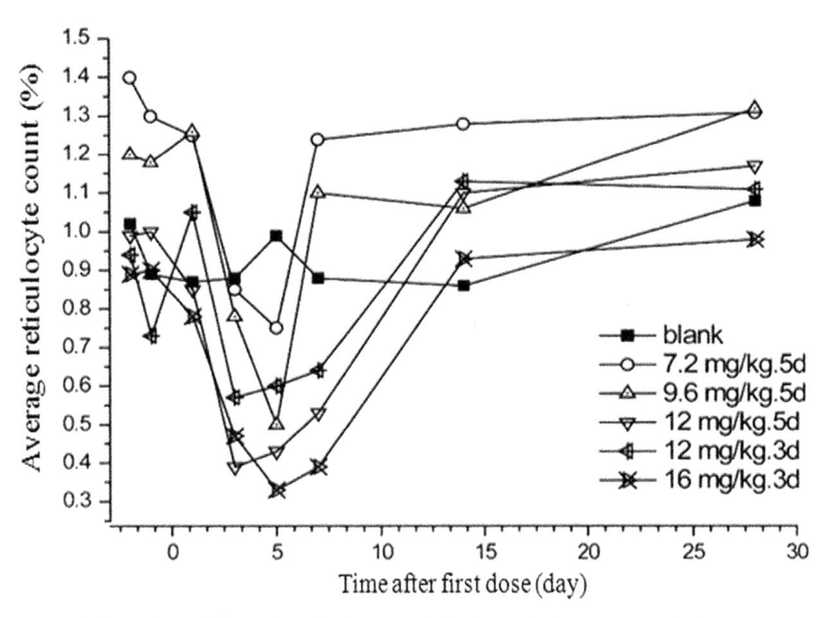

FIGURE 6.6 Changes in reticulocyte levels before and after artemether injections.

3. Increase in Transaminase Activity

No abnormalities in AST and ALT activity were seen in the 7.2 mg/kg, 5-day group and the 9.6 mg/kg, 5-day group. Two out of four cases in the 12 mg/kg, 5-day group had slightly raised AST on Day 7. On Day 14, one case still had elevated AST, but this returned to normal on Day 28. Of the six cases in the 12 mg/kg, 3-day group, AST was raised in two cases on Day 1, and in five cases on Day 3. They returned to normal on Day 7. One case had slightly elevated ALT on Day 3, which returned to normal on Day 7. Four of the five cases in the 16 mg/kg, 3-day group had raised AST on Day 3. One had elevated ALT. By Day 7, high AST level persisted in one case, but recovered on Day 14.

Apart from the above three reactions, the volunteers did not display other side effects or abnormal indicators. The results of the Phase I trial showed that the 480 mg, 5-day regimen (9.6 mg/kg) could be put forward for Phase II trials.

B. PHASE II CLINICAL TRIALS[19]

Patients had the clinical symptoms of malaria and were aged 7–65 years. They had malaria symptoms within 5 days of admission, with asexual parasitemia at 1000 per μL or above, or large ring-form density of 100 per μL or above. They had not used other antimalarials or drugs with antimalarial effects, such as tetracycline or sulfonamides. Pregnant women, those with severe nausea and vomiting, and those aged below 7 or over 65 were not included.

1. Dose-Finding Study of Artemether-Oil Injections in Treating Falciparum Malaria

Based on the Phase I clinical trial, the efficacies of a 9.6 mg/kg, 5-day regimen and a 7.2 mg/kg, 5-day regimen were compared. The 9.6 mg/kg group and 7.2 mg/kg group had 10 and 7 cases of falciparum malaria, respectively. All experienced parasite clearance in 6 days. After the first dose, the average time taken to achieve 95% parasite clearance was 20 h in both groups. Defervescence time was 22.2 ± 8.1 h for the 9.6 mg/kg group and 26.1 ± 7.4 h for the 7.2 mg/kg group. Parasite clearance time was 76.5 ± 16.5 h for the former and 77.6 ± 17 h for the latter. The differences between both groups were not significant. Nine cases in the 9.6 mg/kg group and five cases in the 7.2 mg/kg group were followed for 28 days; there was no recrudesecence.

Previously, a 3- or 4-day course of 600–640 mg artemether (equivalent to 12–12.8 mg/kg) was used to treat adults with falciparum malaria, but the recrudescence rate was 7.8% after a 1-month follow-up. Therefore, a 5-day, 9.6 mg/kg regimen was recommended for further Phase II trials. For ease of administration, 9.6 mg/kg was equivalent to 480 mg in adults, with 160 mg on the first day and 80 mg on Days 2–5.

2. Intramuscular Artemether Injections and Piperaquine Phosphate Tablets in Treating Falciparum Malaria[20]

Falciparum malaria patients were divided randomly into groups and given either intramuscular artemether injections or piperaquine phosphate tablets. Each group had 30 cases. For the 5-day, 480 mg regimen in adults, see Table 6.25 below.

TABLE 6.25 Artemether Injection Regimens in Treating Falciparum Malaria

Age (years)	Total Dose (mg)	Dose Per Day (mg)				
		D1	D2	D3	D4	D5
≥16	480	160	80	80	80	80
11–15	360	120	60	60	60	60
7–10	240	80	40	40	40	40
<7	9.6/kg	3.2/kg	1.6/kg	1.6/kg	1.6/kg	1.6/kg

For piperaquine phosphate, adults were given 1500mg piperaquine base: 600mg in the initial dose, 300mg after 6h and 600mg after 24h. Doses for patients aged 7–15 years were scaled down based on weight.

Average defervescence and parasite clearance times were 20.6 and 75.9h, respectively, in the artemether group, significantly faster than 35.2 and 92.8h, respectively, in the piperaquine phosphate group. The recrudescence rates for both drugs were 4.2% (1/24) for artemether and 10.3% (3/29) for piperaquine. In the piperaquine group, two recrudescence cases were treated with artemether injections and observed for 28 days; no recrudescence occurred.

One case in the artemether group experienced defervescence after treatment but displayed one bout of transient fever thereafter. This response may have been caused by the drug. Before treatment, four cases in the artemether group and three cases in the piperaquine group had raised ALT. Two cases in the former and one case in the latter still showed elevated levels 3 days after treatment. On Day 6, four artemether and six piperaquine cases had elevated ALT. This was not conclusively proven to be drug related.

3. Comparing Intramuscular Artemether Injections and Pyronaridine Phosphate Combination in Treating Falciparum Malaria

Falciparum malaria patients were randomly grouped into intramuscular artemether injection and oral pyronaridine phosphate combination therapy groups. The former had 34 cases and the latter had 33 cases. The regimen used for the artemether injections was the 480mg, 5-day course. The pyronaridine combination therapy included 300mg pyronaridine base, 500mg sulfadoxine, and 25mg pyrimethamine, administered at 0, 6, and 24h. Doses for those aged 7–15 years were calculated based on age.

Average defervescence times were 22.1h for artemether and 21.3h for the pyronaridine combination therapy; average parasite clearance times were 74.1 and 71.1h, respectively. The differences between both groups were not significant. After a 28-day follow-up, 2 out of 30 cases in the artemether group experienced recrudescence (6.7%), but no recrudescence occurred in the 33 cases in the pyronaridine phosphate combination therapy group.

After treatment, one case in the artemether group displayed defervescence but went on to have two bouts of transient fever. No other clinical side effects were seen. For the pyronaridine phosphate combination group, three cases reported gastrointestinal reactions (9.1%), including one case of diarrhea, two cases of nausea, and one case of vomiting. Symptoms disappeared without treatment. Reticulocyte count decreased in 10 cases in the artemether group, falling slightly below normal levels. No such abnormal changes were seen in the pyronaridine group. All other test indicators were normal in both groups.

In the blood chemistry panels, two cases in the artemether group had raised ALT before treatment. This returned to normal after treatment. Another two cases had normal levels before treatment, but this increased 6 days after medication. Before treatment, two cases had elevated AST, and one recovered after treatment. Another two cases were normal, but AST increased on Day 6.

The pyronaridine phosphate combination therapy group had one case with increased ALT on Day 3, which returned to normal on Day 6. Two cases had elevated levels before treatment, recovering on Day 6. Three had raised AST before treatment, which recovered thereafter. High ALT and AST could have been caused by the disease.

C. PHASE III CLINICAL TRIAL OF ARTEMETHER-OIL INJECTIONS[19]

Artemether injections were provided by the Kunming Pharmaceutical Factory, batch numbers 850615-26, 860502, and 860520. Each 1 mL ampule contained 100 mg artemether. The total dose over 5 days was 480 mg. Adults were given 80 mg intramuscular injections a day, with the dose doubled on the first day. Children's regimens are shown in Table 6.25.

Body temperature was checked before medication and every 4 h thereafter until the patients were discharged on Day 7. Defervescence was defined as a reduction in temperature to normal levels, with no further rise within 24 h. If the temperature rose again after 24 h and other causes were excluded, the fever could be drug related.

Before the initial dose, thick blood films were obtained to calculate the number of asexual parasites per microliter. After medication, further checks were conducted at 8 a.m. and 4 p.m. every day. Once parasites were absent in three consecutive films, blood was tested once a day until Day 7. Parasite clearance was taken to be the absence of asexual parasites in the entire film. The time taken to reach this was the parasite clearance time.

1. Falciparum Malaria

A total of 234 falciparum malaria cases were treated, of whom 141 were from malaria-endemic areas (60.3%) and 93 were not (39.7%). Males made up 171 of the cases (73.1%); 63 cases were female (26.9%). In terms of age, 54 cases were under 16 years (23.1%), and 180 cases were 16 and above (76.9%). Nine (3.8%) were in the defervescence stage on admission, while the other 225 had fever. Parasitemia was 7097–70,131 per μL. Thirteen had critical malaria, including seven with cerebral malaria. Of these, five had common cerebral malaria, one was severe, and one was very severe (parasitemia 1,332,840 per μL, erythrocytic infection rate 43.2%, severe damage to liver and kidney function, hemolysis, gastrointestinal hemorrhage, and cerebral edema).

The 221 noncritical cases—including two in late pregnancy—and 13 critical cases were given artemether injections. Clinical symptoms were rapidly controlled and parasites eliminated (Table 6.26). In the two cases of late pregnancy, no premature births, stillbirths, or other adverse reactions were reported. The seven cerebral malaria cases regained consciousness within 7–48 h.

Fourteen cases experienced recrudescence. The recrudescence rate in the local population was 3.8%, as opposed to 11.8% for migrants (Table 6.27). It indicated that the recrudescence rate was linked to immunity.

Plasmodium falciparum from 30 cases was cultured in vitro and tested for susceptibility to chloroquine. Resistance was found in 56.7% cases, showing that artemether injections were effective against chloroquine-resistant falciparum malaria.

2. Side Effects

Out of the 234 cases treated, the body temperature in four cases returned to normal but then showed a transient increase. This could be related to the drug's side effects. Three cases had mild diarrhea, four had abdominal pain, and one had tinnitus. All these symptoms disappeared spontaneously before the termination of treatment. Therefore, it could not be confirmed if they were drug related.

Reticulocytes were checked in 76 cases. Nine cases showed a mild decline on Day 3, as opposed to eight cases on Day 6. ECGs were run on 39 cases. On Day 3, bradycardia appeared in three cases and atrioventricular block in one. This increased to five and two cases, respectively, on Day 6. By the second week, all ECGs were normal.

Routine blood tests were run on 226 patients before treatment and on Days 3 and 6. Apart from this, 43 cases were examined for alkaline phosphatase, bilirubin, creatinine, ALT, and AST, and routine urine tests were performed on 37 cases. None showed obvious abnormal changes.

3. Summary

The Phase I clinical trials indicated that a safe dose for intramuscular artemether injections in healthy adults was 9.6 mg/kg, equivalent to a 5-day course of 480 mg. This 5-day course was used to treat 308 falciparum malaria cases in the Phase II and III clinical trials, including 13 cases of critical malaria; all

TABLE 6.26 Efficacy of Artemether Injections in Treating 234 Cases of Falciparum Malaria

				Recrudescence After 28 d Follow-Up		
Research Unit	**No. of Cases**	**Defervescence Time (M ± SD)**	**Parasite Clearance Time (M ± SD)**	**No. of Cases**	**Recrudescence Cases**	**Recrudescence Rate (%)**
Guangzhou University of Chinese Medicine	57	21.1 ± 10.8	76.2 ± 14.3	48	2	4.2
People's Hospital, Dongfang County	90	23.4 ± 16.5	70.9 ± 16.6	56	4	7.1
Jianfengling Staff Hospital, Hainan	31	21.3 ± 12.8	69.3 ± 13.4	31	0	0
Yunnan Institute for Malaria Prevention and Treatment	27	27.9 ± 12.2	31.9 ± 13.3	27	2	7.4
Xishuangbanna Autonomous Prefecture Hospital, Yunnan	29	29.7 ± 11.2	53.8 ± 23.5	28	6	21.4
Total	234	20–30	48–76	190	14	7.4

TABLE 6.27 Recrudescence Rate in Local and Migrant Populations After Treatment With Artemether Injections

Population	No. of Cases	Recrudescence Cases	Recrudescence Rate (%)
Local	105	4	3.8
Migrant	85	10	11.8

were cured. Defervescence time was 20–30h, and parasite clearance time was 32–76h. A total of 253 cases were followed up for 28days, with 17 cases of recrudescence, a rate of 6.7%.

The treatment dose produced mild adverse reactions and was relatively safe. Artemether injections were highly effective, had rapid action and low toxicity, and were suitable for treating severe malaria.

6.4 DIHYDROARTEMISININ TABLETS

Guo Xingbo, Li Guoqiao
Institute of Tropical Medicine, Guangzhou University of Chinese Medicine, Guangzhou, Guangdong Province, China

A. PHASE I CLINICAL TRIAL[21,22]

This double-blind trial included 24 healthy adult volunteers: 13 males and 11 females. They were randomly assigned to groups and hospitalized for 5days (Days 0–4). Follow-up checks were conducted on Days 7, 14, and 28. ECGs, hematology tests, blood chemistry panels, and routine urine and fecal tests were run on Days −4, −2, 1, 3, 7, 14, and 28. The regimens are shown in Table 6.28. Doses were given once a day for 3days, with the first dose doubled.

No adverse reactions were seen in any of the dose groups. Apart from reticulocyte count, the ECGs, hematology tests, blood chemistry panels (including serum protein, glucose, bilirubin, blood urea nitrogen, creatinine, cholesterol, triglycerides, potassium, sodium, calcium, phosphorus, chloride, alkaline phosphatase, cholinesterase, ALT, and AST), and routine urine and fecal tests showed no abnormal changes. On Days 1 and 3, mean reticulocyte levels decreased slightly in the 9 and 11mg/kg groups but were still within normal range. One case in the 11mg/kg had levels on the lower limit of normal; all returned to premedication levels on Day 7. It was concluded that the maximum tolerated 3-day dose of dihydroartemisinin in healthy people was 11mg/kg (equivalent to 550mg in adults).

TABLE 6.28 Doses Used in the Phase I Clinical Trial of Dihydroartemisinin Tablets (mg/kg)

Total Dose	Dose on Day 1	Dose on Day 2	Dose on Day 3	Increase over Previous Dose (%)	Medicated Cases	Control Cases
0.5	0.2	0.15	0.15			
1.5	0.6	0.45	0.45		2	0
3.0	1.2[a]	0.90	0.90	100.0	2	1
5.0	2.0	1.50	1.50	66.7	5	2
7.0	3.0	2.00	2.00	40.0	3	1
9.0	4.0	2.50	2.50	28.6	3	1
11.0	5.0	3.00	3.00	22.2	3	1

[a] *Predicted treatment dose.*

B. PHASE II CLINICAL TRIALS

1. Dose-Finding Study of Dihydroartemisinin in Treating Falciparum Malaria[23]

Research data on 3-day courses of artesunate injections and tablets showed that there was no significant difference in efficacy between once-daily doses and multiple doses per day. The recrudescence rate was around 50%. To raise efficacy and reduce recrudescence, the regimen had to be extended and the total dose increased. Therefore, a 240 mg, 3-day course and a 360 mg, 5-day course of dihydroartemisinin were compared, with doses administered once daily (Table 6.29). Children's doses were scaled down based on age.

Each group had 30 falciparum malaria patients, with 22 adults and 8 children; all were clinically cured. Defervescence times were 18.5 ± 8.9 and 23.4 ± 15.1 h for the 3- and 5-day groups, respectively. Parasite clearance times were 68.6 ± 11.8 and 64.2 ± 14.8 h for the 3- and 5-day groups, respectively. The 28-day recrudescence rates in the 3- and 5-day groups were 52% and 5.6%, respectively, a significant difference ($P < .01$). A recrudescence rate of 5% was comparable to the high international standards on the treatment of falciparum malaria. However, the number of cases was limited. The trial had to be expanded, including comparison with a 7-day regimen.

In another trial, seven adults were given single oral doses of 120 mg dihydroartemisinin. Time taken to achieve 95% parasite clearance was 16 h. This was comparable to the parasite clearance speeds of intramuscular and intravenous artesunate injections (see above).

2. Comparison of 5- and 7-Day Dihydroartemisinin Tablets Regimens and Piperaquine Phosphate Tablets[24]

A 5-day, 360 mg course and a 7-day, 480 mg course of dihydroartemisinin were used to treat 50 cases of falciparum malaria each. The doses in the first 5 days of the 7-day course were the same as those of the 5-day course, with the addition of 60 mg on Days 6 and 7. The control group adopted

TABLE 6.29 Regimens Used in Two Dose Groups

Regimen	No. of Cases	Total Dose (mg)	Daily Dose (mg)				
			D1	D2	D3	D4	D5
3d	30	240	120	60	60	–	–
5d	30	360	120	60	60	60	60

the then-standard piperaquine phosphate treatment, with an initial dose of 600 mg piperaquine base, followed by 300 mg at 6 h and 600 mg at 24 h. This was given to 51 cases. The 5-day group had 36 adults and 14 children; the 7-day group had 35 adults and 15 children; and the control group had 37 adults and 14 children.

Defervescence times were 16.3 ± 6.6 h for the 5-day course and 19.7 ± 13.2 h for the 7-day course. Parasite clearance times were 64 ± 12 h for the former and 66.5 ± 10.4 h for the latter. The difference was not significant. The recrudescence rate in the 5-day group was 6.3% (2/32), as opposed to none in the 7-day group (0/42). In the piperaquine group, clinical symptoms were not controlled in 11 cases, with either no significant decrease or even an increase in parasitemia (R_{III} resistance). Symptoms were controlled in 13 cases, but the parasites were not eliminated by Day 7 (R_{II} resistance). Another 10 cases had recrudescence within 28 days (R_I resistance). Only 17 cases (33.3%) were cured (sensitive). The 5- and 7-day courses of dihydroartemisinin were significantly more effective than piperaquine. Local strains of *P. falciparum* had clear resistance toward piperaquine (Table 6.30).

During dihydroartemisinin treatment, two cases in the 7-day group presented with a measles-like rash, with no itching or pain. Symptoms were mild. The rash spontaneously disappeared in one case after 12 h. For the other case, the rash subsided on Day 9, 2 days after termination of treatment.

On Day 3, 13 cases in the 5-day group (26.5%) and 19 cases in the 7-day group (38.8%) had slightly reduced reticulocyte levels. All the cases in the former group returned to normal on Day 6, but three in the latter group had not. However, the average value had returned to pretreatment levels.

Six days after the first dose, 1 out of 29 tested cases in the 5-day group had sporadic premature atrial contractions. One out of 32 tested cases in the 7-day group presented with sporadic premature ventricular contractions on Days 3 and 6. Three cases also had first-degree atrioventricular block on Day 6, one of whom recovered on Day 14 and the other two after a month. In the piperaquine group, one case out of 31 tested had frequent, and another case had sporadic premature atrial contractions on Day 6. One of these cases also presented with second-degree atrioventricular block on Day 3. Further research in expanded trials is needed to confirm if these changes were drug related.

There was no significant difference between the recrudescence rates of the 5- and 7-day courses. However, the rate of the 5-day group was significantly lower than that of the 3-day group. The similarity between the 5- and 7-day groups could therefore be due to the small sample size. In addition, even when the dose was increased to 480 mg and the course of treatment extended to 7 days, there were still no obvious side effects. Based on the artesunate and artemether trials, which showed that longer courses could improve the cure rate, the 7-day, 480 mg regimen was put forward for the Phase III clinical trial.

TABLE 6.30 Dihydroartemisinin Regimens and Piperaquine Phosphate in Treating Falciparum Malaria

Group	No. of Cases	Defervescence Time (h) M ± SD	Parasite Clearance Time (h) M ± SD	28d Follow-Up					
				No. of Cases	S	R_I	R_{II}	R_{III}	Cure Rate (%)
Dihydroartemisinin, 5d	50	16.3 ± 6.6	64.0 ± 12.0	32	30	2	0	0	93.7
Dihydroartemisinin, 7d	50	19.7 ± 13.2	66.5 ± 10.4	42	42	0	0	0	100.0
Piperaquine phosphate	51	37.3 ± 16.8[a]	104.6 ± 16.8[b]	51	17	10	13	11	33.3

[a] *$P < .01$ when compared to dihydroartemisinin groups.*
[b] *Not including 24 cases with either no parasite clearance by Day 7 or with treatment failure.*

C. PHASE III CLINICAL TRIALS

From 1988 to 1989, 189 cases of falciparum malaria were given the 7-day, 480 mg course of dihydroartemisinin in the Sanya and Dongfang regions of Hainan, where chloroquine- and piperaquine-resistant falciparum malaria was endemic.[21,25–27] Of these, 119 cases were adults (including two in the fifth to sixth months of pregnancy), and 70 were children.

1. Efficacy

In all cases, clinical symptoms were rapidly controlled and parasites were cleared. The recrudescence rate was low (Table 6.31).

2. Side Effects

On Day 2, one case presented with a measles-like rash on the chest, abdomen, and lower limbs. This subsided on Day 4 after antiallergy treatment. No other clinical side effects were reported. The two pregnant women did not experience abortion or other adverse reactions.

Reticulocyte counts were run on 113 cases. Ten (8.8%) had slightly depressed levels on Day 3 but had basically returned to pretreatment levels on Day 6. This showed that a transient decline in reticulocytes was mainly related to the double dose given at the first administration. The subsequent maintenance doses did not have obvious effects on reticulocytes.

ECGs were conducted on 43 cases, of which four had sinus bradycardia on Days 3 and 6. Before the first dose, two cases had premature atrial contractions, which increased to five cases on Days 3 and 6. In the Phase II trials, the piperaquine control group had two cases with premature atrial contractions on Days 3 and 6. Further study is needed to determine if the increase in the incidence and frequency of premature contractions during treatment is related to the two drugs.

Routine hematology tests were run on 178 cases before treatment and on Days 3 and 6. ALT, AST, alkaline phosphatase, bilirubin, and creatinine levels were measured in 39 cases, and the urine of 50 cases was analyzed. No abnormal changes were seen.

D. CONCLUSION

The Phase I clinical trial showed that the maximum tolerated oral dose of dihydroartemisinin in healthy adults was 11 mg/kg, equivalent to 550 mg over 3 days. In areas where drug-resistant falciparum malaria was endemic, dihydroartemisinin was highly effective and had rapid action. A 360 mg, 5-day regimen had a 28-day cure rate of 90% and above, and a 480 mg, 7-day regimen had a cure rate of over 98%.

Out of all the cases given the 5- and 7-day regimens, 104 were given ECGs. Two (1.9%) presented with premature atrial contractions before treatment, and one (1%) with first-degree atrioventricular block. After treatment, four cases

TABLE 6.31 Efficacy of Dihydroartemisinin in Treating 189 Cases of Falciparum Malaria

Research Unit	No. of Cases	Defervescence Time (h)	Parasite Clearance Time (h)	Recrudescence in 28d	
				No. of Cases	Recrudescence Rate (%)
Guangzhou University of Chinese Medicine	69	19.5 ± 10.2	69.2 ± 16.1	63	1.6
Guangba Farm Hospital	44	24.7 ± 13.0	66.4 ± 9.5	34	2.9
Nanxin Farm Hospital	30	18.8 ± 12.9	65.3 ± 10.0	22	0
Tian'an Health Center	46	14.5 ± 7.6	64.9 ± 15.9	44	4.5

(3.8%) had premature atrial contractions, and one (1%) had premature ventricular contractions. Three cases (2.9%) had first-degree atrioventricular block; one recovered on Day 14 and two after a month. Based on the existing data, it was difficult to conclude if dihydroartemisinin had cardiac side effects. The dose in the Phase I trial in healthy volunteers was significantly higher than those in the Phase II and III trials, and no adverse cardiac reactions were found. Moreover, similar cardiac reactions appeared in the piperaquine group, including two cases with premature atrial contraction (6.5%) and one with first-degree atrioventricular block (3.2%). Since cardiac side effects occurred both before treatment and in the control group, and since dihydroartemisinin did not cause obvious cardiac changes in healthy people, the abnormal ECGs could be due to the disease itself. Other effects included skin rash (0.9%) and a slight decline in reticulocyte levels, but these rapidly returned to normal. This indicated that dihydroartemisinin was relatively safe.

The dihydroartemisinin regimens were easy to administer and highly effective, could act rapidly, and could successfully treat chloroquine- and piperaquine-resistant falciparum malaria. Side effects were mild. A 7-day, 480mg regimen could be used in a clinical setting.

6.5 EVALUATION OF PARASITE CLEARANCE SPEEDS AND EXPLORATION OF REGIMENS

Li Guoqiao, Guo Xingbo
Institute of Tropical Medicine, Guangzhou University of Chinese Medicine, Guangzhou, Guangdong Province, China

A. PARASITE CLEARANCE RATE AND THE CONCEPTS OF R FEVER AND T FEVER, AND R COMA AND T COMA

To evaluate an antimalarial's asexual parasite clearance rate, the parasite's developmental stage and morphology must be considered. What stage of the 48-h developmental stage was the parasite in before medication? After medication, at what stage did development stop? By examining the stage at which development terminated, it would be possible to accurately ascertain the stage targeted by existing antimalarials and those under study, such as chloroquine, quinine, febrifugine (from Chinese medicinal plant *Dichroa febrifuga*), pyronaridine, sulfadoxine-pyrimethamine (Fansidar), artemisinin, piperaquine, mefloquine, artesunate, artemether, dihydroartemisinin, etc. Combining the morphologic information and the four-hourly parasitemia calculation, it would be possible to evaluate clearance rate of antimalarials used.

In the 48-h cycle, when parasites develop into the schizont stage at 26–40h, large ring forms mature into late-stage trophozoites whose nuclei divide to form early schizonts. The schizonts continue to mature and rupture at the 48th hour. From 26 to 48h, late-stage trophozoites and schizonts are sequestrated on capillary walls and cannot normally be detected in peripheral blood smears. Due to this sequestration, from 28 to 32 h, the density of large ring forms in peripheral blood declines sharply over a 4- to 6-h period, until they disappear completely. However, sequestrated parasites can be detected on intradermal blood smears. Correspondingly, Li Guoqiao at Guangzhou University of Chinese Medicine defined two fever types and two coma types based on the morphology of the major brood of parasites detected in the peripheral or intradermal blood smears. "R fever" occurs when the major parasite brood is in tiny to large ring forms, and coma in this period is termed "R coma"; "T fever" occurs when late trophozoites and schizonts are sequestrated on capillary walls; and corresponding coma is termed "T coma."

Since 1972, Li's group has used intradermal blood smears to observe the developmental stages of the parasites from 26 to 48 h of the cycle. Therefore, it was possible to establish if slow-acting antimalarials caused parasitic maturation to cease at the late-trophozoite, early-schizont or late-schizont stage. Fig. 6.7 shows the results of four-hourly blood smear changes conducted before and after the administration of various antimalarials.[28]

In Fig. 6.7, all antimalarials (in oral, injectable, or suppository form) were given 6 h after the first fever appeared, when the major brood parasites were at the tiny ring-form stage of the developmental cycle, corresponding to the R_1 phase of the R fever period. Peripheral blood smears were obtained every 4 h until hour 26 after drug administration; then changed to four-hourly intradermal blood smears thereafter. If the drug was ineffective against late-stage trophozoites, intradermal blood smears were checked every 2 h after hour 38 to observe when mitosis ceased. This protocol made it possible to observe the stage at which parasitic development terminated. In terms of clearance rate, various antimalarials could be roughly divided into three categories.

1. Slow-Acting Antimalarials

A single dose of three sulfadoxine-pyrimethamine tablets (Fansidar) were administered at the tiny ring-form stage (R_1). The parasites continued to develop normally to the small ring-form (R_2) and large ring-form (R_3) stages. From hour 28 to 32 of the developmental cycle, they continued to mature into late-stage trophozoites and were sequestrated in the capillaries. The second febrile episode then occurred (T fever), and the parasites continued to develop into early schizonts. Each nucleus divided into four to six nuclei but could not proceed further to the late-schizont stage. In patients with high parasitemia, when the parasites were eliminated in large numbers, another febrile episode ensued, which was considered as drug fever.

These slow-acting drugs should not be used to treat nonimmune falciparum malaria patients. In patients with an erythrocytic infection rate of 6% or above, the parasites could still develop into late-stage trophozoites the day after medication and coma and other symptoms of severe malaria could result. This type of coma, Li termed T coma, is related to the sequestration of late trophozoites

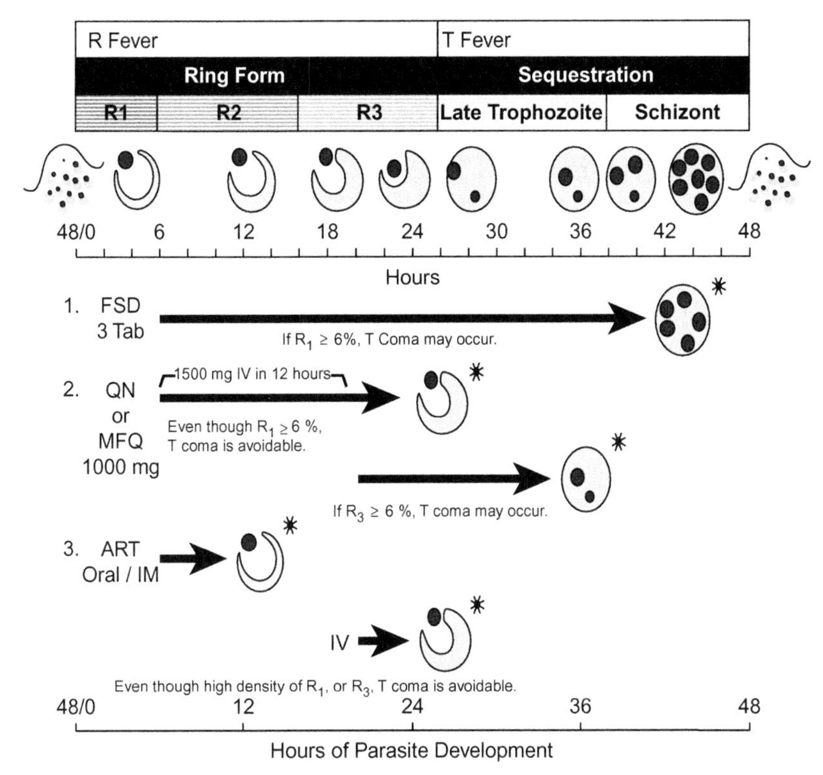

FIGURE 6.7 Clinical and morphological effects of antimalarials on asexual *Plasmodium falciparum*. *ART*, artemisinin and derivatives; *FSD*, Fansidar; *MFQ*, mefloquine; *QN*, quinine; *, stage when asexual parasites cease development.

which causes capillary blockage. In pregnant women, this often results in stillbirth and miscarriage.

2. Intermediate-Term Antimalarials

Chloroquine, quinine, febrifugine (from plant *D. febrifuga* used in traditional Chinese medicine), and mefloquine fall into this category. A full first dose of quinine (1500 mg intravenous infusion over 12 h, i.e., 500 mg every 4 h and not every 8 h), or a single dose of four mefloquine tablets (total 1000 mg), could halt parasitic development in the large ring-form stage (R_3). The parasites did not progress to late-stage, large trophozoites. Therefore, coma in patients with high parasitemia (erythrocytic infection rate 6% and above) may be avoided through use of these drugs. The parasite clearance rates of chloroquine (on nonresistant strains), piperaquine, and pyronaridine are slightly faster than those of quinine and mefloquine. A full first dose terminates parasitic development at the small to large ring-form stage. However, in patients in whom large ring forms are already present at the time of diagnosis and with parasitemia of 6% and above, an immediate administration of a quinine drip or a full oral dose of chloroquine, piperaquine, or mefloquine will not prevent further development into late-stage trophozoites. In such cases, T coma is unavoidable.

3. Fast-Acting Antimalarials

Artemisinins are fast-acting drugs. Regardless of whether the parasites are in the small or large ring-form stage, development ceases 2h after medication. Morphologically, tiny ring forms do not develop into small ring forms (R_2), and large ring forms (R_3) do not become late-stage trophozoites. The following observation was made on the drugs' effects on early schizonts of *P. vivax*: "Oral huanghaosu/artemisinin was administered as the parasitic nucleus began to split into two to three nuclei. From 2 to 7h after medication, blood smears were obtained every hour. Most of the parasites had halted at the two- to three-nuclei stage. A few had produced four to five nuclei. This indicated that, 2h after medication, the development of *P. vivax* schizonts could be controlled."[29] Damage to the membrane system of rodent and simian malaria was observed via electron microscopy 2h after medication.[30,31]

Artesunate injections were introduced in 1980. Because these injections were very safe and the entire initial 120mg dose could be delivered in 5min, a first administration was sufficient to halt parasitic development within 2h. Previously, patients with a high density of large ring forms ($R_3 \geq 6\%$) would experience further parasitic development, with these large ring forms (R_3) becoming late trophozoites (T) in a few hours. The patients would then enter a coma, known as T Coma (described above), which was very difficult to treat. An immediate intravenous infusion of quinine could not act sufficiently quickly to alleviate these circumstances. With intravenous artesunate, however, this development could be halted and coma avoided.[5]

B. EXPLORATION OF ARTEMISININS' TREATMENT REGIMENS

Before 1985, 3-day regimens were used for artemisinin tablets, injections, and suppositories in the treatment of falciparum malaria. The recrudescence rate was 50%–85% for oral preparations, as opposed to 10%–25% for oil- or water-suspension intramuscular injections. This was clearly related to the drug's retention time in the body. Impressed by the long treatment course used for quinine, 3-, 5-, and 7-day courses of water-soluble artesunate injections and oral preparations of other artemisinin derivatives were compared. The 28-day recrudescence rates of the 5- and 7-day courses were 6% and 3%, respectively. It showed that artemisinin had similar properties to quinine, which also had a short half-life. A 7-day regimen was needed to achieve a high cure rate. This could be a characteristic shared by all drugs with short half-lives. The cure rate was not 100%, since a recrudescence rate of around 5% remained in inverse proportion to the level of immunity. Given an identical course of treatment and dose, the recrudescence rate was higher in nonimmune patients (e.g., children) than in patients with partial immunity (e.g., adults in highly endemic areas).

6.6 ARTEMISININS IN TREATING OTHER PARASITIC DISEASES

Li Ying
Shanghai Institute of Materia Medica, Chinese Academy of Sciences, Shanghai, China

The ability of artemisinins to powerfully eliminate erythrocytic *Plasmodium* led scientists to speculate if they would exert the same effects on other parasites in the blood or other organs. Beginning in the 1980s, Chinese researchers made important discoveries in this area.

A. SCHISTOSOMIASIS PREVENTION

Schistosomiasis is a zoonotic disease that also affects humans and is endemic to tropical and subtropical countries. *Schistosoma japonicum* is widespread in China; historically, the incidence of schistosomiasis was around one million cases per annum. Some 60–70 million people are at risk of the disease. Currently, praziquantel is the sole treatment for this disease, but its efficacy has declined in recent years. Resistance to oxamniquine has also emerged.

In the 1980s, Chinese scientists began researching the effects of artemisinin on *S. japonicum* in animals.[32,33] It was discovered that the antiparasitic effects of artemether were relatively high.[34] In an experiment on the susceptibility of *S. japonicum* at different life stages, it was found that juvenile parasites—the stage reached 7 days after mice were infected by cercariae—were most sensitive to artemisinins. The drugs were also effective against juvenile parasites on the 35th day of development.[35] By contrast, praziquantel had preventive effects only 2–8 h after administration.[36] Attention then turned from treatment to prevention.

The National Institute of Parasitic Diseases, Chinese Center for Disease Control and Prevention, together with the Institute of Parasitic Diseases of the Zhejiang Academy of Medical Sciences, conducted large-scale clinical trials with the local schistosomiasis prevention units in Jiangxi, Hunan, Anhui, and Yunnan. The subjects were divided into two groups. The medicated group received 6 mg/kg artemether or artesunate 1 week after exposure to the infective agent in contaminated water. Artemether was administered every 2 weeks, whereas artesunate was given weekly. The control group received a starch placebo. One more dose was given after exposure to the infective agent ceased. Fecal tests were conducted a month after the final dose. The incidence of parasitic ova was markedly different in both groups,[37] showing that artemether and artesunate were highly effective in preventing *S. japonicum* infection.[38]

In 1996, artemether and artesunate were approved for schistosomiasis prophylaxis in China. This drew the attention of the WHO and other international scientists. Laboratory tests found that artemether and artesunate were also effective against juvenile *Schistosoma mansoni* and *Schistosoma haematobia*. Clinical trials conducted in Thailand, Egypt, Tanzania, Kuwait, Morocco, and other countries showed that both drugs had broad-spectrum prophylactic effects against schistosomiasis.[39–42]

Research into the mechanism of action by which artemisinins prevent schistosomiasis must also be emphasized. Artemisinin affects glucose metabolism in the parasite. Artemether not only inhibits the absorption of glucose through its integument, disrupting energy metabolism in the parasite, but also affects intestinal digestion of glycogen and suppresses alkaline phosphatase activity.[43]

Later it was discovered that artemisinins had different effects on juvenile and adult parasites, indicating that different mechanisms of action were present at various life stages. Recent in vitro experiments showed that a combination of artemether and haemin could kill the *Schistosoma* parasite. This did not occur when either substance was used in isolation.[44]

B. PIROPLASMOSIS

The *Babesia* and *Theileria* parasites, which target the red blood cells of domestic animals, have similar developmental stages to those of *Plasmodium*. In the 1980s, to find an effective treatment for *Theileria annulata*, Northwest A & F University used fresh *Artemisia annua* L. to successfully treat 18 cows.[45] The young leaves and twigs of *A. annua* were sliced thinly and soaked in cold water for 30–60 min. Each cow was fed 2–3 kg of this preparation daily, until erythrocytic parasitemia fell to 1% and below. Before medication, the cows' body temperatures were 40°C and above, some reaching 42°C. They returned to normal after 12 days of treatment. After 2–6 days of medication, erythrocytic parasitemia fell to 1% and below. Changes to the morphology and color of the gametocytes were observed after 2–4 days of treatment, and they became smaller in size. Up until 1986, this treatment was used on 66 cows infected with *T. annulata*, and 12 cows infected with *Babesia bigemina*. Six cows infected with *T. annulata* were treated with shade-dried *A. annua*. The total cure rate was 97.6%.

Subsequently, 134 cases with *T. annulata* and three cases with *B. bigemina* were given oral artesunate. Two cases of *B. bigemina* received artemether. The total cure rate was 99.3%, with all indicators significantly superior to those of the fresh *A. annua* regimen.[46]

The morbidity and mortality of *T. annulata* infection in milk cows are high. When 131 infected milk cows were given 35 mg/kg oral artesunate, the cure rate was 99.2% for a regimen of only 3.47 ± 0.75 days.[47] Artesunate and artemether also produced very good results in the treatment of theileriosis in sheep.[48]

C. PARAGONIMIASIS

The consumption of raw seafood can lead to food-borne paragonimiasis, a zoonotic disease caused by the lung fluke. It has been reported that artesunate, when used together with praziquantel, had a certain level of efficacy in 16 cases of paragonimiasis.[49]

D. CLONORCHIASIS

Clonorchiasis is a zoonotic hepatobiliary disease caused by *Clonorchis sinensis*, otherwise known as the liver fluke, which targets the bile ducts. Infection is mainly caused by the consumption of undercooked seafood. The disease is relatively widespread in China. The antiparasitic activity of artemisinins was tested in rats infected with *C. sinensis*. Some of the derivatives were several times more effective than artemisinin itself.[50]

E. TOXOPLASMOSIS

Toxoplasmosis is a zoonotic, parasitic disease. Most patients are carriers, but acute symptoms can occur when immunity is lowered. Current treatments, such as pyrimethamine, sulfadiazine, and spiramycin, inhibit the growth of *Toxoplasma gondii* during the proliferation phase. Reports from China have shown that artemisinins also suppress the development of *T. gondii*. The lowest effective in vitro concentration of artemether against *T. gondii* was 0.1 g/mL. At 4 g/mL, morphological changes were seen in most of the parasites after 48 h, with vacuoles and particles appearing in the cytoplasm. At the same concentration, the spiramycin control was not as efffective.[51]

Several research groups examined the effects of artemether[52] and dihydroartemisinin[53] in mice with acute toxoplasmosis. Both drugs produced similar results to the sulfadiazine control. Overseas studies have also reported on the inhibitory effects of artemisinins on toxoplasmosis in rodents.[54,55]

Doctors in Jiangsu used dihydroartemisinin to treat 18 cases of symptomatic toxoplasmosis.[56] Each adult dose was 40 mg, administered twice daily over 6 days (total dose 480 mg). This regimen was repeated after 5–7 days and was given to all 18 patients, apart from one case who received two courses of treatment. After 2 months, IHA and ELISA tests in 12 cases were negative. Another four cases showed total clearance after 4 months. The cure rate was 88.9%.

REFERENCES

1. Collaborative Research Group on Artemisinin Research. Research into a new antimalarial, artemisinin. *Chin Pharm Bull* 1979;**14**(2):49–53. [青蒿研究（全国）协作组, 抗疟新药青蒿素的研究. 药学通报. 1979, 14(2): 49–53.]
2. National Malaria Treatment Study Group. Certification for artemisinin. In: *Evaluation conference on research into artemisinin as an antimalarial.* Held at the Center for the History of Medicine, Peking University; November 28, 1970. [《青蒿素鉴定书》, 1970年11月28日全国疟疾防治研究领导小组主持"青蒿素（黄花蒿素）治疗疟疾科研成果鉴定会".]

3. Collected data on the use of artemisinin suppositories to treat 463 cases of malaria. In: Guo XB, Li GQ, editors. *Clinical data in support of the new-drug application for artemisinin suppositories*. Held at the Center for the History of Medicine, Peking University; 1994. [郭兴伯, 李国桥整理. 青蒿素栓剂治疗疟疾463例资料综合(青蒿素栓新药申报资料临床研究部分, 1994).]
4. Guo XG, Fu LC, Jian HX, et al. Tolerance and adverse reactions of intravenous artesunate injections in healthy people: phase I clinical trial. *Tradit Chin Drug Res Clin Pharmacol* 1991;**2**(2):29–31. [郭兴伯, 符林春, 简华香等. 健康人静脉注射青蒿琥酯的耐受性和毒副反应研究——I期临床试验.中药新药与临床药理. 1991, 2(2): 29–31.]
5. Li GQ, Guo XB, Jin R, et al. Clinical studies on treatment of cerebral malaria with qinghaosu and its derivatives. *J Tradit Chin Med* 1982;**2**(2):125.
6. Li GQ, Guo XB, Fu LC, et al. Summary of clinical research into intravenous artesunate injections in treating malaria. *Collect Clin Res Artemisinins* 1990;**8**:80–4. [李国桥, 郭兴伯, 符林春等. 青蒿琥酯静脉注射治疗疟疾临床研究总结. 《青蒿素类药临床研究专辑》. 1990, 8: 80–84.]
7. Guo XB, Fu LC, Liu GP, et al. Intravenous artesunate injections in the treatment of 173 cases of falciparum malaria. *J Guangzhou Univ Chin Med* 1988;**5**(1):9–12. [郭兴伯, 符林春, 刘光平等. 青蒿琥酯静脉注射治疗恶性疟疾173例. 广州中医药大学学报. 1988, 5(1): 9–12.]
8. Guo XB, Arnold K, Fu LC, et al. Double-blind dose-finding study of mefloquine-sulfadoxine-pyrimethamine in children with acute falciparum malaria. *Trans R Soc Trop Med Hyg* 1988;**82**:538–40.
9. Guo XB, Fu LC, Fan TT, et al. Clinical report of intravenous artesunate injections in treating 33 cases of cerebral malaria. *Collect Clin Res Artemisinins* 1990;**8**:28–30. [郭兴伯, 符林春, 范太涛等. 青蒿琥酯静脉注射治疗脑型疟33例临床报告. 《青蒿素类药临床研究专辑》. 1990, 8: 28–30.]
10. Guo XB, Fu LC, Li GQ. Observations on a seven-day course of intravenous artesunate injections in treating falciparum malaria. *J Guangzhou Coll Tradit Chin Med* 1989;**6**(1):39–41. [郭兴伯, 苻林春, 李国桥. 青蒿琥酯静脉注射7天疗程治疗恶性疟疗效观察. 广州中医学院学报. 1989, 6(1): 39–41.]
11. Fu LC, Guo XB, Fu YX, et al. A study on a regimen of intramuscular artesunate injections in treating falciparum malaria, and its comparison with piperaquine. *Collect Clin Res Artemisinins* 1990;**8**:33–8. [符林春, 郭兴伯, 符永新等. 青蒿琥酯肌肉注射治疗恶性疟疗程探索与哌喹的随机比较. 《青蒿素类药临床研究专辑》. 1990, 8: 33–38.]
12. Zhang JC, Guo XB, Qian BS, et al. Tolerance and adverse effects of artesunate tablets in healthy individuals. *J N Drugs Clin Remedies* 1992;**11**(2):70–2. [张俊才, 郭兴伯, 钱本顺等. 健康人对青蒿琥酯片的耐受性和不良反应. 新药与临床. 1992, 11(2): 70–72.]
13. Guo XB, Fu YX, Chen PQ, et al. Clinical observations of artesunate tablets in treating 100 cases of falciparum malaria. *Natl Med J China* 1989;**69**(9):515–6. [郭兴伯, 符永新, 陈沛泉等. 青蒿琥酯片治疗恶性疟疾100例的临床观察. 中华医学杂志. 1989, 69(9): 515–516.]
14. Fu LC, Li GQ, Guo XB, et al. A study on the dose of artesunate tablets in treating falciparum malaria. *J Guangzhou Univ Chin Med* 1998;**15**(2):81–3. [符林春, 李广谦, 郭兴伯等. 青蒿琥酯片治疗恶性疟的剂量再探索. 广州中医药大学学报. 1998, 15(2): 81–83.]
15. Li GQ, Fu YX, Bian WX, et al. Comparison of different regimens of artesunate tablets in treating falciparum malaria. *China J Integr Med* 1997;**17**(3):143. [李广谦, 符永新, 卞维秀等. 青蒿琥酯片不同疗程治疗恶性疟疾疗效比较. 中国中西医结合杂志. 1997, 17(3): 143.]
16. Bunnag D, Viravan C, Looareesuwan S, et al. Clinical trial of artesunate and artemether on multidrug resistant falciparum malaria in Thailand: a preliminary report. *Southeast Asian J Trop Med Public Health* 1991;**22**(3):380–5.

17. Li GQ, Guo XB, Wang WL, et al. Clinical study of artemether in treating falciparum malaria. *Natl Med J China* 1982;**62**(5):293–4. [李国桥, 郭兴伯, 王文龙等. 蒿甲醚治疗恶性疟疾的临床研究. 中华医学杂志1982, 62(5): 293–294.]
18. Gu HM, Liu MZ, Lü BF, et al. Comparison of artemether and piperaquine phosphate in treating falciparum malaria. *Acta Pharmacol Sin* 1981;**2**:138–44. [顾浩明, 刘明章, 吕宝芬等. 蒿甲醚治疗恶性疟疾与磷酸哌喹的比较.中国药理学1981, 2: 138–144.]
19. Guo XB, Fu LC, Li GQ, et al. Summary of clinical research on artemether-oil injections in treating falciparum malaria. *Collect Clin Res Artemisinins* 1990;**8**:85–90. [郭兴伯, 符林春, 李国桥等. 蒿甲醚油注射液治疗恶性疟疟疾临床研究总结. 《青蒿素类药临床研究专辑》. 1990, 8: 85–90.]
20. Guo XB, Fu LC, Fu YX, et al. Comparison of artemether and piperaquine phosphate in treating falciparum malaria. *J New Drugs Clin Remedies* 1989;**8**(4):248–50. [郭兴伯, 符林春, 符永新等. 蒿甲醚治疗恶性疟疾与磷酸哌喹的比较. 新药与临床. 1989, 8(4): 248–250.]
21. Guo XB, Fu LC, Li GQ. Summary of clinical research into dihydroartemisinin tablets in treating falciparum malaria. *Collect Clin Res Artemisinins* 1990;**8**:91–4. [郭兴伯, 符林春, 李国桥. 口服双氢青蒿素片治疗恶性疟疾临床研究总结. 《青蒿素类药临床研究专辑》. 1990, 8: 91–94.]
22. Xiyuan Hospital, China Academy of Chinese Medical Sciences. *Phase I clinical trials of dihydroartemisinin tablets. Application data for dihydroartemisinin tablets, 1991, p. 25*. Held at the Center for the History of Medicine, Peking University. [中国中医研究院西苑医院.双氢青蒿素片剂I期临床试验.双氢青蒿素片剂申报材料. 1991, p. 25.]
23. Fu LC, Wang WL, Fu YX, et al. Dose-finding study of oral dihydroartemisinin in treating falciparum malaria. *Collect Clin Res Artemisinins* 1990;**8**:52–7. [符林春, 王文龙, 符永新等. 口服双氢青蒿素治疗恶性疟剂量探索. 《青蒿素类药临床研究专辑》. 1990, 8: 52–57.]
24. Guo XB, Fu LC, Fu YX, et al. A randomized comparative study of dihydroartemisinin and piperaquine phosphate in treating falciparum malaria. *Natl Med J China* 1993;**73**(10):602–4. [郭兴伯, 符林春, 符永新等. 双氢青蒿素与磷酸哌喹治疗恶性疟的随机比较.中华医学杂志. 1993, 73(10): 602–604.]
25. Zhang JC, Yao DJ, Lin SP, et al. Clinical evidence for oral dihydroartemisinin in treating 44 cases of falciparum malaria. *Collect Clin Res Artemisinins* 1990;**8**:58–9. [张俊才, 姚邓居, 林少平等. 口服双氢青蒿素治疗恶性疟44例临床验证. 《青蒿素类药临床研究专辑》. 1990, 8: 58–59.]
26. Wang WL, Zhang ZG, Zheng XZ, et al. Clinical report on dihydroartemisinin tablets in treating 30 cases of falciparum malaria. *Collect Clin Res Artemisinins* 1990;**8**:61–2. [王文龙, 张祖光, 郑心正等. 双氢青蒿素片剂治疗恶性疟疾30例临床报告. 《青蒿素类药临床研究专辑》. 1990, 8: 61–62.]
27. Fu YX, Fu LC, Xie DC, et al. Clinical observation of dihydroartemisinin tablets in treating 59 cases of falciparum malaria. *Collect Clin Res Artemisinins* 1990;**8**:63–5. [符永新, 符林春, 谢带媌等. 双氢青蒿素片治疗恶性疟59例临床观察. 《青蒿素类药临床研究专辑》. 1990, 8: 63–65.]
28. Li GQ, Guo XB, Fu LC, et al. Clinical trials of artemisinin and its derivatives in the treatment of malaria in China. *Trans R Soc Trop Med Hyg* 1994;**88**(Suppl. 1):S5–6.
29. Li GQ, et al. Summary of artemisinin in treating 18 malaria patients. Internal document *Collect Data Res Malar Prev Treat* 1975:30. [李国桥等. 黄蒿素治疗疟疾18例总结. 广东地区《疟疾防治研究资料汇编》（内部资料）. 1975:30.]
30. Pharmacology Laboratory at the Institute of Chinese Materia Medica, China Academy of Chinese Medical Sciences. Pharmacology research into *Qinghao. N Med J* 1979;**1**:23–33. [中医研究院中药研究所药理研究室. 青蒿素的药理研究. 新医药杂志. 1979, (1):23–33.]

31. Li R, et al. Effects of artesunate on the ultrastructure of simian malaria. *Chin Tradit Herb Drugs* 1981;**4**:31. [李锐等. 青蒿琥酯钠对猴疟原虫超微结构的影响. 中草药. 1981, (4):31.]
32. Chen DJ, Fu LF, Shao PP, et al. Research into artemisinin as an experimental treatment for *Schistosoma japonicum* in animals. *Natl Med J China* 1980;**60**(7):422–5. [陈德基, 傅丽芳, 邵萍萍等. 青蒿素实验治疗动物血吸虫病的研究. 中华医学杂志, 1980, 60(7): 422–425.]
33. Yue WJ, You JQ, Mei JY, et al. The effects of some artemisinins on *Schistosoma japonicum*. *Acta Pharmacol Sin* 1981;**16**(8):561–3. [乐文菊, 尤纪清, 梅静艳等. 一些青蒿素衍生物的抗日本血吸虫作用. 药学学报. 1981, 16(8): 561–563.]
34. Yue WJ, You JQ, Yang YQ, et al. Experimental study on artemether in treating schistosomiasis in animals. *Acta Pharmacol Sin* 1982;**17**(3):187–93. [乐文菊, 尤纪清, 杨元清等. 蒿甲醚治疗动物血吸虫病的实验研究. 药学学报. 1982, 17(3): 187–193.]
35. Xiao SH, You JQ, Jiao PY, et al. Efficacy of artemether as an early treatment for schistosomiasis in mice. *Chin J Parasitol Parasit Dis* 1994;**12**(1):7–12. [肖树华, 尤纪清, 焦佩英等. 蒿甲醚早期治疗小鼠血吸虫病的疗效. 中国寄生虫学与寄生虫病杂志. 1994, 12(1): 7–12.]
36. Xiao SH, You JQ, Yang YQ, et al. Experimental studies on early treatment of schistosomal infection with artemether. *Southeast Asian J Trop Med Public Health* 1995;**26**(2):306–18.
37. Wu LJ, Xu PS, Xuan YX, et al. Experimental study on early treatment of *Schistosoma japonicum* with artesunate. *Chin J Schistosomiasis Control* 1995;**7**(3):129–33. [吴玲娟, 徐潘生, 宣尧仙等. 口服青蒿琥酯早期治疗预防日本血吸虫病的实验研究.中国血吸虫病防治杂志. 1995, 7(3): 129–133.]
38. Song Y, Xiao SH, Wu W, et al. Observations on artemether as a schistosomiasis prophylaxis in flood relief workers. *Chin J Parasitol Parasit Dis* 1997;**15**(3):133–7. [宋宇, 肖树华，吴伟等. 蒿甲醚预防抗洪抢险人群感染血吸虫病的观察. 中国寄生虫学与寄生虫病杂志.1997, 15(3):133–137.]
39. Utzinger J, N'Goran EK, N'Dri A, et al. Oral artemether for prevention of *Schistosoma mansoni* infection: randomized controlled trial. *Lancet* 2000;**355**(9212):1320–5.
40. Utzinger J, Xiao S, Keiser J, et al. Current progress in the development and use of artemether for chemoprophylaxis of major human schistosome parasites. *Curr Med Chem* 2001;**8**(15):1841–60.
41. De Clercq D, Vercruysse J, Kongs A, et al. Efficacy of artesunate and praziquantel in schistosoma haematobium infected schoolchildren. *Acta Trop* 2002;**82**(1):61–6.
42. Utzinger J, Keiser J, Xiao SH, et al. Combination chemotherapy of schistosomiasis in laboratory studies and clinical trials. *Antimicrob Agents Chemother* 2003;**47**(5):1487–95.
43. Xiao S, Chollet J, Utzinger J, et al. Artemether administered together with haemin damages schistosomes in vitro. *Trans R Soc Trop Med Hyg* 2001;**95**(1):67–71.
44. Xiao SH, Wu YL, Tanner M, et al. Schistosoma japonicum: in vitro effects of artemether combined with haemin depend on cultivation media and appraisal of artemether products appearing in the media. *Parasitol Res* 2003;**89**(6):459–66.
45. Ding JC, Yue ZQ, Zhang FJ, et al. Summary of *A. annua* L. as a prophylaxis and treatment for *Theileria annulata*. *J Northwest Agric Coll* 1981;**1**:89–92. [丁景昌, 岳治权, 张仿杰等. 青蒿防治牛环形泰勒焦虫病的试验简报. 西北农学院学报. 1981, (1): 89–92.]
46. Yue ZQ. Clinical veterinary use of *A. annua* L. preparations. *Chin J Tradit Vet Sci* 1989;**1**:6–8. [岳治权. 黄花蒿制剂在兽医临床上的应用. 中兽医学杂志. 1989, (1): 6–8.]
47. Liu ZH, Li CC, Yang WS, et al. Study on artesunate in treating bovine *Theileria annulata*. *Chin J Tradit Vet Sci* 1990;**5**:1–2. [刘宗汉, 李春成, 杨伟松等. 青蒿酯治疗奶牛环形泰氏焦虫病的试验. 中兽医医药杂志. 1990, (5): 1–2.]

48. Ma SL, Dong YH, Zhang SZ, et al. Comparative study on the therapeutic efficacy of artesunate and berenil against theileriosis in lambs. *Chin J Tradit Vet Sci* 1997;**6**:7–8. [马绍林, 董玉华, 张世珍等. 青蒿琥酯与贝尼尔对羔羊血孢子病防治效果的比较观察. 中兽医医药杂志. 1997, (6): 7–8.]
49. Wang XY, Fan ZL, Wang GD, et al. Preliminary observations on the treatment of paragonimiasis with artesunate and praziquantel. *J Pract Parasit Dis* 1997;**5**(4):187. [望西玉, 樊中丽, 王过渡等. 青蒿琥酯与吡喹酮治疗肺吸虫病的初步观察. 实用寄生虫病杂志. 1997, 5(4): 187.]
50. Chen RX, Qu ZQ, Zeng MA, et al. Effects of artemisinins on *Clonorchis sinensis* in rats. *Chin Pharm Bull* 1983;**18**(8):410–1. [陈荣信, 屈振骐, 曾明安等. 青蒿素及其衍生物驱大鼠华支睾吸虫的效果观察. 药学通报. 1983, 18(7): 410–411.]
51. Liu Y, Zhang YH, Li Z, et al. Observations on the therapeutic effects of artemether on rabbit toxoplasmosis. *Acta Parasitol Medica Entomol Sinica* 1996;**3**(2):84–8. [刘杨, 张永浩, 李哲等. 蒿甲醚对实验兔弓形虫病的疗效观察. 寄生虫与医学昆虫学报, 1996, 3(2): 84–88.]
52. Liu CM, Ouyang K. Artemether in treating experimental toxoplasmosis. *Hunan Med J* 1998;**15**(5):264–6. [刘翠梅, 欧阳颗. 蒿甲醚治疗实验型弓形虫病. 湖南医学, 1998, 15(5): 264–266.]
53. Yan L, Gan SB, Qi ZQ. Further observations on the therapeutic efficacy of dihydroartemisinin in mice with acute toxoplasmosis. *Acta Parasitol Medica Entomol Sinica* 2002;**9**(7):70–5. [严笠, 甘绍伯, 齐志群. 双氢青蒿素治疗急性弓形虫感染小鼠疗效的进一步观察. 寄生虫与医学昆虫学报. 2002, 9(7): 70–75.]
54. D'Angelo JG, Bordón C, Posner GH, et al. Artemisinin derivatives inhibit *Toxoplasma gondii in vitro* at multiple steps in the lytic cycle. *J Antimicrob Chemother* 2009;**63**(1):146–50.
55. Hencken CP, Jones-Brando L, Bordón C, et al. Thiazole, oxadiazole, and carboxamide derivatives of artemisinin are highly selective and potent inhibitors of *Toxoplasma gondii*. *J Med Chem* 2010;**53**(9):3594–601.
56. Wang CG, Wu J, Zhou YH, et al. Preliminary observations on artemisinin in treating toxoplasmosis and its clinical efficacy. *Chin J Zoonoses* 1997;**13**(1):79–80. [王崇功. 青蒿素治疗弓形虫病的研究和临床治疗效果的初步观察.中国人兽共患病杂志. 1997, 13(1): 79–80.]

Chapter 7

Artemether and Lumefantrine Tablets (Coartem)

Ning Dianxi[1], Zhong Jingxing[1], Zeng Meiyi[2]
[1]*Institute of Microbiology and Epidemiology, Academy of Military Medical Sciences, Beijing, China;* [2]*Institute of Chinese Materia Medica, China Academy of Chinese Medical Sciences, Beijing, China*

Chapter Outline

"Artemether and Lumefantrine" is a fixed-dose combination of two new antimalarials, an innovation of the Institute of Microbiology and Epidemiology at the Academy of Military Medical Sciences (AMMS). The project was initiated in 1982 and was later developed jointly with the Kunming Pharmaceutical Corporation (formerly the Kunming Pharmaceutical Factory). It was registered as a new drug in 1992, under the name "Compound Artemether Tablets" approved by the China Food and Drug Administration. In October 1994, with the approval of the State Science and Technology Commission, CITIC Technology led the AMMS and Kunming Pharmaceutical Corp in signing a license and development agreement with Switzerland's Ciba-Geigy (later Novartis). They would develop

Artemisinin-Based and Other Antimalarials. https://doi.org/10.1016/B978-0-12-813133-6.00007-X

this artemisinin-based combination therapy (ACT) to international standards and market the drug globally. The Sino-Swiss joint research agenda focused on the pharmacodynamics and pharmacokinetics of lumefantrine, as well as clinical pharmacokinetics and multicenter international trials for this ACT.

As this ACT of artemether and lumefantrine combination became more widely used, it gained the attention of scientists who, in turn, initiated clinical trials in different regions. This deepened and broadened research into the drug and in 2002 it was included in WHO's List of Essential Medicines as "Artemether and Lumefantrine" and marketed as Coartem or Riamet. In 2006, it was also included in the WHO's *Guidelines for the Treatment of Malaria*, recommended as a first-line ACT in hyperendemic areas in Africa and in the treatment of highly drug-resistant falciparum malaria in Southeast Asia. As of 2009, the drug has been registered in 86 countries and regions, including the United States. Six hundred million person-doses had been sold by 2013, saving millions of lives.

More than 200 research papers have been published on this ACT worldwide. However, China's early research material on lumefantrine and Compound Artemether Tablets (artemether-lumefantrine combination) had not been made public due to confidentiality and patent applications. This chapter outlines the early phase Chinese development of this ACT and its new-drug application in China and the later international cooperation.

7.1 LUMEFANTRINE

A. CHEMICAL SYNTHESIS

Under the leadership of Professor Deng Rongxian, the antimalarial research group of the AMMS' Institute of Microbiology and Epidemiology synthesized several hundred chemical compounds. The group found that fluoreneamino-ethanols had relatively good antimalarial activity and therefore used them as leading compounds in designing new chemicals.[1]

On this principle, more than 60 compounds were synthesized. Compound 76028 was eventually chosen and developed as a new blood schizontocide. It was named benflumetol but subsequently renamed lumefantrine, Chemical Abstract Service (CAS) name 2-(dibutylamino)-1-[(9Z)-2,7-dichloro-9-(4-chlorobenzylidene)-9H-fluoren-4-yl] ethanol (Fig. 7.1).[2] Its preclinical and clinical trials showed that lumefantrine had extremely low toxicity and was well tolerated, with no significant cross-resistance with chloroquine.[3] Class A drug certification was granted in 1987 and a Class I State Invention Award followed in 1990.

1. Design Concept

Arylamino-ethanols may be considered a new type of compound derived from the classic antimalarial quinine. Quinine was still relatively effective against chloroquine-resistant falciparum malaria but had significant adverse reactions. Work on modifying the structure of quinine was ongoing. Quinine has a quinoline ring and *N*-heterocyclic side chain connected by a hydroxymethylene

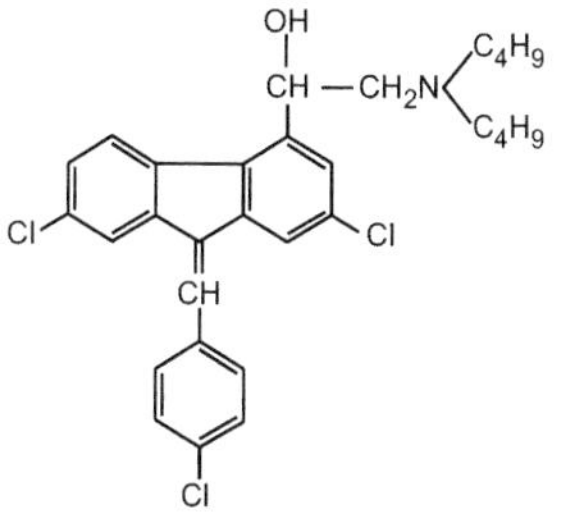

FIGURE 7.1 Chemical structure of lumefantrine.

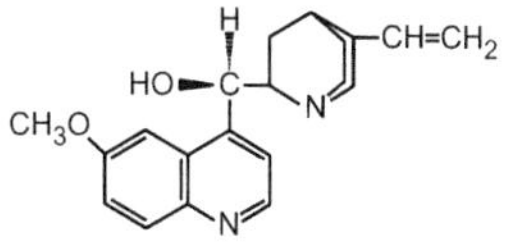

FIGURE 7.2 Chemical structure of quinine.

(–CH(OH)–) group, with two carbon atoms between the hydroxy group and the nitrogen atom in the side chain (Fig. 7.2).

Therefore, various new compounds were conceptualized by replacing the quinoline ring with other rings and retaining the fundamental characteristics of the side chains. But only a few studies on compounds with fluorene rings were published.[4] Fluoreneaminoethanols with different substituents were thus designed as potential new antimalarials for chloroquine-resistant falciparum malaria.[1] First, some 7-Br-, 1,6-Cl_2-, and 2,7-Cl_2-substituted fluorene compounds were synthesized, in which R_1 and R_2 might be mono- or dialkyl (carbon number C_3–C_{10}), pyrrolidinyl or cyclic alkylamino group (Fig. 7.3).[2]

The antimalarial activity of the first 40 or so of these compounds was studied. Mice infected with *Plasmodium berghei* were given subcutaneous injections of these drugs. Many achieved a 100% inhibition rate at doses of 2.5–5.0 mg/kg. Of these, the compounds with 2,7-dichloro fluorene performed relatively well. Aliphatic amine side chains (NR_1R_2) were also more effective, whereas efficacy fell or disappeared with *N*-heterocyclic side chains. However, for oral administration, medium-length alkyl side chains showed higher antimalarial activity.

Further structure modification was carried out to strengthen these compounds' antimalarial activity in oral formulations. Hence, fluoreneaminoethanols with three substituents at positions 2, 7, and 9 were synthesized,[5] with various NR_1R_2 chains such as $-N(C_4H_9)_2$, $-N(C_5H_{11})_2$, $-N(C_9H_{19})_2$, and $-NHC_8H_{19}$. X and Y were H, 2(4)-Br, 2(4)-Cl, 2,4-Cl_2, 3(4)-OCH_3, and 3,4-OCH_2O. In total, 24 new compounds were created (Fig. 7.4).

Antimalarial activity was highest in the compounds with $-N(C_4H_9)_2$ substituents at the side chain and –H, 2-Br, 4-Cl, and 2,4-$(Cl)_2$ at X and Y on the benzene ring. In a 3-day suppressive test, five of these compounds

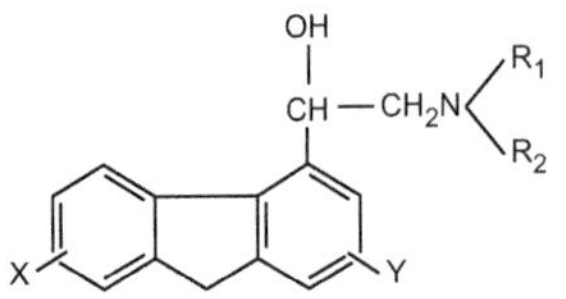

FIGURE 7.3 Fluoreneaminoethanols.

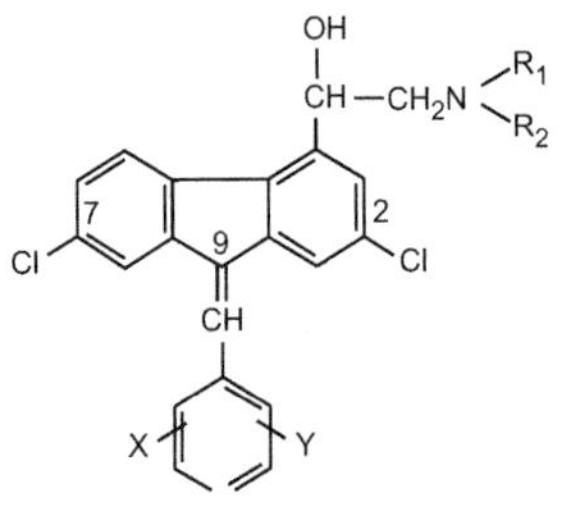

FIGURE 7.4 Substituted fluoreneaminoethanols with substitution at positions 2, 7, and 9.

had a 100% inhibition rate against NK65 strain *P. berghei* at oral doses of 6.25 mg/kg. Eventually, 2-(dibutylamino)-1-[(9Z)-2,7-dichloro-9-(4-chlorobenzylidene)-9H-fluoren-4-yl] ethanol, or lumefantrine, was selected for further study.

2. Process of Synthesis

Two different processes were used for the synthesis of lumefantrine. At first, an existing published method for the synthesis of its analog was followed.[1] This was a 12-step process with anthranilic acid as starting material.[2] This process was not only long, but costs were also high and impact on the environment severe. It was considered not appropriate for mass production.

Subsequently, fluorene was found to be a by-product of coal tar, making it easy to obtain at low cost. A new five-step process was developed with fluorene as starting material. The intermediates and final product were compared against those of the previous method and completely validated via HPLC and NMR, infrared, and mass spectrometry. It showed that the new process was viable.[6]

In the five-step route, fluorene (**2**) is used as starting material. In the presence of ferric chloride, chlorine gas is bubbled through the solution of fluorene in glacial acetic acid to form 2,7-dichlorofluorene (**3**). With anhydrous aluminum chloride as a catalyst, **3** is acylated in dichloroethane using chloroacetyl chloride, producing 2,7-dichlorofluorene-4-chloroacetyl fluorene (**4**). This is reduced with potassium borohydride to form 2-(2,7-dichloro-9H-fluoren-4-yl) oxirane (**5**). Condensation of **5** with dibutylamine produces 2-dibutylamino-1-(2,7-dichloro-9H-fluoren-4-yl) ethanol (**6**), which is then condensed with p-chlorobenzaldehyde to make lumefantrine (**1**) (Fig. 7.5). This process of synthesis was granted a national invention patent.[7]

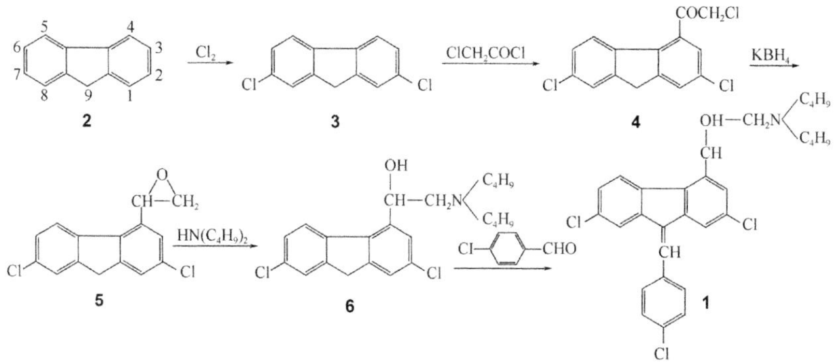

FIGURE 7.5 Synthesis of lumefantrine.

3. Formulations

Lumefantrine is a yellow crystalline powder with a melting point of 130–132°C. It is practically insoluble in water and oil but soluble in unsaturated fatty acids, such as oleic and linoleic acid. Linoleic acid soft capsules are used in a clinical setting; their bioavailability is twice that of the powder form.[8]

The lumefantrine–linoleic acid solution could produce a certain degree of esterification, which was dependent on temperature. At 80°C, the rate of esterification was $2.77 \pm 0.15\%$. Therefore, production should take place under controlled conditions.[9]

4. Structural Characteristics and Antimalarial Effects

The structure of lumefantrine was determined using single-crystal X-ray diffraction and compared to that of quinine.[10] It confirmed the three-dimensional characteristics of arylamino-alcohol antimalarials. The angle between the fluorene ring and dichlorophenyl planes was 52.8 degrees. The N–O distance in the lumefantrine molecule was 2.709 Å, versus 3.04 Å in the quinine molecule. The N–C–C–O dihedral angle in the lumefantrine side chain was 47.6 degrees and that of quinine was 61 degrees.

Since the lumefantrine molecule had a double bond, Z- or E-form geometric isomers should theoretically exist. Single-crystal diffraction showed that lumefantrine was a Z-form isomer. Its crystal structure contained an intramolecular hydrogen bond. Moreover, due to its asymmetric carbon atom, the molecule also had S and R enantiomers. Usually, synthesis yielded a racemic product, and it was determined that both enantiomers and the racemate were equally effective against *Plasmodium falciparum*.[11] This was not the case for some antimalarials, such as quinine and mefloquine, which had rings with terminal nitrogen atoms. In lumefantrine, this consistent efficacy could be due to the large degree of flexibility and space in the side chain.

B. PHARMACODYNAMICS

1. Formulation Tests on *Plasmodium berghei*

Peters' 1965 4-day suppressive test was used with white mice infected with NK65 strain *P. berghei*.[12] Three formulations were used: lumefantrine in linoleic acid, in Tween polysorbate 80 aqueous suspension, and in sodium carboxymethyl cellulose suspension. Each test was repeated three times and the results averaged (Table 7.1). The linoleic acid formulation was significantly better than the other two and was selected for further study.

2. Efficacy Against Drug-Resistant Strains of *Plasmodium berghei*

White mice were infected with normal (N), mildly chloroquine-resistant (NS) and highly chloroquine-resistant (RC) strains of K173 *P. berghei*. Chloroquine and mefloquine were the controls. Each test was repeated three times and the results averaged. The resistance index (I_{50}) was the ratio of the ED_{50} for the resistant strain to the ED_{50} for the normal strain. The results (Table 7.2) showed that the ED_{50} of the three drugs were similar for the normal strain. Both lumefantrine and mefloquine had an effect on the NS strain, but there was some degree of cross-resistance with the RC strain.

3. Parasite Clearance Time of *Plasmodium berghei*

Mice were infected with normal strain K173 *P. berghei*. When parasitemia reached around 10%, a single large oral dose ($ED_{90} \times 20$) was given. After medication, blood smears were obtained every 6 h until full parasite clearance occurred. The time taken for 50%, 90%, and 100% clearance was calculated. This was compared across 10 mice given lumefantrine in linoleic acid, and another 10 given chloroquine in aqueous solution. For lumefantrine, the time taken for 50% and 90% clearance were 74.9 and 90.1 h, respectively. Full

TABLE 7.1 ED_{50} of Three Lumefantrine Formulations Against NK65 *P. berghei*

Formulation	ED_{50} (M ± SD)	ED_{90} (M ± SD)	ED_{90} Ratio
Lumefantrine in linoleic acid	0.58 ± 0.12	1.28 ± 0.15	1.0
Lumefantrine in polysorbate 80 aqueous suspension	2.35 ± 0.50	3.55 ± 0.05	2.7
Lumefantrine in carboxymethyl cellulose suspension	3.80 ± 0.53	7.07 ± 1.24	5.5

clearance occurred at 108 h. The corresponding times for chloroquine were 38.2, 46.2, and 78 h, respectively. It showed that lumefantrine had a slower parasite clearance than chloroquine.

4. Curative Effects on *Plasmodium berghei*

The 5-day medication, 28-day observation protocol from Raether et al.[13] was used to compare the efficacy of lumefantrine and chloroquine against normal strain K173 *P. berghei*. A cure was taken to be full parasite clearance with no recrudescence. The 50% curative dose (CD_{50}) of the lumefantrine in linoleic acid formulation was 11.56 mg/kg, the CD_{90} was 23.19 mg/kg, and the CD_{100} was 30 mg/kg. The corresponding doses for chloroquine were 6.66, 17.49, and 30 mg/kg. Both drugs were effective, but a higher dose of lumefantrine was needed to achieve the same results.

5. Curative Effects on *Plasmodium knowlesi*

Davidson et al.'s 1979 protocol was used, with 7 days of treatment and 105 days of observation.[14] It included three dose groups of the lumefantrine in linoleic acid formulation, with three monkeys in each: 48, 24, and 12 mg/kg. All the animals in the 48 mg/kg group were cured, with no recrudescence at 105 days. In the other two groups, two out of the three monkeys in each group experienced recrudescence within 30 days. For parasite clearance time, see Table 7.3.

6. Antimalarial Activity on Isolates of *Plasmodium falciparum* In Vitro

From the time the AMMS' Institute of Microbiology and Epidemiology began working with Ciba-Geigy in 1992, supplementary studies on the pharmacodynamics of lumefantrine were carried out in the following areas.

TABLE 7.2 ED_{50} of Lumefantrine and Mefloquine on Chloroquine-Resistant K173 *P. berghei*

	ED_{50} (mg/kg)				
Drug	**N**	**NS**	**RC**	**I_{50} (NS)**	**I_{50} (RC)**
Lumefantrine in linoleic acid	1.02	0.61	18.69	0.6	18.2
Mefloquine aqueous suspension	1.43	0.98	28.48	0.7	19.9
Chloroquine aqueous solution	1.35	2.80	72.80	2.1	53.9

TABLE 7.3 *P. knowlesi* Parasite Clearance Using Oral Lumefantrine

Dose (mg/kg)	Average Time Taken for % Reduction in Parasitemia (h)		Average Parasite Clearance Time (h)
	50%	90%	
48	16.07	40.7	100
24	13.37	38.2	100
12	47.40	70.6	108

Antimalarial Activity of Enantiomers and Racemate of Lumefantrine

It had been previously noted that lumefantrine has a racemate and S and R enantiomers. In 1998, Wernsdorfer et al. tested these enantiomers and the racemate on 29 *P. falciparum* strains newly isolated from the Muheza region of Tanzania.[11] Their suppressive effects on the maturation of schizonts were assessed with the WHO standard microtest.[15] The mean EC_{50} for the S and R enantiomers and racemate were 8.87, 9.71, and 12.44 nmol/L, respectively; the EC_{90} were 42.39, 41.12, and 46.98 nmol/L, respectively. The antimalarial activity of all three substances was very similar. The racemate could be used in a clinical setting without separation of the enantiomers.

Antimalarial Activity on Regional Strains of P. falciparum

1. Cameroon. Basco et al. conducted isotopic microtests on lumefantrine against 60 fresh clinical isolates of *P. falciparum* from Yaounde, Cameroon.[16] The geometric mean 50% inhibitory concentration (IC_{50}) was 11.9 nmol/L. The IC_{50} against 15 chloroquine-sensitive and 23 chloroquine-resistant strains were 12.4 and 10.2 nmol/L, respectively; this was not a significant difference ($P > .05$). As reported by Bindschedler, the average peak plasma concentration of lumefantrine in healthy people after a single dose was 0.38 μg/mL after fasting and 5.10 μg/mL after a high-fat meal.[17] It was far higher than the IC_{50}. Therefore, the authors suggested that the in vitro antimalarial activity of lumefantrine against this regional strain was comparable to that of mefloquine, which was in line with the results of the preliminary lumefantrine clinical trials.
2. Tanzania and Thailand. Wernsdorfer et al. used microtests to assess the effects of lumefantrine on *P. falciparum* isolates from Muheza, a hyperendemic area in northeastern Tanzania (1992), and from Mae Hong Son, a hypoendemic region in northwestern Thailand (1997).[18] For Muheza, 64 isolates were tested over concentrations of 10–3000 nmol/L lumefantrine, yielding an EC_{50} and EC_{90} of 9.8 and 38.0 nmol/L, respectively. At 30 nmol/L, more

TABLE 7.4 In Vitro Susceptibilities of 158 Senegalese Isolates of *P. falciparum* to Lumefantrine, Chloroquine, Quinine, Amodiaquine, Pyronaridine, Artemether, and Pyrimethamine

	Chloroquine-Susceptible Isolates ($n=80$)		Chloroquine-Resistant Isolates ($n=78$)		
Drug	IC_{50} (nM/L)[a]	95% CI	IC_{50} (nM/L)[a]	95% CI	*P*
Chloroquine	35.7	30.2–40.9	245	221–269	
Lumefantrine	63.4	53.1–73.7	47.9	38.4–57.9	<0.025
Quinine	157	128–186	302	131–473	<0.001
Amodiaquine	9.6	6.1–13.1	14.3	11.9–16.7	<0.025
Pyronaridine	2.9	2.3–3.5	4.9	3.8–6.0	<0.002
Artemether	5.3	4.1–6.5	5.2	3.6–6.8	<0.45
Pyrimethamine	279	0–1485	347	0–1710	<0.45

[a]*Geometric mean IC_{50}. The threshold IC_{50} for chloroquine resistance is >100 nmol/L.*
Data from Pradines, Tall, Fusai, et al. *Antimicrob Agents Chemother* 1999;**43**(2):418–20.

than 80% of the strains were fully suppressed. For Mae Hong Son, 72 isolates were tested and all were fully inhibited at 300 nmol/L. The EC_{50} and EC_{90} were 29.7 and 128.8 nmol/L, respectively. The Thai isolates were less sensitive to lumefantrine than the Tanzanian ones, perhaps due to the lower sensitivity to arylamino-alcohol antimalarials displayed by strains from that region.

3. Senegal. Pradines et al. studied the effects of lumefantrine on 158 *P. falciparum* isolates from the Fatick and Pikine regions of Senegal, including chloroquine-resistant and chloroquine-sensitive strains.[19] The mean IC_{50} was 55.1 nmol/L. Ten isolates (6%) showed significantly lower sensitivity to lumefantrine ($IC_{50} > 150$ nmol/L). Of all the strains, 78 were considered to be chloroquine-resistant (threshold value: $IC_{50} > 100$ nmol/L). The mean IC_{50} of lumefantrine against chloroquine-resistant and chloroquine-susceptible isolates were 47.9 and 63.4 nmol/L, respectively ($P < .025$). The results are shown in Table 7.4.

The IC_{50} of lumefantrine against all the strains was 12.5–240 nmol/L. Its level of antimalarial activity was two to five times higher than that of chloroquine. It was more effective against chloroquine-resistant strains than chloroquine-sensitive ones, whereas the reverse was the case for quinine, pyronaridine, and amodiaquine. Artemether was equally effective against both strain types.

Comparisons With Other Antimalarials

From 1994 to 1997, Wernsdorfer et al. ran WHO in vitro microtests (Mark II)[15] with commercially available chloroquine, quinine, quinidine, mefloquine, halofantrine, atovaquone, artemisinin, lumefantrine, and pyronaridine in Mae Hong Son, northwestern Thailand.[20] The tests, which measured the drug's suppressive effects on schizont maturation, involved more than 400 fresh isolates. The EC_{50} and EC_{90} of the drugs were calculated, as well as Pearson and Spearman correlation coefficients. There was a clear positive correlation between the effects of lumefantrine and mefloquine ($n=71$, $P<.0001$ at EC_{50} and EC_{90}) and lumefantrine and quinine ($n=20$, $P<.005$ at EC_{50} and EC_{90}). Such a correlation existed between lumefantrine and halofantrine at the EC_{50} level ($n=20$, $P<.001$), but not at the EC_{90} level. More surprising was the high degree of correlation between artemisinin and lumefantrine ($n=49$, $P<.001$ at EC_{90}). There was no positive correlation between lumefantrine and chloroquine or pyronaridine.

Basco et al.'s Cameroon study mentioned above also compared lumefantrine with artesunate, halofantrine, and mefloquine.[16] Artesunate and halofantrine had the highest level of in vitro antimalarial activity, with a mean IC_{50} of 1.28 and 1.60 nmol/L, respectively. Lumefantrine and mefloquine were similar, with an IC_{50} of 11.9 and 11.7 nmol/L, respectively. The average IC_{50} of lumefantrine was 12.4 nmol/L against chloroquine-sensitive strains and 10.2 nmol/L against chloroquine-resistant ones.

Basco et al. then used nonlinear regression analysis with a significance level of .05 to calculate the correlation coefficients of lumefantrine and the two other amino alcohols, mefloquine ($r=.688$) and halofantrine ($r=.677$), as well as artesunate ($r=.420$). They were all positively correlated, suggesting that the three drugs could have latent in vitro cross-resistance.

Similarly, Pradines et al.'s Senegal study also compared lumefantrine to artemether, quinine, pyronaridine, amodiaquine, chloroquine, and pyrimethamine.[19] There was insufficient evidence to suggest cross-resistance between lumefantrine and any of these drugs (see Table 7.4 above).

C. GENERAL PHARMACOLOGY

The impact of the lumefantrine in linoleic acid formulation on the nervous, cardiovascular, and respiratory systems of mice, rats, rabbits, and cats was observed.

A dose of 10 times the ED_{90} for *P. berghei* was used for mice, namely 148 mg/kg (3.7 mg/kg/d × 4d × 10). Two control groups received linoleic acid or normal saline. Each group had 20 mice given one intragastric dose of the drug or a control. The impact of lumefantrine on spontaneous activity, amphetamine stimulation, strychnine and picrotoxin toxicity, and pentobarbital and chloral hydrate sedation was observed. No significant differences were seen between the lumefantrine and control groups, showing that the drug did not have obvious effects on the nervous systems of mice.

For body temperature experiments, the Paget surface-area extrapolation method was used to calculate the dose for rats.[21] Large-dose (10 times ED_{90}), small-dose (ED_{90}), and control groups were set up with 10 rats in each. One intragastric dose was administered and body temperature measured 1, 2, 4, 6, and 8h after medication. There was no statistically significant difference between the groups.

The Paget method was also used to calculate doses for rats, rabbits, and cats, to study the effects of lumefantrine on blood pressure, heart rate, electrocardiograms (ECGs), and respiration. Each animal model had a large-dose, small-dose, and control group. Rats were divided into groups of 10, whereas rabbits and cats were in groups of five. The lumefantrine in linoleic acid formulation had no impact on all the above parameters.

D. TOXICOLOGY

1. Acute Toxicity

Mice were given aqueous suspension lumefantrine in a single dose of 10g/kg orally, or 10, 5, or 2g/kg via intraperitoneal injection. The solvent and a normal saline buffer were used as controls. Each dose group had 40 mice, with equal numbers of males and females. After 2 weeks' observation, the mice were autopsied. All the mice in the 10g/kg oral group and the 2 and 5g/kg injection groups were normal, with no deaths. In the 10g/kg injection group, 14 mice died, a rate of less than 50%. In another test, mice were given lumefantrine in linoleic acid in a single dose of 10g/kg orally, or 5 or 2.5g/kg via subcutaneous injection. There were 60 mice in each group, with equal numbers of males and females. No deaths or symptoms of toxicity occurred during the two-week observation period. The oral LD_{50} was more than 10g/kg in mice; for intraperitoneal injections, it was more than 5g/kg.

Rats were given 10g/kg lumefantrine in linoleic acid, alongside solvent and negative control groups. Each group had 30 rats. The experiment was repeated after 2 weeks' observation. No symptoms of toxicity were seen and no deaths occurred. The LD_{50} was greater than 10g/kg.

Deichmann and LeBlanc's protocol was used to observe the acute toxicity of oral lumefantrine in linoleic acid in dogs.[22] Dogs were given 1.0, 1.5, 2.2, 3.3, and 5.0g/kg doses, with each dose progressively increasing by 50%. Each dose group had one dog. Clinical signs were observed over 2 weeks. Blood (seven parameters), comprehensive metabolic panels (13 parameters), urine (eight parameters), and other tests were run. The dog given the 2.2g/kg dose vomited 8.5h after medication, while the dog given 5g/kg displayed licking and retching behaviors 5–10h after medication. The blood tests, comprehensive metabolic panels, urine tests, and ECGs showed no abnormalities. The maximum tolerated dose was 2.2–5.0g/kg.

The LD_{50} of oral lumefantrine in linoleic acid formulation was >10g/kg for both mice and rats, at which no symptoms of toxicity were observed. For dogs, the maximum tolerated dose was >2.2 g/kg. No deaths occurred at 5 g/kg. In respect to toxicity grading standards, lumefantrine was graded as essentially nontoxic to slightly toxic.

2. Long-Term Toxicity

Rats were given oral doses of 250, 500, and 1000 mg/kg/day lumefantrine in linoleic acid, alongside solvent and negative control groups. Each group had 24 rats, half male and half female. The drug was administered once daily for 13 consecutive weeks, after which 16 rats (eight male and eight female) in each group were killed. The remaining eight rats were observed for further 4 weeks and then killed. Tests included physiological markers (seven parameters), blood (seven parameters), comprehensive metabolic panels (13 parameters), fundus examinations, and ECGs. Twenty-seven tissue blocks were taken for histology examination, and a bone marrow differential cell count was performed.

The results showed that slight intoxication occurred in the 1000 mg/kg/day group. The male rats had slightly elevated AST and ALT levels and a decrease in blood calcium. One male rat displayed slight suppression in bone marrow hematopoiesis, while three had degenerated spermatogonia and spermatocytes. Eleven rats showed a thickening of the glomerular walls. In the 500 mg/kg/day group, male rats had a slight decrease in blood calcium and mild glomerular degeneration. No abnormalities were found in the 250 mg/kg/day group. Autopsies on the rats killed 28 days after medication showed one rat in the 1000 mg/kg/day group with pathologies in the testicles and glomeruli. All other rats were normal.

Therefore, the test found that a 1000 mg/kg/day dose was somewhat toxic and 500 mg/kg/day was slightly toxic. A dose of 250 mg/kg/day showed no toxicity and could be considered safe. Tissue damage occurred mainly in the kidneys with liver biochemical abnormalities of AST and ALT levels as indicators. The damage was reversible.

In another experiment, groups of six dogs (three male, three female) were given one oral dose of 120, 240, or 480 mg/kg/day lumefantrine in linoleic acid daily, alongside solvent and negative control groups. Medication was given for 28 days. On Day 29, four dogs in each group were killed. The other dogs were observed for a 28-day recovery period and then killed. The same tests were run as with the rat experiment above. In the 480 mg/kg/day group, the dogs experienced diarrhea and slight and reversible pathologies in the glomeruli, but other indicators were normal. Sporadic diarrhea occurred in the 120 and 240 mg/kg/day groups as well, but there were no other symptoms of toxicity and all test indicators were within normal range. The 480 mg/kg/day dose was considered mildly toxic and the 240 mg/kg/day dose as safe.

The tests showed that the toxicity of lumefantrine in linoleic acid was quite low, with no significant, species-specific differences between rats and dogs.

Tissue damage occurred mostly to the glomeruli; the main biochemical changes being elevated AST and ALT levels and lowered blood calcium in male rats. However, such changes only occurred at doses of 500 mg/kg/day (rats) and 480 mg/kg/day (dogs).

3. Mutagenicity

An Ames test[23] was run with lumefantrine concentrations of up to 1000 μg/dish. The results were negative. In the chromosomal aberration test, 2.0 g/kg (250 times the clinical dose) lumefantrine in linoleic acid had no significant impact on the chromosomes in mouse bone marrow. For the micronucleus test, mice were given oral doses of 100 mg/kg, equivalent to 12.5 times the clinical dose. There was no effect on the frequency of micronuclei in polychromatic erythrocytes. Finally, in the dominant lethal test, mice were given 800 mg/kg lumefantrine in linoleic acid, or 100 times the clinical dose. It had no obvious effect on the rate of conception and malformations, or number of implants, live fetuses, and early- or late-stage fetal deaths. All these negative results indicated that the drug was safe for clinical use.

4. Teratogenicity

Wilson's protocol was used,[24] with pregnant rats given 150 or 300 mg/kg/day oral lumefantrine in linoleic acid on Days 6–15 of gestation. A 15 mg/kg intraperitoneal injection of cyclophosphamide was used for the positive control. The rats were killed on Day 20 and the number of implants and living, dead, and reabsorbed fetuses were recorded. The appearance, skeletons, and soft tissues of living fetuses were examined for deformities. The proportion of live fetuses was 99.1% and 97.3% for the 150 and 300 mg/kg/day groups, respectively. No deformities were seen and the fetuses developed normally. The proportion of live fetuses in the positive control group was only 79.3%, all of which had deformities in their appearance, skeletons, and internal organs.

5. Male Reproductive Toxicity

Groups of 12 male rats were given 250, 500, or 1000 mg/kg/day oral lumefantrine in linoleic acid for 60 days. Solvent and negative control groups were included. The rats were mated from Day 60. During this period, medication continued for the male rats. After 13 weeks, eight rats in each group were killed and the reproductive organs (testicles, epididymis, vas deferens) taken for histology examination. The remaining four rats were killed 28 days after medication ceased, to observe their recovery from any drug-induced abnormal changes. Half of the pregnant rats for each group were killed on Days 12 and 20 of gestation. The fetuses were examined using the same parameters as in the teratogenicity test.

In the 1000 mg/kg/day group, three of the eight rats killed after 13 weeks had focal "loosening" and swelling of the cytoplasm in the spermatogonia and primary and secondary spermatocytes in the seminiferous tubules. Nuclear microvacuolar degeneration appeared. These pathologies were also seen in one of the four rats killed after recovery. No changes were found in the testicles of the rats in the 250 and 500 mg/kg/day groups.

Among the medicated and control groups, there were no obvious differences in the rate of conception, implantation index, proportion of live fetuses, and their average weight. Of the 10 litters from the 1000 mg/kg/day group, with 51 fetuses in total, two rats had unilateral testicular dysgenesis and one had unilateral cryptorchidism. They were from different litters. The 500 mg/kg/day group had 10 litters and 49 fetuses; two rats also showed unilateral cryptorchidism, but no dysgenesis was found. No such symptoms were seen in the 250 mg/kg/day and control groups. A dose of 1000 mg/kg/day was reproductively toxic to male rats, affecting spermatocyte structure and testicular development in fetal rats. The effects of a 500 mg/kg/day dose were slight. From a reproductive toxicity perspective, a 250 mg/kg/day dose was nontoxic.

E. NON-CLINICAL PHARMACOKINETICS

The area under the curve (AUC) of 125 mg/kg oral lumefantrine in linoleic acid in mice was 145.5 μg·h/mL. For oral lumefantrine in sodium carboxymethyl cellulose suspension, the corresponding AUC was 51.8 μg·h/mL. The bioavailability of the former was three times higher, and its ED_{90} was five times higher. Therefore, the linoleic acid formulation was chosen.

TLC and isotopic labeling were used to investigate the in vivo pharmacokinetics of oral ^{3}H-lumefantrine in linoleic acid in mice. Oral absorption was slow, reaching peak plasma concentration only after an average of 8 h. In the dose range tested—30.8–277.5 mg/kg—C_{max} increased with the dose, but the drug entered erythrocytes slowly. The ratio of the drug in erythrocytes versus in plasma was only 1.3 24 h after medication. However, the concentration in tissues was four to nine times higher than in plasma. The protein binding rate was 90% and above.

The drug was widely distributed in the body, mainly in the lungs, liver, spleen, and kidneys, peaking 6–8 h after medication. It was also eliminated slowly, with a plasma half-life of 78–187 h. The drug was excreted mostly via feces, at a low excretion rate. Only 32%–40% of the dose was excreted in 5 days, mostly in the first 24 h. More than 10% was excreted through bile 24 h after medication, reaching 12.5% in 72 h. Little was excreted through urine, with only 0.2%–1.3% of the dose excreted in 5 days. Mostly, the original drug was excreted via feces and bile, while a metabolite was excreted in the urine.

F. CLINICAL TRIALS OF LUMEFANTRINE CAPSULES

1. Phase I Clinical Trials

Safety and Tolerability

This randomized, double-blind test was conducted with 16 healthy adult male volunteers, who received two types of capsules once daily for 4 days. The first type was a 100 mg lumefantrine in linoleic acid capsule. The second type, the placebo, was a linoleic acid capsule with an identical appearance. Sixteen capsules were given on the first day, followed by eight per day on Days 2–4. The volunteers were divided into dose groups, with lumefantrine capsules replaced by placebos for those in the lower dose groups. The experiment proceeded with two batches at a time, with incremental doses: 500 mg (one subject), 1000 mg (one subject), 2000 mg (four subjects), 3000 mg (two subjects), and 4000 mg (four subjects) and placebo (four subjects). Subjects were hospitalized for 9 days' observation and followed up again after 14 days. Clinical and vital signs were observed, and ECGs, blood tests (seven parameters), comprehensive metabolic panels (16 parameters), and routine urine (eight parameters) and fecal tests were run.

No adverse reactions were seen. One subject each in the 1000 and 2000 mg groups had slightly increased AST and ALT levels 14 days after medication, but this returned to normal on Day 21. One subject in the 3000 mg group and two in the 4000 mg group had slightly elevated triglyceride levels on Days 2 and 7 after medication, which recovered on Day 14. No significant changes occurred in the other parameters. It was concluded that lumefantrine capsules had few side effects and an adult dose of 4000 mg over 4 days was well tolerated.

Clinical Pharmacokinetics

Eight healthy male adult volunteers were given a single oral dose of 800 mg lumefantrine capsules. HPLC was used to determine plasma concentration over time. The drug was detected in plasma around 1.5 h after medication, rising rapidly and reaching C_{max} in about 5 h. Peak concentration was 2000–4000 ng/mL, with a maximum in one subject of 7000 ng/mL and minimum in another subject of 1300 ng/mL. The half-life in plasma was around 24.7–72.1 h, averaging at 46 h. When fitted into a concentration–time curve, the results showed that the drug's in vivo distribution conformed to a two-compartment model. There was a clear lag time of around 1.5 h, with a minimum of 1 h and maximum of 2 h. The drug was rapidly distributed. The distribution half-life was mostly 2–3 h (9.2 h in one individual), at an average of 4 h.

In a separate trial, two healthy adult male volunteers were given four doses of 400 mg lumefantrine capsules over 4 days, with an initial dose of 800 mg. The drug was administered 1 h after a meal. Blood was drawn before medication, 6 h afterward and 24 h after the final dose to determine plasma concentration. One of the volunteers was 47 years old, weighing 71 kg; the other was 27, weighing 50 kg. Therefore, the results obtained from each volunteer differed markedly. The elimination half-life was 38.5 and 9.0 h, the AUC was 511.6 μg·h/mL and 578.0 μg·h/mL, and the accumulation factor was 2.85 and 1.19, respectively. Because the sample size was so small, more research was needed on the impact of multiple doses on drug absorption. Further studies took place during the international cooperation phase of the artemether–lumefantrine ACT (see below). They showed that food intake, particularly of fatty foods, significantly increased absorption of lumefantrine.

2. Phase II and III Trials

Lumefantrine in linoleic acid capsules clinical trials were carried out from 1981 to 1982 in Hainan and Yunnan, where chloroquine-resistant falciparum malaria was endemic.

Dose and Regimen for Falciparum Malaria

Parallel trials on 4- and 6-day lumefantrine regimens were carried out in Hainan and Yunnan. Adults were given 400 mg doses daily, with an initial dose of 800 mg. The 4-day regimen yielded a total dose of 2.0 g; the 6-day regimen, 2.8 g. In Yunnan, 19 cases were given the 6-day regimen and 20, the 4-day one. All patients were cured, with no recrudescence during the 28-day follow-up. Defervescence and parasite clearance times were similar for both regimens. In Hainan, each regimen was given to 20 patients. All subjects were clinically cured, with no difference in defervescence and parasite clearance times for both regimens. There were two cases of recrudescence for the 4-day regimen, and one case for the 6-day regimen (Table 7.5).

Due to the small number of cases, it was difficult to settle on one regimen. Further comparative clinical trials were run in Hainan and Yunnan, together with the standard chloroquine regimen of 1.5 g over 3 days. The 4- and 6-day lumefantrine regimens showed no clear differences but were significantly better than chloroquine (Table 7.6). Because both regimens had similar therapeutic efficacies, the 4-day one was selected as it had a shorter course and a lower total dose.

Four-Day Regimen for Falciparum Malaria

The 4-day lumefantrine regimen was given to 208 falciparum malaria cases in Yunnan and Hainan, of whom 191 (91.8%) came from malaria-endemic regions. There were 112 males and 96 females; 86 were children under 15. The cure rate (Table 7.7) was more than 95%, and no obvious drug-related adverse reactions were observed. Nine cases of drug-resistant falciparum malaria were also treated (six chloroquine-resistant and three multidrug-resistant), with parasitemia between 48,965/mm^3 and 253,344/mm^3. All patients were cured. Average defervescence time was 34.6 h, and average parasite clearance time was 56.6 h. No recrudescence occurred during the 28-day follow-up.

TABLE 7.5 Clinical Efficacy of Two Lumefantrine Regimens on Falciparum Malaria

Region	Regimen	Number of Cases	Defervescence Time (M ± SD)	Parasite Clearance Time (M ± SD)	% Reduction in Parasitemia in 24h	Number of Recrudescence Cases
Yunnan	6-day	19	42.3 ± 13.1	66.5 ± 11.3	53.4	0
	4-day	20	41.8 ± 12.2	67.6 ± 9.7	59.2	0
Hainan	6-day	20	43.9 ± 18.7	66.4 ± 22.4	66.9	1
	4-day	20	43.2 ± 19.2	58.6 ± 17.4	60.2	2

TABLE 7.6 Clinical Efficacy of Lumefantrine and Chloroquine on Falciparum Malaria

Region	Drug/Regimen	No. of Cases	Clinical Efficacy					Fever Reduction Time (M ± SD)	Parasite Clearance Time (M ± SD)	% Reduction in Parasitemia in 24h
			S	R_I	R_{II}	R_{III}	Cure Rate (%)			
Yunnan	Lumefantrine 6-day	25	25	–	–	–	100	38.6 ± 19.8	63.4 ± 9.4	61.5
	Lumefantrine 4-day	25	25	–	–	–	100	39.7 ± 19.6	64.3 ± 3.5	61.8
	Chloroquine 3-day	25	8	4	9	4	32	53.8 ± 14.8	89.3 ± 20.9	68.9
Hainan	Lumefantrine 6-day	32	31	1	–	–	96.8	34.5 ± 17.8	62.1 ± 15.2	73.6
	Lumefantrine 4-day	32	31	1	–	–	96.8	31.9 ± 19.3	58.6 ± 10.6	74.6
	Chloroquine 3-day	32	8	5	8	1	56.2	37.5 ± 18.5	61.5 ± 15.2	74.3

R_I, level I resistance; R_{II}, level II resistance; R_{III}, level III resistance; *S*, sensitive.

TABLE 7.7 Clinical Efficacy of a 4-Day Lumefantrine Regimen on Falciparum Malaria

Region	No. of Cases	Clinical Efficacy					Defervescence Time (h) M ± SD	Parasite Clearance Time (h) M ± SD
		S	R_I	R_{II}	R_{III}	Cure Rate (%)		
Yunnan	133	130[a]	3	–	–	97.7	46.5 ± 12.8	68.21.9
Hainan	75	72[b]	3	–	–	96.0	38.2 ± 18.3	61.2 ± 16.2

[a]*Five cases of* P. vivax *occurred during follow-up.*
[b]*Four cases of* P. vivax *occurred during follow-up.*

Conclusion

A total of 305 falciparum malaria cases received the 4-day lumefantrine in linoleic acid regimen in Yunnan and Hainan, where chloroquine-resistant *P. falciparum* was endemic. All were clinically cured, with a defervescence time of 31–46 h and a parasite clearance time of 62–72 h. Parasitemia fell by 53%–69% 24 h after medication. The 28-day cure rate was 90%–100%, which was superior to chloroquine.

No adverse reactions were seen in the 305 cases given the 4-day regimen or the 96 cases receiving the 6-day regimen nor were significant changes found in the 54 patients given blood, liver, and kidney function tests and ECGs. Based on the preclinical pharmacodynamics and toxicology tests and the clinical trials themselves, lumefantrine was found to have low toxicity, no significant side effects, the ability to thoroughly eliminate parasites and a low recrudescence rate. It was clearly effective against chloroquine-resistant falciparum malaria. However, the drug was slow in controlling symptoms, such as fever, and in clearing parasites. A 4-day regimen of lumefantrine in linoleic acid capsules, totaling 2.0 g, was considered suitable for clinical use.

7.2 COMPOUND ARTEMETHER TABLETS (ARTEMETHER–LUMEFANTRINE COMBINATION)

A. PHARMACODYNAMICS

This research focused on finding an optimal ratio for combining artemether and lumefantrine and comparing the efficacy of the combination to that of its components.

1. Optimized Dosage Ratios for Rodent Malaria

Orthogonal and 4-day suppressive tests were used to compare the efficacy of different ratios of artemether and lumefantrine on white mice infected with K173 strain *P. berghei*. The results showed that a range of 0.25–4.0 mg/kg artemether and 0.38–1.5 mg/kg lumefantrine had lower ED_{50} and ED_{90} than the two drugs used in isolation. Further experiments revealed that the optimal ratio of artemether to lumefantrine was 2:0.75, which had ED_{50} and ED_{90} synergistic indices of >5 and >6 (Table 7.8). Peters' 1968 additive linear graphical method[25] confirmed that artemether and lumefantrine had synergistic effects.

In the 4-day suppressive test, the intragastric ED_{50} of artemether was established as 0.88 mg/kg, and its ED_{90} was 2.79 mg/kg. For intragastric lumefantrine, the ED_{50} and ED_{90} were 0.99 and 2.61 mg/kg, respectively. An artemether–lumefantrine combination, at a ratio of 2:0.75, had an ED_{50} and ED_{90} of 0.44 and 0.89 mg/kg, respectively, lower than those of the two drugs used singly. Hence, both drugs had synergistic effects.

TABLE 7.8 Synergistic Indices of Different Ratios of Artemether and Lumefantrine on *P. berghei*

	Synergistic Index of ED_{50}				Synergistic Index of ED_{90}			
Drug	**1:1**	**2:1**	**2:0.75**	**2:0.5**	**1:1**	**2:1**	**2:0.75**	**2:0.5**
Artemether	1.59	1.23	1.35	0.70	1.90	1.52	1.73	0.29
Lumefantrine	1.74	2.66	3.88	0.43	2.00	3.13	4.79	1.19
Combination	3.33	3.89	5.23	1.23	3.90	4.65	6.52	1.48

Synergistic index = Dosage of the single drug at ED_{50} or ED_{90}/Dosage of the single drug in combination at ED_{50} or ED_{90}.

2. Curative Effects on Rodent Malaria

A 4-day suppressive test was carried out, with blood smears obtained every day after medication for 30 days to observe for parasitemia. Two dose groups each were used for the ACT and its component drugs, with 10 mice per group. At a dose of 13.56 mg/kg, the cure rate of artemether was 0%. For lumefantrine, a dose of 7.05 mg/kg yielded a cure rate of 76%. For the corresponding ACT group at 10.8 mg/kg (when 7.85 mg/kg artemether was combined with 2.95 mg/kg lumefantrine, a ratio of 2:0.75), the cure rate increased to 100%. The combination clearly increased the efficacy and decreased the required dose of the individual drugs.

3. Parasite Clearance Speed on Rodent Malaria

Mice were infected with K173 strain *P. berghei*. When parasitemia increased to around 10%, the mice were given a single intragastric dose of 20 times the ED_{90} of the ACT or its component drugs. Each dose group had 20 mice. Blood smears were obtained every 8 h after medication to determine the time at which 50% and 90% parasite clearance occurred. For artemether, the average time taken for 50% parasite clearance was 23.5 h. Parasitemia increased before 90% clearance was achieved. For lumefantrine, the average time taken for a 50% decrease in parasitemia was 54.3 h. A 90% decrease took 64.3 h. The artemether–lumefantrine combination, at a ratio of 2:0.75, took 28.5 and 49.7 h to achieve 50% and 90% parasite clearance, respectively. The combination had a faster parasite clearance speed than lumefantrine and was comparable to artemether.

4. Delaying Emergence of Drug Resistance in *Plasmodium berghei*

Merkli's dose escalation method was used to cultivate resistance.[26] Three drug groups were included: artemether, lumefantrine, and the 2:0.75 combination. Their respective ED_{90} were the starting dose. Medication was given for three consecutive days each week. Every 7 days, blood smears were obtained to examine successive passages of the parasite. Peters' 4-day suppressive test[12] was run every five passages to determine the ED_{90}. This was compared against the ED_{90} of the first passage to calculate the resistance index (I_{90}). A total of 30 passages were cultivated over 210 days (Table 7.9).

The results indicated that resistance could be cultivated against the component drugs and their combination. For the individual drugs, however, mild to moderate resistance emerged at the 10th passage, but there was still susceptibility to the combination. The combination could delay the onset of resistance. There was resistance to the combination in later passages but at a significantly lower level than resistance toward lumefantrine.

TABLE 7.9 Cultivated Resistance to Artemether, Lumefantrine, and Their Combination in Rodent Malaria

	10th Passage		20th Passage		30th Passage	
Drug	ED_{90} (mg/kg)	I_{90}	ED_{90} (mg/kg)	I_{90}	ED_{90} (mg/kg)	I_{90}
Artemether	18.6	3.1	34.7	6.4	34.2	6.3
Lumefantrine	>100	>40	>500	>205	>1000	>410
Artemether–Lumefantrine (2:0.75)	4.7	1.6	8.4	2.8	57.7	19.3

5. Efficacy on *Plasmodium knowlesi*

Rodent, simian, and human malaria display different, species-specific sensitivity to drugs. Prior animal-model and clinical trials had shown that human malaria was far less sensitive to lumefantrine than rodent malaria was (Table 7.10). A 6-day regimen of 3600 mg lumefantrine capsules, administered in 12 doses totaling 72 mg/kg, had previously been recommended for clinical use.

The optimal ratio of artemether to lumefantrine for rodent malaria was 2:0.75. Recalculated for human malaria, the equivalent ratio was (2×9.6/5.3):(0.75×72/2.7) or 1:5.5. Since the drug sensitivity of simian malaria was closer to that of human malaria, the efficacy of the ratio was tested on simian malaria.

Exploring Various Ratios of Component Drugs

Rhesus macaques infected with *P. knowlesi* were given the drugs intragastrically when parasitemia reached 3%–5%. Blood smears were obtained every 12 h after the first dose and every day after parasite clearance. After 15 days, they were carried out every other day. The total observation time was 105 days. The 1:5.5 artemether to lumefantrine ratio was recalibrated upward and downward to produce alternative ratios of 1:4 and 1:6.

The 1:4 ratio combined 2 mg/kg artemether with 8 mg/kg lumefantrine, administered four times over 3 days. The three monkeys in this group were all cured, with 90% parasite clearance at 19.0 ± 1.1 h and average full parasite clearance at 48 h. For the 1:6 ratio, 2 mg/kg artemether was combined with 12 mg/kg lumefantrine, with the same regimen as the previous ratio. All three monkeys in this group were also cured. The time taken for 90% parasite clearance was 20 ± 1.9 h, and full clearance took 56 h. Both ratios were satisfactory.

TABLE 7.10 Sensitivity of Human and Rodent Malaria to Artemether and Lumefantrine

Drug	ED_{90} for *P. berghei* (mg/kg)	Treatment Dose for *P. falciparum* (mg/kg)	Rodent:Human Dose Ratio
Artemether	5.3	9.6	1:1.8
Lumefantrine	2.7	72.0	1:26.7

Comparing Therapeutic Effects of Combination and Component Drugs

Alongside the previous test, 2 mg/kg doses of artemether or 12 mg/kg doses of lumefantrine were given to groups of three *Rhesus macaques* monkeys. In the artemether group, the time taken for a 90% parasite decrease was 21 ± 10 h. None of the monkeys had full clearance and parasitemia increased 5–7 days after medication ceased, eventually surpassing premedication levels. Two of the monkeys in the lumefantrine group were cured, with one experiencing recrudescence on the 15th day after parasite clearance. Time taken for 90% parasite clearance was 52 ± 18 h, and full clearance took 96 h on average. The combination was clearly more fast-acting than artemether and had a higher cure rate than lumefantrine.

Comparing Various Treatment Regimens

The artemether–lumefantrine combination with a ratio of 1:4 was then formulated in two different regimens: four doses over 3 days or four doses over 4 days. The total dose was the same: 8 mg/kg artemether and 32 mg/kg lumefantrine. Each regimen group had three monkeys. Those in the 4-day group had slower parasite clearance and one monkey experienced recrudescence after 9 days. All moneys in the 3-day group were cured.

6. In Vitro Antimalarial Activity

In 1987, Trager and Jensen's microculture method[27] was used to test the in vitro efficacy of the 1:6 ratio and the component drugs on a strain (Fcc-1) isolated from Hainan. Since the maximum soluble concentration of lumefantrine, 20 ng/mL, had no antimalarial efficacy, its IC_{90} could not be determined. The IC_{90} of artemether was 10.67 ± 9.57 ng/mL and that of the combination was 43.11 ± 9.08 ng/mL. Although the combination's total dose was four times higher than artemether's, given the 1:6 ratio, its artemether content was only 6 ng/mL, lower than that of artemether in isolation.

The IC_{90} of the combination showed evidence of synergistic efficacy due to the addition of lumefantrine.

In 1999, Alin et al. used in vitro cultures to investigate the efficacy of artemether, lumefantrine, and their combination against a multidrug resistant strain from Thailand (T-996) and a chloroquine-resistant strain from India (LS-21). This would also reveal the synergistic effects of both drugs.[28] When the test was run with a concentration range of 0.001–1000 nmol/L erythrocytic medium mixture (EMM) (i.e., 0.0003–298 ng/mL EMM), artemether's IC_{50} against T-996 was 0.28 nmol/L (0.08 ng/mL) EMM. Its IC_{90} against that strain was 35.45 nmol/L (10.58 ng/mL) EMM. Against LS-21, artemether's IC_{50} and IC_{90} were 0.18 nmol/L (0.05 ng/mL) and 7.11 nmol/L (2.12 ng/mL) EMM, respectively.

The test was also done with lumefantrine at a concentration range of 0.002–200 nmol/L (0.0011–106 ng/mL) EMM. The drug's IC_{50} against T-996 was 2.30 nmol/L (1.2 ng/mL) EMM; its IC_{90} was 293.03 nmol/L (154.70 ng/mL) EMM. Against LS-21, lumefantrine's IC_{50} and IC_{90} were 1.91 nmol/L (1.01 ng/mL) and 95.61 nmol/L (50.48 ng/mL) EMM, respectively. LS-21 was more sensitive to both drugs than T-996 was.

Artemether at concentrations of 0.01, 0.03, 0.10, 0.30, 1.00, 3.00, 10.00, 30.00, and 100.00 nmol/L (0.003, 0.009, 0.030, 0.090, 0.298, 0.895, 2.984, 8.952, and 29.840 ng/mL) EMM was combined with lumefantrine at concentrations of 0.01, 0.03, 0.10, 0.30, 1.00, 10.00, 30.00, and 100.00 nmol/L (0.0053, 0.0158, 0.0528, 0.1584, 0.5279, 5.4374, 15.837, and 52.79 ng/mL) EMM. The different combinations achieved up to 100% inhibition at lower concentrations than with the individual drugs. For instance, artemether achieved a 99.4% suppression rate against the LS-21 strain at a concentration of 1000 nmol/L (298 ng/mL) EMM. Lumefantrine had a 98.5% suppression rate against that strain at 20.0 nmol/L (5.3 ng/mL) EMM. However, artemether– and lumefantrine–EMM combinations with molar ratios of 10:0.03, 30:0.01 (high artemether, low lumefantrine), or 0.1:100 (low artemether, high lumefantrine) could all achieve 100% growth inhibition against LS-21. For T-996, synergistic effects could be seen in the IC_{50} of artemether– and lumefantrine–EMM ratios between 10:1 and 1:30. For LS-21, the same effects were observed at the IC_{50} at the ratios between 3:1 and 1:30. Clear synergistic effects were also seen in the IC_{90} and IC_{99} of ratios between 100:1 to 1:100, regardless of the parasitic strain. At the IC_{99}, i.e., the threshold for determining clinical parasitologic response in nonimmune individuals, the observed activity in strain T-996 was an average of 72 times higher than that expected at an additive response to the two monocomponent drugs. The results are presented in Table 7.11.

B. GENERAL PHARMACOLOGY

The artemether–lumefantrine ACT was given intragastrically to mice, rats, and cats, to observe its effects on the nervous, cardiovascular, and respiratory systems.

TABLE 7.11 Growth Inhibition (%) of *P. falciparum* Strains T-996 and LS-21 by Mixtures of Artemether and Benflumetol (Lumefantrine) at Molar Ratios of 1:100 to 100:1[a, 28]

Artemether nmol/L of EMM (ng/mL of EMM)	Benflumetol (Lumefantrine) nmol/L of EMM (ng/mL of EMM)							
	0.01 (0.005)	**0.03 (0.016)**	**0.10 (0.053)**	**0.30 (0.158)**	**1.00 (0.528)**	**10.00 (5.279)**	**30.00 (15.838)**	**100.00 (52.794)**
T-996								
0.01 (0.003)	11.5	14.0	22.5	30.0	39.5	39.0	49.5	93.0
0.03 (0.009)	20.0	26.0	36.5	38.0	47.5	50.0	64.5	97.5
0.10 (0.030)	44.5	45.5	37.5	50.0	65.5	64.0	74.5	100.0
0.30 (0.090)	51.5	55.0	52.0	63.5	77.5	82.0	85.5	100.0
1.00 (0.298)	62.5	64.0	70.5	79.5	83.5	84.0	90.0	100.0
3.00 (0.895)	75.5	67.0	79.5	85.0	94.0	92.0	94.0	100.0
10.00 (2.984)	95.5	97.5	98.5	97.5	100.0	94.5	100.0	100.0
30.00[b] (8.951)	100.0	100.0	100.0	100.0	100.0	100.0	100.0	100.0
LS-21								
0.01 (0.003)	9.5	13.5	15.5	29.5	36.0	37.5	63.0	96.0
0.03 (0.009)	25.0	27.0	32.5	39.0	43.0	49.0	79.0	97.0
0.10 (0.030)	37.0	40.0	49.5	44.0	54.5	51.5	86.5	100.0

Continued

TABLE 7.11 Growth Inhibition (%) of *P. falciparum* Strains T-996 and LS-21 by Mixtures of Artemether and Benflumetol (Lumefantrine) at Molar Ratios of 1:100 to 100:1[a], [28]—cont'd

Artemether nmol/L of EMM (ng/mL of EMM)	Benflumetol (Lumefantrine) nmol/L of EMM (ng/mL of EMM)							
	0.01 (0.005)	**0.03 (0.016)**	**0.10 (0.053)**	**0.30 (0.158)**	**1.00 (0.528)**	**10.00 (5.279)**	**30.00 (15.838)**	**100.00 (52.794)**
0.30 (0.090)	41.5	55.5	57.0	57.0	70.5	63.0	84.0	100.0
1.00 (0.298)	65.5	73.5	67.5	71.5	81.5	79.0	95.0	100.0
3.00 (0.895)	86.0	89.0	87.0	93.5	90.5	88.5	97.5	100.0
10.00 (2.984)	95.5	100.0	98.0	95.5	100.0	98.0	100.0	100.0
30.00[b] (8.951)	100.0	100.0	100.0	100.0	100.0	100.0	100.0	100.0

[a]*Values are the means of eight readings. The inhibition by various ratios of artemether and benflumetol can be derived from the appropriate diagonals. EMM, erythrocyte-medium mixture.*
[b]*At 100.00 nmol/L of EMM; 100% growth inhibition throughout.*

The dose was 112 mg/kg, 10 times a single clinical dose for humans and 20 times the ED_{90} against rodent malaria. In mice, the drug had no significant effects on spontaneous activity, body temperature and responses to amphetamine stimulation, strychnine and picrotoxin toxicity, below-threshold pentobarbital and chloral hydrate sedation, and heat- or acetic-acid-induced pain. No clear impact was seen in the cardiovascular and respiratory systems of anesthetized cats.

C. TOXICOLOGY

1. Acute Toxicity

The acute toxicity of oral and intraperitoneal injected artemether–lumefantrine ACT in mice is shown in Table 7.12.

At the same time, the acute toxicity of both component drugs was also studied, using Smyth's simple additive method to estimate combined toxicity.[29] This, in turn, was used to assess the actual toxicity of the ACT. If the ratio between estimated and actual toxicity was 0.5–2.6, it was additive; if the ratio was >2.7, it was synergistic; if the ratio was <0.4, it was antagonistic. The oral LD_{50} of the ACT was 651 mg/kg and that of artemether was 929 mg/kg. The ratio was 1.43, making the ACT's toxicity additive and not synergistic (Table 7.13).

The LD_{50} of the 1:6 artemether–lumefantrine ACT in mice was 4555 mg/kg. Its LD_5 was 2124 mg/kg. Its chemotherapeutic index was 871, and its safety coefficient around 375. In respect of standard toxicity rankings, the ACT was in the low-toxicity group. The LD_{50} and LD_5 were 81 and 38 times the curative dose for simian malaria (14 mg/kg × 4, total 56 mg/kg), respectively. This indicated that there was a significant difference between the toxic and therapeutic doses, including the estimated recommended clinical dose (12 mg/kg × 4, total 48 mg/kg). The ACT's combined toxicity was slightly additive, whereas the drug had strong synergistic effects. It also reduced the dose of the individual drugs. Therefore, it was considered safe for clinical use.

2. Long-Term Toxicity in Rats

The 1:6 artemether–lumefantrine combination was formulated into large-, medium-, and small-dose groups: 1,792, 896, and 448 mg/kg/day. A saline control group was also used. Each group had 24 Wistar rats, with an equal number of males and females. The drug was administered intragastrically once a day for 14 days.

Clinical signs were recorded daily and 24 h after the final dose, 16 rats in each group (eight male, eight female) were sacrificed by femoral bloodletting. Hematology tests (seven parameters) and blood biochemistry (12 parameters) were run. Eleven organs were weighed and 33 tissue samples taken for histopathology examination. The remaining rats were observed for 28 days and then sacrificed to evaluate the reversibility of damage.

TABLE 7.12 LD (mg/kg), Gradient, and Correlation Coefficient of the Artemether–Lumefantrine ACT in Mice

Route of Administration	LD_5	LD_{50}	LD_{95}	*b*	*R*
Oral	2124 (2013–2240)	4555 (3843–5399)	9423 (8933–9940)	5.0838	0.9920
Intraperitoneal	1031 (945–1126)	1554 (1423–1697)	2341 (2144–2556)	9.2428	0.9909

All rats displayed anorexia across the dose groups, the degree of which correlated positively with the dose. Compared to the control group, the decrease in food consumption was 2–3 g/day for the small-dose, 3–4 g/day for the medium-dose, and 7–8 g/day in the large-dose group. This was the same for both male and female rats. Weight gain was significantly slower in the large- and middle-dose groups when compared to the control group ($P<.01$). The small-dose group showed no abnormalities in weight gain. All other clinical signs were normal after the 14-day regimen and at the end of the 28-day recovery period. Blood tests indicated that erythrocyte and reticulocyte counts decreased in the large-dose group, in comparison to the control ($P<.05$). However, levels returned to normal 28 days after medication ceased. The blood test results of the rats in the small- and medium-dose groups were all within normal range (Tables 7.14 and 7.15).

No visible abnormalities were seen by the naked eye during autopsy and after weighing of the main organs. During histological examination, slight swelling in the glomeruli of the kidneys could be seen in the large-dose group, with a small amount of protein leaking into the Bowman's capsule, detachment of the epithelial cells of the convoluted tubules with pyknotic nuclei. Small amounts of proteinlike exudate were found in the lumen. The liver cells were slightly swollen, and some of the animals had vacuolar change of hepatocytes, sinusoidal congestion, and mild infiltration of lymphocytes into the portal area. Apart from slight changes to the renal convoluted tubules, all the damage had reversed 28 days after medication ceased. No pathologies were found in the small- and medium-dose groups.

3. Long-Term Toxicity in Dogs

The 1:6 artemether–lumefantrine combination was formulated into capsules. Three dose groups of 24 beagles each—12 male, 12 female—were used: 1,000, 556, and 112 mg/kg/day, as well as a control. The drug was administered once daily for 14 days, after which 2/3 of the dogs in each group were sacrificed and examined. The remaining dogs were sacrificed 28 days after medication ceased.

TABLE 7.13 Toxicity of the Artemether–Lumefantrine ACT and Its Component Drugs in Mice

Administration Route	Artemether LD_{50} (mg/kg) (Measured)	Lumefantrine LD_{50} (mg/kg) (Measured)	Artemether–Lumefantrine LD_{50} (mg/kg) (Measured)	Artemether–Lumefantrine LD_{50} (mg/kg) (Estimated)	Artemether–Lumefantrine LD_{50} Ratio (Estimated/Measured)	Combined Toxicity Type
Oral	929	>10,000	4555	6497	1.43	Additive
Intraperitoneal	328	>10,000	1554	1917	1.23	Additive

TABLE 7.14 Effects of the Artemether–Lumefantrine ACT on Erythrocyte Count in Rats (M ± SD)[a]

Dose Group (mg/kg/day)	Sex	Erythrocyte Count on Day 14 ($\times 10^4/mm^3$)	Erythrocyte Count 28 Days After Ceasing Medication ($\times 10^4/mm^3$)
448	Female	657.4 ± 80.7	586.0 ± 20.1
	Male	667.3 ± 119.1	538.5 ± 10.2
896	Female	561.0 ± 49.6	508.7 ± 17.9
	Male	636.3 ± 72.7	510.3 ± 18.5
1792	Female	501.8 ± 55.5[a]	544.0 ± 58.1
	Male	540.6 ± 55.2[a]	557.3 ± 25.6
Control	Female	669.0 ± 86.7	534.5 ± 38.0
	Male	695.0 ± 71.9	542.5 ± 27.4

[a]*$P < .05$ when compared to control.*

TABLE 7.15 Effects of the Artemether–Lumefantrine ACT on Reticulocyte Count in Rats (M ± SD)[a]

Dose Group (mg/kg/day)	Sex	Reticulocyte Count on Day 14 ($\times 10^4/mm^3$)	Reticulocyte Count 28 Days After Ceasing Medication ($\times 10^4/mm^3$)
448	Female	17.7 ± 4.4	14.8 ± 4.8
	Male	19.1 ± 6.8	13.0 ± 3.4
896	Female	16.6 ± 5.1	12.5 ± 4.4
	Male	13.8 ± 8.9	11.5 ± 7.1
1792	Female	8.8 ± 7.0[a]	13.0 ± 3.6
	Male	8.0 ± 10.2[a]	11.3 ± 6.1
Control	Female	17.3 ± 9.1	15.3 ± 3.2
	Male	19.6 ± 5.2	14.8 ± 1.7

[a]*$P < .05$ when compared to control.*

Apart from the observation of clinical signs, ECGs (nine leads), hematology (10 parameters), blood biochemistry panels (17 parameters), routine urine tests (8 parameters), eye fundus examination, bone marrow, and other tissue histopathology examinations (37 tissue samples) were carried out. Clinical signs were recorded daily. ECGs, blood tests, comprehensive metabolic panels, and routine urine tests were conducted 7 days before medication, on the day itself, on Days 7 and 14 of medication, and 14 and 28 days after medication ceased. Electroretinography (ERG) was performed on the 1st and 14th day of medication. Bone marrow and other tissue pathology examinations were carried out on tissue sampled at autopsy. All numerical data was compiled into tables to compare the values before and after medication, as well as among dose groups. Statistical analysis was carried out.

The ECGs showed no significant changes in heart rate, rhythm, and wave shape. There was a lengthening of the Q–T interval in the 1000 mg/kg/day group on the 7th and 14th day of medication, a clear change from before. No obvious changes occurred in the 556 and 112 mg/kg/day groups (Table 7.16).

The reticulocyte count in the 1000 mg/kg/day group showed a downward trend during medication, but the difference was not significant when compared to before medication and the count was within normal range. Bone marrow slides from the 1000 and 556 mg/kg/day groups revealed that the ratio of polychromatic to basophilic erythroblasts was lower than that of the control, but there was no clear difference between the two dose groups. Other blood test parameters showed no significant changes.

TABLE 7.16 Effects of the Artemether–Lumefantrine ACT on Q–T Interval in Dogs (s)

Time of Test		Control Group	Dose Group (mg/kg/day)		
			112	556	1000
Before dosing	D−7	0.195 ± 0.012	0.198 ± 0.012	0.203 ± 0.020	0.198 ± 0.004
	D0	0.192 ± 0.013	0.205 ± 0.008	0.200 ± 0.013	0.200 ± 0.011
During dosing	D+7	0.200 ± 0.011	0.200 ± 0.011	0.208 ± 0.010	0.215 ± 0.010[a]
	D+14	0.200 ± 0.014	0.198 ± 0.012	0.208 ± 0.010	0.218 ± 0.004[b]
Recovery period	D+28	0.200 ± 0.002	0.200 ± 0.014	0.200 ± 0.080	0.205 ± 0.007
	D+42	0.195 ± 0.007	0.195 ± 0.007	0.198 ± 0.025	0.190 ± 0.014

[a] *$P < .05$ compared to before medication.*
[b] *$P < .01$ compared to before medication.*

The 1000 mg/kg/day group had elevated AST and ALP levels 7 days after medication. This difference was significant when compared to the control ($P<.05$). It returned to normal on Day 14. All other blood biochemistry panel indicators were within normal range and no obvious changes were seen. Three dogs in the 1000 mg/kg/day group and one dog in the 556 mg/kg/day group had increased urobilinogen levels 14 days after medication, reaching 4–12 mg/mL. No other changes were found. For the ERG, no changes were found in the amplitude of the a and b waves and the peak time.

"Loosening" of the cytoplasm in the hepatic cells, as well as mild vacuolar change, could be seen in the 1000 mg/kg/day group. Two dogs had partial atrophy, degeneration, and leakage in the glomeruli. The spaces in the Bowman's capsules were enlarged, with slight infiltration of lymphocytes. Similar changes were found in the other two dogs, but to a smaller extent. No obvious pathologies were found in the 112 and 556 mg/kg/day groups as well as the control. All parameters were normal 28 days after medication ceased and the damage to the liver and kidneys had recovered.

These subacute toxicity tests indicated that a dose of 1792 mg/kg/day in rats and 1000 mg/kg/day in dogs was mildly to moderately toxic. A safe dose was 448 mg/kg/day for rats and 556 mg/kg/day for dogs. The target organs were the bone marrow, liver, and kidneys, with reticulocyte count and ALT and AST levels as clinical indicators. The damage was reversible.

D. EARLY CLINICAL TRIALS OF COMPOUND ARTEMETHER TABLETS (ARTEMETHER–LUMEFANTRINE COMBINATION) IN CHINA

Because both artemether and lumefantrine components of the combination had been approved as new drugs and marketed in China, the combination tablet itself "Compound Artemether Tablets" when approved in 1992 was designated a Class III new drug under Chinese regulations. Hence, there was no need for Phase I trials. The "Compound Artemether Tablets" is a fix-dose formulation of artemether–lumefantrine, containing artemether 20 mg and lumefantrine 120 mg in each tablet (a 1:6 ratio.) In the period of 1989–91, from the early stage of clinical optimum ratio determination to the later stage of comparative and extended trials of this artemether-lumefantrine combination "Compound Artemether Tablets" in China, the AMMS' Institute of Microbiology and Epidemiology carried out clinical trials in areas of Hainan and Yunnan where chloroquine-resistant falciparum malaria was endemic.

1. Patient Inclusion Criteria and Observation Method

The subjects were falciparum malaria patients aged 18–50, with females showing no signs of pregnancy. They had not taken any antimalarial medication, or medication with inhibitory effects on the parasite, half a month before the trial.

They were within the first week of the illness, with febrile symptoms. Asexual parasitemia was 5000/mm^3 and above. Patients had no complications; no heart, liver, or kidney disease; and no history of drug allergy.

The WHO's 1973 protocol for evaluating in vivo chloroquine susceptibility was used.[30] This involved a 28-day observation period, and efficacy was assessed based on defervescence and parasite clearance times, parasite clearance in 24 h and recrudescence rate during follow-up.

Before medication, blood smears were obtained and examined under the microscope to confirm the parasitic species. Parasitemia per cubic millimeter of blood was calculated based on the ratio of parasites to leukocytes. Blood smears were also evaluated at 8-h intervals after the first dose, until the parasite was completely cleared (no parasites detected in 300 fields in three successive thick smears). The time of the first negative result was taken as the parasite clearance time. Body temperature was measured before medication and every 4 h afterward, until it returned to normal (axillary temperature 37°C and below); 24 h thereafter, body temperature was measured every morning and afternoon. Patients were closely observed during the medication period and discharged 2 days after parasite clearance and defervescence. The nature, duration, onset, and termination of adverse reactions were recorded. After discharge, patients were followed up once a week for 4 weeks. Body temperature and blood smears were taken, and patients were asked about their symptoms.

2. Selection of Component Drug Ratio and Treatment Regimen

Optimum Component Drug Ratio Study

The experiment with *P. knowlesi* indicated that artemether–lumefantrine ratios of 1:4 and 1:6 allowed both drugs to work synergistically and raised the cure rate. In this trial, ratios of 1:5 and 1:6 were selected and formulated into tablets. For the first ratio, tablets containing 20 mg artemether and 100 mg lumefantrine were used. For the second, the tablets comprised 20 mg artemether and 120 mg lumefantrine. Patients with falciparum malaria were hospitalized and randomly assigned to either group. Each group had 20 patients. Both groups were given four tablets in the initial dose and four tablets 8, 24, and 48 h later. This was a total of 16 tablets. The total dose for the first group was 1920 mg and that of the second group was 2240 mg.

The first group had a 94.2 ± 6.5% reduction in parasitemia in 24 h and a parasite clearance time of 36.0 ± 6.5 h. Defervescence time was 22.4 ± 10.8 h, and there were four cases of recrudescence during the 28-day follow-up. The cure rate was 80%. The second group saw a 96.3 ± 8.1% decrease in parasitemia in 24 h. Parasite clearance and defervescence times were 34.8 ± 5.3 h and 23.2 ± 8.4 h, respectively. No recrudescence occurred during the 28-day follow-up, and the cure rate was 100%. Since the 1:6 ratio had a significantly higher cure rate than the 1:5 ratio, it was selected for further clinical trials.

TABLE 7.17 Clinical Efficacy of Different Regimens of Compound Artemether Tablets (Artemether–Lumefantrine Combination)

Regimen	No. of Cases	Average Reduction in Parasitemia Over 24h (%)	Average Parasite Clearance Time (h)	Average Defer-vescence Time (h)	28-Day Cure Rate (%)
2-day, three doses	20	97.8±8.4	28.8±6.7	20.6±5.1	80.0
3-day, four doses	24	96.5±8.0	41.0±8.0	23.2±6.9	95.9
3-day, three doses	22	93.6±10.1	40.7±7.3	26.8±7.3	72.7

Treatment Regimen Study

The 1:6 ratio was then used to compare different treatment regimens. The first involved four doses over 3 days, with four tablets given at 0, 8, 24, and 48 h (total 16 tablets). The second had four doses in 2 days, with four tablets administered at 0, 8, 24, and 32 h (total 16 tablets). The third had three doses in 3 days, with four tablets at 0, 24, and 48 h (total 12 tablets). The results are shown in Table 7.17. Short-term clinical efficacy was satisfactory for all three regimens, with no significant differences among them. However, the 28-day cure rate was clearly superior in the 3-day, four-dose regimen than in the others. Hence, this regimen was chosen for expanded clinical trials.

3. Comparative Clinical Trial of Combination With Component Drugs

Three drug groups were used in this trial: Compound Artemether Tablets (1:6 artemether plus lumefantrine), 20 mg artemether tablets, and 120 mg lumefantrine powder capsules. The same 3-day, four-dose regimen was used for all groups (see above). Each group had 20 falciparum malaria patients, who were followed up for 28 days. The results can be seen in Table 7.18. The Compound Artemether and artemether group had similar defervescence and parasite clearance times, which were clearly faster than those in the lumefantrine group. However, the cure rate of Compound Artemether was significantly higher than the two individual drugs. The trial indicated that the ACT combined the strengths of the individual drugs and that both drugs had complementary synergistic effects. It was highly effective and fast-acting, with a high cure rate.

The research group had previously carried out clinical trials on artemether in the same region of Hainan. A 3-day, six-dose regimen of 480 mg oral artemether was given to 35 falciparum malaria patients. The parasite clearance and defervescence times were 21.9±4.6 h and 22.1±15.4 h, respectively. The 28-day cure rate was 55.7%. Another trial involved 108 falciparum malaria patients given a

TABLE 7.18 Clinical Efficacy of Compound Artemether Tablets (Artemether–Lumefantrine Combination) and Its Component Drugs

Drug	Cases	Average Reduction in Parasitemia Over 24h (%)	Average Parasite Clearance Time (h)	Average Defervescence Time (h)	28-Day Cure Rate (%)
Artemether–lumefantrine (1:6)	20	97.3±3.8	35.6±7.8	23.8±5.4	95
Artemether	24	95.1±5.2	38.7±8.2	19.7±8.4	45
Lumefantrine	22	74.5±15.7	68.4±14.7	40.2±13.9	65

6-day, 12-dose regimen of lumefantrine capsules, total dose 3600 mg. Parasite clearance time was 72.0 ± 20.4 h, and defervescence time was 45.4 ± 16.2 h. The 28-day cure rate was 88.5%. These clinical trials confirmed that further use of the ACT was justified. It had high efficacy, rapid action, and a high cure rate.

4. Comparative Trials of Compound Artemether (Artemether–Lumefantrine Combination) With Other Antimalarials

Piperaquine Phosphate

This comparative trial took place from July to November 1989, in an area of Hainan where chloroquine-resistance falciparum malaria was endemic. Patients in the first group were given the Compound Artemether Tablets (1:6 artemether–lumefantrine) in the 3-day, four-dose regimen (total dose 2240 mg). Those in the second group received piperaquine phosphate tablets (each containing 150 mg piperaquine base) in a 3-day, four-dose regimen. Four piperaquine tablets were administered in the first dose, followed by two tablets at 8, 24, and 48 h (a total of 10 tablets or 1500 mg piperaquine base).

The compound artemether group had 40 patients, all of whom were clinically cured. Average parasite clearance time was 40.0 h, and average defervescence time was 23.7 h. Three cases of recrudescence occurred during the 28-day follow-up, with a cure rate of 92.5%. Of the 30 cases in the piperaquine phosphate group, six saw no parasite clearance in a week. There was a reduction in fever over 48 h, but the temperature did not return to normal (R_{II} level resistance). Two cases maintained temperatures of 39°C and above after 48 h and there was no clear drop in parasitemia (R_{III} level resistance). There were 11 cases of recrudescence during the 28-day follow-up. The cure rate was 36.7% (Table 7.19).

No abnormalities were found in blood and routine urine tests, as well as in ECGs. Two cases in the ACT group had elevated ALT and AST levels before medication. The piperaquine phosphate group had two cases with elevated ALT and one case with elevated ALT and AST. All laboratory values returned to normal on Days 3 and 6. Other patients' ALT and AST levels were within normal range both before and after medication.

Piperaquine phosphate and its ACTs had been used in Hainan as prophylaxis and treatment since the 1970s. The presence of chloroquine-resistant falciparum malaria there had been confirmed. Like chloroquine, piperaquine was also an aminoquinoline. Using the standard therapeutic dose of piperaquine, this trial indicated that piperaquine resistance had also emerged in Hainan. It also showed that the artemether–lumefantrine ACT was effective against piperaquine-resistant falciparum malaria.

Chloroquine Phosphate

These trials were carried out from August to October 1989, on adult falciparum malaria patients in Sanya's South Island Farm Hospital, Hainan. The 3-day, four-dose regimen of the Compound Artemether Tablets (1:6 artemether–lumefantrine)

TABLE 7.19 Clinical Efficacy of Compound Artemether (Artemether–Lumefantrine Combination) and Piperaquine Phosphate on Falciparum Malaria

Drug	No. of Cases	Average Reduction in Parasitemia After 24h (%)	Average Parasite Clearance Time (h)	Average Defervescence Time (h)	Curative Efficacy				
					S	R_I	R_{II}	R_{III}	Cure Rate (%)
Compound Artemether	40	94.5±9.3	40.0±7.7	23.7±3.3	37	3	0	0	92.5
Piperaquine phosphate	30	52.6±20.1	73.4±18.7[a]	41.8±8.0[a]	11	11	6	2	36.7

[a]*Body temperature did not return to normal within 1 week for eight cases and there was no parasite clearance. This is the average for the remaining 22 cases.*

was used, total dose 2240 mg. A 3-day, four-dose regimen was also used for chloroquine phosphate, total dose 1500 mg chloroquine base. The artemether–lumefantrine ACT group had 35 patients; average parasite clearance time was 37.8 h and defervescence time was 24.2 h. There was one case of recrudescence, yielding a 28-day cure rate of 97.1%.

There were 22 patients in the chloroquine phosphate group. Two cases had fever of 39°C and above 48 h after the first dose. There was no clear decrease in parasitemia; in fact, parasite count increased (R_{III} level resistance). Six cases saw no parasite clearance in a week (R_{II} level resistance), and there were five cases of recrudescence during the 28-day follow-up (R_{I} level resistance). Only nine cases were cured. The 28-day cure rate was 40.9% (Table 7.20). No adverse reactions were found in both groups.

5. Expanded Clinical Trials of Compound Artemether Tablets (Artemether–Lumefantrine Combination)

From July to November 1989, the health and epidemic prevention team of the Chengdu Military District's Logistics Department carried out expanded clinical trials of the Compound Artemether Tablets in the border region between western Yunnan and Burma. The subjects were falciparum malaria patients older than 7 years, with parasitemia above 5000 per mm^3 blood. The 3-day, four-dose Compound artemether regimen with a total dose of 2240 mg was used. It was scaled down for children: Those aged 13–15 were given individual doses of 3.5 tablets, those aged 10–12 received three tablets, and those aged 7–10 years got 2.5 tablets. Piperaquine phosphate was the control, with the same 1500 mg base regimen used as in the previous comparative trial.

The Compound artemether group had 61 patients. Average decrease in parasitemia after 24 h was 99.8 ± 4.0%; average parasite clearance and defervescence times were 23.2 ± 5.1 h and 21.2 ± 8.5 h, respectively. There was no recrudescence during the 28-day follow-up and the cure rate was 100%. No significant adverse reactions were seen. ECGs and liver and kidney function tests showed no significant abnormalities.

For the 30 patients in the piperaquine phosphate group, the average fall in parasitemia after 24 h was 81.6 ± 12.4%. Average parasite clearance and defervescence times were 35.1 ± 14.3 h and 32.2 ± 18.7 h, respectively. There was also no recrudescence during the 28-day follow-up. The cure rate was 100% and no significant adverse reactions were observed. Since most of the patients had been infected in Burma, where piperaquine had not been used, there was still susceptibility to the drug. However, the ACT had shorter defervescence and parasite clearance times.

With the above results in mind, the South Island and Ledong Baoguo Farm Hospitals in Sanya, Hainan, continued with expanded Compound artemether trials in 1990. These were regions where chloroquine and piperaquine resistance were endemic. South Island Farm Hospital took in 167 falciparum malaria

TABLE 7.20 Clinical Efficacy of Compound Artemether (Artemether–Lumefantrine ACT) and Chloroquine Phosphate on Falciparum Malaria

Drug	No. of Cases	Average Parasite Clearance Time (h)	Average Defervescence Time (h)	Curative Efficacy				
				S	R_I	R_{II}	R_{III}	Cure Rate (%)
Compound Artemether	35	37.8 ± 4.2	24.2 ± 10.1	34	1	0	0	97.1
Chloroquine phosphate	22	87.3 ± 24.7[a]	56.5 ± 23.7[a]	9	5	6	2	40.9

[a]*Body temperature did not return to normal in 1 week for eight cases. This is the average for 14 cases.*

patients aged 7–50. Of these, 150 were locals and 17 were from elsewhere, with no history of malaria. Parasitemia was over 5000 per mm^3, with 35 cases over 50,000 per mm^3. The AST, ALT, and BUN levels of 20 cases were checked before and after medication. Reticulocyte tests were carried out for 50 cases before and after medication. All 167 patients had different degrees of fever; 27 cases had body temperatures over 40°C. Febrile symptoms cleared 8–40 h after treatment, with average defervescence time at 20.4 ± 8.4 h. The decrease in parasitemia after 24 h was 97.4%, and full clearance followed within 24–48 h. The average was 37.9 ± 7.8 h. All patients were clinically cured. During the 28-day follow-up, 10 cases had recrudescence of the asexual parasite between Days 21 and 28. The 28-day cure rate was 94.1%. No significant adverse reactions were found and the results of the reticulocyte counts and ALT, AST, and BUN tests were normal.

Ledong Baoguo Farm Hospital had 33 patients, including 25 who were from elsewhere and had no immunity. Nine were aged 7–15, with the rest aged 16–45. The level of parasitemia in 18 of the patients was 30,000–100,000 per mm^3. The 3-day, four-dose artemether–lumefantrine ACT regimen, total dose 2240 mg, was used. The dose was scaled down for children based on age. All patients were given blood tests before medication; 30 were given ECGs and their AST, ALT, and BUN levels were checked. All 33 cases were clinically cured. The decrease in parasitemia after 24 h was 96.2 ± 4.3%. Average parasite clearance time was 29.0 ± 7.9 h, and average defervescence time was 25.7 ± 16.8 h. The 28-day cure rate was 100%, with no recrudescence. The comprehensive metabolic panels, blood tests, routine urine tests, reticulocyte counts, ECGs and AST, ALT, and BUN tests all showed no abnormal changes before and after medication.

In 1990, the Hainan Institute of Tropical Diseases used the Compound artemether to treat vivax malaria in Ledong, where the disease was endemic. Satisfactory results were obtained. In total, 48 vivax malaria patients were taken in, with parasitemia between 1000 and 60,000 per mm^3. They were aged 16–50 years, apart from six patients aged 6–15. The 3-day, four-dose regimen of the ACT was used, with a total dose of 2240 mg. All patients were clinically cured. Average defervescence time was 13.6 ± 6.9 h, and average parasite clearance time was 22.8 ± 9.5 h. There were four cases of recrudescence between Days 21 and 28 of the follow-up. No drug-related adverse reactions were seen during the observation period, and blood and urine tests showed no abnormal changes.

From June to November 1996, a parallel comparative trial took place in Hainan with the Compound artemether tablets and lumefantrine capsules. The standard 2240 mg ACT regimen was used for the former. For the latter, a 4-day, four-dose regimen was used with the first dose double that of the subsequent ones. The total dose was 2400 mg. A total of 100 falciparum malaria patients were randomly assigned to either drug group and hospitalized for observation for 28 days. The ACT group had 51 cases and the lumefantrine group, 49 cases. All subjects were older than 13 years. Both drugs were effective against

TABLE 7.21 Efficacy of Compound Artemether Tablets (Artemether–Lumefantrine Combination) and Lumefantrine Capsules on Falciparum Malaria

Drug	No. of Cases	Average Parasite Clearance Time (h)	Average Defervescence Time (h)	28-Day Cure Rate (%)
Compound artemether tablets	51	17.1 ± 8.6	29.7 ± 8.9	98.0
Lumefantrine capsules	49	29.4 ± 24.9	54.7 ± 17.4	95.9

falciparum malaria, but the ACT had a superior parasite clearance time and was better at controlling symptoms (Table 7.21).

In 1996, the standard 2240 mg Compound artemether tablets regimen was also used to treat 22 cases of falciparum and 24 cases of vivax malaria in the area between the western border of Yunnan and Burma, where drug-resistant *P. falciparum* was coendemic with *Plasmodium vivax*. The patients were aged 5–68 and were migrants from hypoendemic or nonendemic areas who had moved in for 1–3 years. In the falciparum malaria cases, average defervescence and parasite clearance times were 26.2 ± 9.1 h and 33.4 ± 8.3 h, respectively. There was one case of recrudescence on Day 24 of the observation period. Before medication, four cases tested positive for gametocytes. These were cleared half a month after treatment. In the vivax malaria cases, average parasite clearance time was 28.9 ± 8.3 h, and average defervescence time was 22.9 ± 7.1 h. Within 9 months, there were 21 cases of recrudescence, occurring within 86.6 days on average. Nevertheless, reinfection could not be ruled out. The trial showed that the ACT Compound artemether was fast-acting and highly effective in treating falciparum and vivax malaria. Adverse reactions were mild. For vivax malaria, however, it should be combined with primaquine for a more complete treatment.

The AMMS' Institute of Microbiology and Epidemiology also sent a team to Somalia in 1988. There, it treated 24 cases of falciparum malaria with the ACT. Average parasite clearance time was 29.9 h, and average defervescence time was 25.7 h. The 28-day cure rate was 100%. No significant adverse reactions were observed.

From 1989 to 1996, Compound artemether tablets had been used to treat a total of 453 falciparum malaria patients in Hainan and Yunnan, where drug-resistant strains of the disease were endemic. A 3-day, four-dose regimen totaling 16 tablets and 2240 mg was used. All patients were clinically cured. Across all the tests, seven treatment groups of patients had a 28-day cure rate of 92.5%–98.0%, and two had a rate of 100%.

E. INTERNATIONAL MULTICENTER CLINICAL TRIALS OF ARTEMETHER AND LUMEFANTRINE TABLETS (COARTEM)[31]

After the 1990–94 joint Chinese-Swiss research and development period, artemether–lumefantrine tablets, Coartem was registered in Switzerland in 1995, listed in the 2013 edition of the International Pharmacopoiae as "artemether and lumefantrine tablets." Twenty international, multicenter clinical studies were organized by Novartis between 1994 and 2007 under the names of "CGP 56697," "Co-artemether," "artemether and lumefantrine tablets," "Coartem Tablets" or "Coartem." These were conducted in countries where falciparum malaria was endemic, such as China, Thailand, Vietnam, India, and Bangladesh in Asia; Gambia, Tanzania, Kenya, Nigeria, Benin, Mali, Mozambique, Senegal, and Cameroon in Africa; and Brazil in Latin America. Travelers from nonendemic countries, and who thus had no immunity, were also included in the trials. These patients were from Switzerland, Germany, France, Italy, the Netherlands, the United Kingdom, and Columbia. A total of 3599 cases were included, of whom 1572 were adults (over 16 years) and 2027 were children (16 years or below). They included Asian, Black, Caucasian, and Hispanic patients. A few of the important clinical trials are outlined below. The 28-day cure rates were values with Polymerase Chain Reaction (PCR) adjustment. The median was used to calculate defervescence and parasite clearance times.

1. Comparative Trial of Coartem and Its Component Drugs (AB/M02)

From 1994 to 1996, based on the Institute of Microbiology and Epidemiology's existing research, scientists from the Institute and Novartis carried out a GCP-compliant comparative trial on Coartem 's standard regimen and its two component drugs. To prevent reinfection, all patients were under strict observation in a single center, the Naval Hospital in Sanya, Hainan, China for 28 days. The doses of artemether and lumefantrine were based on their ratios in Coartem. The total Coartem dose was 2240 mg, that of artemether was 320 mg, and that of lumefantrine was 1920 mg.

Without PCR adjustment, the 28-day cure rate for Coartem was 100% (50 cases). The cure rates of artemether and lumefantrine were 54.5% (44 cases) and 92.2% (51 cases), respectively. Decrease in parasitemia after 24 h was 99.3% for Coartem, 99.9% for artemether, and 78.2% for lumefantrine. Defervescence time was 24 h for Coartem, 21 h for artemether, and 60 h for lumefantrine. Parasite clearance time was 30 h for Coartem, 30 h for artemether, and 54 h for lumefantrine. The results indicated that Coartem was superior to its component drugs and were consistent with those of the previous trials conducted by the Institute.

2. Comparative Trial of Four- and Six-Dose Regimens (A025)

Novartis treated adults and youths from the United Kingdom and India with the 3-day, four-dose Coartem regimen in 1995–96. Without PCR correction, the 28-day cure rates were 100% (12 cases) and 95.4% (89 cases). However, three trials involving adults and youths in Thailand, where drug resistance is severe, produced uncorrected cure rates of 69.3% (126 cases), 82.1% (309 cases), and 76.5% (87 cases). The cure rate for children was only 60.8% (111 cases). In the Gambian and Tanzanian trials, moreover, the cure rates in children were 70.9% (60 cases) and 63.6% (130 cases). It showed that the regimen had different efficacies in various regions.

Therefore, in 1996–97, researchers from Novartis designed two new six-dose regimens. The first was a 60-h, six-dose regimen, with the drug given at 0, 8, 24, 36, 48, and 60 h (total dose 3360 mg). The second was a 96-h, six-dose regimen, with medication at 0, 8, 24, 48, 72, and 96 h (total dose 3360 mg). These were compared with the standard four-dose regimen (total dose 2240 mg) in a trial, A025, in Bangkok and Mae La, Thailand. The 28-day cure rates of the 60-h, six-dose regimen and the 96-h, six-dose regimen were 96.9% (96 cases) and 99.1% (106 cases), respectively. Defervescence time was 35 h for the former and 21.8 h for the latter. Parasite clearance time was 43.6 h for both. These regimens were significantly more effective than the original four-dose one, which had a cure rate of 83.3% (102 cases). Of the two, the 60-h, six-dose regimen had better patient adherence. The results showed that in Thailand, where drug resistance is high, the dose of the Coartem had to be increased to 3360 mg for adults to achieve satisfactory levels of efficacy.

3. Efficacy of the Six-Dose Regimen in Thailand (A026, A028)

To prevent regional differences in the regimen used, the WHO proposed that Novartis apply one standard regimen across the board. From 1996 to 1997, Novartis carried out clinical, pharmacokinetic, and safety trials of the 60-h, six-dose regimen on adults and children aged two and above in Thailand (A026). Then, in 1998–99, Novartis commenced another trial (A028), using the same protocol to conclusively validate the efficacy of Coartem in the formulation intended for marketing. In this trial, adults were those aged 16 and above; children were those aged below. Mefloquine and artesunate (MAS) was used as the control in both trials. Children's doses were scaled down adjusted to body weight. Four weight scales were used: 5–15 kg (one tablet per dose), 15–25 kg (two tablets per dose), 25–35 kg (three tablets per dose), and above 35 kg (four tablets per dose).

The 28-day cure rate was 97.7% in Trial A026, which had 133 cases. Of these, 38 were children, who had a cure rate of 94.7%. For Trial A028, the 28-day cure rate was 96.1%. Of the 154 cases treated, 15 were children, who had a cure rate of 93.3%. Defervescence times were 22 h for A026 and 29 h for A028.

Parasite clearance times were 48 h for A026 and 29.3 h for A028, and the recrudescence rate was 1.3% and 2.4%, respectively. The 28-day cure rate for MAS in both studies was 100%; in A026 (47 cases, including 17 children) and A028 (53 cases, including 12 children). Defervescence times were 22.2 and 23 h, respectively, and parasite clearance times were 48 and 31 h, respectively. There was no recrudescence.

4. Efficacy of the Six-Dose Regimen in Infants and Children in Africa (A2403, B2303)

After the Thai trials, in 2002–03, Novartis and the WHO collaborated to test the six-dose Coartem regimen in Kenya, Nigeria, and Tanzania. The trial, A2403, assessed the regimen's safety and efficacy in children weighing 5–35 kg. Half the patients were aged 2 months to 3 years and weighed 5 to <10 kg. They were an extremely vulnerable group, with no immunity and high mortality. The 28-day cure rate was 96.7% (299 cases); for infants weighing 5 to <15 kg, the cure rate was 96.9% (256 cases). Defervescence time was 7.8 h and parasite clearance time was 24 h. Of the 42 cases of treatment failure, 32 experienced recrudescence of *P. falciparum*.

At the same time, Novartis also conducted trials on the six-dose regimen with infants and children in eight centers in sub-Saharan Africa: Benin (one center), Kenya (three centers), Mali (one center), Mozambique (one center), and Tanzania (two centers). Patients in the trial, B2303, were aged below 12 years and weighed 5–35 kg; 60.8% weighed between 5 and 15 kg. The tablets used were either dispersible tablets specially developed by Novartis, or crushed normal tablets. The 28-day cure rate was relatively high: 97.8% (412 cases). In infants weighing 5–15 kg, the cure rate was 98.0% (244 cases). Defervescence time was 7.8 h and parasite clearance time was 34.9 h.[32]

Table 7.22 consolidates Makanga and Krudsood's summary of Novartis' six trials on the six-dose regimen.[33] Coartem's efficacy in different regions, in different centers of the same region, and in various ethnicities and age groups can be clearly seen. Trial A2401, which ran from 2001 to 2005, took place in Switzerland (five centers), Germany (four centers), France (four centers), the Netherlands (one center), and Columbia (one center). The patients were mostly Caucasians who had traveled to falciparum-malaria endemic areas.

5. Non-Novartis Trials

Makanga and Krudsood summarized more than 40 independent clinical trials on Coartem.[33] They were mostly comparative trials between the artemether–lumefantrine ACT (Coartem) and existing antimalarials and other ACTs, which were carried out after the WHO had recommended that the ACT be used in areas with drug-resistant malaria. This increased the number of countries using Coartem. Trials involving Coartem in Uganda, Ethiopia, Tanzania,

TABLE 7.22 Clinical Studies Evaluating the Efficacy and Safety of the Six-Dose Regimen of Artemether-Lumefantrine

Study Number	A025[10]	A026[11]	A028[12]	A2403[13]	B2303[14]	A2401 [15]
Design	Randomized double-blind	Randomized open-label	Randomized open-label	Open-label	Randomized investigator blind	Open-label
Comparator	Four-dose regimen	MAS[a]	MAS[a]	–	Dispersible formulation	–
Patients	Adults and children (>2 years)	Adults and children (≥2 years)	Adults and adolescents (>12 years)	Infants and children (5–25 kg)	Infants and children (5 to <35 kg)	Adult nonimmune travellers
N AL/total	120/359[b]	150/200	164/219	310	452/899	165
Geography	Thailand	Thailand	Thailand	Kenya Tanzania Nigeria	Kenya Tanzania Mall Benin Mozambique	EU Colombia
28-day PCR-corrected cure rate (evaluable pts)	96.9%	97.7%	95.5%	96.7%	97.8%	96.0%
Median time to fever clearance (h) (mITT pop[c])	35 (n=59)	22 (n=87)	29 (n=76)	7.8 (n=309)	7.8 (n=311)	36.5 (n=100)
Median time to parasite clearance (h) (mITT pop[c])	43.6 (n=118)	48 (n=149)	29.3 (n=164)	24.0 (n=310)	34.9 (n=452)	41.8 (n=162)

AL, artemether/lumefantrine; *MAS*, mefloquine + artesunate.
[a]*Study was not designed to compare AL with mefloquine + artesunate.*
[b]*AL over 60 h.*
[c]*28-day cure rate was a secondary endpoint.*
Data from Makanga and Krudsood. The clinical efficacy of artemether/lumefantrine (Coartem). *Malar J* 2009; **8** (Suppl. 1):S5.

TABLE 7.23 Results of a Meta-Analysis Comparing the Efficacy of Several Antimalarial Combinations

Treatment Combination	28-Day, PCR-Corrected Cure Rate (%)
Artemether–lumefantrine	97.4
Mefloquine + artesunate	96.9
Amodiaquine + artesunate	88.5
Amodiaquine + sulfadoxine–pyrimethamine	85.7
Sulfadoxine–pyrimethamine + artesunate	82.6
Chloroquine + sulfadoxine–pyrimethamine	72.1
Chloroquine + artesunate	45.3

Data from Makanga M, Krudsood S. The clinical efficacy of artemether/lumefantrine (Coartem®). *Malar J* 2009;**8**(Suppl. 1):S5.

Zanzibar, Sudan, Zambia, Mozambique, Ghana, Nigeria, Senegal, Mali, Gambia, Burkina Faso, Rwanda, Angola, Burundi, the Republic of Congo, Thailand, Laos, Indonesia, Bangladesh, and Nepal are presented below.

A meta-analysis was conducted on the artemether–lumefantrine ACTand other ACTs, based on Jansen et al. summary of 32 studies conducted primarily in Africa, South America, and Asia and using a Bayesian random effects model.[34] It showed that Coartem was one of the most effective, with a 28-day cure rate of 97.4% (Table 7.23).

6. Effect on Gametocytes

During the clinical trials, the Institute of Microbiology and Epidemiology and Novartis discovered that Coartem could eliminate gametocytes. Sutherland et al. reported that the drug could reduce malaria transmission. In a randomized controlled trial, they treated falciparum malaria patients aged 1–10 years with chloroquine and sulfadoxine–pyrimethamine (SP) ($n=91$) or the Co-artemether (Coartem) ($n=406$). Gametocytes were checked on Days 7, 14, and 28. On Day 28 after treatment, the children in the ACT group had a gametocyte rate of only 7.94% ($P<.001$). The gametocyte rate in the chloroquine and SP group was 48.8% ($P<.001$).[35]

7. Tolerance and Safety

Based on extensive data from the clinical trials, involving 1869 patients and including 243 children aged 5–12 years and 368 children aged below five,

Bakshi et al. evaluated the tolerability and safety of artemether–lumefantrine (Coartem).[36] They also compared it to other antimalarials. The most common side effect of artemether–lumefantrine in adults involved the gastrointestinal system, such as abdominal pain (11.6%), anorexia (13.2%), nausea (6.3%), vomiting (2.4%), and diarrhea (3.5%). The central nervous system was also affected, including headaches (20.7%), dizziness (16.1%), and sleep disturbance (12.5%). Skin reactions included itching (2.4%) and rashes (1.3%). Other side effects were fatigue (6.1%), weakness (13.9%), heart palpitations (8.3%), myalgia (11.6%), arthralgia (10.5%), and coughing (11.3%). In children, common side effects were abdominal pain (11%), vomiting (6.7%), diarrhea (5.4%) anorexia (9.5%), coughing (11.3%), and headaches (8.7%).

More than 90% of the patients who reported adverse reactions also had overlapping symptoms of acute malaria, and most of these side effects were mild or moderate. The side effects of chloroquine (vomiting and itching), mefloquine (dizziness, nausea, and vomiting), SP (drowsiness), and quinine (vomiting and dizziness) were comparatively more severe. Coartem had no serious or persistent neurological side effects and did not cause any clinical changes in blood, liver, and kidney tests. A series of ECGs from 713 patients showed an increased incidence of prolonged QT intervals, like those seen in chloroquine, mefloquine, and artesunate–mefloquine patients. However, it occurred less frequently than with quinine or halofantrine. All the patients with prolonged QT intervals did not report adverse cardiac symptoms. It indicated that Coartem could be safely used to treat acute, uncomplicated falciparum malaria in children and adults.

After 16 trials and more than 3400 patients—1427 adults and 1992 children, taking in both the four-dose and the six-dose regimens—Novartis could draw certain conclusions about Coartem 's adverse reactions.[31] The most frequent side effects in adults were headaches, weakness, dizziness, and anorexia, which often occurred within 1–3 days of the illness. These could be symptoms of the disease relating to fever. The six-dose regimen had a 5.4% rate of adverse reactions. For children, fever, vomiting, and anorexia were the most common side effects with the six-dose regimen. As in adults, coughing could be a result of respiratory infection. These reactions were mild or moderate. Therefore, Coartem was seen as well-tolerated and safe, and suitable for a broad range of patients.

It had previously been reported that artemisinin-type drugs displayed neurotoxicity in animals. The experiments by McLean et al. also showed that dihydroartemisinin was the most neurotoxic, while the toxicity level of artemether was much lower.[37] However, there was little clinical evidence of neurotoxicity with artemisinin-type drugs. The authors believed that many of the early animal-model experiments involved slow-release formulations, which were administered via intramuscular injection. In a clinical setting, however, the drugs were given orally or as suppositories, which were rapidly cleared from the body and reduced the amount of time in which the nervous system was exposed to the drugs. This lowered the risk of toxicity.

Moreover, Toovey et al. found that the artemether–lumefantrine ACT (Co-artemether) could cause asymptomatic hearing loss.[38] In Novartis' 16 trials, however, hearing loss was reported in only 0.7% of the cases (16 out of 2318). In other antimalarials, this proportion was 6% (23 out of 1223 cases). Since acute malaria could also cause temporary hearing loss, the authors proposed that further tests on the neurotoxicity of artemisinin were needed. Later, such tests were performed on over 200 patients, many had received multiple courses of an artemisinin, and no effect on neurophysiological parameters was found which included no hearing loss.[39]

Falade et al. analyzed the safety and tolerability of Coartem in more than 6300 patients, taken from the trials by Novartis and other independent researchers.[40] Most reported side effects were very mild, mostly involving the gastrointestinal and nervous systems. Many were also classic symptoms of malaria, and an increase in the number of doses did not add to the severity of these side effects. They did not cause any safety concerns for the heart, nervous system, and blood chemistry, including hearing. The authors considered this very important for children in hyperendemic areas, since they may require repeated courses of antimalarials over a lifetime. Novartis further investigated the effects of Coartem on hearing.[31] One case experienced severe urticaria, which was related to Coartem. Falade also drew on Bakshi's study to analyze the ACT's effects on anemia and thrombocytopenia in 15 clinical trials, since these symptoms were common in malaria patients. It was found that both conditions normalized or improved significantly after treatment with the artemether–lumefantrine ACT.[36]

In Zambia, Manyando et al. carried out exploratory research on the effects of the artemether–lumefantrine ACT and SP in 1001 pregnant women and 933 newborns.[41] A total of 495 pregnant women and 466 newborns were given the ACT; 506 pregnant women and 467 newborns were given SP. There were no clinical differences between both groups in terms of the rate of perinatal and neonatal mortality, stillbirths, physical defects, and preterm deliveries. The WHO states that the artemether–lumefantrine ACT can be used in the second and third trimesters. In the absence of further data, it should not be used in the first trimester. Nevertheless, the treatment of pregnant women should not be neglected despite these ambiguities. For example, quinine is safe for use in the first trimester but, due to its long treatment duration, exposure may continue into the second and third trimesters, increasing the risk of hypoglycemia. If ACTs are on hand, they should be given in the first trimester. Further data on the drug's effects on infant development will provide a firmer foundation for the WHO's recommendations.[42]

Lumefantrine and halofantrine are homologous. The therapeutic dose of halofantrine has cardiotoxic effects, since it prolongs ventricular repolarization. This is reflected in a lengthened QTc interval in ECGs. Lengthened QTc intervals were also seen in dogs and rats given intramuscular injections of artemether. Thus, van Vugt et al. conducted a trial with 150 patients given the six-dose regimen of artemether–lumefantrine ACT, and another 50 receiving artesunate–mefloquine.[43] ECGs were conducted before and after medication to assess changes in the QTc interval and their relationship to the dose. None of the patients showed prolonged QTc intervals, and there was no relationship to the plasma concentration of lumefantrine. Bindschedler et al. also compared the effects of Co-artemether and halofantrine on the QTc interval of 13 healthy subjects.[44] All subjects who received halofantrine had prolonged QTc intervals, the largest being 28 ms. There was also a dose–effect relationship. The QTc intervals in the subjects given Co-artemether remained unchanged. These results were statistically significant.

F. PHARMACOKINETICS

After the signing of the international cooperation agreement, Novartis and the Institute of Microbiology and Epidemiology conducted studies on the clinical pharmacokinetics of the ACT and its component drugs. These would provide a firm basis for the ACT's clinical regimens. Population pharmacokinetics studies also elucidated the synergistic effects of artemether and lumefantrine. They demonstrated that artemether, with its short half-life, was responsible for rapid defervescence and parasite clearance. Lumefantrine, with its longer half-life, raised the cure rate and prevented recrudescence.

1. Pharmacokinetics of a Single Oral Dose in Healthy Volunteers

Pharmacokinetic Parameters

Lefèvre et al. studied the pharmacokinetics of the oral artemether and lumefantrine (Riamet) in healthy people, as well as of artemether, its dihydroartemisinin metabolite, and lumefantrine.[45] The plasma concentrations of artemether and dihydroartemisinin were measured using reversed phase HPLC with reductive electrochemical detection, a method proposed by Navaratnam et al.[46] and van Agtmael et al.[47] For benflumetol (lumefantrine), reversed phase HPLC with UV detection was used, as proposed by Zeng et al. from the Institute,[48] as well as by Mansor.[49]

Twelve healthy Caucasian volunteers were fasted and given a single oral dose of Co-artemether comprising 80 mg artemether and 480 mg lumefantrine. Blood samples were collected at 0, 0.5, 1, 1.5, 2, 3, 4, 6, 8, 10, 12, 15, 24, 34, 48, 72, 96, and 168 h. The pharmacokinetics of ACT, its dihydroartemisinin metabolite and lumefantrine were measured (Table 7.24).

The Influence of Food

A crossover trial was run with 16 healthy Chinese volunteers, with Co-artemether given after fasting and 15–25 min after a high-fat breakfast. The same oral ACT dose and blood sampling schedule were as with the previous trial. The pharmacokinetics of artemether, dihydroartemisinin, and lumefantrine are shown in Table 7.25.[45]

TABLE 7.24 Mean (± SD) Pharmacokinetic Parameters of Artemether, Dihydroartemisinin (DHA) and Lumefantrine in 12 Healthy Caucasian Volunteers After a Single Oral Dose of Co-artemether (80/480 mg)

Parameters	Artemether	DHA	Lumefantrine
C_{max}	34.2 ± 26.6 μg/L	55.0 ± 31.6 μg/L	0.50 ± 0.74 mg/L
T_{max}[a]	0.75 h	1.0 h	4.0 h
AUC_{0-t}	AUC_{0-15h} 98.1 ± 78.1 μg·h/L	AUC_{0-15h} 200 ± 144 μg·h/L	AUC_{0-168h} 10.3 ± 15.9 mg·h/L
$T_{½\beta}$	3.0 ± 1.5 h	2.0 ± 0.9 h	30.9 ± 27.6 h

[a]*Median value.*
Data from Lefèvre G, Thomsen MS. Clinical pharmacokinetics of artemether and lumefantrine (Riamet®). *Clin Drug Investig* 1999;**18**(6):467–80.

TABLE 7.25 Mean (± SD) Pharmacokinetic Parameters of Artemether, Dihydroartemisinin (DHA) and Lumefantrine in 16 Healthy Chinese Individuals After a Single Dose of Co-artemether (80/480 mg) Under Fasted and Fed Condition

	Fasted	After High-Fat Meal	Ratio
Artemether			
C_{max} (μg/L)	52.3 ± 44.8	104 ± 53	2.0
T_{max} (h) (median)	1.25	2.0	
$AUC_{0\to16h}$ (μg·h/L)	143 ± 110	338 ± 175	2.4
$T_{½\beta}$ (h)	2.2 ± 0.9	2.3 ± 1.2	
DHA			
C_{max} (μg/L)	33.3 ± 21.2	49.7 ± 23.3	1.5
T_{max} (h) (median)	1.5	2.0	
$AUC_{0\to16h}$ (μg·h/L)	90.5 ± 49.8	169 ± 57.1	1.9
$T_{½\beta}$ (h)	2.5 ± 1.1	3.1 ± 1.6	
Lumefantrine			
C_{max} (mg/L)	0.38 ± 0.15	5.10 ± 1.90	13.3
T_{max} (h) (median)	6.0	6.0	
$AUC_{0\to168h}$ (mg·h/L)	6.8 ± 3.3	108 ± 47	15.8
$T_{½\beta}$ (h)	35.1 ± 19.1	71.3 ± 20.7	

Data from Lefèvre G, Thomsen MS. Clinical pharmacokinetics of artemether and lumefantrine (Riamet®). Clin Drug Investig 1999;**18**(6):467–80.

The increase in AUC and C_{max} was statistically significant in all three compounds after food ($P<.05$). The oral bioavailability (based on AUC) of artemether was more than twofold higher in the volunteers given a high-fat breakfast than in the fasted volunteers. A similar trend was seen with dihydroartemisinin. The bioavailability of lumefantrine was 16 times higher after the meal, and its elimination half-life $T_{½\beta}$ also doubled from 35.1 to 71.3h. In fact, this was not just the time in which the compound was eliminated but also in which it was absorbed. Actual elimination time could not be calculated due to the lack of a parenteral formulation.

2. Population Pharmacokinetics of Oral Coartem in Falciparum Malaria Patients

Pharmacokinetic Parameters of Artemether, Dihydroartemisinin, and Lumefantrine

Van Agtmael et al. conducted a study on 88 falciparum malaria patients in Hainan.[47] Of these, 48 received artemether and 40 received Co-artemether. The population pharmacokinetic parameters of artemether were determined, as well as the impact of lumefantrine on these parameters. The patients had uncomplicated falciparum malaria, had not been treated with antimalarials within 28 days, were not pregnant or breastfeeding, and had not received any prophylactic, therapeutic, or other unknown medication in the past 14 days. All patients were given four-dose regimens, with medication administered at 0, 8, 24, and 48h. Twelve of the patients in the Co-artemether group received four tablets (20mg artemether, 120mg lumefantrine in each tablet) in each dose. Another 12 in the artemether group were given four 20mg artemether tablets in each dose. Patients were fasted 2h before medication and half an hour afterward. Blood samples were taken 0.5, 1.0, 1.5, 2.0, 4.0, and 8.0h after the first dose; 1.0, 1.5, 8.0, and 16.0h after the second; 1.0, 1.5, 8.0, and 24.0h after the third dose; and 1.0, 1.5, 2.0, 4.0, 8.0, 16.0, and 32.0h after the fourth dose.

The remaining patients in each group received the same four-dose regimen but could eat small amounts before and after medication. Blood samples were only taken at 1.5 and 8.0h (i.e., at the lowest and highest concentration points). Drug concentration in plasma was measured using reversed phase HPLC with electrochemical detection, as proposed by van Agtmael.

The artemether concentration–time curves in both groups clearly showed that C_{max} and AUC rose and fell with the dose. On average, peak concentration after the first dose was three times higher than after the last dose. The C_{max1}/C_{max4} ratios were 3.2 in the artemether group and 2.6 in the Co-artemether group. This trend was reversed for dihydroartemisinin. Its C_{max} and AUC rose slightly after the second dose and, in the Co-artemether group, its peak concentration was significantly higher after the fourth dose than after the third. The C_{max1}/C_{max4} ratios of dihydroartemisinin were 0.7 in the artemether group and 1.2 in the Co-artemether group. Although the concentrations of both

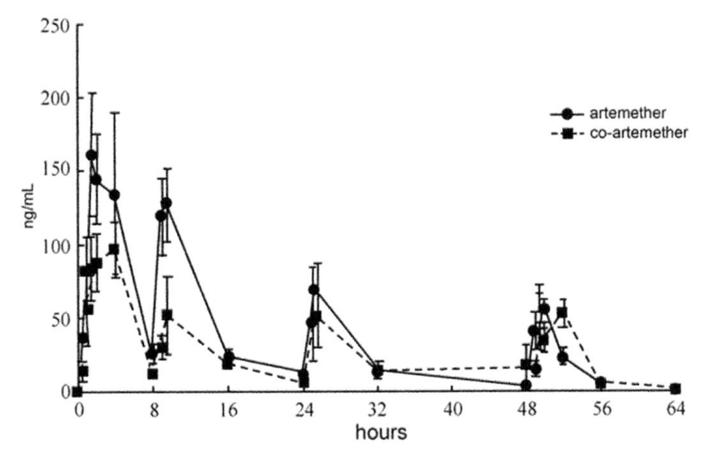

FIGURE 7.6 Mean (and standard error) artemether plasma concentration in 24 patients given oral artemether or co-artemether at 0, 8, 24, and 48 h.[50]

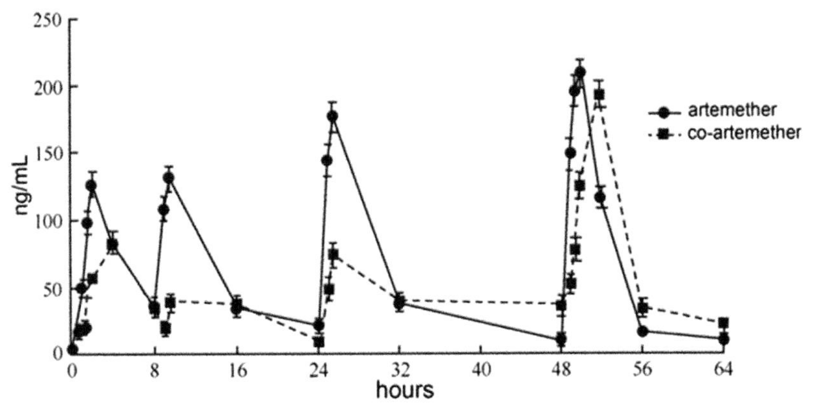

FIGURE 7.7 Mean (and standard error) dihydroartemisinin plasma concentration in 24 patients given oral artemether or co-artemether at 0, 8, 24, and 48 h.[50]

compounds rose or fell with the dose, there was marked variation across individuals (Figs. 7.6 and 7.7).

In both groups, the $AUC_{0\to\infty}$ for dihydroartemisinin (2926 and 3135 ng·h/mL in the artemether and Co-artemether group, respectively) was significantly higher than the $AUC_{0\to\infty}$ for artemether (1834 and 2035 ng·h/mL in the artemether and Co-artemether group, respectively). This was mainly due to the dihydroartemisinin peaks being higher than the artemether peaks after the third and fourth doses. The C_{max3} of dihydroartemisinin was 260 and 168 ng/mL in the artemether and Co-artemether group, respectively, and the C_{max4} was 283 and 171 ng/mL, respectively. For artemether, the C_{max3} in the artemether group was 85 ng/mL, whereas it was 116 ng/mL in the Co-artemether group. Its C_{max4}

was 60 ng/mL in the artemether group and 58 ng/mL in the Co-artemether group. These results corresponded to dihydroartemisinin's lower $V_{d/f3}$ and $V_{d/f4}$ values. Simultaneously, the bioavailability of artemether decreased over time, whereas that of dihydroartemisinin increased. The authors believed that this phenomenon may be due to enzymatic autoinduction in the intestinal mucosa or liver, which accelerates the first-pass metabolism of artemether over time. Since both artemether and dihydroartemisinin have antimalarial effects, changes in their concentration will not influence the efficacy of artemether.

The results showed that a single oral dose of artemether had a mean lag time of 0.48 h. Plasma concentration then rose rapidly (T_{abs}=0.82 h, T_{max}= 1.73 h, C_{max}= 157 ng/mL) and fell quickly (T_{half}= 1.16 h). For dihydroartemisinin, the parameters were T_{lag}=0.35 h, C_{max}= 161 ng/mL, T_{max}= 1.86 h, and T_{half}=0.95 h. In vitro studies showed that the EC_{50} of artemether was around 1 ng/mL and that of dihydroartemisinin was around 0.5 ng/mL. Here, all the C_{max} values exceeded the EC_{50} in 1–6 h. Although the range of average C_{max} values in this study was relatively large (16–372 ng/mL), all patients achieved drug concentrations necessary for parasite clearance. Average parasite clearance time was 30 h.

At the AMMS, Lefèvre et al. conducted a randomized, double-blind pharmacokinetics test on 36 Chinese patients out of a group of 144.[45] The patients were divided into three groups of 12. The first group received four doses of artemether–lumefantrine (Riamet, Co-artemether), with 80 mg artemether and 480 mg lumefantrine for each dose; the second, four doses of 80 mg artemether for each dose; and the third, four doses of 480 mg lumefantrine for each dose. The drugs were administered at 0, 8, 24, and 48 h. There were no regulations as to food or drink intake, but 23% of the patients did not eat during the first two doses. By the latter two doses, all the patients had eaten. Blood samples were taken before medication and at 0.5, 1, 1.5, 2, 4, 8, 9, 9.5, 16, 24, 25, 25.5, 32, 48, 49, 49.5, 50, 52, 56, 64, 80, 96, 120, 168, 336, 504, and 672 h after the first dose. The remaining 108 patients had five blood samples taken before medication and 1.5 and 8 h after each dose. The parameters for the 36 patients are in Table 7.26 and Fig. 7.8.

These results indicated that lumefantrine concentration gradually increased with successive doses, regardless of whether it was given in single-drug or ACT form. The reverse was true for artemether: Concentration decreased with each successive dose, and its C_{max} after the fourth dose was only half or a third of the value after the first dose. However, the concentration of its dihydroartemisinin metabolite gradually rose, with its C_{max} after the fourth dose being 1.4 (in the artemether group) or two (in the Co-artemether group) times higher than the value after the first dose.

This was also seen in the blood samples obtained from the remaining 108 patients. Whether this was due to autoinduction of artemether, or the influence of lumefantrine on artemether metabolism, will be discussed in the metabolism section below. Based on the AUC, the total bioavailability of artemether was higher in the single drug than with the ACT, with a ratio of 0.72. However, the

TABLE 7.26 Mean (± SD) Pharmacokinetic Parameters of Artemether, Dihydroartemisinin (DHA) and Lumefantrine Following Four Doses of 80 mg Artemether, 480 mg Lumefantrine, or Combination of Both Drugs (Co-artemether 80/480 mg) in 36 Chinese Patients (12 Per Treatment Group) With Malaria. Doses Were Given at 0, 8, 24, and 48 h

		Artemether		DHA		Lumefantrine	
	Dose	Co-artemether	Artemether	Co-artemether	Artemether	Co-artemether	Artemether
C_{max} (μg/L)	I	150 ± 58	225 ± 177	105 ± 42	154 ± 71	3.10 ± 3.16[a]	4.16 ± 3.54[a]
	II	56 ± 90	134 ± 87	53 ± 43	139 ± 82	8.93 ± 3.91[a]	9.41 ± 4.01[a]
	III	65 ± 92	83 ± 62	105 ± 86	186 ± 132	10.7 ± 4.9[a]	12.8 ± 3.6[a]
	IV	84 ± 62	69 ± 38	218 ± 66	221 ± 119	13.4 ± 5.7[a]	13.8 ± 6.7[a]
T_{max} (h) (median)	I	2.0	2.0	4.0	2.0	8.0	8.0
	II	1.5	1.5	8.0	1.5	8.0	8.0
	III	1.5	1.25	1.5	1.5	8.0	8.0
	IV	2.0	2.0	4.0	2.0	8.0	8.0
$AUC_{(0-672h)}$ (mg·h/L)		1.7 ± 0.77	2.3 ± 1.2	6.6 ± 12.4	3.7 ± 1.2	925 ± 403	955 ± 385

[a]*mg/mL.*
Data from Lefèvre G, Thomsen MS. Clinical pharmacokinetics of artemether and lumefantrine (Riamet®). *Clin Drug Investig* 1999;**18**(6):467–80.

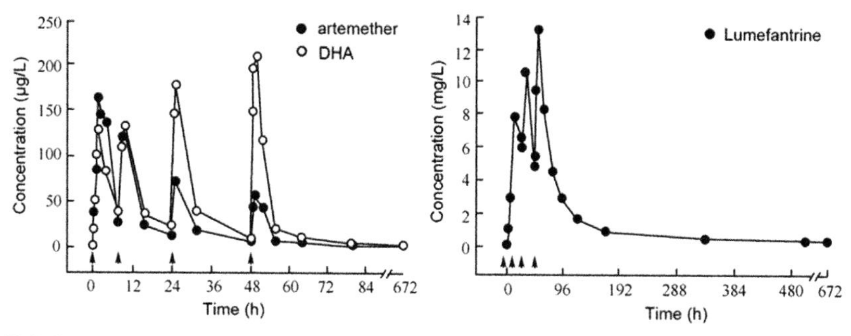

FIGURE 7.8 Mean plasma profiles of artemether, dihydroartemisinin (DHA) and lumefantrine during four doses of co-artemether (80/480 mg) at 0, 8, 24, and 48 h (see *arrows*) in 12 Chinese patients with malaria.[45]

TABLE 7.27 Cure Rate, Parasite Clearance Time (PCT) and Fever Resolution Time (FRT) After Co-artemether Compared With Either Component Given Alone in Chinese Patients With Malaria. Doses Were Given at 0, 8, 24, and 48 h

	Co-artemether 4 × 80/480 mg	Artemether 4 mg × 80 mg	Lumefantrine 4 mg × 480 mg
28-Day cure rate	100% (n=50)	54.5% (n=44)	92.2% (n=51)
Median PCT (h)	30 (n=53)	30 (n=52)	54 (n=52)
Median FRT (h)	24 (n=36)	21 (n=30)	60 (n=38)

Data from Lefèvre G, Thomsen MS. Clinical pharmacokinetics of artemether and lumefantrine (Riamet®). *Clin Drug Investig* 1999;**18**(6):467–80.

total bioavailability of dihydroartemisinin was 40% higher with the ACT than with the single drug. This had no real impact on therapeutic efficacy, since there was a statistically significant difference between artemether and the ACT in terms of cure rate and parasite clearance and defervescence times (Table 7.27).

Relationship Between Efficacy of Artemether–Lumefantrine Combination and the AUCs of Component Drug Artemether and Lumefantrine

The treatment objective for malaria is the thorough clearance of parasites from the patient, to avoid recrudescence and obtain a complete cure. Therefore, two indicators of therapeutic efficacy are the cure rate and parasite clearance time. The four-dose Coartem regimen was very effective in Africa, India, and China, but less so in areas of Thailand where drug resistance was severe. Clinical trials indicated that higher artemether or dihydroartemisinin AUC levels could markedly reduce parasite clearance times. However, although artemether had a rapid parasite clearance time, its elimination half-life was also short. In a study of patients given different artemether–benflumetol (lumefantrine) regimens, Na-Bangchang et al. found a robust marker of efficacy: Patients who were cured had significantly higher plasma concentrations of lumefantrine than in patients whose treatment had failed.[51]

By comparing three different regimens on falciparum malaria patients in Thailand, Ezzet et al. examined the population pharmacokinetics of lumefantrine[52] to provide further basis for the application of artemether–benflumetol in that country.[53] From September 1996 to February 1997, Ezzet et al. conducted a study in Bangkok and the northwest border area of Mae La. The objectives were to estimate the population pharmacokinetics of lumefantrine, ascertain the factors influencing plasma concentration of the drug, and increase the dose of artemether–benflumetol to enhance clinical efficacy.[52] In this double-blind test, 266 uncomplicated falciparum malaria patients were given artemether–benflumetol in three different regimens. These were a four-dose, 48-h regimen (total 1920 mg lumefantrine); a six-dose, 60-h regimen (total 2880 mg lumefantrine); and a six-dose, 96-h regimen (total 2880 mg lumefantrine).

The patients were observed for five consecutive days, and then at 7-day intervals beginning on Day 7 and ending on Day 63. If the parasites reappeared during this period, it was considered recrudescence. PCR was used to distinguish between recrudescence and reinfection. Blood samples were obtained at the following times and stored at −70°C following treatment:

- Regimen A (four doses, 48 h): Before medication and at 4, 8, 24, 28, 32, 44, 48, 60, 72, 80, 120, 168, and 240 h.
- Regimen B (six doses, 60 h): Before medication and at 8, 24, 36, 44, 48, 60, 64, 72, 96, 108, 120, 168, and 240 h.
- Regimen C (six doses, 96 h): Before medication and at 8, 24, 32, 48, 52, 64, 72, 80, 96, 108, 120, 168, and 240 h.

Plasma concentration of lumefantrine was determined using the reversed-phase HPLC with UV detection method established by Zeng et al.[48] Detailed observations were made on 51 hospitalized adults in Bangkok. The remaining data were obtained from 215 patients in Mae La, who were of various ages.

The results of Ezzet's study can be summarized as follows:

1. *Relationship between lumefantrine's AUC and cure rate of drug regimens.* In areas where malaria is endemic, patients with immunity often have higher cure rates than those without, such as children. In these cases, Regimen A could achieve a high level of efficacy. Therefore, the authors considered this a satisfactory regimen for use in most endemic areas. Immunity in the western border region of Thailand was relatively lower, as the level of transmission was less intensive and more erratic. The average AUC of lumefantrine in Regimens A, B, and C were 356, 561, and 712 ng·h/mL, respectively.

These values approximated the level of bioavailability. The bioavailability of lumefantrine in Regimens B and C were therefore 60% and 100% higher than in Regimen A, making them more reliable. The 28-day cure rates of regimens A, B, and C were 83%, 97%, and 99%, respectively. The results demonstrated a close relationship between lumefantrine's AUC and cure rate of various artemether–benflumetol (lumefantrine) regimens.

2. *Relationship between lumefantrine's minimum inhibition concentration and recrudescence*. Prior dose-related research indicated that if the plasma concentration of lumefantrine could be maintained at 280 ng/mL and above for 7 days after medication, treatment failure (i.e., recrudescence) could be avoided. This critical concentration was close to the in vivo minimum inhibition concentration ($MIC \approx EC_{99}$). The longer the concentration could be kept above this threshold, the lower the chances of recrudescence and the higher the AUC value would be.[52] This trial showed that lumefantrine concentration dropped below 280 ng/mL in 204 h (8 days) for Regimen A, 252 h (10 days) for Regimen B, and 298 h (12 days) for Regimen C. Regimens B and C could maintain the threshold concentration for 25% and 50% longer than Regimen A.
3. *Effects of food intake on lumefantrine's bioavailability (AUC)*: Lumefantrine is a lipophilic compound. During the acute phase of malaria, patients have poor appetite. Hence, the bioavailability of the drug is relatively low. Lefèvre et al. had previously discovered that oral bioavailability increased 16 times in healthy subjects after a high-fat meal. Ezzet et al. compared 13 patients who ate normally before medication and 24 who took only liquids. Oral bioavailability in the first group was 336% times that of the latter. In another experiment, patients who ate small or normal amounts had 48% and 108% higher oral bioavailability, respectively, than patients who drank only liquids. Those who ate normally had 42% higher bioavailability than those who ate light meals.
4. *Relationship between lumefantrine's bioavailability concentration (AUC), the baseline parasitemia, and cure rate*. In the Thai study, it was found that the bioavailability fraction values of the first to the fourth dose F_1, F_2, F_3, and F_4 were lower among the Bangkok patients with higher levels of parasitemia.[53] At the same time, there was a link between interpatient variability in F_1 values, baseline parasitemia, and body temperature. After plotting average baseline parasitemia over time, it was discovered that 12 of the 15 recrudescence cases had high levels of parasitemia and shorter periods in which lumefantrine concentration was maintained above 280 ng/mL. For patients in high, medium, and low parasitemia groups (the top, middle, and bottom lines in the graph), cure rates were 50%, 74%, and 95%, respectively, if lumefantrine concentration could be maintained at 280 ng/mL and above. If the concentration fell below this threshold, cure rates fell to 33%, 56%, and 71%, respectively (Fig. 7.9).

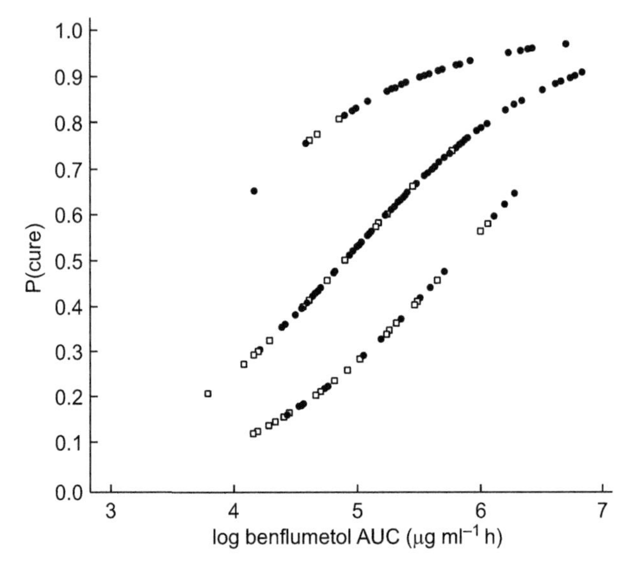

FIGURE 7.9 Estimated probability of cure against benflumetol (lumefantrine) log [$AUC_{(0,\infty)}$], (μg/mL·h).[53] Note: (a) ●, cured; □, recrudescence; (b) top, middle and bottom curves correspond to parasite counts in the categories of <5000/μL, 5000–50,000/μL, and >50,000/μL.

Artemether and dihydroartemisinin had no clear impact on cure rates, since their AUC did not influence curative frequency. Lumefantrine's AUC also had no statistically significant effect on parasite clearance time.

These population-based results showed that the pharmacokinetics and pharmacodynamics of lumefantrine and artemether were very different in terms of their antimalarial effects. Artemether was rapidly absorbed and effectively cleared parasites throughout its short half-life. Lumefantrine's AUC was critical for maintaining the cure rate. When used with artemether, it could prevent recrudescence. Its antimalarial effects were in line with its longer half-life.

White et al. summarized the artemether–lumefantrine clinical trials that had taken place in China, Africa, India, and Southeast Asia since 1992.[54] With the 4-day, 48-h regimen, 90% of patients experienced defervescence and 53% saw parasite clearance in 24h. On average, parasitemia fell by 99% in less than 24h, and parasite clearance time was 36h. This was mainly due to artemether. In Tanzania, where the rate of transmission was high and stable, the 7- and 14-day cure rates in children aged 1–5 were 94% and 86%, respectively. In Gambia, where transmission was much lower, the rates were 100% and 93%, respectively. The 28-day cure rate of this regimen was 85% in Thailand, where transmission was low but drug resistance was the most severe worldwide. The rate was only 69% in the northwestern border of Thailand. With six-dose, 60-h or six-dose, 96-h regimens, however, this rose to 97% and 99%, respectively.

White et al.[54] and Bindschedler et al.[17] evaluated the ACT's strengths, stating that artemether eliminated parasites and relieved clinical symptoms rapidly, including fever, which contributed to patient well-being. Because both component drugs had completely different modes of action, artemether had already eliminated most of the parasites by the time lumefantrine took effect. Artemether was also swiftly eliminated from the body, which reduced the chance of the remaining parasites developing selective resistance toward the drug and dihydroartemisinin. Lumefantrine alone eliminated a parasitic biomass of 10^5 at most, which also minimized the emergence of selective resistance and extended the effective lifespan of the drug. Koram et al. were the first to report on this synergistic and complementary relationship between artemether and lumefantrine.[55] Artemether–lumefantrine or Co-artemether also markedly reduced gametocytes, as well as the emergence and transmission of drug resistance. Therefore, it could minimize the spread of resistant strains of falciparum malaria.

3. Metabolism and Drug Interactions

CYP Enzymes and the Metabolism of Artemether and Lumefantrine

Giao et al. outlined the metabolism of artemether and lumefantrine, as well as potential drug interactions.[56] As we have seen, an oral dose of artemether was rapidly absorbed, reaching C_{max} in around 2 h. It underwent extensive and thorough first-pass metabolism in the liver, and perhaps in the intestinal tract as well, to form dihydroartemisinin. This, in turn, reached C_{max} in 2–6 h. At least 90% of artemether was bound to proteins, mainly to alpha-1-acid glycoprotein and albumin but also to various lipoproteins.

In healthy people, the bioavailability of artemether could increase if the drug was taken with grapefruit juice. This suggested that artemether was a substrate of CYP3A4 in the intestines, since grapefruit juice can inhibit this enzyme. If CYP3A4 catalyzes the breakdown of artemether into dihydroartemisinin, when the enzyme is inhibited by grapefruit juice, the metabolism of artemether is itself suppressed, raising its plasma concentration and therefore its bioavailability. However, this effect disappeared as the dose frequency increased, which indicated that the CYP3A4-mediated metabolic process was a saturable one.

Thus, intestinal CYP3A4 played a role in the metabolism of artemether. However, an examination of the ratio of 6β-hydroxycortisol to cortisol in urine showed that artemether had no effect on the activity of *hepatic* CYP3A4. Neither did the hepatic enzyme play a role in the decrease in artemether plasma concentration over time. CYP2D6 and CYP2C19 were also not involved in the metabolism of artemether. Based on the in vitro research by Zimmerlin et al., Lefèvre concluded that CYP1A1, CYP1A2, CYP2B6, CYP2C9, and CYP2C19 could all catalyze the biotransformation of artemether. This was especially the case for CYP2B6, which might act as a time-dependent catalyst. Apart from

CYP3A4, therefore, other supplementary metabolic pathways existed and it was to be expected that artemether would interact with other drugs. Still, many enzymes might affect first-pass metabolism and hence bioavailability, or influence the elimination rate and metabolic catalysis. More research was required.

Lumefantrine seems to bind completely (>99%) to proteins, mainly lipoproteins. In vitro, lumefantrine is metabolized to form desbutyl-lumefantrine, which is excreted via bile and feces. CYP3A4 is involved in the metabolic process. The 6β-hydroxycortisol to cortisol ratio in the urine of healthy volunteers given the oral ACT was measured. As mentioned above, it showed that both artemether and lumefantrine did not affect hepatic CYP3A4 activity.[57] Some papers found a degree of CYP2D6 inhibition, but the clinical significance is unclear.[54]

Interactions With Mefloquine

In a clinical setting, Co-artemether may be used when mefloquine prophylaxis or therapy has failed to deal with falciparum malaria. Since artemether, lumefantrine, and mefloquine metabolism are catalyzed by CYP3A4, Novartis' Lefèvre et al. studied the pharmacodynamics of Co-artemether when taken in tandem with mefloquine.[57]

The six-dose Co-artemether regimen was used and administered 24, 32, 48, 60, 72, and 84 h after the mefloquine was given. Each Co-artemether dose had four tablets containing 80 mg artemether and 480 mg lumefantrine. The mefloquine dose was 1000 mg; to prevent vomiting, it was divided into three doses at 12-h intervals. A total of 42 healthy volunteers were allocated to three groups: Co-artemether group, a mefloquine group, and a Co-artemether-mefloquine group. Because the elimination half-life of mefloquine lasted several weeks, a parallel instead of a crossover test design was used. The subjects ate a standard diet, with no caffeine or grapefruit juice. Blood samples were obtained at set intervals for pharmacokinetic analysis.

Mefloquine was a substrate of CYP3A4. If it was administered together with another substrate or an inhibitor, drug interactions may occur. Even though artemether and lumefantrine metabolism were catalyzed by CYP3A4, the study suggested that mefloquine interacted only with lumefantrine. The C_{max} of lumefantrine in the ACT-only group was 28.3 ± 13.6 μg/mL; when administered with mefloquine, the C_{max} was 20.0 ± 8.3 μg/mL. However, because lumefantrine was mostly influenced by food intake, this difference was not clinically significant. Also because of artemether's wide therapeutic index, it was unlikely that its efficacy would be affected by drug interactions.

The authors also pointed out that, in the Co-artemether group, the dihydroartemisinin-to-artemether ratio, C_{max} and AUC increased with the number of doses. This also occurred in the Co-artemether-mefloquine group. Artemether and lumefantrine, or lumefantrine alone, could be acting as CYP3A4 inducers, leading to increased metabolism of artemether.

Interactions With Quinine

Coartem may also be used after treatment with quinine has failed. Quinine was cardiotoxic, the effects of which may be enhanced by Coartem. Moreover, quinine metabolism was primarily catalyzed by hepatic CYP3A4. Its elimination was accelerated by rifampicin, which was a strong CYP3A4 inducer. Conversely, the CYP3A4 inhibitor cimetidine could lower its elimination rate and lengthen its half-life.

Lefèvre et al. thus carried out a drug interaction study with the same design as with the mefloquine experiment.[58] Intravenous quinine was given after the sixth and final artemether–lumefantrine (Riamet) dose. The quinine dose was 10 mg/kg at a speed of 20 mL/h, total dose 40 mL over 2 h. Blood was drawn at set intervals for pharmacokinetic analysis. Coadministration of artemether–lumefantrine and quinine produced a small increase in cardiotoxicity, and the plasma concentrations of artemether and dihydroartemisinin were relatively lower. Apart from that, there were no other pharmacokinetic or metabolic effects. These changes had no obvious clinical significance.

The drug interaction tests indicated that, in a clinical context, artemether and lumefantrine were not mutually inhibiting. The inverse relationship between artemether and the dose—unlike that of dihydroartemisinin, which increased with the dose—could be due to autoinduction. Both artemether and dihydroartemisinin were active compounds. Hence, in clinical practice, these metabolic changes had no impact on therapeutic efficacy. It seemed that the metabolism of dihydroartemisinin was less linked to CYP enzymes. Not much was known about lumefantrine as an effector.

Interactions With Ketoconazole

Lefèvre et al. studied the effects of ketoconazole on the pharmacokinetics of artemether–lumefantrine (Riamet), as well as on the ECGs of healthy people.[59] Ketoconazole was a powerful inhibitor of hepatic and intestinal CYP3A4 and of intestinal P-glycoprotein. Sixteen subjects were randomly assigned to crossover groups and given either a single artemether–lumefantrine dose or the ACT combined with ketoconazole (400 mg on Day 1, followed by 200 mg for four additional days). Blood samples were taken and the plasma concentrations of artemether, dihydroartemisinin, and lumefantrine were measured.

The results showed that ketoconazole could influence the pharmacokinetics of artemether, dihydroartemisinin, and lumefantrine. The $AUC_{0-\infty}$ increased from 320 to 740 ng·h/mL (2.4 times) for artemether, from 331 to 501 ng·h/mL (1.7 times) for dihydroartemisinin, and from 207 to 333 ng·h/mL (1.7 times) for lumefantrine. The C_{max} increased 2.2, 1.4, and 1.3 times, respectively. The elimination half-lives of artemether and dihydroartemisinin rose but that of lumefantrine remained unchanged. However, these increases were much smaller than the impact of food on the ACT, namely a 16-fold increase for lumefantrine.

Neither were there any effects on adverse reactions and ECGs. This suggested that no adjustment in dose was needed for patients taking the ACT together with ketoconazole or other strong CYP3A4 inhibitors.

Plasma Protein Binding

Colussi et al. used ultrafiltration and erythrocyte partitioning to measure the protein binding rate of artemether and lumefantrine.[60] This rate was very high: 95%–98% for artemether and >99% for lumefantrine. At normal protein concentrations, 33% of artemether in the blood was bound to alpha-1-acid glycoprotein, 17% to albumin, 12% to high density lipoproteins (HDLs), 9.3% to low density lipoproteins (LDLs), and 12% to very low density lipoproteins (VLDLs). For lumefantrine, 77% was bound to HDLs, 7.3% to LDLs, and 6.6% to VLDLs. The erythrocytic binding rate was 10% for both drugs. The apparent volume of distribution was 1 L/kg for artemether and 2 L/kg for lumefantrine, indicating that there was broad tissue binding in both compounds. This was in line with their lipophilic qualities.

G. EFFECTIVENESS AND COMPLIANCE OF ARTEMETHER AND LUMEFANTRINE TABLETS (COARTEM)

The effectiveness of a drug in clinical practice was determined by its efficacy, cost, availability, deliverability, and acceptability.[61] Since malaria often affected the poor, each aspect of effectiveness had to be explored and refined for Coartem to achieve its full potential. Clinical trials had already addressed the drug's efficacy. In terms of cost-effectiveness, Novartis marketed this ACT under two brand names: Coartem, sold at low cost in developing countries, and Riamet, sold at a higher price in the private-sector markets of developed countries. Coartem was also marketed at two price points, one for sale to the public sector and the other, to the private. As for deliverability and availability, the WHO obtained and distributed Coartem to ensure supply.[62]

Compliance depended on a clear understanding of the correct usage of a drug among patients or guardians, especially those with a low level of literacy. It also depended on the feasibility of a particular regimen in clinical practice. These, in turn, ensured that maximum therapeutic efficacy could be achieved. Preventing improper drug administration—such as failure to take the drug on time, to finish the treatment course, or to consume high-fat foods after each dose to maintain lumefantrine at effective plasma concentration levels—would minimize any impact on efficacy and avoid drug resistance. Children's doses should be clear and practicable. The safety and efficacy of the drug for pregnant women should also be made explicit.

According to Fogg et al., basic factors enabling or threatening drug adherence must be assessed before the introduction of a new drug.[63] They studied patients with uncomplicated falciparum malaria in Mbarara, southwestern Uganda, who had been prescribed artemether–lumefantrine. When the drug was dispensed, the use of the drug was explained; the authors conducted home visits, during which the drug's packaging was checked for unused tablets. Patients were asked questions to determine if they had followed the regimen. Out of 210 patients whose data could be analyzed, 21 (10%) either definitely or probably had not adhered to the regimen. The 16 nonadherent patients had lower lumefantrine plasma concentrations than in adherent patients ($n = 171$); however, this was not statistically significant. The study found that the ACT was very effective in Mbarara. Patients received clear explanations, as well as pictorial instructions for children weighing 10–14 kg. The key factor affecting adherence was low educational levels.

Also in Mbarara, Piola et al. compared the therapeutic outcomes of 313 cases given direct instructions from a doctor with that of 644 outpatient cases who only received nutritional advice from a doctor.[64] The 28-day cure rate for the first group was 97.7% (296/303) and that of the second group was 98.0% (603/615). Contact was lost with 39 patients. There was no difference between both groups and it was found that outpatient cases could comply with the artemether–lumefantrine regimen, which was very effective. However, cost was still an issue. Other ACTs could be used if treatment failure occurred after 2 weeks, or for pregnant women.

ACKNOWLEDGMENTS

We would like to thank G. Lefèvre, M.S. Thomsen, F. Ezzet, R. Mull, C.J. van Boxtel for permission to use their published figures and tables; and H. Rietveld of Novatis for his kind support.

REFERENCES

1. Deng RX, Yu LB, Zhang HB, et al. Antimalarial research: synthesis of α-(alkylaminomethyl)-halogenated-4-fluorenemethanols. *Acta Pharm Sin* 1981;**16**(12):920–4. [邓蓉仙，余礼碧，张洪北，等. 抗疟药的研究α-(烷氨基甲基)-卤代-4-芴甲醇类化合物的合成. 药学学报 1981, 16(12): 920–924.]
2. Zhao DC, Zhong JX, Geng RL, et al. Antimalarial research II: synthesis of α-alkylaminomethyl-1,6-dichloro-4-fluorenemethanols. *Acta Pharm Sin* 1982;**17**(1):28–31. [赵德昌，钟景星，耿荣良，等，抗疟药的研究 II.α-(烷氨基甲基)-1,6-二氯-4-芴甲醇类化合物的合成. 药学学报 1982, 17(1): 28–31.]
3. Deng RX, Zhong JX, Zhao DC, et al. Synthesis and antimalarial activity of fluorenemethanols. *Acta Pharm Sin* 1997;**32**(11):874–8. [邓蓉仙，钟景星，赵德昌，等. 芴甲醇类化合物的合成及抗疟作用. 药学学报 1997, 32(11): 874–878.]
4. Atkinson Edward R, Granchelli Felix E. Antimalarials. 3. Fluorenemethanols. *J Med Chem* 1974;**17**(9):1009–21.
5. Deng R. Recent progress in antimalarial research in China. *Chin J Pharm* 1989;**20**(8):372–6. [邓蓉仙. 我国近几年抗疟药新进展. 中国医药工业 1989, 20: 372–376.]
6. Deng RX, Zhong JX, Zhao DC, et al. New route for the synthesis of the antimalarial lumefantrine. *Acta Pharm Sin* 2000;**35**(1):22–5. [邓蓉仙，钟景星，赵德昌，等. 抗疟药本芴醇的合成新路线. 药学学报 2000, 35(1): 22–25.]

7. Deng RX, Zhong JX, Zhao DC. *New technique for the synthesis of the antimalarial lumefantrine*. CN1029680. 1999. [邓蓉仙，钟景星，赵德昌. 抗疟新药本芴醇的合成新工艺：中国 CN1029680. 1999.]
8. Wang YL, Ding DB, Ding JX. Increasing the bioavailability of hydrophobic fluorenemethanols via oleic acid capsules. *Chin Pharm Bull* 1982;**17**(1):4–7. [王云玲，丁德本，丁建新. 用油酸胶囊剂提高疏水性芴甲醇抗疟药的生物利用度. 药学通报 1982, 17(1): 4–7.]
9. Wang YL, Ding JX, Geng RL. Studies on the stability of fluorenemethanol antimalarials in linoleic acid capsules. *Chin J Pharm Anal* 1984;**4**(2):84–7. [王云玲，丁建新，耿荣良. 芴甲醇类抗疟药亚油酸软胶囊剂的稳定性研究. 药物分析杂志 1984, 4(2): 84–87.]
10. Zhong JX, Deng RX, Wang J, et al. Crystal and molecular structure of the antimalarial α-(dibutylaminomethyl)-2,7-dichloro-9-(p-chlorobenzylidene)-4-fluorenemethanol (lumefantrine). *Acta Pharm Sin* 2000;**35**(1):824–9. [钟景星，邓蓉仙，王俭，等. 抗疟药α-(二正丁氨基甲基)-2,7-二氯-9-对氯亚苄基-4-芴甲醇的晶体和分子结构. 药学学报 2000, 35(1): 824–829.]
11. Wernsdorfer WH, Landgraf B, Kilimali VA, et al. Activity of benflumetol and its enantiomers in fresh isolates of *Plasmodium falciparum* from East Africa. *Acta Trop* 1998;**70**(1):9–15.
12. Peters W. Drug resistance in *Plasmodium berghei* Vincke and Lips, 1948. I. Chloroquine resistance. *Exp Parasit* 1965;**17**:80–9.
13. Raether W, Fink E. Antimalarial activity of floxacrine (HOE 991) I. Studies on blood schizontocidal action of floxacrine against *Plasmodium berghei*, *P. vinckei* and *P. cynomolgi*. *Ann Trop Med Parasitol* 1979;**73**(6):505–26.
14. Davidson Jr. DE, Johnsen DO, Tanticharoenyos P, et al. Evaluating new antimalarial drugs against trophozoite induced *Plasmodium cynomolgi* malaria in rhesus monkeys. *Am J Trop Med Hyg* 1976;**25**(1):26–33.
15. Wernsdorfer WH, Payne D. Drug sensitivity tests in malaria parasites. In: Wernsdorfer WH, McGregor IA, editors. *Malaria: principles and practice of malariology*. Edinburgh: Churchill Livingstone; 1988. p. 1765–800.
16. Basco LK, Bickii J, Ringwald P. In vitro activity of lumefantrine (benflumetol) against clinical isolates of *Plasmodium falciparum* in Yaoundé, Cameroon. *Antimicrob Agents Chemother* 1998;**42**(9):2347–51.
17. Bindschedler M, Degen P, Lu ZL, et al. Comparative bioavailability of benflumetol after administration of a single oral dose of co-artemether under fed and fasted condition to healthy subjects. In: *Abstract P-01-96 from the XIV international congress for tropical medicine and malaria*. 1996. p. 346. Nagasaki, Japan.
18. Wernsdorfer WH, Kilimali VAEB, Prajakwong S, et al. Sensitivity to benflumetol of *Plasmodium falciparum* from Muheza, Northeastern Tanzania, and Mae Hong Son, northwestern Thailand. In: *2nd European congress on tropical medicine of the federation of European societies for tropical medicine and international health, and 4th residential meeting of the royal society of tropical medicine and hygiene, Liverpool, UK*. 1998. p. 35.
19. Pradines B, Tall A, Fusai T, et al. In vitro activities of benflumetol against 158 senegalese isolates of *Plasmodium falciparum* in comparison with those of standard antimalarial drugs. *Antimicrob Agents Chemother* 1999;**43**(2):418–20.
20. Wernsdorfer G, Parajakwong S, Noedl H, et al. Sensitivity to benflumetol of *Plasmodium falciparum* and its relation to the response to other antimalarial drugs. In: *2nd European congress on tropical medicine of the federation of European societies for tropical medicine and international health and 4th residential meeting of the royal society of tropical medicine and hygiene, Liverpool, UK*. 1998. p. 36.
21. Xu SY, et al. *Experimental methodology in pharmacology*. Beijing: People's Medical Publishing House; 1982. p. 1184. [徐淑云等主编，《药理实验方法学》，人民卫生出版社, 1982, p. 1184.]

22. Deichmann WB, LeBlanc TJ. Determination of the approximate lethal dose with about six animals. *J Ind Hyg Toxicol* 1943;**25**(9):415–7.
23. Ames BN, McCann J, Yamasaki E. Methods for detecting carcinogens and mutagens with the *Salmonella*/mammalian-microsome mutagenicity test. *Mutat Res* 1975;**31**(6):347–64.
24. Wilson JG. Methods for administering agents and detecting malformations in experimental animals. In: Wilson JG, Warkany J, editors. *Teratology: principles and techniques*. Chicago: University of Chicago Press; 1965. p. 262–75.
25. Peters W. The chemotherapy of rodent malaria, V. The action of some sulphonamides alone or with folic reductase inhibitors against malaria vectors and parasites, part 2: schizontocidal action in the albino mouse. *Ann Trop Med Parasitol* 1968;**62**(4):488–94.
26. Merkli B, Richle R. *Plasmodium berghei*: diet and drug dosage regimens influencing selection of drug-resistant parasites in mice. *Exp Parasitol* 1983;**55**(3):372–6.
27. Trager W, Jensen JB. Human malaria parasites in continuous culture. *Science* 1976;**193**(4254):673–5.
28. Alin MH, Björkman A, Wernsdorfer WH. Synergism of benflumetol and artemether in *Plasmodium falciparum*. *Am J Trop Med Hyg* 1999;**61**(3):439–45.
29. Smyth Jr. HF. An exploration of joint toxic action: twenty-seven industrial chemicals intubated in rats in all possible Pairs. *Toxicol Appl Pharmacol* 1969;**14**(2):340.
30. Chemotherapy of malaria and resistance to antimalarials. Report of a WHO scientific group. *World Health Organ Tech Rep Ser* 1973;**529**:1–121.
31. Cousin M, Kummerer S, Lefèvre G, et al. *Coartem® (artemether/lumefantrine) tablets for the treatment of malaria in patients with acute, uncomplicated infection due to Plasmodium falciparum or mixed infection including P. falciparum*. NDA; 2008. p. 22–68.
32. Abdulla S, Sagara I, Borrmann S, et al. Efficacy and safety of artemether-lumefantrine dispersible tablets compared with crushed commercial tablets in African infants and children with uncomplicated malaria: a randomized, single-blind, multicentre trial. *Lancet* 2008; **372**(9652):1819–27.
33. Makanga M, Krudsood S. The clinical efficacy of artemether/lumefantrine (Coartem®). *Malar J* 2009;**8**(Suppl. 1):S5.
34. Jansen FH, Lesaffre E, Penali LK, et al. Assessment of the relative advantage of various artesunate-based combination therapies by a multi-treatment Bayesian random-effects meta-analysis. *Am J Trop Med Hyg* 2007;**77**(6):1005–9.
35. Sutherland CJ, Ord R, Dunyo S, et al. Reduction of malaria transmission to *Anopheles* mosquitoes with a six-dose regimen of co-artemether. *PLoS Med* 2005;**2**(4):0338–45. e92.
36. Bakshi R, Hermeling-Fritz I, Gathmann I, et al. An integrated assessment of the clinical safety of artemether-lumefantrine: a new oral fixed-dose combination antimalarial drug. *Trans R Soc Trop Med Hyg* 2000;**94**(4):419–24.
37. McLean WG, Ray DE, Smith SL, et al. Neurotoxicity of artemisinin derivatives: clinical hazard or experimental artefact?. In: *2nd European congress on tropical medicine of the federation of European societies for tropical medicine and international health and 4th residential meeting of the royal society of tropical medicine and hygiene, Liverpool, UK*. 1998. p. 381.
38. Toovey S, Jamieson A. Audiometric changes associated with the treatment of uncomplicated falciparum malaria with co-artemether. *Trans R Soc Trop Med Hyg* 2004;**98**(5):261–7.
39. Kissinger E, Hien TT, Hung NT, et al. Clinical and neurophysiological study of the effects of multiple doses of artemisinin on brain-stem function in Vietnamese patients. *Am J Trop Med Hyg* 2000;**63**(1, 2):48–55.
40. Falade C, Manyando C. Safety profile of Coartem®: the evidence base. *Malar J* 2009;**8** (Suppl. 1):S6.

41. Manyando C, Mkandawire R, Puma L, et al. Safety of artemether-lumefantrine in pregnant women with malaria: results of a prospective cohort study in Zambia. *Malar J* 2010;**9**:249.
42. World Health Organization. *Guidelines for the treatment of malaria*. Geneva: World Health Organization; 2006. p. 32–3.
43. van Vugt M, Ezzet F, Nosten F, et al. No evidence of cardiotoxicity during antimalarial treatment with artemether-lumefantrine. *Am J Trop Med Hyg* 1999;**61**(6):964–7.
44. Bindschedler M, Lefèvre G, Degen P, et al. Comparison of the cardiac effects of the antimalarials co-artemether and halofantrine in healthy participants. *Am J Trop Med Hyg* 2002;**66**(3):295–300.
45. Lefèvre G, Thomsen MS. Clinical pharmacokinetics of artemether and lumefantrine (Riamet®). *Clin Drug Investig* 1999;**18**(6):467–80.
46. Navaratnam V, Mansor SM, Chin LK, et al. Determination of artemether and dihydroartemisinin in blood plasma by high-performance liquid chromatography for application in clinical pharmacological studies. *J Chromatogr B* 1995;**669**(2):289–94.
47. van Agtmael MA, Butter JJ, Portier EJ, et al. Validation of an improved reversed phase high performance liquid chromatography assay with reductive electrochemical detection for the determination of artemisinin derivatives in man. *Ther Drug Monit* 1998;**20**(1):109–16.
48. Zeng MY, Lu ZL, Yang SC, et al. Determination of benflumetol in human plasma by reversed-phase high-performance liquid chromatography with ultraviolet detection. *J Chromatogr B* 1996;**681**(2):299–306.
49. Mansor SM, Navaratnam V, Yahaya N, et al. Determination of a new antimalarial drug, benflumetol, in blood plasma by high-performance liquid chromatography. *J Chromatogr B* 1996;**682**(2):321–5.
50. van Agtmael MA, Shan QC, Jiao XQ, et al. Multiple dose pharmacokinetics of artemether in Chinese patients with uncomplicated falciparum malaria. *Int J Antimicrob Agents* 1999; **12**:151–58.
51. Na-Bangchang K, Karbwang J, Tasanor U, et al. Pharmacokinetics of benflumetol given as a fixed combination artemether-benflumetol (CGP 56697) in Thai patients with uncomplicated *falciparum* malaria. *Int J Clin Pharmacol Res* 1999;**19**(2):41–6.
52. Ezzet F, van Vugt M, Nosten F, et al. Pharmacokinetics and pharmacodynamics of lumefantrine (benflumetol) in acute falciparum malaria. *Antimicrob Agents Chemother* 2000;**44**(3):697–704.
53. Ezzet F, Mull R, Karbwang J. Population pharmacokinetics and therapeutic response of CGP 56697 (artemether + benflumetol) in malaria patients. *Br J Clin Pharmacol* 1998;**46**(6):553–61.
54. White NJ, van Vugt M, Ezzet F. Clinical pharmacokinetics and pharmacodynamics of artemether-lumefantrine. *Clin Pharmacokinet* 1999;**37**(2):105–25.
55. Koram KA, Abuaku B, Duah N, et al. Comparative efficacy of antimalarial drugs including ACTs in the treatment of uncomplicated malaria among children under 5 years in Ghana. *Acta Trop* 2005;**95**(3):194–203.
56. Giao PT, de Vries PJ. Pharmacokinetic interactions of antimalarial Agents. *Clin Pharmacokinet* 2001;**40**(5):343–73.
57. Lefèvre G, Bindschedler M, Ezzet F, et al. Pharmacokinetic interaction trial between co-artemether and mefloquine. *Eur J Pharm Sci* 2000;**10**(2):141–51.
58. Lefèvre G, Carpenter P, Souppart C, et al. Interaction trial between artemether-lumefantrine (Riamet) and quinine in healthy subjects. *J Clin Pharmacol* 2002;**42**(10):1147–58.
59. Lefèvre G, Carpenter P, Souppart C, et al. Pharmacokinetics and electrocardiographic pharmacodynamics of artemether-lumefantrine (Riamet) with concomitant administration of ketoconazole in healthy subjects. *Br J Clin Pharmacol* 2002;**54**(5):485–92.

60. Colussi D, Parisot C, Legay F, et al. Binding of artemether and lumefantrine to plasma proteins and erythrocytes. *Eur J Pharm Sci* 1999;**9**(1):9–16.
61. Newman RD, Parise ME, Slutsker L, et al. Safety, efficacy and determinants of effectiveness of antimalarial drugs during pregnancy: implications for prevention programmes in *Plasmodium falciparum*-endemic Sub-Saharan Africa. *Trop Med Int Health* 2003;**8**(6):488–506.
62. World Health Organization. *Procurement of artemether/lumefantrine (Coartem) through WHO*. 2007.
63. Fogg C, Bajunirwe F, Piola P, et al. Adherence to a six-dose regimen of artemether-lumefantrine for treatment of uncomplicated *Plasmodium falciparum* malaria in Uganda. *Am J Trop Med Hyg* 2004;**71**(5):525–30.
64. Piola P, Fogg C, Bajunirwe F, et al. Supervised versus unsupervised intake of six-dose artemether-lumefantrine for treatment of acute, uncomplicated *Plasmodium falciparum* malaria in Mbarara, Uganda: a randomized trial. *Lancet* 2005;**365**(9469):1467–73.

Chapter 8

Artemisinin–Naphthoquine Phosphate Combination (ARCO)

Chapter Outline

The artemisinin–naphthoquine phosphate tablet (ARCO) is an artemisinin combination therapy (ACT) created by the antimalarial research group from the Institute of Microbiology and Epidemiology at the Academy of Military Medical Sciences (AMMS). In 1992, the necessary research results regarding naphthoquine phosphate as an antimalarial were collected and a new-drug certification was issued in 1993. A State Award for Inventions, Second Class followed in 1996.

To safeguard the new drug and to delay the onset of drug resistance, research on naphthoquine phosphate ACT began in 1989. After a series of drug-combination studies, an ACT comprising naphthoquine and artemisinin at a ratio of 1:2.5 was settled on. A patent application was made in 1997 and granted in 2001. Preclinical studies were completed in 1998 and clinical trials approved that same year. These were funded by the People's Liberation Army's Initiatives 85 and 95 and by the Ministry of Science and Technology's program for new drug research and manufacture.

Following 10 years of research into this ACT, it was given new-drug certification in 2003 and transferred to Kunming Pharmaceutical Corporation in 2004. International registration was undertaken after production permits were granted in 2005. As of May 2013, ARCO had undergone studies in 13 international

Artemisinin-Based and Other Antimalarials. https://doi.org/10.1016/B978-0-12-813133-6.00008-1

multicenter clinical trials and been registered or licensed in 22 countries. Currently, it is the only ACT worldwide that can cure drug-resistant falciparum malaria with a single oral dose.

The following is an account of the preclinical and clinical research into naphthoquine phosphate and its ACT, ARCO.

8.1 NAPHTHOQUINE PHOSPHATE RESEARCH

Li Fulin[†]*, Zhang Zhixiang*

Institute of Microbiology and Epidemiology, Academy of Military Medical Sciences, Beijing, China

A. CHEMICAL SYNTHESIS

1. Design

Compound **1** (naphthoquine phosphate, Fig. 8.1), with the chemical name 4-[(7-chloro-4-quinolinyl)amino]-2-[[(1,1dimethylethyl)amino]methyl]-5,6,7,8-tetra-hydro-1-naphthalenol diphosphate dihydrate, was a new antimalarial synthesized by Li Fulin and others at the Institute of Microbiology and Epidemiology at the AMMS.

In 1978, Schmidt et al. had found that 6-phenyl-4-tertbutyl-2-diethyl-amino-o-cresol (Compound **2**) possessed only a weak antimalarial effect. However, if the 2-diethylamine group and 4′-hydrogen atom at 6-phenyl were replaced with *t*-butylamino and chloride, respectively, the antimalarial effect of the resulting compound (Compound **3**) increased significantly. Compound **3** was highly effective against the multidrug-resistant Smith strain of *Plasmodium falciparum*

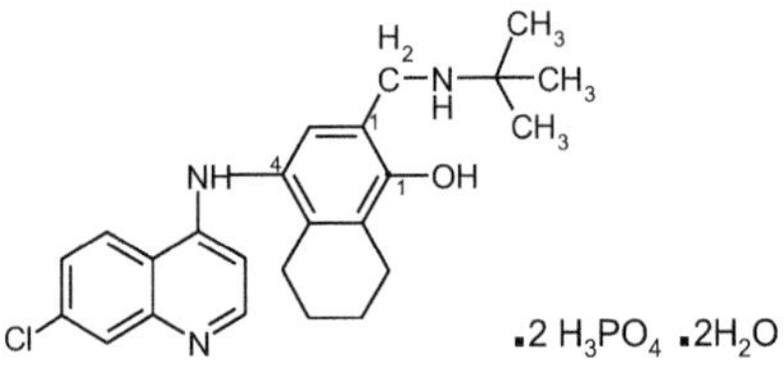

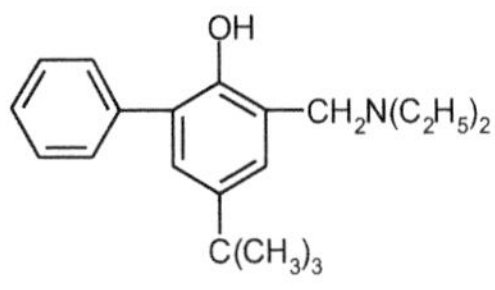

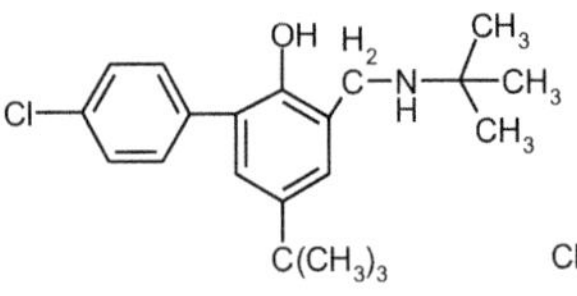

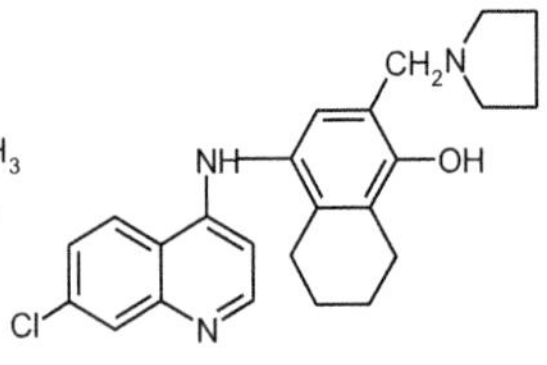

FIGURE 8.1 Structures of naphthoquine phosphate and its derivatives.

[†]Deceased.

from Vietnam.[1] This indicated that *t*-butyl groups in amino-substituted phenols had a particular impact on antimalarial efficacy.

Zhang Mingli et al. synthesized 4-[(7-chloro-4-quinolinyl)amino]-2-(pyrrolidinyl-1-methyl)-5,6,7,8-tetrahydro-1-naphthol (Compound **4**). The compound not only showed an increase in antimalarial efficacy but also had some effect on chloroquine-resistant *Plasmodium berghei* in mice. It could inhibit chloroquine-sensitive and chloroquine-resistant strains of *P. berghei* at doses of 2.5 and 20 mg/kg, respectively.[2]

In 1981, Li Fulin and others designed and synthesized a series of compounds to discover new antimalarials with higher efficacy and no cross-resistance to chloroquine. They replaced the pyrrolidinyl in Compound **4** with tertbutyl-amino, using quinoline, acridine, phenazine, and other aromatic heterocyclics as a scaffold. Dihydrochloride salt of naphthoquine yielded the best results. Given subcutaneously, the ED_{50} of this compound was 0.33 mg/kg on normal strains of rodent malaria, 3–4 times lower than the ED_{50} of the chloroquine group (1.12 mg/kg). On chloroquine-resistant strains, its ED_{50} was 3.86 mg/kg, as opposed to 34.5 mg/kg for chloroquine.[3] Thereafter, Li et al. also synthesized a monohydrochloride salt of naphthoquine and naphthoquine diphosphate dihydrate (Compound **1**). Because Compound **1** had high yields, a stable hydrated form, and good bioavailability, further research was conducted on its synthesis, chemical analysis, preparation, pharmacology, toxicity, and clinical efficacy.

2. Process of Synthesis

Naphthoquine phosphate was synthesized via a five-step process using α-naphthalenol as starting material (see Fig. 8.2).[4]

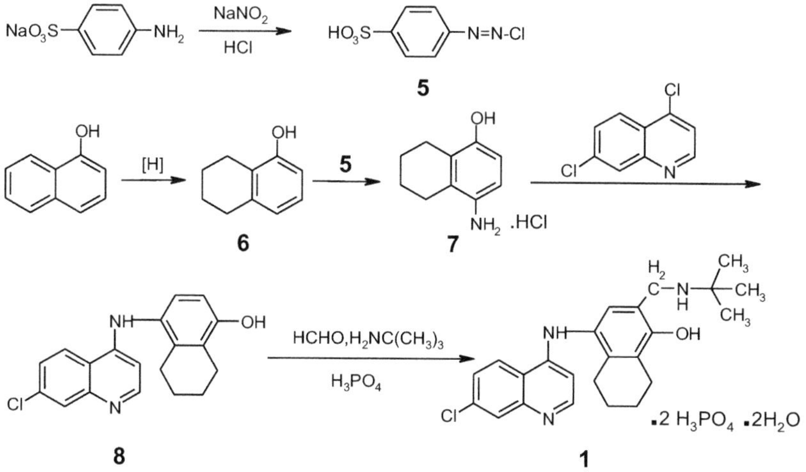

FIGURE 8.2 Procedure for synthesizing naphthoquine phosphate.

First, sodium-p-aminobenzenesulfonate undergoes a diazotization reaction to produce its diazonium salt (Compound **5**). At the same time, α-naphthalenol is reduced using raney nickel to obtain Compound **6**. Intermediates **5** and **6** are coupled and reduced using sodium dithionite without further purification, yielding 4-amino-5,6,7,8-tetrahydronaphthalen-1-ol hydrochloride (Compound **7**). This is then condensed with 4,7-dichloroquinoline to produce Intermediate **8**, which undergoes a Mannich reaction with tertbutylamine and formaldehyde. Finally, it undergoes salification with phosphoric acid to produce naphthoquine phosphate (Compound **1**). This processing method was convenient, materials were available domestically, and costs were low. Total yield was about 30%.[4]

Naphthoquine phosphate is a yellow powder, odorless, and bitter. It changes color when exposed to light and has a melting point of 266–272°C. It is slightly soluble in water; insoluble in chloroform, anhydrous ethanol, diethyl ether, and benzene; and soluble in 50% ethanol.

The structure of naphthoquine phosphate was confirmed by elemental, infrared, and ultraviolet spectrum analyses, nuclear magnetic resonance (^{1}H-NMR and ^{13}C-NMR) and mass spectrum analyses, and further validated using single-crystal X-ray diffraction.[5]

B. PHARMACODYNAMICS

Jiao Xiuqing
Institute of Microbiology and Epidemiology, Academy of Military Medical Sciences, Beijing, China

Pharmacology research into naphthoquine phosphate included pharmacodynamics, pharmacokinetics, general pharmacology, and method of action studies. Toxicity research included general and special toxicity studies.

1. Pharmacodynamics[6]: Rodent Malaria

Suppression Effects on Plasmodium berghei

In vivo studies were conducted with normal (N) and chloroquine-resistant (RC) strains of K173 *P. berghei* in Kunming outbred Swiss mice. Eight other antimalarials, including chloroquine, were used as parallel controls. Evaluation was based on the Peters 4-day suppressive test,[7] with results shown in Tables 8.1 and 8.2.

Table 8.1 shows that naphthoquine phosphate had the best suppressive effect against the N strain of *P. berghei*, with a potency 2–103 times that of the other eight antimalarial drugs. Table 8.2 shows similar results against the RC strain, with the potency of naphthoquine phosphate 15–33 times higher than that of the other eight drugs. Turning to the I_{90} resistance index ($I_{90}=ED_{90}$ of RC strain/ ED_{90} of N strain), the I_{90} of naphthoquine phosphate was 1.7, as opposed to 3.4–224.0 for the other drugs. It indicated that the RC strain was more sensitive to naphthoquine phosphate than to the other drugs.

TABLE 8.1 Antimalarial Efficacy of Naphthoquine Phosphate and Other Drugs Against Normal (N) Strain *P. berghei*

Drug	ED_{50} (mg/kg/d)	ED_{90} (mg/kg/d)	ED_{90} Ratio[a] (Valence Ratio)
Naphthoquine phosphate	0.40 ± 0.04	0.69 ± 0.06	1.0
Piperaquine phosphate	1.09 ± 0.10	1.53 ± 0.15	2.2
Amodiaquine	1.29 ± 0.03	2.25 ± 0.07	3.3
Benflumetol	1.56 ± 0.19	2.44 ± 0.34	3.5
Mefloquine	1.43 ± 0.01	2.94 ± 0.05	4.3
Chloroquine phosphate	1.35 ± 0.01	3.25 ± 0.13	4.7
Artemether	2.06 ± 0.33	5.42 ± 0.56	7.9
Artemisinin	18.78 ± 2.21	56.59 ± 7.26	82.0
Quinine hydrochloride	35.10 ± 0.75	71.50 ± 4.75	103.6

ED_{50} is the 50% effective naphthoquine dose. ED_{90} is the 90% effective naphthoquine dose. The value is mean ± standard deviation.
[a]*The ED_{90} of naphthoquine was used as a baseline of 1.0, on which the ratios of the other drugs were calculated. The potency of naphthoquine is a multiple of the other drugs.*

TABLE 8.2 Antimalarial Efficacy of Naphthoquine Phosphate and Other Drugs Against Chloroquine-Resistant (RC) Strain of *P. berghei*

Drug	ED_{90} (mg/kg/d)		I_{90}[a]
	RC Strain	N Strain	
Naphthoquine phosphate	1.20	0.69	1.7
Artemether	18.30	5.42	3.4
Quinine hydrochloride	367.40	71.50	5.7
Artemisinin	401.20	56.59	7.1
Amodiaquine	18.60	2.25	8.3
Mefloquine	182.52	2.94	62.1
Benflumetol	182.70	2.44	74.9
Chloroquine phosphate	291.60	3.25	89.7
Piperaquine phosphate	342.73	1.53	224.0

[a]*I_{90}, ED_{90} of RC strain/ED_{90} of N strain. In 1980, Merkli introduced rubrics for different levels of resistance in rodent malaria. An $I_{90} \leq 2$ indicates sensitivity (S); from 2 to 10, light resistance (R+); from 10 to 100, moderate resistance (R++); >100, high resistance (R+++).*

Curative Effects on Plasmodium berghei

Chloroquine phosphate, commonly regarded as the standard drug at the time, was used as the control. Both drugs were administered in seven dose groups of 15 randomly allocated mice for each group. Raether's 1979 experimental method was used,[8] with the drug administered for 5 days and observation carried out for 28 days. The results showed that the CD_{50} and CD_{90} of naphthoquine phosphate were 3.54 and 5.97 mg/kg/d, respectively, as opposed to 6.16 and 10.40 mg/kg/d for chloroquine phosphate. It indicated that the curative effect of naphthoquine phosphate was greater than that of chloroquine phosphate for normal strain *P. berghei*.

Speed of Parasite Clearance on Plasmodium berghei

With chloroquine phosphate as control, 10 times the ED_{90} dose for the normal (N) strain was administered over a 3-day period. When parasitemia reached 3%–5% of erythrocytes following infection with the N strain, the drugs were administered once daily for 3 days. Caudal blood films were obtained every 12 h after the first dose and examined under the microscope using Giemsa stain, until parasites were cleared (i.e., no parasites found in 20,000 erythrocytes). Taking parasitemia at first dose as 100%, the percentage reduction at subsequent times was calculated. The time taken for parasitemia to decrease by 50%, 90%, and 100% was determined.

The results in Table 8.3 show that naphthoquine phosphate took 10 h less than chloroquine phosphate to achieve 50% and 90% reductions in parasitemia, and 17 h less for full clearance. It indicated that naphthoquine phosphate had a greater rate of parasite clearance than chloroquine phosphate ($P<.05$).

Prophylaxis Against Plasmodium berghei

Piperaquine phosphate, the drug with the longest duration of efficacy at the time, was the control. Both drugs were administered in two dose groups of

TABLE 8.3 Speed of Parasite Clearance on Normal (N) Strain *P. berghei*

	Time Taken to Achieve Percentage Reduction in Parasitemia (M ± SD)		Full Clearance Time (h) (M ± SD)
Drug	50%	90%	
Naphthoquine phosphate	14.4 ± 41.7	25.9 ± 2.1	52.8 ± 6.6
Chloroquine phosphate	25.7 ± 7.7	35.9 ± 3.1	69.0 ± 6.0

Dose was N strain $ED_{90} \times 10$. Time taken for 50% and 90% reduction and full clearance were compared, with $P<.05$.

150 mice each. Drugs were given once orally. One, two, three, and four weeks after administration, 24–31 mice at a time were infected with normal (N) strain *P. berghei* and tested for parasitemia. Full prophylaxis was taken to be the absence of parasites in all the animals in a group. Absence in only some of the animals was taken to be partial prophylaxis. Prophylactic efficacy was calculated based on the interval between drug administration and infection.

The results (Table 8.4) showed that at doses of 100 and 40 mg/kg, full or partial prophylaxis lasted 1 week longer for naphthoquine phosphate than for piperaquine phosphate.

Effects on *Plasmodium yoelii* Sporozoite Infectivity

In vivo studies were conducted using the By265 strain of *P. yoelii* on pure-bred black C57BL/6Jax mice, via Hor strain *Anopheles stephensi* mosquitoes. Primaquine phosphate was the control. The minimum effective dose (MED, 3/5 negative) and full minimum effective dose (FMED, fully negative) were calculated for subcutaneous and oral administration. As shown in Table 8.5, naphthoquine phosphate could prevent infection in mice at the 5 mg/kg dose. Although its efficacy was worse than that of pyrimethamine, it was significantly better than chloroquine and primaquine phosphate.

Long-Term Causal Prophylaxis Against *Plasmodium yoelii* Sporozoite Infectivity

The 1980 Peters protocol for long-term causal prophylaxis[9] was used in two steps. The first was a 3-day long-term prophylaxis experiment, and the second, a comparison of long-term causal prophylactic efficacy. Piperaquine phosphate was the control. Each mouse was given one dose of the drug a week. After 6 weeks, sporozoites were introduced. Blood films were examined 7 and 14 days later with the number of mice uninfected taken as the measure of efficacy.

The results of the 3-day experiment (Table 8.6) showed that naphthoquine phosphate had a clear long-term causal prophylactic effect on sporozoite infectivity. The comparative study (Table 8.7) also showed that, at the same dose, the prophylactic effect of naphthoquine phosphate lasted 1–2 weeks longer than that of piperaquine phosphate.

Causal Prophylaxis Versus Residual Effects on *Plasmodium yoelli*

An experiment was conducted with naphthoquine phosphate, primaquine phosphate, and chloroquine to distinguish between causal prophylaxis and residual effects. The improved 1970 Gregory and Peters protocol[10] was used, with *P. yoelii*, C57BL/6Jax mice, and *A. stephensi* mosquitoes. It was found that naphthoquine phosphate showed significant persistent effects at doses of 5 mg/kg and above. Primaquine phosphate had a causal prophylactic effect. Chloroquine had neither.

TABLE 8.4 Prophylactic Efficacy of Naphthoquine and Piperaquine Phosphate

Drug	Dose (mg/kg)	Number of Infections (/Total Number)				Full Prophylactic Duration (week)	Partial Prophylactic Duration (week)
		1 week	2 week	3 week	4 week		
Naphthoquine phosphate	100	0/28	0/28	0/30	9/29	3	4
	40	0/27	0/28	24/30		2	3
Piperaquine phosphate	100	0/28	0/29	23/24		2	3
	40	0/27	25/31			1	2

TABLE 8.5 Effects of Naphthoquine Phosphate on *P. yoelii* Sporozoite Infectivity

	MED (mg/kg)		FMED (mg/kg)	
Drug	**Subcutaneous**	**Oral**	**Subcutaneous**	**Oral**
Naphthoquine phosphate	5.0	5.0	10.0	5.0
Chloroquine phosphate			100 (ineffective)	100 (ineffective)
Primaquine phosphate	10~20	20.0	20~40	>40
Pyrimethamine	0.625	0.625	1.25	1.25

MED, 3/5 negative by microscopy; *FMED*, full negative by microscopy.

TABLE 8.6 Oral Naphthoquine Phosphate in 3-Day Long-Term Prophylaxis Against Sporozoite Infectivity

		Number of Infections (/Total Number)	
Drug	**Dose (mg/kg/d)**	**7 days**	**14 days**
Naphthoquine phosphate	80	0/10	0/10
	40	0/10	0/10
	20	0/10	0/10
	10	0/10	0/10
No-treatment control		10/10	10/10

Effects on Plasmodium yoelii in the Tissue Phase

The study used the Nigeria strain of *P. yoelii*, *A. stephensi*, and Wistar rats. Primaquine phosphate and chloroquine were the controls. The livers of all three rats in the nonmedicated control group showed exoerythrocytic parasites 47–48 h after sporozoite infection. Merozoites formed and matured. Exoerythrocytic schizonts were circular or oval. Merozoite nuclei were purplish-red or purplish-black, 0.4–0.5 μm in diameter, very numerous, and difficult to calculate. Size and distribution of nuclei were even. At 80 mg/kg chloroquine, the density, size, and shape of the parasites in the liver were basically similar to the exoerythrocytic parasites in the nonmedicated group and had also developed mature schizonts.

TABLE 8.7 Naphthoquine and Piperaquine Phosphate in 6-Week Long-Term Prophylaxis Against Sporozoite Infectivity

		Number of Infections (/Total Number)					
Drug	**Dose mg/kg**	**1 week**	**2 week**	**3 weeks**	**4 weeks**	**5 weeks**	**6 weeks**
Naphthoquine phosphate	250	0/10	0/10	0/10	0/10	0/10	2/10
	100	0/10	0/10	2/10	5/10	10/10	10/10
	40	0/10	4/10	8/10			
Piperaquine phosphate	250	0/10	0/10	0/10	0/10	8/10	10/10
	100	0/10	2/10	10/10			
	40	4/10	10/10				
No-treatment control		20/20					

At 10mg/kg primaquine phosphate vacuoles appeared in the exoerythrocytic schizonts, as well as irregularities in the morphology of the parasite. However, some developing merozoites could still be seen. Only when the dose was increased to 20mg/kg did the exoerythrocytic schizonts show serious degeneration.

The exoerythrocytic schizonts in the animals given 20mg/kg naphthoquine phosphate developed mostly normally, with only slight differences. At 40mg/kg, changes occurred in most of the schizonts. Some developed obvious vacuoles, with the shape of the parasite becoming mostly oval. Smaller vacuoles also appeared in others, but merozoites were still able to form. At 80mg/kg, all the schizonts were altered and it was hard to distinguish if individual parasites contained merozoites.

Effects on Plasmodium yoelii Gametocytes and Sporozoites

The By265 strain of *P. yoelii*, Kunming outbred Swiss mice, C57BL/6Jax purebred black mice, and Hor strain *A. stephensi* were used to observe the effects of naphthoquine phosphate on *P. yoelii* gametocytes and sporozoites.

For gametocytes, mice were tested 2.5 to 3days after infection. Those with lower gametocyte densities were removed and three mice were selected from each group. The drug was administered twice after fasting, with an interval of 4h between doses. Further observations were carried out 12h after the final dose. The number of gametocytes per 100 leukocytes was calculated and compared to before medication (100%), to derive the rate of suppression and the ED_{50} dose. The ED_{50} of naphthoquine phosphate against *P. yoelii* gametocytes was found to be 26mg/kg, seven times higher than that of primaquine phosphate (ED_{50}=3.7mg/kg).

For sporozoites, mosquitoes were fasted for 24h before consuming rodent blood containing gametocytes. They were then fed liquid forms of the drug at different concentrations. Eight to nine days after infection, some mosquitoes were selected and their stomach dissected. Using microscopy, the number of oocysts was calculated and their development or degeneration observed. The results showed that naphthoquine phosphate could reduce oocyst density at 0.01% concentration. But when mosquitoes from the same group fed on C57BL/6Jax mice 14days after infection, all mice showed parasitemia after 7days.

Therefore, several experiments proved that naphthoquine phosphate had some effect on *P. yoelii* gametocytes but was inferior to primaquine phosphate. It had no effect on sporozoites.

2. Pharmacodynamics: Simian Malaria

Antimalarial Effects on Plasmodium cynomolgi

An in vivo study was run on Rhesus macaques (*Macaca mulatta*) infected with C strain *P. cynomolgi* via blood transfusion, with chloroquine as control.

TABLE 8.8 Curative Efficacy of Oral Naphthoquine Phosphate and Chloroquine Phosphate on *P. cynomolgi*

Drug	Dose (mg/kg/d)			Minimum Short-Term Cure Dose (mg/kg/d)	Minimum Cure Dose (mg/kg/d)
	10.00	5.00	3.16		
Naphthoquine phosphate	C	C	C		
	C	C	C	3.16	5.00
	C	C	CS (40)		
Chloroquine phosphate	C	MS (27)	MS (27)		
	C	MS (27)	MS (21)	10.00	>10.00
	MS (27)				

C, cure; *CS*, complete suppression; *MS*, marked suppression. Number in parentheses is the day on which recrudescence occurred.

The 1976 Davidson protocol[11] was used, with 7 days of treatment and 105 days of observation. After medication, regular tests were run. Speed of parasite clearance was based on reductions in parasitemia of 50%, 90%, and 100%. The efficacy of the drug was evaluated based on whether recrudescence occurred. The benchmarks were Ineffective (I), no obvious difference in parasitemia before and after medication, or even an increase; Slight Suppression (SS), parasitemia suppressed temporarily but later increased; Marked Suppression (MS), parasites eliminated for at least 20 days but with recrudescence in 30 days; Complete Suppression (CS), no recrudescence 30–105 days after parasite clearance; Cure (C), no recrudescence 105 days after clearance.

Although naphthoquine phosphate was 3–18 h faster than chloroquine in achieving 50%, 90%, and 100% reductions, this was not statistically significant ($P > .05$). Table 8.8 shows that the minimum curative dose of naphthoquine phosphate was 5 mg/kg/d, whereas that of chloroquine phosphate was more than 10 mg/kg/d. It indicated that the curative efficacy of naphthoquine phosphate was better than that of chloroquine phosphate.

Antimalarial Effects on Plasmodium knowlesi

Rhesus monkeys were infected with *P. knowlesi* Nuri via blood transfusion. When parasitemia reached 2%–5%, the drug was administered once a day for 1, 2, or 3 days. The same benchmarks were used as with the *P. cynomolgi* experiment. The curative efficacy of the 3-day regimen against *P. knowlesi* is shown in Tables 8.9 and 8.10. At the same dose, naphthoquine phosphate took 3–4 h less than chloroquine phosphate in reducing parasitemia by 50% and 90%. It took 12–36 h less than chloroquine phosphate for full clearance.

TABLE 8.9 Oral Naphthoquine Phosphate and Chloroquine Phosphate: Speed of Parasite Clearance Against *P. knowlesi*

Drug	Dose (mg/kg)	Number of Animals	Time Taken for Percentage Reduction in Parasitemia (M ± SD) (h)		Full Clearance Time (M ± SD) (h)
			50%	90%	
Naphthoquine phosphate	10.00	3	5.4 ± 1.6	14.5 ± 3.5	60 ± 12.0
	6.00	3	8.9 ± 2.0	20.3 ± 2.2	56 ± 6.9
	3.16	3	7.6 ± 2.7	15.7 ± 4.9	60 ± 0.0
Chloroquine phosphate	10.00	3	9.5 ± 3.0	20.6 ± 3.5	72 ± 20.7
	6.00	3	11.9 ± 1.3	20.7 ± 13.9	76 ± 6.9
	3.16	2	47.8 ± 2.2	58.8 ± 2.4	96 ± 16.9

TABLE 8.10 Oral Naphthoquine Phosphate and Chloroquine Phosphate: Curative Efficacy Against *P. knowlesi*

Drug	Dose (mg/kg/d)			Minimum Short-Term Cure Dose (mg/kg/d)	Minimum Cure Dose (mg/kg/d)
	10.00	6.00	3.16		
Naphthoquine phosphate	C	MS (22)	C		
	C	C	MS (13)	3.16	10.00
	C	MS (24)	MS (13)		
Chloroquine phosphate	C	MS (9)	MS (6)		
	C	MS (8)	MS (6)	10.00	>10.00
	MS (19)	MS (9)			

C, cure; *MS*, marked suppression. Number in parentheses indicates day on which recrudescence occurred.

The minimum curative dose for naphthoquine phosphate was 3.16 mg/kg/d, as opposed to 10 mg/kg/d for chloroquine phosphate (Table 8.10). Table 8.11 shows the efficacies of the 1- and 2-day oral regimens. Both the 10 mg/kg 2-day regimen and the single 20 mg/kg dose took similar time to achieve parasite clearance (44–50 h). At 20 mg/kg and above, both 1- and 2-day regimens could

TABLE 8.11 Curative Efficacy of Single- and Double-Dose Oral Naphthoquine Phosphate Against *P. knowlesi*

Dose (mg/kg/d)	Treatment Time	Number of Animals	Full Clearance Time (M ± SD) (h)	Number Cured/Total Number
10	1	3	44 ± 13.8	1/3
10	2[a]	3	44 ± 13.8	3/3
20	1	3	48 ± 20.8	3/3
30	1	4	50 ± 15.1	4/4

[a]*Once a day for 2 days.*

cure *P. knowlesi*. Therefore, certain doses of naphthoquine phosphate could still yield good therapeutic results with a shorter course of treatment.

Causal Prophylaxis Against Plasmodium cynomolgi

Schmidt's 1963 protocol[12] was used to observe the causal prophylactic effects of naphthoquine phosphate on C and B strain *P. cynomolgi* sporozoites. Four other antimalarial drugs, including primaquine phosphate, and a no-treatment group were the controls. The drugs were administered in a single dose 1 day and 2 h before sporozoite inoculation. Medication was continued for 5 days and blood tests were conducted 7–105 days after infection.

The results (Table 8.12) showed that 10 mg/kg naphthoquine phosphate had a prophylactic effect on C strain *P. cynomolgi* sporozoites. At 3.16 mg/kg, it only delayed the appearance of parasites. Its efficacy was inferior to those of primaquine phosphate and pyrimethamine, but better than those of piperaquine phosphate and chloroquine phosphate. At 10 mg/kg piperaquine phosphate, parasites were found in the venous blood of both monkeys 8–9 days after sporozoite infection, and parasite clearance was observed from Days 3 to 105. This may have been due to the drug's residual effects in vivo or a deficiency in that strain's secondary exoerythrocytic stage.

Naphthoquine phosphate had no prophylactic effect on B strain *P. cynomolgi* even at a dose of 31.6 mg/kg/d. Both experimental monkeys showed parasitemia 40–60 days after infection. On the other hand, primaquine phosphate could prevent infection at 0.316 mg/kg/d. Therefore, it could be deduced that naphthoquine phosphate was not ideal for eliminating sporozoites and schizonts. Due to its lasting effects in vivo, however, it still held promise as a long-term prophylaxis.

TABLE 8.12 Causal Prophylactic Effect of Naphthoquine Phosphate on C and B Strain *P. cynomolgi*

Strain	Drug	Dose (mg/kg/d)	Number Immune/ Total Number	Postinfection Days When Blood Smears Positive	Days Delayed Parasite Emergence
C strain	Naphthoquine phosphate	31.6	1/1		105(–)[a]
		10.0	3/3		105(–)[a]
		3.16	0/3	15–19	6–20
	Pyrimethamine	2.5	2/2		105(–)[a]
	Primaquine phosphate	0.316	2/2		105(–)[a]
	Piperaquine phosphate	10.0	0/2	8–9	0[b]
	Chloroquine phosphate	31.6	0/1	76	68
	Control		0/2	8–9	
B strain	Naphthoquine phosphate	31.6	0/2	40–60	31–51
	Primaquine phosphate	0.316	2/2		105(–)[a]
	Control		0/2	9	

[a]*Blood smears remained negative on Day 105.*
[b]*Parasitemia turned negative from Days 3 to 105 for both monkeys in the piperaquine phosphate group.*

Long-Term Curative Efficacy Against Plasmodium cynomolgi

Davidson's 1981 protocol[13] was used to test curative efficacy against B and C strain *P. cynomolgi*. The controls for B strain study were chloroquine phosphate, naphthoquine phosphate + primaquine phosphate, and chloroquine phosphate + primaquine phosphate, and for C strain piperaquine phosphate and chloroquine phosphate. The drugs were administered 7–9 days after sporozoite infection when venous parasitemia reached 1.2%–4.8%. Medication was given once daily for 3 days, and parasitemia observed every 12 h until full clearance. Thereafter, regular tests were run until Day 105. Total elimination of parasitemia was considered a cure. Recrudescence was taken to mean the reappearance of parasites within 20 days. If parasites reappeared between Days 21 and 105, it was considered recurrence.

Table 8.13 shows that at 10 mg/kg/d, naphthoquine phosphate could cure C strain *P. cynomolgi*. Chloroquine and piperaquine experienced recurrence at the same dose. Nevertheless, naphthoquine phosphate did not achieve ideal curative efficacy against the B strain. At 20 mg/kg/d, it only delayed the reappearance of the parasite. Both naphthoquine and chloroquine phosphate had to be combined with primaquine phosphate to cure the B strain. Both curative and prophylactic efficacy depended on the strain of the parasite.

3. In Vitro Efficacy Study Against *Plasmodium falciparum*

The Fcc-1/HN strain of *P. falciparum* was cultured in RPMI1640 for 48 h according to the Trager–Jensen protocol.[14] Parasites in the no-treatment group developed well, with parasites at every stage and reproduction at around 300%. Schizonts exceeded 5%. In terms of dose and efficacy, naphthoquine phosphate had a higher suppression rate than chloroquine phosphates. At a high concentration (4.5 nmol/L), it was able to thoroughly eliminate the parasite. The IC_{50}, IC_{90}, and IC_{95} (50%, 90%, and 95% inhibitory concentrations) of both drugs were calculated via simple regression. The values for naphthoquine phosphate were 1.62, 3.209, and 3.90 nmol/L, respectively, 2–3 times smaller than those of chloroquine phosphate. Naphthoquine phosphate was more effective than chloroquine phosphate, consistent with the results from the in vivo studies.

4. In Vivo Efficacy Study Against Cultivated Naphthoquine Phosphate-Resistant *Plasmodium berghei*

An in vivo study was conducted with normal (N) strain *P. berghei* (K173) and Kunming outbred Swiss mice. Merkli's 1983 method of progressive doses[15] was used to cultivate a naphthoquine-resistant strain of the parasite (RNQ/K173). The initial dose was equivalent to the ED_{90} of the N strain. A new passage was produced each week and resistance was tested every 5–10 passages. Once a certain level of resistance was reached, the drugs were stopped and

TABLE 8.13 Curative Efficacy of Naphthoquine Phosphate Against B and C Strain *P. cynomolgi*

Strain	Drug	Dose (mg/kg/d)	Recrudescence/Total Cases	Day of Recrudescence
B	Naphthoquine phosphate	20	3/3	57, 61, 67
		10	3/3	38, 42, 38
		5	2/2	24, 27
	Chloroquine phosphate	20	2/2	24, 24
	Naphthoquine phosphate + primaquine phosphate	10.0 + 1.5	0/2	105(–)
	Chloroquine phosphate + primaquine phosphate	10.0 + 1.5	0/2	105(–)
C	Naphthoquine phosphate	31.6	0/2	105(–)
		10.0	0/4	105(–)
		3.16	2/3	20, 30
	Piperaquine phosphate	10.0	2/2	21, 25
	Chloroquine phosphate	31.6	0/2	105(–)
		10.0	2/2	25, 102

TABLE 8.14 Cultivation of Naphthoquine Phosphate-Resistant K173 Strain of *P. berghei*

Passage	ED_{90} (mg/kg/d)	I_{90}
5	0.59	0.9
10	2.30	3.3
15	8.02	11.6
20	24.94	36.1
25	34.18	49.5
30	57.66	83.6
35	136.11	197.3
40	113.50	164.5
50	133.40	193.2
60	125.04	181.2

further passages monitored for a reduction in resistance. Sensitivity tests were conducted with other antimalarials to determine the degree of cross-resistance.

From Table 8.14, it can be seen that slight resistance emerged from the 10th passage of the normal K173 strain, reaching severe resistance at the 35th passage. As resistance grew, the dose tolerated by the parasites increased as well, from 0.7 mg/kg/day at the start to 24.94 mg/kg/day at the 20th passage (36.1-fold), and 164.5-fold at the 40th passage. Thereafter, tolerance did not increase markedly, and parasitic tolerance fluctuated within a definite range as the degree of resistance stabilized. As resistance became established, the parasites' malarial pigment diminished.

Stability of resistance is shown in Table 8.15. The resistance of the strongly naphthoquine-resistant RNQ/K173 strain disappeared completely in 15 passages after medication ceased. Of the nine other antimalarials tested, the RNQ/K173 strain was sensitive to sulfanilamide and chloroquine phosphate but showed slight cross-resistance to the other drugs (Table 8.16).

5. Pharmacodynamics: Conclusion

The experimental results of naphthoquine phosphate on the reproductive cycles of rodent malaria, transmission of simian malaria via blood and sporozoite transmission, and in vitro cultivation of human malaria parasites showed that the drug had a definite effect. It had high and persistent efficacy as well as fast action. It had an impact on the tissue phase and gametocytes of some parasitic

TABLE 8.15 Stability of Resistance in Naphthoquine Phosphate-Resistant Strain of *P. berghei*

Passage After Drug Withdrawal	ED_{90} (mg/kg/d)	I_{90}
5	110.90	160.7
10	11.44	16.6
15	0.64	1.0

TABLE 8.16 Sensitivity of Naphthoquine-Resistant Strain to Common Antimalarials

Drug	ED_{90} (mg/kg/d)	I_{90}	Resistance Level
Naphthoquine phosphate	125.04	181.2	R+++
Amodiaquine	5.73	2.5	R+
Chloroquine phosphate	7.16	2.2	R+
Quinine hydrochloride	359.33	5.0	R+
Piperaquine phosphate	5.86	3.8	R+
Benflumetol (renamed as lumefantrine)	18.02	4.2	R+
Artemisinin	142.58	2.5	R+
Artemether	12.76	2.4	R+
Sulfanilamide	0.96	0.6	S
Primaquine phosphate	3.05	1.2	S

strains but was ineffective against sporozoites. Finally, it produced good therapeutic outcomes against chloroquine-resistant parasites. Although resistance could be easily cultivated to serious levels, sensitivity was quickly regained once medication ceased.

C. GENERAL PHARMACOLOGY[16]

All the drugs in the general pharmacological studies were prepared in normal saline. Two dose groups were used based on the CD_{90} of naphthoquine phosphate on rodent malaria: small ($CD_{90} \times 1$) and large ($CD_{90} \times 10$). The effective dose was taken to be a 5-day oral course totaling $CD_{90} \times 5$. The Paget body surface area method[17] was used to calculate oral doses for rats, rabbits, and cats, from which weight-based doses were then derived (mg/kg). Intravenous doses were set at a 10th of the oral dose. Small doses were administered over 1 min and large doses over 10 min, at constant speed via peristaltic pump. The laboratory was kept quiet for central nervous system experiments. For cardiovascular or respiratory system experiments, rats and cats were anesthetized using intraperitoneal urethane. Procaine was used as local anesthesia on the rabbits.

Spontaneous Activity in Mice

Kunming outbred Swiss mice were divided into three dose groups (large, small, and control) of 10 mice each. These were further divided into five subgroups of two mice each. Forty-five minutes after medication, the mice in each subgroup were placed in a three-photobeam observation cage to measure spontaneous activity. The frequency of spontaneous activity was measured over 10 min and averaged over the subgroups. The results showed that there was no obvious difference in spontaneous activity across the groups and there were no outward abnormalities. Naphthoquine did not affect spontaneous activity in mice.

Effects on Amphetamine Stimulation in Mice

Four experimental groups were used with 10 mice each: Control, control plus amphetamine, large-dose group plus amphetamine, small-dose group plus amphetamine. These were divided into five subgroups of two mice each. Forty-five minutes after medication, the mice were injected with 4 mg/kg amphetamine. Normal saline was used for the control group. After 15 min, the frequency of spontaneous activity was observed. The groups injected with amphetamine showed obvious increases in spontaneous activity compared to the normal saline control group ($P < .01$). However, there were no obvious differences among the amphetamine groups, indicating that naphthoquine did not affect the stimulating effects of amphetamine on the central nervous system.

Effects on Strychnine and Picrotoxin Convulsions in Mice

There were six experimental groups: Control plus trychnine, control plus picrotoxin, large dose plus strychnine, large dose plus picrotoxin, small dose plus strychnine, and small dose plus picrotoxin. Forty-five minutes after naphthoquine phosphate was administered, 1 mg/kg strychnine or 7.5 mg/kg picrotoxin were injected subcutaneously. There were no obvious time differences among the groups regarding the onset of convulsions. Naphthoquine phosphate did not influence strychnine and picrotoxin convulsions.

Effects on Sleep Onset With Sodium Pentobarbital and Chloral Hydrate in Mice

Six experimental groups were used with 10 mice in each: Control plus sodium pentobarbital, control plus chloral hydrate, large dose plus sodium pentobarbital, large dose plus chloral hydrate, small dose plus sodium pentobarbital, small dose plus chloral hydrate. Forty-five minutes after administration of naphthoquine phosphate, the mice were given 20mg/kg sodium pentobarbital or 190mg/kg chloral hydrate via intraperitoneal injection. Cessation of the righting reflex was taken as the moment of sleep and the number of animals falling asleep within 15min was observed. There were no obvious differences between the dose and control groups.

The same experimental arrangement was used with the mice given higher doses—30mg/kg sodium pentobarbital or 300mg/kg chloral hydrate—via intraperitoneal injection 45min after receiving naphthoquine phosphate. Absence and reemergence of the righting reflex were taken as indicators of sleep and wakefulness. The onset and duration of sleep were observed. The control and dose groups were basically similar, showing that naphthoquine phosphate did not influence onset and duration of sleep with sodium pentobarbital or chloral hydrate.

Effects on Heat-Induced Pain in Mice

Female mice with pain response time of 5–30s were divided into four groups of 10 mice: control, large dose, small dose, and positive control (pethidine). The mice were placed in beakers at constant temperature (55±0.5°C). Licking of hind paws or kicking of hind legs was taken as the onset of pain response. The pain response was tested before medication and at 1 and 2h thereafter. There were no obvious differences in the onset of pain response before or after drug administration, as well as compared to the control. Subcutaneous injection of pethidine in the positive control group (20mg/kg) had an obvious analgesic effect. Naphthoquine phosphate did not affect heat-induced pain.

Effects on Acetic Acid-Induced Pain in Mice

Four experimental groups of 10 mice each were used: control plus acetic acid, large dose plus acetic acid, small dose plus acetic acid, aspirin 625mg/kg plus acetic acid. Forty-five minutes after administration of naphthoquine phosphate, the mice were given intraperitoneal injections of 0.2mL 1% acetic acid. A writhing response induced by acetic acid irritation was taken as the indicator, and each group was monitored over 15min for this response. Mice in the aspirin group showed significantly less writhing, clearly differing from the control plus acetic acid group. The other two groups did not differ from the control group, indicating that naphthoquine phosphate did not influence acetic acid pain.

Effects on Body Temperature in Rats

Wistar rats were divided into four groups of five animals: No-treatment control, large-dose, small-dose, and chlorpromazine. A needle-type semiconductor temperature gauge was inserted 1.0–1.5cm into the muscle of the hind leg, and the temperature was read off the scale. Three readings were taken and averaged before medication, with further readings at 30, 90, 150, 210, and 270 min thereafter. There were no obvious differences between the naphthoquine phosphate groups and the control, whereas the drop in body temperature of the chlorpromazine group was statistically significant ($P<.05$).

Effects on Cardiovascular and Respiratory Systems in Rats, Rabbits and Cats

Wistar rats, Japanese white rabbits, and cats were used in this study and separated into large- and small-dose groups of five animals each. Naphthoquine phosphate was administered intravenously. Blood pressure, heart rate, electrocardiograms (ECGs), and breathing were monitored before medication and at 15, 30, 45, and 60 min thereafter. In both dose groups, all three types of animal showed no significant changes to these indicators before and after medication. Naphthoquine phosphate had no clear effect on the blood pressures, heart rates, ECGs, and breathing of rats, rabbits, and cats.

Conclusion

Naphthoquine phosphate had no obvious effect on the nervous systems of mice and rats at the $CD_{90}\times 10$ dose. No external behavioral abnormalities were observed. This dose also did not have an impact on the cardiovascular and respiratory systems of rats, rabbits, and cats, with no clear differences before and after medication. Naphthoquine phosphate had no significant effect on nervous, cardiovascular, and respiratory systems.

D. NON-CLINICAL PHARMACOKINETIC STUDIES IN ANIMALS[18]

Pharmacokinetics of ^{3}H-Naphthoquine Phosphate via Intragastric Administration in Mice

Male and female Kunming outbred Swiss mice, weighing 25–30 g, were fasted for 12–18 h. They were split into three ^{3}H-naphthoquine phosphate dose groups: 10.5 mg/120 μCi/kg, 42 mg/480 μCi/kg, and 168 mg/1920 μCi/kg. The drug was administered intragastrically. Blood was drawn orbitally at different times after medication, with heparin as an anticlotting agent. Plasma was isolated and extracted with ethyl acetate dried over a water bath. When the amount of ^{3}H-naphthoquine phosphate was at 100 ng/mL plasma, its average recovery ratio was 93.7%. The amount of naphthoquine phosphate contained was isolated and measured using the TLC method (silica gel 60F254 plate, E. Merck, eluent n-BuOH:acetic acid:water = 60:15:25). Five mice were tested at a time.

The pharmacokinetic parameters are shown in Table 8.17. The blood concentration of the drug followed a two-compartment open kinetic model. Low,

TABLE 8.17 Pharmacokinetic Parameters of Intragastric ^{3}H-Naphthoquine Phosphate in Mice

Parameter	10.5 mg 120 µCi/kg	42 mg 480 µCi/kg	168 mg 1920 µCi/kg
A (ng/mL)	0.253 45	1.321 17	7.312 98
B (ng/mL)	0.094 48	0.128 81	0.149 72
α (d^{-1})	7.563 46	7.752 74	7.069 82
β (d^{-1})	0.208 12	0.121 90	0.054 37
Ka (d^{-1})	574.984 19	77.816 66	109.347 25
$T_{1/2\alpha}$ (d)	0.091 64	0.089 41	0.098 04
$T_{1/2\beta i}$ (d)	3.330 49	5.685 99	12.748 56
$T_{1/2}Ka$ (d)	0.001 21	0.008 91	0.006 34
K_{21} a (d^{-1})	2.224 39	0.866 43	0.204 57
K_{10} (d^{-1})	0.707 71	1.090 70	1.879 04
K_{12} (d^{-1})	4.839 98	5.91 74	5.040 58
$AUC_{0\sim120}$ (ng·h/mL)	0.486 87	1.208 44	3.719 87
CL(s) (mL/d)	21.566 24	34.755 42	45.162 81
T_{max} (min)[a]	15	30	30
C_{max} (ng/mL)	0.363 00	1.017 00	6.667 00
V/F(c) (mL)	30.473 34	31.862 62	24.039 99

[a]*Actual value.*

medium, and high doses peaked at 15, 30, and 30 min on average, reaching concentrations of 0.36–6.67 ng/mL. The $T_{1/2\alpha}$ values were 0.09164, 0.08941, and 0.09804 days, respectively, and the $T_{1/2\beta}$ values were 3.33, 5.69, and 12.75 days, respectively. It indicated that naphthoquine phosphate was absorbed quickly in the alimentary canal and eliminated slowly.

Distribution in Tissues in Mice

Mice were given 42 mg/480 µCi/kg intragastric ^{3}H-naphthoquine phosphate. The radioactivity in different tissues was measured at various times. Table 8.18 shows that naphthoquine phosphate was relatively widely distributed in the tissues. Concentration was highest in the liver. Two hours after administration, radioactivity in the liver was 6 times that of the stomach, 7 times that of the lungs, and 25 times that of the spleen. It was still the highest at the 336th hour.

TABLE 8.18 Organ/Tissue Distribution (%) in Mice After Receiving 42 mg/480 μCi/kg Intragastric ^{3}H-Naphthoquine Phosphate (Mean ± SD, $n = 5$)

Tissue	Time After Administration			
	2 h	4 h	168 h	336 h
Heart[a]	0.042 ± 0.004	0.042 ± 0.011	0.010 ± 0.000	0.000 ± 0.000
Liver[a]	3.886 ± 0.567	0.378 ± 0.133	0.880 ± 0.084	0.582 ± 0.049
Spleen[a]	0.158 ± 0.037	0.308 ± 0.039	0.080 ± 0.041	0.060 ± 0.000
Lung[a]	0.556 ± 0.009	0.864 ± 0.078	0.150 ± 0.041	0.080 ± 0.007
Kidney[a]	0.642 ± 0.060	0.588 ± 0.151	0.134 ± 0.031	0.054 ± 0.015
Brain[a]	0.116 ± 0.017	0.084 ± 0.009	0.020 ± 0.000	0.000 ± 0.000
Stomach[a]	0.696 ± 0.265	0.118 ± 0.008	0.026 ± 0.005	0.015 ± 0.007
Duodenum[a]	0.244 ± 0.044	0.076 ± 0.011	0.016 ± 0.005	0.000 ± 0.000
Genitals[a]	0.050 ± 0.014	0.060 ± 0.023	0.038 ± 0.034	0.030 ± 0.007
Eyeballs[a]	0.010 ± 0.000	0.010 ± 0.000	0.002 ± 0.001	0.000 ± 0.000
Fat[b]	0.543 ± 0.082	0.336 ± 0.085	0.200 ± 0.034	0.158 ± 0.013
Bone marrow[b]	0.126 ± 0.057	0.082 ± 0.029	0.013 ± 0.005	0.000 ± 0.000
Muscle[b]	0.274 ± 0.022	0.140 ± 0.016	0.045 ± 0.007	0.000 ± 0.000

[a] *Percentage of radioactivity in the entire organ to total radioactivity in drug administered.*
[b] *Percentage of radioactivity per gram of tissue to total radioactivity in drug administered.*

TLC conducted on homogenate obtained from the liver showed a peak in radiation similar to the reference Rf value, indicating that the radioactivity accumulated in the liver came mostly from the original drug.

Distribution of ^{3}H-Naphthoquine Phosphate in Red Cells and Plasma in Mice

Mice were given three different intragastric doses of the drug and blood was drawn orbitally at various times. Plasma and erythrocytes (washed twice in saline) were separated and extracted with ethyl acetate. TLC was conducted on the plasma and erythrocytes to monitor the presence of the drug. Five mice were tested at a time. Table 8.19 shows that the distribution of the drug varied over time. Half an hour after medication, the ratio of the drug present in erythrocytes versus in plasma was mainly less than 1 at all three doses. From 1 h to 7 days after medication, this ratio rose above 1. This was especially notable at the medium and high doses.

TABLE 8.19 Distribution of ^{3}H-Naphthoquine Phosphate in Red Cells and Plasma (cpm/mL)

	10.5 mg/ 120 μCi/kg			42.0 mg/ 480 μCi/kg			168 mg/ 1920 μCi/kg		
Time (h)	Blood Cells	Plasma	Cells/ Plasma	Blood Cells	Plasma	Cells/Plasma	Blood Cells	Plasma	Cells/ Plasma
0.083	3685	7450	0.49	9,415	8,090	1.17	53,675	87,290	0.62
0.250	5950	9200	0.65	18,340	17,180	1.07	73,850	87,570	0.84
0.500	10,315	7805	1.33	21,920	25,795	0.85	150,530	169,160	0.89
1	9065	6910	1.31	29,625	23,510	1.26	170,965	167,030	1.02
2	7730	5295	1.46	22,815	21,610	1.06	179,295	148,125	1.21
4	6845	4075	1.68	20,860	14,980	1.39	93,750	50,235	1.87
8	4920	3035	1.62	18,835	4,895	3.85	43,775	18,470	2.37
12	4870	2590	1.88	11,665	4,365	2.68	27,730	12,495	2.22
24	2180	1710	1.27	7,595	2,925	2.60	22,790	12,264	1.86
72	1540	1345	1.14	5,490	1,980	2.77	6,825	2,681	2.55
168	865	570	1.52	3,670	1,940	1.89	4,170	2,473	1.69
336	–	–	–	1,178	562	2.10	3,325	1,839	1.81
504	–	–	–	–	–	–	2,225	1,246	1.79

Excretion in Urine and Feces in Mice

Mice were given three different intragastric doses of naphthoquine phosphate. Their urine and feces were collected at various times, undergoing extraction with ethyl acetate and then TLC to determine the concentration of the drug. The results showed that the main excretion pathway was via urine (40% over 7 days) as opposed to feces (around 23%). The three different doses displayed no obvious differences in proportion.

Bile Excretion in Rats

Wistar rats weighing between 250 and 300 g were fasted for 12–18 h. They were anesthetized using 2% sodium pentobarbital and a cannula inserted into the bile duct. Twelve hours after consciousness was regained, 40 mg/80 μCi/kg of ^{3}H-naphthoquine phosphate was administered. Bile was collected every 6 h and radioactivity measured on the clear media prepared by using the homogenous method. The drug was excreted in the bile mainly in the first 24 h. Thereafter excretion decreased, totaling 27.61% in 3 days. In the next test, bile containing a high concentration of ^{3}H-naphthoquine phosphate obtained above was injected into the duodenum of a rat. Twelve hours later, the liver, blood, and bile were extracted and tested for radioactivity. Homogenate from the liver was concentrated and separated. TLC of the homogenate showed the same Rf value as the reference, indicating that the original drug was being excreted.

Plasma Protein Binding Rate (Equilibrium Dialysis) in Rats

Different concentrations of ^{3}H-naphthoquine phosphate were mixed with fresh rat plasma and placed in dialysis tubing (10 mm × 6.4 dia, molecular-weight cutoff 12,000–14,000) with 1/15 mol/L phosphate buffer (pH 7.4). The mixture was shaken at a constant temperature of 37°C for 17 h (150–200 times per minute). The concentration of the drug inside and outside the tube was measured. The drug's plasma protein binding rate was measured at 87%–89%. Plasma was then loaded onto a cellulose acetate membrane and electrophoresis used to determine different bands of radioactivity (100 V at 25°C for 50 min). The drug was found to bind mostly with albumin (above 60%), followed by γ- and α-2 globulins.

Conclusion

Isotopic labeling and TLC were used to observe the absorption, distribution, and excretion of naphthoquine phosphate in vivo. They showed that oral ^{3}H-naphthoquine phosphate was absorbed quickly, peaking at 15–30 min and at a concentration of 0.36–6.67 ng/mL. Around 50% was absorbed in the alimentary canal within 1 h. The drug was broadly distributed throughout the tissues but mostly in the liver, followed by the kidneys, lungs, and spleen as well as the brain. Sixty minutes after medication, the drug was concentrated more in erythrocytes than in plasma, reaching ratios of 2.55–3.85. It was eliminated slowly, with a half-life of 3.3–12.7 days. The drug was excreted mainly in urine

(44.3%–44.7%), then in feces (23.8%–27.8%) and bile (24%), also remaining in the enterohepatic circulation. The plasma protein binding rate was 87%–89%.

E. NON-CLINICAL PHARMACOKINETICS: ACTION MECHANISM

The pharmacodynamic studies showed that naphthoquine phosphate was effective, fast-acting, and long-lasting, highly capable of eliminating chloroquine-resistant strains. It acted on multiple levels and deserved further study due to these outstanding characteristics.

Pharmacokinetic Mechanism of Action on Normal Mice[19]

Kunming outbred Swiss mice weighing 25±2g were randomly divided into 17 groups of three. They were given 10mg/kg intragastric naphthoquine phosphate. Blood was drawn orbitally before medication and after, at 0.17, 0.33, and all the way to 144h. Plasma and erythrocytes were separated according to Zou Jing's 1985 protocol, and extraction was performed using Lu Zhiliang's 1988 method.[20] The concentration of naphthoquine phosphate was determined via HPLC. Table 8.20 shows that drug concentration in erythrocytes and plasma peaked 2h after medication and then decreased gradually. It could be detected at 144h. This indicated that the drug was concentrated more in erythrocytes than in plasma, consistent with the findings of the isotopic labeling experiments. Table 8.21 sets out the pharmacokinetic parameters calculated using the 3p87 software.[21]

The $T_{1/2}Ka$ of naphthoquine phosphate absorbed via the intestinal tract was 0.51h, indicating that the drug was absorbed quickly. $K_{12}>K_{21}>K_{10}$, showing that the drug crossed faster from the central to the peripheral compartment than from peripheral to central, resulting in wide distribution in the tissues. Elimination from the central compartment was slow, therefore the drug's long half-life ($T_{1/2\beta}=199$h).

Pharmacokinetic Studies on Infected Mice

Forty-eight mice infected with K173 strain *P. berghei* were selected when parasitemia reached 2%–5%. They were randomly divided into 16 groups of three and given one intragastric 10mg/kg dose. Blood was collected orbitally both before medication and after, at 0.33, 0.75, 1.0, and up to 168h.

Table 8.22 shows that naphthoquine phosphate levels peaked in the plasma 2h after medication, then fell gradually. Drug concentration increased again at 96h and was still detected in the plasma at 168h. Concentration in erythrocytes peaked 36h after medication, dropping to its lowest point at 72h before rising again to another peak at 144h. It was still very high at 168h. The concentration in erythrocytes was far greater in infected mice than in normal mice, up to 150 times higher at 144h.

TABLE 8.20 Naphthoquine Phosphate in Erythrocytes and Plasma of Normal Mice

	Concentration (ng/mL)		
Sampling Time (h)	Plasma	Red Cells	Red Cells/Plasma
0	0	0	–
0.17	63.10	79.13	1.25
0.33	86.20	122.45	1.42
0.50	91.83	90.13	0.98
0.75	169.17	208.59	1.23
1.00	271.07	198.57	0.73
2.00	300.84	273.29	0.91
4.00	197.42	209.46	1.06
6.00	198.97	231.03	1.16
12.00	159.32	197.73	1.24
24.00	100.89	211.43	2.10
36.00	105.55	137.48	1.30
48.00	79.68	79.91	1.00
72.00	51.11	101.49	1.99
96.00	45.30	78.24	1.73
120.00	42.15	55.59	1.32
144.00	39.69	38.96	0.98

The pharmacokinetic parameters are shown in Table 8.23. The absorption half-life, $T_{1/2}Ka$, was 0.245 h, indicating that naphthoquine phosphate could be absorbed quickly via the intestinal tract. The elimination half-life, $T_{1/2\beta}$, was long at 474 h. The volume of distribution $V/F(c)$ was 83 L/kg, $K_{12} >> K_{21} > K_{10}$, indicating that the drug was widely distributed in the tissues and was eliminated from the central compartment slowly. There were obvious differences in drug concentration in the blood of normal and infected mice, signaling that pharmacokinetic processes changed after infection.

An experiment was then run to analyze the increase in erythrocytic drug concentration in infected mice, as well as to make up for the lack of observations on the parasite itself in the previous study. Table 8.24 shows a correlation between erythrocytic drug concentration and parasitemia. At insufficient drug

TABLE 8.21 Pharmacokinetic Parameters of Naphthoquine Phosphate in Plasma of Normal Mice

Parameter	Unit	Value
A	ng/mL	190.90515
α	h	0.05010
B	ng/mL	63.49218
β	h	0.00349
Ka	h	1.36637
$V/F(c)$	mL	0.04045
$T_{1/2\alpha}$	h	13.83429
$T_{1/2\beta}$	h	198.61946
$T_{1/2}Ka$	h	0.50729
K_{21}	h	0.01543
K_{10}	h	0.01133
K_{12}	h	0.02683
$AUC_{0\sim120}$	ng·h/mL	21,817.72270
CL(s)	mL/day	0.00046
T_{max}	h	2.72000
C_{max}	ng/mL	223.28975

concentrations, the parasite was not fully controlled. Therefore, it may be seen that infected erythrocytes could induce higher drug concentrations.

To verify this relationship, 24 infected mice were randomly divided into two groups and given the drug intragastrically 4 and 6 days after infection. Blood was drawn orbitally 1, 2, 4, and 8 h after medication to monitor parasitemia and drug concentration in plasma and erythrocytes. Drug concentration in erythrocytes increased in tandem with parasitemia, whereas concentration in plasma decreased as parasitemia rose (Table 8.25).

Mechanism of Action on Parasite Ultrastructure[22]

Ten mice infected with normal strain K173 *P. berghei* were selected, with parasitemia at 25%–35%. They were randomly divided into two groups, one given naphthoquine phosphate and the other, chloroquine phosphate. The dose was $ED_{90} \times 10$, delivered once intragastrically. Parasitic ultrastructures were observed before medication and at 1, 2, 4, 8, 16, 24, and 48 h after.

TABLE 8.22 Naphthoquine Phosphate in Erythrocytes and Plasma of Infected Mice

	Concentration (ng/mL)		
Sampling Time (h)	Plasma	Red Cells	Red Cells/Plasma
0	0	0	–
0.33	61.50	185.4	3.01
0.75	87.60	260.9	2.98
1.00	119.62	364.0	3.04
2.00	156.04	495.8	3.18
4.00	65.23	719.9	11.04
6.00	53.29	1271.3	23.86
12.00	80.19	1881.8	23.47
24.00	13.73	1870.8	136.26
36.00	16.15	2979.2	184.47
48.00	12.94	1624.7	125.56
72.00	11.05	558.3	50.52
96.00	21.79	1439.3	66.05
120.00	17.79	4099.9	230.47
144.00	17.03	5851.7	343.61
168.00	9.20	2133.6	231.91

In the naphthoquine group, the parasites showed a decrease in malarial pigment grains 1 h after medication. Gaps in the compound membranes of a few of the parasites also widened. At 2 h, structural changes were seen. Food vacuoles fused and formed lacunae, compound membranes became blurred and swelling occurred in the mitochondria. At 4–16 h, structural changes gradually became more obvious as the food vacuole disappeared and the nuclear membrane swelled and formed vacuoles. The nucleoplasm also changed. At 24–48 h, the trophozoite collapsed, only the outline of the nucleus could be seen and the nucleoplasm disintegrated.

Changes in the chloroquine group occurred 2–4 h later than in the naphthoquine phosphate group. The food vacuole was first to be altered, with the entire structure following at 16–24 h after medication. The changes were obvious after 48 h.

TABLE 8.23 Pharmacokinetic Parameters of Naphthoquine Phosphate in Infected Mice

Parameter	Unit	Value
A	ng/mL	111.18596
α	h	0.14474
B	ng/mL	14.81020
β	h	0.00146
Ka	h	2.82774
Lag time	h	0.07423
V/F(c)	L/kg	83.13
$T_{1/2\alpha}$	h	4.78897
$T_{1/2\beta}$	h	473.88019
$T_{1/2}Ka$	h	0.24512
K_{21}	h	0.01909
K_{10}	h	0.01109
K_{12}	h	0.11602
$AUC_{0\sim120}$	ng·h/mL	10,848.84280
CL(s)	mL/d	0.00092
T_{max}	h	1.20000
C_{max}	ng/mL	104.04564

Conclusion

There were clear in vivo differences in the pharmacokinetics of naphthoquine phosphate between normal and infected mice. Drug concentration was significantly higher in the erythrocytes of infected mice than in those of normal mice. It increased in proportion to parasitemia, indicating that infected erythrocytes had a concentrating effect on the drug. Judging by its effects on the parasite's ultrastructure, the drug acted not only on the digestive system but also on the membranes. Once damaged, the membranes were more permeable to the drug. Naphthoquine phosphate's efficacy, speed, and ability to circumvent chloroquine resistance could be the result of this concentrating effect and disruption of membranes.

The elimination half-life in the plasma of normal mice was 198 h, which increased to 473 h for infected mice. It was distributed widely in the tissues at concentrations many times higher than in plasma and remained in the enterohepatic circulation. These contributed to its long-lasting effects.

TABLE 8.24 Parasitemia and Drug Concentrations in Erythrocytes of Infected Mice

Sampling Time After Administration (h)	Average Parasitemia (%)	Average Concentration in Red Cells (ng/mL)
2.0	0.5	156.4
4.0	1.2	367.5
6.0	2.1	819.2
12.0	1.9	780.6
24.0	4.2	1528.4
36.0	4.1	1515.2
48.0	0.8	214.9
72.0	3.5	610.3
96.0	5.8	1192.4
120.0	6.5	1713.8

F. TOXICOLOGY

1. General Toxicity

These studies included general and special toxicity studies. General toxicity was divided into single (acute) and multiple (long-term) doses. Special toxicity took in reproductive toxicity (reproduction in males, teratogenicity in females) and mutagenesis (mutation, micronucleus, chromosomal aberration).

Acute Toxicity[23]

Mice were divided randomly into two groups of 20 based on weight, with an equal number of males and females. The drug was administered orally or via intraperitoneal injection, and the mice were observed every day after medication for 2 weeks. Clinical features and deaths were recorded. Finney's probit analysis method[24] was used to calculate the LD_{50}, LD_5, and LD_{95}, their respective 95% confidence intervals and *b* values.

Six hours after oral naphthoquine phosphate, hyperactivity and convulsions appeared in the mice, correlated positively with the dose. $LD_{50} \pm SD$ was 794 ± 27 mg/kg. There were no obvious symptoms among the injected mice: $LD_{50} \pm SD$ was 127 ± 3 mg/kg. Both dead and surviving mice were autopsied and examined with the naked eye. Blood congestion was observed to different degrees in the liver and kidneys.

More female mice died than male in the oral administered group. There was no obvious sex difference in the injected group. Deaths occurred in the oral

TABLE 8.25 Parasitemia and Drug Concentrations in Erythrocytes and Plasma of Infected Mice

	Medication on Day 4 After Infection				Medication on Day 6 after Infection			
Time After Drug (h)	% Parasitemia	Plasma Conc. (ng/mL)	Red Cell Conc. (ng/mL)	Red Cell/Plasma	% Parasitemia	Plasma Conc. (ng/mL)	Red Cell Conc. (ng/mL)	Red Cell/Plasma
1.0	10.9	45.4	461.1	11.1	18.7	66.2	1,531.6	23.1
2.0	32.5	32.4	6911.6	213.3	38.3	48.5	11,502.4	237.1
4.0	16.7	79.6	5905.3	74.0	29.3	44.5	5,705.8	128.1
8.0	10.1	45.0	2286.2	50.8	12.6	28.3	3,810.0	134.9

group from 6h after medication. As the dose increased, onset of death lengthened to 4 days. In the injection group, death occurred within 1 h.

The same protocol was used for rats. The LD_{50} of oral naphthoquine phosphate was 1018 ± 102 mg/kg. Rats with naphthoquine poisoning showed marked anorexia, gradually increasing sluggishness, weight loss, fewer feeding behaviors, and loose stool. Those severely intoxicated had bloody secretions around the muzzle and delayed reactions or paralysis in the hind legs. These became more obvious 1–3 days before death. Death occurred mainly 3–7 days after medication. In most of the dead rats—as well as in a small number of surviving rats in the large-dose group—autopsies showed visible blood congestion and nodular necrosis in the liver, with focal distribution and a definite link to the degree of intoxication.

For dogs, Deichmann's and LeBlanc's 1943 method of calculating the lethal dose based on six animals[25] was used. Four oral doses of naphthoquine phosphate were chosen—100, 70, 55, and 33 mg/kg—and given to one randomly assigned dog each. The dogs were fasted before medication and fed 1 h afterward. They were observed for 14 days. Clinical signs were recorded daily; appetite and body weight were measured once every 7 days. Dogs were autopsied on the day of death, with surviving dogs killed and dissected on Day 14, and observations were made with the naked eye. The approximate lethal dose (ALD) was defined as the lowest dose able to induce death, whereas the approximate maximum dose (AMD) was taken to be the highest dose without clear clinical signs of intoxication.

Nine hours after 100 mg/kg oral naphthoquine phosphate, the dog displayed tonic spasms lasting 15 min. Symptoms eased and disappeared after 72 h. At 70 mg/kg, the dog also experienced tonic spasms 6 h after medication, as well as opisthotonus and salivation lasting 20 min. These developed into spastic tics and the dog died 10 h after medication. No signs of intoxication were seen at the 55 and 33 mg/kg doses. Energy levels, appetite, and activity were normal.

The leukocyte count of the dog given the 100 mg/kg dose was elevated, but its composition was normal. Serum glutamic oxaloacetic transaminase (SGOT) levels were slightly raised after medication, but returned to normal on Day 7. For the dog given the 55 mg/kg dose, the number of eosinophil granulocytes was raised on Days 7 and 14, with no obvious aberrations in other areas. No abnormalities were found in the dog given the 33 mg/kg dose.

The dog given 70 mg/kg naphthoquine phosphate died. The autopsy revealed mild blood congestion in the liver, stomach, and intestines. There were no abnormalities found in the other dogs.

In summary, the $LD_{50} \pm SD$ of naphthoquine phosphate for mice was 794 ± 27 mg/kg for oral doses and 127 ± 3 mg/kg for intraperitoneal injection. For rats, the $LD_{50} \pm SD$ for an oral dose was 1018 ± 102 mg/kg. The oral ALD and AMD for dogs were 70 and 55 mg/kg, respectively. Regarding toxicity ratings, naphthoquine phosphate was slightly toxic. Because of its high antimalarial efficacy, with an ED_{50} of 0.4 mg/kg and ED_{95} of 0.69 mg/kg against

P. berghei, its therapeutic index (LD_{50}/ED_{50}) was 1985 and its safety index (LD_5/ED_{95}), 726. The autopsies conducted as part of the acute toxicity studies showed signs of liver damage. Therefore, attention should be focused on the liver as a target organ.

Long-Term Toxicity[26]

Wistar rats were randomly divided into four dose groups: 80, 40, 20 mg/kg/d, and control. Each group had 24 rats, with an equal number of males and females. The drug was administrated once daily for 14 days. On Day 15, eight rats per group (four male and four female) were randomly selected for observation over 28 days. Blood was drawn orbitally from two-thirds of the rats for blood tests (9 parameters), comprehensive metabolic panels (16 parameters), and pathology tests. Along with the autopsy, nine of the organs were weighed and 42 tissue blocks for histology examinations undertaken.

At the 80 mg/kg/d dose, the rats were markedly anorexic and lost weight. A few displayed thinning fur and diarrhea during the medication period. One female rat died after the 10th dose. The rats given 40 mg/kg/d also ate less and gained weight more slowly than the control. The 20 mg/kg/d group was basically the same as the control in terms of feeding and weight gain.

Blood tests showed a slight elevation in leukocyte and neutrophil count at the 80 mg/kg/d dose. There was a similar upward tendency at the 40 mg/kg/d dose, while the 20 mg/kg/d dose group was normal.

At the 80 mg/kg/d dose, increases in glutamic-oxaloacetic transaminase (GOT), glutamic-pyruvic transaminase (GPT), and lactate dehydrogenase (LDH) could be seen in the comprehensive metabolic panel, alongside a reduction in blood sugar. This was more marked in the female rats, who also displayed higher phosphocreatine kinase (CK) levels. Slight increases in GOT and LDH and a small dip in blood sugar was seen at the 40 mg/kg/d dose. The 20 mg/kg/d dose was normal.

The rats were killed after medication ceased. Nine organs were weighed and ratios calculated based on body mass, as relative organ mass. The 80 mg/kg/d dose group showed higher liver, lung, and brain relative mass than the control. The thymus of female rats had less relative mass than those in the control. Liver and lung mass was also slightly higher in the 40 mg/kg/d dose group. Autopsies revealed enlarged livers with obvious blood congestion in both these groups. Some of the livers had suppurative nodular necrosis. A small number of rats displayed blood congestion in the brain, enlarged spleens, and atrophied thymuses. Organ mass and appearance were similar in the 20 mg/kg/d dose group and the control.

Histology examinations showed damage to the liver in the form of vesicular fatty change, ballooning degeneration, patchy and focal necrosis, formation of abscesses, and congested blood in liver sinusoids and hepatic veins. Kidney damage included swelling, blood congestion, and degeneration of the tubular epithelium. Small changes were seen in the heart muscle. There was slight bone marrow suppression, reduction in the epithelial cells of the thymus, and fewer lymphocytes in the lymph nodes. Such changes were obvious and frequent at the 80 mg/kg/d dose but slight at the 40 mg/kg/d dose. The 20 mg/kg/d group was basically normal. In all dose groups, foamy type II pneumocytes could be seen in the alveolar spaces and walls, and in subpleural lung tissues. However, there was no obvious dose–effect relationship.

The remaining rats were killed 28 days after medication ceased to see if the pathologies were reversible. Blood and comprehensive metabolic panel tests were within normal limits for all dose groups. Histology examinations demonstrated a slight increase in type II pneumocytes in the lungs. In the 80 mg/kg/d group, the liver and kidneys were recovering but had not yet returned to normal. All other pathologies had returned to normal and no new ones were seen.

For dogs, naphthoquine phosphate was formulated into tablets of 0.1 g naphthoquine base each. The dogs were randomly divided into four groups: 5, 10, 20 mg/kg/d, and control. Each group had three females and three males. The drug was administrated once daily for 14 days. Fifteen days after ending medication, two-thirds of the dogs were killed, leaving one-third (one female and one male) for observation over 28 days. Clinical signs were recorded daily, starting 7 days before examination and ending 28 days after medication ceased (Day 42). The dogs were weighed every 7 days. The following were carried out 7 days before medication, on the day of medication, and at 7-day intervals thereafter: blood tests (11 parameters), including tests on the bone marrow of both dead and surviving dogs; comprehensive metabolic panels (16 parameters); routine urine tests (seven parameters); and nine-lead ECGs. Fundus examinations were carried out on the 20 mg/kg/d and control groups once before medication and again before termination. Pathology examinations were performed on days 15 and 42. Ten organs were weighed and histology examination done on 44 tissue blocks from the nervous, circulatory, digestive, respiratory, urinary, reproductive, glandular, skeletal, muscular, and lymphatic systems.

In the 20 mg/kg/d group, four of the six dogs displayed tremors throughout the whole body, but chiefly in the hind legs, 6–10 days after medication. This continued for 5–9 days and disappeared after medication was stopped. Similar signs appeared in two of the dogs in the 10 mg/kg/d group, but the presentation was not as severe. No such symptoms were seen in the 5 mg/kg/d group and control. After medication, no abnormalities were seen in feeding, body weight, rectal temperature, respiration, and heart rate.

At 10 mg/kg/d and above, the Q–T interval was prolonged and two out of three dogs had flattened T waves. No obvious differences were seen between the 5 mg/kg/d group and control.

GPT and GOT levels rose in the 20 mg/kg/d group 7 and 14 days after medication, with the rise in GPT especially marked. This was also seen in the 10 mg/kg/d group, but the values were within normal bounds. There were no obvious differences between the 5 mg/kg/d group and control. Blood and comprehensive metabolic panel tests showed no changes. Bone marrow smears showed no obvious changes in the bone marrow hematopoietic system of the 20mg/kg/d group. No obvious abnormalities were seen in the fundus examinations of that group.

Swelling of hepatocytes, reduction in cytoplasmic density, vacuolar degeneration, and lymphocytic infiltration into the portal area were observed in those dogs in the 20mg/kg/d group that were killed 15days after medication. Lipid and bile stains were negative. No drug-related pathological signs were seen at 10mg/kg/d and below.

During the recovery period, medication ceased and observation was carried out for 28days. Across all groups, all indicators displayed no obvious changes. Pathological liver changes were reversed and no new injuries seen.

In summary, the severely toxic dose of oral naphthoquine phosphate for rats was 80mg/kg/d. A moderately toxic dose was 40mg/kg/d for rats and 20mg/kg/d for dogs, whereas an essentially safe dose was 20 and 5mg/kg/d, respectively. At 10mg/kg/d, dogs experienced only slight and transient abnormalities.

Chloroquine groups were set up in parallel with each long-term toxicity study. It was found that 80mg/kg/d was the lethal oral chloroquine dose in rats, and damage to the heart muscle occurred at 20mg/kg/d. Four out of five dogs died after a second oral dose of 15mg/kg/d chloroquine. Therefore, naphthoquine was less toxic than chloroquine.

In clinical trials, chloroquine was administrated four times over 3days, totaling 1500mg (i.e., 30mg/kg). The dose on the first day was 900mg, or 18mg/kg. For naphthoquine, a clinical dose of 800–1000mg was proposed, with an initial dose of 600mg or 12mg/kg. Both this dose and the duration of treatment were lower than those of chloroquine. Therefore, naphthoquine had a broader therapeutic window.

The target organ in naphthoquine toxicity was mainly the liver, followed by the heart and kidneys. However, such damage was reversible and recovery was generally possible in a short time. Although the drug was given orally, no symptoms appeared in the gastrointestinal tract. More significant were the lengthening of the Q–T interval in ECGs, and the elevation of GPT and GOT levels in plasma. These indicators deserved more clinical attention.

The rat study noted an increase in type II pneumocytes in the lungs. The mechanism for this was unclear, but it was known that such an increase was a common, nonspecific, adaptive tissue-repair response to toxicity in the lungs. It was an adverse reaction, but one not seen in the dog study. Thus, Phase I clinical trials should focus on indicators of pulmonary function.

2. Special Toxicity: Reproductive[27]

Teratogenesis

Wilson's protocol[28] was used to determine if naphthoquine phosphate was teratogenic or toxic to mouse embryos. Mice were divided into four dose groups of 20 pregnant mice each: 100, 50, 25mg/kg/d, and a saline control. The drug was administrated once daily 6–15 days after fertilization, during organogenesis. The mice were killed on the 17th day of gestation to determine the total number of implantations and living or dead embryos, and to check for abnormalities. The skeletal and soft tissue development of live embryos was studied.

The rate of living, dead, and reabsorbed embryos showed no significant differences between all the dose groups and the control. No abnormalities were seen in the appearance and skeletons of the embryos, although cleft palate and cryptorchidism appeared in the 100 mg/kg/d group. The total rate of abnormality was 10.87%, clearly different from the control ($P < .05$).

Oral doses of 100, 50, and 25 mg/kg/d were equivalent to 1/10, 1/20, and 1/40 of the LD_{50}, respectively. Naphthoquine phosphate was not toxic to embryos at these doses. At 100 mg/kg/d (equivalent to 10 times the clinical dose), soft tissue deformities could be seen, primarily in the higher incidence of cleft palate. One mouse in this group died at the sixth dose. Others had lower body weights and smaller and lighter fetuses than in the control, low-dose, and medium-dose groups. Therefore, it may be assumed that the higher incidence of soft tissue deformity was related to toxicity to the mother. The 50 mg/kg/d dose (five times the clinical dose) was regarded as safe.

Reproductive Toxicity in Male Mice

Dose groups of 2.5, 5.0, 10.0 mg/kg/d, and normal saline control were set up with 20 male mice each. The drug was administrated once daily for 13 weeks. Eight weeks after medication, the mice were mated with nonmedicated mature females. Fertilization was determined based on the presence of a vaginal plug and 20 pregnant mice were obtained per group. The male mice were medicated until Week 13, when two-thirds (14 mice) were killed to obtain the testicles, epididymis, vas deferens, and prostate for histopathology examination. The rest were observed for another 4 weeks. Half of the pregnant mice were killed on the 12th day of pregnancy; half the remaining number of mice were killed on the 17th day to ascertain the rate of conception and implantation, as well as living, reabsorbed, or dead embryos. The length and weight of the fetuses was measured. The phenotypes, skeletons, and soft tissues of living fetuses were examined.

In all dose groups, the rate of fertilization and implantation and number of living embryos—as well as average fetal length and weight—showed no obvious differences with the control. Naphthoquine phosphate did not influence reproductive function in male mice.

The 2.5, 5.0, and 10.0 mg/kg/d groups produced 121, 140, and 134 fetuses, respectively. The proportion of females was 44.6%, 41.4%, and 49.3%, respectively, with males at 55.4%, 58.6%, and 50.7%. Control group fetuses were 43.2% female and 56.8% male; therefore there was no clear difference in sex ratio. Fetal appearance and skeletal and soft tissue development were normal.

The testicles were weighed on the 91st day of medication and 28 days after medication ceased, and the epididymis, vas deferens, and prostate were examined. All dose groups were the same as the control.

3. Special Toxicity: Mutagenesis[29]

Reverse Mutation

The 1975 Ames test[30] was used, with four histidine auxotrophic *Salmonella typhimurium* strains (TA97, TA98, TA100, and TA102) and four naphthoquine phosphate dose groups (0.1, 1.0, 10.0, and 100.0 μg/dish). Distilled water was the negative control. Known mutagens such as NTG (*N*-methyl-*N*′-nitro-*N*-nitrosoguanidine), 9-aminoacridine, mitomycin C, and aflatoxin B1(AFTB1) were the positive controls. The strains were cultivated both with and without mammalian liver S9 fraction. The number of reverse-mutated colonies was then calculated. A positive result was defined as a dose-dependent increase of at least 100% in the number of revertant colonies, as compared to the control. All other outcomes were considered negative.

There was no obvious mutagenic effect on all four bacteria strains, as opposed to the obvious mutagenesis in the known-mutagen groups.

Micronucleus

Adult male Kunming outbred Swiss mice weighing 25–35 g were divided into groups of five. They were given doses of 10, 50, or 100 mg/kg. The mice were killed at different times (16, 24, 48, and 72 h). A solvent control and mitomycin C-positive control were also set up.

The thigh bone was obtained after termination and the bone marrow washed twice with calf serum and centrifuged to extract the precipitate, prepared as a smear and stained. From each specimen, 1000 intact polychromatic erythrocytes were examined using oil immersion microscopy for the presence of micronucleoli. The number of cells with either single or multiple micronucleoli were counted and expressed as micronucleated cell rate per 1000 cells. A positive result was defined as a dose-dependent, statistically significant increase in the frequency of micronucleated cells when compared to the controls.

There were no clear differences in the frequency of micronucleated cell rate between the three dose groups and the solvent control, regardless of when the samples were obtained.

Chromosomal Aberration

A Chinese hamster lung cell line (CHL) was used, with three naphthoquine phosphate dose groups: 2.5, 5.0, and 10.0 μg/mL, the highest being the IC_{50} for cell proliferation. Saline was set as the negative control; mitomycin C, cyclophosphamide (CP), and S9 + CP as the positive controls. The cells were cultured for 24 and 48 h. The rate of chromosomal aberration was less than 2% at the three doses and at both times, including under different conditions. This was less than the control (2.5%) and normal saline (3%) groups. It showed that, under these conditions, naphthoquine had no effect on chromosomal aberration in CHL cells.

G. NAPHTHOQUINE PHOSPHATE TABLETS CLINICAL TRIALS[31]

Phase I trials were carried out at the test bed for national new-drug clinical trials: the PLA's 307 Hospital. Phases II and III were organized by Chongqing Medical University, where national clinical and pharmacological trials for antiparasitic drugs are based. Five medical research units took part in the trials.

1. Phase I Clinical Trials

The aim of the Phase I trials was to determine the dose tolerance and pharmacokinetic parameters of naphthoquine phosphate tablets in healthy people, as a basis for designing safe and effective clinical protocols.

Dose Safety and Tolerability

The selection criteria for trial subjects were:

1. At least middle-school education, volunteer, male, weighing between 55 and 70kg;
2. Normal results for chest X-rays, ECGs, respiratory function tests, routine urine tests, and comprehensive metabolic panels;
3. Good nutritional status; normal blood pressure, respiration and pulse; normal nervous system; normal results on fundus examination; no history of drug allergy; no medical symptoms or alcoholism.

Twenty volunteers were chosen as to the above criteria. Six dose groups and a placebo group were set up, with two to four subjects randomly assigned to each.

Each naphthoquine phosphate trial tablet contained 0.1 g naphthoquine base. Placebo tablets had identical shape, color, and taste. The trial was double-blind; tablets were given before food and their consumption monitored. Doses and times are shown in Table 8.26.

The subjects were hospitalized 2days before medication, with the day of medication set as Day 0. Observation was carried out for 14days, with total time under hospital supervision at 16days. It was carried out as follows:

1. Daily ward rounds were made before and for a week after medication. Subjects were asked about adverse reactions. The type, degree, and duration of these reactions were recorded. Body temperature, blood pressure, respiration, and pulse were measured. After a week, these rounds were made once every 2days.

TABLE 8.26 Dose Tolerance of Naphthoquine Phosphate Tablets

Total Dose (g)	Dose at Different Time (Number of Tablets)			Increment Rate (%)	Number of Volunteers
	0 h	8 h	24 h		
0.4	4	–	–	–	2
0.6	6	–	–	50	2
0.8	6	–	2	35	4
1.0	6	–	4	25	4
1.2	6	3	3	20	2
1.4	6	4	4	17	2
Placebo	6	4	4	–	4

The shortfall of six tablets at 0 h and of four tablets at 8 and 24 h were made up with placebos. Each naphthoquine phosphate tablet contained 0.1 g naphthoquine base.

2. Laboratory tests. Comprehensive metabolic panel with 17 parameters: Total protein, albumin, blood sugar, bilirubin, urea nitrogen, creatinine, cholesterol, triglyceride, phosphocreatine kinase, alkaline phosphatase, glutamic-pyruvic transaminase, glutamic-oxaloacetic transaminase, lactate dehydrogenase, methemoglobin, Ca^{2+}, Na^{+}, K^{+}. Blood tests: hemoglobin, erythrocytes, leukocytes and platelet count, leukocyte classification, reticulocytes. Urine tests: Specific gravity, pH, protein, sugar, acetone bodies, urobilirubin, microscopic examination of sediment in urine. Fecal tests: Occult blood, presence of parasitic ova. Tests were done before medication and 2, 4, 7, and 14 days after.
3. Chest X-rays and respiration tests were done before medication and 14 days afterward. ECGs were conducted at the same time as the comprehensive metabolic panel. If abnormalities were detected 14 days after medication, the corresponding indicators were tested every week until it returned to normal. Due to safety considerations, the trial started with the lowest dose. The next-dose trial would be stopped if obvious adverse reactions or abnormal indicators emerged.

During the observation period, two subjects—one in the control group, the other in the 0.8 g group—showed no increase in weight. All other subjects gained 1–2.3 kg, averaging at 1.4 kg. Blood pressure, respiration, pulse, fundus examination, and nervous system were normal.

Sixteen subjects were given naphthoquine phosphate. One in the 0.4 g group, one in the 0.8 g group, and one in the 1.2 g group experienced slight fullness in the upper abdomen, with increased flatulence. Besides this, one subject in the

1.0 g group also had slight hiccups. All these subjects presented with a small increase in intestinal sounds. Treatment was not required and appetite was not affected. Symptoms disappeared in 2–4 h.

One subject in the 0.8 g group and one in the control group experienced abnormally high levels of phosphocreatine kinase 14 days after medication. They returned to normal within a week. In addition, a subject in the 1.2 g group and another in the 1.4 g group had increased glutamic-pyruvic transaminase (GPT) and glutamic-oxaloacetic transaminase (GOT) levels 7 and 14 days after medication, recovering within 1–2 weeks. All other indicators were normal.

Blood tests showed no abnormalities before and after medication in all dose groups. One subject in the 0.8 g group and one in the 1.4 g group showed significant levels of protein in the urine at Days 4 and 14, which returned to normal in a day.

One subject in the 0.6 g group had abnormal pulmonary function before medication and at Day 14. This presented as a lower than normal FEV1 and FVC. Chest X-rays and ECGs were normal before and after medication.

This Phase I study showed that 1.0 g was a safe dose of naphthoquine phosphate tablets. Preclinical toxicity studies had indicated that the liver was the main target organ. The two large-dose groups in the Phase I trials had temporary increases in GPT and GOT levels. Next-level clinical trials should therefore focus on hepatic function indicators.

Clinical Pharmacokinetics[32]

Each naphthoquine phosphate tablet in this study contained 0.1 g naphthoquine base, with 2,6-bis((tert-butylamino)methyl)-4-((7-chloroquinolin-4-yl)amino) phenol as internal standard. Fourteen healthy volunteers were recruited. They were male, aged 18–37 years, weighed 54–68 kg, and had passed a physical examination.

Six tablets totaling 0.6 g were given orally, taken with 50 mL warm water. Eight subjects were fasted, with the other six medicated on half-empty stomachs. The subjects had breakfast 1 h after medication.

Between 2.5 and 3.0 mL of blood was drawn from an arm vein before medication and 0.33, 0.67, 1, 2, 3, 4, 6, 8, 10, 12, 18, 24, 36, 48, 72, 96, 120, and 168 h thereafter. The plasma was separated using 1 mol/L potassium oxalate anticoagulant and refrigerated. One milliliter of plasma was added to 125 ng internal standard, 0.6 mL phosphate buffer (pH 8) and 2.5 mL solvent (petroleum ether: dichloroethane:isobutanol = 10:7:1). This was shaken and centrifuged to obtain the organic components. Another 2.8 mL solvent was added to the aqueous component, shaken, and centrifuged, then combined with the organic component. The mixture was washed in 0.1 Eq hydrochloric acid (150 μL), the organic component discarded and HPLC conducted on the acid layer.

The HPLC column was 300 mm × 3.9 mm (internal diameter), with YWG-C6H5 adsorbent. Eluent was methanol:triethylamine:water:phosphoric acid at 50:1:50:0.65. Flow rate was 0.5 mL/min, chart speed 25 cm/min, detector wavelength 340 mm. Under these conditions, the lowest detection concentration was 5 ng/mL, with a linear relationship from 7.8 to 500 ng/mL at a coefficient of 0.9999. Average recovery rate was more than 95%. Within-day or day-to-day variation coefficient was about 5%. The data were analyzed using 3p87 software.

Plasma drug concentration peaked in the fasted subjects 2.4 h after medication, at 98.89–245.19 ng/mL. This dropped gradually to 7.56 ng/mL at 120 h. Table 8.27 shows the main pharmacokinetic parameters. The concentration–time relationship followed a two-compartment model. The drug was absorbed quickly, with an average half-life of 1.21 h. It was eliminated slowly, with a half-life of 40.93 h. Distribution was rapid, at a half-life of 2.71 h. The K_{12} constant (0.85680 h) was significantly higher than K_{21} (0.06772 h) and K_{10} (0.30263 h), showing that the drug rapidly entered the tissues.

For the subjects medicated on half-empty stomachs, peak time was 2 h at 96.37 ng/mL, falling gradually to about 5 ng/mL at 168 h. The 3p87 software fitted

TABLE 8.27 Pharmacokinetic Parameters of Naphthoquine Phosphate in Fasted Volunteers ($n = 8$)

Parameter	Unit	Value
A	ng/mL	180.766914
α	h	0.255401
B	ng/mL	39.3141182
β	h	0.016935
Ka	h	0.572743
Lag time	h	0.702543
$V/F(c)$	L/kg	0.698749
$T_{1/2\alpha}$	h	2.713955
$T_{1/2\beta}$	h	40.930374
$T_{1/2}Ka$	h	1.210225
K_{21}	h	0.067722
K_{10}	h	0.302633
K_{12}	h	0.856804
$AUC_{0\sim120h}$	ng·h/mL	2837.3567
CL(s)	mL/h	0.211464
T_{max}	h	2.510000
C_{max}	ng/mL	127.216255

TABLE 8.28 Pharmacokinetic Parameters of Naphthoquine Phosphate in Volunteers With Half-Empty Stomachs ($n=6$)

Parameter	Unit	Value
A	ng/mL	119.11567
α	h	0.15272
B	ng/mL	29.60708
β	h	0.01207
Ka	h	0.52416
Lag time	h	0.32816
$V/F(c)$	L/kg	5.29399
$T_{1/2\alpha}$	h	4.53875
$T_{1/2\beta}$	h	57.43484
$T_{1/2}Ka$	h	1.32239
K_{21}	h	0.04796
K_{10}	h	0.03843
K_{12}	h	0.07840
$AUC_{0\sim120h}$	ng·h/mL	2949.51
CL(s)	mL/h	0.20342
T_{max}	h	4.18
C_{max}	ng/mL	74.65965

the data into the two-compartment model. Table 8.28 shows the main pharmacokinetic parameters. The elimination half-life was 57.43 h and distribution half-life, 4.53 h. Four hours after medication, drug concentration in erythrocytes was 3–4 times higher than in plasma. The plasma protein binding rate was 89%–98%.

2. Phase II Clinical Trials

Phase II trials included tests to determine an effective dose, as well as comparative studies.

Plasmodium falciparum Dose Study

Patients with falciparum malaria were chosen, with parasitemia of no less than 5000 asexual parasites per 1 μL blood. They had to be within 5 days of infection and have no comorbidities, obvious heart, lung, or liver disease, or history of drug allergies. Subjects had to be older than 16 years. Both sexes were included, except for pregnant women. The patients had taken no other antimalarials–or drugs with antimalarial effects, such as sulfonamide—at least half a month before the study.

To ensure patients' safety, observation was halted and other antimalarials given if the following symptoms occurred: Vomiting within half an hour of medication; clear adverse reactions during treatment; and no reduction in parasitemia 48 h after medication.

Sixty patients were randomly divided into three dose groups: 0.6, 0.8, and 1.0 g. Each naphthoquine phosphate tablet contained 0.1 g naphthoquine base. The low dose group was given a single oral dose. The middle- and high-dose groups received an initial dose of 0.6 g, then 0.2 or 0.4 g, respectively, after 24 h. The trial began with the lowest dose, going higher if the anticipated efficacy was not reached.

Using the WHO's 1973 protocol for in vivo chloroquine-sensitivity tests,[33] therapeutic efficacy was based on fever reduction, parasite clearance, and 28-day recrudescence. The criteria were as follows:

- Cure (S): No parasites found by Day 7 and no recrudescence by Day 28;
- Clinical cure (RI): No parasites by Day 7, but with recrudescence by Day 28;
- Effective (RII): Significant drop in the number of parasites, but no clearance in 1 week;
- Failure (RIII): Equal or increased parasitemia 48 h after medication.

The parasites were examined before medication to determine their developmental stage. The number of parasites per 1 μL blood was calculated based on the ratio of parasites to leukocytes. After the first dose, the blood was checked for parasites every 8 h until three consecutive negative results were obtained (no parasites in thick blood smear in 300 fields). The time of the first negative result was taken as the parasite clearance time. Blood tests were done 7, 14, 21, and 28 days after medication to check on recrudescence.

Temperature was measured before medication and then every 4 h afterward until it returned to normal (no rise in temperature within 24 h). The time when temperature returned to normal was recorded. Then, temperature was taken twice daily, once in the morning and again in the afternoon.

Any adverse reactions were recorded daily until the patient left hospital, including the type of reaction, duration, and onset. ECGs and blood, routine urine, and liver function tests were run before medication and 3, 7, and 14 days afterward. If abnormalities were detected, tests continued until the indicator returned to normal.

The three dose groups had 12, 18, and 30 patients, respectively, aged between 18 and 52 years; all had a history of malaria in the last 2 years. At the time of the study, they were in the first to fifth days of illness and had not taken any other antimalarials for the past half-month. None had drug allergies or heart, liver, or kidney disease. Body temperature was 38.5–40.4°C, parasite density 8000–62,000 per μL. The patients in each group had roughly equal body temperatures, immunity, and had similar parasite counts.

TABLE 8.29 Efficacy of Different Doses of Naphthoquine Phosphate Tablets on *P. falciparum*

Dose (g)	Number of Patients	Parasite Clearance Time (h) (M ± SD)	Defervescence Time (h) (M ± SD)	28-Day Cure Rate (%)
0.6	12	73.5 ± 13.7[a]	36.4 ± 11.7	75.0
0.8	18	68.3 ± 24.5[b]	33.6 ± 18.2	94.4
1.0	30	65.7 ± 16.5	37.1 ± 20.4	100.0

[a]*Mean value of 11 subjects.*
[b]*Mean value of 17 subjects.*

No adverse reactions appeared in all the patients after medication, and all indicators were normal. Of the 12 patients in the low-dose group, nine were cured (75%), two clinically cured, and one unaffected. It was not a suitable therapeutic dose. Higher cure rates were achieved in the middle- and high-dose groups (Table 8.29). However, one patient in the 0.8 g group did not show parasite clearance in a week. All patients in the high-dose group were cured. Thus, 1.0 g naphthoquine phosphate was set as the dose for further clinical studies.

Comparative Study on Plasmodium falciparum

In 1989, clinical trials were held in Dongfang county, Hainan to evaluate the therapeutic efficacy of naphthoquine phosphate tablets against *P. falciparum*, as well as any adverse reactions that resulted. Piperaquine phosphate was the control. The age range of patients was extended to include 7-year-olds. All other criteria, indicators, and protocols were the same as those of the dose study.

Each naphthoquine phosphate tablet contained 0.1 g base. Adults took 10 tablets: Six in the initial dose, followed by another four 24 h later. Each piperaquine phosphate tablet contained 0.15 g base. Adults took 10: Six in the initial dose, followed by another two after 6 h and four after 24 h. Children received 20 mg/kg naphthoquine and 30 mg/kg piperaquine, scaled down from the adult doses.

The naphthoquine phosphate and piperaquine phosphate groups had 30 *P. falciparum* patients each. The naphthoquine group contained 26 males and four females. Among these, 22 were locals with partial immunity to malaria. Eight were from elsewhere. Microscope examination showed *P. falciparum* with parasitemia at 27,000 per μL. The piperaquine phosphate group had 22 males and 8 females. Nine were nonlocal and 21 were locals with partial immunity. Average parasitemia, observed under microscope, was 31,000 per μL.

All 30 in the naphthoquine phosphate group were cured. Defervescence and parasite clearance times were 31.2 and 90.3 h, respectively. No recrudescence occurred within 28 days. For the piperaquine phosphate group, 15 of the patients

TABLE 8.30 Therapeutic Efficacy of Naphthoquine Phosphate Tablets and Piperaquine Phosphate Tablets in the Treatment of *P. falciparum*

Drug	Number of Patients	Defervescence Time (h) (M ± SD)	Parasite Clearance Time (h) (M ± SD)	28-Day Cure Rate (%)
Naphthoquine Phosphate	30	31.2 ± 18.0	90.3 ± 21.1	100.0
Piperaquine Phosphate	30	36.9 ± 17.5[a]	107.8 ± 16.1[b]	30.0

[a]*Mean value of 18 subjects.*
[b]*Mean value of 15 subjects.*

experienced no parasite clearance within a week, indicating that drug resistance was at RII–RIII levels. The other patients were observed for 28 days and there were six cases of recrudescence. Therefore, the 28-day cure rate was 30%. It proved that the therapeutic efficacy of naphthoquine phosphate was better than that of piperaquine phosphate (Table 8.30).

There was one case each of nausea, dizziness, and headache in the piperaquine phosphate group. Apart from this, no other adverse reactions were noted in both groups. One patient in the naphthoquine phosphate group had a slight, temporary increase in alanine aminotransferase (ALT). No other aberrations were seen in the blood tests, comprehensive metabolic panels, routine urine tests, and ECGs of all other patients, including the six in the naphthoquine phosphate group who had elevated ALT levels before medication.

3. Phase III Clinical Trials

The purpose of these studies was to expand on existing trials and on the number of clinical observations. Chloroquine phosphate was used as the control, to evaluate the local level of resistance and the efficacy of naphthoquine phosphate on drug-resistant malaria. The drug's efficacy against *P. vivax* and its prophylactic potential were also investigated.

Comparative Study of Naphthoquine Phosphate Tablets and Chloroquine Phosphate Tablets on P. falciparum

Each naphthoquine phosphate tablet contained 0.1 g naphthoquine. Adults were given 10: Six in the initial dose and four 24 h later. Each chloroquine phosphate tablet contained 0.15 g chloroquine. Adults took 10: Four at the initial dose, and two tablets 8, 24, and 48 h later. Patients were randomly assigned to the groups. The selection criteria, therapeutic standards, and test indicators were the same as those of the Phase II trials.

TABLE 8.31 Therapeutic Efficacy of Naphthoquine Phosphate Tablets and Chloroquine Phosphate Tablets Against *P. falciparum*

Drug	Number of Patient	Defervescence Time (h) (M ± SD)	Parasite Clearance Time (h) (M ± SD)	Cure Rate (%)
Naphthoquine phosphate tablets	40	35.9 ± 16.7	61.7 ± 18.1	100.0
Chloroquine phosphate tablets	37	37.5 ± 28.5	63.5 ± 15.2	56.8

All 40 patients in the naphthoquine phosphate group were cured. Average defervescence and parasite clearance times were 35.9 and 61.7 h, respectively, and no recrudescence occurred in 28 days. Of the 37 patients in the chloroquine phosphate group, 28 were clinically cured. Nine experienced no parasite clearance and seven cases of recrudescence occurred in 28 days. The 28-day cure rate was 56.8% (Table 8.31). It indicated that the local level of drug resistance was relatively high and that naphthoquine phosphate tablets were sufficiently effective in areas of high drug resistance. No adverse reactions or abnormal indicators were noted in both groups.

Clinical Observations on 205 P. falciparum Patients Given Naphthoquine Phosphate Tablets

This study investigated the efficacy and toxicity of a 1.0 g regimen, split into two doses. Each naphthoquine phosphate tablet contained 0.1 g naphthoquine. Hainan's Institute of Parasitic Diseases and the South Island Farm Hospital conducted the expanded clinical trials. Patient selection, protocol, and therapeutic standards were the same as with the Phase II trial. All 104 patients at the South Island Farm Hospital were clinically cured, with no recrudescence over 28 days. The clinical cure rate was 100%. Of the 101 patients at the Institute of Parasitic Diseases, 98 were clinically cured. Three experienced no parasite clearance after 1 week; however, checks at 14, 21, and 28 days showed no parasites. One case of recrudescence occurred at Day 17. The clinical cure rate was 99% (Table 8.32).

There were no significant adverse reactions among all the patients. Some complained of headache and dizziness, but symptoms eased as body temperature returned to normal and parasites were cleared. Blood and routine urine tests were normal. Before medication, 80 patients were tested for liver function and

TABLE 8.32 Therapeutic Efficacy of Naphthoquine Phosphate Tablets on 205 *P. falciparum* Patients [1 g regimen]

Hospital	Number of Patients	Defervescence Time (h) (M ± SD)	Parasite Clearance Time (h) (M ± SD)	Cure Rate (%)
Institute of Parasitic Diseases	101	50.1 ± 38.8[a]	58.1 ± 21.2	99
South Island Farm	104	37.6 ± 18.4	66.7 ± 19.6	100

[a]*Average of 98 patients.*

10 were found to have elevated ALT levels. Tests at Days 3 and 7 showed that ALT levels had returned to normal in nine of the patients. One patient's ALT level was still high but returned to normal on Day 14. Patients with normal liver function showed no abnormalities at Days 3 and 7. Urine tests and ECGs were normal before and after medication.

Therapeutic Efficacy of Naphthoquine Phosphate Tablets 0.8 g Regimen on P. falciparum

In the Phase II trial, 1 patient out of the 18 in the 0.8 g naphthoquine group experienced no parasite clearance in 7 days. All 30 in the 1.0 g group showed parasite clearance in 7 days, with no recrudescence in 28 days. This was therefore set as the dose in further trials. However, because expanded trials revealed that some patients had no parasite clearance in 7 days even at the 1.0 g dose, it was decided to investigate the efficacy of the 0.8 g naphthoquine dose further.

Adults were given eight naphthoquine phosphate tablets, each containing 0.1 g naphthoquine. The initial dose was six tablets, followed by another two after 24 h. Children's doses were scaled down based on age. Patient selection, test indicators, and therapeutic standards were the same as in the Phase II trials.

The trial involved 100 *P. falciparum* patients, 71 male and 29 females, aged between 6 and 50 years. Thirty patients were less than 15 years old. Seventy were locals with partial immunity to malaria; the other 30 were from elsewhere. All patients were in the first to fifth day of the illness and had fever of varying degrees. Eleven cases had body temperature of less than 38°C; 22, of 38–39°C; 47, of 39.1–40°C; 20, of more than 40°C. Microscopic examination showed nine cases with parasitemia of less than 10,000 per μL; 48 cases with 10,000–50,000 per μL; 30 cases with 51,000–100,000 per μL; 13 cases with more than 100,000 per μL.

All patients showed no adverse reactions after medication, and test indicators were normal. Average parasite density fell by 75% 24h after medication. Defervescence and parasite clearance times were 46.2 ± 21.6h and 63.4 ± 17.2h, respectively. No recrudescence was observed within 28days and all patients were cured. The results indicated that the cure rate of 0.8g naphthoquine and 1.0g doses were basically identical.

Comparative Study of Naphthoquine Phosphate and Chloroquine Phosphate Tablets on P. vivax

Naphthoquine phosphate tablets containing 0.1g naphthoquine were given once orally. The adult dose was 0.6g (six tablets). The chloroquine phosphate tablets contained 0.15g chloroquine. Adults were given 10 tablets, four at the initial dose and two after 8, 24, and 48h. Child doses were scaled down based on age. Groups were assigned randomly after patients were hospitalized.

Subjects had parasitemia of more than 3000 per μL; all were within 7days of infection. Observation was carried out over 8weeks. The other parameters were the same as in the Phase II trials.

The 35 subjects in each group were all clinically cured. Average defervescence and parasite clearance times for naphthoquine phosphate tablet group were 22.7 and 46.4h, respectively, with no recrudescence within 8weeks. Average defervescence and parasite clearance times for chloroquine phosphate group were 19.1 and 41.6h, respectively. Recrudescence occurred in nine patients, yielding a 56-day cure rate of 74.3% (Table 8.33).

Expanded Trials on Plasmodium vivax

The purpose of the expanded trials was to increase the number of observations and evaluate the efficacy of naphthoquine phosphate tablets in single dose, 0.6g naphthoquine regimen on *P. vivax*. Child doses were scaled down based on age.

Of the 81 patients recruited, 72 came from malaria-endemic areas and had partial immunity. Nine were from elsewhere. There were 56 males and 25 females, aged between 5 and 55years. Eleven were younger than 10years and 23 were aged 10–16. Sixteen patients were in the dormant phase when hospitalized and had no fever. The other 65 had fever of varying degrees: 17 had a body temperature lower than 38°C; 41, from 38 to 40°C; 7, higher than 40°C. Fifty-nine patients had parasitemia of less than 10,000 per μL; 21, of 10,000–30,000 per μL; 1, of more than 30,000 per μL.

All patients were clinically cured, with average parasitemia falling 92.5% after 24h. Defervescence and parasite clearance times were 19.7 ± 9.4h and 47.7 ± 11.1h, respectively. During the 6-week follow-up, one patient tested positive for *P. vivax* on Day 42. The 42-day clinical cure rate was 98.8%.

Malaria Prophylaxis Using Naphthoquine Phosphate Tablets

In the prophylaxis test on mice, it was found that naphthoquine phosphate had a prophylactic effect 1–2weeks longer than corresponding doses of piperaquine

TABLE 8.33 Therapeutic Efficacy of Naphthoquine Phosphate Tablets and Chloroquine Phosphate Tablets on *P. vivax*

Drug	Number of Patients	Defervescence Time (h) (M ± SD)	Parasite Clearance Time (h) (M ± SD)	Results Over 56 days	
				Recurrence	Cure Rate (%)
Naphthoquine phosphate tablets	35	22.7 ± 7.4	46.4 ± 9.3	0	100
Chloroquine phosphate tablets	35	19.1 ± 8.3	41.6 ± 9.2	9	74.3

TABLE 8.34 Thirty-Day Prophylaxis With Naphthoquine Phosphate Tablets

Village	Number of Volunteers	Dose (g)	Number Infected	Incidence Rate in 30 Days (%)
Qianjin	96	0.4	0	0
Zhatao	60	0.5	0	0
Zhaban	52	0.5	0	0
Qianjin[a]	66	0	13	19.7
Zhatao[a]	41	0	3	7.3
Zhaban[a]	33	0	6	18.2

[a]*Control.*

phosphate. From August to October 1991, a study of its prophylactic effects on human malaria was conducted in suburban areas of Sanya, Hainan.

The trial took place on the farms in Sanya's South Island. There, malaria was endemic throughout the year, with falciparum malaria being the most common. The incidence of disease peaked from June to October. *Anopheles dirus* was the main vector. The local population was engaged in rice and rubber planting. Three villages and two rubber planting teams were selected for observation. Most of the villagers were of the Li ethnic group with partial immunity to malaria, whereas the members of the rubber planting teams were from elsewhere and had no immunity. Both groups took few malaria-control measures.

Two prophylactic doses were used. The villages and rubber planting teams were divided into units. Each unit was given one of two doses. Two-thirds of the members in each unit were randomly selected as the no-drug control. Before medication, blood tests were run to determine infection.

The villages had two dose groups: 96 subjects given naphthoquine phosphate tablets totaling 0.4 g naphthoquine and 112 subjects given 0.5 g. None of the subjects were infected during the 30-day observation period. The incidence of infection (Table 8.34) in the controls was 19.7% (Qianjin), 7.3% (Zhatao), and 18.2% (Zhaban).

In the rubber-planting teams, 54 subjects were given 0.4 g and 50 given 0.5 g. They were observed over 45 days. No infections occurred within 30 days; the 45-day incidence rate was 1.9% and 4.0%, respectively. The incidence rates for the two controls were 17.6% and 20.0% (Table 8.35).

The results showed that a single oral dose of 0.4 or 0.5 g naphthoquine had at least a 30-day prophylactic effect during peak seasons in malaria-endemic areas.

Expanded Prophylaxis Trials

Based on the results of the 1991 study, a single dose of naphthoquine phosphate tablets that amounted to 0.4 g naphthoquine was selected for the expanded

TABLE 8.35 Forty-Five-Day Prophylaxis With Naphthoquine Phosphate Tablets

Planting Team	Number of Volunteers	Dose (g)	Number Infected	Incidence Rate in 45 days (%)
Fuping group	54	0.4	1[a]	1.9
Fuhe group	50	0.5	2[a]	4.0
Fuping group	34	0	6	17.6
Fuhe group	35	0	7	20.0

[a] *Infection appeared after 40–43 days.*

trial. From July 10 to August 25, 1994, a single-blind trial was conducted with the residents of Dongfang and Baoting counties in Hainan, where falciparum malaria is endemic. Adults were given a single oral dose of 0.4 g naphthoquine. Children's doses were scaled down based on age. The control group was given a placebo. Subjects were observed for 45 days after medication.

After a single oral dose of 0.4 g, no infections occurred in both locations within a month. The 45-day infection rate was less than 1%. Compared to the control group, the difference was extremely significant ($P < .01$). The expanded trials further confirmed that naphthoquine phosphate tablets had prophylactic effects (Table 8.36).

H. NAPHTHOQUINE PHOSPHATE TABLETS: KEY CLINICAL FINDINGS, DISCUSSION, AND CONCLUSION

The emergence of resistance to chloroquine and other antimalarial drugs has threatened the global fight against malaria. Research into naphthoquine phosphate was initiated to treat and prevent drug-resistant malaria. The Phase II and III clinical trials confirmed that local strains of falciparum malaria were relatively resistant to chloroquine and piperaquine. RI to RII was 36.7%–40.5%; RIII was 2.7%–33.3%. Nevertheless, the cure rate of naphthoquine phosphate in areas of high drug resistance was 99%–100%. It showed that the drug had good therapeutic outcomes against chloroquine- and piperaquine-resistant *P. falciparum*. This was related to its ability to act on the parasite's digestive system and membranes, whereas chloroquine affected the digestive system alone.

The parasite clearance speed of naphthoquine phosphate was lower than that of artemisinin, but it could thoroughly eliminate parasites. Its cure rate was significantly higher than that of artemisinin. Across the various studies, a total dose of 0.8–1.0 g naphthoquine for adults, taken in two doses, achieved a cure rate of 99%–100% in 423 falciparum malaria patients. The drug's therapeutic efficacy was equal to that of mefloquine, but with significantly lower toxicity.

TABLE 8.36 Expanded Trials on the Prophylactic Effect of Naphthoquine Phosphate Tablets

County	Group	Number of Volunteers	Results Within 30 Days		Results Within 45 Days	
			Number of Infected Patients	Infection Rate (%)	Number of Infected Patients	Infection Rate (%)
Dongfang	Drug group	551	0	0	2	0.36
	Control group	546	16	2.9	52	9.5
Baoting	Drug group	540	0	0	1	0.19
	Control group	540	20	3.7	60	11.1

Hainan was selected as the test site for trials on *P. vivax*. Going by the geographical distribution of *P. vivax* strains, the type present in Hainan may be the Chesson strain, which has a short incubation period, frequent recurrence without a lengthy dormant phase, and higher drug resistance than in other strains. A single dose of naphthoquine phosphate containing 0.6 g naphthoquine was given to 35 *P. vivax* patients. Another 35 received a chloroquine phosphate control. Both drugs had good immediate therapeutic effects, with no significant differences in defervescence and parasite clearance times. However, an 8-week follow-up showed no recrudescence in the naphthoquine phosphate group and nine cases of recrudescence in the chloroquine phosphate group (25.7%). It showed that the long-term efficacy of naphthoquine was clearly better than that of chloroquine ($P<.01$). Naphthoquine's biological half-life was at least three times that of chloroquine. Its long-lasting efficacy could be due to its slow elimination, which produced a suppressive effect.

Chloroquine had already been proven to be unable to thoroughly cure *P. vivax*. However, preclinical studies showed that naphthoquine could eliminate the parasite in some animals. Further studies were needed to see if it could do the same for human *P. vivax*. Although it could not target the tissue phase in human malaria and therefore was not a complete cure, naphthoquine's efficacy against *P. vivax* was still superior to chloroquine's, requiring a smaller dose administered only once.

The prophylaxis trial was conducted in an area where drug-resistant falciparum malaria was endemic, and at the peak malaria season. Subjects with partial and no immunity were randomly allocated and compared, which made the trial results especially reliable. Previous tests had shown that naphthoquine phosphate had long-term prophylactic efficacy, 1–2 weeks longer than a corresponding dose of piperaquine. The prophylactic efficacy of piperaquine had been established at 20–30 days. Naphthoquine phosphate at 0.4 and 0.5 g doses of naphthoquine could fully prevent infection within 30 days, with an efficacy rate of 96%–99% in 45 days. Although the number of subjects was limited, the results were reliable due to the time and location of the observations and the rigorous experimental design.

Regarding toxicity, an adult dose of 1.0 g naphthoquine was used to treat 305 falciparum malaria patients; a dose of 0.8 g on 118 falciparum malaria patients; and a dose of 0.6 g on 116 vivax malaria patients. No obvious adverse reactions were seen. Blood and routine urine tests were normal in 151 cases. Liver function and ECGs were normal in 110 cases. Apart from one patient in the 1.0 g group experiencing a slight increase in ALT, no aberrations were found in any of the other indicators. In addition, the rise in ALT may not have been related to the drug. Because the parasite developed in the liver during the tissue phase, this was often accompanied by liver damage, leading to elevated ALT levels. Of the 110 cases tested for liver function, 16 had high ALT levels before medication. This did not increase further after medication, but slowly returned to normal instead. Although preclinical studies showed that the liver was the main target organ in naphthoquine toxicity, there was a strong dose–effect relationship. Moreover, Phase I and pharmacokinetic tests showed no abnormal indicators in the 1.0 g groups. Therefore, 1.0 g may be regarded as a safe dose.

In conclusion, a regimen of naphthoquine phosphate tablets of 0.8–1.0 g naphthoquine—at an initial dose of 0.6 g, followed by 0.2–0.4 g 24 h later—had good therapeutic outcomes against falciparum malaria, including chloroquine- and piperaquine-resistant strains. The cure rate was 99% and above. A single dose of 0.6 g was effective against *P. vivax*, with a 6- to 8-week cure rate of 98.8%–100%. Also, a single 0.4 g dose could efficiently prevent infection within a month.

Note: The following were responsible for pharmacology, toxicology, and clinical research into naphthoquine phosphate.

Chief researcher: Jiao Xiuqing
Pharmacodynamics: Ding Deben, Shi Yunlin, Gao Xusheng, Li Guofu, Wang Shufen
General pharmacology: Ye Hongjin
Pharmacokinetics, animal models: Zu Jing
General toxicology: Wu Boan, Ning Dianxi, Wu Zengdan
Reproductive toxicology: Shi Xiaochun
Mutagenic toxicity: Wang Zhiqiao, Gao Peiyong
Clinical research: Liu Guangyu, Wang Shufan, Pang Xuejian, Fu Linchun, Chen Jifeng
Clinical pharmacokinetics: Lu Zhiliang

8.2 ARTEMISININ–NAPHTHOQUINE PHOSPHATE COMBINATION

Wang Jingyan, Ding Deben, Li Guofu
Institute of Microbiology and Epidemiology, Academy of Military Medical Sciences, Beijing, China

A. PRECLINICAL PHARMACOLOGY STUDIES

1. Pharmacodynamics

Partner Drug Dosage Compatibility on Rodent Malaria

Several doses of naphthoquine phosphate and artemisinin were used in an in vivo study on K173 strain *P. berghei*. The compatible doses were determined via an orthogonal test design. Mice were given the drugs intragastrically. The ED_{90} of each combination was calculated based on the Peters standard 4-day suppressive test.[7] This also established the dose at which synergistic effects occurred. An additive graphical model was used to assess each combination.

The results showed that the ED_{90} of naphthoquine phosphate and artemisinin were 0.63 and 42.76 mg/kg, respectively. A synergistic dose (where the ED_{90} of the combination was smaller than the ED_{90} of each drug) ranged between 0.2 and 0.4 mg/kg naphthoquine phosphate and 10–20 mg/kg artemisinin. The ED_{90} of the different combinations were all close to or below the regression line, showing that both drugs had synergistic effects.

Optimized Dosage Ratio for Rodent Malaria

An in vivo study was conducted with chloroquine-sensitive and chloroquine-resistant (RCQ) *P. berghei* strains. Mice were given 0.2–0.4 mg/kg naphthoquine phosphate and 10–20 mg/kg artemisinin, with the ED_{90} and ED_{50} measured using the 4-day suppressive test. The synergistic index was calculated based on the ED_{50} or ED_{90} of the individual drugs compared to those of the combination. The ideal ACT would have a synergistic index of higher than two.

The results showed that a ratio of 1:50 naphthoquine phosphate to artemisinin had ED_{50} and ED_{90} synergistic indices above two. In other words, the ideal naphthoquine phosphate-to-artemisinin ratio for rodent malaria was 1:50. Its ED_{90} synergistic index was 4.2 for the sensitive strain (Table 8.37) and 8.2 for the resistant strain (Table 8.38).

Parasite Clearance Rate

Mice were infected with *P. berghei*. When parasitemia reached 5%–10%, they were given one dose intragastrically. All three groups were given 20 times the ED_{90} of naphthoquine phosphate, artemisinin, or the 1:50 naphthoquine phosphate–artemisinin combination. Caudal blood smears were obtained every 6 h to measure parasite density, and the time taken for parasitemia to fall by 90% was recorded.

The time taken for a 90% decrease in parasitemia was 25.9 h for the combination group. Compared to the artemisinin group, at 24.1 h, the difference was not statistically significant ($P > .1$), but the difference with the naphthoquine phosphate group, at 31.1 h, was obvious ($P < .001$). Parasite clearance time was 49.8 h for the combination group, with no significant difference from the artemisinin group (51.4 h, $P > .05$), but a clear difference from the naphthoquine group (65.3 h, $P < .001$). The rate of parasite clearance was 100% in the combination group and 80% in the other two groups (Table 8.39).

Cure Rate on Rodent Malaria

Mice were infected and given the drugs once intragastrically and observed for 30 days. Caudal blood smears were examined for the parasite. If no parasites were seen after 30 days, this was considered a complete cure. The combination, naphthoquine phosphate, and artemisinin groups were further divided into different dose groups: 5, 10, 20, and 30 times the respective drug's ED_{90}. The combination and naphthoquine phosphate groups achieved a 100%

TABLE 8.37 Naphthoquine Phosphate (NQ), Artemisinin (QHS), and Their Combinations on a Sensitive Strain of *P. berghei*

Drug Ratio	Drug	$ED_{50} \pm SD$ (mg/kg/d)	$ED_{90} \pm SD$ (mg/kg/d)	Synergistic Index	
				ED_{50}	ED_{90}
	QHS	17.33±5.23	34.57±3.48	–	–
	NQ	0.39±0.10	0.76±0.18	–	–
NQ+QHS					
1:100	QHS	10.58±1.95	23.29±1.17	1.6	1.5
	NQ	0.11±0.02	0.23±0.01	3.5	3.3
1:50	QHS	8.35±3.84	17.31±3.56	2.1	2.0
	NQ	0.17±0.08	0.34±0.06	2.3	2.2
1:25	QHS	7.56±3.48	11.72±2.76	2.3	2.9
	NQ	0.31±0.15	0.48±0.12	1.3	1.6

TABLE 8.38 Naphthoquine Phosphate (NQ), Artemisinin (QHS), and Their Combinations on a Chloroquine-Resistant Strain of *P. berghei* (RCQ)

Drug Ratio	Drug	$ED_{50} \pm SD$ (mg/kg/d)	$ED_{90} \pm SD$ (mg/kg/d)	Synergistic Index	
				ED_{50}	ED_{90}
	QHS	74.76±21.55	280.18±29.37	–	–
	NQ	0.75±0.15	9.69±2.77	–	–
NQ+QHS					
1:100	QHS	29.23±5.24	155.21±57.76	2.6	1.8
	NQ	0.28±0.07	1.55±0.58	2.7	6.3
1:50	QHS	22.31±1.57	94.05±4.34	3.4	3.0
	NQ	0.45±0.04	1.88±0.08	1.7	5.2
1:25	QHS	15.94±3.40	64.86±4.43	4.7	4.3
	NQ	0.84±0.15	2.37±0.14	0.9	4.1

TABLE 8.39 Parasite Clearance Rate of Naphthoquine Phosphate (NQ), Artemisinin (QHS), and Their Combination (NQ + QHS) on a Sensitive Strain of *P. berghei*

Drug	$20 \times ED_{90}$ (mg/kg)	Time Taken for 90% Decrease in Parasitemia (h) (M±SD)	Parasite Clearance Time (h) (M ± SD)	Rate of Parasite Clearance (%)
QHS	691.4	24.1 ± 2.8	51.4 ± 5.2	80.0
NQ	15.2	31.1 ± 2.8	65.3 ± 4.1	80.0
NQ + QHS	6.8 + 346.2	25.9 ± 2.9	49.8 ± 7.1	100.0

30-day cure rate at the $30 \times ED_{90}$ dose. The rate for the artemisinin group was 0%. At the $20 \times ED_{90}$ dose, the cure rate was 75% for the combination and 79% for the naphthoquine phosphate group. It indicated that the combination and naphthoquine phosphate cure rates were similar (Table 8.40). However, the combination achieved the same efficacy at only half the naphthoquine phosphate dose.

Cross-Resistance Study in Plasmodium berghei

A 4-day suppressive test was run on different strains of *P. berghei*: sensitive, highly chloroquine-resistant, highly naphthoquine phosphate-resistant, and mildly artemisinin-resistant. The resistance index (I_{90}) was assessed based on the ratio of the ED_{90} of the resistant strain to the ED_{90} of the sensitive strain ($I_{90} < 2$, sensitive; $2 < I_{90} < 10$, slightly resistant; $10 < I_{90} < 100$, moderately resistant; $I_{90} > 100$, highly resistant). The results are in Table 8.41.

Out of all the drugs, the combination had the lowest resistance index against the chloroquine-resistant strain. Therefore, the combination could reduce the degree of cross-resistance between naphthoquine phosphate, artemisinin, and chloroquine. The resistance index of naphthoquine phosphate on the naphthoquine phosphate-resistant strain was 312.3, a very high level. However, the index for the compound was lowered to 3.2, a drastic difference when compared to the single-drug group.

The resistance indices of the following eight drugs were measured against chloroquine-resistant *P. berghei* (RCQ): Naphthoquine phosphate, artemisinin, the naphthoquine phosphate–artemisinin combination, chloroquine phosphate, amodiaquine, piperaquine phosphate, artemether, and lumefantrine. This would determine the combination's degree of cross-resistance to chloroquine-resistant malaria. The results can be seen in Table 8.42. RCQ was a highly

TABLE 8.40 Cure Rates of Naphthoquine Phosphate (NQ), Artemisinin (QHS), and Their Combination (NQ+QHS) on *P. berghei*

Drug	Dose (mg/kg)	Multiples of ED_{90} Dose Equivalence	Number of Mice	Number of Mice Cured	Cure Rate in 30 days (%)
QHS	1037.1	30	24	0	0
	691.4	20	24	0	0
	345.7	10	24	0	0
	172.9	5	24	0	0
NQ	22.8	30	24	24	100.0
	15.2	20	24	19	79.2
	7.6	10	24	16	66.7
	3.8	5	24	0	0
NQ+QHS	10.2+519.3	30	24	24	100.0
(1:50)	6.8+346.2	20	24	18	75.0
	3.4+173.1	10	24	0	0
	1.7+86.6	5	24	0	0

TABLE 8.41 Resistance Indices of Naphthoquine Phosphate (NQ), Artemisinin (QHS), and Their Combination (NQ+QHS) on Chloroquine-, Artemisinin-, and Naphthoquine Phosphate-Resistant Strains of *P. berghei*

Drug	Resistance Index (I_{90})		
	Chloroquine-Resistant Strain	Artemisinin-Resistant Strain	Naphthoquine-Resistant Strain
QHS	8.1	6.1	2.4
NQ	12.8	1.1	312.3
NQ+QHS	5.4	2.3	3.2

chloroquine-resistant strain (I_{90}>170) and had a moderate resistance to naphthoquine phosphate (I_{90}=25.4). However, it showed milder resistance to the combination (I_{90}=3.0), demonstrating that the combination could reduce naphthoquine phosphate's degree of cross-resistance to chloroquine-resistant malaria.

TABLE 8.42 Resistance Indices for Chloroquine-Resistant Strain of *P. berghei*

Drug	ED_{90} for Sensitive Strain (mg/kg)	ED_{90} for Chloroquine-Resistant Strain (mg/kg)	I_{90}
Chloroquine phosphate	2.12 ± 0.26	>360	>170
NQ + QHS	19.72 ± 1.07	59.83 ± 4.43	3.0
Naphthoquine phosphate	0.60 ± 0.05	15.21 ± 5.07	25.4
Artemisinin	43.36 ± 3.57	184.90 ± 21.14	4.3
Amodiaquine	1.67 ± 0.02	51.95 ± 3.21	31.1
Piperaquine phosphate	1.67 ± 0.18	6.27 ± 1.51	3.8
Artemether	4.21 ± 0.18	23.29 ± 1.05	5.5
Lumefantrine	2.40 ± 0.19	23.16 ± 0.58	9.7

Delaying Emergence of Drug Resistance in Plasmodium berghei

A drug-resistant strain of *P. berghei* was cultivated by progressively increasing small doses of the drug over multiple passages of the parasite. Parallel tests were then run with the individual drugs and the combination to determine the speed and degree of resistance. The experiment was run over 700 days, with 100 passages of the parasites cultivated. At the 20th passage, the parasites were still susceptible to the combination (I_{90} = 1.5), slightly resistant to artemisinin (I_{90} = 4.4), and moderately resistant to naphthoquine phosphate (I_{90} = 32.8). The combination could delay the onset of resistance (Table 8.43). By the 100th passage, the resistance index of the combination was lower than those of the individual drugs, proving that the combination lowered the degree of resistance.

Optimized Dosage Ratio for Simian Malaria

The dose in this study was based on the findings that an ideal synergistic dose of naphthoquine phosphate to artemisinin for mice was 1:50. It was known that the human dose of naphthoquine phosphate was 20 mg/kg, 7.9 times the therapeutic dose in mice (2.52 mg/kg) and that the human dose of artemisinin was 50 mg/kg, 0.29 times the dose in mice (171.04 mg/kg). Based on the 1:50 ratio, it was deduced that the ideal dose in humans was (1 × 7.9):(50 × 0.29), or approximately 1:2.

The fact that the pharmacodynamics of human and simian malaria was more closely aligned, the simian malaria test used a combination ratio of 1:2.

TABLE 8.43 Naphthoquine Phosphate (NQ), Artemisinin (QHS), and Their Combination (NQ + QHS) on *P. berghei* With Cultivated Resistance

	Resistance to NQ (RNQ)		Resistance to QHS (RQHS)		Resistance to NQ + QHS (RNQ + RQHS)	
Passage	ED_{90}	I_{90}	ED_{90}	I_{90}	ED_{90}	I_{90}
0	0.76	–	34.57	–	17.71	–
20	24.94	32.8	153.32	4.4	26.51	1.5
50	133.40	175.5	100.00	2.9	104.19	5.9
100	152.23	200	194.85	5.6	77.29	4.4

However, because human and simian malaria were caused by different parasitic species, the 1:2 ratio was adjusted downward to 1:1 to yield another potential dose. A third dose group was initially set at 1:3, but the proportion of artemisinin was judged to be too high and the ratio reduced to 1:2.5. This would ensure the success and reliability of the test.

Three dose groups were created using 10 mg/kg naphthoquine phosphate with 10, 20, and 25 mg/kg artemisinin, respectively. The individual-drug groups were given 10 mg/kg naphthoquine phosphate or 30 mg/kg artemisinin. Each group had three rhesus monkeys infected with *P. knowlesi.* They were given a single oral dose and observed for 105 days.

After 24 h, there was no significant difference in the rate of parasite reduction among the three combination groups or between the combination and single-drug groups. Parasite clearance times were similar in the three combination groups, which were 13–16 h shorter than the naphthoquine phosphate group. The artemisinin group saw no parasite clearance (Table 8.44). Over 105 days, two monkeys in the 1:1 and 1:2 groups were cured, as opposed to all the monkeys in the 1:2.5 group. None of the monkeys in the single-drug groups were cured. Even at less than the effective dose, the combination had synergistic effects. A suitable combination dose ratio was 1:2.5.

2. General Pharmacological Tests

The study focused on the effects of the naphthoquine phosphate–artemisinin combination on the nervous, cardiovascular, and respiratory systems of mice, rats, and anesthetized cats.

For mice, the dose was derived from the ED_{90} of naphthoquine phosphate established in the 4-day suppressive test, which was 0.76 mg/kg/day.

TABLE 8.44 Therapeutic Effects of Single-Dose Naphthoquine Phosphate (NQ), Artemisinin (QHS), and Their Combination on *P. knowlesi*

Drug	Dose (mg/kg)	Drug Ratio	Decrease in Parasitemia Over 24h (%) (M ± SD)	Clearance Time (h) (M ± SD)	Number Cured in 105 days/Total
NQ+QHS	10+10	1:1	98.7 ± 1.1	56.0 ± 16.0	2/3
	10+20	1:2	98.7 ± 1.0	53.3 ± 4.6	2/3
	10+25	1:2.5	99.0 ± 0.7	56.0 ± 8.0	3/3
NQ	10	–	95.2 ± 3.7	69.3 ± 4.6	0/3
QHS	30	–	96.6 ± 3.4	No clearance	0/3

The total dose of naphthoquine phosphate was calculated as 0.76 mg/kg/d × 4 = 3.04 mg/kg. The artemisinin dose was 2.5 times that at 7.6 mg/kg. The study included two combination-dose groups with both partner drugs at a small or large dose. The small-dose group had 3.04 mg/kg naphthoquine phosphate and 7.6 mg/kg artemisinin, respectively, based on the ED_{90} of naphthoquine phosphate. The large-dose group contained 10 times the ED_{90} of naphthoquine phosphate, or 30.4 mg/kg naphthoquine phosphate, and 76 mg/kg artemisinin, respectively. The combination-dose groups were set at 106.4 and 10.64 mg/kg. A control group consumed an identical volume of water. The effects of these doses on spontaneous activity and heat-induced pain were studied.

Based on body surface area,[17] the dose values for mice were extrapolated to rats and cats and converted to mg/kg. The combination doses for rats were 74.6 and 7.46 mg/kg. Body temperature tests were carried out on the rats. For cats, the two combination doses were 31.6 and 3.16 mg/kg. ECGs and blood pressure, heart rate, and respiratory tests were conducted on the anesthetized cats before medication and at various intervals thereafter.

These studies showed that a single oral dose of the combination drug, at 10 times the ED_{90} of naphthoquine phosphate, had no impact on nervous, cardiovascular, and respiratory systems.

3. Acute Toxicity

Acute Toxicity in Mice

Mice were given a single oral or intraperitoneally injected dose of the naphthoquine phosphate–artemisinin combination. Its median lethal dose (LD_{50}) was

TABLE 8.45 LD of the Naphthoquine Phosphate-Artemisinin Combination in Mice (mg/kg)

Administration Method	LD_{50}	LD_5	LD_{95}
Oral	2907	1718	4745
	(2670~3166)	(1467~2163)	(3950~5700)
Intraperitoneal	428	239	768
	(381~482)	(192~297)	(617~957)

Twenty mice per group, equal number of males and females. Data in brackets at 95% confidence interval.

2907 mg/kg (2670–3166 mg/kg) for the oral and 428 mg/kg (381–482 mg/kg) for the injectable dose (Table 8.45). By toxicity rating standards, the combination's toxicity was relatively low. Its oral chemotherapeutic index was 341. Therefore, it had a broad therapeutic window.

The LD_{50} of naphthoquine and artemisinin were 897 mg/kg (828–972 mg/kg) and 4475 mg/kg (4022–4979 mg/kg), respectively (Table 8.46). Their estimated combined toxicity was calculated using the Finney formula:[34]

$$\frac{1}{\text{calculated } \mathbf{LD_{50}}} = \frac{\text{a}}{LD_{50}\,(\text{A})} + \frac{\text{b}}{LD_{50}\,(\text{B})} + \cdots$$

Drugs A and B are used in a combination at percentages a and b, respectively. The ratio of naphthoquine phosphate to artemisinin in this combination was 1:2.5. The LD_{50} of naphthoquine (Drug A) was 879 mg/kg. That of artemisinin (Drug B) was 4475 mg/kg. The percentages of the two drugs were 28.6% ($a = 1/3.5$) and 71.4% ($b = 2.5/3.5$), respectively.

$$\frac{1}{\text{calculated } \mathbf{LD_{50}}} = \frac{28.6\%}{879\,(\text{NQ})} + \frac{71.4\%}{4475\,(\text{QHS})} + \cdots = 0.2857/897 + 0.7143/4475$$
$$= 0.0003185 + 0.0001596 = 0.0004781$$

The estimated LD_{50} of the compound: 1/0.0004781 = 2092 mg/kg. Actual LD_{50} of the compound: 2907 mg/kg.

The actual and estimated LD_{50} values were compared. If the ratio was higher than 2.7, it meant that both drugs had synergistic toxicity. A ratio of less than 0.4 indicated antagonistic toxicity. Ratios of 0.5–2.6 were considered additive toxicity. In this case, the actual-to-estimated LD_{50} ratio was 0.71, showing that the compound had only a slight increase in toxicity over the individual drugs.

TABLE 8.46 LD of Oral Naphthoquine Phosphate (NQ) and Artemisinin (QHS) in Mice (mg/kg)

Drug	LD_{50}	LD_5	LD_{95}
NQ	897	583	1379
	(828~972)	(491~691)	(1182~1609)
QHS	4475	2592	7727
	(4022~4979)	(2062~3257)	(6111~9769)

Twenty mice per group, equal number of males and females. Data in brackets at 95% confidence interval.

Acute Toxicity in Rats

Rats were given one oral dose of the combination. The LD_{50} (Table 8.47) was 3197.1 mg/kg (2816.9–3628.6 mg/kg). The main symptoms were tremors and other aberrations of the nervous system.

Acute Toxicity in Dogs

The method of Deichmann et al. of progressive increases from the 50% dose was used.[25] Six beagles were randomly divided into six dose groups, given single oral doses, and observed over 14 days. Symptoms and time of death were recorded. The maximum tolerated dose (MTD) was taken as the highest dose at which no signs of toxicity were seen. The ALD was the lowest dose at which death occurred.

The MTD and ALD of the combination were 160 and 550 mg/kg, respectively. The first signs of intoxication appeared in the gastrointestinal tract, in the form of vomiting. The vomitus consisted mainly of mucus and stomach contents. At higher doses, neurological symptoms also appeared in the form of serious tonic spasms. Death occurred mainly after aberrations in the nervous system, with the seizures causing respiratory failure. Autopsies showed no visible abnormalities in the main organs.

4. Long-Term Toxicity

Long-Term Toxicity in Rats

Three dose groups of the combination and a solvent control were used. The doses were calculated based on the long-term naphthoquine phosphate toxicity study—80, 40, and 20 mg/kg/d—as well as the optimal ratio of the naphthoquine phosphate–artemisinin combination, 1:2.5. Therefore, the dose of artemisinin was set at 2.5 times that of naphthoquine phosphate. The combination doses were derived by adding up the corresponding doses of the two component

TABLE 8.47 LD of the Oral Naphthoquine Phosphate–Artemisinin Combination in Rats (mg/kg)

Sex	LD_5	LD_{50}	LD_{95}
Female	2229.9	3616.7	5866.1
	(1708.5~2910.3)	(3123.3~4188.1)	(4490.8~7662.6)
Male	1228.3	2796.5	6366.7
	(738.6~2042.8)	(2258.7~3462.4)	(4204.5~9641.0)
Female+Male	1622.4	3197.1	6300.2
	(1250.9~2104.3)	(2816.9~3628.6)	(4932.5~8047.1)

Twenty rats per group, equal number of males and females. Data in brackets at 95% confidence interval.

drugs: 280, 140, and 70 mg/kg/d. Wistar rats were randomly allocated to these groups. Each group had 24 rats, with an equal number of males and females. The drugs were administered intragastrically every day over 14 days. The control group was given an equal amount of buffer. On Day 15, two-thirds of the rats of each group were killed (eight males, eight females). The remainder were observed over a recovery period and then killed at Day 42.

Clinical signs were recorded daily over 42 days. Blood tests (seven parameters), comprehensive metabolic panels (12 parameters), pathology examinations (gross anatomy, weighing of 11 organs, histology examination of 33 tissue blocks), and bone marrow tests (smear and biopsy) were conducted on Days 15 and 42. An additional reticulocyte count was done on Day 7.

On Day 4, behavioral symptoms of a reduction in activity and weakness appeared in the 280 and 140 mg/kg/d groups. The rats in the large-dose group developed diarrhea, anorexia, and weight loss. One male rat died on Day 12. No abnormalities appeared in the 70 mg/kg/d and control groups. Blood tests showed that the reticulocyte count had fallen in the large-dose group ($P<.01$), with an increase in white blood cells and neutrophilic granulocytes. All other indicators were normal in the high- and medium-dose groups. There were no significant differences between the low-dose and control groups. The comprehensive metabolic panel showed that the high-dose group had elevated AST, ALP, and cholesterol levels. AST and cholesterol levels were also high in the middle-dose group. However, there were no other significant changes, as well as no clear differences between the low-dose and control groups.

Autopsies revealed that the livers of the rats in the 280 mg/kg/d group were yellow and greasy. Some had abscesses and patchy or focal necrosis. The relative weights of the brain, lungs, liver, spleen, and kidneys had increased. The weight of the thymus also increased in the 140 mg/kg/d group. The low-dose and control groups showed no abnormalities. Histology tests indicated that 9 out of the 16 rats (56.3%) in the high-dose group had confluent bronchopneumonia. A few cases had purulent foci, some of which had merged, alongside an increase in foam cells. All 16 showed inflammation and degeneration of the liver cells, including cell enlargement and ballooning degeneration. Four had large areas of necrotic foci, six had a slight increase in reticular cells in the thymus and lymph nodes, and five had mild interstitial nephritis. These symptoms appeared in the medium-dose group but to a smaller degree. No pathologies appeared in the low-dose and control groups.

Bone marrow smears showed a clear impact on red cell maturation and mitotic indices ($P<.001$) in the high-dose group. Nuclei were faded, fragmented, and shrunken. Significant damage appeared in the myeloid cells at every stage ($P<.001$), in the form of swollen, spinous, fragmented, and faded nuclei and hydropic change. There were no abnormalities in the other groups. All groups showed no change to the lymphocytes, reticular cells, megakaryocytes, plasma cells, and monocytes. Bone marrow biopsies showed small sites of necrosis in one rat in the 140 mg/kg/d group and two rats in the 280 mg/kg/d group, including loosening of the bone marrow tissue and reduction in hematopoietic cells.

One rat in the 280 mg/kg/d group died on Day 3 after medication ceased, and another on Day 6. Six days after medication ceased, the surviving animals in that group and those in the 140 mg/kg/d group showed no signs of intoxication. Food intake and weight started to increase. On Day 42, blood tests and comprehensive metabolic panels showed no significant differences with the control. There was marked recovery of the lung, liver, spleen, kidney, heart, and bone marrow damage in the high-dose group, although they had not returned to normal. All other indicators were normal and no new damage observed.

The lethal oral dose of the naphthoquine phosphate–artemisinin combination was taken as 280 mg/kg/d over 14 days. The 140 mg/kg/d dose produced moderate intoxication, and 70 mg/kg/d was basically safe. The target organs were the liver and bone marrow. AST, ALT, and cholesterol levels, as well as reticulocyte count, could serve as indicators of intoxication. The damage produced was reversible.

Long-Term Toxicity in Dogs[35]

Dogs were divided into three dose groups and one control. The previous long-term toxicity test for naphthoquine phosphate in dogs indicated that the safe dose was 5 mg/kg/d. Therefore, this was taken as the low dose. Using the 1:2.5 ratio, the dose of artemisinin was set at 12.5 mg/kg/d. The combination's low-dose group was thus 17.5 mg/kg/d. This was multiplied by eight to yield a high dose of 140 mg/kg/d. The middle dose was 87.5 mg/kg/d. Each group had six beagles, with an equal number of males and females. They were given a single oral dose daily for 14 days. On Day 15, two-thirds were killed for pathology examination. The other third was observed over 28 days and terminated thereafter.

The behavioral symptoms of the dogs were observed. Blood tests (eight parameters), comprehensive metabolic panels (13 parameters), urine tests (eight parameters), and ECGs (nine leads) were conducted 7 days before medication, on the day of medication itself and at 7, 14, 28, and 42 days. Fundus examinations were done before and after medication. At autopsy, 12 organs were weighed and 43 tissue block samples—including bone marrow—taken for histology tests.

From the day of medication to Day 10, five dogs in the 140 mg/kg/d group displayed nausea, vomiting, and anorexia. Three dogs experienced paroxysmal tonic spasms from Day 1 to Day 3, which lasted 2–3 min and eased spontaneously. All dogs showed anorexia on Days 7 and 14 ($P<.05$), with weight loss on Day 14. One dog in the 87.5 mg/kg/d group had nausea and vomiting after medication. This group also had anorexia at Days 7 and 14 ($P<.05$). No obvious adverse reactions occurred in the 17.5 mg/kg/d group. The combination had no significant effect on body temperature, heart rate, and respiration across all groups.

In the 140 mg/kg/d group, white blood cell count rose on Days 14 and 28. Neutrophilic granulocytes increased on Days 7 and 14. All other indicators fluctuated within normal range.

On Days 7 and 14, ALT and AST levels in the 140 and 87.5 mg/kg/d groups showed a marked increase. Other indicators were within normal bounds. A prolonged Q–T interval was seen on Days 7 and 14 in the 140 mg/kg/d group ($P<.05$). Other indicators were normal.

One dog in the 140 mg/kg/d group showed hematopoietic suppression at every stage, with a marked decrease in cell count. The nuclei of early erythroblasts were unevenly pigmented and had partly degenerated, while those of later erythroblasts ruptured. Some myeloblasts had faded nuclei, which formed vacuoles. Later myeloblasts degenerated, with dense, purple, toxic granules in the cytoplasm. Neutrophil nuclei were shrunken and spinous; some were swollen and discolored. Vacuoles formed in the cytoplasm of the band granulocytes. Reticular cells were swollen, with faded nuclei, and phagocytosed large numbers of nucleated erythrocytes, nucleus fragments, and fat droplets. Black, refractive, round grains of varying sizes appeared in plasma cell cytoplasm. Slight swelling occurred in megakaryocyte nuclei. In the 87.5 mg/kg/d group, the nuclei of red blood cells showed slight dissipation and shrinkage. Mild vacuolar degeneration was seen in the nuclei of myeloblasts, but without clear toxic granules in the cytoplasm. Phagocytosis by reticular cells of erythrocytes, lymphocytes, and nuclear fragments was less evident. No significant changes occurred in the 17.5 mg/kg/d and control groups.

Both superficial and cross-sectional examination of the livers of the dogs in the 140 mg/kg/d group showed yellow grease and white, spotlike nodules. The lungs of a few of the dogs had dark red spots or patches. Liver, spleen, and kidney mass increased. The same yellow grease was noted in the livers of the dogs in the 87.5 mg/kg/d group, though to a lesser extent. Liver mass also increased. Histological examination found obvious pathological changes in the livers of all dogs in the 140 mg/kg/d group. Liver cell swelling was widespread, with foci or areas of nodular necrosis in the liver lobules. All dogs had slight interstitial inflammation in the lungs, where foam cells could be seen. They also had vacuolar degeneration and a small amount of protein in the epithelial cells of the convoluted tubules, where the renal cortex and medulla meet. Pathologies of the immune system were also marked in all the dogs of that group, presenting as enlargement of germinal centers in the lymphoid follicles and abnormal proliferation of monocytes and macrophages. The tissues of the thymus were disordered, with a reduction in the cortex. These pathologies lessened after medication ceased but had not returned to normal in 28 days. Scattered necrotic loci could still be seen in the liver lobules, and slight abnormalities persisted in the lymph nodes and lungs. The 87.5 mg/kg/d group experienced similar pathologies but to a lesser degree.

The livers of the dogs had basically returned to normal after 28 days, with occasional foci of nodular necrosis. Damage to the immune system had markedly improved. No obvious pathologies were seen in the 17.5 mg/kg/d and control groups. Clotting time, routine urine tests, and fundus examinations were normal.

For the naphthoquine phosphate–artemisinin combination, a severely toxic dose was 140 mg/kg/d delivered orally over 14 days. A similar regimen of 87.5 mg/kg/d was moderately toxic, whereas 17.5 mg/kg/d was basically safe. The target organs were the liver and bone marrow, and AST and ALT levels could be used as indicators. Furthermore, the resulting pathologies could be reversed and no new damage was seen.

B. CLINICAL TRIALS

1. Phase I Clinical Trial on Tolerance and Safety

China's General Administration of Food and Drug approved clinical trials for the artemisinin–naphthoquine phosphate (ACT) tablets in 1998. Each artemisinin–naphthoquine phosphate ACT tablet was 175 mg, containing 50 mg naphthoquine and 125 mg artemisinin. These were conducted at the clinical pharmacology test center of the PLA's 307 Hospital. Twenty-eight healthy male volunteers were randomly divided into four dose groups: 350 mg (two tablets), 700 mg (four tablets), 1400 mg (eight tablets), and 2100 mg (12 tablets). There were four volunteers in the first group and eight in all the others. A single oral dose was administered. The trial began with the lowest dose, proceeding to the next dose after a week. Clinical observations and routine tests were performed daily. Comprehensive metabolic panels (17 parameters), blood tests (13 parameters), urine tests (eight parameters), and ECGs (nine leads) were run before medication and 24 and 72 h thereafter. The trial proceeded with the next dose only if the previous one was safe. Each volunteer received only one dose.

TABLE 8.48 Biodata of Volunteers in the Oral Single Dose Artemisinin–Naphthoquine Phosphate Tablets Trial

		M±SD			
Dose Group	Number of Volunteers	Age (year)	Height (m)	Weight (kg)	Body Mass Index (kg/m²)
350mg	4	36.0±3.2	1.64±0.02	59.3±3.6	22.0±1.4
700mg	8	32.9±3.1	1.69±0.04	61.5±5.0	21.5±1.4
1400mg	8	37.5±2.0	1.67±0.07	60.4±6.6	21.7±1.3
2100mg	8	35.9±2.4	1.71±0.07	63.8±7.7	21.8±1.3
Total	28	35.6±1.9	1.68±0.03	61.2±1.9	21.7±0.2
	Range	30~39	1.60~1.78	51~75	20.2~24.0

The volunteers had BMIs of 19–24, were aged 30–40 years, and had no history of smoking. They were given a comprehensive physical examination—including liver and kidney function, routine blood and urine tests, clotting function, ECGs, and serological tests—to establish that all indicators were normal. Written declarations of informed consent were signed. The details of the volunteers are presented in Table 8.48.

All subjects completed the trial to plan, allowing for safety assessments to be made. Out of the adverse reactions observed, two may be linked to the ACT. One member of the 350mg dose group experienced a slight increase in ALP and AST levels on the first and third days after medication. This was only temporary, as the levels returned to normal on Day 7. One subject in the 2100mg group had elevated ALP on Day 3 after medication, which also returned to normal on Day 7. This could be related to the naphthoquine phosphate in the artemisinin–naphthoquine phosphate tablets, but no dose–response relationship was observed and recovery was spontaneous.

In all dose groups, the average body temperature, heart rate, respiration, and blood pressure did not change significantly before and after medication. No clinical aberrations were seen on the ECGs before medication and at 1, 24, and 72h afterward. In sum, no clinical changes were seen in the blood and urine tests, comprehensive metabolic panels, ECGs, and vital signs of healthy subjects after a single oral dose of 350–2100mg artemisinin–naphthoquine phosphate tablets. The combination was safe and well-tolerated.

2. Pharmacokinetics of Single Oral Dose in Healthy Volunteers

The Phase I pharmacokinetics trial was carried out at the same center as the safety trial. Forty healthy male volunteers were selected for this artemisinin–naphthoquine phosphate ACT single-dose open trial. Thirty of them were divided into three dose groups of 10 each: 700, 1400, and 2100 mg. The volunteers were fasted for 10 h. After emptying their bladders, they were given a single oral dose of the ACT with 250 mL lukewarm water. Before medication and at various times afterward, 5 mL blood was drawn from an arm vein to determine the concentration of naphthoquine phosphate and artemisinin. Urine was also collected at intervals to establish drug concentration. The subjects drank water 2 h after medication and were given low-fat meals 4 h after medication.

The remaining 10 volunteers were given 1000 mg artemisinin and 400 mg naphthoquine phosphate orally to determine the pharmacokinetics of the individual drugs. The trial was designed as a single-dose, 2-week crossover study. Blood and urine samples were obtained and tested based on the same protocols as the artemisinin–naphthoquine phosphate groups.

Blood was taken from an arm vein before medication and 15, 30, and 45 min, and 1, 1.5, 2, 3, 4, 6, 8, 12, 18, 24, 48, 72, 120, 168, and 216 h afterward. It was placed with heparin in a test tube. The plasma was separated via centrifuge and stored at −20°C. Urine was collected before medication and 0–2, 2–4, 4–8, 8–12, 12–24, 24–48, 48–72, 72–120, 120–168, and 168–216 h later and stored at low temperature. The total volume of urine at each interval was recorded. Samples from each interval were obtained and stored at −20°C.

Measurements were carried out using the QTRAP triple quadrupole mass spectrometer, equipped with ESI and Analyst 1.3.2 data-processing software from Applied Biosystems; Agilent's 1100 HPLC pump and auto-sampler; and Agilent's ZORBAX Extend-C18 chromatography column (4.6 mm × 150 mm, 5 μm particle size). Established analytic methods were used to test specificity, sensitivity, precision, accuracy, recovery, calibration standards, stability, and quality control. If these were acceptable, the samples were analyzed. Tables 8.49 and 8.50 show the pharmacokinetic parameters of naphthoquine phosphate and artemisinin after a single oral dose of the ACT. The data for the individual drugs are in Table 4.51.

The original drug could be detected in the urine after administration of the ACT. The amount excreted increased together with the dose. The results are presented in Tables 8.52 and 8.53. Data for the individual drugs are in Tables 8.54 and 8.55.

The blood concentration of naphthoquine phosphate showed a strong linear relationship (r = .9956) from 0.50 to 500 μg/L. The lowest concentration was 0.50 μg/L. At concentrations of 5, 50, and 400 ng/mL, the recovery rates (n = 3) were 78.3 ± 4.99%, 86.0 ± 2.72%, and 87.9 ± 1.85%, respectively. The within-day or day-to-day variation was less than 15%. For artemisinin, the drug's blood concentration also had a strong linear relationship (r = .9990) from 4.00 to 1000 μg/L. The lowest concentration was 4.00 μg/L. Recovery rates (n = 3) were 93.3 ± 9.81%, 86.1 ± 5.29%, and 89.2 ± 3.50% at blood concentrations of 10, 100, and 800 ng/mL, respectively. Variation coefficients were also less than 15%.

The urine concentration of naphthoquine phosphate had a strong linear relationship (r = .9991) from 0.50 to 500 μg/L. The lowest concentration

TABLE 8.49 Clinical Pharmacokinetic Parameters of Naphthoquine Phosphate at Various Oral Doses of Artemisinin–Naphthoquine Phosphate Tablets ($n = 10$)

Parameter	700 mg	1400 mg	2100 mg
Ke (h^{-1})	0.0035 ± 0.0015	0.0029 ± 0.0013	0.0046 ± 0.0029
$T_{1/2}$ (h)	256.44 ± 179.37	276.35 ± 107.50	233.27 ± 190.74
C_{max} (μg/L)	11.40 ± 4.45	27.44 ± 16.21	59.83 ± 20.03
T_{max} (h)	3.50 ± 5.23	3.03 ± 1.92	2.50 ± 1.05
AUC_{0-216h} (μg·h/L)	479.95 ± 153.94	955.03 ± 352.05	1874.99 ± 257.57
$AUC_{0-\infty}$ (μg·h/L)	1002.34 ± 381.95	2011.28 ± 958.69	3709.33 ± 1860.28
MRT (h)	499.55 ± 359.04	674.03 ± 534.38	366.85 ± 159.39
CL/F (L/h)	324.91 ± 120.50	182.51 ± 71.40	107.41 ± 56.29
V/F (L)	107,439.17 ± 53,932.20	65,553.64 ± 20,670.12	26,760.64 ± 9254.23

TABLE 8.50 Clinical Pharmacokinetic Parameters of Artemisinin at Various Oral Doses of Artemisinin–Naphthoquine Phosphate Combination Tablets ($n = 10$)

Parameter	700 mg	1400 mg	2100 mg
Ke (h^{-1})	0.18 ± 0.03	0.19 ± 0.03	0.16 ± 0.04
$T_{1/2}$ (h)	4.01 ± 0.63	3.69 ± 0.60	4.90 ± 1.85
C_{max} (μg/L)	427.30 ± 143.01	697.70 ± 246.51	892.60 ± 219.78
T_{max} (h)	2.20 ± 1.06	2.38 ± 1.09	2.10 ± 1.63
AUC_{0-24h} (μg·h/L)	2642.42 ± 1293.25	4746.70 ± 2085.43	6736.12 ± 2551.60
$AUC_{0-\infty}$ (μg·h/L)	2708.87 ± 1320.48	4842.57 ± 2172.61	6820.63 ± 2524.43
MRT (h)	6.86 ± 1.02	6.73 ± 1.21	8.31 ± 1.46
CL/F (L/h)	148.23 ± 81.56	91.92 ± 61.64	71.52 ± 70.11
V/F (L)	847.09 ± 444.78	476.83 ± 315.64	515.55 ± 631.24

TABLE 8.51 Clinical Pharmacokinetic Parameters of Artemisinin Tablets or Naphthoquine Phosphate Tablets After Single Oral Doses ($n = 10$)

Parameter	Artemisinin 1000 mg	Naphthoquine Phosphate 400 mg
Ke (h^{-1})	0.2036 ± 0.0488	0.0042 ± 0.0025
$T_{1/2}$ (h)	3.58 ± 0.86	298.55 ± 294.57
C_{max} (μg/L)	466.50 ± 120.15	13.41 ± 6.37
T_{max} (h)	2.15 ± 0.91	3.43 ± 3.16
AUC_{0-t} (μg·h/L)[a]	2785.00 ± 847.84	423.75 ± 162.74
$AUC_{0-\infty}$ (μg·h/L)	2825.37 ± 866.56	854.00 ± 528.39
MRT (h)	5.89 ± 0.80	248.03 ± 347.96
CL/F (L/h)	381.29 ± 102.97	622.24 ± 294.39
V/F (L)	1929.95 ± 618.93	188,762.21 ± 90,230.48

[a]*Artemisinin: AUC_{0-24h}; naphthoquine phosphate: AUC_{0-216h}.*

was 0.50 μg/L. Relative recovery rates were 84.3 ± 3.31%, 81.9 ± 5.53%, and 84.7 ± 4.42%. The within-day and day-to-day variation coefficients were less than 15%. A strong linear relationship was also observed in the urine concentration of artemisinin ($r = .9975$) from 4.00 to 1000 μg/L. The lowest concentration was 4.00 μg/L. Relative recovery rates were 80.2 ± 4.19%, 88.6 ± 2.78%, and 88.5 ± 4.89%, and variation coefficients were also less than 15%.

For the artemisinin–naphthoquine phosphate combination, the correlation coefficients between the dose and the C_{max}, $AUC_{0\sim216h}$, and $AUC_{0\sim\infty}$ of naphthoquine were 0.9815, 0.9835, and 0.9894, respectively. The corresponding values for artemisinin were 0.9956, 0.9998, and 0.9997, respectively. It indicated that the C_{max} and AUC of naphthoquine phosphate and artemisinin in plasma increased in tandem with the dose, with a strong linear relationship. There were no significant changes to the $T_{1/2}$ and T_{max} of naphthoquine and artemisinin as the dose increased.

In the crossover study, the 10 volunteers were given 1000 mg artemisinin on an empty stomach. A week after clearance, they received 400 mg naphthoquine phosphate on an empty stomach. For naphthoquine phosphate tablets, the C_{max} was 13.41 ± 6.37 μg/L; AUC_{0-216h} was 423.75 ± 162.74 μg·h/L; $AUC_{0\sim\infty}$ was 854.00 ± 528.39 μg·h/L; $T_{1/2}$ was 298.55 ± 294.57 h; T_{max} was 3.43 ± 3.16 h; CL/F was 622.24 ± 294.39 L/h; V/F was 188,762.21 ± 90,230.48 L. The values for artemisinin were C_{max} 466.50 ± 120.15 μg/L; $AUC_{0\sim24h}$ 2785.00 ± 847.84 μg/L/h; $AUC_{0\sim\infty}$ 2825.37 ± 866.56 μg/L/h; $T_{1/2}$ 3.58 ± 0.86 h; T_{max} 2.15 ± 0.91 h; CL/F 381.29 ± 102.97 L/h; V/F 1929.95 ± 618.93 L.

TABLE 8.52 Amount of Naphthoquine Phosphate Excreted in Urine After Single-Dose Oral Artemisinin–Naphthoquine Phosphate Tablets

Dose	700 mg		1400 mg		2100 mg	
Time (h)	Amount (mg)	Percentage (%)	Amount (mg)	Percentage (%)	Amount (mg)	Percentage (%)
0–2	0.0067	0.0034	0.0049	0.0024	0.0868	0.0058
2–4	0.0077	0.0038	0.1695	0.0422	0.1614	0.0108
4–8	0.0329	0.0165	0.2450	0.0645	1.0754	0.0717
8–12	0.0695	0.0347	0.1585	0.0582	0.6895	0.0460
12–24	0.1402	0.0701	0.9490	0.2245	2.4532	0.1635
24–48	0.1199	0.0600	0.3312	0.0773	1.0078	0.0672
48–72	0.2193	0.1096	0.3798	0.1092	1.0382	0.0692
72–120	0.2461	0.1231	0.7570	0.1946	1.8950	0.1263
120–168	0.4396	0.2198	0.7274	0.1754	2.3513	0.1568
168–216	0.2457	0.1229	0.2239	0.0425	0.6201	0.0413

TABLE 8.53 Amount of Artemisinin Excreted in Urine After Single-Dose Oral Artemisinin–Naphthoquine Phosphate Tablets

Dose	700 mg		1400 mg		2100 mg	
Time (h)	Amount (mg)	Percentage (%)	Amount (mg)	Percentage (%)	Amount (mg)	Percentage (%)
0–2	0.0052	0.0010	0.0167	0.0017	0.0237	0.0016
2–4	0.0097	0.0019	0.0279	0.0028	0.0520	0.0035
4–8	0.0070	0.0014	0.0179	0.0018	0.0267	0.0018
8–12	0.0049	0.0010	0.0074	0.0007	0.0133	0.0009
12–24	0.0168	0.0034	0.0108	0.0011	0.0165	0.0011

TABLE 8.54 Amount of Artemisinin Excreted in Urine After 1000 mg Oral Artemisinin Tablets

Time (h)	Amount (mg)	Percentage (%)
0–2	0.0112	0.0011
2–4	0.0153	0.0015
4–8	0.0157	0.0016
8–12	0.0066	0.0007
12–24	0.0049	0.0005

TABLE 8.55 Amount of Naphthoquine Phosphate Excreted in Urine After 400 mg Naphthoquine Phosphate Tablets

Time (h)	Amount (mg)	Percentage (%)
0–2	0.0148	0.0037
2–4	0.0377	0.0094
4–8	0.1018	0.0254
8–12	0.1754	0.0438
12–24	0.3066	0.0767
24–48	0.8730	0.2183
48–72	0.8218	0.2054
72–120	0.5748	0.1437
120–168	0.3191	0.0798
168–216	0.0790	0.0198

3. Phase II Clinical Trials

Phase II trials for the artemisinin–naphthoquine phosphate combination tablets were designed by Chongqing Medical University, at its clinical test center for parasitic diseases. They were carried out by Hainan's Tropical Disease Research Institute, South Island Farm Hospital, and Dongfang Tianan Hospital. Laboratory tests were conducted by the Third People's Hospital and Dongfang People's Hospital in Hainan. The trials began in May 1998 and ended in July 2000.

The trials were randomized, open, and comparative. A suitable therapeutic dose would be determined based on the ACT's safe dose on falciparum malaria. The safety and efficacy of this dose would be compared against the individual drugs in the ACT. This would provide clinical data to justify use of the artemisinin–naphthoquine phosphate ACT tablets. Multicenter trials taking in a larger number of falciparum malaria patients, as well as trials on vivax malaria, would help to establish a therapeutic regimen for both malaria types.

Dose-Finding Study

The study used 175mg artemisinin–naphthoquine phosphate ACT tablets containing 50mg naphthoquine and 125mg artemisinin. Three dose groups were set up: 700, 1400, and 2100mg. The groups had 30 falciparum malaria patients each. Drugs were given in a single dose and patients hospitalized for 3–7days for observation. Patients' symptoms were recorded before medication and once a day thereafter, until they left hospital. The patients returned for tests if they experienced fever or discomfort. Follow-ups were conducted for 28days. Blood tests (nine parameters), comprehensive metabolic panels (six parameters), and urine tests (eight parameters) were performed before medication and 3, 7, and 14days afterward. ECGs (12 leads) were run before medication and 1, 3, 7, and 14days afterward. The patients' conditions are presented in Table 8.56. Their ages, course of infection, body temperatures, and parasite densities were similar across the groups ($P > .05$).

The clinical efficacy of the three dose groups can be seen in Table 8.57. All subjects were clinically cured in the short term, with rapid defervescence and parasite clearance. The 28-day cure rates were 96.7%, 93.3%, and 83.3% for the high-, middle-, and low-dose groups, respectively. There was one case of recrudescence in the high-dose group, two in the middle-dose group, and five in the low-dose group. The cure rate of the low-dose group differed markedly from those of the high- and middle-dose groups ($P < .01$), but there was no significant difference between the high- and middle-dose groups ($P > .05$).

One to two days after medication, patients in the various groups displayed differing degrees of headache, dizziness, general weakness, and loss of appetite. However, these symptoms eased on their own as the illness improved. No meaningful changes were seen in the blood tests, comprehensive metabolic panels, ECGs, and routine urine tests.

The results showed that all three doses of the artemisinin–naphthoquine phosphate ACT were both effective against falciparum malaria and safe. The 28-day cure rate increased with the dose, with a clear dose-efficacy relationship. Cure rates for the middle- and high-dose groups exceeded 90%, and there was no statistical significance between the two groups. Therefore, 1400mg (eight tablets) was chosen as the clinical dose for adults.

Comparative Trial of Artemisinin–Naphthoquine Phosphate Combination and Its Component Drugs[36]

This study included three groups. The first was given 1400mg artemisinin–naphthoquine phosphate combination tablets; the second, 1000mg naphthoquine tablets, with an initial dose of 600mg and another 400mg after 24h; the

TABLE 8.56 Falciparum Malaria Patients' Conditions Before Receiving Artemisinin–Naphthoquine Phosphate Tablets

Dose	Number of Patients	Average Age (years)	Malaria History		Time Since Infection (days)	Body Temperature (°C)	Parasite Density (10,000 per μL)
			Yes	No			
700	30	21.6 ± 7.3	21	9	3.6 ± 1.9	38.3 ± 1.4	3.02 ± 3.68
1400	30	28.0 ± 12.7	18	12	3.3 ± 1.5	38.1 ± 1.3	3.31 ± 3.16
2100	30	24.6 ± 9.1	22	8	3.4 ± 1.5	38.0 ± 1.2	3.03 ± 3.73

TABLE 8.57 Efficacy of Artemisinin–Naphthoquine Phosphate Tablets Against Falciparum Malaria

Dose (mg)	Number of Patients	Defervescence Time (h)	Decrease in Parasitemia Over 24 h (%)	Clearance Time (h)	Cases of Recrudescence	Cure Rate in 28 days (%)
700	30	17.1 ± 12.4	97.7 ± 7.5	32.1 ± 10.6	5	83.3[a]
1400	30	13.5 ± 7.4	99.0 ± 2.4	29.6 ± 9.2	2	93.3[b]
2100	30	11.8 ± 7.8	99.2 ± 2.9	30.0 ± 10.3	1	96.7

[a]*Difference between low-dose group and middle- and high-dose groups at P<.01.*
[b]*Difference between middle- and high-dose groups at P>.05.*

third, 2500mg artemisinin tablets, with 1000mg in the first dose and 500mg 8, 24, and 48h later. Patients were randomly allocated at a ratio of 3:3:1. The details can be seen in Table 8.58.

The results are presented in Table 8.59. The artemisinin–naphthoquine phosphate ACT's defervescence time, 24-h parasite clearance rate, and full parasite clearance time were the same as those of artemisinin. However, its 28-day cure rate was better than artemisinin's and the same as naphthoquine phosphate's. The ACT tablets combined the rapid action of artemisinin with the longer efficacy and high cure rate of naphthoquine phosphate. It also overcame the weaknesses of both individual drugs.

No drug-related adverse reactions were observed in the three groups. Blood test parameters were in the normal range. No abnormalities were seen in the reticulocyte count, routine urine tests, and ECGs.

The composition of the combination was based on a 1:2.5 ratio of naphthoquine phosphate and artemisinin. For artemisinin, a total dose of 2500mg was split into four oral doses. It was fast-acting but, due to its high recrudescence rate, patients' tolerance had to be considered and only 30 cases were included in this group. Naphthoquine phosphate with a total dose of 1000mg naphthoquine was used over two doses. It had persistent effects and could clear parasites thoroughly. Cure rates were high but the drug was slow to act. The 1400mg combination tablets comprised 1000mg artemisinin and 400mg naphthoquine, administered once. It lowered the required dose of the individual drugs, shortened the course of treatment, and combined the strengths of its components while avoiding their weaknesses. At a significantly reduced dose—1/2.5 that of the individual drugs—the combination was highly effective, fast-acting, and persistent. This confirmed the ACT's synergistic effects. Clinical tests showed that this did not extend to synergistic toxicity. It justified further use of the artemisinin–naphthoquine phosphate combination.

4. Phase III Trials[37]

Phase III trials were conducted by Hainan's Tropical Disease Research Institute, South Island Farm Hospital, and Dongfang Tianan Hospital. These expanded trials were based on a 1400mg oral dose of the artemisinin–naphthoquine phosphate combination. The three hospitals recruited a total of 220 falciparum malaria patients (129 male and 91 female), aged 26.4 ± 11.3 years. Of these, 173 patients had a history of malaria. Average course of infection was 2.7 ± 1.3 days, body temperature 39.0 ± 1.3°C, and parasitemia 28,107 ± 31,411 per μL. Average defervescence time was 17.4 ± 10.8h, the 24-h parasite clearance rate was 95.3 ± 15.1%, and full clearance time was 29.8 ± 8.8h. Five patients experienced recrudescence at Days 10, 17, 18, 19, and 24, respectively. The 28-day cure rate was 97.7%. No drug-related adverse reactions were observed and the drug was well-tolerated.

TABLE 8.58 Condition of Falciparum Malaria Patients in the Comparative Trial

Drug Group	Total Dose (mg)	Number of Patients	Average Age (years)	Time Since Infection (day)	Body Temperature (°C)	Parasite Density (10,000 per μL)
NQ-QHS	1400	100	27.7 ± 11.5	3.1 ± 1.5	38.7 ± 1.2	3.54 ± 3.90
NQ	1000	100	29.0 ± 10.7	2.7 ± 1.3	38.2 ± 3.2	2.73 ± 3.05
QHS	2500	30	23.5 ± 8.9	3.3 ± 1.1	38.3 ± 1.1	2.66 ± 3.19

TABLE 8.59 Therapeutic Efficacy of Artemisinin–Naphthoquine Phosphate Combination (NQ-QHS), Naphthoquine Phosphate (NQ), and Artemisinin (QHS) on *P. falciparum*

Drug Group	Dose (mg)	Number of Doses	Defervescence Time (h)	Decrease in Parasitemia Over 24h (%)	Clearance Time (h)	Cure Rate in 28 days (%)
NQ-QHS	1400	1	17.5 ± 12.3	96.9 ± 12.9	30.0 ± 8.8	97.0
NQ	1000	2	32.7 ± 17.7	81.4 ± 17.6	45.5 ± 10.0	100
QHS	2500	4	18.1 ± 9.7	98.7 ± 2.3	29.1 ± 6.0	66.7

At the same time, 109 patients with vivax malaria were recruited. There were 83 males and 26 females, aged 23.02 ± 9.0, 100 of whom had a history of malaria. Time since infection was 2.6 ± 1.4 days, body temperature 37.9 ± 1.5°C, and parasitemia 6782 ± 7861 per μL. The subjects were given a single oral dose of 1400 mg artemisinin–naphthoquine phosphate ACT. Defervescence time was 13.0 ± 5.5 h, the 24-h parasite clearance rate was 100% with a parasite clearance time of 18.1 ± 5.7 h. Eleven cases of recrudescence occurred in 56 days, between Days 47 and 56. In the short term, 98 patients were cured, yielding a corresponding cure rate of 90%. No drug-related symptoms were observed over 56 days and the drug was well tolerated. The defervescence time, 24-h parasite clearance rate, and full parasite clearance time were faster for vivax than for falciparum malaria. However, the 11 vivax recrudescence cases showed that the ACT was not a complete cure. Therefore, reinfection could not be ruled out in endemic areas.

In summary, the Phase II and III trials confirmed that a suitable dose of the artemisinin–naphthoquine phosphate ACT was 1400 mg, administered once orally. All 320 falciparum malaria patients and 109 vivax malaria were clinically cured, with no significant drug-related adverse reactions. The 28-day cure rate for falciparum malaria was 97.5%, while the 56-day cure rate for vivax malaria was 90% (Table 8.60). Compared to its component drugs, the ACT—using a naphthoquine phosphate-to-artemisinin ratio of 1:2.5—combined their strengths and overcame their weaknesses, obtaining synergistic effects. The ACT greatly reduced the required doses of the component drugs, which not only lowered toxicity but also mitigated and delayed the onset of drug resistance. In terms of safeguarding the new drug, this was very promising.

The artemisinin–naphthoquine phosphate ACT had high efficacy, rapid action, a high cure rate, and low toxicity. Because its regimen required only one dose, it was a new-model, fixed-ratio ACT with good patient compliance and acceptability.

5. International Multicenter Clinical Trials[38]

The artemisinin–naphthoquine phosphate ACT was given the trade name ARCO. Kunming Pharmaceuticals Corporation was responsible for 13 international multicenter clinical trials. As of May 2013, 2037 patients were included, with 1153 given ARCO and 884 in the control. The average 28-day cure rate for ARCO was 96% for falciparum malaria (Table 8.61). The drug was safe and effective across patients of different ethnic groups. ARCO has been approved or registered in 22 countries (Table 8.62). Its single oral dose greatly improves compliance and serves both as a model and a stimulus for future research worldwide.

Note: The following were involved in artemisinin–naphthoquine phosphate combination research.

TABLE 8.60 Therapeutic Efficacy of Single Dose 1400 mg Artemisinin–Naphthoquine Phosphate Tablets in Multicenter Clinical Trials

Parasite	Number of Patients	Defervescence Time (h)	Parasite Clearance Time (h)	Cure Rate (%)[a]
P. falciparum	320	18 ± 11	30 ± 9	97.5
P. vivax	109	13 ± 6	18 ± 6	90

[a] *28-day cure rate for falciparum malaria; 56-day cure rate for vivax malaria.*

TABLE 8.61 Results of International Multicenter Clinical Trials on Artemisinin–Naphthoquine Phosphate Tablets (ARCO)

Name of Clinical Trial	Country	ARCO Group/ Control Group, ARCO Cure Rate
Comparative randomized clinical trials of artemisinin–naphthoquine and Dihydroartemisinin–Piperaquine in the treatment of falciparum and vivax malaria[39]	Indonesia	201/200, 97%
Comparative clinical trials of artemisinin–naphthoquine combination (ARCO) and Chloroquine–Fansidar in the treatment of falciparum malaria[40]	Papua New Guinea	51/49, 100%
Comparative clinical trials of artemisinin–naphthoquine (ARCO), Artemether, and Artemether–Fansidar in the treatment of falciparum malaria[41]	Sudan	129/65/71, 100%
Comparative clinical trials of ARCO and Coartem in the treatment of falciparum malaria[42]	Uganda	107/105, 100%
Safety and efficacy study of ARCO and Coartem in the treatment of falciparum malaria (Age range: 4 months–16 years)[43]	Uganda	87/77, 100%
Clinical trials of artemisinin and naphthoquine combination in the treatment of falciparum malaria[44]	Nigeria	121, 96%
Comparative clinical trials of ARCO, Coartem and artesunate–amodiaquine in the treatment of falciparum malaria	Nigeria	56/56/50, 100%

TABLE 8.61 Results of International Multicenter Clinical Trials on Artemisinin–Naphthoquine Phosphate Tablets (ARCO)—cont'd

Name of Clinical Trial	Country	ARCO Group/ Control Group, ARCO Cure Rate
Comparative clinical trials of ARCO and Coartem in the treatment of falciparum malaria (Age range: 5–14 years)	Nigeria	55/55, 90%
Comparative clinical trials of artemisinin–naphthoquine and artemether–lumefantrine in the treatment of falciparum malaria[45]	Cote d'Ivoire	60/60, 100%
Clinical efficacy of artemisinin–naphthoquine phosphate in the treatment of falciparum malaria[46]	Burma	55, 98.1%
Clinical efficacy of artemisinin–naphthoquine (ARCO) in the treatment of falciparum malaria[47]	Benin	50, 100%
Comparative clinical trials of ARCO and Artemether tablets in the treatment of falciparum malaria in children	Benin	93/96, 97.85%
Clinical efficacy of artemisinin–naphthoquine phosphate (ARCO) in the treatment of falciparum malaria in adults[48]	Cameroon	88, 100%

TABLE 8.62 Countries Where Artemisinin–Naphthoquine Phosphate Tablets (ARCO) is Registered or Approved

Cambodia	Benin
Congo (Brazzaville)	Nigeria
Cote d'Ivoire	Chad
Uganda	Kenya
Togo	Burma
Liberia	Cameroon
Papua New Guinea[a]	Central African Republic
Samoa[a]	Congo (Kinshasa)
Solomon Islands[a]	Gabon
Vanuatu[a]	Tanzania
Fiji[a]	Zambia

[a] *ARCO was approved in the South Pacific.*

Pharmacodynamics tests in mice: Li Guofu
Pharmacodynamics tests in monkeys: Ding Deben
Acute toxicity tests in mice: Wu Boan
Acute and long-term toxicity tests in rats and dogs were supervised by Wang Jingyan
General pharmacology tests were supervised by Wang Jingyan
Phase I clinical trials in the PLA's 307 Hospital: Liu Zeyuan
Phase II and III clinical trials in Hainan were supervised by Wang Jingyan.
International clinical trials and registration were organized by Kunming Pharmaceuticals Corporation.

REFERENCES

1. Schmidt LH, Crosby R. Antimalarial activities of WR-194, 965, an α-Amino-o-Cresol derivative. *Antimicrob Agents Chemother* 1978;**14**(5):672–9.
2. Zhang LM, Shen JH, Wang YL, et al. Synthesis and antimalarial activity of 2-dialkyl ammonium methyl-4-heterocyclic-amino-5,6,7,8-tetra-hydrogen naphthol derivatives. *Pharm Ind* 1985;**16**(2):8–12. [张立明，沈季华，王云玲，等. 2-二烷氨甲基-4-杂环氨基-5,6,7,8-四氢萘酚衍生物的合成及抗疟活性. 医药工业, 1985, 16(2): 8–12.]
3. Li FL, Wang LH, Ding DB, et al. The synthesis of 4-arylamino-tertbutylaminomethyl phenol antimalarials. *Acta Pharm Sin* 1982;**17**(1):77–9. [李福林，王丽华，丁德本，等. 疟疾治疗药物——4-芳胺基-2-特丁胺甲基酚类化合物的合成. 药学学报, 1982, 17(1): 77–79.]
4. Li FL, Li YL, He GH, et al. *Preparation method for naphthoquine salts* China Patent CN 1093356 A, published. October 12 , 1994. [李福林，李玉兰，何闺华，等. 萘酚喹盐类的制备方法: CN 1093356A, 公开日期1994年10月12日.]
5. Naphthoquine Phosphate Research Group. *Materials for new-drug application (certificate and manufacturing), Dossier 2: confirmation of the chemical structure of naphthoquine phosphate.* Archives of Institute V, AMMS; 1992. [军事医学科学院微生物流行病研究所磷酸萘酚喹课题组. 磷酸萘酚喹化学结构确证. 磷酸萘酚喹申报材料之二.1992.]
6. Naphthoquine Phosphate Research Group. *Materials for new-drug application (certificate and manufacturing), Dossier 5: main pharmacodynamic test data.* Archives of Institute V, AMMS; 1992. [磷酸萘酚喹课题组. 申报新药(证书、生产)资料之五，主要药效学试验资料及文献资料.内部资料, 1992, 资料保存在军事医学科学院五所档案室.]
7. Peters W. Drug resistance in *Plasmodium berghei* Vincke and Lips, 1948. I. Chloroquine resistance. *Exp Parasit* 1965;**17**:80–9.
8. Raether W, Fink E. Antimalarial activity of floxacrine (HOE 991) I. Studies on blood schizontocidal action of floxacrine against *Plasmodium berghei*, *P. vinckei* and *P. cynomolgi*. *Ann Trop Med Parasitol* 1979;**73**(6):505–26.
9. Peters W, Ramkaran AE. The chemotherapy of rodent malaria, XXXII. The influence of *p*-aminobenzoic acid on the transmission of *Plasmodium yoelii* and *P. berghei* by *Anopheles stephensii*. *Ann Trop Med Parasitol* 1980;**74**(3):275–82.
10. Gregory KG, Peters W. The chemotherapy of rodent malaria. IX. Causal prophylaxis. I. A method for demonstrating drug action on exo-erythrocytic stages. *Ann Trop Med Parasitol* 1970;**64**(1):15–24.
11. Davidson Jr DE, Johnsen DO, Tanticharoenyos P, et al. Evaluating new antimalarial drugs against trophozoite induced *Plasmodium cynomolgi* malaria in rhesus monkeys. *Am J Trop Med Hyg* 1976;**25**(1):26–33.

12. Schmidt LH, Rossan RN, Fisher KF. The activity of a repository form of 4,6-diamino-1-(p-chlorophenyl)-1,2-dihydro-2,2-dimethyl-*s*-triazine against infections with *Plasmodium cynomolgi*. *Am J Trop Med Hyg* 1963;**12**:494–503.
13. Davidson Jr DE, Ager AL, Brown JL, et al. New tissue schizontocidal antimalarial drugs. *Bull World Health Organ* 1981;**59**(3):463–79.
14. Trager W, Jensen JB. Human malaria parasites in continuous culture. *Science* 1976;**193**(4254):673–5.
15. Merkli B, Richle R. *Plasmodium berghei*: diet and drug dosage regimens influencing selection of drug-resistant parasites in mice. *Exp Parasitol* 1983;**55**(3):372–6.
16. Naphthoquine Phosphate Research Group. *Materials for new-drug application (certificate and manufacturing), Dossier 6: pharmacological test data*. Archives of Institute V, AMMS; 1992. [磷酸萘酚喹课题组. 申报新药(证书、生产)资料之六，一般药理试验资料及文献资料.内部资料, 1992,资料保存在军事医学科学院五所档案室.]
17. Xu SY, Bian RL, Chen X, editors. *Experimental methodology of pharmacology*. Beijing: People's Medical Publishing House; 1982. p. 1184. [徐叔云，卞如濂，陈修， 药理实验方法学，北京，人民卫生出版社 1982: 1184.]
18. Naphthoquine Phosphate Research Group. *Materials for new-drug application (certificate and manufacturing), Dossier 15: animal-model pharmacokinetic test data*. Archives of Institute V, AMMS; 1992. [磷酸萘酚喹课题组. 申报新药(证书、生产)资料之十五，动物药代动力学试验资料及文件资料.内部资料, 1992, 资料保存在军事医学科学院五所档案室.]
19. Tian JG, Lu ZL, Jiao XQ. *Pharmacokinetic study of naphthoquine in healthy and infected mice*. Archives of Institute V, AMMS; 1996. [田建广，卢志良，焦岫卿. 磷酸萘酚喹在正常和感染疟原虫小鼠体内药代动力学研究. 内部资料, 1996, 资料保存在军事医学科学院五所档案室.]
20. Both these tests were based on internal protocol.
21. 3p87 is a program that runs on IBM computers and contains different algorithms for the calculation of pharmacokinetic parameters.
22. Shi YL, Kong WW, Jiao XQ. *The effects of naphthoquine on the ultrastructure of Murine malaria*. Archives of Institute V, AMMS; 1992. [时云林，孔惟惟，焦岫卿. 磷酸萘酚喹对鼠疟原虫超微结构的影响. 内部资料, 1992, 资料保存在军事医学科学院五所档案室.]
23. Naphthoquine Phosphate Research Group. *Materials for new-drug application (certificate and manufacturing), Dossier 7: animal-model acute toxicity test data*. Archives of Institute V, AMMS; 1992. [磷酸萘酚喹课题组. 申报新药(证书、生产)资料之七，动物急性毒性试验资料及文献资料. 内部资料, 1992,资料保存在军事医学科学院五所档案室.]
24. Finney DJ. *Statistical method in biological assay*. 3rd ed. London: Charles Griffin & Co.; 1978.
25. Deichmann WB, LeBlanc TJ. Determination of the approximate lethal dose with about six animals. *J Ind Hygiene Toxicol* 1943;**25**(9):415–7.
26. Naphthoquine Phosphate Research Group. *Materials for new-drug application (certificate and manufacturing), Dossier 8: animal-model long-term toxicity test data*. Archives of Institute V, AMMS; 1992. [磷酸萘酚喹课题组. 申报新药(证书、生产)资料之八，动物长期毒性试验资料及文献资料. 内部资料, 1992,资料保存在军事医学科学院五所档案室.]
27. Naphthoquine Phosphate Research Group. *Materials for new-drug application (certificate and manufacturing), Dossier 12: reproductive toxicity test data*. Archives of Institute V, AMMS; 1992. [磷酸萘酚喹课题组. 申报新药(证书、生产)资料之十二，生殖毒性试验资料及文献资料. 内部资料, 1992, 资料保存在军事医学科学院五所档案室.]
28. Wilson JG. Methods for administering agents and detecting malformations in experimental animals. In: Wilson JG, Warkany J, editors. *Teratology: principles and techniques*. Chicago: University of Chicago Press; 1965. p. 262–75.

29. Naphthoquine Phosphate Research Group. *Materials for new-drug application (certificate and manufacturing), Dossier 11: mutagenesis test data*. Archives of Institute V, AMMS; 1992. [磷酸萘酚喹课题组. 申报新药(证书、生产)资料之十一，致突变试验资料及文献资料. 内部资料, 1992,资料保存在军事医学科学院五所档案室.]
30. Ames BN, McCann J, Yamasaki E. Methods for detecting carcinogens and mutagens with the salmonella/mammalian-microsome mutagenicity test. *Mutat Res* 1975;**31**(6):347–64.
31. Naphthoquine Phosphate Research Group. *Materials for new-drug application (certificate and manufacturing), Dossier 25: clinical trial data*. Archives of Institute V, AMMS; 1992. [磷酸萘酚喹课题组. 申报新药(证书、生产)资料之二十五，临床研究试验资料及文献资料. 内部资料, 1992,资料保存在军事医学科学院五所档案室.]
32. Naphthoquine Phosphate Research Group. *Materials for new-drug application (certificate and manufacturing), Dossier 20: pharmacokinetic test data in healthy volunteers*. Archives of Institute V, AMMS; 1992. [磷酸萘酚喹课题组. 申报新药(证书、生产)资料之二十，健康人药代试验资料及文献资料. 内部资料, 1992,资料保存在军事医学科学院五所档案室.]
33. World Health Organization. Chemotherapy of malaria and resistance to antimalarials: report of a WHO scientific group. *World Health Organ Tech Rep Ser* 1973;**529**:1–121.
34. Yang RD. A method for determining combined toxicity in mixed preparations. *Agrochemicals* 1981;**2**:36–7. [杨瑞典. 混合制剂联合毒力测定方法. 农药, 1981, 2: 36–37.]
35. Wang JY, Wu BA, Dong ZH, et al. Long-term toxicity of naphthoquine act in beagles. *Bull Acad Mil Med Sci* 2003;**27**(3):196–9. [王京燕，邬伯安，徐在海，等. 复方磷酸萘酚喹对比格犬的长期毒性.军事医学科学院院刊, 2003, 27(3) 196–199.]
36. Wang JY, Shan CQ, Fu DD, et al. Clinical study of naphthoquine act and its individual drugs in the treatment of falciparum malaria. *Acta Parasitol Med Entomol Sin* 2003;**21**(3):131–3. [王京燕，单成启，符大东，等. 复方磷酸萘酚喹及其组份单药治疗恶性疟的临床研究. 中国寄生虫学与寄生虫病杂志, 2003, 21(3): 131–133.]
37. Wang JY, Shan CQ, Fu DD, et al. Clinical study of naphthoquine act in the treatment of falciparum malaria. *Chin J Parasit Dis Con* 2003;**16**(3):134–6. [王京燕，单成启，符大东，等. 复方磷酸萘酚喹片治疗恶性疟的临床研究.中国寄生虫病防治杂志, 2003, 16(3): 134–136.]
38. For an overview of some of these studies, see: Hombhanje FW, Huang QY. Artemisinin-naphthoquine combination (ARCO®): an overview of the progress. *Pharmaceuticals* 2010;**3**(12):3581–93.
39. Tjitra E, Hasugian AR, Siswantoro H, et al. Efficacy and safety of artemisinin-naphthoquine versus dihydroartemisinin- piperaquine in adult patients with uncomplicated malaria: a multi-center study in Indonesia. *Malar J* 2012;**11**:153.
40. Hombhanje FW, Linge D, Saweri A, et al. Artemisinin-naphthoquine combination (ARCO™) therapy for uncomplicated falciparum malaria in adults of Papua New Guinea: a preliminary report on safety and efficacy. *Malar J* 2009;**8**:196.
41. Bakri YMN, Naser Mahmoud MH, Ali Babekir H, et al. Efficacy and safety of artemisinin-naphthoquine (ARCO®) in the treatment of uncomplicated *Plasmodium falciparum* among Sudanese adults. *Glob Adv Res J Med Med Sci* 2014;**3**(1):001–7.
42. Mworozi EA, Maganda A, Masembe V, et al. An open label randomized study comparing ARCO—a new act—and Coartem in treatment of uncomplicated malaria in adults in Uganda. *Int J Infect Dis* December 2008;Suppl. 1:e313.
43. Rujumba J, Mworozi EA, Maganda AK, et al. A comparative study of ARCO® and coartem in the treatment of uncomplicated malaria in patients aged 4 months to 16 years attending Mulago hospital, Kampala, Uganda. *Int J Infect Dis* 2010;**14**:e332.

44. Meremikwu M, Odey F, Alaribe A, et al. Open-label randomised trial of a fixed dose combination of artemisinin and naphthoquine for treating uncomplicated falciparum malaria in Calabar, Nigeria. [MIM Nairobi, 16205673]. In: The 5th Multilateral Initiative on Malaria (MIM) Pan-Africa Conference, November 2–6, 2009, Nairobi, Kenya.
45. Offianan AT, Louis KP, Jean-Didier Y, et al. A comparative, randomized clinical trial of artemisinin/naphthoquine twice daily one day versus artemether/lumefantrine six doses regimen in children and adults with uncomplicated falciparum malaria in Côte d'Ivoire. *Malar J* 2009;**8**:148.
46. Tun T, Tint HS, Lin K, et al. Efficacy of oral single-dose therapy with artemisinin-naphthoquine phosphate in uncomplicated falciparum malaria. *Acta Trop* 2009;**111**(3):275–8.
47. Kinde-Gazard D, Massougbodji A, Capo-Chichi L, et al. *Etude de l'efficacité et de la tolérance de la combinaison fixe Artémisinine-Naphthoquine (ARCO®) en prise unique versus Arthemether-Luméfantrine (COARTEM®) pour la prise en charge du paludisme non compliqué à Plasmodium falciparum au Bénin*. Unpublished data from Kunming Pharmaceutical Corporation. 2009.
48. Same-Ekobo A, Huang QY, Kuete T, et al. *Etude de l'efficacité et de la tolérance de ARCO® (Artémisinine-Naphthoquine Phosphate) dans le traitement du paludisme simple à Plasmodium falciparum au Cameroun*. Unpublished data from Kunming Pharmaceutical Corporation. 2011.

Chapter 9

Artemisinins and Pyronaridine Phosphate Combination

Chen Chang

National Institute of Parasitic Diseases, Chinese Center for Disease Control and Prevention, Shanghai, China

Chapter Outline

Artemisinin-Based and Other Antimalarials. https://doi.org/10.1016/B978-0-12-813133-6.00009-3

The National Project 523 Leading Group tasked the Chinese Center for Disease Control and Prevention's National Institute of Parasitic Diseases with studying a new antimalarial drug with no cross-resistance to chloroquine. Based on existing research the Institute designed and synthesized a new antimalarial in 1970. This drug, pyronaridine phosphate—pyronaridine for short—targets the erythrocytic stage of the malarial parasite, is highly effective, and has low toxicity and no cross-resistance with chloroquine. It can be administered orally, via intramuscular injection or intravenously to treat various species of malaria, including severe falciparum cases such as cerebral malaria.

Following the WHO/TDR's recommendation that artemisinins be used in combination with other antimalarial drugs, various artemisinin derivatives were administered with pyronaridine to treat normal and drug-resistant falciparum malaria. This combination therapy overcomes the shortcomings of each drug when administered individually and achieves a higher level of efficacy.

9.1 CHEMICAL STRUCTURE AND SYNTHESIS OF PYRONARIDINE PHOSPHATE

In the history of antimalarials, chloroquine had emerged as the best orally administered drug for treatment of malaria. The appearance of chloroquine-resistant strains of *Plasmodium falciparum*, however, meant that the drug lost its initial effectiveness. Other antimalarials followed, such as amodiaquine, cycloquine, and amopyroquine (Fig. 9.1). The relationship between structure and therapeutic efficacy of these antimalarials had previously been studied comparing them against chloroquine. With this information a new antimalarial was designed and synthesized.

A. DESIGNING THE CHEMICAL STRUCTURE OF THE NEW ANTIMALARIAL

The effective doses of chloroquine, amodiaquine, and cycloquine for parasite clearance in 50% of mice infected by *Plasmodium berghei* (ED_{50}) were 15.8±4.1 mg/kg, 45.0±3.4 mg/kg, and 4.6±1.2 mg/kg, respectively.[1] In the treatment of children with *Plasmodium vivax*, the minimal effective dose for parasite clearance was 2.5 mg/kg for chloroquine, >5 mg/kg for amodiaquine, and 5 mg/kg for cycloquine.[1] The results suggested that cycloquine was more effective than amodiaquine, but the clinical efficacy of both drugs was lower than that of chloroquine. The structure of cycloquine contains a double diethylamine side chain, whereas amodiaquine has only one similar side chain (Fig. 9.1). This showed that a double-side-chain configuration was superior to a single-chain one.

The intragastric ED_{50} of amopyroquine against *P. berghei* in mice and its lethal dose in 50% of normal mice (LD_{50}) were 13.7±3.4 mg/kg and 781±98 mg/kg, respectively. By contrast, the corresponding doses of chloroquine were 42.9±15.4 mg/kg and 478±95 mg/kg, respectively.[2] This showed that the efficacy

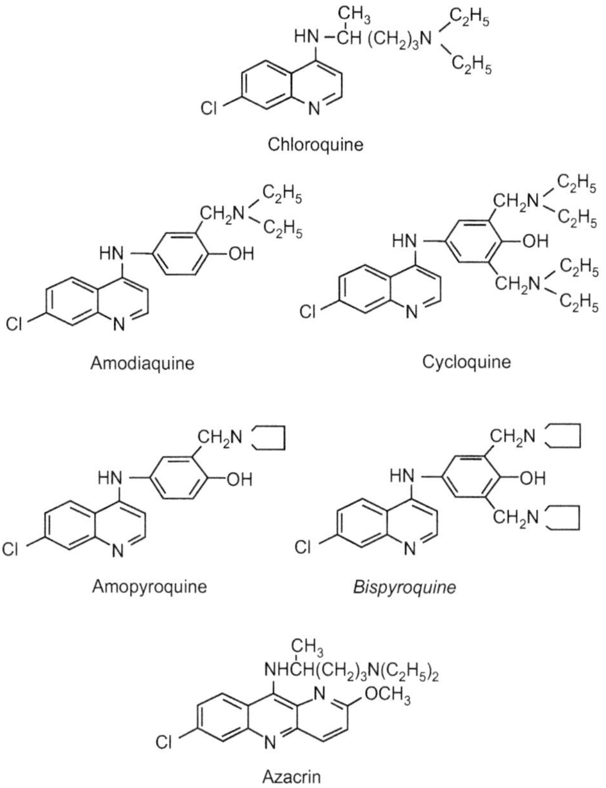

FIGURE 9.1 Chemical structures of chloroquine and related antimalarials.

of amopyroquine in the treatment of rodent malaria was better than that of chloroquine. Intramuscular injections to treat falciparum and vivax malaria showed an average parasite clearance time of 31.3 and 31.4 h for amopyroquine, as opposed to 34.1 and 34.2 h for chloroquine. However, in the 1962 article by Hoekenga, when an intramuscular injection of 40 mg of chloroquine was given to a 6-year-old child, and when a similar injection of 120 mg chloroquine was administered to an adult, both individuals died after a few minutes.[3] By contrast, intramuscular injections of amopyroquine were well tolerated and amopyroquine was therefore recommended as a drug of choice for intramuscular administration, with a single dose of 90–150 mg to treat malaria in adults.[3] The studies demonstrated that the toxicity of amopyroquine was far less than that of chloroquine and that its clinical effectiveness was comparable. In the study of children with *P. vivax* mentioned above, the efficacy of amodiaquine and cycloquine was inferior to that of chloroquine. Amopyroquine has a single cyclic pyrrolidine side chain (Fig. 9.1). It indicated that this cyclic side chain resulted in higher efficacy than the acyclic diethylamine side chains of cycloquine and amodiaquine.

In a test of efficacy against *Plasmodium lophurae* in chicks, amopyroquine—with its cyclic pyrrolidine side chain—proved the most effective when compared to analogs in which other cyclic side chains were substituted, e.g., with piperidine, hexamethylenimine, or 4-methylpiperazine.[4] Again, this demonstrated that pyrrolidine would be a good choice for the double side chains.

Given the experimental results above—in which a double side chain was better than a single one, and a cyclic pyrrolidine side chain superior to an acyclic diethylamine one—the Institute of Parasitic Diseases (then under the Chinese Academy of Medical Sciences) synthesized a new compound in 1964. This new compound, M6407, had two cyclic pyrrolidine side chains and was named bispyroquine (Fig. 9.1).[5,6] The curative doses of bispyroquine in 50% of mice infected with *P. berghei* (CD_{50}) were 17.4 ± 1.4 mg/kg for intragastric administration and 7.3 ± 1.0 mg/kg for intramuscular injection. In comparison, the corresponding doses of chloroquine were 54.0 ± 1.4 mg/kg and 52.8 ± 1.2 mg/kg, respectively. The intragastric and intramuscular LD_{50} of bispyroquine were 664 ± 24 mg/kg and 128 ± 7.0 mg/kg in mice, while those of chloroquine were 488 ± 14 mg/kg and 71 ± 5.0 mg/kg, respectively. The intragastric therapeutic index (LD_{50}/CD_{50} or ED_{50}) of bispyroquine was 37, as opposed to 9 for chloroquine; the intramuscular therapeutic index was 17.5 for bispyroquine and 1.3 for chloroquine. In the treatment of rodent malaria, bispyroquine was superior to chloroquine both in efficacy and toxicity.[7] Oral administration of bispyroquine in treating vivax malaria resulted in a similar level of clinical efficacy to that of chloroquine, but its side effects were much lower.[5]

Subsequently, oral administration of bispyroquine was used to treat vivax and falciparum malaria in southern Henan province, with chloroquine as a benchmark. The results showed that the efficacy of bispyroquine was similar to that of chloroquine in treating both forms of malaria. However, its side effects were markedly lower than those of chloroquine.[8] This proved that, in comparison to chloroquine, bispyroquine's clinical efficacy was indeed reliable and its toxicity was less.

The intragastric cross-resistance of bispyroquine and amopyroquine in relation to chloroquine was compared. It was found that the ED_{50} against chloroquine-sensitive *P. berghei* in mice was 15.4 mg/kg for bispyroquine and 65.3 mg/kg for amopyroquine. Against chloroquine-resistant *P. berghei*, the ED_{50} was 56.5 mg/kg for bispyroquine and 306 mg/kg for amopyroquine. The cross-resistance indices (ED_{50} against chloroquine-resistant strain/ED_{50} against chloroquine-sensitive strain) were 3.67-fold for bispyroquine and 4.69-fold for amopyroquine. This demonstrated that bispyroquine, with its double pyrrolidine side chain, was more effective than amopyroquine and its single pyrrolidine side chain. The aforementioned indices also showed that bispyroquine was better than amopyroquine at reducing cross-resistance to chloroquine.

However, these indices showed that bispyroquine and amopyroquine had some cross-resistance with chloroquine. This may have been due to the quinoline nucleus common to the chemical structures of both drugs and of chloroquine itself (Fig. 9.1). Substituting this nucleus with another could allow the new compound's cross-resistance with chloroquine to be reduced even further, or be

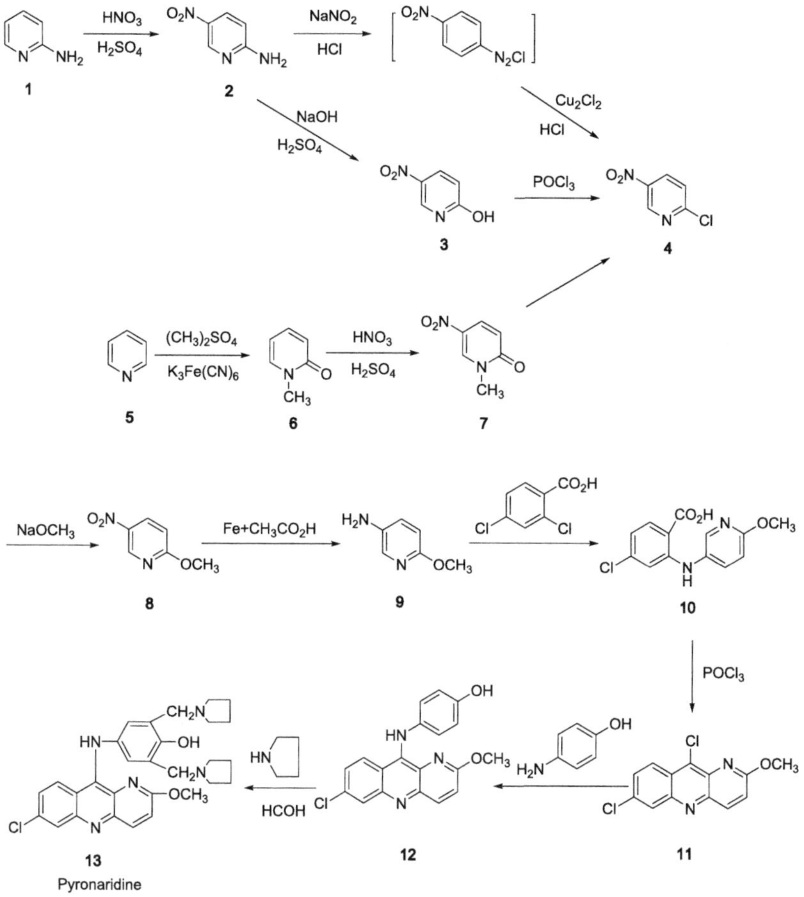

FIGURE 9.2 Synthesis of pyronaridine.

eliminated entirely. Therefore, it was decided to replace the quinoline nucleus. After comparing the various nuclei of different antimalarial compounds, the benzonaphthyridine nucleus of azacrin was selected (Fig. 9.1). After combining this with a phenol and adding double pyrrolidine side chains, a new compound was created with the chemical structure—10-[3,5-bis(pyrrolidinyl-*N*-methyl)-4-hydroxyphenyl]-amino-7-chloro-2-methoxybenzo[*b*]1,5-naphthyridine (Compound **13**, Fig. 9.2). The compound, number 7351, was named pyronaridine.

B. SYNTHESIS OF PYRONARIDINE

The synthesis of pyronaridine started with 2-aminopyridine or pyridine (Fig. 9.2).[9–11] 2-Aminopyridine (Compound **1**) underwent mononitration at position 5 to obtain 2-amino-5-nitropyridine (Compound **2**). Following hydrolysis, 2-hydroxy-5-nitropyridine (Compound **3**) was synthesized. Chlorination then yielded a 2-chloride Compound **4**, which could also be obtained by diazotization of Compound **2** followed by treatment with cuprous chloride (Cu_2Cl_2). Compound **4** could also be formed through the oxidization of pyridine (Compound **5**), producing *N*-methylpyridine-2-one (Compound **6**), which was then nitrated to Compound **7** and subsequently chlorinated to derive Compound **4**.

Compound **4** was then reacted with sodium methoxide to obtain a 2-methoxy Compound **8**, which was reduced to yield a 5-amino Compound **9**. This was condensed with 2,4-diclorobenzoic acid to produce Compound **10** and subsequently cyclized to create Compound **11**. The latter was condensed with p-aminophenol to form 2-methoxy-7-chloro-10-(4′-hydroxyphenyl)-aminobenzo(b)1,5-naphthyridine (Compound **12**), and finally underwent a Mannich reaction to obtain pyronaridine (Compound **13**). Pyronaridine was subsequently treated with phosphoric acid to yield pyronaridine tetraphosphate, or pyronaridine phosphate.

Several analyses and spectroscopic tests confirmed that the chemical structure of the synthesized pyronaridine was in accordance with the structure initially designed. Its molecular formula was $C_{29}H_{32}ClN_5O_2 \cdot 4H_3PO_4$, in which the base content was 56.89%.

Going further, various pyronaridine derivatives and analogs were produced. Among them were three compounds synthesized at the recommendation of WHO/TDR/CHEMAL. After tests on *P. berghei* in mice, however, they were found to be less effective than pyronaridine in treating rodent malaria.[10,12–14]

In 1975, after the completion of a draft on quality standards and specifications for pyronaridine phosphate, the National Project 523 Leading Group convened a conference in Shanghai to appraise the compound. It was at this conference that pyronaridine was assessed. With cooperation from the Hangzhou Minsheng Pharmaceutical Factory, the technical aspects of pyronaridine manufacture were evaluated in 1979. Finally, in 1985, pyronaridine phosphate, pyronaridine phosphate tablets, and pyronaridine phosphate injections were approved and registered as antimalarial drugs. From 1990 onward, pyronaridine phosphate, its tablets and injections have been listed in the *Pharmacopoeia of the People's Republic of China*, with the foreign name "malaridine phosphate."

In the various experiments and clinical therapies described below, the dosages of pyronaridine phosphate and other controls refer to the drugs' bases.

9.2 PHARMACOLOGICAL STUDIES OF PYRONARIDINE PHOSPHATE

A. THE PHARMACODYNAMICS OF PYRONARIDINE PHOSPHATE

1. In Vivo Tests of Pyronaridine Against Malaria Parasites[15]

Efficacy on Rodent Malaria

Mice were given a single intragastric dose of pyronaridine on the fourth day after infection with *P. berghei*. On Day 7, a blood test was conducted to determine

TABLE 9.1 Efficacy of Chloroquine and Pyronaridine on Normal *P. berghei*

Drug	Route of Administration	$ED_{50} \pm SD$ (mg/kg)	Therapeutic Index (TI)
Pyronaridine	Intragastric	6.8±1.4	201.5
Chloroquine	Intragastric	45.6±6.2	14.5
Pyronaridine	Intramuscular injection	4.9±0.6	50.4
Chloroquine	Intramuscular injection	30.9±5.8	2.9

the ED_{50} of the drug in comparison to a control group given chloroquine. The results are shown in Table 9.1 and demonstrate that the therapeutic efficacy of pyronaridine on rodent malaria is higher than that of chloroquine.

WHO/TDR/CHEMAL established that, for subcutaneous injections of pyronaridine, the dosage to effect a complete cure of *P. berghei* rodent malaria was 20mg/kg. Before this, tests in China showed that a complete cure of KSP II strain *P. berghei* in mice was brought about by dosages of 10mg/kg (intragastric administration) and 5mg/kg (subcutaneous injection). By contrast, the ED_{50} of intragastrically administered pyronaridine in mice infected by mosquito-borne *Plasmodium yoelii* was 2.67mg/kg. This indicated that the drug could effectively eliminate *P. yoelii* in the erythrocytic cycle but had no clear impact on gametocytes.[16]

Persistent Effects on Rodent Malaria in the Erythrocytic Stage

Fifteen days before mice were inoculated with *P. yoelii*, they were given intragastric doses of 100mg, 10mg, or 5mg/kg pyronaridine. The duration of efficacy or protection, in which erythrocytic invasion by the parasite was inhibited, was ≥15, ≥7 and ≥3days, respectively. Intramuscular injection increased the duration of protection by 2–3days compared to intragastric administration.

In another test, mice were given intragastric doses of 10mg/kg pyronaridine 6, 5, and 3days before being infected with *P. berghei*. Subsequently, five out of nine, six out of nine, and all eight mice in the three test groups, respectively, showed no parasitized erythrocytes. A similar intragastric administration of 67mg/kg chloroquine, delivered just 2days before infection, resulted in only one in nine mice presenting with no parasitized erythrocytes. A control group of nine mice untreated with any drugs before infection all showed parasitic invasion of erythrocytes. These experiments showed that the persistent effects of pyronaridine in inhibiting *P. berghei* in the erythrocytic phase lasted longer than those of chloroquine.[17]

Efficacy on Chloroquine-Resistant Strains of P. berghei

Mice were given an intragastric dose of pyronaridine 4h before being infected with a chloroquine-resistant strain of *P. berghei*. On the second day and third day they were again fed one dose each, with a blood test being conducted on Day 8. The same was performed on a control group of mice infected with a chloroquine-sensitive (i.e., normal) strain of *P. berghei*. For the ED_{50} of pyronaridine on the chloroquine-resistant and chloroquine-sensitive strains, see Table 9.2.

The results contained in Table 9.2 suggest that pyronaridine and chloroquine did not show cross-resistance.

Using a regimen of 3-day subcutaneous injections of pyronaridine, WHO/TDR/CHEMAL determined that the dose suppressing 90% of the parasite (SD_{90}) was 2.34 mg/kg for the chloroquine-sensitive strain and 3.0 mg/kg for the chloroquine-resistant strain. The resistance index was 1.28. The corresponding SD_{90} for chloroquine was 2.7 and 200 mg/kg, respectively, and the resistance index was 74.1. Another experiment with a 4-day regimen showed that the ED_{50} and ED_{90} of pyronaridine on a chloroquine-resistant strain (*P. yoelii* NS) were 0.6 and 1.2 mg/kg, respectively. For chloroquine, they were 2.4 and 56.0 mg/kg. The ED_{50} and ED_{90} of a control group infected with a chloroquine-sensitive strain were 0.4 and 0.9 mg/kg for pyronaridine, as opposed to 1.8 and 3.1 mg/kg for chloroquine. The resistance index (I_{90}) for pyronaridine was 1.3, as against 18.1 for chloroquine.[18] These results showed that there was no cross-resistance between pyronaridine and chloroquine.

Experiments using 4-day therapeutic regimens of intragastric pyronaridine against sensitive and resistant strains (ANKA and K173)[19–21] demonstrated that pyronaridine showed no cross-resistance to chloroquine, mefloquine, and hydroxypiperaquine. The resistance index of pyronaridine was significantly lower than those of artemisinin derivatives and mefloquine, which were tested at the same time.

TABLE 9.2 Efficacy of Pyronaridine and Chloroquine on Chloroquine-Resistant and Chloroquine-Sensitive Strains of *P. berghei* in Mice

Drug	$ED_{50} \pm SD$ (mg/kg) of Sensitive Strains	$ED_{50} \pm SD$ (mg/kg) of Resistant Strains	Resistance Index (I_{50})
Pyronaridine	2.01 ± 0.19	1.99 ± 0.27	0.99
	2.40 ± 0.26	2.66 ± 0.36	1.11
	2.61 ± 0.44	2.88 ± 0.48	1.1
Chloroquine	15.90 ± 4.2	840[a]	>52

[a]*When dosage reached 840 mg/kg, all mice died without parasite clearance.*

Efficacy on Simian Malaria

Rhesus monkeys (*Macaca mulatta*) were infected with *Plasmodium inui*, *Plasmodium cynomolgi*, and *Plasmodium knowlesi* via blood passage. Pyronaridine was administered orally, through intramuscular injection or intravenous drip, to examine its efficacy on simian malaria. For the intravenous drip, pyronaridine was diluted in saline to produce a 0.05% solution and delivered at a constant rate over 1 h. A control group was given chloroquine. The results of each group are displayed in Table 9.3. Pyronaridine had a definite effect on *P. inui*, *P. cynomolgi*, and *P. knowlesi*. By increasing the dosage of pyronaridine to two 10 mg/kg intramuscular injections, five monkeys testing positive for *P. cynomolgi* all tested negative for the parasite 24–72 h after first administration of the drug. Observation continued for 3 months and all five showed no parasite recrudescence (Table 9.3) and were cured.[22]

An owl monkey (*Aotus trivirgatus*) infected with *P. falciparum* was given two 40 mg/kg intramuscular injections of pyronaridine at an interval of 6 h. Three days after the first dose, the parasite was eliminated. A subsequent examination 3 months later showed no recrudescence.[23]

Rhesus monkeys were injected with *P. cynomolgi* sporozoites and fed pyronaridine on Day 7. The drug inhibited or blocked further development of the ring forms. No trophozoites were observed within 10–20 days or even longer, unlike with a control group of monkeys given chloroquine. Therefore, pyronaridine inhibits or stops the development of the ring forms into schizonts, thus exerting a persistent effect on schizonts in the erythrocytic phase.[24] However, it has no effects on schizonts in the tissue phase.

2. In Vitro Testing of Drug-Resistant *Plasmodium falciparum*

A 48-h in vitro assay was used to assess the inhibitory concentration (IC_{50}) of pyronaridine on 50% of chloroquine-sensitive *P. falciparum* strains extracted on location from Anhui, Jiangsu, and Hainan. This IC_{50} was determined to be 15.0 ± 1.7 nmol/L, whereas the IC_{50} against chloroquine-resistant *P. falciparum* strains from Hainan and Cambodia was 15.4 ± 2.5 nmol/L. The resistance index (I_{50}) was 1.02. The corresponding IC_{50} of chloroquine in these cases was 56.0 ± 7.8 nmol/L and 322 ± 69 nmol/L, respectively, with an I_{50} of 5.75. This showed that there was no cross-resistance between pyronaridine and chloroquine.[25]

In another in vitro assay, the IC_{50} of pyronaridine on chloroquine-resistant *P. falciparum* strains extracted from four patients in Yunnan was measured at 1.25 pmol/well, as against 6.42 pmol/well for chloroquine. The ratio between the two was 1:5.1. The Yunnan data showed that the four cases of falciparum malaria were resistant to chloroquine but still susceptible to pyronaridine.[26] Furthermore, the prevalence of chloroquine-resistant *P. falciparum* in the area was measured at 97%–100%. Chloroquine-resistant *P. falciparum* strains obtained from 24 cases in Yunnan, when tested with pyronaridine, showed an

TABLE 9.3 Efficacy of Pyronaridine on Various Types of Simian Malaria

Drug	Dose (mg/kg)	Route of Administration	Type of Malaria	Parasite Clearance Time (hour)	Recrudescence Time (day)
Pyronaridine	28 × 1 (3 monkeys)	PO	*P. inui*	4–48	11–33
	20 × 1 (3 monkeys)	PO	*P. inui*	168[a]	11
	4.5 × 2[b] (4 monkeys)	IM	*P. cynomolgi*	48–120	15–18
	3.0 × 2 (4 monkeys)	IM	*P. cynomolgi*	72–120	10–15
	3.0 × 1 (3 monkeys)	IM	*P. cynomolgi*	73–97	7–8
	1.5 × 1 (1 monkey)	IM	*P. cynomolgi*	72	4
	10.0 × 2 (5 monkeys)	IM	*P. cynomolgi*	24–72	>3 months without recurrence
Chloroquine	4.5 × 2 (2 monkeys)	IM	*P. cynomolgi*	73, 96	10, 14
	3.0 × 2 (2 monkeys)	IM	*P. cynomolgi*	96, 120	10, 14
	3.0 × 1 (3 monkeys)	IM	*P. cynomolgi*	93–120	1–2
Pyronaridine	5.0 × 1 (5 monkeys)	IV drip	*P. knowlesi*	48–72	7–11
	2.0 × 1 (2 monkeys)	IV drip	*P. knowlesi*	48, 72	4, 8
	1.5 × 2 (1 monkey)	IM	*P. knowlesi*	72	11
	1.5 × 1 (1 monkey)	IM	*P. knowlesi*	72	4

IM, Intramuscular injection; *IV drip*, Intravenous drip; *PO*, Oral.
[a] *Two monkeys showed no parasite clearance.*
[b] *×2 indicates a twice-daily dose at an interval of 4–6 h.*

IC_{100} of 3.26 pmol/well on average. By comparison, the IC_{100} of pyronaridine on chloroquine-sensitive strains in this case was ≤4.0 pmol/well. This proved that pyronaridine could act on chloroquine-resistant *P. falciparum*.[27]

WHO/TDR/CHEMAL conducted in vitro experiments on a clone of a chloroquine-sensitive strain of *P. falciparum* D-6 and on a clone of a chloroquine-resistant strain W-2. It was found that the ED_{50} of pyronaridine on D-6 was 0.51 ng/mL as compared to 3.42 ng/mL for chloroquine. Similarly, the ED_{50} of pyronaridine on W-2 was 0.45 ng/mL, compared to 15.58 ng/mL for chloroquine. The resistance index of pyronaridine was less than 1, as opposed to the index of 4.55 for chloroquine. This showed that there was no cross-resistance between pyronaridine and chloroquine.

Pyronaridine was tested against chloroquine, the two enantiomers of chloroquine, and desethylchloroquine to compare their effects on a chloroquine-sensitive strain of *P. falciparum* F_{32} obtained from Tanzania. Similar tests were conducted involving KI, a chloroquine-resistant strain from cases in Thailand. It was found that pyronaridine and all the other four drugs were active on chloroquine-sensitive F_{32}, but only pyronaridine was effective on chloroquine-resistant KI. Hence, pyronaridine was more effective than the other four drugs ($P<.05$), showing that it can be used to treat chloroquine-resistant falciparum malaria.[28]

In Thailand, where chloroquine-resistance is endemic, a strain was isolated from 43 out of 73 patients infected with resistant *P. falciparum*. To determine the resistance of pyronaridine, an in vitro test was conducted while further samples from the remaining 30 patients were kept in continuous culture. These two experiments showed the resistance rate of pyronaridine was 0%, while that of chloroquine was 100%. The resistance rates for mefloquine were 7% and 3% for the two groups. It showed that the resistant local strain was susceptible to pyronaridine but not to chloroquine and mefloquine, and that there was no cross-resistance between pyronaridine and chloroquine or mefloquine.[29]

A multidrug-resistant *P. falciparum* strain was also obtained from patients in eastern Thailand, where resistance is endemic. An in vitro experiment determined that the IC_{50} of chloroquine, quinine, and pyrimethamine was 316, 388, and 11,800 nmol/L, respectively. That of pyronaridine was 8.40 nmol/L. By way of contrast, a strain taken from northern Thailand—an area where *P. falciparum* shows less resistance—was also tested in vitro. Here, the IC_{50} of pyronaridine was 10.1 nmol/L and 167, 248, and 1980 nmol/L for the other three drugs, respectively. This showed a degree of resistance to the three drugs but not to pyronaridine.[30]

Further, two resistant strains of *P. falciparum* from the same area of eastern Thailand were obtained from cases in which mefloquine and enpiroline treatment had failed. The in vitro experiment showed that there was resistance to mefloquine and enpiroline, but the IC_{50} of pyronaridine was 9.59 and 8.78 nmol/L, respectively. Thus, there was no cross-resistance between pyronaridine and either mefloquine or enpiroline. Pyronaridine could be considered as a treatment for multidrug-resistant malaria.[30]

In vitro tests were also conducted with strains of *P. falciparum*, obtained from Cambodia, which were resistant to chloroquine, pyrimethamine, cycloguanil, and mefloquine. They all showed susceptibility to pyronaridine.[31]

Similar tests were performed on a chloroquine-resistant strain of *P. falciparum* obtained from 31 patients and a chloroquine-susceptible strain from 12 patients. All these patients were from Africa. It showed that the IC_{50} of pyronaridine on these two strains was 7.7 and 8.4 nmol/L, respectively, proving that there was no cross-resistance between pyronaridine and chloroquine. Furthermore, it demonstrated that pyronaridine had a greater effect on chloroquine-resistant *P. falciparum* in different regions. It was thought to be a promising potential drug for the treatment of drug-resistant falciparum malaria.[32]

3. Mode of Action of Pyronaridine on the Plasmodium Parasite

Observations on the Ultrastructure of a Chloroquine-Susceptible Strain of Plasmodium berghei *in the Erythrocytic Stage*[33]

Mice infected with *P. berghei* were given a single dose of 6 mg/kg intragastric pyronaridine. After 15–30 min, it was found that pyronaridine acted first on the parasite's food vacuole. It caused several changes, including fusion and enlargement of the food vacuole, nuclear membrane, and mitochondrial swelling, and aggregation of the pigment. In addition, pyronaridine acted on the pellicular complexes of the erythrocytic trophozoite, making them swell and form multilamellate whorls which increased progressively. Four to 16 h after the drug was administered, the trophozoites disintegrated. Similar changes could be seen in schizonts 1–2 h after pyronaridine was given, and it was difficult to detect the parasite after 16 h.

By contrast, chloroquine also affected the food vacuole first, causing it to undergo the aforementioned changes, with disintegration of the parasite's structure after 16 h. However, it had no clear effect on the pellicular complexes of the trophozoite.

Observations on the Ultrastructure of a Chloroquine-Resistant Strain of Plasmodium berghei *in the Erythrocytic Stage*[34]

Mice infected with chloroquine-resistant *P. berghei* were given one dose of 6 mg/kg pyronaridine intragastrically. Fifteen minutes to 24 h later, the pellicular complexes of the trophozoites swelled up and progressively formed multilamellate whorls. Later only small, fragmented dead parasites were seen, but there were no significant changes in their food vacuoles. This showed that pyronaridine may be able to kill chloroquine-resistant strains by damaging the structure and physiological functions of the pellicular complexes.

A control group of mice was infected with a resistant strain and given 400 mg/kg of chloroquine. No significant changes in the ultrastructure of the parasite in its erythrocytic stage were found, demonstrating that chloroquine had no clear effect on both the food vacuole and pellicular complexes. This may be why pyronaridine has no cross-resistance with chloroquine-resistant strains.

Observations on the Morphology of Plasmodium falciparum[35]

Following in vitro tests of cultured human *P. falciparum*, the effects of pyronaridine on the food vacuole and pellicular complexes were soon detected. Approximately 8h after the drug took effect, some dead and damaged parasites appeared. Chloroquine and mefloquine affected the food vacuole but had no obvious effect on the pellicular complexes. These results may indicate why pyronaridine can treat chloroquine-resistant and multidrug-resistant falciparum malaria.

Observations on the Morphology of Plasmodium falciparum *in an Owl Monkey*[23]

An owl monkey (*A. trivirgatus*) infected with *P. falciparum* received 20mg/kg of pyronaridine via two intramuscular injections at an interval of 6h. Thirty minutes after the drug was administered, significant changes occurred in the food vacuoles of the trophozoites and schizonts, causing the denaturation of hemoglobin. Swelling in the pellicular complexes of the merozoites was also observed, which showed changes in multilamellae.

Effects on Multidrug-Resistant Plasmodium falciparum[36]

In vitro tests were conducted using the chloroquine- and pyrimethamine-resistant strain of *P. falciparum* KI. It was found that pyronaridine could inhibit DNA topoisomerase II in KI. The concentration at which complete suppression occurred was 11 μmol/L. This showed that pyronaridine may be targeting DNA topoisomerase II in drug-resistant *P. falciparum.*

Effects on the Phagocytic Function of Macrophages[37]

Mice were given three 2mg intragastric doses of pyronaridine to observe the changes in macrophage phagocytosis in the abdominal cavity. These were contrasted with in vitro tests. The results of in vivo and in vitro tests both showed that the average phagocytic rate and phagocytic index of pyronaridine were significantly higher than those of the known positive control drug ($P<.01$–$.001$). It demonstrated that pyronaridine could nonspecifically enhance macrophage phagocytosis. Moreover, it was observed that the parasite was digested within 2h after being engulfed by the macrophages, leaving only debris and pigment. One could conclude that the therapeutic effect of pyronaridine may be related to the promotion of macrophage phagocytosis.

4. Pharmacokinetics of Pyronaridine

Absorption of Pyronaridine in Rabbits[38]

Two groups of three rabbits were given single doses of 30 and 60mg/kg, respectively. Plasma concentration was measured using the spectrofluorometric method. The results showed that the drugs were absorbed relatively quickly after administration. The peak time for each group was 1.38 ± 0.22 h and 1.62 ± 0.37 h, respectively, with peak concentrations at 768±17.0ng/mL and 1514±376.0ng/mL. Bioavailability in both groups was 34.6%. The elimination half-life was 56.0±7.0h and 54.0±4.0h, respectively.

Four rabbits were given a single dose of 6mg/kg pyronaridine each via intramuscular injection. The drug was absorbed quickly and completely, with peak time at 0.75±0.44h and peak concentration at 1534.0ng/mL. Bioavailability was high, up to 100%. The half-life of intramuscular injection was 52±8h.

Subsequently, four rabbits were given a single dose of 6mg/kg of pyronaridine each via intravenous injection. The half-life of the drug was measured at 59.0±10.0h. It could be seen from the pharmacokinetic parameters of intravenous injection that the distribution of pyronaridine in blood components was heterogeneous but stabilized from 72–96h after injection. The concentration of pyronaridine was significantly higher in red blood cells than in plasma. The red blood cell to plasma pyronaridine concentration ratios were 3.36, 6.01, and 6.07 at 3, 72, and 96h, respectively.

Distribution and Excretion of ^{3}H-Pyronaridine in Mice[39]

Two groups of three mice were given a single dose of 30mg/kg ^{3}H-pyronaridine, via intragastric administration or subcutaneous injection. Using the same methods described above, the average radiation concentration was 103–128μCi per mouse for intragastric administration and 72–95μCi per mouse for subcutaneous injection. At designated time points, radioactive activity of various organs was measured, and the percentage of organ radioactivity to radioactivity of drug administered calculated. Twenty-four hours after intragastric administration, average radiation concentration was highest in the liver, at 18% of the total radioactive dose in drug administered. Small and large intestines and their contents contained 1.7%, kidneys 1.4%, lungs 1.1%, and heart 0.16%. There were only trace amounts in the brain and eyes and the proportion concentrated in the body was 12%. The average proportion excreted in urine and feces was 21.3% and 33.6%, respectively.

Similar results were obtained 24h after subcutaneous injection, except that the proportion in the liver and body was 20% each. Excretion in urine was 21%–22% and 17.9% in feces. Pyronaridine was excreted slowly and lasted for a relatively long time in vivo. After 96h, a small amount of radioactive matter was progressively discharged. Radioactivity could still be found in the urine and feces on day 11. A study mentioned earlier showed that pyronaridine had persistent effects on the erythrocytic cycle of rodent *P. yoelii*. This may be related to the slow excretion of pyronaridine in mice.

Whole-Blood Distribution of Pyronaridine in Rabbits[40]

The whole-blood distribution of pyronaridine in rabbits was measured using HPLC. A rabbit was given 20mg/kg intragastric pyronaridine. Six hours later the plasma concentration of pyronaridine was 0.21μg/mL; whole-blood concentration was 1.61 μg/mL, showing that pyronaridine was concentrated mainly in red blood cells.

HPLC-Electrochemical Detection Assay for Plasma Concentration in Monkeys[41]

On three consecutive weeks, a rhesus monkey was given a single dose of 160 mg pyronaridine via intramuscular injection. The monkey did not show adverse reactions at this high dosage. An hour after administration, the concentration of pyronaridine in the whole blood reached its peak at 1983 ng/mL, then decreased rapidly to 1319 ng/mL after 2 h, 701 ng/mL after 8 h, 290 ng/mL after 1 day, 155 ng/mL after 3 days, 71 ng/mL after 7 days, and 27 ng/mL after 14 days. After both 21 and 28 days, the number fell below 20 ng/mL, which meant that the concentration of the drug had reached the limit of detection. The elimination half-life of pyronaridine was 64 h. Twenty-four hours after injection, drug concentration was 5.72 μg/mL in urine. In the absorption peaks found after testing the urine, one extremely faint absorption peak was found which could possibly be a pyronaridine metabolite, but the concentration was too low to be measured. The researchers considered that the HPLC method was suitable for studying the clinical pharmacokinetics of this promising drug.

Pyronaridine Pharmacokinetics in Malaria Patients[42]

A study was conducted on 10 patients with malaria: three of falciparum and seven of vivax malaria. Of these, four were given an intramuscular injection of 204 mg (3.8 mg/kg) of pyronaridine. Using the spectrofluorometric method, it was found that pyronaridine was absorbed quickly and completely. Drug concentration reached a peak 0.66 ± 0.2 h after injection, with an average peak concentration of 525 ± 104 ng/mL and half-life of 63 ± 5 h. Another three cases were given a 0.6 g (11.1 mg/kg) dose of pyronaridine in enteric-coated tablets. Peak absorption was reached after 14.0 ± 0.3 h, with an average peak concentration of 130 ± 32 ng/mL, half-life was 65 ± 6 h. The final three cases were given a 0.6 g (10.1 mg/kg) dose of pyronaridine in capsules. Peak absorption came after 4.72 ± 0.26 h, with an average peak concentration of 255 ± 144 ng/mL and half-life of 63 ± 6 h. With intramuscular injection as a reference, the relative bioavailability of the two oral formulations was 19 ± 7% and 32 ± 7%, respectively.

Pyronaridine Pharmacokinetics in the Human Body

In 1998, a new, sensitive HPLC fluorescence detection method was used to measure the pharmacokinetic parameters of pyronaridine in humans (WHO/TDR/TDP, Proposed meeting material, September 1998, Geneva). Ten healthy male subjects in Thailand participated. Five received oral pyronaridine in aqueous solution, and five consumed pyronaridine gelatin capsules. The dose was the same throughout, at 6 mg/kg. After administration, drug tolerance in both groups was good.

The pharmacokinetic parameters of the different dosage forms were similar, with no statistical difference ($P > .05$). Peak absorption of the aqueous solution was reached in 6.3 ± 7.3 h and peak concentration was 154.2 ± 85.6 ng/mL. Peak time for the capsules was 11.9 ± 13.3 h, with a peak concentration of 120.4 ± 41.3 ng/mL. Elimination half-life for the aqueous solution was 160.3 ± 59.4 h. The capsules had a half-life of 191.2 ± 49.7 h, three times longer than the half-life reported in domestic studies. The domestic studies carried out observations for only 72 h, but this experiment prolonged it to 336 h because the concentration of pyronaridine still stood at 50 and 60 ng/mL 2–6 days after administration. Therefore, it is worth considering a longer observation time; in addition, differences could have been produced by limitations in the different methods of detection.

HPLC in Determining the Clinical Pharmacology of Pyronaridine[43]

HPLC fluorescence detection was used to conduct a clinical pharmacokinetic study on a healthy adult male volunteer. The subject was given a single dose of 400 mg (8 mg/kg) oral pyronaridine via capsule. Peak plasma concentration was reached at 0.5 h. Peak concentration was 495.8 ng/mL. The elimination half-life of pyronaridine was measured at 251 h. This method of analysis had high specificity, selectivity, and sensitivity and was simple and quick to use. In Thailand, it has become the standard technology to measure pyronaridine in plasma and for conducting clinical pharmacokinetic studies.

B. TOXICOLOGY OF PYRONARIDINE

1. Acute Toxicity Tests[15]

Acute Toxicity in Mice and Rats

The median lethal dose (LD_{50}) in mice given intragastric pyronaridine was 1277 ± 16.2 mg/kg, 1369 ± 234.2 mg/kg, or 1729.4 mg/kg. The LD_{50} of a chloroquine control group was 663.4 ± 76.7 mg/kg. For intramuscular injection, the LD_{50} of pyronaridine was 250.6 ± 33.1 mg/kg and 89.7 ± 34.0 mg/kg for chloroquine. In rats, the LD_{50} for intragastric administration was 1281 mg/kg.

Acute Toxicity in Rabbits

Three groups of rabbits—the first with five, the second with four, and the third with five subjects—were given 20, 40, and 80 mg/kg intramuscular injections of pyronaridine, respectively. Only in the 80 mg/kg group did one rabbit out of the five die 40 min after administration. The remaining 13 rabbits across all groups survived. A further group of three rabbits given 80 mg/kg chloroquine all died after 25–72 min. Of the other controls, one out of four rabbits given 40 mg/kg chloroquine died after 18 min. One out of five rabbits given 20 mg/kg chloroquine died in 24 h. In dosages of pyronaridine above 40 mg/kg, electrocardiograms (ECGs) showed a slowing in heart rate, lengthening of P–R, QRS, and Q–T, and other reversible changes. As the dosage increased, premature ventricular contraction and even ventricular fibrillation could be seen. Such changes appeared at 20 mg/kg chloroquine, with severe arrhythmia at 20 and 40 mg/kg.

Acute Toxicity in Dogs

Oral pyronaridine 120 and 240 mg/kg was fed to two groups of two dogs. The drug was administered four times in 3 days: Twice on Day 1 and once on Days 2 and 3. Vomiting occurred in all the dogs, but there were no other adverse reactions. A blood test showed no abnormalities.

In another experiment, 60, 40, and 20 mg/kg pyronaridine was given via intramuscular injection to three groups of dogs—the first with five, the second with five, and the third with three subjects. Of five given 60 mg/kg, one dog died 23 min after injection. Another vomited and, 20 min later, developed paroxysmal spasms that lasted 1–2 min and subsided after 15 min. It returned to normal the next day. The serum glutamic pyruvic transaminase (SGPT) level in a third dog rose from 22.5 to 85 units and returned to normal in a week. The remaining two dogs in this group, as well as the 40 and 20 mg/kg groups, displayed no adverse reactions. ECGs and routine blood tests showed no abnormalities. The surviving dogs were tested and observed for a month, also with no abnormal findings.

In another test, 5, 10, and 20 mg/kg of chloroquine were given to three other groups of dogs via intramuscular injection; these groups had two, three, and three subjects, respectively. All three dogs in the 20 mg/kg group died within half an hour. Of the 10 mg/kg group, one died and the other two displayed trembling, nausea, foaming at the mouth, and an inability to stand. They recovered gradually after 1–4 h. Similar reactions were seen in the 5 mg/kg group, but they recovered more rapidly. The ECGs of the dogs given chloroquine showed mainly a delay in conduction which returned to normal in 2 h.

All the above tests proved that the toxicity of pyronaridine was significantly lower than that of chloroquine.

Acute Toxicity in Monkeys

A dose of pyronaridine of 240 mg/kg was fed to four monkeys. The drug was administered four times in 3 days: twice on Day 1 and once on Days 2 and 3. The SGPT level in one of the monkeys rose from 25 to 107 units, but fell to 20 units in a week. The remaining monkeys showed no adverse reaction. ECGs and biochemical examination were normal and drug tolerance was good.

2. Subacute Toxicity Tests[15]

Effects on the Growth of Rats

Two groups of 15 rats were given 200 and 40 mg/kg intragastric pyronaridine, respectively, to observe their growth and development. The drug was administered once a day over 14 days. Seven days after the start of the regimen, the average weight of the rats given the higher dose had increased by 6 g. After 14 days, the average increase was 22 g; three rats had died in that time. Rats in the low dosage group had an average increase in weight of 27 and 45 g after 7 and 14 days, respectively, and one rat died on Day 13. Nonmedicated rats fed normally had a corresponding weight increase of 22 and 46 g. This showed that the large dosage group experienced suppressed growth, but that there was no effect in the low dosage group.

A control study was conducted on an identical number of rats fed 100 and 20 mg/kg chloroquine. During the periods mentioned above, rats in the large dosage group had an average weight gain of 4 g and a weight loss of 2 g, respectively, and three of the rats died in 14 days. The effects were obviously greater than those of pyronaridine. Of the low dosage group, the rats' weight increased by an average of 22 and 45 g, showing no effects. Dissection 24 h after administration showed that both drugs had no obvious influence on the major organs.

Subacute Toxicity in Rabbits

Five rabbits were given 10 mg/kg intravenous pyronaridine. The drip was administered twice on Day 1 with a 6-h interval, full administration taking an hour each time. Subsequently, the drip was delivered once a day for 7 days in total. The rabbits had no adverse reactions and were observed over a month. Regular ECGs and biochemical measurements were normal. In a similar group of five rabbits given chloroquine, one died 20 min after the fifth infusion. After dissection, atrial enlargement with congestion, diffuse emphysema, and pathological changes in the lungs could be seen with the naked eye.

Subacute Toxicity in Dogs

Two groups of two dogs were given 12 and 24 mg/kg intragastric pyronaridine, respectively. This was administered once a day for 30 days. The dogs showed no adverse reactions and ECGs and biochemical tests were also normal. In a control, two dogs given 24 mg/kg chloroquine displayed intermittent tremors and both died in 2 weeks. Two dogs given 12 mg/kg chloroquine showed reduced food intake, trembling, foaming at the mouth, and other reactions after the fourth dose. The SGPT level of one of the dogs increased from 16 to 62.5 units. These dogs also died in 30 days. One dog showed severe heart failure before death.

3. Cardiovascular Toxicity of Rapid Intravenous Injections[15]

Toxicity in Anesthetized Rabbits

An incision was made on the necks of 10 anesthetized rabbits. A trachea cannula and carotid cannula were inserted to record respiration, arterial blood pressure, and ECGs. Every 2 min, a 2 mL solution of pyronaridine (containing 4 mg) was injected into the rabbits' femoral vein at high speed. The cumulative toxic dose was taken to be that at which blood pressure dropped to 40 mmHg and obvious arrhythmia appeared. When the toxic dose was reached and blood pressure decreased to 40 mmHg, ECGs showed a merger of T and P waves. P–R and Q–T were extended and the QRS widened. Subsequently, this developed into a 2:1 atrial–ventricular conduction blockage, or sinus bradycardia, and premature ventricular beats. This progressed to idioventricular rhythm and death, which was then recorded as the average cumulative lethal dose. Although similar changes occurred in a chloroquine control group of 10 rabbits at the corresponding cumulative doses, there were still significant differences between both groups (see Table 9.4.).

Toxicity in Anesthetized Dogs

Using the method described above, five anesthetized dogs were injected with a 5 mL (10 mg) solution of pyronaridine every 2 min. When blood pressure dropped to 40 mmHg, the ECG showed similar changes to those of the rabbits. The average cumulative toxic and lethal doses were noted. A control group of five dogs was given chloroquine. There were significant differences in average cumulative toxic and lethal doses in the two groups (see Table 9.4).

These experiments show that the cardiovascular toxicity of pyronaridine and its effects on blood pressure are significantly lower than those of chloroquine.

4. Safety of Intravenous Injections and Drips in a Simulated Clinical Setting

Intravenous Injection in Anesthetized Rabbits

Using the experimental methods described above, changes in the rabbits' respiration, blood pressure, and ECGs were observed during administration of the

TABLE 9.4 Toxicity of Intravenous Pyronaridine in Anesthetized Animals

		Average Cumulative Dose (mg/kg)		
Drug	Animal (Number)	Onset of Serious Arrhythmia	Blood Pressure Below 40 mmHg	Lethal Dose (mg/kg)
Pyronaridine	Rabbit (10)	23.3 ± 4.0	39.8 ± 9.3	64.6 ± 10.1
Chloroquine	Rabbit (10)	11.8 ± 1.9 ($P < .05$)	11.4 ± 3.1 ($P < .025$)	20.7 ± 3.8 ($P < .001$)
Pyronaridine	Dog (5)	–	97.8 ± 18.7	116.0 ± 22.3
Chloroquine	Dog (5)	–	28.8 ± 4.4 ($P < .01$)	47.4 ± 6.7 ($P < .025$)

drug and 1 h afterward. The pyronaridine and chloroquine groups each had five anesthetized rabbits, which were given the drug at a constant speed with a rate-regulated injector. A dose of 1 mg/kg intravenous pyronaridine solution was administered per minute, with full dose 15 mg/kg delivered in 15 min. During this time, the blood pressure of the rabbits in the pyronaridine group decreased by varying degrees. The blood pressure of two of the five rabbits fell below 40 mmHg, but then rose. When the cumulative dose reached 15 mg/kg, the blood pressure in all five rabbits had recovered or was close to their original levels. None died. Before the cumulative doses were reached in the chloroquine group, the blood pressure of all five rabbits fell below 40 mmHg. The blood pressure of two rabbits dropped rapidly to zero and they died. The other three rabbits' blood pressure returned to normal or close to their original levels.

Slow Intravenous Infusion in Anesthetized Rabbits

Using the same method as the aforementioned tests on cardiovascular toxicity, rabbits were given the drug at a constant speed with a rate-regulated injector via the auricular vein. Five rabbits were given pyronaridine and five given chloroquine equivalent to 30 mg/kg administered slowly over 1 h at 4.5 mL diluted solution per minute. The blood pressure of the rabbits in the pyronaridine group did not change significantly or else decreased slightly. ECGs showed only a merger in T and P waves and extension in the P–R and QRS. There were no abnormal rhythms and no animals died.

Of the chloroquine group, the blood pressure of two rabbits dropped gradually after 3 min. After 7–9 min of infusion, the blood pressure of one of these rabbits fell by 25%, then rose almost to its original level and recovered quickly when administration of the drug stopped. The blood pressure of the other rabbit dropped to zero 32 min after drug administration and it died. At the 10-min mark, the ECG of this rabbit showed a prolonged P–R interval and a significantly widened QRS. The width and depth of the S wave almost doubled, heartbeat gradually slowed, and blood pressure dropped to zero. An extension in the P–R and QRS was also found in the four surviving rabbits, but no abnormal rhythms were seen.

Intravenous Drip in Rabbits

Five healthy unanesthetized rabbits were given a total dose of 80 mg/kg intravenous pyronaridine via the auricular vein. A control group was given an identical dose of chloroquine. The drugs were diluted in saline to form a 0.1% solution, with 10 mg/kg delivered each time via intravenous drip over the course of an hour. This was administered twice on Day 1, at an interval of 6 h, and then once a day. The total course was 7 days. During this time and 48 h afterward, the five rabbits in the pyronaridine group showed no adverse reactions, with no abnormal changes in ECGs and blood tests. The rabbits were observed for a month and regular ECGs and blood tests also recorded no changes. In the chloroquine group, one rabbit in the five died 20 min after the fifth infusion.

After dissection, atrial enlargement with congestion, diffuse emphysema, and pathological changes in the lungs could be seen with the naked eye. The other rabbits showed no abnormalities.

The surviving rabbits in both groups were observed after the drugs were stopped. After 20 days, the weight of the rabbits in both groups showed an increase. ECGs before and 10, 20, and 30 days after the drugs were stopped showed no abnormalities in both groups.

These experiments showed that the cardiovascular toxicity of pyronaridine depended not only on the dosage but also on the speed of drug administration. Rapid intravenous injection of 10 mg/kg could lead to death. Yet when the dosage was raised to 30 mg/kg and the period of delivery extended to an hour, no rabbits died. While the drug was administered, there was no change in blood pressure and only small and reversible changes were seen in ECGs. It indicated that intravenous delivery of pyronaridine at a slow rate not only preserved its therapeutic efficacy but increased its safety. It also showed that parenteral administration was safe.

5. Specific Toxicity Tests

Mutagenicity[44]

An Ames test was run using five strains of histidine-deficient *Salmonella typhimurium*: TA-100, TA-98, TA-1535, TA-1537, and TA-1538. The dose of pyronaridine was 100–1000 μg/dish. Results showed that pyronaridine caused mutation in TA-1537 alone. At this dose, the number of revertants showed a correlation to increases in dosage. Chloroquine produced reversions in TA-1537 at 100–250 μg/dish. Pyronaridine caused no mutagenicity on the other four strains.

Teratogenicity

Four groups of pregnant rats were given doses of 1100, 330, 165, and 84 mg/kg intragastric pyronaridine, respectively, 7 days into pregnancy. The large 1100 mg/kg dose was administered in a single feeding, whereas the three lower doses were given once daily over 3 days. A known teratogenic drug was used on a control group. Twenty days into pregnancy, the rats were dissected to observe the embryo.

Results showed that pyronaridine could cause an increase in the rate of early-term absorption of the embryo, which correlated with higher doses. It could delay the ossification of the sternum and occipital bone of the fetal rats, and had a certain level of embryotoxicity. However, compared to the teratogenic drug, the embryotoxicity of pyronaridine was clearly lower, at significant or very significant levels of difference ($P<.05–.001$). There were no malformations in the visceral and other skeletal structures of the fetuses in the pyronaridine groups.[45] To further investigate the embryotoxicity of pyronaridine on rats, an experiment was run on the effects of pyronaridine on three generations of rats. Female and male parent rats were fed 10 and 20 mg/kg pyronaridine daily for 60 days. The first generation produced from these rats, as well as the second generation produced from the first, both showed no outward or skeletal malformations. There was also no effect on the growth and development of the fetal rats.[46]

Dominant Lethal Test[47]

Three groups of 10 male rats were fed 343, 171.5, and 86 mg/kg pyronaridine, respectively, once a day for 3 days. After the rats were mated, the fertilization rate, average number of implantations, average number of live births, and post-implantation mortality rate were measured. The experimental results showed nothing abnormal, proving that pyronaridine had no obvious influence on the maturation and activity of male sperm cells.

Phototoxicity

Two groups of mice were given 600 and 800 mg/kg intragastric pyronaridine, respectively, with 20 mice in the first group and 18 in the second. They were then irradiated with UV light for 24 h. No phototoxic reaction appeared. However, 300 mg/kg intragastric chloroquine produced a mild phototoxic reaction in mice.[48] In another experiment, 10 mice were fed 300 mg/kg pyronaridine, with the results also showing that pyronaridine produced no phototoxic reaction and hence no phototoxicity.[49]

9.3 PYRONARIDINE CLINICAL TRIALS

Preclinical pharmacology and toxicology tests showed that pyronaridine phosphate had a significant effect on rodent malaria and in vitro cultures of human *P. falciparum* and that its toxicity was lower than that of chloroquine. Because the doses of chloroquine used to treat vivax and falciparum malaria were 1.2 and 1.5 g, respectively, administered over 3 days, it was proposed that the clinical dose of oral pyronaridine should be 1.2 g for normal and 1.6 g for severe malaria. This would be given four times in 3 days, with two doses on Day 1 at an interval of 4–6 h, and one each on Days 2 and 3.

Severe malaria could be treated with an intravenous drip of 2–6 mg/kg or with an intramuscular injection of 2–4 mg/kg. In both cases the drug would be administered twice a day, with an interval of 8–12 h. For the intravenous drip, the pyronaridine injection could be mixed with 150–500 mL of 5% glucose or 5% glucose saline solution, agitated, and then infused over 1–3 h. As a control, the standard chloroquine therapy of 1.5 g in 3 days was used (i.e., 1.5 g administered in four doses, with two doses at a 4- to 6-hour interval on Day 1, the first dose being double that of the second, and one dose each on Days 2 and 3). The total dosage of chloroquine by intramuscular injection would be 2 mg/kg, given either once or over two injections at an interval of 4–6 h.

Standard clinical trial protocol demanded that patients be examined once every 7 or 10 days after parasite clearance, for 28 days or a month follow-up.

Any side effects after administration of the drug would be observed, and ECGs, blood, urine, and other routine tests conducted before and after treatment.

In the early 1970s, before clinical trials were conducted, researchers ran an experiment on healthy volunteers under hospital and doctor supervision following the proposed regimen mentioned above. The result was that individual subjects experienced dizziness or abdominal discomfort, which were relatively mild and disappeared after the drug was stopped. Side effects of intramuscular injection and intravenous drip were even milder. Before and after the trial, ECGs and blood and urine tests showed no significant changes. This indicated that the proposed drug dosages were safe. Thereafter, tests carried out in Zhejiang and Hainan on malaria patients further proved that this dosage was safe and effective.

With the aid of research units in Hainan, Yunnan, Zhejiang, Henan, Anhui, Shandong, Sichuan, Jiangsu, Hubei, and other provinces, clinical or expanded field trials were run during the malaria season. The conclusions are listed below.

A. EFFICACY ON VIVAX MALARIA

The efficacy of pyronaridine on vivax malaria is shown in Table 9.5.[50–56] Oral pyronaridine was given to 244 patients and injections to 324; all showed reduction in fever and parasite clearance. Average time taken for fever reduction and parasite clearance with oral and injected pyronaridine was the same as those of chloroquine. After 28-day or 1-month follow-up, the recrudescence rate was about 10%. However, because patients were tracked during the malaria season reinfection could not be ruled out. The adverse clinical reactions in pyronaridine patients were significantly lower than those in the chloroquine group. Eleven vivax malaria patients in late pregnancy were all cured with pyronaridine, without any side effects.

B. EFFICACY ON FALCIPARUM MALARIA

The efficacy of pyronaridine on 651 cases of falciparum malaria is shown in Table 9.6.[50–58] Between the pyronaridine and chloroquine control groups, there were no significant differences in the average time for fever reduction and parasite clearance. The pyronaridine group included a 4-month-old infant, a 77-year-old man, four pregnant women, and 12 cases of cerebral malaria displaying convulsions, high fever, coma, and other symptoms; all were cured. The recrudescence rate was about 11% during the follow-up period, but as this was during the malaria season, reinfection could not be ruled out. Patients whose symptoms did not abate with chloroquine and who showed no parasite clearance were treated with pyronaridine instead; all showed reduction in fever and parasite clearance, with no parasite recrudescence 1 month later.

The side effects of oral pyronaridine included dizziness, headache, nausea, vomiting, abdominal discomfort, and loose stool. Reactions were mild and disappeared on their own after the drug was stopped. The rate of occurrence of side

TABLE 9.5 Efficacy of Pyronaridine on 568 Cases of Vivax Malaria

Route of Administration	Total Dose	Length of Treatment (Day)	Number of Cases	Average Fever Reduction Time (Hour)	Average Parasite Clearance Time (Hour)	Recrudescence Number/Follow-Up Number
Oral	1.2 g	3	194	18–36	40–64	7/94
Intramuscular injection	1.6 g	3	50	13–18	–	–
	2 mg/kg	1	37	18–30	47.5 ± 17.2	3/16
	4 mg/kg	1	260	17–48	28–43	4/29
Intravenous drip	2 mg/kg	1	14	20.3 ± 7.8	60.1 ± 29.3	2/10
	6 mg/kg	1	13	21.3 ± 9.8	50.7 ± 21.2	0/7

TABLE 9.6 Efficacy of Pyronaridine on 649 Cases of Falciparum Malaria

Administration Route	Total Dose	Days Treated	Cases	Average Fever Reduction Time (h)	Average Parasite Clearance Time (h)	Recrudescence Cases/Follow-Up
Oral	0.8 g[a]	2	32	27.0 ± 14.1	57.2 ± 10.2	0/32
	1.2 g	2	64	30.2 ± 43.5	57.9 ± 41.4	6/60
	1.6 g	3	57	41.2 ± 32.9	58.8 ± 48.3	9/54
Intramuscular injection	2 mg/kg	1	46	26.4 ± 14.8	53.3 ± 15.6	4/30
	3 mg/kg	1	17	24.5 ± 15.5	51.3 ± 15.3	–
	4 mg/kg	1	329	19–34	32–56.7	3/44
	6 mg/kg	2	11	36	53	–
	9 mg/kg	2	11	26	41	–
Intravenous drip	2 mg/kg	1 (1 dose)	22	27.2 ± 16.0	72.1 ± 25.5	4/6
	4 mg/kg	1 (2 doses)	18	19.4	53.6	–
	6 mg/kg	1 (2 doses)	42	27.8 ± 17.8	58.1 ± 14.7	0/4

[a]*Ordinary tablets administered twice, with 0.5 g on Day 1 and 0.3 g on Day 2. All other oral preparations were enteric-coated tablets.*

effects was significantly lower than that of chloroquine. There were no obvious reactions in most of the intramuscular injection cases, with the few side effects being slight dizziness, headache, nausea, vomiting, and individual cases of heart palpitation or rash that disappeared 2 days after symptomatic treatment. The intravenous drip had no obvious side effects. ECGs and blood, urine, and other routine tests were run before and after oral and parenteral administration of the drug. They showed no significant changes and the drugs were easily tolerated.

C. EFFICACY ON CEREBRAL AND OTHER FORMS OF SEVERE MALARIA

Intramuscular injections of 4 mg/kg or intravenous drips of 4–6 mg/kg pyronaridine were given in several doses to 121 severe cases of malaria, including cerebral and gastrointestinal malaria, malaria with jaundice and high fever with convulsions, and serious malaria in late pregnancy leading to coma; all were cured.[50–52,54,55,59–64] The malaria prevention research unit at the Guangzhou University of Chinese Medicine saved 14 cases of cerebral malaria with pyronaridine. The average time for fever reduction was 46.6 h. Parasite clearance took 77.4 h on average. It was thought that the drug acted quickly on the ring forms of *P. falciparum*, with no significant side effects in the process.

D. EFFICACY ON FALCIPARUM MALARIA IN AREAS WITH ENDEMIC CHLOROQUINE RESISTANCE

The first case of enteric-coated pyronaridine tablets being used to treat multidrug-resistant falciparum malaria took place in 1981. After undergoing various drug regimens, a woman returning to France from a holiday in Indonesia was identified by French doctors as a patient with chloroquine-resistant and Fansidar-resistant falciparum malaria. She was given 1.2 g enteric-coated pyronaridine tablets over 3 days. The trophozoites disappeared on Days 3–4, and 6 months of follow-up showed no recrudescence. No side effects were seen during treatment.[65]

In an area of Thailand where multidrug-resistant falciparum malaria is endemic, enteric-coated pyronaridine tablets were used to treat 32 cases of uncomplicated falciparum malaria. The total dosage was increased to 1.8 g, divided into six doses, with two doses on Day 1 and one dose daily for 5 days. Average fever reduction time was 81.4 ± 52.2 h and average parasite clearance time was 86.7 ± 24.4 h. After 28 days of follow-up, recrudescence occurred in 3 out of 26 of the cases. Side effects were mild. Another 69 cases of falciparum malaria were given a total dose of 1.2 g using the 3-day regimen; all showed fever reduction and parasite clearance, with no significant difference with the group given 1.8 g. However, the recrudescence rate was higher in the 1.2 g group. Pyronaridine was considered a very promising drug in the treatment of multidrug-resistant falciparum malaria in Thailand, and a drug of choice for combination therapy.[66]

In 1986–87, in an area of Zanzibar where falciparum malaria is endemic, patients were given 4 mg/kg pyronaridine via deep intramuscular injection. This was repeated after 6 h. In those patients where parasite clearance did not occur after 48 h, a third injection was given. Standard chloroquine therapy was used as a control. Pyronaridine was given to 140 patients, aged from 6 months to 65 years, with falciparum malaria. Of these, 117 were children and 23 adults. They were all treated while hospitalized and laboratory examinations run before and after administration. The fever reduction time of pyronaridine group was 22.1 ± 4.3 h and the parasite clearance time was 39.8 ± 4.8 h. In the control group of 30 children with falciparum malaria given conventional chloroquine treatment, the fever reduction and parasite clearance times were 66 ± 7.9 h and 180 ± 23.7 h, respectively. Hence, the efficacy of pyronaridine was higher ($P < .01$) and its clinical side effects were small.

ECGs and routine urine analysis were conducted on all cases, while 70 cases were tested for liver function. There were no obvious abnormalities. Eight patients who did not respond to chloroquine were also treated with pyronaridine. In the pyronaridine group, there were 10 cases of cerebral malaria and four cases with second-trimester pregnancies; all were cured. The pregnant women were followed for almost a year. There were no abnormalities in mothers and infants and the infants showed good development.[63]

In a hospital in Mali, 3–6 mg/kg pyronaridine was given via intravenous drip. This was repeated after 4–6 h. The symptoms abated in 1–2 days and all 19 cases of falciparum malaria were cured.[64]

While treating cases of falciparum malaria in Yaounde, Cameroon, French researchers randomly assigned patients to pyronaridine and chloroquine groups. A total dose of 32 mg/kg pyronaridine in enteric-coated tablets were given four times over 3 days, with two doses on Day 1 and one dose each on Days 2 and 3. Forty patients were in this group. Average fever reduction time was 33.5 ± 14.3 h and parasite clearance time was 76.8 ± 14.6 h. During a 14-day follow-up, there was no recrudescence. For chloroquine, a total dose of 25 mg/kg was administered once daily over 3 days: 10 mg/kg on Days 1 and 2 and 5 mg/kg on Day 3. This group had 41 cases. Apart from those cases which did not respond to chloroquine, fever reduction time was 40 ± 20.8 h and parasite clearance time was 70 ± 14.9 h. The recrudescence rate was 41% during the 14-day follow-up.[67]

In vitro tests run on the strains of *P. falciparum* isolated from unresponsive cases showed that these patients to have an RI–RIII level of chloroquine resistance. It was thought that pyronaridine could act rapidly on African falciparum malaria without severe side effects. Patients' tolerance was good, making it a potential drug for the treatment of chloroquine-resistant falciparum malaria.[67]

E. EFFICACY ON FALCIPARUM MALARIA IN AFRICAN CHILDREN

In 1996, French researchers in Yaounde treated 88 cases of falciparum malaria in children aged 5–15. These were randomly divided into pyronaridine and chloroquine groups. Forty-four received a total dose of 32 mg/kg pyronaridine in enteric-coated tablets over 3 days. Fever reduction time was 35.71 ± 5.3 h and parasite clearance time 72.6 ± 15.6 h. For 14 days, 41 cases were tracked, showing no recrudescence. Another 44 children received a total dose of 25 mg/kg chloroquine orally over 3 days. Fever reduction time was 31.3 ± 15.2 h and parasite clearance time 70.9 ± 16.1 h. Forty cases were tracked for 14 days, which yielded 16 cases of recrudescence.[68]

In vitro tests on the *P. falciparum* strains obtained from 40 patients in the pyronaridine group showed that 23 of these patients had chloroquine-resistant strains. The side effects of pyronaridine were mild and transient. It was thought that children could tolerate pyronaridine well. In Africa, where chloroquine-resistant falciparum malaria has been confirmed, pyronaridine was a safe and effective antimalarial drug for treating children.[68]

In treating cases of falciparum malaria, clinical trials showed that pyronaridine was effective in patients with RI- to RIII-level chloroquine resistance, or where chloroquine therapies had failed.[52,54,55,61–64]

F. EFFICACY ON *PLASMODIUM OVALE*- AND *PLASMODIUM MALARIAE*-INFECTED PATIENTS

French researchers in Yaounde treated eight cases (aged seven to 38) infected with *P. ovale* and 14 cases (aged eight to 45) infected with *P. malariae* using enteric-coated pyronaridine tablets. The total dose was 32 mg/kg over 3 days. Average fever reduction time for both groups was 44.6 and 49.8 h, respectively. Average parasite clearance time was 58.3 and 60.9 h, respectively. After a 14-day follow-up, there was no recrudescence in both groups. Nine of the 14 *P. malariae* and 3 of the 8 *P. ovale* patients had mild abdominal discomfort (abdominal pain, diarrhea) and itching, but these side effects disappeared spontaneously after medication ceased. Clinical side effects were mild. Parasites were obtained for in vitro analysis, which showed sensitivity to chloroquine and pyronaridine. Pyronaridine could be an effective alternative drug for treating patients infected with *P. falciparum*, *P. ovale*, and *P. malariae* in sub-Saharan Africa.[69]

9.4 COMBINATION OF ARTEMISININ DERIVATIVES AND PYRONARIDINE IN TREATING FALCIPARUM MALARIA

Using antimalarial drugs in combination not only improves their efficacy but may also delay or inhibit the emergence of drug resistance. Fansimef, a combination of mefloquine and Fansidar (sulfadoxine–pyrimethamine), can retard the development of mefloquine resistance.[70] Pyronaridine, with its long half-life, has been combined with Fansidar, which also has a long half-life. Pharmacology experiments showed that the drugs with this combination had an additive effect against *P. berghei* schizonts, also in terms of additive toxicity in mice and rats. The combination could also delay the development of drug resistance. Clinical trials involving falciparum malaria indicated that the combination increased efficacy, shortened the course of treatment and produced minimal recrudescence. Side effects were mild and patients tolerated it well.[71]

WHO/TDR repeatedly stressed that artemisinin derivatives should not be used in isolation when treating falciparum malaria. It advocated for artemisinin-based combination therapies (ACTs), in which artemisinin or its derivatives are combined with other antimalarials. Therefore, clinical trials were conducted combining various artemisinin derivatives with pyronaridine.

A. EFFICACY OF VARIOUS ARTEMISININS AND PYRONARIDINE ON FALCIPARUM MALARIA

In 1996 and 1997, trials were conducted in areas where multidrug-resistant falciparum malaria is endemic: Ledong in Hainan, Mengla in Yunan, and the Yuanjiang Autonomous County of Hani, Yi, and Dai. Dihydroartemisinin, artemether, and artesunate were each combined with pyronaridine in oral treatment for falciparum malaria.

Total dosages of 200mg dihydroartemisinin, 300mg artemether, or 300mg artesunate were combined with 800mg pyronaridine. These dosages were administered over two doses, with an interval of 12–24h. Both medicines in each combination were taken orally at the same time. Before and after medication, routine ECGs and blood and urine tests were conducted. Patients were followed up at 28 and 40days, showing no recrudescence. The results can be seen in Table 9.7. Side effects were light, with individual patients experiencing temporary mild abdominal pain, nausea, and stomach discomfort, but these symptoms disappeared spontaneously. No abnormalities were found in the ECGs and blood and urine tests. The combination of artemether and pyronaridine was used to treat a woman at the sixth month of pregnancy. No adverse reactions occurred after medication.

In the group receiving dihydroartemisinin and pyronaridine, a gametocyte density of 3–158 per μL was found in three of the cases. Three days after medication, the patient with the lowest density showed no gametocytes. For the other patients, similar results were obtained 14 and 20days after medication, respectively. The same test was carried out for the artemether and pyronaridine group, in which three patients had a gametocyte density of 50–265 per μL. Two of the patients showed no gametocytes 4 and 5days after medication, respectively. The one remaining case still had gametocytes 5days after medication. In the artesunate and pyronaridine group, one patient had a gametocyte density of 345 per μL, which fell to zero 4days after treatment.

The combination of pyronaridine with any artemisinin derivative could reduce the required dosage of each individual drug, shorten the course of treatment, and improve cure rates. Adverse reactions were slight. To overcome the long duration of treatment necessary and high recrudescence of artemisinin derivatives when used in isolation, combinations with pyronaridine are ideal for treating drug-resistant falciparum malaria.[72,73]

TABLE 9.7 Efficacy of Various Artemisinins With Pyronaridine on Falciparum Malaria

Drug Combination	Cases	Average Fever Reduction Time (h)	Average Parasite Clearance Time (h)	Recrudescence Number/ Follow-Up Number
DHA+PR	10	23.1±12.7	34.4±10.0	0/10[a]
ATM+PR	10	22.2±13.7	35.9±12.9	0/9
ATS+PR	9	15.2±6.8	29.9±9.6	0/6[b]

ATM, artemether; *ATS*, artesunate; *DHA*, dihydroartemisinin; *PR*, pyronaridine.
[a]*Seven of the cases were followed for 40 days.*
[b]*These were followed for 30 days, with the rest of the cases followed for 28 days.*

B. EFFICACY OF DIHYDROARTEMISININ AND PYRONARIDINE ON FALCIPARUM MALARIA

In Ledong, Hainan, where drug-resistant falciparum malaria is endemic, a trial was run combining dihydroartemisinin with pyronaridine. Dihydroartemisinin and pyronaridine separately were used as controls. Eighty-one cases of falciparum malaria were randomly assigned to these three groups, with the results shown in Table 9.8. All the patients experienced reduction in fever and parasite clearance. A 28-day follow-up was conducted on all the cases: One patient in the dihydroartemisinin group experienced recrudescence of *P. falciparum* and accompanied by a temperature of 39.6°C, and another four cases showed recrudescence of vivax malaria. There was no recrudescence in the combination therapy group and the pyronaridine group. It showed that combination therapy could overcome the high recrudescence rate of dihydroartemisinin. At the same time, drug dosages and the course of treatment could be reduced. Adverse reactions were slight, and only individual cases had mild headache and dizziness which spontaneously diminished in 1–2 days.

After administration of the combination therapy, rate of occurrence of gametocytes was 20%. Average gametocyte density was 4 per μL and average gametocyte clearance time was 5.7 days. The corresponding rate for the dihydroartemisinin group was 16.7%, with an average density of 3 per μL and average clearance time of 3.5 days. For the pyronaridine group, the rate was 60.9%, average density was 12 per μL, and the average clearance time 11.5 days. The rate of occurrence of gametocytes in the pyronaridine group was higher than that of the combination and dihydroartemisinin groups, with a statistically significant difference ($P<.01$).[74]

TABLE 9.8 Efficacy of Dihydroartemisinin With Pyronaridine on Falciparum Malaria

Drug/Total Dose (mg)	Length of Treatment (Day)	Cases	Fever Reduction Time (h)	Parasite Clearance Time (h)	Recrudescence Number/ Follow-Up Number
DHA (200) + PR (800)	3[a]	32	35.7 ± 24.7	23.8 ± 10.1	0/32
DHA (480)	7[b]	24	52.6 ± 38.9	22.9 ± 6.5	5/24[c]
PR (1600)	3[d]	25	35.8 ± 16.5	49.4 ± 20.3	0/25

[a]*One dose of DHA and PR each were given on the first administration. After 4–6 h, PR was given alone. On Days 2 and 3, one dose of DHA and PR each were administered together.*
[b]*Two doses were administered on the first day.*
[c]*One case of recrudescence accompanied by a temperature of 39.6°C, and the other four cases were* P. vivax.
[d]*Two doses were given on Day 1 and once daily thereafter.*

C. EFFICACY OF DIHYDROARTEMISININ AND PYRONARIDINE ON CHLOROQUINE-RESISTANT FALCIPARUM MALARIA IN AFRICA

In 2000, falciparum malaria patients in Tessenei, Eritrea—where chloroquine-resistant falciparum malaria is endemic—were given a combination of dihydroartemisinin and pyronaridine. They were observed in hospital, with pyronaridine used in isolation as the control. A 2-week WHO in vivo test was conducted to determine chloroquine resistance. Cases which failed to respond to the standard 3-day regimen of oral chloroquine within 3–14 days (the asexual form of the parasite was not cleared or there was recrudescence within 14 days) were used as subjects in this trial. The results are in Table 9.9.

Both the combination and pyronaridine groups of chloroquine-resistant cases were effectively treated, with no recrudescence in 14 days. Adverse reactions were mild. Gametocytes were still present in both groups. Two patients (15.4%) in the combination group still carried gametocytes, as opposed to five cases (55.6%) in the pyronaridine group, higher than in the combination group. The combination of two drugs represented a potential treatment for falciparum malaria in areas of Africa with chloroquine resistance. To promptly eliminate the source of infection and control transmission, an appropriate amount of primaquine was added in both groups to kill the gametocytes.[75]

In sum, the various artemisinin derivatives in combination with pyronaridine were tested in domestic and international clinical trials. They proved to be ideal combination therapies worthy of further study.

WHO/TDR supports combination therapies in the treatment of malaria. Already in 2001, the TDR joined with Shin Poong Pharmaceutical Co. in South Korea to develop the pyronaridine–artesunate product[76] Pyramax. This obtained WHO certification in 2012.

TABLE 9.9 Efficacy of Dihydroartemisinin With Pyronaridine on Chloroquine-Resistant Falciparum Malaria in Africa

Total Dose (mg)	Length of Treatment (Day)	Cases	Average Asexual Parasite Clearance Time (Hour)
DHA (200) + PR (800)	2[a]	13	27.69
PR (1200)	3[b]	11	50.18

[a] *Drugs were given once daily.*
[b] *Two doses were given on Day 1 at an interval of 4–6 h, and one dose each on Days 2 and 3.*

REFERENCES

1. Ren DX, Sun JL, Huang WZ, et al. *Comparative study on the effects of cycloquine, amodiaquine and chloroquine*. Annual Report of the Institute of Parasitic Diseases, Chinese Academy of Medical Sciences; 1962. p. 216. [任道性，孙金琳，黄文洲，等. 环喹啉、氨酚喹啉与氯喹疗效的比较研究. 中国医学科学院寄生虫病研究所年报，1962:216.]
2. Ren DX, Pan YR, Huang ZS. *Experimental study on a new drug, M6303, an 4-aminoquinoline I. Observation of the effects against Plasmodium berghei and toxicity in mice*. Annual Report of the Institute of Parasitic Diseases, Chinese Academy of Medical Sciences; 1963. p. 122. [任道性，潘玉蓉，黄再松. 4-氨基喹啉类新药M6303的实验研究I.对伯氏疟原虫(Plasmodium berghei)和对小鼠毒性观察. 中国医学科学院寄生虫病研究所年报，1963:122.]
3. Hoekenga MT. Intramuscular amopyroquin for acute malaria. *Am J Trop Med Hyg* 1962;**11**(1):1–5.
4. Nobles WL, Tietz RF, Koh YS, et al. Antimalarial agents VIII. Synthesis of amopyroquine. *J Pharm Sci* 1963;**52**(6):600.
5. Huang LS, Zheng XY, Chen C, et al. Synthesis of M6407, a new antimalarial agent. *Acta Pharm Sin* 1979;**14**(9):561. [黄兰孙，郑贤育，陈昌，等. 新抗疟药M6407的合成. 药学学报, 1979, 14(9): 561.]
6. Zheng XY, Chen C, Huang LS. Synthesis of antimalarial bispyroquine and its analogues. *Chin J Pharm* 1990;**21**(5):200. [郑贤育，陈昌，黄兰孙. 抗疟药双咯喹及其类似物的合成. 中国医药工业杂志, 1990, 21(5): 200.]
7. Ren DX, Huang LS, Zheng XY, et al. Synthesis, toxicity and antimalarial effects of bispyroquine. *Acta Pharm Sin* 1983;**4**(1):69. [任道性，黄兰孙，郑贤育，等. 双咯喹的合成、毒性与抗疟作用. 中国药理学报, 1983, 4(1): 69.]
8. Huang ZS, Shen DY, Deng MH, et al. Effects of bispyroquine in the treatment of 42 malaria cases. *J N Drugs Clin Remedies* 1987;**6**(4):244. [黄再松，沈大勇，邓明华，等. 双咯喹治疗42例疟疾的疗效. 新药与临床, 1987, 6(4): 244.]
9. Zheng XY, Xia Y, Gao FH, et al. Synthesis of 7351, a new antimalarial drug. *Acta Pharm Sin* 1979;**14**(12):736. [郑贤育，夏毅，高芳华，等. 抗疟新药7351的合成. 药学学报, 1979, 14(12): 736.]
10. Zheng XY, Chen C, Gao FH, et al. Synthesis of a new antimalarial drug, pyronaridine, and its analogues. *Acta Pharm Sin* 1982;**17**(2):118. [郑贤育，陈昌，高芳华，等. 抗疟新药咯萘啶及其类似物的合成. 药学学报, 1982, 17(2): 118.]
11. Chen C. Synthesis of a new antimalarial, pyronaridine phosphate. *Chin J Pharm* 1981;**9**:12. [陈昌. 抗疟新药磷酸咯萘啶的合成. 医药工业, 1981, (9): 12.]
12. Chen C, Zheng XY, Zhu PE, et al. Studies on new antimalarials – synthesis of heterocyclic compounds with double Mannich basic chains of p-aminophenol. *Acta Pharm Sin* 1982;**17**(2):112. [陈昌，郑贤育，朱佩萼，等. 抗疟新药的研究——具有对-氨基酚双Mannich 碱的杂环化合物的合成. 药学学报, 1982, 17(2): 112.]
13. Chen C, Gao FH, Zhu PE, et al. Studies on new antimalarials – synthesis of derivatives of benzo[b]1,5-naphthyridine. *Acta Pharm Sin* 1982;**17**(5):344. [陈昌，高芳华，朱佩萼，等. 抗疟新药的研究——苯并[b]1,5-萘啶衍生物的合成. 药学学报, 1982, 17(5): 344.]
14. Chen C, Zheng XY, Guo HZ. Synthesis of pyronaridine related compounds and comparison of antimalarial activities. *Acta Pharm Sin* 1993;**28**(8):594. [陈昌，郑贤育，郭惠珠. 抗疟药咯萘啶有关化合物的合成及抗疟活性比较.药学学报, 1993, 28(8): 594.]
15. New Drug Group of the Former Department of Malaria Research. Experimental studies on the efficacy and toxicity of a new antimalarial, 7351. *Acta Pharm Sin* 1980;**15**(10):630. [原疟疾研究室新药组. 抗疟新药7351的疗效和毒性的实验研究. 药学学报, 1980, 15(10): 630.]

16. Shao BR, Ye XY, Zheng H. The effects of pyronaridine and three other antimalarials on the proliferation of *P. yoelii* gametocytes and sporozoites. *Chin J Parasitol Parasit Dis* 1984;**2**(1):28. [邵葆若，叶秀玉，郑浩. 咯萘啶等4种抗疟药对约氏鼠疟原虫配子体与孢子增殖的影响. 寄生虫学与寄生虫病杂志, 1984, 2(1): 28.]
17. Shao BR, Ye XY, Zheng H. Sustained effects of pyronaridine on erythrocytic rodent malaria schizonts. *Chin J Parasitol Parasit Dis* 1984;**2**(4):232. [邵葆若，叶秀玉，郑浩. 咯萘啶对鼠疟原虫红内期裂殖体的持效作用. 寄生虫学与寄生虫病杂志, 1984, 2(4): 232.]
18. Peters W, Robinson BL. The chemotherapy of rodent malaria. XLVII. Studies on pyronaridine and other Mannich base antimalarials. *Ann Trop Med Parasitol* 1992;**86**(5):455.
19. Dai ZR, Ma ZM, Chen L. Development of a hydroxy-piperaquine-resistant line of *Plasmodium berghei* ANKA strain. *Acad J Second Mil Med Univ* 1986;**7**(4):265. [戴祖瑞，马志明，陈林. 伯氏疟原虫ANKA株抗羟基哌喹系的培育. 第二军医大学学报, 1986, 7(4): 265.]
20. Li GD. Development of a piperaquine-resistant line of *Plasmodium berghei* K173 strain. *Acta Pharm Sin* 1985;**20**(6):412. [李高德. 伯氏疟原虫K173株对喹哌抗药性的实验研究. 药学学报, 1985, 20(6): 412.]
21. Li GD, Qu FY, Chen L. Development of a piperaquine-resistant line of *Plasmodium berghei* ANKA strain. *Chin J Parasitol Parasit Dis* 1985;**3**(3):189. [李高德，瞿逢伊，陈林. 伯氏疟原虫ANKA株抗喹哌系的培育. 寄生虫学与寄生虫病杂志, 1985, 3(3): 189.]
22. Ye XY, Shao BR. Tissue schizontocidal action and acute toxicity of trifluoroacetyl primaquine. *Acta Pharmacol Sin* 1990;**11**(4):359. [叶秀玉，邵葆若. 三氟乙酰伯氨喹杀疟原虫组织期的作用及其毒性. 中国药理学报, 1990, 11(4): 359.]
23. Satoru K, Shigeyuki K, Chen C. The effects of pyronaridine on the morphology of *Plasmodium falciparum* in *Aotus trivirgatus*. *Am J Trop Med Hyg* 1996;**55**(2):223.
24. Zhang JX, Lin BY, Chen KY, et al. Sustained activity of pyronaridine against *Plasmodium cynomolgi* in rhesus monkeys. *Chin J Parasitol Parasit Dis* 1986;**4**(2):129. [张家埙，林宝英，陈克涌，等. 咯萘啶对食蟹猴疟原虫持效作用的探讨. 寄生虫学与寄生虫病杂志, 1986, 4(2): 129.]
25. Li J, Huang WJ. Effects of artesunate, pyronaridine, and hydroxypiperaquine on chloroquine-sensitive and chloroquine-resistant isolates of *Plasmodium falciparum* in vitro. *Acta Pharmacol Sin* 1988;**9**(1):83. [李军，黄文锦. 青蒿酯钠、咯萘啶和羟基哌喹对恶性疟原虫氯喹敏感株和抗性虫株的体外效应. 中国药理学报, 1988, 9(1): 83.]
26. Che LG, Huang KG, Yang HL. Determining the sensitivity of pyronaridine against a chloroquine-resistant strain of *Plasmodium falciparum*. *Chin J Parasitol Parasit Dis* 1984;**2**(4):280. [车立刚，黄开国，杨恒林. 抗氯喹恶性疟原虫株对咯萘啶敏感性的测定. 寄生虫学与寄生虫病杂志, 1984, 2(4): 280.]
27. Yang HL, Yang PF, He H, et al. In vitro sensitivity of *Plasmodium falciparum* to pyronaridine and artesunate. *Chin J Parasit Dis Control* 1989;**2**(3):169. [杨恒林，杨品芳，何慧，等. 恶性疟原虫对咯萘啶、青蒿琥酯敏感性现场体外测定. 中国寄生虫病防治杂志, 1989, 2(3): 169.]
28. Fu S, Björkman A, Wåhlin B, et al. In vitro activity of chloroquine, the two enantiomers of chloroquine, desethylchloroquine and pyronaridine against *Plasmodium falciparum*. *Br J Clin Pharmacol* 1986;**22**(1):93.
29. Tang LH, Sucharit P. In vitro study on susceptibility of *Plasmodium falciparum* to pyronaridine, chloroquine, and mefloquine in Thailand. *Chin J Parasitol Parasit Dis* 1988;(Suppl.):16. [汤林华，Sucharit P. 泰国恶性疟原虫对咯萘啶、氯喹及甲氟喹的敏感性研究. 中国寄生虫学与寄生虫病杂志, 1988, (增刊): 16.]
30. Childs GE, Hausler B, Milhous W, et al. In vitro activity of pyronaridine against field isolates and reference clones of *Plasmodium falciparum*. *Ann J Trop Med Hyg* 1988;**38**(1):24.

31. Basco LK, Le Bras J. In vitro susceptibility of Cambodian isolates of *Plasmodium falciparum* to halofantrine, pyronaridine and artemisinin derivatives. *Ann Trop Med Parasitol* 1994;**88**(2):137.
32. Basco LK, Le Bras J. In vitro activity of pyronaridine against African strains of *Plasmodium falciparum*. *Ann Trop Med Parasitol* 1992;**86**(5):447.
33. Wu LJ. Effects of pyronaridine on the ultrastructure of the erythrocytic stage of *Plasmodium berghei* in mice. *Acta Pharmacol Sin* 1985;**6**(4):280. [吴莉菊. 咯萘啶对伯氏鼠疟原虫红内期超微结构的影响. 中国药理学报, 1985, 6(4): 280.]
34. Wu LJ. Observations on the effects of pyronaridine on the ultrastructure of the erythrocytic stage of chloroquine-resistant *Plasmodium berghei*. *Chin J Parasitol Parasit Dis* 1986;**4**(4):263. [吴莉菊. 咯萘啶对伯氏疟原虫抗氯喹系红内期作用的超微结构观察. 寄生虫学与寄生虫病杂志, 1986, 4(4): 263.]
35. Wu LJ, Rabbege JR, Nagasawa H, et al. Morphological effects of pyronaridine on malarial parasites. *Am J Trop Med Hyg* 1988;**38**(1):30.
36. Chavalitshewinkoon P, Wilairat P, Gamage S, et al. Structure-activity relationship and modes of action of 9-anilinoacridines against chloroquine-resistant *Plasmodium falciparum* in vitro. *Antimicrob Agents Chemother* 1993;**37**(3):403.
37. Huang MF, Wu GW, Cheng DX. Effects of pyronaridine on the phagocytosis of macrophages in peritoneal exudates of mice. *Chin J Parasitol Parasit Dis* 1989;**7**(3):218. [黄美芳，吴光武，程道新. 咯萘啶对小鼠腹腔巨噬细胞吞噬功能的影响. 中国寄生虫学与寄生虫病杂志, 1989, 7(3): 218.]
38. Feng Z, Jiang NX, Wang CY, et al. Pharmacokinetics of pyronaridine, an antimalarial, in rabbits. *Acta Pharm Sin* 1986;**21**(11):801. [冯正，江乃雄，王翠英，等. 抗疟药咯萘啶在兔体内的药代动力学. 药学学报, 1986, 21(11): 801.]
39. Feng Z, Wu ZF, Wang CY, et al. Distribution and excretion of ^{3}H-pyronaridine in mice. *Acta Pharm Sin* 1988;**23**(8):629. [冯正，吴祖帆，王翠英，等. 3H-咯萘啶在小鼠体内的分布和排泄. 药学学报, 1988, 23(8): 629.]
40. Liu XR, Wu RJ. HPLC assay for pyronaridine and primaquine in plasma. *J China Pharm Univ* 1989;**20**(4):240. [刘星荣，吴如金. HPLC法测定咯萘啶和伯喹的血药浓度. 中国药科大学学报, 1989, 20(4): 240.]
41. Wages SA, Patchen LC, Churchill FC. Analysis of blood and urine samples from *Macaca mulatta* for pyronaridine by high-performance liquid chromatography with electrochemical detection. *J Chromatogr* 1990;**527**(1):115.
42. Feng Z, Wu ZF, Wang CY, et al. Pharmacokinetics of pyronaridine in malaria patients. *Acta Pharm Sin* 1987;**8**(6):543. [冯正，吴祖帆，王翠英，等. 咯萘啶在疟疾病人体内的药物动力学. 中国药理学报, 1987, 8(6): 543.]
43. Jayaraman SD, Ismail S, Nair NK, et al. Determination of pyronaridine in blood plasma by high-performance liquid chromatography for application in clinical pharmacological studies. *J Chromatogr B Biomed Sci Appl* 1997;**690**(1–2):253.
44. Ni YC, Xu YQ, Shao BR. Mutagenicity of a new antimalarial drug, pyronaridine, in the microsome system of *Salmonella typhimurium*. *Acta Pharmacol Sin* 1982;**3**(1):51. [倪奕昌，徐月琴，邵葆若. 抗疟药咯萘啶在鼠伤寒沙门氏菌/微粒体系统中的诱变性. 中国药理学报, 1982, 3(1): 51.]
45. Ni YC, Zhan CQ, Ha SH, et al. The embryotoxicity of a new antimalarial pyronaridine in rats. *Acta Pharm Sin* 1982;**17**(6):401. [倪奕昌，湛崇清，哈淑华，等. 抗疟新药咯萘啶对大鼠胚胎毒性的观察. 药学学报, 1982, 17(6): 401.]
46. Shao BR, Zhan CQ, Ha SH. Influence of pyronaridine phosphate on three generations of rats. *Acta Pharmacol Sin* 1985;**6**(2):131. [邵葆若，湛崇清，哈淑华. 磷酸咯萘啶对大鼠三代生殖的影响. 中国药理学报, 1985, 6(2): 131.]

47. Shao BR, Zhan CQ, Ha SH, et al. Dominant lethal test with pyronaridine on male mice. *Chin J Parasitol Parasit Dis* 1983;**1**(2):121. [邵葆若，湛崇清，哈淑华，等. 咯萘啶对雄性小鼠的显性致死试验. 寄生虫学与寄生虫病杂志, 1983, 1(2): 121.]
48. Shao BR, Zhan CQ, Ha SH. Evaluation of the phototoxicity of five antimalarial agents and praziquantel in mice. *Acta Pharmacol Sin* 1986;**7**(3):273. [邵葆若，湛崇清，哈淑华. 五个抗疟药与吡喹酮的小鼠光毒试验. 中国药理学报, 1986, 7(3): 273.]
49. Yuan BJ, Li BC, Chen NC. Evaluation of the phototoxic effects of chloroquine and eleven other antimalarials in mice. *Acta Pharmacol Sin* 1986;**7**(5):468. [袁伯俊，李保春，沈念慈. 氯喹等12种抗疟药对小鼠的光毒检测. 中国药理学报, 1986, 7(5): 468.]
50. Huang ZS, Liu DQ, Wang YC, et al. Observations on the efficacy of the antimalarial pyronaridine. *Natl Med J China* 1985;**65**(6):366. [黄在松，刘德全，王元昌，等. 咯萘啶抗疟效果观察. 中华医学杂志, 1985, 65(6): 366.]
51. Xu YX, Liu DQ, Wang YC, et al. Clinical observations on the efficacy of pyronaridine phosphate injections in treating malaria. *Natl Med J China* 1982;**62**(11):686. [许永湘，刘德全，王元昌，等. 磷酸咯萘啶针剂治疗疟疾的临床观察. 中华医学杂志, 1982, 62(11): 686.]
52. Xu YX, Wang YC, Liu DQ, et al. Clinical observations on pyronaridine phosphate intravenous drip for the treatment of malaria. *Chin J Intern Med* 1982;**21**(11):655. [许永湘，王元昌，刘德全，等. 磷酸咯萘啶静脉滴注治疗疟疾的临床观察. 中华内科杂志, 1982, 21(11): 655.]
53. Kang WM, Yang RZ, Huang RZ, et al. Observations on the efficacy of pyronaridine on vivax malaria. *Chin J Parasitol Parasit Dis* 1984;**2**(4):263. [康万民，杨仁政，黄瑞芝，等. 咯萘啶治疗间日疟的疗效观察. 寄生虫学与寄生虫病杂志, 1984, 2(4): 263.]
54. Weng QL, Liu YC. Efficacy of pyronaridine in the treatment of 30 falciparum malaria cases. *Chin J N Drugs Clin Remedies* 1987;**6**(4):246. [翁其琳，刘运彩. 咯萘啶治疗恶性疟30例的疗效. 新药与临床, 1987, 6(4): 246.]
55. Pyronaridine Joint Research Group. Efficacy of pyronaridine on 510 acute malaria cases. *Chin J Intern Med* 1985;**24**(11):646. [咯萘啶研究协作组. 磷酸咯萘啶治疗510例疟疾现症病人的观察. 中华内科杂志, 1985, 24(11): 646.]
56. Shao BR. A review of antimalarial drug pyronaridine. *Chin Med J* 1990;**103**(5):428.
57. Huang ZS, Feng Z, Meng F, et al. Comparison of the therapeutic effects of normal and enteric-coated pyronaridine tablets in falciparum malaria patients. *Chin J Parasitol Parasit Dis* 1989;**7**(1):19. [黄在松，冯正，蒙锋，等. 磷酸咯萘啶普通片与肠溶片对恶性疟疗效的比较. 中国寄生虫学与寄生虫病杂志, 1989, 7(1): 19.]
58. Chen L, Dai ZR, Qian YL, et al. Observations on the efficacy of the combined use of several new antimalarials for the treatment of falciparum malaria in Hainan province. *Chin J Parasitol Parasit Dis* 1989;**7**(2):81. [陈林，戴祖瑞，钱永乐，等. 几种新抗疟药联合应用治疗海南恶性疟的效果观察. 中国寄生虫学与寄生虫病杂志, 1989, 7(2): 81.]
59. Zhang JS, Xu YX, You BY, et al. Emergency treatment of three cases of cerebral malaria. *Chin J Parasitol Parasit Dis* 1983;**1**(1):59. [张家埙，许永湘，尤伯英，等. 三例危重脑型恶性疟的救治. 寄生虫学与寄生虫病杂志, 1983, 1(1): 59.]
60. Luo ZH, Gao J, Dai XZ, et al. Successful emergency treatment of a case with cerebral malaria and renal failure. *Chin J Parasitol Parasit Dis* 1992;**10**(2):158. [罗振辉，高健，戴祥章，等. 一例抢救成功的脑型恶性疟疾伴肾衰病例. 中国寄生虫学与寄生虫病杂志, 1992, 10(2): 158.]
61. Wei G. Experiences from the emergency treatment of 60 cerebral malaria cases. *Hainan Med J* 1991;**2**(4):6. [韦珪. 抢救脑型疟疾60例的体会. 海南医学, 1991, 2(4): 6.]
62. Cai XZ, Pang XJ, Chen WJ. Pyronaridine phosphate in the treatment of 10 chloroquine-resistant falciparum malaria cases. *Chin J Parasitol Parasit Dis* 1987;**5**(3):228. [蔡贤铮，庞学坚，陈文江. 磷酸咯萘啶治疗抗氯喹恶性疟10例. 中国寄生虫学与寄生虫病杂志, 1987, 5(3): 228.]

63. Xiong RJ. Efficacy of pyronaridine in 140 cases of falciparum malaria. *J Nanjing Med Univ* 1988;**8**(3):208. [熊人杰. 磷酸咯萘啶治疗恶性疟疾140例临床疗效观察. 南京医学院学报, 1988, 8(3): 208.]
64. Hu WX. Clinical analysis of 2,102 cases of falciparum malaria. *Chin J Parasit Dis Control* 1992;**5**(3):215. [吴文仙. 2102例恶性疟疾临床分析. 中国寄生虫病防治杂志, 1992, 5(3): 215.]
65. Lapierre J. Paludisme à *Plasmodium falciparum*, polychimiorésistant, traité avec succès par la benzonaphthyridine. *La Nouv Presse Méd.* 1982;**11**(9):673.
66. Looareesuwan S, Kyle DE, Viravan C, et al. Clinical study of pyronaridine for the treatment of acute uncomplicated falciparum malaria in Thailand. *Am J Trop Med Hyg* 1996;**54**(2):205.
67. Ringwald P, Bickii J, Basco L. Randomised trial of pyronaridine versus chloroquine for acute uncomplicated falciparum malaria in Africa. *Lancet* 1996;**347**(8993):24.
68. Ringwald P, Bickii J, Basco LK. Efficacy of oral pyronaridine for the treatment of acute uncomplicated falciparum malaria in African children. *Clin Infect Dis* 1998;**26**(4):946.
69. Ringwaid P, Bickii J, Same-Ekobo A. Pyronaridine for treatment of *Plasmodium ovale* and *Plasmodium malariae* infections. *Antimicrob Agents Chemother* 1997;**41**(10):2317.
70. Merkli B, Richle R, Peters W. The inhibitory effect of a drug combination on the development of mefloquine resistance in *Plasmodium berghei*. *Ann Trop Med Parasitol* 1980;**74**(1):1.
71. Shao BR, Huang ZS, Shi XH, et al. Antimalarial and toxic effect of triple combination of pyronaridine, sulfadoxine and pyrimethamine. *Southeast Asian J Trop Med Pub Health* 1989;**20**(2):257.
72. Cai XZ, Chen C, Zheng XY, et al. Preliminary study of dihydroartemisinin combined with pyronaridine for the treatment of falciparum malaria. *J Pract Parasit Dis* 1999;**7**(3):104. [蔡贤铮，陈昌，郑贤育，等. 双氢青蒿素与咯萘啶联用治疗恶性疟的初步研究. 实用寄生虫病杂志, 1999, 7(3): 104.]
73. Cai XZ, Che LG, Chen C, et al. Clinical efficacy of artemether or artesunate combined with pyronaridine in the treatment of falciparum malaria. *J Pract Parasit Dis* 1999;**7**(4):156. [蔡贤铮，车立刚，陈昌，等. 蒿甲醚或青蒿琥酯与咯萘啶联用治疗恶性疟的临床效果观察. 实用寄生虫病杂志, 1999, 7(4): 156.]
74. Liu DQ, Lin SG, Feng XP, et al. Study on the treatment of multidrug-resistant falciparum malaria with a combination of dihydroartemisinin and pyronaridine. *Chin J Parasitol Parasit Dis* 2002;**20**(4):193. [刘德全，林世干，冯晓平，等. 双氢青蒿素与咯萘啶伍用治疗抗性恶性疟的研究. 中国寄生虫学与寄生虫病杂志, 2002, 20(4): 193.]
75. Shang LY, Han QX. Observations on the efficacy of pyronaridine phosphate and its dihydroartemisinin combination in the treatment of malaria. *Chin J Parasit Dis Control* 2001;**14**(3):171. [尚乐园，韩庆霞. 磷酸咯萘啶及其合并双氢青蒿素治疗疟疾效果观察. 中国寄生虫病防治杂志, 2001, 14(3): 171.]
76. Kanyok T. Support for the development of two antimalarial combinations: the pyronaridine-artesunate project for treatment of uncomplicated malaria. *TDR News* 2001;**65**:10.

Chapter 10

Dihydroartemisinin and Artemisinin in Combination With Piperaquine (Artekin, Artequick); Primaquine and Malaria Transmission; and Malaria Elimination*

Chapter Outline

* Fast Elimination of Malaria by Source Eradication (FEMSE).

Artemisinin-Based and Other Antimalarials. https://doi.org/10.1016/B978-0-12-813133-6.00010-X

10.1 PIPERAQUINE AND PIPERAQUINE PHOSPHATE

A. SYNTHESIS OF PIPERAQUINE AND PIPERAQUINE PHOSPHATE

Zhang Xiuping[1], *Zhang Zhixiang*[2]
[1]*State Institute of Pharmaceutical Industry, Shanghai, China;* [2]*Institute of Microbiology and Epidemiology, Academy of Military Medical Sciences, Beijing, China*

Laboratory studies had previously shown that intramuscular injections of cycloguanil pamoate (CI501) had antimalarial effects lasting half a year or more, but the drug could cause severe localized reactions as well.[1] Since antimalarials with long-lasting efficacy were of great importance to troop strength and public health in malarial regions, the State Institute of Pharmaceutical Industry (SIPI) in Shanghai (simply called Shanghai Institute of Pharmaceutical Industry below) and Second Military Medical University made it a key research goal.

In 1965, a new type of compound with piperazinyl side chains and bis-(7-chloro-4-aminoquinoline) was found to have long-lasting suppressive effects on *Plasmodium berghei*.[2] Two of these—13228 RP and 12494 RP—were combined at a ratio of 3:1 to form a combination drug, Code 16126 RP. A trial in West Africa showed that the compound had persistent prophylactic effects of up to 3 weeks.[3] However, this was not followed by other clinical reports on the drug.

In 1966, SIPI synthesized 13228 RP and named it piperaquine (**1**). Its chemical name was 1,3-bis[4-(7-chloroquinolinyl-4)piperazinyl-1]propane, with a formula of $C_{29}H_{32}Cl_2N_6$ and CAS registration number 4085-31-8. The Institute subsequently synthesized piperaquine tetraphosphate tetrahydrate, or piperaquine phosphate (PQP) for short (**2**). Its molecular formula was $C_{29}H_{52}Cl_2N_6O_{20}P_4$, with Chemical Abstract Service (CAS) registration number 915967-82-7.

1. Properties

Piperaquine (**1**) is a white or pale-yellow crystalline powder, odorless, and tasteless. Its color changes readily in light. It has a melting point of 199–204°C, at which it decomposes. It is highly soluble in chloroform, slightly soluble in ethanol, and almost insoluble in water. Piperaquine is also highly fat-soluble ($\log P = 6.16$) and has a logarithmic dissociation constant (pK_a) of 8.92.[4]

PQP (**2**) is a white or pale-yellow crystalline powder, odorless, and with a slightly bitter taste. It is hygroscopic and changes color in light. It is soluble in water above 90°C, slightly soluble in cold water, and almost insoluble in anhydrous ethanol and chloroform. It decomposes at a melting point of 246–252°C.

2. Preparation

Many methods of synthesizing piperaquine and its intermediates have been published both in China and abroad,[2,5–9] all with their own shortcomings: low yield, cost, or difficulty of obtaining raw materials, complexity, and impact on the environment or on health. Based on its research, the Shanghai Institute of Pharmaceutical Industry (SIPI) chose the process outlined in Fig. 10.1.

First, 4,7-dichloroquinoline was condensed with excess piperazine hexahydrate to get 4-piperazinyl-7-chloroquinoline. The leftover piperazine was retrieved. The product was then reacted in ethanol with 1-bromo-3-chloropropane and sodium carbonate to yield piperaquine. This was salified with phosphoric acid and refined to obtain PQP.

Compared to the process published by the French company Rhone-Poulenc S.A.,[2] the above method made three improvements. First, four to five times the volume of piperazine hexahydrate was used in the first step to minimize side products from the substitution reaction. Subsequently, 40%–50% of the piperazine could be recycled. Second, 1-bromo-3-chloropropane and ethanol were used in the second step instead of costly dibromopropane and butanone (which had to be imported at the time). Using sodium carbonate instead of triethylamine as a neutralizer also reduced costs. Third, the salification and refining process had originally been divided into two steps. Now, one step was used, namely salification and recrystallization in water. This saved a large amount of isobutanol, again lowering costs. In 1970, the Shanghai 14th Pharmaceutical Factory began using this process to manufacture piperaquine.[10] After the drug was included in the *Pharmacopoeia of the People's Republic of China* in 1977, more factories also commenced production.

In 2007, Zhang Zhixiang et al. from the Academy of Military Medical Sciences (AMMS) further refined the method of producing 4-piperazinyl-7-chloroquinoline. 4,7-Dichloroquinoline was reacted with excess anhydrous piperazine and anhydrous sodium carbonate in anhydrous ethanol. Sodium carbonate was

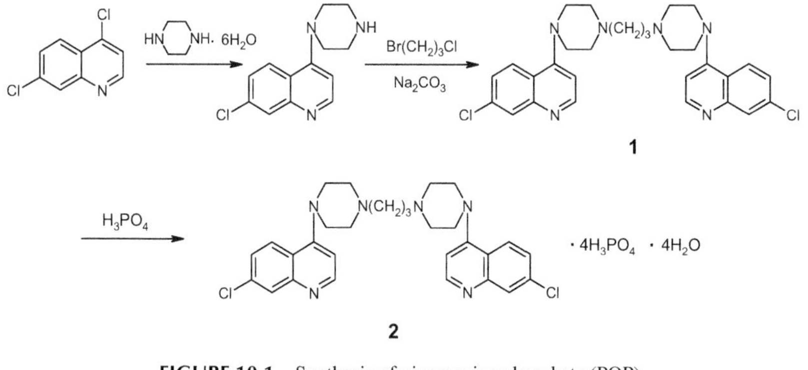

FIGURE 10.1 Synthesis of piperaquine phosphate (PQP).

filtered out and the ethanol recycled. The rest of the product was washed with water; the oil then obtained could be easily processed to form light-yellow crystals of 4-piperazinyl-7-chloroquinoline. Yield and purity both markedly increased.

3. Analysis of Impurities

Using TLC, the Shanghai Institute of Pharmaceutical Industry (SIPI) measured the purity of piperaquine. Points indicative of impurities were found both above and below piperaquine. The higher point (Impurity **2** in Fig. 10.2) was known to be 1,4-bis-(4,7-dichloroquinolinyl)piperazine, the result of a reaction between 4,7-dichloroquinoline and an intermediate. During this process, most of Impurity **2** may be removed via acid–base extraction or ethanol. When PQP was used to formulate Malaria Prevention Tablets Number 3 unless the impurities were removed toxicity reactions such as headache, dizziness, fatigue, nausea, and abdominal pain and facial numbness occurred.[11] Therefore, these toxic effects could be associated with Impurity **2**.

It had previously been reported that active ingredient PQP contained four main impurities (Fig. 10.2). In 2007, Dongre et al. used mass spectrometry (MS, MS/MS) and other spectral methods to determine the structure of PQP and deduce those of Impurities **1**, **2**, **3**, and **6**.[12] Impurity **1** was presumed to be 7-chloro-4-(piperazinyl)quinoline, in a small amount, an unreacted intermediate in piperaquine synthesis. Impurity **3** was the residue of a reaction between an intermediate and 1-bromo-3-chloropropane. Impurity **6** was a position isomer of piperaquine, the product of Impurity **4** in the starting material (4,5-bichloroquinoline) and Impurity **5**. Lindegardh et al.[13] showed that the NMR spectrum

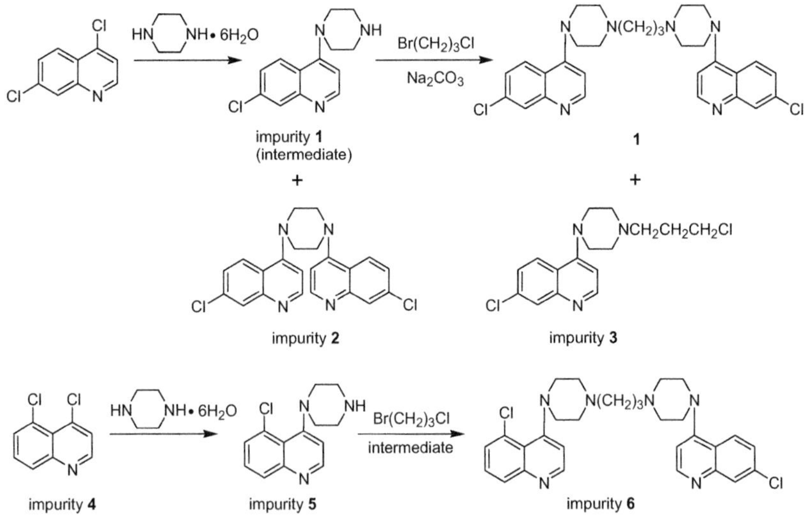

FIGURE 10.2 Impurities in the synthesis process of piperaquine.

and HPLC-MS revealed that clinical use of a dihydroartemisinin–piperaquine combination contained up to 1.5%–5% of Impurity **6**.

This combination of dihydroartemisinin (DHA)–PQP (later developed as an artemisinin combination therapy (ACT) Artekin, Duo-Cotecxin) was widely tested in Southeast Asia. Its satisfactory therapeutic efficacy and good safety record showed that Impurity **6** produced no obvious toxic reactions. However, the presence of the impurity had to be strictly monitored and curbed to adhere to GDP quality control standards and FDA regulations. Stringent control over the amount of 4,5-bichloroquinoline in the starting material should resolve the issue.

Dongre et al.[12] conducted gradient reversed-phase chromatography on Impurities **1**, **2**, **3**, and **6** and PQP. The relative retention times of the peaks were 0.20, 0.85, 0.90, and 1.10 min for Impurities **1**, **2**, **3**, and **6**, respectively.

ACKNOWLEDGMENTS

The authors of this section would like to express their gratitude to Chen Wenbin, who participated in the early work on piperaquine synthesis.

B. ANTIMALARIAL ACTIVITY AND TOXICITY OF PIPERAQUINE[14–16]

Chen Lin
Antimalarial Laboratory, Institute of Parasitic Diseases, Second Military Medical University, Shanghai, China

1. Effect on *Plasmodium berghei* and *Plasmodium cynomolgi*

The ED_{50} of piperaquine on the ANKA strain of *P. berghei* was 1.71 (1.59–1.83) mg/kg. The ED_{50} for chloroquine was 3.09 (2.77–3.44) mg/kg and for ATS, it was 9.13 (7.72–10.80) mg/kg. Rhesus monkeys infected with *P. cynomolgi* were given either piperaquine or chloroquine. There were no significant differences in the rate or time of parasite clearance.

2. Effect on ANKA Strain of *Plasmodium berghei*: Ultrastructure Changes[17]

After piperaquine was used to treat ANKA strain of *P. berghei*, clear electron microscopic changes appeared in the membranes of the late trophozoite's food vacuoles (fv), showing that this was the drug's initial target. Round or oval spots of agglutinated pigment (p) could be seen in the food vacuoles, obviously different from the classic crystal structures caused by chloroquine. Piperaquine crystallization was clearly different from that of chloroquine.

There were no obvious effects on the early trophozoites (tiny ring form), schizonts, and gametocytes in infected mice 24 h after intragastric administration of piperaquine. Marked changes in appearance occurred among late trophozoites (large ring form).

One hour after medication, a small proportion of the late trophozoites had markedly swollen food vacuoles with wavy outlines, as opposed to the no-treatment control (Fig. 10.3O). The outlines of hemoglobin in the food vacuole were clear. Electron density was uniform (Fig. 10.3A) and pigment grains still appeared as individual rectangular crystals (Fig. 10.3B).

Two hours after medication, the membranes of the food vacuole continued to swell, forming multilayered whorls. The pigment grains took on varying forms, mostly as spindles, and began to agglutinate (Fig. 10.3C).

Four hours after medication, most of the trophozoite food vacuoles showed whorled membranes (Fig. 10.3D). The mitochondria (m) were swollen, some becoming mushroom-shaped, and electron density fell (Fig. 10.3E). One or two bodies with multilamellate membranes (mb) appeared (Fig. 10.3F). The gap between the nuclear (n) membranes grew wider (Fig. 10.3D). Agglutinated round or oval pigment grains were seen in the food vacuole (Fig. 10.3F). Each trophozoite had up to several such pigment clumps, surrounded by a single membrane and containing 10–20 grains. The electron density of the grains increased (Fig. 10.3G).

Eight hours after medication, the above changes persisted. However, the pigment in the trophozoites' residual body continued to present as rectangular crystals (Fig. 10.3H).

Twelve hours after medication, most of the trophozoites were seriously damaged. The nuclear membranes were swollen, with wide gaps in between. The chromatin had agglutinated in star-shaped clusters. Mitochondria were swollen and electron density decreased significantly (Fig. 10.3I). Some of the parasites had disintegrated, leaving only parts of the membrane. Autophagic vacuoles appeared in others, some of which had been exocytosed into the erythrocyte (Fig. 10.3J).

Sixteen hours after medication, damage to most trophozoites became more serious. Spots of agglutinated pigment joined to form obvious clusters, encased in a single membrane. The electron density of the pigment grains increased noticeably (Fig. 10.3K) and more autophagic vacuoles appeared in the cytoplasm.

In 20–24 h, all the trophozoites disintegrated, leaving only membrane parts and vacuoles of different sizes. No obvious changes were seen in the early ring forms, schizonts, and gametocytes.

After 36–48 h, the parasite was hard to detect. The few individuals found had mostly disintegrated, with only membrane parts and vacuoles remaining (Fig. 10.3L).

As a comparison, the effects of chloroquine on ANKA strain *P. berghei* in the erythrocytic phase were observed. Four hours after medication, the food vacuoles in a few of the trophozoites had enlarged membranes and the pigment grains had agglutinated into rectangular crystals (Fig. 10.3M). After 8 h, more severe damage was seen in the food vacuole membranes of most of the parasites. In 12 h, autophagic vacuoles had formed in most of the trophozoites. Some had been exocytosed. Membranous residues also appeared in the parasite (Fig. 10.3N).

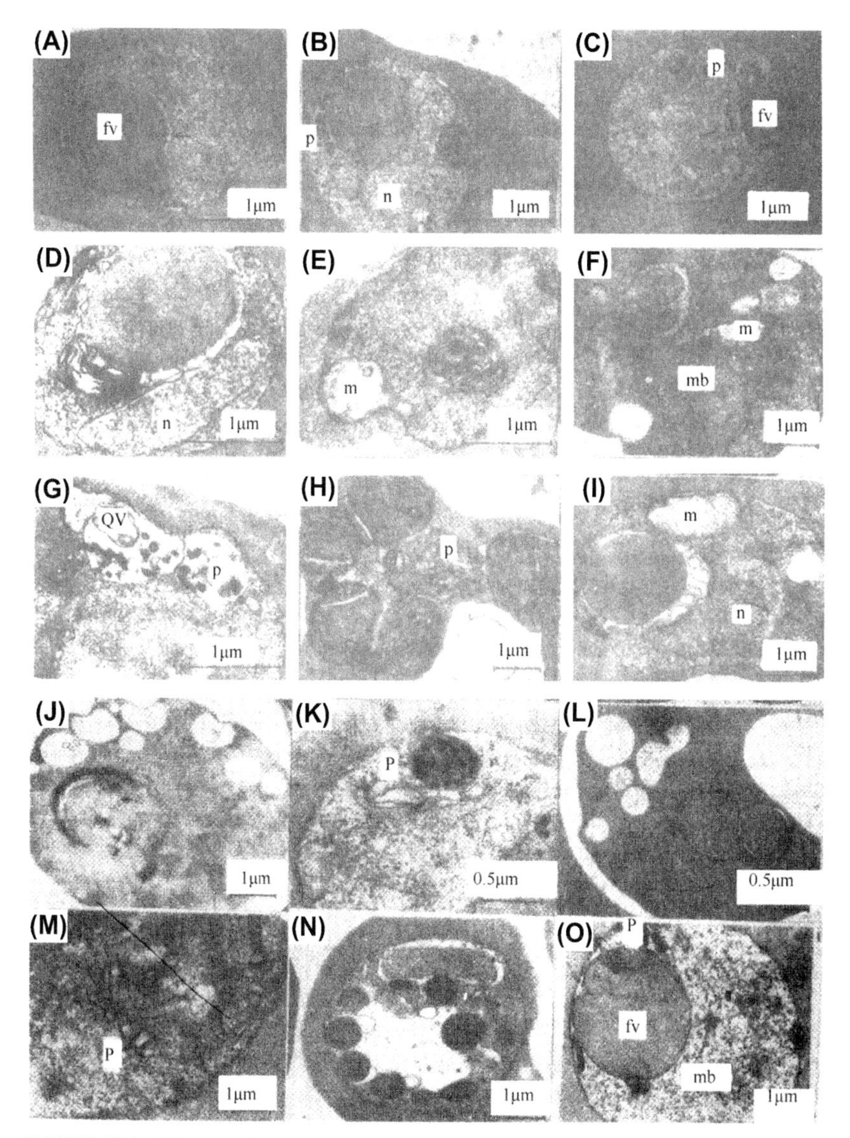

FIGURE 10.3 Electron micrographs of ANKA strain *P. berghei* trophozoites in the blood of mice given 6.4 mg/kg intragastric piperaquine (A–L), 40 mg/kg intragastric chloroquine (M and N) or no treatment (O). (A and B) Trophozoites after 1 h: swelling of food vacuole (fv) membrane (A) and normal pigment (p) bars (B); (C) Trophozoites after 2 h: formation of multilamellate whorls and abnormal pigment grains; (D–G) Trophozoites after 4 h: swelling of mitochondrion (m, E), proliferation of multilamellate membrane body (mb, F), and abnormal shape of the clumped pigment grains with a very dense appearance (G); (H) Schizont after 8 h: normal clumped pigment bars in the residual body; (I and J) Trophozoites after 12 h: swollen intermembranous space between the outer and inner nuclear membranes (I) and disintegrated parasite with some membranous residues (J); (K) Trophozoite after 16 h: digestive vacuole with clumped round or oval pigment grains which were abnormally electron-lucent in appearance; (L) Trophozoite after 36 h: conspicuous whorls of membrane within the cytoplasm; (M) Trophozoite after 4 h: normal scattered pigment bars; (N) Trophozoite after 12 h: membranous digestive residues.

However, the shape of the parasite was unaffected and the organelles could be discerned. After 16h the parasites had disintegrated, leaving autophagic vacuoles of different sizes or membranous residues. Schizonts and gametocytes showed no abnormalities. The parasite fragments were hard to detect 24h after medication.

Comparing the effects of piperaquine and chloroquine on the parasitic ultrastructure, both caused changes in the membranes of the food vacuole, neither had an impact on the early ring forms, schizonts, and gametocytes. With piperaquine, pigment grains became round or oval and agglutinated; with chloroquine, they agglutinated into rectangular crystals, indicating that the components of the crystals may be different. Piperaquine had obvious effects on mitochondria, whereas no abnormalities were seen in the mitochondria with chloroquine. Generally chloroquine could eliminate parasites faster than piperaquine.

These results showed that the main effects of piperaquine on the ultrastructure of late-stage trophozoites were changes in the food vacuole membranes and mitochondria, as well as abnormalities in the pigment. The gap between the inner and outer membranes of the food vacuole became wider. The membranes were swollen and formed whorls, possibly due to disruption of the membranous enzymes and hence the function of the membranes. Mitochondria were also swollen and deformed, undermining their physiological function. The proliferation of mitochondria and the appearance of multilamellate membrane bodies within them could be a compensatory response to structural damage. Howells maintains this was a way to overcome the effects of the drug. No significant changes were seen in the parasite's plasma membrane and erythrocytes, suggesting that piperaquine acted selectively. Regarding the pigment, which gradually changed from individual rectangular crystals dispersed in the cytoplasm to agglutinated round or oval grains, further research was needed.

3. Effects on Piperaquine-Resistant ANKA Strain of *Plasmodium berghei*: Ultrastructure Changes[18]

Piperaquine was an effective prophylactic and therapeutic drug, particularly against chloroquine-resistant falciparum malaria. It had been widely used in Hainan, Guangdong, and Yunnan for many years, as well as in some Southeast Asian countries where chloroquine-resistant falciparum malaria was endemic. Good results were obtained in all these areas.

In 1984, Chen Lin produced a piperaquine-resistant version of ANKA strain *P. berghei* (PR line), observing the ultrastructures of its erythrocytic stage. This line (PR/ANKA) was cultivated from 1983 to 1984 by the antimalarial research laboratory of the Shanghai Second Military Medical University, using small dose increments. The resistance was relatively stable and could be transmitted via *Anopheles stephensi*. Five healthy Kunming mice, weighing 20 ± 2 g, were infected with parasites cultivated in rodent blood via intraperitoneal injection. The resistance index of the PR/ANKA strain was 197.2 and 1×10^7 infected erythrocytes were injected. The mice were checked after 7 days and three mice with parasitemia of more than 20% were chosen. Blood was drawn orbitally and placed in a flat-bottomed flask, each mouse providing three to four drops. Once the blood had congealed, 2 mL of 3% glutaraldehyde (pH 7.25) was immediately added. After 5 min, the sample was cut into 1 mm^3 pieces, left for an hour and then washed three times in buffer for 15 min each time. It was then transferred to a 1.5% osmic acid solution for an hour, washed in buffer and distilled water for another hour, then progressively dehydrated with 50%–100% acetone. A temperature of 4°C was maintained throughout. The sample was embedded in Quetol 812 epoxy and placed in a constant temperature oven (35°C for 24 h, 45°C for another 24 h, and 60°C for 48 h). It was sectioned with an LKB Ultrotome V, stained with uranyl acetate and lead citrate, and observed with an H600 transmission electron microscope.

The ultrastructure morphology of all stages of its erythrocytic phase (ring forms, late trophozoites, immature and mature schizonts, male and female gametocyte of the normal ANKA strain of *P. berghei*) was documented. The malaria pigment appeared as rectangular crystals (Fig. 10.4). Contrasting with the above, the 226 specimens of the PR/ANKA strain had the following characteristics.

Although the structures of PR/ANKA trophozoites were intact, the number of food vacuoles had increased to three or four (Figs. 10.5 and 10.6), with markedly swollen and wavy membranes. Multilamellate whorls appeared in some of the membranes. The outlines of erythrocytic cytoplasm could clearly be seen in the food vacuoles. Their electron density was the same as that of erythrocytes. The nuclear membranes were also swollen, with a wider gap between the inner and outer membranes. The chromatin had agglutinated into spot-like and star-shaped clusters (Fig. 10.7). The number of mitochondria increased to six or seven; their membranes were swollen and electron density declined significantly. Many took on a dumbbell shape (Figs. 10.5 and 10.8) and some contained multilamellate membrane bodies. Compared to the normal strain, the

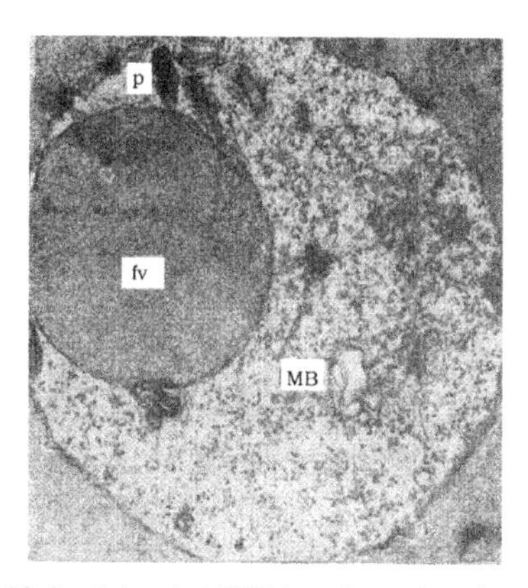

FIGURE 10.4 *P. berghei* ANKA strain trophozoite (×15,000).

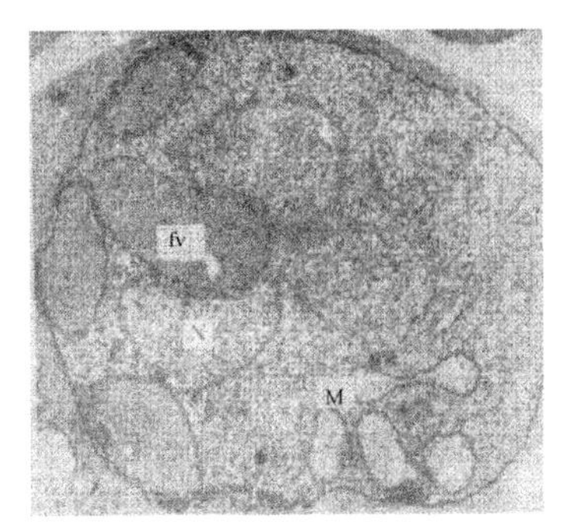

FIGURE 10.5 Increased food vacuoles (fv) and mitochondria (M) (×12,0000).

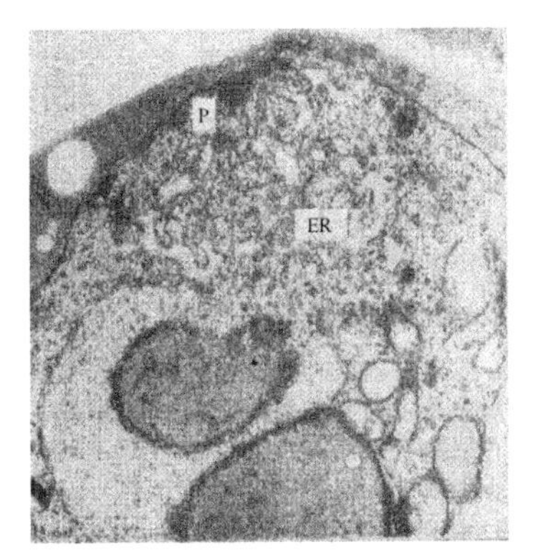

FIGURE 10.6 Endoplasmic reticulum (ER) (×15,000).

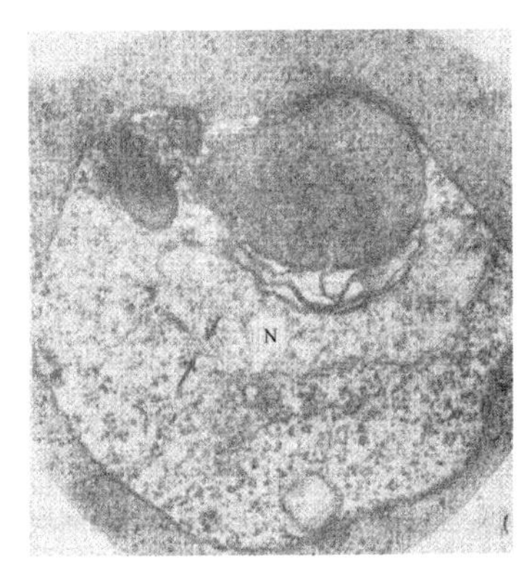

FIGURE 10.7 Swelling of nuclear membrane and spotlike agglutination of chromatin (×20,000).

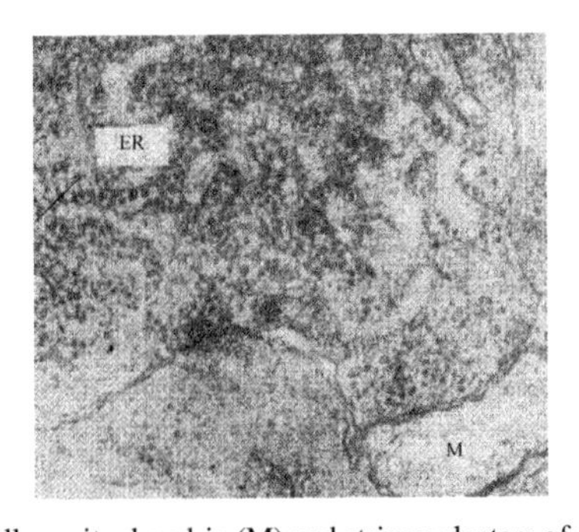

FIGURE 10.8 Swollen mitochondria (M) and stringy clusters of ribosomes (×35,000).

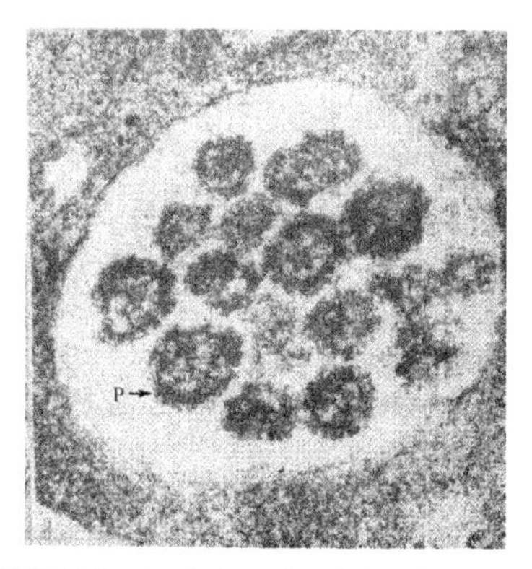

FIGURE 10.9 Agglutinated malarial pigment grains.

endoplasmic reticulum was more densely packed, with wormlike clusters of ribosomes (Fig. 10.8). The malarial pigment was mostly round or oval, agglutinated, with 10 or more pigment grains. Their electron density had increased in some specimens. Others had lower densities. They were enveloped in a single-layered phagocytic vacuole (Fig. 10.9). A few pigment grains were shaped like long spindles and there was a tendency to cluster.

Mature schizonts were basically normal. Each mature schizont had 6–10 merozoites and residual bodies. The former appeared in pairs, as rods or tubules. The gametocytes were also basically normal. Nuclei and mitochondria could be clearly seen. The rough endoplasmic reticulum was visible.

Parts of the parasite had disintegrated, leaving only membrane residues. Vacuoles of different sizes appeared. These characteristics had not been seen in the normal ANKA strain of *P. berghei*. There was significant effect on the ultrastructures of the erythrocytic stage. The ultrastructural changes contrasted with the 1968 paper by Howells et al., on the reduction or absence of pigment in chloroquine-resistant *P. berghei*. Such a difference could be due to varying digestive functions and degradation and metabolic pathways for hemoglobin in piperaquine- and chloroquine-resistant strains. In addition, there were similarities to the 1965 Peters paper on chloroquine-resistant *P. berghei*, in which the number of mitochondria increased significantly. It indicated that metabolic and respiratory functions were markedly stronger in resistant strains.

4. Toxicity

Acute Toxicity

The LD_{50} of piperaquine was 800 mg/kg for intragastric and 2500 mg/kg for subcutaneous administration. Its ED_{50} was 4 mg/kg/day, as opposed to 11 mg/kg/day for chloroquine. Piperaquine's efficacy against *P. berghei* was almost three times that of chloroquine. For a 5-day intravenous regimen, the LD_{50} of piperaquine was 95 mg/kg in total, while that of chloroquine was 35 mg/kg. Its toxicity was three times lower than that of chloroquine.[19]

The research data on Malaria Prevention Tablets Number 3 (piperaquine and sulfadoxine) compiled between 1966 and 1972 by the Shanghai Institute of Pharmaceutical Industry (SIPI) and other national collaborative working groups demonstrated that the intragastric LD_{50} of piperaquine in mice was 1098.5 mg/kg. The ED_{50} was 4.0 mg/kg. For chloroquine, the LD_{50} was 437.0 mg/kg and the ED_{50}, 10.5 mg/kg. The therapeutic index (LD_{50}/ED_{50}) of piperaquine was 274.6, significantly higher than that of chloroquine (41.9).

Four dogs were given 200 mg/kg oral piperaquine, among which two were also given 8 mg/kg sulfadoxine. One died. Another nine dogs were given 100 mg/kg oral piperaquine, among which eight also received 10 mg/kg sulfadoxine. None of the dogs in this group died. In another experiment, one monkey was given 200 mg/kg oral piperaquine. Two others were given 100 mg/kg doses. Again, none died.

The toxic effects of chloroquine on the cardiovascular systems of rabbits were closer to those observed in a clinical setting than similar reactions in dogs. Therefore, rabbits were chosen for the cardiovascular toxicity experiment. Every 2 min, equal doses of piperaquine or chloroquine were injected intravenously. The doses which caused serious drops in blood pressure (40 mmHg) and death (0 mmHg) were compared. The results showed that the corresponding doses for piperaquine were 42.1 ± 8.5 mg/kg and 45.1 ± 8.7 mg/kg, respectively. Those for chloroquine were 16.2 ± 3.1 mg/kg and 23.0 ± 3.2 mg/kg. The difference was significant and indicated that piperaquine was less toxic to the cardiovascular system of rabbits than chloroquine was.

Long-Term Toxicity

Rats and dogs were given intragastric doses of 25 or 50 mg/kg daily over 35 days. The dogs' weights were stable and no adverse reactions, including vomiting, were seen. All blood and clotting tests were normal. Apart from the presence of basophilic granules in the various blood cells and organs, such as the liver, kidneys, adrenal glands and lungs, the rats and dogs showed good long-term drug tolerance. It was supposed that these granules were the intracellular form of a piperaquine metabolite. The affected cells showed no other abnormalities.[19]

The research data collected by SIPI between 1966 and 1972 also included an experiment in which four dogs were given 100 mg/kg enteric coated (antiemetic formulation), piperaquine tablets every 7 days. This continued for 14 weeks, with a total dose of 1400 mg/kg. Liver function was checked 2 days after each dose. Three dogs had consistently normal ALT levels, but one had significantly higher levels after the second and 12th doses, which quickly recovered after medication ceased. It showed that the drug's impact on the liver was slight and reversible. Postmedication blood tests and monthly renal function checks were basically normal. The dogs were killed 3 months after medication ceased and the major organs examined. No drug-related pathologies were seen.

Another long-term toxicity experiment on Malaria Prevention Tablets Number 3 involved six dogs given 25 mg/kg oral PQP base and 2.5 mg/kg sulfadoxine once a week for half a year. A total of 26 doses were given. The main toxic side effect was liver damage. The ALT liver function indicator increased temporarily, but this was not a sustained rise. It could still return to normal levels during medication. There was a reduction in leukocyte count and the proportion of neutrophils, but this was not serious and no bone marrow suppression was found. The same drug combination was given to two monkeys once a week for 3 months, a total of 13 doses. Another three monkeys were given double the dose once every 2 weeks for half a year, a total of 10 doses. Liver function tests were normal and mild pathological changes were seen on liver tissue. Blood pressure and electrocardiograms (ECGs) were normal. The same drop in leukocyte count and neutrophil percentage were observed, which gradually reversed after medication ceased. There was no bone marrow suppression.

The clinical dose of oral piperaquine—12 mg/kg on the first day and 9 mg/kg on the second and third days—was equivalent to 1/8 of the canine dose of 100 mg/kg per week for 14 weeks.

C. PHARMACOKINETICS OF PIPERAQUINE

The absorption, distribution, and excretion of intragastric PQP and piperaquine in mice were examined using ^{14}C as a tracer. After absorption in the gastrointestinal tract, PQP rapidly accumulated in the liver, kidneys, lungs, spleen, and other organs. Eight hours after medication, around a quarter of the dose was accumulated in the liver. The drug was eliminated slowly, with a half-life of about 9 days. The main elimination pathway could be through bile. Therefore, it may exist in the hepatic circulation. This may be a key reason why the drug persisted in vivo.

Tests on mice suggested that the amount of pure piperaquine accumulated in the liver was less than with PQP. The speed of accumulation was also delayed. However, distribution and rate of elimination were the same 1 day after medication.

D. CLINICAL STUDIES OF PIPERAQUINE

1. Clinical Sensitivity of *Plasmodium falciparum* to Piperaquine[20]

Test sites in Hainan Island in a region where malaria was highly prevalent where it was hard to avoid the possibility of reinfection were used. Therefore, only the 7-day sensitivity test, recommended by the WHO, could be used. Forty-three cases of falciparum malaria were given piperaquine and observed on site. Body temperature returned to normal 3 days after medication. Full parasite clearance occurred in 24 cases 34–48 h after medication. The other 19 cases had full clearance in 58–72 h. No parasites were observed 7 days after medication.

2. Prophylaxis Trials[20]

Studied Population and Methods

From 1972 to 1974, large-scale field trials on the long-term prophylactic effects of PQP were conducted over the summer in a county in Hainan. They included 4000 subjects with partial or no immunity. The local dry season lasted from January to April, and the rainy season from May to September. Nine test sites were established, all in forested areas where the malaria vector, *Anopheles dirus*, had been identified. The parasite rate at each test site was ascertained before medication. *P. falciparum* was the primary strain of malaria, with a few vivax cases. The parasite carrier rate among the residents was around 10%, with some as high as 20% and above. Out of 157 people sampled, 39.5% had swollen spleens. Three to four years before the trial, some of the sites had implemented twice-yearly regimens of chloroquine plus primaquine (PMQ), administered over 4 days, or pyrimethamine plus PMQ. Outpatient records showed malaria morbidity was 16%–28% per year. Infections occurred year-round but peaked from June to September.

Most of the subjects were nonlocals who had been in the region for less than 3 years, or were children 6 years old and above who were born on the island. Each group was handled separately.

Tablets containing 0.3 g piperaquine were used. Adults were given two tablets; children aged 6–9 received one; those aged 10–15 took 1.5 tablet. Medication took place over the 4-month peak season, with one dose every 30 days, administered in the first 10 days of each month. The subjects were forestry workers. The drugs were given en masse, except for pregnant women and those with heart, liver, or kidney disease. Tablets were administered individually and administration of the drugs was observed.

The observation period began when the first drug dose was given and ended 30 days after the fourth dose was taken. Thick blood smears were obtained from those subjects experiencing fever of over 37.5°C. The absence of parasites in the entire smear was considered a negative result. If asexual parasites were found and the species identified, a diagnosis of malaria would be made and included in the morbidity rate. No antimalarials were given before confirmation, but the subject was closely observed. Treatment was promptly given after diagnosis. Subjects who contracted malaria after the observation period and without further medication were included in the nonmedicated morbidity rate.

At selected test sites, thick and thin blood smears were obtained from the subjects every 30 days, at the same time when the once monthly medication was given. Blood tests were again conducted 30 days after the fourth dose. These five sets of blood smears per subject would establish changes in the parasite rate over the course of treatment.

The WHO's 7-day sensitivity test was also used. Subjects were chosen who had clinical symptoms of malaria, more than 1000 per μL asexual *P. falciparum* in the blood and who had received no medication for half a month before infection. Those aged 16 and above were given six piperaquine tablets (total 1.8 g); those aged 5–15 received five. The medication was given in two doses, at an interval of 8–24 h. Thick and thin blood smears were obtained from each subject on the day of medication before the drugs were received, and 1–7 days afterward. The thick smears were examined for 100 fields. If no asexual parasites were seen, this was considered a negative result. Depending on the subject's condition, body temperature was measured one to four times daily. Beginning 24 h after medication, individual subjects were surveyed over 3 days for any side effects. The incidence of such effects was recorded.

Changes in Malaria Morbidity Rate in the Two Trial Periods 1972 and 1973

Results of the following two major trial periods demonstrated the prophylactic efficacy of piperaquine evidenced by changes in the malaria morbidity rate in the population studies.

1. The 1972 trial period from June to September; included more than 1000 people medicated and observed over 4 months, a total of 4958 person-months. Malaria morbidity was 97 person-months, with an average incidence rate of 1.96% (1.43%–2.31%). Of these, 1128 person-months involved piperaquine produced with the ethanol process (Batch No.720405). Morbidity in this group was 26 person-months, with an average incidence rate of 2.30% (2.01%–3.07%). In addition, 1122 person-months used piperaquine produced with the acid–base process, yielding a morbidity of 16 person-months and an average incidence rate of 1.43% (0.51%–2.11%). At the same time, 416 person-months were nonmedicated. Morbidity was 51 person-months and the average incidence rate was 12.26% (9.79%–19.11%). This showed that piperaquine had a clear prophylactic effect.

2. The 1973 trial period from June to September included more than 3000 people and 12,094 person-months. Adults accounted for 10,282 person-months and children, for 1842 person-months. The full course of medication and complete data collection was achieved for 6052 person-months.

At each site, adults and children were handled separately and randomly assigned to four groups: A, B, C, and D. Group A was given piperaquine produced with the acid–base process, batch number 730401. Group B received piperaquine produced with the ethanol process, batch number 730402. Group C used "refined" piperaquine, batch number 730404. Group D was given Malaria Prevention Tablets Number 3, each containing 0.15 g piperaquine base and 0.05 g sulfadoxine. In this group, adults received four tablets per dose, children aged 6–9 received two and those aged 10–15 received three. Children under six and those unable to receive medication were the controls.

A chi-squared test was run on the data from Groups A, B, and D. The χ^2 value of Groups A and B was 1.918 and that of Groups A and D was 1.157. The P values were more than 0.05, showing that the difference in morbidity-person-months was not significant. Comparing these groups to the control, however, yielded an χ^2 value of 15.337, where $P<.01$. The difference was very significant, showing that the preparations all had clear prophylactic effects (Table 10.1).

Table 10.1 adds the χ^2 values of similar data trends and the degrees of freedom to compare the overall results from each test site. The differences were not significant. A comparison of the three processing methods also found no significant difference ($P>.05$), whether in adults or children. Neither was there any significant difference in the data for ethanol and acid–base processed piperaquine ($P>.05$) across the test sites.

Therefore, there was no qualitative difference in prophylactic efficacy between piperaquine and Malaria Prevention Tablets Number 3, or among the three processing methods. However, the incidence rate of Malaria Prevention Tablets Number 3 was slightly lower than those of the three types of piperaquine. A monthly breakdown of the incidence rates can be seen in Table 10.2.

The table shows that the morbidity rate in the medicated groups stabilized and fell month-on-month, from 2.8% in June to 1.4% in September. Morbidity rose in the nonmedicated group every month, reaching 10.3% in September. Although monthly morbidity for the three types of piperaquine fluctuated slightly, their rates by September were approximately equal. A similar situation could be seen with Malaria Prevention Tablets Number 3. Three months before the medication period, the morbidity rate for the tablets was lower than those of the piperaquine groups, but the rates had equalized by September.

Duration of Effective Prophylaxis[20]

In the four-group study in the 1973 trial period in Hainan above, time at which infection occurred was calculated, with 2 days after medication as the starting point. Nonmedicated subjects were excluded who had contracted malaria, as well as those whose compliance was uncertain. Group A had 39 cases of infection, with the disease emerging after 18.5 ± 8.4 days; Group B had 29 cases, average time 18.8 ± 6.8 days; Group C had 35 cases, average time 19.4 ± 7.8 days; Group D had 19 cases, average time 19.1 ± 9.6 days. There was no clear difference among the groups. When the time of infection was broken down into 10-day intervals, it was found that around half of the cases became ill on Days 21–30. It indicated that the duration of effective prophylaxis for piperaquine and Malaria Prevention Tablets Number 3 was around 20 days.

Piperaquine was tested in areas of Guangxi where falciparum and vivax malaria are endemic, as well as in areas of Hubei, Henan, Shandong, and Jiangsu where vivax malaria is prevalent. The results showed that piperaquine also had persistent prophylactic effects on vivax malaria.

TABLE 10.1 Malaria Incidence in Six Test Sites After a Four-Month Drug Regimen

	Number of Subjects				Malaria Incidence (Morbidity)			
Site Code	Group A	Group B	Group C	Group D	Group A	Group B	Group C	Group D
Adults								
01	264	284	300	300	7 (2.7)	5 (1.8)	5 (1.7)	6 (2.0)
02	400	412	396	396	10 (2.5)	6 (1.5)	6 (1.5)	4 (1.0)
03	48	40	64	48	0	0	1 (1.6)	1 (2.1)
04	160	160	136	148	2 (1.3)	2 (1.3)	3 (2.2)	0
05	192	176	204	180	8 (4.2)	3 (1.7)	6 (2.9)	1 (0.6)
06	60	56	52	68	1 (1.7)	1 (1.8)	0	1 (1.5)
Total	1124	1128	1152	1140	28 (2.5)	17 (1.5)	21 (1.8)	13 (1.1)
Children								
07	112	104	96	104	3 (2.7)	4 (3.8)	5 (5.2)	1 (1.0)
08	136	124	124	112	5 (3.7)	2 (1.6)	0	0
09	8	12	12	4	0	1 (8.3)	1 (8.3)	1 (1/4[a])
Total	256	240	232	220	8 (3.1)	7 (2.9)	6 (2.6)	2 (0.4)

Of the 1218 nonmedicated person-months, morbidity was 95, with a rate of 7.8%.
[a]*One infection out of four subjects given malaria prevention tablets number 3.*

TABLE 10.2 Monthly Malaria Incidence Rates From June to September

	Number of Subjects				Incidence				Monthly Morbidity (%)			
Group	Jun	Jul	Aug	Sept	Jun	Jul	Aug	Sept	Jun	Jul	Aug	Sept
A	550	529	507	470	18	14	13	5	3.3	2.6	2.6	1.1
B	544	539	508	472	18	4	7	7	3.3	0.7	1.4	1.5
C	441	419	415	387	13	3	13	6	2.9	0.7	3.1	1.6
D	420	422	404	402	5	1	5	6	1.2	0.2	1.2	1.5
Total	1955	1909	1834	1731	54	22	38	24	2.8	1.2	2.1	1.4
No drug	264	309	326	319	21	18	23	33	8.0	5.8	7.1	10.3

TABLE 10.3 Monthly Population Parasite Carrier Rate in a Four-Month Course of Piperaquine

		Number Testing Positive					Parasite Rate (%)				
			30 days After Medication					30 days After Medication			
Site	Number Given Drug	Before Drug	Jun	Jul	Aug	Sept	Before Drug	Jun	Jul	Aug	Sept
011											
Adult	126	38	36	31	25	8	30.2	28.6	24.6	19.8	6.3
Child	70	24	28	21	20	15	34.3	40.0	30.0	28.6	21.4
012											
Adult	180	65	48	10	14	17	36.1	26.7	5.6	7.8	9.4
Total	376	127	112	62	59	40	33.8	29.8	16.5	15.7	10.6

Parasite Carrier Rate[20]

From June to September 1974 in Hainan, piperaquine was administered over 4 months. The population's parasite rate was checked 30 days after each dose (Table 10.3).

Table 10.3 shows that the 30-day postmedication parasite rate in adults at both test sites fell every month. This was especially marked at Site 012. However, the decline was not as great in children. After 4 months of medication, the parasite rate had dropped to 10.6%, down from 33.8% before the trial. This was a significant decrease.

Adverse Effects

Trials in the 1973 trial period above using three types of piperaquine and Malaria Prevention Tablets Number 3 included 9347 adult person-months and 1831 child person-months.

The main nervous system side effect was dizziness, followed by headache and drowsiness. A few experienced facial numbness; this happened most frequently with Malaria Prevention Tablets Number 3 (8.4%) but usually disappeared spontaneously in half a day to a day. Side effects on the gastrointestinal tract included nausea, vomiting, abdominal pain, and diarrhea. A small number had heart palpitations and isolated cases experienced serious side effects such as tightness in the chest and shortness of breath. These abated after a period of rest and did not affect normal activity. Side effects occurred between 1 h and 2 days after medication. A minority of women had irregular menstruation, which generally returned to normal in 2 months.

The incidence of side effects with the three types of piperaquine versus Malaria Prevention Tablets Number 3 was compared. The χ^2 value of the adult group was 140.4 ($P<.01$), a very significant difference; that of the child group was 10.3 ($0.01<P<.05$), also a significant difference. It suggested that Malaria Prevention Tablets Number 3 produced more side effects than piperaquine did, both in adults and in children. The same calculation for the three types of piperaquine alone yielded an χ^2 value of 2.85 ($P>.05$), indicating no significant difference.

The incidence of side effects was lower in children than in adults. This was statistically significant and could be due to underreporting in children.

Conclusion

After 3 years of large-scale field trials, piperaquine was found to be a relatively good inhibitory prophylactic drug for both falciparum and vivax malaria. It had persistent effects and could effectively lower the morbidity rate. Although its effective duration was around 20 days, one dose a month was sufficient to produce satisfactory reductions in morbidity. It could also be used therapeutically. Side effects were minor and the drug did not have a bitter taste, making it easily acceptable.

3. Early Studies and Resistance

According to the malaria laboratory at the National Institute of Parasitic Diseases, of the Chinese Center for Disease Control and Prevention, *P. berghei* had cross-resistance to piperaquine and to chloroquine. In 1976, however, Zhou Mingxing, Chen Lin, Zhou Yuanchang, et al. used piperaquine to treat drug-resistant falciparum malaria cases and the therapeutic results were satisfactory.[21] From 1978 chloroquine resistance to *P. falciparum* was spreading widely in Hainan and Yunnan, all areas in China where falciparum malaria was endemic stopped using chloroquine and switched to piperaquine. In the first few years, the therapeutic efficacy of piperaquine on falciparum malaria was good. However, its efficacy declined progressively each year until 1988, when significant piperaquine resistance emerged in *P. falciparum*.

10.2 DIHYDROARTEMISININ–PIPERAQUINE PHOSPHATE TABLETS (ARTEKIN)

Dihydroartemisinin–piperaquine phosphate (DHA–PQP) ACT was called Artekin while still in development and when it was first marketed. It has since been renamed Duo-Cotecxin for the international market. In August 2009, it was in the eighth edition of the WHO's list of prequalified antimalarials.

DHA is a derivative of artemisinin that retains the latter's peroxide bridge which is responsible for its efficacy. It is the pharmaceutical ingredient for synthesizing artemether and artesunate (ATS), and the active metabolite of these two drugs. Therefore, it has a lower cost of production. Its antimalarial efficacy is also higher than that of artemisinin. Out of all known antimalarials, artemisinin and derivatives have a particularly rapid action. When used in ACTs, they can greatly reduce the incidence of severe malaria, thereby decreasing morbidity and mortality. They have few side effects and are very safe. In addition, these antimalarials have the advantage of inhibiting *P. falciparum* gametocytes, hence reducing the spread of the disease. However, artemisinin-type drugs have a short half-life, requiring a 7-day regimen to obtain a high cure rate. To achieve the same results with a short treatment course, they must be combined with a drug possessing a long half-life.

PQP was the first antimalarial product to be developed in China. In the late 1960s, it was used as long-term prophylaxis, and in 1978 it replaced chloroquine in falciparum-malaria-endemic areas nationwide. By now, it has been widely used in China for over 30 years. PQP and chloroquine are 4-aminoquinolines, but PQP had a higher efficacy and markedly lower toxicity. A 3-day regimen produces cure rates of 90% and above. When used in isolation, however, it easily leads to drug resistance. It also acted more slowly than artemisinin-type antimalarials.

A. PHARMACOLOGY AND TOXICOLOGY OF DIHYDROARTEMISININ–PIPERAQUINE PHOSPHATE COMBINATION (ARTEKIN)

Li Guoqiao
Institute of Tropical Medicine, Guangzhou University of Chinese Medicine, Guangzhou, Guangdong Province, China

1. Clinical Basis for Ratio Design of Artekin

The DHA–PQP dose ratio was the result of many years' clinical experience. Since the switch to PQP in 1978 (see above), there was ample data to show that a total dose of 1500 mg base PQP—600 mg as the first dose, 300 mg 6 h later and 300 mg each on the second and third days—could achieve high cure rates against falciparum malaria. The initial 600 mg dose can cause side effects in the central nervous system and gastrointestinal tract, although these were milder than with chloroquine. DHA provided fast action, allowing the PQP regimen to be changed to four doses over 2 days. This reduced the initial dose by 37.5%, lowering side effects but still maintaining efficacy.

The use of DHA was also supported by the experience accumulated since 1989 when clinical trials were completed. Its regimen for falciparum malaria was 480 mg over 5 days. The initial dose of 60 mg alone produced 95% parasite clearance in 24 h. To ensure rapid action for patients with no immunity, each dose of the DHA in the combination was set at 80 mg, given four times over 2 days (total 320 mg). Clinical practice showed that the combination of DHA and PQP possessed the efficacy of both drugs and overcame their weaknesses. The resulting ACT, Artekin, was superior to both DHA and PQP (Table 10.4).

Artekin comprises DHA and PQP base at a ratio of 1:4.8, which was set in 1999 and validated in 2000 after experiments on rodent malaria. It combined DHA's fast action, low toxicity and low potential for resistance with PQP's short yet effective treatment duration. This synergy avoided the shortcomings of both drugs in the following ways:

1. *Shorter treatment duration and better cure rates: The combination of* DHA and PQP reduced their respective regimens—seven days for DHA and 3 days for PQP—to 2 days. The cure rate was higher than those of the individual drugs.
2. *Rapid action*: Artekin had DHA's rapid action, which minimized the occurrence of severe malaria and significantly reduced mortality. PQP alone does not have this advantage.
3. *Increased efficacy without increased toxicity*: $ED_{50/90}$ experiments with rodent malaria showed that DHA and PQP had synergistic effects. Animal models also indicated that the ACT's toxicity was additive, not synergistic. Since DHA produced almost no toxic side effects in a clinical setting, the only side effect observed during Artekin trials was the mild gastrointestinal

TABLE 10.4 Characteristics of Artekin and Its Components

	Advantage					Disadvantage		
Drug	Fast-Acting	Short Treatment Duration	High Cure Rate	Low Toxicity	Not Susceptible to Resistance	Slow-Acting	Long Treatment Duration	Suscept. to Resistance
DHA	+			+	+		+	
PQP		+	+	±		+		+
Artekin	+	+	+	+	+	−	−	−

reactions caused by PQP. The incidence of nausea was less than 3% and that of vomiting was less than 2%.

4. *Delayed drug resistance*: PQP had a long half-life, allowing the parasite to develop resistance more easily. When resistance was cultivated in rodent malaria, the I_{90} resistance index of the parasite rose to 63.23 by the 30th passage. DHA's half-life was short and, despite its widespread use over 20 years in China, no drug resistance had emerged. The same mouse model showed that, by the 30th passage, the I_{90} was still below 10. The I_{90} of Artekin was 2.8 after 30 passages, indicating that the ACT had avoided PQP's susceptibility to resistance.

2. Pharmacodynamics of Artekin in Rodent Malaria

Experimental Ratio Study

An orthogonal test design was used, together with the Peters 4-day suppressive test protocol.[22] A combination of 0.08–0.8 mg/kg DHA and 0.1–1.2 mg/kg PQP was found to be an effective range against the K173 strain. Further experiments refined these ratios. The best ratio was 1:5 DHA:PQP which had a synergistic index of more than 2. The $ED_{50/90}$ of this ratio against K173 was 1.24 and 1.83 mg/kg/day for 4 days, respectively. The synergistic index, calculated using the Berenbaum method, was close to 1, showing that both drugs had additive effects in suppressing rodent malaria.

Parasite Clearance

Artekin's speed of parasite clearance was slightly faster than those of DHA and PQP. Using the time taken for 90% clearance (CT_{90}) as a benchmark, Artekin took 27.57 h, 3 h faster than DHA (30.68 h) and 5 h faster than PQP (32.75 h).

28-Day Efficacy

The $CD_{50/90}$ of Artekin on the K173 strain was 2.83 and 6.88 mg/kg/day for 5 days, respectively. Its therapeutic efficacy was markedly higher than those of its component drugs: at least 100 times that of DHA and double that of PQP.

Efficacy on RC Strain

Peters' 4-day suppressive test[22] was used. The ED_{50} and ED_{90} of Artekin on a chloroquine-resistant (RC) strain were 2.2 and 3.6 mg/kg, respectively. The $I_{50/90}$ was 1.8 and 1.2, respectively, showing that Artekin was effective against RC strains.

Resistance

Artekin-, PQP- and DHA-resistant strains of K173 *P. berghei* were cultivated using progressive dose increments. The starting dose was four times the ED_{90} of each drug, increasing 50% every three to five passages to a total of 30 passages.

The resistance index was measured every five passages. At the 20th and 30th passages, the indices for PQP were 21.2 and 63.2, respectively; for DHA, they were 4.5 and 9.9; for Artekin, they were 3.8 and 2.8. It showed that the ACT could reduce the degree of resistance to each individual drug.

3. Toxicology

Central Nervous System

Mice were given 13.36 mg/kg (the ED_{90} dose) or 133.6 mg/kg (10 times ED_{90}) Artekin intragastrically. The Irwin behavioral test and cage-shaking did not reveal any obvious side effects on the nervous systems of the mice. An experiment with below-threshold doses of pentobarbital showed that, at 10 times the ED_{90}, Artekin tended to prolong sleep, indicating that the ACT had a calming effect at high doses.

Cardiovascular and Respiratory Systems

Beagles were given 7.97 mg/kg (the curative dose for simian malaria) or 23.9 mg/kg (three times the same curative dose) Artekin intragastrically. There were no significant effects on heart rate, blood pressure, ECGs, and breathing.

Acute Toxicity in Mice

Mice were randomly assigned to parallel tests on the toxicity of Artekin, DHA, and PQP. Each drug had six dose groups. The drugs were administered once intragastrically except for Artekin, which also included groups given a single intraperitoneal injection. Acute toxicity reactions and deaths were observed for 14 days. The Bliss method was used to calculate the LD_{50} of each drug, and the Finney formula to estimate toxicity[23] for Artekin. The LD_{50} of Artekin was 961.7 mg/kg for oral administration and 527.6 mg/kg for intraperitoneal injection. The ratio of the Finney-estimated LD_{50} to the actual LD_{50} of Artekin was 0.83, showing that its toxicity was additive.

Acute Toxicity in Beagles

Deichmann's method of ascertaining the approximate lethal dose (ALD) from six animals was used.[24] The dose of Artekin was increased by 50% for each animal. The dogs were given 33, 50, 75, 253, or 378 mg/kg oral Artekin. The ALD was 253 mg/kg and the maximum tolerated dose (MTD) was 33 mg/kg.

Long-Term Toxicity in Wistar Rats

Doses of 41.62, 83.25, 166.49, and 332.98 mg/kg/day, as well as a control group, were used. The drug was administered over 14 days. A dose of 83.25 mg/kg/day was found to be basically safe, with 166.49 mg/kg/day being mildly toxic and 332.98 mg/kg/day being moderately to severely toxic. The main target organ was the liver, with AST levels as an indicator. The damage was reversible.

Long-Term Toxicity in Beagles

The test used 6.93, 13.88, 27.75, and 55.49 mg/kg/day doses, as well as a control group. Medication took place over 14 days. The 6.93 and 13.88 mg/kg/day doses produced no toxic reactions. The 27.75 mg/kg/day dose was mildly to moderately toxic; the 55.49 mg/kg/day dose was severely toxic. The main target organ was the liver and reticulocyte count as an indicator for a hematology change. The changes were reversible.

Mutagenicity, Ames Test

The TA97, TA98, TA100 and TA102 histidine-auxotrophic strains of *Salmonella typhimurium* were used.[25] At concentrations of 0.5–5000.0 μg/dish, Artekin was no different from the buffer control in its impact on the number of revertants, regardless of the presence of S9. It showed that Artekin did not have a mutagenic effect on the test strains.

Mutagenicity, Chromosomal Aberration

Chinese hamster lung cells (CHL) were chosen to test the effects of Artekin on the chromosome. Concentrations of 1.25, 2.5, 5, and 8 μg/mL were used. The highest concentration inhibited cell growth by 50%. Nonactivated CHL was cultured in the medium for 24 or 48 h. CHL activated with S9 was cultured for 6 h. The Artekin was then removed and the cells further cultivated for 24 or 48 h, after which the chromosomes were examined. Regardless of activation, Artekin had no teratogenic effects on CHL chromosomes.

Mutagenicity, Micronucleus Test

Micronucleus effects of Artekin on mouse bone marrow over time were studied. Kunming mice were given 480 mg/kg Artekin intragastrically: 12, 24, 36, 48, and 72 h afterward, there was no clear difference in the frequency of micronuclei among polychromatic erythrocytes in the bone marrow when compared to the buffer control group ($P > .05$).

Twenty-four hours after mice were given a single intragastric dose of 480, 192, or 48 mg/kg Artekin, there was also no clear difference in the frequency of micronuclei ($P > .05$, Dunnett t test). There was also no obvious decrease in the ratio of polychromatic to normochromatic erythrocytes (PCE/NCE) when compared to the buffer control ($P > .05$). Artekin did not increase the frequency of micronuclei in polychromatic erythrocytes in mouse bone marrow, and hence had no clear toxicity to bone marrow cells.

4. Clinical Pharmacokinetics of the Piperaquine Component in Artekin

Te-Yu Hung et al.[26] conducted pharmacokinetic studies on piperaquine in Cambodia, involving 38 adults and 47 children with malaria. They were given

Artekin following the four-dose, 2-day regimen, which included total dose 32–35 mg/kg piperaquine. The drug's half-life was 19–28 days in adults, average 23 days; for children, it was 10–18 days, average 14 days.

B. ARTEKIN CLINICAL TRIALS[38]

Li Guoqiao, Guo Xingbo
Institute of Tropical Medicine, Guangzhou University of Chinese Medicine, Guangzhou, Guangdong Province, China

1. Drugs and Doses

1. *Artekin*: Each tablet contained 40 mg DHA and 320 mg PQP. At 0, 8, 24, and 32 h, adults were given two tablets, a total of eight. Those aged 11–15 took 3/4 of the dose; those aged 7–10 took 1/2 of the dose.
2. *Artekin + Trimethoprim (TMP)*: Each tablet contained 32 mg DHA, 320 mg PQP, and 90 mg TMP. The same regimen was used as with Artekin.
3. *Artesunate + Mefloquine*: 4 mg/kg ATS and 10 mg/kg mefloquine was given at 0, 24, and 48 h.

2. Comparative Trials[38]

Trial in Kampong Speu, Cambodia

Artekin and Artekin + TMP groups each had 50 cases of uncomplicated falciparum malaria. Defervescence and parasite clearance times were similar. There were no cases of recrudescence during the 28-day follow-up (Table 10.5).

Both drugs had a low incidence of side effects. There were no adverse reactions in the Artekin group; for the Artekin + TMP group, there were three cases of nausea and one case each of vomiting, dizziness, deafness, and itching.

TABLE 10.5 Efficacy of Artekin and Artekin + TMP on Falciparum Malaria, Kampong Speu, Cambodia

Drug	Number of Cases	Average Parasite Clearance Time (h) (M ± SD)	Average Defervescence Time (h) (M ± SD)	Cure Rate in 28 days (%)
Artekin	50	31.7 ± 9.0	12.7 ± 7.2	100
Artekin + TMP	50	32.8 ± 8.8	16.5 ± 7.9	100

Trial in Battambang, Cambodia

Multidrug-resistant falciparum malaria was widespread in Battambang province. In 2001, a randomized trial was conducted using Artekin and Artekin+TMP. Each group had 25 cases of falciparum malaria, and defervescence and parasite clearance times were similar (Table 10.6). In the Artekin group, 18 completed the 28-day follow-up and there was no recrudescence. For the Artekin+TMP group, 20 completed the follow-up and there was one case of recrudescence on Day 14. The incidence of side effects was low for both drugs (Table 10.7).

Trial in Hainan

Chloroquine-resistant falciparum malaria was endemic in Hainan. Clinical trials on Artekin and Artekin+TMP were conducted there in 2001. Each drug group included 30 cases. Defervescence and parasite clearance times were similar and both groups each had one case of recrudescence over the 28-day follow-up (Table 10.8). The incidence of side effects was low for both drugs.

TABLE 10.6 Efficacy of Artekin and Artekin + TMP on Falciparum Malaria, Battambang, Cambodia

Drug	Number of Cases	Average Parasite Clearance Time (h) M ± SD	Average Defervescence Time (h) M ± SD	Cure Rate in 28 days (%)
Artekin	25	36.3 ± 19.9	41.6 ± 25.2	100
Artekin + TMP	25	35.8 ± 17.0	31.1 ± 20.5	95

TABLE 10.7 Observed Side Effects of Artekin and Artekin + TMP

Side Effect	Artekin (n = 24)	Artekin + TMP (n = 25)	*P* value
Nausea	2 (8.3%)	3 (12%)	>0.05
Vomiting	1 (4.2%)	0	>0.05
Abdominal pain	2 (8.3%)	3 (12%)	>0.05
Loss of appetite	1 (4.2%)	1 (4%)	>0.05
Rash	0	1 (4%)	>0.05
Diarrhea	0	1 (4%)	>0.05
Itching	0	1 (4%)	>0.05

TABLE 10.8 Efficacy of Artekin and Artekin + TMP in Hainan

Drug	Number of Cases	Average Parasite Clearance Time (h) (M ± SD)	Average Defervescence Time (h) (M ± SD)	Cure Rate (%)
Artekin	30	56.7 ± 16.3	20.2 ± 11.3	96.7
Artekin + TMP	30	58.4 ± 18.6	22.8 ± 1.05	96.7

TABLE 10.9 Side Effects Observed

Side Effect	Artekin (n = 104)	Artekin + TMP (n = 105)
Nausea	3 (2.9%)	8 (7.6%)
Vomiting	1 (1.0%)	2 (1.9%)
Loss of appetite	1 (1.0%)	1 (1.0%)
Abdominal pain	2 (1.9%)	3 (2.9%)
Diarrhea	0	1 (1.0%)
Itching	0	2 (1.9%)
Rash	0	1 (1.0%)
Dizziness	0	2 (1.9%)
Headache	0	1 (1.0%)
Deafness	0	1 (1.0%)

Trial in Ho Chi Minh City, Vietnam[39]

In 2001, the Vietnam research team of Oxford University's Tropical Medicine Research Unit conducted a trial in Ho Chi Minh City's Cho Quan Hospital. Uncomplicated falciparum malaria patients were randomly assigned to Artekin, Artekin + TMP, and mefloquine + ATS groups at a ratio of 2:2:1. The Artekin group had 162 cases, the Artekin + TMP group had 156 and the mefloquine + ATS group had 77. Patients were followed for 56 days. The cure rate for the Artekin group was over 97%, the incidence of side effects less than 5%, and there were no obvious abnormalities in liver and kidney function.

Safety Evaluation

Side effects observed during the comparative trials are shown in Table 10.9. Artekin had infrequent and mild side effects, with a slightly lower incidence than with Artekin + TMP.

3. Expanded Trials[38]

Efficacy

The health department of the Burmese Defense Ministry carried out expanded trials on Artekin, involving 84 cases of uncomplicated falciparum malaria; all were adults. Over the 28-day follow-up, the cure rate was 91.7%. There were seven cases of recrudescence (8.3%), although reinfection could not be ruled out in some cases. Average parasite clearance time was 23.2±7.7h; average defervescence time was 19.8±6.3h.

Further trials were conducted by the health department of Cambodia's Defence Ministry. These trials compared Artekin to Artekin+TMP. The Artekin group had 42 cases of falciparum malaria; 11 completed the 28-day follow-up and there was no recrudescence. The Artekin+TMP group had 34 cases, of which 14 completed the follow-up. There was also no recrudescence. Defervescence and parasite clearance times were similar.

Adverse Reactions

Adverse reactions were observed and recorded for the 126 cases in the expanded trials. Nausea (3.2%) and vomiting (2.4%) were the most frequent side effects, both for Artekin and Artekin+TMP. These were transient and eased spontaneously without treatment. Other symptoms were those common during the acute phase of malaria; incidence was low and disease-related causes could not be ruled out.

4. Conclusion

Trials on Artekin's efficacy against falciparum malaria took place in four sites, including China, Cambodia, and Vietnam. It was compared to Artekin+TMP and ATS+mefloquine. A total of 266 cases of uncomplicated falciparum malaria were treated with Artekin: The 28-day cure rate was 96.7%–100%. Artekin+TMP was used on 261 cases, with a cure rate of 95%–100%. Expanded trials were carried out on two groups of patients in Cambodia and Myanmar, totaling 126 cases. Of these, 95 cases completed the follow-up. The 28-day cure rate was 92.6% (88/95) and the recrudescence rate was 7.4%, although reinfection was possible.

The main adverse reactions were nausea and vomiting. The incidence of nausea was 2.88% (6/208), while that of vomiting was 1.44% (3/208). It showed that Artekin was well tolerated.

Multidrug-resistant falciparum malaria is widespread in Cambodia, Vietnam, Myanmar, and Thailand. These are areas in which the frequency of treatment failure and recrudescence are among the highest in the world. Artekin's ability to obtain high cure rates for uncomplicated falciparum malaria in this region showed that local parasite strains were highly sensitive to the drug. Its curative efficacy was the same as that of Artekin+TMP, but it had slightly fewer adverse reactions.

10.3 ARTEMISININ–PIPERAQUINE TABLETS (ARTEQUICK)

Li Guoqiao
Institute of Tropical Medicine, Guangzhou University of Chinese Medicine, Guangzhou, Guangdong Province, China

The idea of combining artemisinin and piperaquine was first considered in 2003 to create an ACT for mass drug administration and to eliminate malaria via rapid removal of its source. This ACT needed to have a high cure rate and low side effects, be easy to administer, and cheap to produce. From the clinical experience accumulated by Project 523 since 1973, a reduction in gastrointestinal side effects could be achieved by removing the phosphate from piperaquine phosphate (PQP) and using the piperaquine amount required in two doses. However, the requisite dose of artemisinin was only determined in 2003, after new clinical trials. It confirmed that the ratio of artemisinin to piperaquine should be 1:6. Each Artequick tablet had 62.5 mg artemisinin and 375 mg piperaquine. The medication was given twice over 24 h (see Section C below).

A. LABORATORY TOXICOLOGY STUDIES

1. Acute Toxicity in Mice and Rats

Acute toxicity test in mice showed that Artequick's toxicity was the same as that of piperaquine. Its LD_{50} was 1053.5 mg/kg, while that of piperaquine was 928.7 mg/kg, and of artemisinin >7500 mg/kg. When artemisinin was combined with piperaquine, the result had synergistic efficacy on rodent malaria, but only additive toxicity.

Artequick was fed to rats for 14 days. The toxicity results showed that a dose of 30 mg/kg/day was basically safe and 115 mg/kg/day was moderately toxic. The main target organ was the liver, with ALT and AST levels as indicators. The damage was reversible.

2. Mutagenicity

In 1981 Li Guangyi, Yang Lixin et al. reported on an Ames test with artemisinin. The TA100, TA98, TA1535, TA1537, and TA1538 strains of *S. typhimurium* were used, both with and without S9. The results proved that artemisinin was not a mutagen. In addition, Yang et al. also published a study on the effects of artemisinin on the frequency of micronuclei in polychromatic erythrocytes in mouse bone marrow. Because micronuclei are the result of chromosomal aberrations, an increase reflects the degree of such aberrations. The experiment used cyclophosphamide as a positive control. Matter and Schmid's protocol was used, with the maximum dose at 1/5 of the LD_{50}. Artemisinin had no effect on the rate of micronuclei, while the frequency of micronuclei increased across all doses of the positive control in proportion to the dose.

These results showed that artemisinin was not a mutagen. Artemisinin was metabolized in the body to form DHA. Only then did it produce antimalarial effects. Micronucleus, CHL chromosomal aberration, and Ames tests on DHA were all negative, further confirming that artemisinin was not mutagenic.

Piperaquine was the base and active ingredient in PQP. Ames, chromosomal aberration, and micronucleus tests on Artekin were all negative (see above), showing that PQP was not mutagenic. Since Artequick's component drugs were not mutagenic, the ACT itself had additive and not synergistic toxicity, and Artekin and Artequick had the same active ingredients, it was believed that Artequick was unlikely to be a mutagen.

3. Reproductive Toxicity in Mice and Rats

These teratogenic tests involved mice and rats. Rats were given daily doses of 1/40–1/25 of the LD_{50} of artemisinin from the 6th to the 15th day of gestation. The pregnant rats were killed on the 20th day. Out of the 200 fetuses in the control group, only four were reabsorbed (2%). The rest were alive. Teratogenic examinations were carried out on the skeletons and soft tissues; all were normal. For artemisinin, the group given 1/40 of the LD_{50} had 35 live fetuses. All other groups had a 100% reabsorption rate. The mothers in the high dose group all had different degrees of vaginal bleeding during medication.

Artemisinin was given at the early (Days 6–8), middle (Days 9–11), and late (Days 12–14) stages of organogenesis to determine which stage was most susceptible to toxicity. The dose was 1/25 of the LD_{50}. At the early stage, 22% of the fetuses were reabsorbed. Out of the 164 live fetuses, 10 (6.1%) had umbilical hernias. The bones and soft tissues of the rest appeared normal. At the middle and late stages, 100% of fetuses were reabsorbed.

A further study was conducted with pregnant rats given 1/25 of the LD_{50} on Days 1–6, Days 7–14, or Days 15–19 of gestation. When artemisinin was administered before organogenesis, on Days 1–6, most of the fetuses were alive. Only 6.4% were reabsorbed. The live fetuses developed normally, with no deformities. Medication in the early stage of pregnancy led to fewer effects on fetal rats. At the middle and late stages, all the fetuses were reabsorbed. The drug had clear toxic effects on the fetus. These experiments were repeated in mice, with basically the same results.

In sum, the tests showed that artemisinin had significant toxicity on the embryos of mice and rats. If administered before organogenesis, it had little influence on fetal development. When the drug was given during early organogenesis, 6.1% of fetal rats had umbilical hernias. Although these results had to be validated further, they demanded close attention, since fetal reabsorption and teratogenesis are on the same spectrum, differing only in the degree of toxicity. The marked ability of artemisinin to cause reabsorption could also mask its teratogenic effects. Teratogenicity tests of PQP on mice, using 1/27 to 1/3 of the LD_{50} administered from Days 9–14 of pregnancy, showed no toxicity to embryos and no teratogenic effects.

Teratogenicity tests had also been run for Artekin, having the same active metabolite DHA with Artequick. Sprague Dawley (SD) rats were divided into three Artekin dose groups—15, 45, and 135 mg/kg—a buffer control group and a cyclophosphamide positive control group. Artekin affected the general condition, weight, and gestational weight gain of maternal rats. It was toxic mainly to embryos. In the 135 mg/kg group, all fetuses were reabsorbed; in the 45 mg/kg group, the reabsorption rate reached 59.1% and the malformation rate, 5.4%. The differences with the buffer group were clear. In addition, although Artekin had no effects on the physical development of the fetus, it significantly delayed skeletal development. Under these conditions, the minimum-toxicity, no effect level (NOEL) of Artekin in SD rats was 15 mg/kg.

The chemical structures and metabolites of Artequick's component drugs artemisinin and PQP were unrelated to known carcinogens. Mutagenicity tests on both individual drugs were negative. Their toxicity was additive, rather than synergistic. Hence, no carcinogenicity tests were run.

B. ALLERGIC AND HEMOLYTIC REACTIONS

Artequick's component drugs artemisinin and piperaquine have been marketed and widely used as antimalarials for many years. From 1973 to 1993, the Shanghai Zhongxi Pharmaceutical Factory produced a total of 217.6 tons of oral piperaquine and PQP, equal to 120 million therapeutic doses (50 million adults and 76 million children). Tens of millions of people worldwide had also used artemisinin and its combinations, Artequick and Artekin. No hemolytic reactions had been reported for both Artekin and Artequick. From the findings of the 1989–94 Phase II and III clinical trials, DHA could occasionally cause mild drug eruption, with an incidence of 0.17%–0.9%. This did not cause obvious discomfort to patients or affect compliance. No treatment was needed and it disappeared spontaneously after medication ceased. No allergic reactions were reported for artemisinin.

C. CLINICAL TRIALS ON ARTEQUICK

These included tolerance tests in healthy subjects (Phase I trials), Phase II dose finding trials, and Phase III expanded clinical trials.

1. Phase I Trials[40]

Oral Artequick was given to 26 healthy volunteers. A randomized, double-blind protocol was used, beginning with the lowest dose. Three dose groups were used, with two in each group receiving a placebo. The test proceeded with the next group if the previous dose group experienced no adverse reactions. Each group's regimen is shown in Table 10.10.

TABLE 10.10 Regimen and Dose Tolerance of Artequick

Dose Group (ART + PQ)	Dose (Tablets)			% Increase in Dose	Test Dose/Predicted Therapeutic Dose	No. of Cases
	0 h	24 h	Total			
1 tablet (62.5 + 375 mg)	1	1	2	−50%	0.50	2[a]
2 tablets (125 + 750 mg)	2	2	4	Predicted therapeutic dose	1.00	8[a]
2½ tablets (156.25 + 937.5 mg)	2½	2½	5	+25%	1.25	6[b]
3 tablets (187.5 + 1125 mg)	3	3	6	+20%	1.50	4[c]

ART, artemisinin, *PQ*, piperaquine.
[a]*First group.*
[b]*Second group.*
[c]*Third group.*

All subjects were followed up for 28 days. The subjects were observed 2, 4, 8, 24, 48, and 72 h after medication on the first and second days. Adverse reactions were recorded. Blood tests, comprehensive metabolic panels, ECGs and routine urine tests were carried out before medication and at Days 3, 7, 14, 21, and 28 thereafter.

Vital signs, blood tests, urine tests, and ECGs showed no abnormalities and no differences with the placebo group. The incidence of adverse reactions was as follows:

1. *Single-tablet group*: Two subjects were given the drug four times. No adverse reactions.
2. *Two tablets (estimated therapeutic dose)*: One case each of slight dizziness and facial numbness.
3. *Two and half tablets:* Two cases of drowsiness, three cases of fatigue, one case of facial numbness, two cases of nausea, one case each of abdominal pain and diarrhea.
4. *Three tablets*: Four cases of dizziness, four cases of drowsiness, three cases of fatigue, two cases of facial numbness, two cases of nausea, one case of vomiting, one case of abdominal pain.

The severity of adverse reactions increased with the dose. The symptoms experienced in the group given 2½ tablets usually lasted for several hours. Those of the three-tablet group lasted a day. However, the symptoms disappeared without treatment.

The results showed that healthy subjects could tolerate a two-tablet dose (i.e., the estimated therapeutic dose). The incidence of adverse reactions was low and such reactions were mild. Side effects were more frequent in the 2½- and 3-tablet groups and symptoms were more severe. Common reactions were dizziness, fatigue, drowsiness, facial numbness, nausea, and vomiting. Laboratory tests and ECGs showed no abnormalities across the doses. Therefore, the recommended dose of Artequick for the Phase II trials was two tablets, followed by another two on Day 2—four tablets in total.

2. Phase II Trials[40]

Dose

Two dose groups were used: 1400 mg (200 mg artemisinin, 1200 mg piperaquine) and 1750 mg (the estimated therapeutic dose of 250 mg artemisinin, 1500 mg piperaquine). Their efficacy and safety were compared to establish a clinical dose for Artequick. Each group included 50 cases of uncomplicated falciparum malaria. The 28-day cure rates were 82% for the 1,400 mg group and 96% for the 1,750 mg group. The latter dose was significantly more effective than the former ($P < .05$).

The drug was well tolerated by 30 pretrial patients receiving 1400 mg Artequick and the 100 patients in the Phase II dose trial. Individual patients had

TABLE 10.11 Adverse Reactions in Patients Given 1400 or 1750 mg Artequick

Side Effect	1400 mg Group (n = 80)	1750 mg Group (n = 50)
Nausea	1	1
Vomiting	0	0
Diarrhea	2	3
Loss of appetite	1	2
Dizziness	1	0
Fatigue	0	1

diarrhea, nausea, dizziness, and other symptoms, all of which were self-limited (Table 10.11). No abnormal changes were seen in the blood tests, comprehensive metabolic panels, and ECGs conducted on Day 0 before medication and on Day 7 after medication.

The experiment showed that the 28-day cure rate of 1400 mg Artequick on uncomplicated falciparum malaria was only 82%. That of 1750 mg Artequick was 96%. This dose was well tolerated. Therefore, a 2-day regimen of 1750 mg Artequick was recommended as the therapeutic dose.

Randomized Trial With Artequick, Artekin, and Coartem

Each Artequick tablet contained 62.5 mg artemisinin and 375 mg piperaquine; four tablets were used. The Artekin tablets had 40 mg DHA and 320 mg PQP; patients received eight tablets. The Coartem tablets contained 20 mg artemether and 120 mg benflumetol (lumefantrine), with patients getting 24 tablets. The drug regimens may be seen in Table 10.12.

Uncomplicated falciparum malaria patients were randomly assigned to the three drug groups at a ratio of 2:1:1. There were 110 patients in the Artequick group, 55 in the Artekin group, and 55 in the Coartem group. Average parasite clearance and defervescence times were similar across all groups. The 28-day cure rates were 95.5% for Artequick, 98.2% for Artekin, and 83.6% for Coartem. Artequick and Artekin had similar cure rates, somewhat higher than Coartem's (Table 10.13). However, the author's own experience as well as a large volume of data indicated that a 5-day regimen of 24 Coartem tablets could achieve a 28-day cure rate of 95% and above, equal to Artekin and Artequick.

The three drugs were well-tolerated, with the incidence of gastrointestinal side effects such as nausea and vomiting within 2%. The symptoms were also self-limited. Blood tests, comprehensive metabolic panels, and ECGs conducted on Day 0 before medication and Day 7 after medication showed no abnormal changes.

TABLE 10.12 Drug Regimens of Artequick, Artekin, and Coartem[a] (unit: tablet)

Drug	Day 0		Day 1		Day 2		Total
	0 h	8 h	24 h	32 h	48 h	56 h	
Artequick	2		2				4
Artekin	2	2	2	2			8
Coartem	4	4	4	4	4	4	24

[a]*Adult dose. Doses were scaled down for those aged 7–15.*

TABLE 10.13 Efficacy of Artequick, Artekin, and Coartem

Drug	No. of Cases	Parasite Clearance Time (h) (M ± SD)	Defervescence Time (h) (M ± SD)	Reinfection in 28 days (Cases)	Recrudescence in 28 days (Cases)	Cure Rate in 28 days (%)
Artequick	110	66.7 ± 21.9	31.6 ± 17.7	8	5	95.5
Artekin	55	65.6 ± 27.3	34.6 ± 21.8	0	1	98.2
Coartem	55	65.3 ± 22.5	36.9 ± 15.4	4	9	83.6

Randomized Trial With Artequick and Artekin

The study took place in central Vietnam, where multidrug-resistant falciparum malaria was endemic. It evaluated the efficacy and safety of Artequick against drug-resistant *P. falciparum*. A total of 103 uncomplicated falciparum malaria cases were included: 52 in the Artequick group (13 aged 7–15, 39 adults) and 51 cases in the Artekin group (13 aged 7–15, 38 adults). After medication, fever and other symptoms rapidly disappeared in both groups. There was no recrudescence during the 28-day follow-up and the cure rate was 100%.

The incidence of dizziness was similar in both groups, but this was not serious and did not affect daily activities. Disease-related causes also could not be ruled out. In addition, the incidence of vomiting was 9.8% in the Artekin group.

Safety of Artequick in the Phase II Trials

During the Phase II clinical trials, the recommended therapeutic dose of Artequick was given to a total of 212 uncomplicated falciparum malaria patients. It was compared to Artekin and Coartem, the most effective antimalarials then on the market. Artekin was used to treat 106 cases, while 55 received Coartem. All three drugs were well tolerated. For adverse reactions, see Table 10.14. The incidence of vomiting was slightly lower for Artequick than for Artekin, but the difference was not significant ($P > .05$).

TABLE 10.14 Side Effects Observed With Artequick, Artekin, and Coartem

Adverse Reaction (Likely Drug-Related)	Artequick (n = 212)	Artekin (n = 106)	Coartem (n = 55)
Loss of appetite	8 (3.8%)	3 (2.8%)	4 (7.3%)
Nausea	4 (1.9%)	7 (6.6%)	1 (1.8%)
Vomiting	2 (0.9%)	7 (6.6%)	0
Diarrhea	6 (2.8%)	3 (2.8%)	2 (3.6%)
Abdominal pain	3 (1.4%)	2 (1.9%)	1 (1.7%)
Headache	5 (2.4%)	3 (2.8%)	3 (5.5%)
Dizziness	8 (3.8%)	5 (4.7%)	3 (5.5%)
Insomnia	4 (1.9%)	3 (2.8%)	1 (1.8%)
Fatigue	2 (0.9%)	1 (0.9%)	1 (1.8%)

The blood tests, comprehensive metabolic panels, and ECGs showed no abnormal changes between Days 0 and 7 across all three drugs.

Summary of Phase II Trials

The results of the Phase II Artequick trials showed that a total adult dose of 1750 mg (250 mg artemisinin and 1500 mg piperaquine), given in two doses over 2 days, had a high cure rate against uncomplicated falciparum malaria. It was convenient to administer and well tolerated. Expanded clinical trials were proposed.

3. Phase III Trials[40]

The expanded trials took place in six test sites in Vietnam, Cambodia, and Indonesia. An adult dose of 1750 mg, taken in four tablets, was used.

Indonesia

Indonesia's National Institute of Health Research and Development ran the trial at the Bangka Island medical center. Artequick was compared to a 3-day regimen of 600 mg ATS and 1500 mg amodiaquine. The study included 145 falciparum malaria and 174 vivax malaria cases.

Of the 145 falciparum malaria cases, 71 were given Artequick and 74, ATS + amodiaquine. The 28-day cure rates were 95.7% (66/69) and 92.5% (62/67), respectively. The difference was not statistically significant. Defervescence and parasite clearance times were relatively rapid in both groups.

The 174 vivax malaria cases were divided equally into Artequick and ATS + amodiaquine groups. The 28-day recrudescence rates were 0% (0/83) and 3.5% (3/85), respectively.

In sum, Artequick was used to treat a total of 158 (71 + 87) patients, while ATS + amodiaquine were used on 161 (74 + 87). After medication, the latter group showed side effects such as dizziness (7/161, 4.4%), nausea (26/161, 16.2%), and vomiting (9/161, 5.6%). For Artequick, the side effects were nausea (6/158, 3.8%) and vomiting (3/158, 1.9%).

Cambodia

The health department of the Defence Ministry carried out a trial in Battambang's Military Region 5 Hospital. Artequick was given to 50 cases of uncomplicated falciparum malaria. All cases were hospitalized for observation for 21 days; 44 cases were observed for 28 days. There was no recrudescence.

The National Malaria Control Center also conducted a trial in Pursat, on the Thai–Cambodian border, where multidrug-resistant falciparum malaria was endemic. A total of 58 uncomplicated falciparum malaria cases were included, 44 of whom completed the 28-day follow-up. There were two cases of recrudescence and the cure rate was 95.5%. No significant adverse reactions were seen. In addition, Artequick was used to treat 62 cases of vivax malaria in Kampong

Speu. Symptoms eased rapidly after medication. The recrudescence rate was 5.6% (3/54) over the 28-day follow-up. There was one case each of dizziness, nausea, and vomiting, all of which were self-limited.

Vietnam

The Ministry of Health's Quy Nhon Institute of Malariology, Parasitology and Entomology carried out a trial in Ninh Hai County, Ninh Thuan Province, where multidrug-resistant falciparum malaria was endemic. Sixty children aged 6–15 with uncomplicated falciparum malaria were given Artequick and followed for 28 days. There was no recrudescence. The incidence of nausea and vomiting was 8.3% (5/60), which could have been linked to the five cases already experiencing nausea and vomiting before medication.

Xuan Loc District General Hospital in Dong Nai Province treated 20 cases of uncomplicated falciparum malaria with Artequick. The patients were hospitalized for observation for 28 days. There was one case of recrudescence (D22), with a rate of 5.0%. One case had nausea and vomiting, an incidence of 5.0%.

Summary of Phase III Trials

In total, 259 patients with uncomplicated falciparum malaria were treated with Artequick during the Phase III trials; all were clinically cured in a short time. Of these, 237 cases were followed up for 28 days. The cure rate was 97.5% (231/237) and the recrudescence rate, 2.5% (6/237). For the 149 cases of vivax malaria, all were clinically cured. Among the 137 cases followed for 28 days, the short-term recrudescence rate was 2.2% (3/137).

Adverse reactions were observed and recorded for the 569 cases in the expanded trials (408 cases for Artequick, 161 cases for ATS + amodiaquine). The incidence of nausea and vomiting for Artequick was 3.4% and 2.7%, respectively. For ATS + amodiaquine, the incidence of nausea was 16.2%, that of vomiting 5.6%, and that of dizziness 4.4%. Artequick was clearly safer than ATS + amodiaquine.

4. Conclusion

Four Artequick tablets given over 2 days—a total dose of 1750 mg (250 mg artemisinin and 1500 mg piperaquine)—were used to treat 471 cases of uncomplicated falciparum malaria. The clinical symptoms were rapidly controlled and the 28-day cure rate was over 95%. For vivax malaria, Artequick was used on 149 cases, with a 28-day recrudescence rate of 2.2%.

The recommended therapeutic dose of Artequick was given to 620 patients in total, and the incidence of nausea and vomiting was within 2%. Artequick could act against drug-resistant *P. falciparum*. It was highly effective, fast-acting, had low toxicity, and was easy to administer. Therefore, it is an effective treatment for uncomplicated malaria.

10.4 DIHYDROARTEMISININ–PIPERAQUINE–PRIMAQUINE–TRIMETHOPRIM (CV8 TABLETS) CLINICAL TRIALS[27]

Li Guoqiao
Institute of Tropical Medicine, Guangzhou University of Chinese Medicine, Guangzhou, Guangdong Province, China

From 1991 to 1992, Vietnam experienced a serious malaria outbreak involving multidrug-resistant *P. falciparum*. There was no ideal first-line treatment. The Guangzhou University of Chinese Medicine's Sanya Tropical Medicine Institute and Cho Ray Hospital's Tropical Diseases Clinical Research Center in Ho Chi Minh City cooperated to develop a quick-acting, highly effective ACT with low toxicity and the ability to kill gametocytes rapidly. This ACT with piperaquine was called CV8. The "C" stood for China and the "V" for Vietnam, and it was the eighth proposed formula that was validated after clinical trials.

CV8 comprises dihydroartemisinin (DHA), piperaquine phosphate (PQP), primaquine phosphate (PMQ), and trimethoprim (TMP). Animal toxicity tests were run in China and Vietnam. Once CV8's toxicity was found to be lower than those of chloroquine and Fansimef (FSM), clinical trials were initiated.

1. Tolerance Test in Healthy Volunteers

Twenty-six healthy adult male volunteers were assigned to three different dose groups and a placebo group by drawing lots. The trial was double-blind. Each volunteer was hospitalized for observation for 9 days (Day −1, Day 0, Days 1–7). Follow-up checks were done on Days 14, 21, and 28. Before treatment (Day 0) and on Days 3, 7, 14, and 28, blood tests, comprehensive metabolic panels, and routine urine tests were run. An ECG was performed on Days 0–3, 7, 14, 21, and 28. The doses and groups are shown in Table 10.15.

TABLE 10.15 CV8 Dose Groups and Number of Volunteers

Dose (Capsules)			% Increase in Dose	Number of CV8 Recipients	Number of Placebo Recipients
Daily Amount	Number of Days	Total			
5[a]	3	15	–	8	2
6	3	18	20.0	6	2
7	3	21	40.0	6	2

[a]*Recommended dose was five capsules once a day. Each CV8 capsule contained 17 mg DHA, 170 mg PQP, 2.7 mg PMQ, and 48 mg TMP.*

The results showed that the side effects of CV8 occurred mainly as gastrointestinal symptoms. Of the eight subjects given the recommended therapeutic dose, two had mild nausea accompanied by fatigue, but this did not affect normal activity. As the dose increased by 20% (six capsules) and 40% (seven capsules), the incidence of such adverse reactions also rose (Table 10.16). The abdominal side effects began 30–60 min after medication, were transient, and disappeared spontaneously without treatment in all groups.

On Days 3, 7, 14, 21, and 28, the 20 medicated subjects showed no abnormal changes in their blood tests and comprehensive metabolic panels. ECGs and routine urine tests also did not reveal any abnormalities. Although there was an increase in side effects among the 12 volunteers given 120% and 140% of the recommended therapeutic dose, their liver and kidney function, ECGs, blood, and other tests showed no abnormalities.

The Phase I trial results showed that healthy people could tolerate 1.2 to 1.4 times the recommended therapeutic dose of CV8. Therefore, the recommended dose was safe.

2. Comparative Studies of CV8 on Falciparum Malaria

CV8 and Fansimef (FSM)

Xuan Loc and Quy Nhon hospitals conducted a randomized, open, comparative trial between CV8 and FSM. A 3-day regimen of CV8 was used on 76 cases of falciparum malaria. FSM was administered once, in three tablets containing 250 mg mefloquine, 500 mg sulfadoxine, and 25 mg pyrimethamine each. Fifty falciparum malaria patients received FSM. CV8's parasite elimination speed and defervescence and parasite clearance times were all faster than those of FSM. The 28-day cure rates were more than 90% in both groups, but the FSM group saw one case each of RII and RIII. All in the CV8 group were rapidly clinically cured (Table 10.17; Fig. 10.10).

CV8 and Artesunate

Quy Nhon Hospital compared the effects of CV8 and ATS on falciparum malaria. A total of 600 mg oral ATS was given over 5 days. The defervescence and parasite clearance times and cure rate of CV8 were significantly superior ($P < .05$) to those of ATS (Table 10.18).

3. Expanded Trials on CV8

Falciparum Malaria

The Tan Phu, Xuan Loc, Lam Dong Second Provincial, and Quy Nhon hospitals used a 3-day course of CV8 to treat 163 cases of falciparum malaria. Average defervescence and parasite clearance times were 19.7–26.0 and 24.7–34.8 h, respectively. A total of 147 cases completed the 28-day follow-up. There were

TABLE 10.16 Side Effects of CV8 in Healthy Volunteers

Group	Number of Cases	Nausea	Vomiting	Abdominal Pain	Loss of Appetite	Dizziness	Fatigue
Placebo	6	0	0	0	0	0	0
5 capsules	8	2	0	0	1	0	2
6 capsules	6	3	1	1	1	2	4
7 capsules	6	5	3	1	2	4	5

TABLE 10.17 Efficacy of CV8 and Fansimef (FSM) on Falciparum Malaria

Drug	Number of Cases	Defervescence Time (h) M ± SD	Parasite Clearance Time (h) M ± SD	28-day Efficacy				
				Cases Tracked	RI	RII	RIII	Cure Rate (%)
CV8	76	23.4 ± 13.4	51.6 ± 18.1	73	4	0	0	94.5
FSM	50	39.0 ± 29.0	59.4 ± 19.8	50	1	1	1	94.0

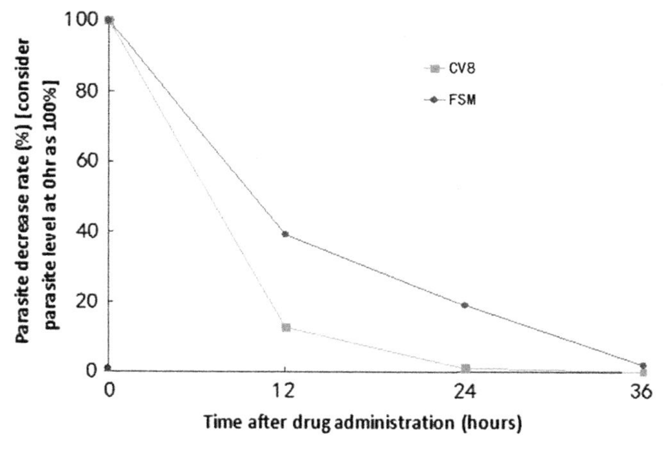

FIGURE 10.10 Parasite elimination rate of CV8 and FSM.

TABLE 10.18 Efficacy of CV8 and Artesunate (ATS) on Falciparum Malaria

Drug	Number of Cases	Defervescence Time (h) M±SD	Parasite Clearance Time (h) M±SD	28-day Recrudescence	
				Cases	%
CV8	33	11.6±6.7	22.9±10.1	1	3.0
ATS	23	24.0±12.0	30.3±10.2	5	21.7

nine cases of recrudescence—although reinfection was not ruled out via a Polymerase Chain Reaction (PCR) test—and the cure rate was 93.9%.

Vivax Malaria

Xuan Loc Hospital treated 21 vivax malaria cases with a 3-day course of CV8. Average defervescence time was 17.7±10.2h; average parasite clearance time was 31.8±20.1h. There was no recrudescence over the 28-day follow-up; however, one patient relapsed after 33days.

Carriers

Cho Ray Hospital's Tropical Diseases Clinical Research Center, and the antimalarial stations of Binh Thuan Province used CV8 on 154 *P. falciparum* carriers in the Suoi Kiet test site, Tanh Linh District. Of these, 134 were observed for 28days. There were seven cases of recrudescence and a cure rate of 94.8%. Apart from this, 78 cases of vivax malaria were treated, among whom 71 were followed for 28days. There were two cases (2.8%) of short-term recrudescence.

TABLE 10.19 In Vivo Presence of *P. falciparum* Gametocytes After Administration of CV8 and Fansimef (FSM)

Drug	Number of Positive Cases	Number of Gametocytes per μL (M ± SD)	Gametocyte Clearance Time (days)	
			Range	M ± SD
CV8	28	373 ± 441	2–6	3.9 ± 1.2
FSM	23	316 ± 539	7 to >28	>17.8 ± 6.4

4. Effects of CV8 on Falciparum Gametocytes

In the CV8-FSM comparative trial, 28 patients in the CV8 group and 23 in the FSM group were found to have *P. falciparum* gametocytes. After medication, the gametocytes remained in vivo for 2–6 days in the CV8 group, an average of 3.9 days; for the FSM group, this was seven to more than 28 days, an average of more than 17.8 days (Table 10.19). In a test village, the *P. falciparum* gametocytes in 33 patients were eliminated 7 days after administration of CV8. Only one case tested positive for gametocytes again on the 28th day. These results suggested that CV8 could rapidly kill *P. falciparum* gametocytes, which was important in blocking malaria transmission and controlling its spread.

5. Safety Evaluation

Side Effects

A total of 216 cases were treated with CV8 and 51 with FSM. Detailed records of side effects were kept for both drugs. The incidence of symptoms involving the central nervous system—such as insomnia, headache and dizziness—was similar in both drugs. Abdominal pain, diarrhea, and anorexia were infrequent. These symptoms could be linked to the disease and not necessarily drug induced.

For a 3-day course of CV8, in which the drug was administered at 0, 24, and 48 h, the incidence of nausea was 16.5% and that of vomiting was 14.9%. The rates were similar for FSM, with an incidence of 15.7% for nausea and 13.7% for vomiting. A different 3-day course of CV8, where the drug was administered at 0, 6, 24, and 48 h at 25% smaller doses, caused the incidence of nausea and vomiting to fall to 9.5% and 5.3%, respectively.

ECGs

Twenty cases were given the 3-day, three-dose regimen of CV8. Another 15 cases received the 3-day, four-dose course. ECGs conducted on Days 0, 3, and 7 showed no abnormal changes.

Laboratory Tests

Erythrocyte and leukocyte checks on 149 patients at Days 0 and 5 or 7 showed no abnormalities. The ALT, AST, bilirubin, and creatinine levels of 40 patients were checked on Days 0, 3, and 7. There were no unusual changes. Those patients with abnormal levels before medication returned to normal or improved significantly on Days 3 and 7, as the illness was being treated.

6. Conclusion

In areas where drug-resistant falciparum malaria was endemic, a total of 426 cases were given a 3-day course of CV8. Of these, 387 cases were tracked for 28 days. The cure rate was 94.6%. CV8 eliminated parasites more rapidly than FSM, had a higher cure rate than a 5-day course of ATS, and could rapidly kill *P. falciparum* gametocytes. Of the 99 cases of vivax malaria given CV8, 92 were followed up for 28 days. The recrudescence rate was 2.2%. Side effects included transient nausea and vomiting, the frequency of which fell to 9.5% and 5.3%, respectively, with a four-dose regimen. This was lower than the incidence in FSM. After medication, no abnormalities were seen in ECGs, blood tests or comprehensive metabolic panels. It showed that CV8 was safe.

10.5 THE ROLE OF PRIMAQUINE IN ACTs

Li Guoqiao

Institute of Tropical Medicine, Guangzhou University of Chinese Medicine, Guangzhou, Guangdong Province, China

To treat falciparum malaria, not only must the asexual parasite be rapidly cleared to control symptoms, the gametocyte must also be eliminated to prevent transmission. Artemisinin was thought to be effective against gametocytes. Therefore, it was assumed that ACTs removed the need for additional drugs to deal with gametocytes. This was not so.

Artemisinin and its derivatives were indeed effective against *P. falciparum* gametocytes. They could eliminate immature, early-stage gametocytes in the bone marrow within 7–10 days. They could also stop preinfectious gametocytes, which had just entered the blood, from maturing and becoming infectious.[28] Therefore, they were effective in killing gametocytes and preventing their maturation. However, many patients already carried infectious gametocytes. Only 14 days after treatment with artemisinin and its derivatives did 100% of the mature gametocytes lose their infectivity (Table 10.20).[29–32] In other words, for 10–13 days after medication, the gametocytes in a patient's blood could still become sporozoites in the mosquito vector if the patient were to be bitten, thus spreading the disease. Therefore, ACTs

TABLE 10.20 Effects of Artemisinin and Derivatives Against *P. falciparum* Gametocytes

Publication Year	Drug Tested	No. of Cases	Cases With Infectious Gametocytes in *Anopheles* Mosquitoes						
			D0	D4	D7	D10	D14	D21	D28
1993[30]	ATS	5	5/5	5/5	3/5	–	0/5	0/5	–
1994[29]	ART	9	9/9	9/9	4/9	–	0/9	–	–
	MFQ	9	9/9	9/9	8/9	–	4/9	1/9	–
	MFQ + PMQ	9	9/9	0/9	0/9	–	0/9	0/9	–
1996[31]	ATS	9	9/9	9/9	7/9	3/9	0/9	–	–
	QN	7	3/7	7/7	7/7	7/7	7/7	2/7	0/7
1999[32]	DHA	6	6/6	6/6	6/6	2/6	0/6	0/6	–
	QN	8	4/8	8/8	8/8	8/8	8/8	2/8	0/8

ART, artemisinin; *ATS*, artesunate; *DHA*, dihydroartemisinin; *MFQ*, Mefolquine; *PMQ*, primaquine; *QN*, quinine.

may not necessarily halt transmission without the addition of PMQ that can quickly eliminate gametocytes.

All 29 cases using artemisinin-type antimalarials lost 100% of their gametocyte infectivity 14 days after drug taking.

PMQ was known to kill gametocytes. However, its routine dose was 22.5–45 mg, which could cause hemolysis in glucose-6-phosphate dehydrogenase (G6PD)-deficient patients. This was difficult to avoid even at doses as low as 15 mg, taken for 4–5 days.

From 1993 to 1996, the author undertook extensive research in Vietnam on the suppressive effects of PMQ on gametocyte infectivity. A single dose of 6–8 mg PMQ could cause gametocytes in the blood to lose infectivity within 24 h (Tables 10.21 and 10.22). Even a dose of 3.75 mg had an impact. PMQ had synergistic effects with artemisinin-type drugs. Hence, 8 mg PMQ was added to the daily ACT dose.

A study in Thailand showed that a dose of 15 mg PMQ, administered for 14 days, caused hemolysis in all 23 G6PD-deficient patients.[33] To explore the relationship between the dose and the side effect, from 2003 to 2006, four G6PD-deficient vivax malaria patients were hospitalized for observation in Cambodia. The 14-day, 15 mg regimen was given and all patients experienced a hemolysis reaction 4–5 days after medication. The dose was reduced to 10 mg for another 13 G6PD-deficient vivax malaria patients, administered for 8–14 days. Only one patient had mild hemolysis on Days 7–8 (Table 10.23). Compared to the results for the 27 cases with 14-day, 15 mg regimen (23 cases reported, 4 described above), it showed that the side effect was clearly related to the dose. When 8 mg PMQ was added to the daily dose of Artequick and administered for 2 days, four G6PD-deficient falciparum malaria cases showed no hemolysis during their 2 weeks in hospital (Table 10.24).

Therefore, a single dose of 8 mg PMQ, combined with artemisinin or an ACT, could cause gametocytes to lose infectivity within 24 h. Given only for 2 days, this dose could greatly reduce the risk of hemolysis. In 1997, CV8 was registered for production in Vietnam. The first day's dose, spread over two administrations, contained 11.36 mg PMQ, while those of the second and third days contained 5.68 mg. The total dose was 22.7 mg. Since 1997, Vietnam has manufactured more than 2 million doses of CV8. It has been used as first-line treatment, with no reports of hemolysis.

From 2005 to 2006, Cambodia implemented mass drug administration with Artequick (see below). To make dispensing easier and ensure rapid loss of gametocyte infectivity, 4 mg PMQ was added to each Artequick tablet. For adults taking two tablets daily, this meant a total daily dose of 8 mg PMQ for 2 days. The rate of G6PD deficiency in Cambodia was 14.7% and 28,000 people took one course of medication a month for 2 months. No hemolysis occurred.

TABLE 10.21 Effects of PMQ + DHA + Piperaquine, PMQ + Piperaquine, and PMQ + DHA on *P. falciparum* Gametocytes

Drug/Dose (mg)			No. of Cases	Number of Carriers Transmitting Infection to *A. dirus*									
				0h		12h		24h		48h		D7	
PMQ	DHA	PQ		O	S	O	S	O	S	O	S	O	S
15.0	64×4	375×4	11	11	11	1	0	0	0	0	0	0	0
11.25	–	375×4	2	2	2	2	2	1	0	0	0	0	0
11.25	80×4	–	2	2	2	0	0	0	0	0	0	0	0
7.5	–	375×4	6	6	6	6	5	0	0	0	0	0	0
7.5	80×4	–	6	6	6	1	1	0	0	0	0	0	0
3.75	–	375×4	5	5	5	5	5	4	3	2	2	1	1
3.75	80×4	–	4	4	4	4	4	1	0	0	0	0	0
Nonmedicated			6	6	6	6	6	6	6	6	6	6	6

DHA, dihydroartemisinin; *O*, oocyst; *PMQ*, primaquine; *PQ*, piperaquine; *S*, sporozoite. 7.5 mg PMQ + 80 mg DHA can cause gametocytes to lose infectivity in 24 h; 3.75 mg PMQ has a mild suppressive effect on gametocytes, which is synergistic with DHA.

TABLE 10.22 Effects on *P. falciparum* Gametocytes of Different Doses of PMQ Added to the Initial Dose of Artequick

Drug/Dose (mg)		No. of Cases	Number of Carriers Transmitting Infection to *A. dirus*											
			0h		12h		24h		48h		D7		D14	
			O	S	O	S	O	S	O	S	O	S	O	S
PMQ	6	7	7	7	4	3	0	0	0		0	0	0	0
PMQ	7.5	3	3	3	2	1	0	0	0	0	0	0	0	0
PMQ	8	32	32	32	7	1	1[a]	0	1[a]	0	1[a]	0	1[a]	0
Control[b]		4	4	4	–	–	–	–	4	4	4	4	4	4

O, oocyst; *PMQ*, primaquine; *S*, sporozoite. Ten cases received an initial dose of Artequick (artemisinin 125 mg, piperaquine 750 mg) with 6 or 7.5 mg PMQ; 32 cases received Artequick and 8 mg PMQ; total 42 cases. *P. falciparum* gametocytes lost infectivity within 24 h for all cases.
[a]*Different patients.*
[b]*Nonmedicated control group: 3/4 tested positive on Day 21 and 2/4 on Day 28.*

TABLE 10.23 RBC Monitoring on 13 Cases of *Plasmodium vivax* Patients With G6PD Deficiency Given Primaquine 10 mg (×8 days: 5 Cases; ×14 days: 8 Cases)

G6PD Deficiency (n = 13)		G6PD Normal (n = 23)	
Day	RBC (X ± SD)	Day	RBC (X ± SD)
D0	4.92 ± 0.69	D0	4.80 ± 0.90
D2	4.65 ± 0.69	D2	4.65 ± 1.03
D4	4.53 ± 0.44	D4	4.81 ± 0.92
D6	4.32 ± 0.53	D6	4.83 ± 0.81
D8	4.18 ± 0.73	D8	4.94 ± 0.90
D10	4.08 ± 0.59[a]		
D12	4.21 ± 0.30		
D14	4.37 ± 0.27		

[a] *1 case of mild hemolysis, RBC declined 40%.*

TABLE 10.24 RBC Monitoring on Four Cases of *Plasmodium vivax* Patients With G6PD Deficiency Given Primaquine 8 mg + Artequick 2 tablets ×2 days

G6PD Deficiency (n = 4)		G6PD Normal (n = 13)	
Day	RBC (X ± SD)	Day	RBC (X ± SD)
D0	5.39 ± 1.03	D0	5.21 ± 0.90
D2	4.69 ± 0.64	D3	4.60 ± 0.77
D4	4.72 ± 0.79	D7	4.88 ± 0.99
D6	4.73 ± 0.79		
D8	4.82 ± 0.72		

In 2007, 36,000 people in Moheli Island in Comoros adopted the same regimen as in Cambodia. The rate of G6PD deficiency in Comoros was 15.0% and again, no hemolytic reactions were seen. Because the dose of PMQ was reduced to 1/4–1/5 of the 30–45 mg routine dose, the likelihood of other side effects occurring was small.

10.6 FAST ELIMINATION OF MALARIA BY SOURCE ERADICATION (FEMSE) USING ACTs

Li Guoqiao
Institute of Tropical Medicine, Guangzhou University of Chinese Medicine, Guangzhou, Guangdong Province, China

In 1991, at the invitation of Cho Ray Hospital in Ho Chi Minh City, Vietnam, the malaria research team of the Guangzhou University of Chinese Medicine shared its expertise on the use of artemisinin to treat cerebral malaria.[34] Between then and 1995, the Xuan Loc Hospital treated 168 cases of cerebral malaria with artesunate (ATS), with a cure rate of 90.5%. The experience of curing severe malaria patients and visits to hospitals in central and southern Vietnam led to the idea that a fundamental reduction in mortality could only be achieved by swiftly lowering the morbidity rate. To do this, it was critical to block the transmission of gametocytes. Thus, research into the effects of artemisinin on *P. falciparum* gametocytes, the source of infection, began. Several years' work showed that artemisinin had a rapid effect on early-stage gametocytes and could block their maturation. However, it was slow to eliminate mature gametocytes in the blood. It took 14 days for 100% of the gametocytes to lose infectivity.

This led to another idea. In 1993, while investigating CV8, the team also considered the ability of primaquine (PMQ) to eliminate gametocytes. Given the hemolytic side effects of PMQ, studies focused on finding an effective and safe dose of the drug. Based on the routine adult dose of 22.5 mg for 3 days, smaller, single doses of 15, 11.25, and 7.5 mg were proposed. All doses could cause *P. falciparum* gametocytes to lose infectivity in 24 h. Even a dose of 3.75 mg had some effects and was synergistic with artemisinin (see above). Eventually, it was decided to include 5.68 mg PMQ (i.e., 10 mg primaquine phosphate) with each ACT dose. Thus, CV8 became the first such combination to include PMQ. In 1997, it was approved and registered by the Ministry of Health in Vietnam, manufactured by Vietnam's Pharmaceutical Factory No. 26. For many years, CV8 was dispensed in community hospitals in areas where falciparum malaria was endemic, given free of charge for the treatment of falciparum malaria patients. It markedly reduced the morbidity and mortality rates for falciparum malaria in Vietnam and inspired the reform of antimalarial measures.

Tropical or subtropical areas where falciparum malaria is endemic have long relied on the spraying of insecticides or insecticide-treated bed nets to reduce infection, or taken medication during the peak malaria season. Since 2000, a new approach was gradually developed based on several decades of research into malaria control measures worldwide, namely to block malaria transmission by eliminating infective parasitic gametocytes in patients and carriers and removing the source of infection. This could be simpler, cheaper, and more manageable than eliminating the mosquito vector.

1. Antimalarial Efforts in Six Countries of the Mekong Region

Malaria prevention and control measures in Hainan of China are among the most effective worldwide. It was predicted that malaria would be eradicated there by 2015. Since 1955, when comprehensive prevention and control work began, annual morbidity in Hainan fell from 276,000 to 14,000 in 1987. The average annual decrease was 9% over 32 years. In that time, the total morbidity was 1.868 million, with 1077 deaths. From 1987 to the eradication of malaria in 2015, it would take 28 years before malaria would be eliminated. It would be a long, hard journey of 60 years.

Vietnam, Thailand, Laos, Cambodia, and Myanmar also embarked on this lengthy path. From WHO statistics, since 1990, these countries achieved similar results to Hainan in their fight against malaria (Table 10.25). The morbidity figures for these countries fell by an average of 5%–11% per annum.

The history lessons of malaria control in these six countries showed that when the mosquito vector was common in the wild, vector control was very difficult. It was a significant limitation to antimalarial policies which focused on eliminating mosquitoes. Therefore, a new approach was needed to reevaluate traditional methods and adopt a policy centered around actively, swiftly, and thoroughly eliminating the source of transmission: the in vivo parasite. This would fulfill the objective of rapidly controlling and eventually eradicating malaria. The parasites could persist in the human body for 1–2 years or more. However, the lifespan of a mosquito was usually only a month and the parasites died along with their mosquito vector. Therefore, prevention measures should target the parasite in the human body.

2. Pilot Fast Elimination of Malaria by Source Eradication (FEMSE) Projects

The lengthy, combined experiences of Hainan, Vietnam, and Cambodia in devising antimalarial policies came to fruition in 2003. At that time, the European Union had given its sixth consecutive year of antimalarial aid to Cambodia. The morbidity rate in these 6 years had only fallen by 22.2%, and an average of 132,500 people still fell victim to the disease annually during this time period. It also coincided with the start of the author's ACT clinical trials in Cambodia. The director of Cambodia's National Center for Malaria Control, Duong Socheat, discussed the issue of rapidly reducing the morbidity rate with the author, and FEMSE was proposed. With the support of the Cambodian Minister of Health and the deputy governor in charge of health in Kampong Speu, it was decided to conduct a study in two highly malarial areas of that province: Aoral (average parasite rate 55.8%) and Sprin. The region had a population of 28,000 and three test sites. After the study, the local parasite rate fell by 95% in a year and the transmission of malaria was swiftly controlled.

In the meantime, research continued into the effects of low doses of PMQ on gametocyte infectivity, and the relationship between the drug's dose and

TABLE 10.25 Malaria Morbidity and Mortality in the Mekong Region Over the 19-Year Period 1990–2008[a]

Country	Malaria Cases		Decrease Over 19years (%)	Average Annual Decrease (%)	Total Morbidity	Total Mortality
	1990	2008				
Thailand	273,880	26,150	90.5	11.0	1,673,843	7551[c]
Vietnam	123,796	11,355	90.8	11.0	1,556,098	137,763
Cambodia	123,796	42,124	66.0	5.5	1,372,055	12,939
Laos	41,048[b]	17,648	57.0	4.6	635,483	6,099
Burma	989,042	411,494	58.4	4.5	8,598,650	31566[c]

[a]*Data from World Health Organization, World Malaria Report 2009.*
[b]*Figures available for the 13-year period 1996–2008.*
[c]*Figure for 1991.*

the incidence of hemolysis in G6PD-deficient patients. A study of gametocyte infectivity in 42 falciparum malaria cases, with *Anopheles* mosquitoes as the vector, showed that a single adult dose of 6–8 mg PMQ, coupled with an ACT, could cause gametocytes to lose infectivity within 24 h. Another study involving 17 G6PD-deficient patients, observed over 2 weeks, indicated that the hemolysis reaction had a strong dose–effect relationship (see above).

Cambodia's experience from 2004 to 2006, which saw the parasite rate fall by 95% in a year as well as rapid control of the disease,[35,36] proved to be a model for the safe application of FEMSE.

In 2007, the African island nation of Comoros instituted a FEMSE study on Moheli Island, with a population of 36,000.[37] Moheli had 25 villages and was a typical highly malarial endemic area. Before FEMSE was carried out, the parasite rate averaged at 23.0%; 10 villages had rates of over 30%. The six villages with the highest parasite rates registered at 94.4%, 81.0%, 65.0%, 58.0%, 52.0%, and 50.0%. After blood smears were obtained randomly from these villages, 182 people tested positive. Of these, 176 were infected with *P. falciparum* (96.7%) and 57 with *Plasmodium malariae* (31.3%). The latter were mostly coinfections with *P. falciparum*. No vivax malaria cases were seen.

Two months after the mass drug administration (MDA) in FEMSE, the population parasite carrier rate decreased by 93.9% (Fig. 10.11) and monthly morbidity fell by 93.2% (Fig. 10.12). Before FEMSE, 3.1% (8/258) of mosquito vectors sampled tested positive for the parasite. Four months later, this had reduced to 0%. The numbers were 0/400, 0/456, and 0/517 at 4, 5, and 6 months after FEMSE, respectively. The rate of G6PD deficiency was 15.0%. A 2-day course of Artequick and 8 mg PMQ was given to the population of 36,000 in two

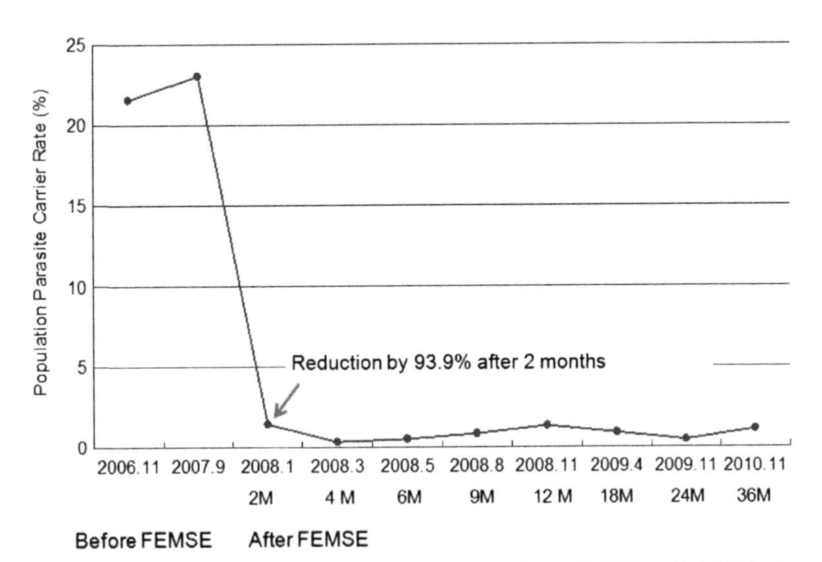

FIGURE 10.11 Population parasite carrier rate before and after FEMSE on Moheli Island.

mass administration exercises. No hemolytic reactions were seen. The island rapidly became an area of low malaria incidence. Four years later, the morbidity and parasite rates remained at relatively low levels.

In early October 2012, Anjouan Island introduced FEMSE among its population of 350,000. The malaria situation before and after FEMSE is shown in Fig. 10.13.

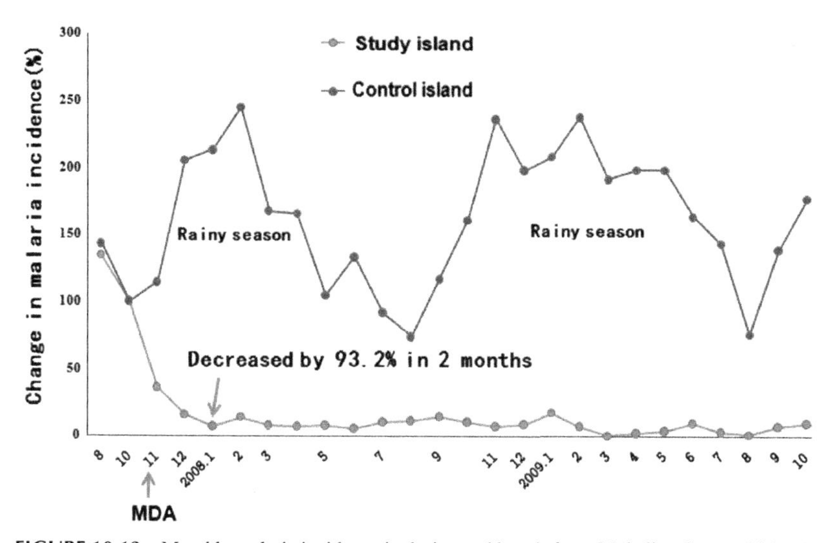

FIGURE 10.12 Monthly malaria incidence in designated hospitals on Moheli and control islands.

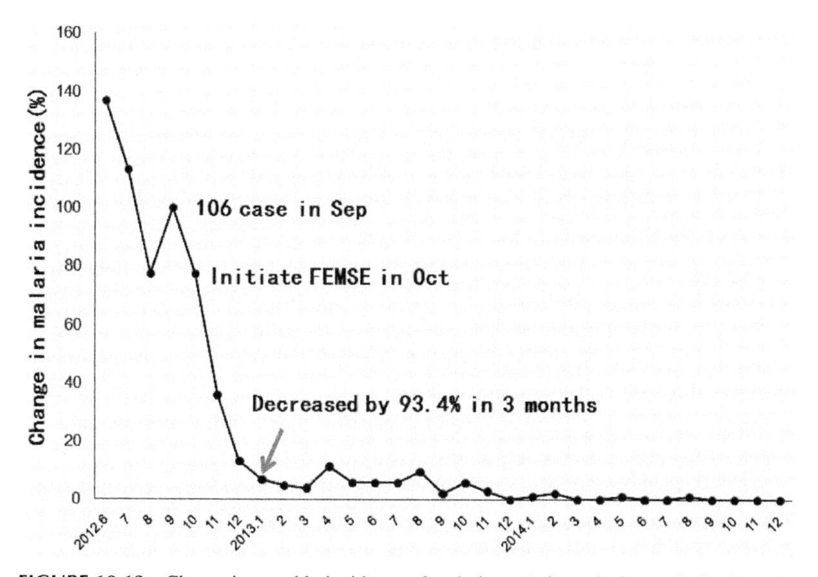

FIGURE 10.13 Change in monthly incidence of malaria on Anjouan before and after FEMSE.

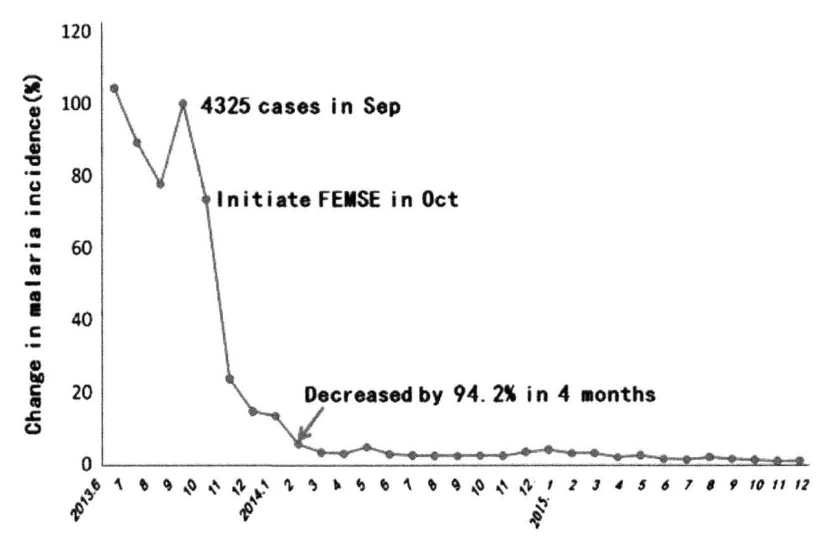

FIGURE 10.14 Monthly incidence of malaria on Grande Comore before and after FEMSE.

In late October 2013, FEMSE was introduced in Grande Comore, population 400,000. The malaria situation before and after FEMSE is shown in Fig. 10.14.

When Anjouan and Grande Comore instituted FEMSE in 2012 and 2013, this greatly reduced the introduction of new sources of infection to Moheli via the mobile population. The opportunity was taken to conduct a polymerase chain reaction (PCR) diagnostic screening on 6000 inhabitants of the villages in Moheli which still had malaria cases. Two hundred villagers were found to be low-density carriers. They were given a 3-day course of six Artequick tablets in total, with 9 mg PMQ added to the first dose. After medication, 92.2%, 93.9%, and 100% of the PCR tests turned negative on Days 5, 7, and 10, respectively. This was another big step for FEMSE on that island. Based on the experience of FEMSE in Comoros, involving three islands and nearly 800,000 people, a three-step policy plan and its expected outcomes was drawn up (Fig. 10.15).

The results achieved in Comoros showed that a change in traditional antimalarial measures, from controlling the mosquito vector to eliminating the source of infection in the human subject, could swiftly transform a hyperendemic region into a hypoendemic one. It could halt mortality and reduce the number of years needed for a country to control and eradicate malaria from several decades to several years. This greatly reduces the cost of antimalarial policies and rapidly eases the economic burden that malaria imposes on society.

History had shown that turning a hyperendemic region into a hypoendemic one took decades. Even then, a low parasite rate could persist for 10–20 years or even longer. This was a serious problem, primarily due to the lack of source eradication.

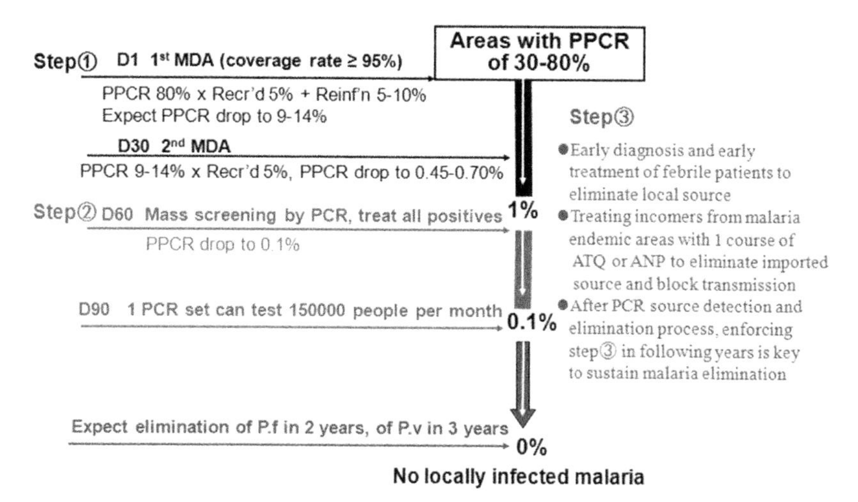

FIGURE 10.15 FEMSE policy plan and expected outcomes. *ANP*, artemisinin–naphthoquine–primaquine combination; *ATQ*, Artequick; *MDA*, mass drug administration; *PCR*, polymerase chain reaction; *PPCR*, population parasite carrier rate.

Past policies focusing on vector control had achieved results under specific conditions. In tropical regions with millions of inhabitants, however, several decades were needed to eradicate malaria via vector-specific methods, especially in areas where mosquitoes were endemic in the wild. In Hainan, for example, from the 1970s to the 1980s, MDA took place in hyperendemic regions in April, July, and October every year. A 3-day course of PMQ and chloroquine (piperaquine from 1978) was given to 70%–80% of the population. This only reduced morbidity and did not eradicate the source of infection. From a report by the Prevention, Treatment and Research into Malaria in China, "In malarial areas where *A. sinensis* (the wild mosquito) is the main vector… the control measures must be based on source eradication… to avoid repeating the lessons that some countries have learnt: that less satisfactory results are obtained in the same amount of time via indoor spraying of insecticides alone."

FEMSE controlled and could eradicated malaria by eliminating the source of infection. The lifespan of a mosquito is only a month. When carrier mosquitoes die, parasite transmission also ceases. A new generation of mosquitoes had to feed on human gametocyte carriers to pass the disease to others. FEMSE renders this impossible, thereby blocking the spread of malaria. In a few months, it reduces the parasite and morbidity rates in hyperendemic areas by more than 95%. However, local incidents or outbreaks could still occur due to the ubiquity of the mosquito vector. This creates lingering sources of infection and allows for undetected carriers to persist. Therefore, after PCR checks are carried out, strong and effective village antimalaria networks are needed. Village malaria workers armed with fast-acting, convenient diagnostic and treatment tools, rapid diagnostic tests (RDTs), PCR, ACTs, and PMQ, could detect and treat malaria cases promptly, thoroughly removing local and incoming sources of infection. With such initiatives, malaria may be eradicated in an area within a few years.

The control and elimination of malaria was also closely associated with regional, social and economic development. In less-developed areas, FEMSE is a shortcut to malaria eradication that saves both money and time. Previously, antimalarial efforts in the six Greater Mekong countries had been limited by socioeconomic factors and the absence of ACT treatments. Medical staff has not realized the overall potential of PMQ, for example, and were discouraged from using it due to its hemolytic side effects at the normally used high dose. Therefore, it was not fully exploited to block malaria transmission.

3. Principles of Fast Malaria Elimination by Source Eradication (FEMSE)

ACT-based FEMSE used in highly malarial areas, in place of traditional methods focusing on mosquito vectors, could rapidly control and eradicate malaria.

As we have seen, the persistence of the malaria parasite in humans—as opposed to the short lifespan of the mosquito vector—make it more critical to block transmission from the human host. This is also easier than controlling the mosquito population. After 10 years of being tested and improved in hyperendemic areas, FEMSE achieved unprecedented results: a dramatic fall in parasite rates in both infected patients and carriers as well as in vectors, also in overall morbidity. Malaria transmission reaches extremely low levels, allowing a hyperendemic country of several hundred thousand to become a hypoendemic one with zero malaria mortality. Together with a robust network of village malarial workers (VMWs) and monthly reporting, malaria could be eradicated in a few years. As of now, FEMSE is the fastest and most cost-effective method for controlling and eradicating malaria. Its main principles are as follows.

1. *Rapidly eliminating the source of infection.*
2. *Recommended regimen*: A 2- or 3-day course of an ACT, with 8 mg PMQ added on the first day alone.
   - **a.** Mass Drug Administration. Artequick is recommended as it is effective, convenient to use, and has few side effects. In a therapeutic context, a 2-day regimen of Artequick is used: two tablets and 8 mg PMQ on the first day, and another two tablets on the second day. Two such courses are needed for mass drug administration under FEMSE, with the second course given 30 days after the first.
   - **b.** Special cases where extra care is needed to prevent lingering sources of infection include severe malaria and possible drug resistance.
     - **i.** Patients who cannot take medication orally should receive artesunate (ATS) or artemether injections. When oral medication

can be administered, the recommended ACT + PMQ course should be given to thoroughly eliminate the parasite.

ii. Chloroquine- or piperaquine-resistance is suspected if malaria occurs twice in the same patient within 2 months, and if treatment with Artequick is followed by an RI (recrudescence in 28 days) or RII (at least 75% parasite clearance in 48 h and symptoms controlled, but no full clearance by Day 7) response. A 3-day regimen of six Artequick tablets plus 8 mg PMQ on the first day should be given to thoroughly eliminate the parasite.

c. PMQ had previously been contraindicated for infants and pregnant women. This was because a high dose and lengthy course of treatment could lead to hemolysis in G6PD-deficient patients. As mentioned above, however, mass drug administration under FEMSE only required a single dose of 8 mg PMQ, which was only 1/5 of the previously recommended dose. No hemolysis was seen during FEMSE in Cambodia and Comoros, further validating its safety. Parasite carrier rates are higher in infants and pregnant women due to their lowered immunity. Mass drug administration could benefit them by eliminating their parasite load. They also make up no small proportion of the population and, if excluded, could represent new sources of infection. For women in the early stages of pregnancy (3 months and below), the ACT should not be given. Instead, they should receive 500 mg piperaquine every day for 3 days, with the addition of 8 mg PMQ on the first day.

3. *PCR Diagnostics:* After 60 days (i.e., two courses of mass drug administration), the population parasite carrier rate usually falls to 1% or less. From Day 60 onward—or, for individuals given only one course of medication, from Day 30 onward—PCR diagnostics should be used on the whole population, with two drops of blood drawn from each person. PCR positive individuals (low-density carriers) should be given a 3-day course of six Artequick tablets, including 8 mg PMQ on the first day, within 24 h. The increased dose suppresses the usual 5% recrudescence rate observed in past antimalarial efforts.
4. *Early diagnosis and treatment*: After the two courses of mass drug administration, a few cases of recrudescence or reinfection will occur. At this eradication stage, prompt control of gametocytes via early diagnosis and treatment is the key to preventing further transmission. Individuals with fever should undergo diagnostic checking with rapid diagnostic kits (RDT) or blood smears within 24 h. Those testing positive should be given a 3-day course of six Artequick tablets, with 8 mg PMQ added on the first day.
5. *Controlling external sources of infection*: People from countries or regions where malaria is endemic may be carriers of the parasite. Therefore, medication points should be set up in airports and harbors to administer A-P tablets containing 160 mg artemisinin and 8 mg PMQ. One tablet per person should be taken, which has no side effects. During their stay, travelers should

take one tablet every 7 days to prevent the spread of gametocytes and further maturation of the asexual parasite. Early diagnosis and treatment should be given, either by VMWs or community clinics, to all incoming travelers who contract malaria.

6. *Vector control*: In the maintenance phase, if economic conditions permit, bed nets may also be used to reduce contact with mosquitoes. This further consolidates the results obtained by FEMSE.

REFERENCES

1. Shen NC, Jiang W, Tang HL, et al. Pre-clinical toxicity studies of new antimalarial compounds II: pre-clinical toxicity studies of piperaquine phosphate and its act, malaria prevention tablets number 3. *J Second Mil Med Univ* 1981;**2**(1):40–6. [沈念慈，姜渭，唐惠兰，等. 新抗疟化合物的临床前毒性研究II. 磷酸哌喹及其复方“防疟片3号”的临床前毒性研究. 第二军医大学学报，1981, 2(1):40–46.]
2. Rhone-Poulenc SA. *Quinoline derivatives* Brit. Patent 991,838. May 12, 1965.
3. Lafaix C, Rey M, Diop Mar I, et al. Essai de traitement curatif du paludisme pour un nouvel antipaludigue de synthese, le 16,126 RP. *Bull Soc Med Afr Noire Langue Fr* 1976;**12**(3):546–51.
4. Davis TM, Hung TY, Sim IK, et al. Piperaquine: a resurgent antimalarial drug. *Drugs* 2005;**65**(1):75–87.
5. Yadav GC, Srinivasan S, Bhovi MG, et al. *Preparation of bisquinoline compounds* US Patent 20060270852. November 30, 2006.
6. Singh T, Stein RG, Hoops JF, et al. Antimalarials. 7-Chloro-4-(substituted amino) quinolines. *J Med Chem* 1971;**14**(4):283–6.
7. Vennerstrom JL, Ager AL, Dorn A, et al. Bisquinolines. 2. Antimalarial. *N,N*-bis (7-chloroquinolin-4-yl)heteroalkanediamines. *J Med Chem* 1998;**41**(22):4360–4.
8. Tripathi RC, Saxena M, Chandra S, et al. Synthesis of 4-arylsulphonyl/piperazinyl-7- chloroquinolines and related compounds as potential antimalarial agents. *Indian J Chem* 1995;**34B**: 164–6.
9. Agrawal VK, Sharma S. Antiparasitic agents: Part VI – synthesis of 7-chloro-4-(substituted phenyl)-amino and 4-(substituted piperazin-1-yl)-quinolines as potential antiparasitic agents. *Indian J Chem* 1987;**26B**:550–5.
10. Shanghai Institute of Pharmaceutical Industry. In: *Piperaquine phosphate (production unit: Shanghai 14th pharmaceutical factory)*. National API Production Compendium; 1980. p. 305–7 (Unpublished material). [上海医药工业研究院. 喹哌磷酸盐(生产单位：上海第十四制药厂). 全国原料药工艺汇编(内部资料)，1980:305–307.]
11. Shanghai Institute of Pharmaceutical Industry, PLA Rear Unit 243, Shanghai 2nd pharmaceutical factory, et al. *The synthesis, pharmacology and prophylactic-therapeutic trials of malaria prevention tablets number 3*. Materials on Synthesis and Clinical Observations, Malaria Research; 1975. p. 1–23 (Unpublished material). [上海医药工业研究院，中国人民解放军后字243部队，上海第二制药厂，等.“防疟片3号”的化学合成、药理及现场防治试验.《疟疾研究》化学合成与临床观察专集(内部资料)，1975:1–23.]
12. Dongre VG, Karmuse PP, Ghugare PD, et al. Characterization and quantitative determination of impurities in piperaquine phosphate by HPLC and LC/MS/MS. *J Pharm Biomed Anal* 2007;**43**(1):186–95.
13. Lindegardh N, Giorgi F, Galletti B, et al. Identification of an isomer impurity in piperaquine drug substance. *J Chromatogr A* 2006;**1135**(2):166–9.

14. Chen L. *Experimental research and clinical applications of antimalarial drugs*. Shanghai: Second Military Medical University Press; 2001. p. 10. [陈林. 抗疟药的实验研究与临床应用. 上海：第二军医大学出版社，2001.10.]
15. Chen L, Dai ZR, Ma ZM, et al. Research into drugs for malaria prevention and treatment ii: screening for causal prophylactic drugs using animal models 1, *P. yoelii* and *A. stephensi*. *Acta Pharm Sin* 1981;**16**(4):260–6. [陈林，戴祖瑞，马志明，等. 疟疾防治药物的研究——II疟疾病因性预防药物筛选试验动物模型1. 约氏鼠疟原虫—斯氏按蚊系统. 药学学报，1981.16(4):260–266.]
16. Dai ZR, Chen L, Li YT, et al. Simian malaria model with *P. cynomolgi* and *A. stephensi*: some biological characteristics and responses to common antimalarials. *Acta Pharm Sin* 1983;**18**(12):881–6. [戴祖瑞，陈林，李裕棠，等. 食蟹猴疟原虫—斯氏按蚊系统猴疟模型的一些生物学特性和对常用抗疟药物的生物效应. 药学学报，1983, 18(12):881–886.]
17. Chen L, Qian YL, Li ZL, et al. The effects of piperaquine on the ultrastructure of ANKA strain *P. berghei* in the erythrocytic stage. *Acta Pharmacol Sin* 1986;**7**(4):p 351. [陈林，钱永乐，李泽琳，等. 哌喹对伯氏疟原虫ANKA株红内期超微结构的影响. 中国药理学报，1986,7(4):351.]
18. Chen L, Qian YL, Li ZL, et al. Observations on the ultrastructure of piperaquine-resistant ANKA strain *P. berghei* in the erythrocytic stage. *J Parasitol Parasit Dis* 1985;**3**(4):p 281. [陈林，钱永乐，李泽琳，等. 伯氏疟原虫ANKA株抗哌喹系红内期超微结构的观察. 寄生虫学与寄生虫病杂志，1985, 3(4):281.]
19. Information provided by Aventis, August 1962.
20. Chen L, Zai FY, Zhou YC, et al. On-site observations in Hainan on the prophylactic effects of a new antimalarial, piperaquine. *Med J Chin PLA* 1979;**4**(2):104–8. [陈林，瞿逢伊，周元昌，等. 抗疟新药哌喹在海南岛现场防疟效果的观察. 解放军医学杂志，1979, 4(2):104–108.]
21. Zhou MX, Chen L, Zhou YC, et al. A report on two cases of chloroquine-resistant falciparum malaria in Hainan. *Med J Chin PLA* 1979;**2**:p 125. [周明行，陈林，周元昌，等. 海南岛某地抗氯喹恶性疟2例报告. 解放军医学杂志，1979, (2):125.]
22. Peters W. Drug resistance in *Plasmodium berghei* Vincke and Lips, 1948. I. Chloroquine resistance. *Exp Parasit* 1965;**17**:80–9.
23. Finney DJ. *Statistical method in biological assay*. 3rd ed. London: Charles Griffin & Co.; 1978.
24. Deichmann WB, LeBlanc TJ. Determination of the approximate lethal dose with about six animals. *J Ind Hygiene Toxicol* 1943;**25**(9):415–7.
25. Ames BN, McCann J, Yamasaki E. Methods for detecting carcinogens and mutagens with the salmonella/mammalian-microsome mutagenicity test. *Mutat Res* 1975;**31**(6):347–64.
26. Hung TY, Davis TM, Ilett KF, et al. Population pharmacokinetics of piperaquine in adults and children with uncomplicated falciparum or vivax malaria. *Br J Clin Pharmacol* 2003;**57**(3):253–62.
27. (a) This information is derived from the "CV8 clinical research summary, 1996" by Li Guoqiao and Cho Ray Hospital's Trinh Kim- Anh and Huang An-Binh. (b) Dossier for New Drug Registration Application CV8 1996; Internal document, Guangzhou University of Chinese Medicine.
28. Jian HX, Chen PQ, Fu LC, et al. Effects of artesunate on *P. falciparum* gametocytes. *J Guangzhou Coll Tradit Chin Med* 1998;**15**(1):31–3. [简华香，陈沛泉，符林春，等. 青蒿琥酯对恶性疟原虫配子体的作用广州中医药大学学报，1998, 15(1):31–33.]
29. Chen PQ, Li GQ, Guo XB, et al. Effects of artemisinin on *P. falciparum* gametocyte infectivity. *Natl Med J China* 1994;**74**(4):209–10. [陈沛泉，李国桥，郭兴伯，等. 青蒿素对恶性疟配子体感染性的影响. 中华医学杂志，1994, 74(4):209–210.]
30. Chen PQ, Li GQ, Guo XB, et al. Observations on the impact of artesunate on *P. falciparum* gametocyte infectivity. *Tradit Chin Drug Res Clin Pharmacol* 1993;**4**(3):p 40. [陈沛泉，李国桥，郭兴伯，等. 青蒿琥酯对恶性疟原虫配子体感染性影响的观察. 中药新药与临床药理，1993, 4(3):40.]

31. Li G, Chen P, Li G, et al. Effect of artesunate on the infectivity of *Plasmodium falciparum* gametocytes (PFG) and a randomized comparative study with quinine. In: *XIVth International congress for tropical medicine and malaria, Nagasaki, Japan*. 1996. p. 211.
32. Chen PQ, Guo XB, Li GQ, et al. Effect of dihydroartemisinin on the development of sexual forms of *Plasmodium falciparum. Tradit Chin Drug Res Clin Pharmacol* 1999;**11**:333–5. [陈沛泉，郭兴伯，李广谦，等.双氢青蒿素对恶性疟原虫有性生殖的影响.中药新药与临床药理. 1999,11:333–335.]
33. Buchachart K, Krudsood S, Singhasivanon P, et al. Effects of primaquine standard dose (15 mg/day for 14 days) in the treatment of vivax malaria patients in Thailand. *Southeast Asian J Trop Med Public Health* 2001;**32**(4):720–6.
34. Fu LC, Wang XH, Guo XB, et al. Clinical therapeutic effects of artesunate in the treatment of cerebral malaria. *J Guangzhou Coll Tradit Chin Med* 1998;**15**(1):27–30. [苻林春，王新华，郭兴伯，等. 青蒿琥酯治疗脑型疟的临床疗效. 广州中医药大学学报，1998, 15(1):27–30.]
35. Song J, Socheat D, Tan B, et al. An experiment in rapidly controlling malaria in a hyperendemic area. *J Guangzhou Coll Tradit Chin Med* 2006;**23**(2):89–94. [宋健平，Duong Socheat，谈博，等. 高度疟疾流行区快速控制疟疾试验. 广州中医药大学学报，2006, 23(2):89–94.]
36. Song J, Socheat D, Tan B, et al. Rapid and effective malaria control in Cambodia through mass administration of artemisinin-piperaquine. *Malar J* 2010;**9**:57.
37. Li G, Song J, Deng C, et al. One-year report on the fast elimination of malaria by source eradication Project in Moheli island of Comoros. *J Guangzhou Coll Tradit Chin Med* 2010;**27**:90–8.
38. Dossier for New Drug Registration Application Artekin 2002 (Internal document, Guangzhou University of Chinese Medicine).
39. Hien TT, Dolecek C, Dung NT, et al. Dihydroartemisinin-piperaquine against multidrug-resistant *Plasmodium falciparum* malaria in Vietnam: randomised clinical trial. *Lancet* 2004;363, 18–22.
40. Dossier for New Drug Registration Application Artequick 2005 (Internal document, Guangzhou University of Chinese Medicine).

Index

Note: Page numbers followed by "f" indicate figures, "t" indicate tables.

B

D

E

F

G

H

I

J

K

L

N

O

P

Q

R

S

T

X

Y

Z

Printed and bound by CPI Group (UK) Ltd, Croydon, CR0 4YY

08/05/2025

01865002-0001